Linking the Gaseous and Condensed Phases of Matter

Condensed Phases of Matter

The Behavior of Slow Electrons

NATO ASI Series

Advanced Science Institutes Series

A series presenting the results of activities sponsored by the NATO Science Committee, which aims at the dissemination of advanced scientific and technological knowledge, with a view to strengthening links between scientific communities.

The series is published by an international board of publishers in conjunction with the NATO Scientific Affairs Division

A	**Life Sciences**	Plenum Publishing Corporation
B	**Physics**	New York and London
C	**Mathematical and Physical Sciences**	Kluwer Academic Publishers
D	**Behavioral and Social Sciences**	Dordrecht, Boston, and London
E	**Applied Sciences**	
F	**Computer and Systems Sciences**	Springer-Verlag
G	**Ecological Sciences**	Berlin, Heidelberg, New York, London,
H	**Cell Biology**	Paris, Tokyo, Hong Kong, and Barcelona
I	**Global Environmental Change**	

Recent Volumes in this Series

Volume 320 —Singular Limits of Dispersive Waves
edited by N. M. Ercolani, I. R. Gabitov, C. D. Levermore, and D. Serre

Volume 321 —Topics in Atomic and Nuclear Collisions
edited by B. Remaud, A. Calboreanu, and V. Zoran

Volume 322 —Techniques and Concepts of High-Energy Physics VII
edited by Thomas Ferbel

Volume 323 —Soft Order in Physical Systems
edited by Y. Rabin and R. Bruinsma

Volume 324 —On Three Levels: Micro-, Meso-, and Macro-Approaches in Physics
edited by Mark Fannes, Christian Maes, and André Verbeure

Volume 325 —Statistical Mechanics, Protein Structure, and Protein Substrate Interactions
edited by Sebastian Doniach

Volume 326 —Linking the Gaseous and Condensed Phases of Matter: The Behavior of Slow Electrons
edited by Loucas G. Christophorou, Eugen Illenberger, and Werner F. Schmidt

Volume 327— Laser Interactions with Atoms, Solids, and Plasmas
edited by Richard M. More

Series B: Physics

Linking the Gaseous and Condensed Phases of Matter

The Behavior of Slow Electrons

Edited by

Loucas G. Christophorou

Oak Ridge National Laboratory
Oak Ridge, Tennessee
and University of Tennessee
Knoxville, Tennessee

Eugen Illenberger

Freie Universität
Berlin, Germany

and

Werner F. Schmidt

Hahn–Meitner Institüt
Berlin, Germany

Springer Science+Business Media, LLC

Proceedings of a NATO Advanced Study Institute on
Linking the Gaseous and Condensed Phases of Matter: The Behavior of Slow Electrons,
held September 5–18, 1993,
in Patras, Greece

NATO-PCO-DATA BASE

The electronic index to the NATO ASI Series provides full bibliographical references (with keywords and/or abstracts) to more than 30,000 contributions from international scientists published in all sections of the NATO ASI Series. Access to the NATO-PCO-DATA BASE is possible in two ways:

—via online FILE 128 (NATO-PCO-DATA BASE) hosted by ESRIN, Via Galileo Galilei, I-00044 Frascati, Italy

—via CD-ROM "NATO Science and Technology Disk" with user-friendly retrieval software in English, French, and German (©WTV GmbH and DATAWARE Technologies, Inc. 1989). The CD-ROM also contains the AGARD Aerospace Database.

The CD-ROM can be ordered through any member of the Board of Publishers or through NATO-PCO, Overijse, Belgium.

Library of Congress Cataloging-in-Publication Data

Linking the gaseous and condensed phases of matter : the behavior of
  slow electrons / edited by Loucas G.Christophorou, Eugen
  Illenberger, and Werner F. Schmidt.
      p.   cm. -- (NATO ASI series. Series B, Physics ; v. 326)
    "Published in cooperation with NATO Scientific Affairs Division."
    "Proceedings of a NATO Advanced Study Institute on Linking the
  Gaseous and Condensed Phases of Matter: the Behavior of Slow
  Electrons, held September 5-18, 1993, Patras, Greece"--T.p. verso.
    Includes bibliographical references and index.
    ISBN 978-1-4613-6083-4   ISBN 978-1-4615-2540-0 (eBook)
    DOI 10.1007/978-1-4615-2540-0
    1. Electronic structure--Congresses. 2. Gases--Congresses.
  3. Condensed matter--Congresses. 4. Electrons--Congresses.
  I. Christophorou, L. G. II. Illenberger, E. (Eugen) III. Schmidt,
  Werner F. IV. North Atlantic Treaty Organization. Scientific
  Affairs Division. V. NATO Advanced Study Institute on Linking the
  Gaseous and Condensed Phases of Matter: the Behavior of Slow
  Electrons (1993 : Patrai, Greece) VI. Series.
  QC176.8.E4L56 1994
  530.4'74--dc20   94-30228
                                                            CIP

ISBN 978-1-4613-6083-4

©1994 Springer Science+Business Media New York
Originally published by Plenum Press in 1994
Softcover reprint of the hardcover 1st edition 1994

PREFACE

The Advanced Study Institute (ASI) on "Linking the Gaseous and Condensed Phases of Matter: The Behavior of Slow Electrons" was held at Patras, Greece, September 5-18, 1993.

The organizers of the Patras ASI felt that the study of the electronic properties of matter in various states of aggregation has advanced to a point where further progress required the interfacing of the phases of matter in order to find out and to understand how the microscopic and macroscopic properties of materials and processes change as we go from low pressure gas to the condensed phase. This approach is of foremost significance both from the point of view of basic research and of applications. Linking the electronic properties of the gaseous and condensed phases of matter is a fascinating new frontier of science embracing scientists not only from physics and chemistry but also from the life sciences and engineering.

The Patras ASI brought together some of the world's foremost experts who work in the field of electronic properties of molecular gases, clusters, liquids, and solids. The thirty five lectures given at the meeting as well as the twenty nine poster papers presented and the formal and informal discussions that took place focused largely on the behavior of slow electrons in matter. Slow electrons and their interactions with matter are exceedingly appropriate probes of the effects of density and structure of the medium on the basic physical and chemical reactions. A wide spectrum of elementary reactions involving directly or indirectly the generation, depletion, scattering or transport of slow electrons in molecular systems in all three states of aggregation including transition between different states were discussed.

The proceedings of the Patras ASI are contained in this volume. They are grouped into nine chapters as follows: Interactions of Slow Electrons as a Function of State; Ionization in Dilute and Condensed Matter; Elementary Processes Induced in Clusters by Electrons and Photons; Electron Motion in Gases and Liquids; Electron Attachment in the Gaseous and the Condensed Phases of Matter; Electron-Ion Recombination in Gases and Liquids; Electron Transfer at Interfaces; Applications; Summary of Discussion Panel. Collectively the material in this volume is an excellent representation of the activity in the field. We hope that it will be useful to the researcher and to the graduate student and that it will guide and accelerate further studies in this broad and interdisciplinary area.

The Patras ASI was a learning, engaging, lively and enjoyable experience. The presentations and the discussions were a real joy as were the beauty of Greece and the hospitality of its people. It can, perhaps, be said, also, that in effect the Patras Institute formalized the establishment of the new field of <u>Interphase Science</u> the aim of which is the unification of knowledge on all phases of matter.

We are grateful to the Scientific Affairs Division of NATO for providing the bulk of the Support for the Institute and to the Office of Health and Environmental Research of the Department of Energy for its generous contribution to the Institute. We also acknowledge with gratitude the contributions of the Hahn-Meitner Institute, the Oak Ridge National Laboratory, The Free University of Berlin, The University of Patras, and The University of Tennessee.

Finally, we thank the members of the International Organizing Committee (A. A. Christodoulides, T. D. Märk, M. N. Pisanias, L. Sanche, S. Süzer, and M. N. Varma), the Scientific Secretary Klaus Lacmann, the lecturers, the session chairmen, and the participants for their outstanding work and cooperation. Their collective contributions and their enthusiasm made this a profitable and indeed a memorable event for everyone.

Loucas G. Christophorou
Oak Ridge National Laboratory/University of Tennessee

Eugen Illenberger
Freie Universität Berlin

Werner F. Schmidt
Hahn-Meitner Institute

November, 1993

CONTENTS

SECTION I: INTERACTIONS OF SLOW ELECTRONS AS A FUNCTION OF STATE

Linking the Gaseous and the Condensed Phases of Matter:
The Slow Electron and Its Interactions
L. G. Christophorou 3

Comparisons Between Low-Energy Electron Scattering from
Gaseous and Condensed-Phase Atoms and Molecules
L. Sanche 31

Anion Formation in Low Energy Electron Impact to
Gaseous and Condensed Molecules
E. Illenberger 49

SECTION II: IONIZATION IN DILUTE AND IN CONDENSED MATTER

Ionization of Atoms or Molecules by Radiation as a Function of Phase
W. F. Schmidt 75

High Energy Ionization in Liquids - The Free Ion Yield
R. A. Holroyd 91

Photo- and Penning Ionization of Molecules in the Gas Phase
and in the Liquid Phase
H. Morgner 103

Positron and Positronium Annihilation in Gases and Liquids
A. G. Khrapak 121

Self Trapping of Light Particles in Fluids: The Path Integral Approach
B. N. Miller, J. Chen, T. Reese, and G. Worrell 141

SECTION III: ELEMENTARY PROCESSES INDUCED IN CLUSTERS BY ELECTRONS AND PHOTONS

Clusters: An Introduction
  E. Illenberger 151

Mechanisms and Kinetics of Electron Impact Ionization of Atoms,
  Molecules, and Clusters
  T. D. Märk 155

Photofragmentation as a Probe of Electron Thermalization in Size-Selected
  Cluster Anions
  D. J. Lavrich, P. J. Campagnola, and M. A. Johnson 183

Multiphoton Ionization Studies of Van der Waals Molecules and Clusters
  J. C. Miller 203

Core Level Excitation in Free Clusters: NEXAFS, EXAFS, and
  Coulomb Explosion
  J. Geiger, S. Rabe, C. Heinzel, H. Baumgärtel, and E. Rühl 217

Reaction in the NO_2-C_2H_4 Van der Waals Complex
  J. C. Loison, C. Dedonder-Lardeux, C. Jouvet, and D. Solgadi 223

Phase Transitions in Clusters: A Bridge to Condensed Matter
  R. S. Berry 231

Exploring Potential Surface Landscapes and How They Govern Dynamics
  R. S. Berry 251

SECTION IV: ELECTRON MOTION IN GASES AND LIQUIDS

Density and Field Dependence of Excess Electron Mobility in High-
  Density Noble Gases
  A. F. Borghesani and M. Santini 259

Excess Electron Localization in High-Density Neon Gas
  A. F. Borghesani and M. Santini 281

Boltzmann Equation for Slow Electron Transport in Gases
  and Liquids
  Y. Sakai 303

Electron Scattering in Dense Gases and Liquids and Related
  Phenomena
  I. T. Iakubov 319

Multiple Scattering of Electrons in Polar Gases-Evidence for Short
  Living Dipole-Bound Electron States in CH_3CN
  Th. Klahn, P. Krebs, and U. Lang 339

Thermodynamics of Electron Injection
  R. Schiller and R. A. Holroyd 347

**SECTION V: ELECTRON ATTACHMENT IN THE GASEOUS AND THE
CONDENSED PHASES OF MATTER**

Electron Attachment to Molecules
  E. Illenberger 355

Effects of the Solid Phase on Resonance Stabilization, Dissociative
  Attachment and Dipolar Dissociation
  L. Sanche 377

Photoinduced Dissociative Electron Capture Processes in Binary
  Ion-Molecule Complexes
  D. M. Cyr and M. A. Johnson 397

Electron Attachment to Excited Molecules
  L. G. Christophorou, L. A. Pinnaduwage, and P. G. Datskos 415

Electron Reactions in Nonpolar Liquids-Pressure Effects
  R. A. Holroyd 443

Thermodynamic Properties of the Electron
  M. Henchman 455

The Theory of Electron Attachment to Molecules
  J. N. Bardsley 461

SECTION VI: ELECTRON-ION RECOMBINATION IN GASES AND LIQUIDS

Electron-Ion Recombination in Dense Molecular Media
  Y. Hatano 467

FALP Studies of Electron-Ion Recombination and Electron
  Attachment
  D. Smith and P. Spanel 487

The Theory of Electron-Ion Recombination
  J. N. Bardsley 495

SECTION VII: ELECTRON TRANSFER AT INTERFACES

Low Energy Electrons for the Investigation of Liquid Surfaces
  H. Morgner 501

Photoelectron Spectroscopy at Liquid Water Surfaces
  M. Faubel and B. Steiner 517

Light-Induced Electron Emission from Surfaces of Organic Liquids
  K. Lacmann, H. Koizumi, and W. F. Schmidt 525

SECTION VIII: APPLICATIONS

Physics of Noble Gas X-Ray Detectors: A Monte Carlo
  Simulation Study
  T.H.V.T. Dias 543

A. Carcinogen-Screening Test Based on Electrons
  G. Bakale 561

SECTION IX: SUMMARY OF DISCUSSION PANEL

Summary of the Discussion Panel on Experimental Techniques
  E. Illenberger 569

Theory: Interactions of Electrons With Dense Media
  R. Schiller 573

The Behavior of Slow Electrons in Molecular Substances
  and Its Significance in Radiation and Life Sciences
  M. Inokuti 577

PARTICIPANTS 581

CONFERENCE PICTURE 589

SUBJECT INDEX 591

SECTION I. INTERACTIONS OF SLOW ELECTRONS AS A FUNCTION OF STATE

LINKING THE GASEOUS AND THE CONDENSED PHASES OF MATTER: THE SLOW ELECTRON AND ITS INTERACTIONS[1]

Loucas G. Christophorou

Atomic, Molecular, and High Voltage Physics Group, Health and Safety Research Division, Oak Ridge National Laboratory, Post Office Box 2008, Oak Ridge, Tennessee 37831-6122, and Department of Physics, The University of Tennessee, Knoxville, Tennessee, 37996

ABSTRACT

The interfacing of the gaseous and the condensed phases of matter as effected by interphase and cluster studies on the behavior of key reactions involving slow electrons either as reacting initial particles or as products of the reactions themselves is discussed. Emphasis is placed on the measurement of both the cross sections and the energetics involved, although most of the available information to date is on the latter. The discussion is selectively focussed on electron scattering (especially the role of negative ion states in gases, clusters, and dense matter), ionization, electron attachment and photodetachment. The dominant role of the electric polarization of the medium is emphasized.

INTRODUCTION

Interphase and Cluster Studies

The interfacing of the gaseous and the condensed phases of matter requires multidisciplinary and systematic investigations as to how the microscopic and the macroscopic properties of materials and the elementary processes involving electrons, photons, ions, and neutral particles change as one makes the transition from a low pressure gas (isolated-particle behavior) to the condensed phase. There have been two complementary approaches in this endeavor: (i) interphase physics/chemistry and (ii) clusters. In the former approach, a given reaction (or property) is studied as a function of the density and the nature of the medium in which it occurs from the low-pressure gas to the liquid or the solid. Actually, traditionally such studies begin at either end of the density range: from the liquid (solid) density to progressively lower densities and from a low density gas (binary collisions)

[1]Research sponsored by the Office of Health and Environmental Reserach, U.S. Department of Energy, under Contract DE-AC05-84OR21400 with Martin Marietta Energy Systems, Inc., and by the Office of Naval Research under Contract N00014-89-J-1990 with the University of Tennessee, Knoxville, Tennessee, 37996.

to progressively higher densities (multiple scattering regime), and to the condensed phase. Both bridge the density gap between low pressure gases and condensed matter (e.g., see Refs. 1-5 and references therein). In the latter approach, the properties and reactions of a given species (atom or molecule) are studied as a function of its size (increased gradually by clustering), cluster[2] shape and cluster composition. A unique feature of clusters is that they allow studies of the transition from large finite clusters to the bulk and thus determination of the minimum cluster size beyond which the cluster properties no longer vary with size but are essentially similar to those of a macroscopic sample of the material (e.g., see Ref. 6 (and references cited therein) and Refs. 7-12).

Why Slow Electrons?

Slow electrons are abundant and reactive species in all phases of matter[13]. They are generated in gaseous and condensed matter by a multiplicity of mechanisms: energy transfer from high-energy particles to atoms and molecules; absorption of light by neutrals or negative ions; collisions of excited and unexcited atoms, negative ions with neutrals, electrons with neutrals or ions; injection from surfaces. They lose their energy and slow-down in matter in elastic and a multiplicity of inelastic collisions[14]. They interact before they are thermalized (i.e., during their slowing down) and also after they have reached thermal equilibrium or steady-state conditions (when an applied electric field E is superimposed on the medium). This distinction is significant since the interactions of slow electrons in a dense medium depend on the "state" of the electron itself.

For a low-pressure gas, the electrons are normally free and the collision (interaction) mean free path l is much longer than the electron de-Broglie wavelength. In a dense medium (high pressure gas, liquid or solid) l is smaller than the de-Broglie wavelength and as the medium density increases the electrons become localized or delocalized into conduction bands. In the former case, their mobilities are low and their kinetic energies are thermal and in the latter case their mobilities are high and their kinetic energies (under an applied electric field) can exceed considerably thermal energies[3].

The reactions of slow electrons in dense matter often differ greatly from those in a low-pressure gas. They are unique in that they help us unravel the structure of atoms and molecules, probe the structural and dynamical changes with the density and the nature of matter, and quantify the energetics and dynamics of basic reactions in matter.

This Lecture

In this lecture we discuss the behavior of key reactions--as studied by interphase and cluster researches--involving slow electrons either as reacting initial particles or as products of the interactions themselves. We emphasize measurement of both cross sections and energetics, although most of the available information to date is on the latter. We selectively focus on electron scattering (especially the role of negative ion states in gases, clusters, and dense matter), ionization, electron attachment and photodetachment and emphasize the dominant role of the electric polarization of the medium on the reaction energetics. The comparison between the gaseous and the liquid phase measurements is restricted to dielectric

---

[2]Normally the term cluster is used to describe finite aggregates of 2 to 10^4 particles (atoms or molecules).

liquids with conduction bands[3] where the excess electrons are quasi-free--not localized as, say, in polar media--and the connection between the electron behavior in the two phases is more apparent. Finally, the general nature of this lecture unavoidably touches on aspects of the theme of this meeting that will be covered by subsequent lecturers. We hope that it will enhance the value of and the anticipation for these upcoming and in-depth lectures.

DIRECT AND INDIRECT ELECTRON COLLISIONS

Slow electrons lose their energy and slow down in matter in <u>elastic</u> and (a multiplicity of) <u>inelastic</u> collisions. Such collisions are either <u>direct</u> or <u>indirect</u> (Fig. 1). In a direct - glancing - collision the electron is scattered at a distance from the target, the duration of the collision is short, and the cross section for the collision--whether elastic or inelastic--is appreciable over a broad range of incident electron energies. In contrast, in indirect collisions the electron is temporarily captured by the target forming a transient anion whose

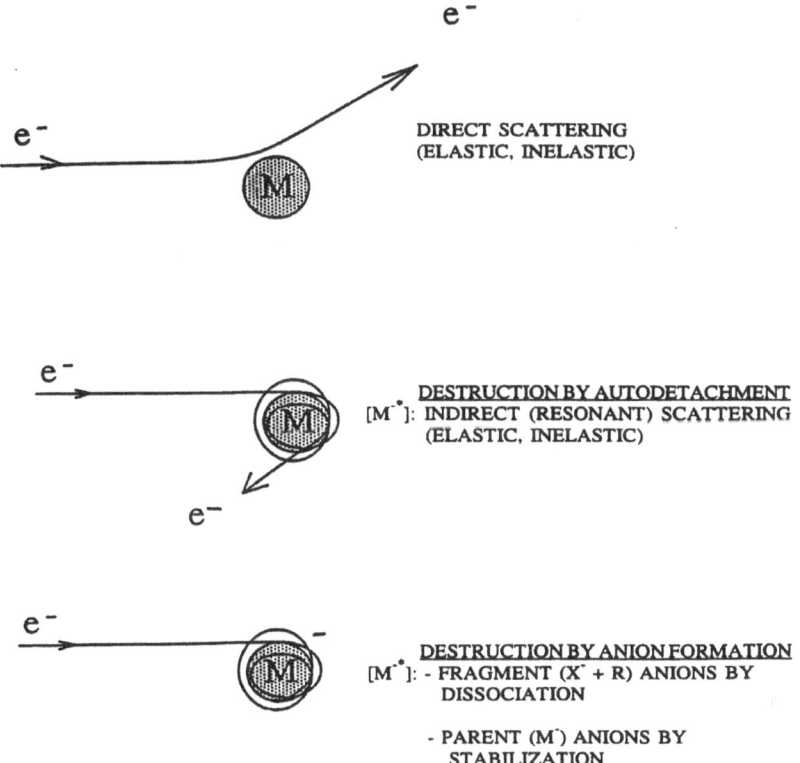

Figure 1. Schematic pictures depicting direct and indirect electron-molecule collisions; M^{-*} indicates a transient negative ion with excess energy denoted by the asterisk.

---

[3]These are dense media whose $V_o < 0$ eV; V_o is the energy of the excess electron at the bottom of the conduction band. It is defined as $V_o = W_M - W_{vac}$ where W_M and W_{vac} are, respectively, the work functions of a metal immersed in the dense medium and in vacuum.

lifetime can vary from $\sim 10^{-15}$ to $> 10^{-3}$ s (Refs. 13-15). Such collisions are resonant, i.e., they occur over a limited energy range--where empty orbitals exist for electrons to enter into and be temporarily retained. Subsequent to its formation the transient anion M^{-*} is destructed by autodetachment (i.e., __indirect__ elastic or inelastic electron scattering) or by "permanent" negative ion formation: fragment anions by dissociation and parent anions by stabilization. These processes are of general occurrence in nature. An example of an indirect collision leading only to elastic and inelastic scattering is shown in Fig. 2 for N_2 (no permanent N_2^- or N^- ions are formed since the electron affinities of N_2 and N are negative). The

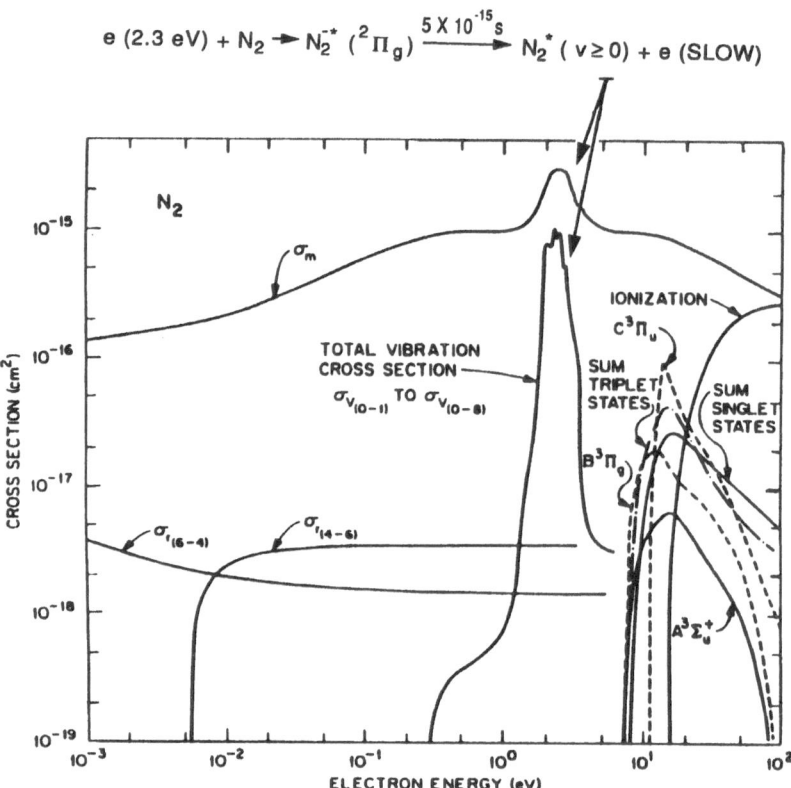

$$e\ (2.3\ eV) + N_2 \rightarrow N_2^{-*}\ (^2\Pi_g) \xrightarrow{5 \times 10^{-15}\,s} N_2^*\ (v \geq 0) + e\ (SLOW)$$

Figure 2. Cross sections for various electron scattering processes in N_2 as a function of electron energy (see the text) (based on Fig. 51 of Ref. 14, Vol. 2, p. 194).

pronounced peak in the momentum transfer cross section σ_m and in the total vibrational cross section at ~ 2.3 eV--pointed to by the arrows in Fig. 2--is due, respectively, to the elastic and the inelastic scattering of electrons via the decay of the transient anion N_2^{-*} formed by the temporary capture by N_2 molecules of ~ 2.3 eV electrons. Examples of the formation of "permanent" fragment negative ions via indirect (resonant) collisions are shown in Fig. 3. Clearly, these negative ion resonances occur abundantly in the energy range below ~ 20 eV and their cross sections increase as their energy positions are lowered.

Indirect collisions forming transient anions occur in condensed matter as well, but as we shall see their "isolated species" properties, energetics, and effects on other processes involving slow electrons are modified in dense matter by the nature and the density of the medium in which they occur. We shall try to understand some of these changes by considering knowledge from selected investigations on condensed phase and clusters.

ELECTRON SCATTERING

In Table 1 are listed examples of the effects of phase on various electron energy loss

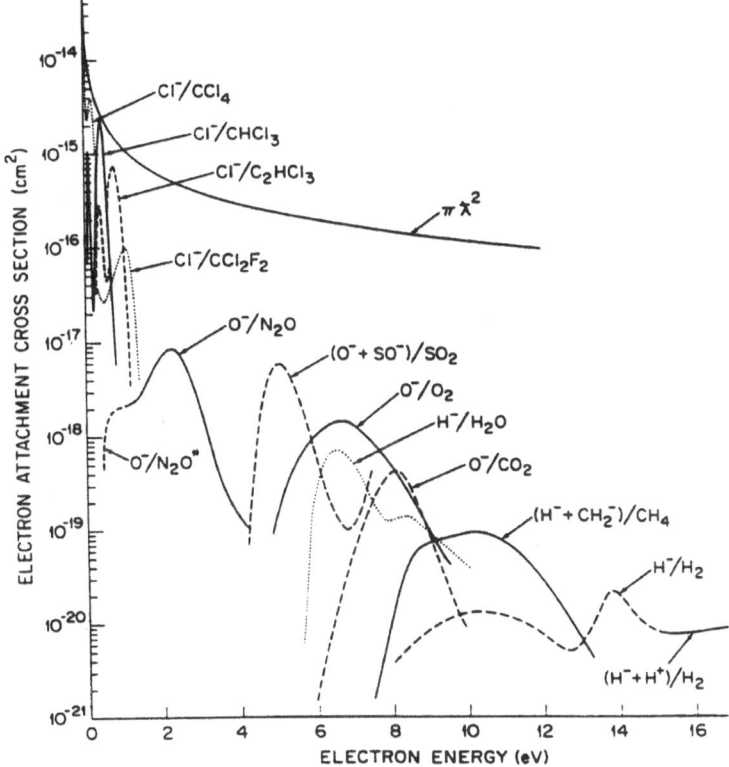

Figure 3. Dissociative electron attachment cross sections as a function of electron energy for a number of molecules (from Ref. 14, Vol. 1, p. 559).

processes and cross sections (see also Refs. 16,17). Let us illustrate some of these by examining the effect of phase on the differential oscillator strength distribution $f(\varepsilon)$ of water shown in Fig. 4. Clearly one sees a general loss of structure and a shift of $f(\varepsilon)$ to higher energies in going from the vapor to the condensed phase. Interestingly, the main difference is between the vapor and the condensed matter and not between the various forms (liquid, hexagonal ice, amorphous ice) of the latter. These phase effects seem to diminish with excitation energy. Important as this information is, it is still largely indirect[18,19] and limited. It should be noted in this connection that total <u>inelastic</u> cross sections estimated[20] for slow (1-20 eV) electron scattering in amorphous ice compare in magnitude (they are, actually,

Table 1. Examples of Condensed Phase Effects on the Oscillator Strength Spectrum

Shift of oscillator strength to higher energies (occurs over wide ranges of excitation energy)

Excitation of special modes of motion (plasmons in metals and excitons in molecular and ionic crystals; occur at specific energies)

Intermolecular vibrations (vibrational and translational phonons)

High Rydberg states (normally absent)

Negative Ion Resonances (exist as in single molecules but perturbed by the medium, e.g., their symmetries, lifetimes, selection rules, energetics)

Ejected electrons undergo further interactions

Scattering affected by correlation (medium structure) effects, nuclear/electronic excitation mode coupling, screening of interaction potential

Dissociation processes influenced by fast radiationless transitions and changes in energetics

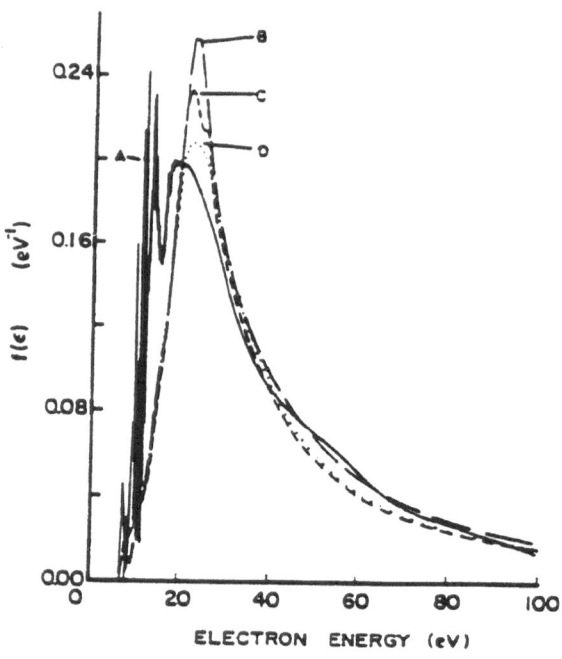

Figure 4. Differential oscillator strength distribution, $f(\varepsilon)$, of water vapor (A), liquid water (B), hexagonal ice (C), and amorphous ice (D) (Refs. 18 and 19).

slightly larger) and general trend with gas phase data (see discussion in Ref. 20); this conclusion, however, requres further scrutiny.

Let us now look at a system, namely xenon, for which scattering cross sections for slow electrons are well-known in the gas[13,14,21] and electron drift velocities as a function of the density, N, reduced electric field E, E/N, have been measured in the low pressure gas[13,14,21], the high pressure gas[2,22], the liquid[2,23], and the solid[2,23]. From the liquid phase electron transport data Sakai et al.[24] deduced two types of scattering cross sections, namely, a cross section for <u>elastic energy loss</u> (energy transfer to acoustic phonons which is independent of the liquid structure)

$$\sigma_o(\varepsilon) = 2\pi \int_0^\pi (1-\cos\,\theta)\,\sigma(\varepsilon,\theta)\,\sin\,\theta\,d\theta, \tag{1}$$

and a cross section for <u>elastic momentum transfer</u> which depends on the liquid structure through the structure factor $S(K)$

$$\sigma_1(\varepsilon) = 2\pi \int_0^\pi (1-\cos\,\theta)\,\sigma(\varepsilon,\theta)\,S(K)\,\sin\,\theta\,d\theta \tag{2}$$

where $\sigma(\varepsilon,\theta)$ is the differential scattering cross section, θ is the scattering angle and ε is the electron energy. These are compared with the low-pressure gas momentum transfer cross section σ_m in Fig. 5. Both $\sigma_o(\varepsilon)$ and $\sigma_1(\varepsilon)$ exhibit a displaced and shallower (relative to the low pressure gas) Ramsauer-Townsend minimum and are much lower in magnitude than $\sigma_m(\varepsilon)$ especially at low ε. As ε increases, $S(K) \rightarrow 1$ and the gas and liquid cross sections converge. A quantitative determination of the scattering cross section in the liquid is still lacking although the lowering of its magnitude in the condensed phase clearly is due to the screening of the scattering potential by the medium.

The indicated changes in the energy dependence and the magnitude of the scattering cross section in going from the low-pressure gas to the liquid explain nicely the much higher drift velocities, w, of slow electrons in gaseous compared to <u>liquid</u> xenon (Fig. 6). At a fixed E/N the w(E/N) of slow electrons in <u>solid</u> xenon are about twice those in <u>liquid</u> xenon[22]. While this is consistent with the lower values of the total electron scattering cross section in solid xenon (σ_s) deduced by Bader et al.[25] (Fig. 7), the behavior of σ_s (Fig. 7) is different at high energy from that of σ_1 or σ_o (Fig. 5) in the liquid, probably indicating different scattering processes.

A behavior similar to that in xenon has been reported[26] for the w(E/N) of room temperature liquids with conduction bands [eg. see, Fig. 8][4].

---

[4]See Stephens[27] for a possible relationship between the mobilities of thermal electrons in different pentane isomers and their respective structure functions.

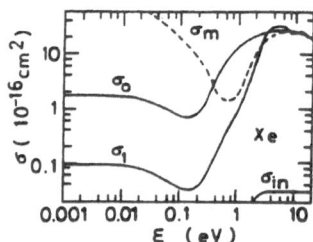

Figure 5. Cross sections $\sigma_o(\varepsilon)$ and $\sigma_1(\varepsilon)$ for liquid xenon and $\sigma_m(\varepsilon)$ for gaseous xenon (see the text and Refs. 3 and 24).

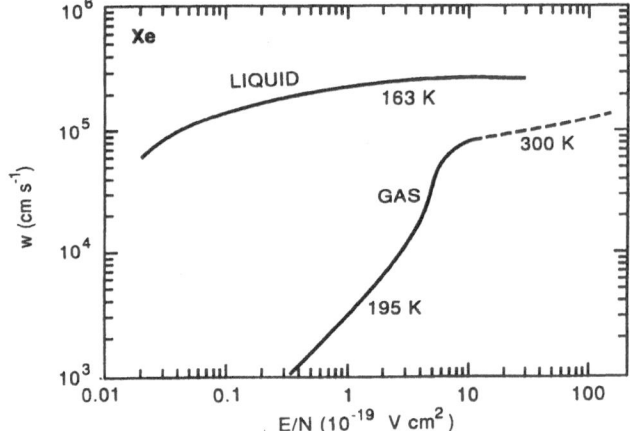

Figure 6. Electron drift velocity w as a function of E/N for liquid[23] and gaseous[21] Xe at the indicated temperatures.

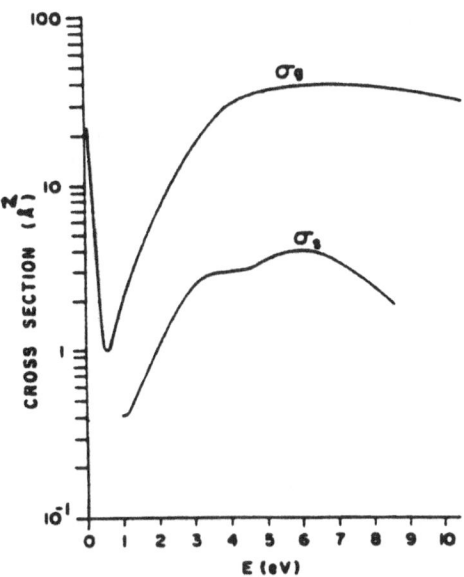

Figure 7. Comparison of the total electron scattering cross section for gaseous, σ_g, Xe and for solid, σ_s, Xe (17 K deposition on Pt.) plotted as a function of the electron energy (note that the zero energy in the solid is shifted by the V_o value of the solid) (from Ref. 25).

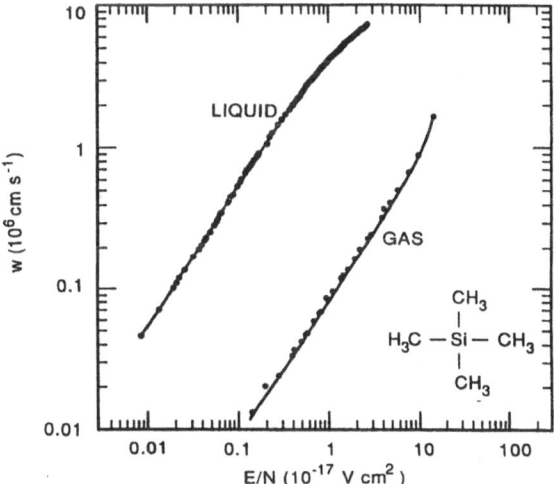

Figure 8. w vs E/N for gaseous and liquid tetramethylsilane (T ≈ 300 K) (based on Ref. 26).

The cross sections in Figs. 5 and 7 are due to direct electron scattering. In condensed matter--as in low pressure gases--indirect scattering of slow electrons occurs abundantly. It manifests itself through a strong enhancement in the scattering cross section over limited energy regions where the negative ion states (NISs) are located. A prototypical example is shown in Fig. 9 for solid films of N_2. The cross section function for excitation of the N_2 molecule in solid N_2 films in the first (v = 1) vibrational level via its lowest, N_2^{-*} ($^2\Pi_g$), negative ion state has--as in the isolated-molecule case--distinct structure which originates from the vibrational levels of the transient N_2^{-*} ($^2\Pi_g$) state. The position of the first peak in Fig. 9 (first broken vertical line) lies ~ 0.7 eV lower than the corresponding position in the low pressure gas (first solid vertical line in Fig. 9). This lowering of the energy position of the N_2^{-*} ($^2\Pi_g$) state is a general characteristic of the changes in the energetics of the NISs of atoms and molecules embedded in dense matter and is due to the polarization of the medium which surrounds the temporarily localized electron on the molecule. The cross section functions for excitation of N_2 in the v = 2, 3, etc. levels exhibit similar behavior[16]. Interestingly, however, while in the solid--as in the isolated molecule--the vibrational excitation increased by ~ 2 orders of magnitude in the region of the resonance, Sanche et al.[16] observed no such enhancement in the elastic scattering channel in the N_2 solid film. This is in sharp contrast to the gas where indirect <u>elastic</u> electron scattering is profoundly enhanced. The <u>formation</u> of the NIS in the condensed phase--as in gases--is localized in space and is dominated by the short-range forces. For this reason in the solid--as in the gas-- the formation of a NIS leads to <u>intra</u>molecular changes that enhance excitation of individual vibrational levels of the molecular constituents. The <u>decay</u> of the NIS in the condensed phase, however, should differ from that in the gas because it is affected by the medium due to polarization. The <u>inter</u>molecular changes lead to continuous accoustic excitations (phonons) which in essence replace the (low pressure) elastic scattering. This process is very probable due to the large number of such low frequency vibrations (see Refs. 2,16,17,28, and 29). It, thus, seems that the formation of negative ion states in dense matter enhances <u>intra</u>molecular vibrational excitation and <u>inter</u>molecular excitation of phonon modes of the lattice. A theoretical understanding of slow electron interactions in condensed matter in a manner analogous to gases, is, however, still lacking.

The electrons that have been slowed down via the NISs cause (see later) profound changes in the observed electron attachment properties of molecules in dense matter and in clusters.

IONIZATION IN DENSE MATTER AND IN CLUSTERS

Electron Impact Ionization

Inspite of its basic and applied significance, there appears to be no <u>direct</u> measurement of the electron impact ionization cross section as a function of the electron energy $\sigma_{ei}(\varepsilon)$ for any system. There is, however, some limited information on the effective ionization cross section of H_2 in H_2 clusters as a function of the cluster size (Fig. 10) which indicates that the $\sigma_{ei}(\varepsilon)$ for closely packed and interacting H_2 molecules is smaller than that for the isolated H_2 molecules[30]. There exists, also, some indirect information on media with $V_o < 0$ eV such as xenon. For this medium the electron impact ionization coefficient $\alpha(E/N)$ does not scale with the density N in going from the gas to the liquid[31]. In Table 2 are compared: the total electron yield G_{te} (e,γ,x) for electrons, γ-rays and x-rays; the energy to produce an electron-ion pair for β particles, W_β; and the ionization threshold energy for the low pressure gas, I_G, and the liquid, I_L. The lowering of W_β (increase in G_{te}) for the liquid can be attributed to the lowering of I in the liquid compared to the gas (see below). Due to the lower I_L, electronic states that would have led to inelastic scattering in the gas lead to ionization in the liquid.

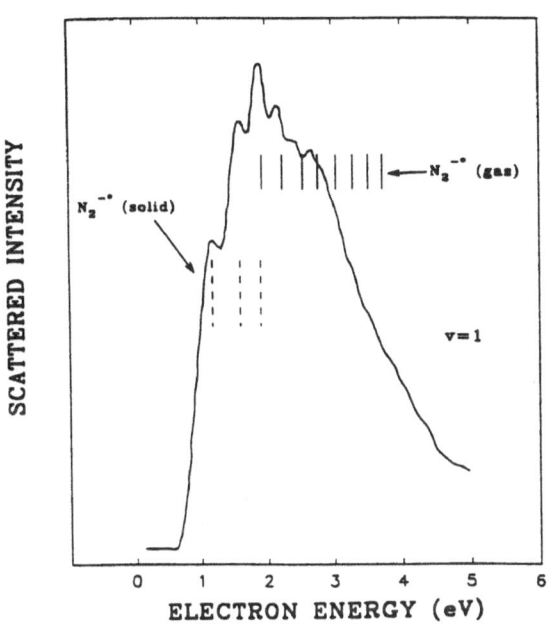

Figure 9. Relative cross section for indirect scattering of slow electrons via the N_2^{-*} ($^2\Pi_g$) negative ion state of N_2 in solid N_2 film leaving N_2 excited in the $v = 1$ vibrational level. Note that the position of the first peak in the cross section function for the solid lies ~ 0.7 eV lower than for the corresponding gas (see the text) (based on Fig. 13b in Ref. 16).

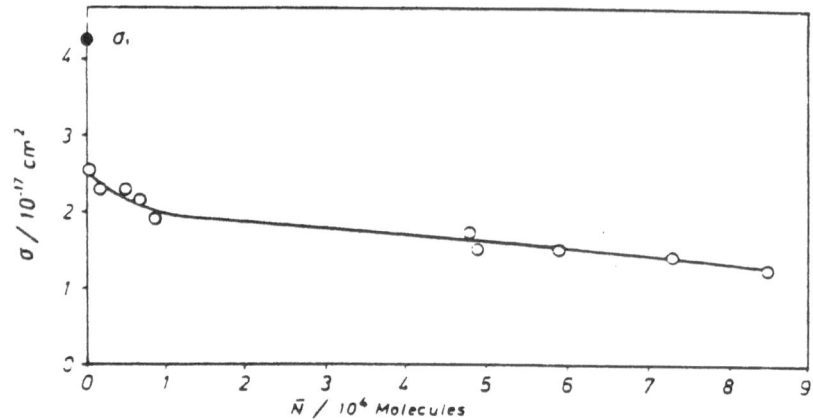

Figure 10. Effective electron impact ionization cross section of cluster beams of H_2 as a function of the mean cluster size N for 500 eV. The solid point shows the value, σ_1, of the cross section for uncondensed H_2 at the same electron energy[30].

Table 2. Comparison of G_{te} (e,γ,x), W_β and I for gaseous and liquid xenon[1]

Quantity	Liquid	Gas
$G_{te}(e,γ,x)$	6.4	4.6 (electrons/100 eV)
W_β	~16	21.9 (eV/i.p.)
I	~9.0	12.13 eV

[1]See the text and Ref. 1.

Single and Multiphoton Ionization of Molecules in Fluids

Here again absolute cross section measurements are scarced. However, many studies on pure liquids and solutions quantified the effect of the medium on the energetics of the reaction

$$M_F + nh\nu \; (n \geq 1) \; \rightarrow \; M_F^+ + e_F \tag{3}$$

where F stands for the fluid medium in which the molecule M is embedded and the reaction takes place. If we neglect the small broadening of the valence electron levels in the fluid due to condensation, for liquids [F = L) the measurements showed[1,2,32,33] that

$$I_L = I_G + V_o + P^+ \qquad (4)$$

In Eq. (4), P^+ is the polarization energy of the positive ion M_F^+ and is normally represented by the Born expression

$$P^+ = -\frac{e^2}{2R}[1 - \frac{1}{\varepsilon_{opt}}] \qquad (5)$$

where R is the effective radius of the positive ion cavity and ε_{opt} is the optical dielectric constant of the medium. For a number of organic molecules in nonpolar liquids[3], $I_L - I_G$ (= $V_o + P^+$) $\approx$ -1 to -3 eV and $<R> \sim$ 2.7 Å. A major uncertainty in the determination of P^+ remains the lack of accurate knowledge of R.

The validity of (4) can indeed be seen from measurements (see Ref. 34, Table 3) of I_L of a certain molecule (azulene) in a number of dielectric liquids with different V_o values. Since in this case we are dealing with the same positive ion, P^+ may be regarded as constant for these nonpolar liquids and, hence, the I_L should correlate with the V_o of the liquids. Indeed, the linear dependence of I_L on V_o implied under such conditions by Eq. (4) is exhibited by the measurements in Table 3 (see Fig. 11).

Table 3: I_L and I_G for Azulene in Dielectric Liquids With Various V_o Values[34]

Liquid	$I_L(eV)$	$I_G(eV)$	$V_o(eV)$
Tetramethyltin (TMT)[1]	5.33	7.42	-0.75
Tetramethylsilane (TMS)[1]	5.45		-0.55
2,2,4,4-Tetramethylpentane (TMP)[1]	5.70		-0.36
n-Pentane (n-Pt)[1]	6.12		+0.01
n-Tridecane	6.28		+0.21

[1]Abbreviations used in Fig. 11.

Similarly, for a fluid of number density N the ionization threshold, $I_F(N)$, of a molecule (atom) embedded in the fluid has been found (e.g., Refs. 35,36) to be related to its isolated-molecule value, I_G, by

$$I_F(N) = I_G + P^+(N) + V_o(N) \qquad (6)$$

i.e., by the N-dependence of P^+ and V_o; the N-dependence of P^+ is dominated by the N-dependence of ε_{opt} while the N-dependence of V_o is more complicated. The $V_o(N)$ has been calculated in a few cases using the Springett, Jortner, and Cohen (SJC) model[35-37]. In Fig. 12 is plotted the I_F of the molecule TMPD (N,N,N',N'-tetramethyl-p-phenylenediamine) in ethane as a function of the ethane density $\rho(M/l)$ from the dilute gas to the liquid. The effect of increasing medium density on I_F is a continuous gradual decrease and a smooth transition to the liquid. The solid lines represent the dependence of I_F on N for the indicated values of the "hard core" radius $<\bar{a}>$ as predicted by the SJC model.

From the experimental investigations to date it appears that the energetics of ionization in dense matter are modified from their values in dilute gases by the charge induced medium polarization energies. The energetics of the process for nonpolar fluids is reasonably well understood and needs to the extended to polar media.

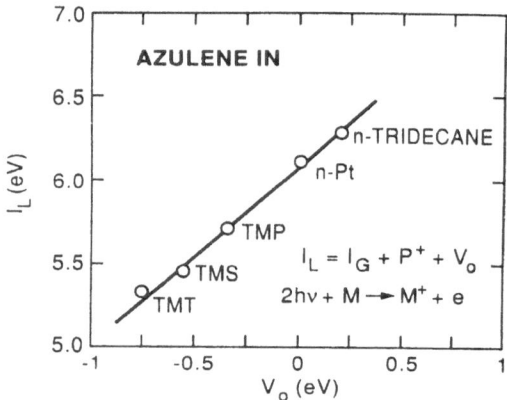

Figure 11. I_L vs V_o for the azulene molecule, M, ionized by two photon absorption ($2\,h\upsilon + M_L \rightarrow M_L^+ + e_L$) in a number of nonpolar liquids (Ref. 34; see Table 3).

Photoionization Energetics in Clusters

Many studies have been devoted recently to this topic and subsequent lectures will deal with the subject in more detail. Here we merely lay down a few aspects of the energetics of the process as a background for the other lectures and, also, to integrate our discussion on the kind of knowledge obtained by the two complementary approaches we mentioned in the Introduction. Let us first refer to Fig. 13 which shows schematically[6] the dependence on the cluster size of a cluster property $x(n)$; n is the number of the cluster constitutents. For small n, $x(n)$ exhibits specific cluster size effects (e.g., shell closure effects). For large n, Jorntner[6] expressed $x(n)$ as

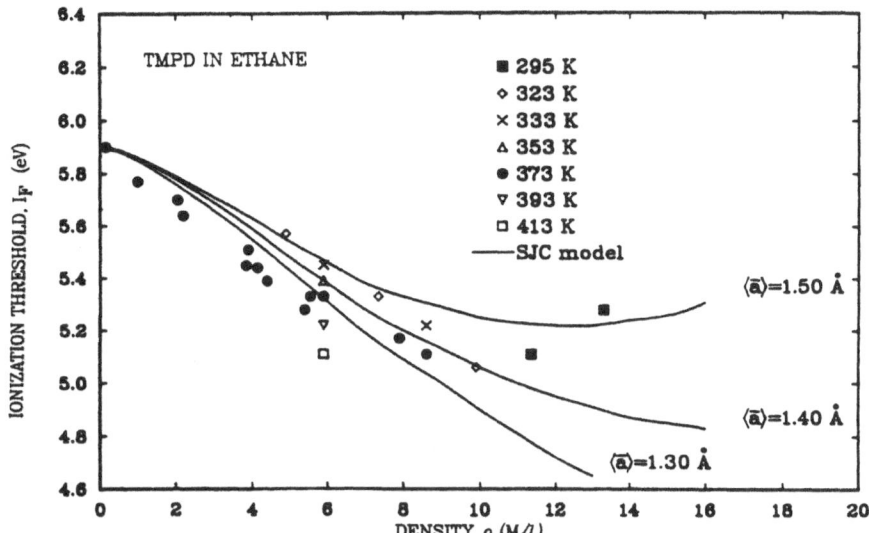

Figure 12. Ionization threshold of TMPD (N,N,N'N'-tetramethyl-p-phenylenediamine) in ethane as a function of the ethane density from a dilute gas to the liquid at the indicated temperatures. The data points ■ are measurements made in liquid ethane at 295 K. The solid lines represent the dependence of I_F on N for the indicated values of the "hard core" radius <ā> as predicted by the Springett-Jortner-Cohen (SJC) model[35,37].

$$x(n) = \chi(\infty) + An^{-\beta} \tag{7}$$

namely, in terms of the corresponding bulk value $x(\infty)$ and a correction term $An^{-\beta}$ ($0 \le \beta \le 1$) which accounts for the modification of the bulk value in the cluster due to the "excluded volume" outside of it. Expressing Eq. (7) in terms of the cluster radius R_c (taken equal to $R_o n^{1/3}$ where R_o is the effective single particle radius determined from the molecular (atomic) volume using the bulk medium density) we have

$$\chi(R_c) = \chi(\infty) + (A\ R_o^{3\beta})\ R_c^{-3\beta} \tag{8}$$

This "cluster-size equation" attempts to provide a quantitative description of the "smooth" cluster size dependence of $x(n)$ for large clusters and it interpolates to the corresponding bulk value $x(\infty)$ in much the same way as Eqs. (4) and (6) we have discussed earlier.

Let us apply (7) and (8) to the case of the ionization potential $I(n)$ of clusters. We can write

$$I(n) = I_A + P^+(R_c) = I_A + P^+(\infty) + C(R_c) = I(\infty) + C(R_c) \tag{9}$$

where I_A is the ionization potential of the isolated species A, $P^+(\infty)$ is the bulk polarization energy and $C(R_c)$ is the correction for the finite cluster size (i.e., the long-range electrostatic

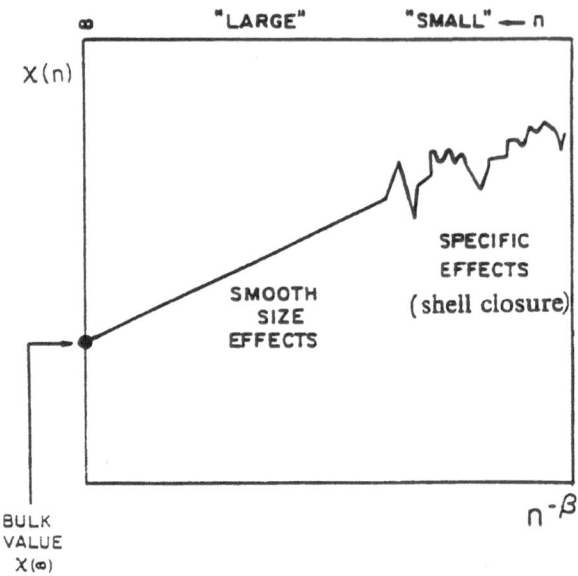

Figure 13. Schematic representation of the cluster size dependence of a cluster property $x(n)$ on the number, n, of the cluster constituents (from Ref. 6; see the text).

interaction of the generated charge with the "excluded region"). If we use for $C(R_c)$ the classical electrostatic expression which gives the energy required to remove an electron from a uniformly conducting sphere having the same dimensions as the cluster, namely,

$$C(R_c) = (e^2/2R_c)(1 - 1/\varepsilon_{opt}) = (e^2/2R_o)(1 - 1/\varepsilon_{opt})n^{-1/3} \tag{10}$$

we have

$$I(n) = I(\infty) + (e^2/2R_o)(1 - 1/\varepsilon_{opt})n^{-1/3} \tag{11}$$

In Eq. (11), $I(\infty) = I_A + P^+(\infty)$ is the bulk photoelectric threshold for molecular clusters, the top of the bulk valence energy for semiconductor clusters, or the bulk metal work function for metallic clusters.

The validity of Eq. (11) is illustrated in Fig. 14 for the rare gas and mercury clusters[8]. The results of Rademann et al.[8] for mercury clusters show in addition the gradual transition from van der Waals-type to metallic properties; on the basis of their data in Fig. 14 this transition occurs in the cluster size range of ~ 70 atoms. In cluster energetics as in interphase studies energetics a major uncertainty in applying relations such as Eqs. (4), (5), and (11) is the proper knowledge of R.

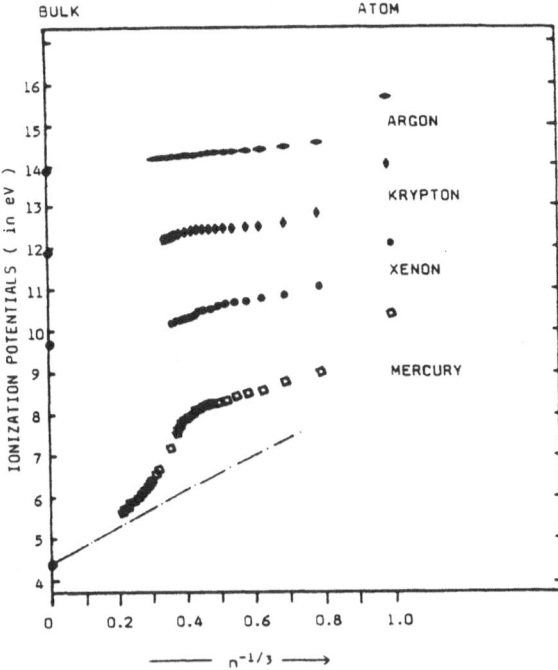

Figure 14. Ionization potential (in eV) vs $n^{-1/3}$ for Ar, Kr, Xe, and Hg clusters (from Ref. 8) (see the text).

ELECTRON ATTACHMENT

Subsequent lectures will--most assuredly--discuss the effect of the state and the nature of the medium on the attachment of slow electrons to molecules. We, also, elaborated on interphase studies of electron attachment and the relevance of electron attachment processes in gases to those in the condensed phase earlier (e.g., Refs. 1-3). We shall, thus, be brief. In Fig. 15, the rate constant for the dissociative electron attachment reaction

$$e + N_2O \rightarrow N_2O^{-*} \rightarrow O^- + N_2 \tag{12}$$

as measured in gaseous argon, $(k_a)_G$, and in liquid argon, $(k_a)_L$, is plotted as a function of respective values of the mean electron energy, $<\varepsilon>_G$ in gaseous argon and $<\varepsilon>_L$ in liquid argon. In spite of the large uncertainty in $<\varepsilon>_L$, the comparison of $(k_a)_G$ $(<\varepsilon>_G)$ with $(k_a)_L$ $(<\varepsilon>_L)$ in Fig. 15 shows three rather distinct features: (i) the dissociative attachment resonant reaction (12) occurs in liquid argon as it occurs in the argon buffer gas, (ii) the position of the dissociative attachment resonance shifts to lower energy as a result of the polarization of the medium around the transient N_2O^{-*}, and (iii) the rate constant for the reaction is higher in the liquid than in the gas largely because of the lowering of the energy of the negative ion state.

Studies of electron attachment to molecules in clusters and in solid films generally showed similar changes: the perturbation from the isolated-molecule case of the formation of the transient anion and its decay channels by the medium. Thus, as is shown in Fig. 16a, the negative ion resonance that produces O^- from O_2 and peaks at ~ 6.7 eV in the gas exists in solid films of mixtures of O_2 in Ar, but it is shifted to lower energies by ~ 0.7 eV. The yield of O^- from O_2 from the solid O_2 film shows (Fig. 16a)--in contrast to the isolated O_2 case--enhanced O^- production at ~ 8 and ~ 14 eV. These enhancements ("peaks") in the O^- yield have been attributed by Sanche and collaborators (e.g., see Ref. 16) to the transitions

$$O_2(^3\Sigma_g^-) + e \rightarrow O_2^{-*}\,(^2\Sigma_g^+) \rightarrow O^- + O \tag{13}$$

and

$$O_2(^3\Sigma_g^-) + e \rightarrow O_2^{-*}\,(^2\Sigma_u^+) \rightarrow O^- + O \tag{14}$$

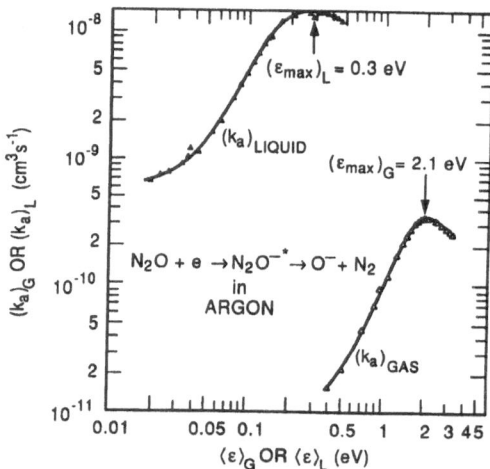

Figure 15. Electron attachment rate constant for N_2O in gaseous $((k_a)_G)$ and liquid $((k_a)_L)$ argon plotted as a function of the respective, $<\varepsilon>_G$ and $<\varepsilon>_L$, mean electron energies (the data for liquid argon are from Ref. 38).

which are symmetry forbidden in the isolated molecule case, but become allowed in the solid because of the perturbation by the medium. Observation of O^- via dissociative electron attachment to O_2 in O_2 clusters as a function of cluster size showed similar evidence. This is indicated by the enhancement in the O^- yield around 8 and 14 eV in Fig. 16c with increasing stagnation pressure from 1 to 3.5 bar (the cluster size distribution increases with increasing stagnation pressure).

From these rather limited studies it can be inferred that in going from the gaseous to the condensed phase additional negative ion states contribute to indirect energy loss and dissociative electron attachment processes.

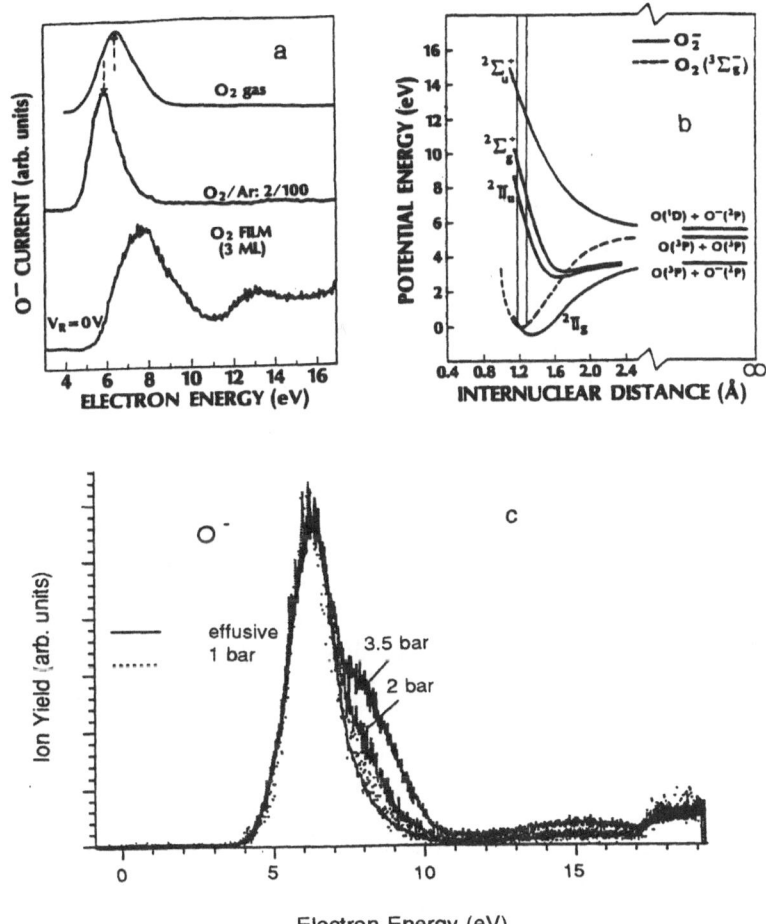

Figure 16. a. O⁻ from O_2 in O_2 gas (upper curve) and in solid films [O_2/Ar: 2/100; pure O_2)] (lower curves).

b. Schematic potential energy curves for $O_2(^3\Sigma_g^-)$ (---) and O_2^- ($^2\Pi_g$, $^2\Pi_u$, $^2\Sigma_g^+$, $^2\Sigma_u^+$) (——).

c. Yield of O⁻ from O_2 clusters at 1, 2, and 3.5 bar stagnation pressure normalized at the lowest maximum to the yield of O⁻ from O_2 at low pressure (effusive ——). (a and b from Ref. 16; c from 39).

21

Besides these changes, the proximity of collision partners within the condensed medium or the cluster will enhance the production of parent negative ions (e.g., O_2^-). Furhermore, the negative ion properties of single molecules or clusters with strong "zero" or "near-zero" nondissociative electron attachment resonances are strongly influenced by inelastic electron scattering by the cluster constituents. This is beautifully illustrated in Fig. 17 where the production of $(O_2)_2^-$ in O_2/N_2 mixed clusters is strongly influenced by the inelastic electron scattering by N_2, especially by the inelastic electron scattering via the 2.3 eV negative ion resonance N_2^{-*} ($^2\Pi_g$). The yield of this ion exhibits a strong resonance at ~ 2.3 eV in the case of the O_2/N_2 clusters (Fig. 17b) which is absent in the homogeneous clusters of O_2 (Fig. 17a). The profound effect of inelastic electron scattering (especially via NIRs) by the cluster constitutents on the intensity and energy dependence of the anions formed in clusters by the capture of "near-zero" energy electrons indicates that such processes can be employed to probe electronic and negative ion states of molecules in clusters.

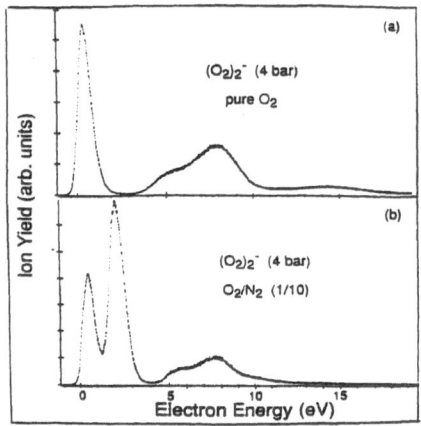

Figure 17. Yield of $(O_2)_2^-$ from (a) pure O_2 clusters and (b) O_2/N_2 clusters (O_2/N_2 gas mixture ratio 1/10). The stagnation pressure was 4 bar[40]. The pronounced additional resonance in (b) at ~ 2.3 eV is due to the formation of $(O_2)_2^-$ by capture of electrons inelastically scattered to near-zero energy via the 2.3 eV N_2^{-*} ($^2\Pi_g$) resonance of N_2.

PHOTODETACHMENT IN GASES, LIQUIDS, AND CLUSTERS

The photodetachment of electrons from negative ions in low-pressure gases is well-understood and the relation of the photodetachment threshold E_{th} to the electron affinity, EA, of the electron attaching species is well-established (e.g., see Ref. 13). Near threshold, the photodetachment cross section, σ_{pd}, is predicted[41,42] to vary as

$$\sigma_{pd}(E) = BE(E-E_{th})^{(2l+1)/2} \alpha k^{2l+1}, \qquad (15)$$

where B is a constant, $E = h\upsilon$ is the photon energy, and k and l are the linear and angular momenta of the ejected electron. For atoms, A, the value of the E_{th} for the process

$$A^- + h\upsilon \rightarrow A + e \qquad (16)$$

is equal to the electron affinity EA(A) of A, which in turn, is equal to the "vertical detachment energy" (VDE). For molecules, M, the relation of E_{th} for the process

$$M^- + h\nu \rightarrow M + e \tag{17}$$

to the electron affinity, EA(M), of M and the VDE is complicated by possible differences in the structural parameters of M and M⁻. If we define the VDE for (17) as the minimum energy required to eject the electron from the negative ion in its ground electronic and molecular state without changing the internuclear separations, then the VDE is related to the EA and E_{th} by

$$E_{th} = VDE = EA + \Delta E, \tag{18}$$

i.e., the VDE for (17) exceeds the EA by ΔE; the magnitude of ΔE depends on the relative positions of the potential energy curves (surfaces) of M and M⁻.

The photodetachment of electrons from negative ions in dense gases, liquids and clusters has been studied--and the energetics of the process have been related to their gas-phase values--for only a limited number of cases (e.g., see Refs. 43-46). Even more limited seem to be the measurements of the photodetachment cross sections. Experimental studies are in progress at the author's laboratory to determine the photodetachment energetics and cross sections as a function of medium density from a low-pressure gas to the liquid. Establishing the energetics and the cross sections for photodetachment as a function of the nature and density of the medium is significant because it gives a direct measure of the stability of the anion in dense matter and because it provides a basic input for understanding electron transfer mechanisms in dense matter.

In dense gases, liquids and clusters, the photodetachment process for molecular negative ions is complicated by the effect of the medium on E_{th} and the potential energy curve (surface) of M⁻. Rewriting reaction (17) for, say, the liquid as:

$$M_L^- + h\nu \rightarrow M_L + e_L \tag{19}$$

and assuming that $(EA)_L = (EA)_G + V_o - P^-$ (Refs. 1,44) we have

$$(E_{th})_L = (VDE)_L = (EA)_L + (\Delta E)_L$$

$$= (EA)_G + V_o - P^- + (\Delta E)_L \tag{20}$$

where V_o and P^- are the polarization energies of the electron and the negative ion in the medium, and $(EA)_L$ and $(EA)_G$ are the values of the electron affinity of M in the liquid and the gas phase.

Among the few studies of photodetachment of negative ions in liquids is that on $C_6F_6^-$ (Refs. 43,44). Let us, then, by way of example, refer to the method of Faidas et al. [44] which utilizes a two-laser photoconductivity technique suitable for photodetachment studies in dense fluids and the results they obtained on $C_6F_6^-$ photodetachment in liquid TMS using this method. A schematic of their experimental arrangement and an outlay of the principle of their technique is shown in Fig. 18. A cell--with appropriate light windows and feedthroughs--contains the molecule (C_6F_6) under study disolved in the liquid (TMS) at an appropriate concentration. Two counter-propagating coaxial laser beams traverse the interaction volume in the liquid cell with a time delay of ~ 5 μs. The first laser beam from

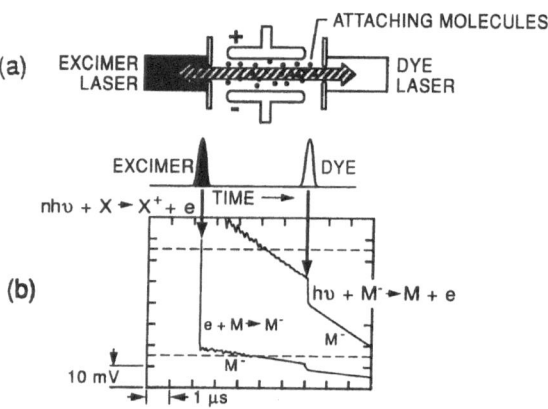

Figure 18. Schematics of the two laser photodetachment technique and an oscillogram of the conductivity signal (see the text and Ref. 44).

an excimer laser (λ = 308 nm; fwhm = 15 ns) ionizes biphotonically the liquid and produces electrons which give rise to a fast signal (initial drop in Fig. 18b identified by $\downarrow$ and the process $nh\upsilon + X \rightarrow X^+ + e$). These electrons quickly (within < 1 ns) attach to C_6F_6 forming $C_6F_6^-$ (process $e + M \rightarrow M^-$ in Fig. 18b); the slow falling portion of the signal (indicated by M^- in Fig. 18b) after the initial steep fall is due to these slow-moving anions. At the preset time delay ($\sim$ 5 μs), the second laser pulse (tunable dye laser, fwhm = 0.6 ns) detaches the electron from $C_6F_6^-$ (when $h\upsilon > (E_{th})_L$) and produces a second transient signal (step drop in Fig. 18b identified with $\downarrow$ and $h\upsilon + M^- \rightarrow M + e$), followed by a slow drop when the detached electrons attach to C_6F_6 again forming slow moving $C_6F_6^-$.

The basis for determining the photodetachment cross section $\sigma_{pd}(E)$ as a function of the photon energy E can be seen by referring to Fig. 19. The first (excimer) laser pulse has an essentially flat intensity profile $I_e(r)$ for distances $r < \alpha$ where α is the radius of the cross sectional area of the interaction volume (Fig. 19a). This pulse generates electrons uniformly in the interaction volume with a density distribution $n_{ei}(r) \propto I_e^2(r)$. Under the experimental conditions employed by Faidas et al. all these electrons were captured by C_6F_6 within 1 ns and the resultant negative ions $C_6F_6^-$ were essentially stationary with an ion density distribution $n_i(r) = n_{ei}(r) \propto I_e^2(r)$ when the second (tunable dye) laser pulse arrived $\sim$ 5 μs later. The intensity profile $I_d(r)$ of the dye laser pulse was Gaussian and lay well within α (Fig. 19c). The density distribution, $n_{ed}(r)$, of the photodetached electrons is

$$n_{ed}(r) = \sigma_{pd}(E)\, n_i(r)\, I_d(r) \tag{21}$$

Under the experimental conditions employed, $n_{ed}(r) << n_i(r)$, the $I_e(r)$, $n_i(r)$, and $I_d(r)$ were virtually constant along the axis of the interaction volume, and

$$\sigma_{pd}(E) = \frac{N_{ed}}{N_i}\,\frac{\pi\alpha^2}{I_t} \tag{22}$$

Thus, $\sigma_{pd}(E)$ can be determined from a measurement of the ratio N_{ed}/N_i of the total number of photodetached electrons, N_{ed}, to the total number of negative ions N_i, a measurement of the total number of photons in the dye laser pulse I_t and a knowledge of α and the intensity profiles of the two laser beams.

In Fig. 20 is shown the $\sigma_{pd}(E)$ for $C_6F_6^-$ in TMS (T $\sim$ 298 K) determined by this technique. The cross section $\sigma_{pd}(E)$ exhibits two well-defined maxima at 2.58 and 3.15 eV due presumably to excited negative ion states of $C_6F_6^-$. The photodetachment threshold $(E_{th})_L$ was determined by fitting the experimental measurements near threshold to

$$(\sigma_{pd}/E)^{1/n} = B[E - (E_{th})_L] \tag{23}$$

with $n = 1/2(2l+1)$. The best fit was obtained for $n = 3/2$ (i.e., $l = 1$, see Eqs. (15) and (23)) and $(E_{th})_L = 1.51$ eV. Using this value, Eq. (20), and literature values of $(EA)_G$ Faidas et al. concluded that the electron affinity of C_6F_6 is lowered in liquid TMS by $\sim$ 0.4 eV compared to its isolated-molecule value.

Finally, photodetachment studies on a number of cluster negative ions aided considerably our understanding of the energetics of the process. Experimental studies on the vertical values of EA (photodetachment thresholds; e.g., see Refs. 45,46) revealed the existence of the expected (Fig. 13, and Refs. 6 and 11) specific size variations and the smooth (increase) transition to the bulk property as the cluster size is increased. This is clearly shown in Fig.

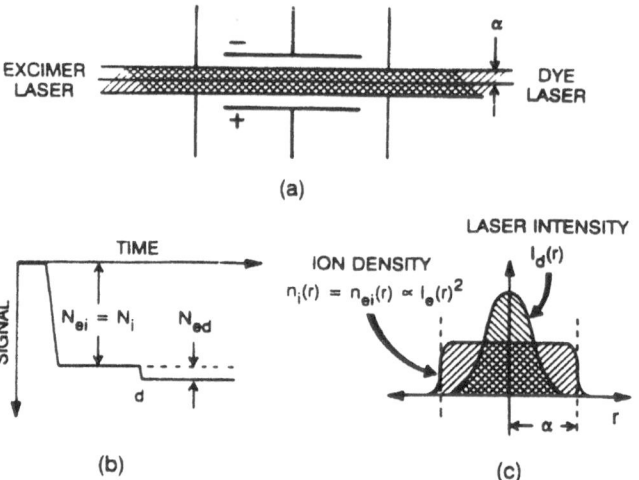

(a)

(b)   (c)

Figure 19. Schematics of the photodetachment technique illustrating: (a) the laser interaction region, (b) the type of signal measured, and (c) the laser intensity and negative ion density profiles (see the text and Ref. 44).

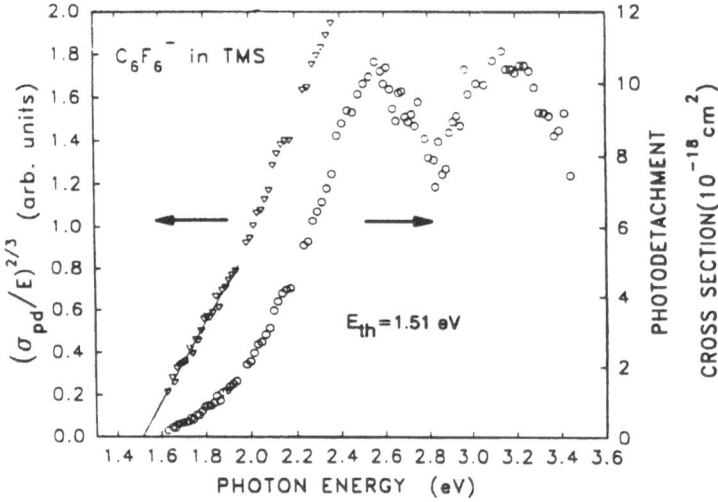

Figure 20. The absolute photodetachment cross section $\sigma_{pd}(E)$ (right-hand side Y axis) of $C_6F_6^-$ in tetramethylsilane as a function of the photon energy, E, and a linear least squares fit of $(\sigma_{pd}/E)^{2/3}$ (left-hand side Y axis) for $E < 1.95$ eV. The intercept with the X axis gives a threshold energy E_{th} of 1.51 eV (Ref. 44).

21 where the measured EA values of Cu_n clusters are plotted as a function of cluster size[45]. Odd copper clusters are seen to have substantially lower EA than adjacent even clusters and this continues at least through Cu_{17}^-. It was pointed out[45] that the LUMO (lowest unoccuppied molecular orbital) of the even clusters would be expected to be strongly antibonding and the corresponding negative ion more weakly bound than if the electron were to be placed in the half-filled nonbinding HOMO of an odd cluster.[5]

The electron affinity, EA(n), of the larger clusters would be predicted to increase as [e.g., see Ref. 6)

$$EA(n) = EA(R_c) = EA(\infty) - a \frac{e^2}{R_c} \qquad (24)$$

where EA(∞) is the bulk band energy and a = 1/2 for the polarization and a = 5/8 for the metal droplet model of negative metal clusters. The predictions of (24)--as in the case of (5), (10), and (20) we have seen earlier--depend on the accuracy of R_c.

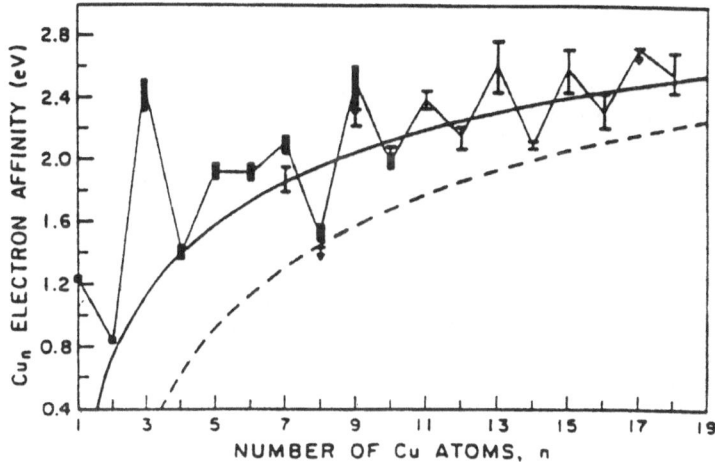

Figure 21. Experimental electron affinities for Cu_n compared with predictions of the classical spherical drop model ■ (Ref. 45); I (Ref. 47); --- and —— predicted using Eq. (24) and a = 5/8, EA (∞) = 4.65 eV, R = $n^{1/3} r_b$ (----) and R = $n^{1/3} r_b + \Delta r$ (——) with r_b (half the average distance between the atoms in the bulk) = 1.41 Å and Δr (correction for orbital overlap among bounded atoms) = 0.53 Å (from Ref. 45).

---

[5]An opposite alternation is exhibited by the ionization threshold energy I (i.e., $I_{odd} < I_{even}$) (e.g., see Ref. 11).

REFERENCES

1. Christophorou, L. G., and Siomos, K., Interphase Physics: Linking Knowledge on Electron-Molecule Interactions in Gases to Knowledge on Such Processes in Condensed Matter, in "Electron-Molecule Interactions and Their Applications," Christophorou, L.G., Ed., *Academic Press*, Orlando, 1984, Vol. 2, Chapt. 4.

2. Christophorou, L. G., Gas/Liquid Transition: Interphase Physics, in "The Liquid State and Its Electrical Properties," Kunhardt, E. E., Christophorou, L. G., and Luessen, L. H., (Eds.), *Plenum Press*, New York, 1988, pp. 283-316.

3. Christophorou, L. G., Radiation Interactions in High-Pressure Gases, in "Physical and Chemical Mechanisms in Molecular Radiation Biology," Glass, W. A. and Varma, N. M., Eds., *Plenum Press*, New York, 1991, pp. 183-230.

4. Kunhardt, E. E., Christophorou, L. G., and Luessen, L. H., (Eds.), "The Liquid State and Its Electrical Properties," *Plenum Press*, New York, 1988.

5. Freeman, G. R., (Ed.), "Kinetics of Nonhomogeneous Processes," *Wiley-Interscience*, New York, 1987.

6. Jortner, J., Cluster Size Effects, *Z. Phys. D* 24:247 (1992).

7. Kappes, M. M., Schär, M., Radi, P., and Schumacher, E., On the Manifestation of Electronic Structure Effects in Metal Clusters, *J. Chem. Phys.* 84:1863 (1986).

8. Rademann, K., Reactivity and Electron Structures of Isolated Clusters, Ber. Bunsenges. *Phys. Chem.* 94:1295 (1990); Photoelectron Spectroscopy and UV/VIS-Photoabsorption Spectroscopy of Isolated Clusters, *Z.Phys. D* 19:161 (1991).

9. Leopold, D. G., Ho, J., and Lineberger, W. C., Photoelectron Spectroscopy of Mass-Selected Metal Cluster Anions. I. Cu_n^-, n = 1-10, *J. Chem. Phys.* 86:1715 (1987).

10. Märk, T. D., Free Electron Attachment to van der Waals Clusters, *Intern. J. Mass Spectr. Ion Proc.* 107:143 (1991).

11. Bonačić-Koutecký, V., Fantucci, P., and Koutecký, J., Quantum Chemistry of Small Clusters of Elements of Groups Ia, Ib, and IIa: Fundamental Concepts, Predictions, and Interpretation of Experiments, *Chem. Rev.* 91:1035 (1991).

12. Illenberger, E., Electron Attachment Reactions in Molecular Clusters, *Chem. Rev.* 92:1589 (1992).

13. Christophorou, L. G., "Atomic and Molecular Radiation Physics," *Wiley-Interscience*, New York, 1971.

14. Christophorou, L. G. (Ed.), "Electron-Molecule Interactions and Their Applications," *Academic Press*, Orlando, 1984, Vols. 1 and 2.

15. Christophorou, L. G., The Lifetimes of Metastable Negative Ions, *Adv. Electr. Electron Phys.* 46:55 (1978).

16. Sanche, L., Primary Interactions of Low Energy Electrons in Condensed Matter, in "Excess Electrons in Dielectric Media," Ferradini, C., and Jay-Gerin, J.-P., Eds., *CRC Press*, Boca Raton, 1991, pp. 1-42.

17. Inokuti, M., How is Radiation Energy Absorption Different Between the Condensed Phase and the Gas Phase?, in "Radiation Effects and Defects in Solids," Gordon and Breach, U.K., 117, (1991), pp. 143-162.

18. LaVerne, J. A. and Mozumder, A., Effect of Phase on the Stopping and Range Distribution of Low-Energy Electrons in Water, *J. Phys. Chem.* 90:3242 (1986).

19. Ashley, J. C., Stopping Power of Liquid Water for Low-Energy Electrons, *Rad. Research* 89:25 (1982).

20. Michaud, M. and Sanche, L., Total Cross Sections for Slow-Electron (1-20 eV) Scattering in Solid H_2O, *Phys. Rev. A* 36:4672 (1987).

21. Hunter, S. R., Carter, J. G., and Christophorou, L. G., Low-Energy Electron Drift and Scattering in Krypton and Xenon, *Phys. Rev. A* 38:5539 (1988).

22. Dmitrenko, V. V., Romanyuk, A. S., Suchkov, S. I. and Uteshev, Z. M., Electron Mobility in Dense Xenon Gas, *Sov. Phys. Tech. Phys.* 28:1440 (1983).

23. Miller, L. S., Howe, S. and Spear, W. E., Charge Transport in Solid and Liquid Ar, Kr, and Xe, *Phys. Rev.* 166:871 (1968).

24. Sakai, Y., Nakamura, S., and Tagashira, H., Drift Velocity of Hot Electrons in Liquid Ar, Kr, and Xe, *IEEE Trans. Electr. Insul.* , EI-20:133 (1985).

25. Bader, G., Perluzzo, G., Caron, L. G., and Sanche, L., Elastic and Inelastic Mean-Free Path Determination in Solid Xenon from Electron Transission Experiments, *Phys. Rev. B*, 26:6019 (1992).

26. Faidas, H., Christophorou, L. G., McCorkle, D. L., and Carter, J. G., Electron Drift Velocities and Electron Mobilities in Fast Room-Temperature Dielectric Liquids and Their Corresponding Vapors, *Nucl. Instr. Meth. Phys. Res.* A294:575 (1990).

27. Stephens, J. A., Comment on Electron Mobility in the Liquid Isomeric Pentanes, *J. Chem. Phys.* 84:4721 (1986).

28. Fano, U., Studies of Slow Electron Action on Condensed Media, *Radiat. Phys. Chem.* 32:95 (1988); Short- and Long-Range Interactions of Slow Electrons in Condensed Matter: Effects of Reflection and Transmission, *Phys. Rev.* B36:1929 (1987).

29. Mills, D. L., Resonant Scattering of Slow Electrons in Molecular Solids: Suppression of the Elastic Beam, *Phys. Rev. B* 45:36 (1992); Resonant Scattering of Electrons in the Presence of Coupling to Phonons, *Phys. Rev. B* 45:13221 (1992).

30. Henkes, W. and Mikosch, F., The Effective Cross Section for Ionization by Electrons of Molecules in Hydrogen Clusters, *Int. J.Mass Spectr. Ion Phys.* 13:151 (1974).

31. Derenzo, S.E., Mast, T. S., Zaklad, H., and Müller, R.A., Electron Avalanche in Liquid Xenon, *Phys. Rev.* A9:2582 (1974).

32. Schmidt, W. F., Electron Conduction Processes in Dielectric Liquids, *IEEE Trans. Electr. Insul.* EI-19:389 (1984).

33. Raz, B. and Jortner, J., Energy of Quasi-Free Electron State in Liquid and Solid Rare Gases, *Chem. Phys. Lett.* 4:155 (1969).

34. Faidas, H. and Christophorou, L. G., Laser Multiphoton Ionization of Aromatic Molecules in Nonpolar Liquids, *Rad. Phys. Chem.* 32:433 (1988).

35. Faidas, H., Christophorou, L. G., Datskos, P. G., and McCorkle, D. L., The Ionization Threshold of N, N, N',N'-Tetramethyl-p-phenylenediamine in Dense Fluid Ethane; Effects of Fluid Density and Temperature, *J. Chem. Phys.* 90:6619 (1989).

36. Steinberger, I. T., Photoconductivity, Conduction Electron Energies, and Excitons in Simple Fluids, in Ref. 4, p. 235.

37. Springett, B. E., Jortner, J., and Cohen, M. H., Stability Criterion for the Localization of an Excess Electron in a Nonpolar Fluid, *J. Chem. Phys.* 48:2720 (1968).

38. Bakale, G., Sowada, U., and Schmidt, W. F., Effect of an Electric Field on Electron Attachment to SF_6, N_2O, and O_2 in Liquid Argon and Xenon, *J. Phys. Chem.* 80:2556 (1976).

39. Jaffke, T., Hashemi, R., Christophorou, L. G., and Illenberger, E., Mechanisms of Anion Formation in O_2, O_2/Ar, and O_2/Ne Clusters; the Role of Inelastic Electron Scattering, *Z. Phys. D* 25:77 (1992).

40. Hashemi, R., Jaffke, T., Christophorou, L. G., and Illenberger, E., Role of Inelastic Electron Scattering by N_2 in the Formation of $(O_2)_n^-$ Anions in Mixed O_2/N_2 Clusters, *J. Phys. Chem.* 96:10605 (1992).

41. Christophorou, L. G., Electron Attachment and Detachment Processes in Electronegative Gases, *Cotrib. Plasma Physics* 27:237 (1987).

42. Geltman, S., Theory of Threshold Energy Dependence of Photodetachment of Diatomic Molecular Negative Ions, *Phys. Rev.* 112:176 (1958).

43. Sowada, U., and Holroyd, R. A., Laser Photodetachment Spectra of $C_6F_6^-$ in Nonpolar Liquids, *J. Phys. Chem.* 84:1150 (1980).

44. Faidas, H., Christophorou, L. G., and McCorkle, D. L., Laser Photodetachment in Liquids: $C_6F_6^-$ in Tetramethylsilane, *Chem. Phys.* 193:487 (1992).

45. Leopold, D. G., Ho, J., and Lineberger, W. C., Photoelectron Spectroscopy of Mass-Selected Metal Cluster Anions I. Cu_n^-, n=1-10, *J. Chem. Phys.* 86:1715 (1987).

46. DeLuca, M. J., Han, C.-C., and Johnson, M. A., Photoabsorption of Negative Cluster Ions Near the Electron Detachment Threshold: A Study of the $(O_2)_n^-$ System, *J. Chem. Phys.* 93:268 (1990).

47. Zheng, L.-S., Karner, C. M., Brucat, P. J., Yang, S. H., Pettiette, C. L., Craycraft, M. J., and Smalley, R. E., Photodetachment Studies of Metal Clusters: Electron Affinity Measurements for Cu_x, *J. Chem. Phys.* 85:1681 (1986).

COMPARISONS BETWEEN LOW-ENERGY ELECTRON SCATTERING FROM GASEOUS AND CONDENSED-PHASE ATOMS AND MOLECULES

Léon Sanche

Groupe CRM en Sciences des radiations
Faculté de médecine, Université de Sherbrooke
Sherbrooke, QC, Canada J1H 5N4

INTRODUCTION

Low-energy electron scattering from atoms and molecules has been studied for more than half a century.[1] Various types of experiments have been performed in order to measure the energy and angular dependence of the magnitude of the cross sections involved in the elastic, inelastic, ionising and dissociative processes induced by the collision of an electron with a gaseous target. During the last decade, such experiments have also been performed with atoms and molecules lying within[2] or at the surface[3] of solids. There exists now considerable information on elastic and inelastic scattering of low-energy electrons as well as on electron attachment in the solid phase. The main purpose of the present article is to show, with appropriate examples, how *the behavior of low energy electron scattered in the gas-phase can be linked to the more recent results obtained in the solid phase.* In another article of this book, attempts are made to do the same between gas and condensed phase results on electron attachment and dipolar dissociation processes. Due to the vast amount of information already available in the gas-phase on low-energy electron interactions, providing a strong link between the two phases should considerably enhance our present understanding of the action of low-energy (< 30 eV) electrons in condensed media.

In the next section, some general considerations about low-energy electron interactions with atoms and molecules are given with reference to concepts developed in the gas-phase. High resolution electron scattering experiments are described in Sec. III. In the subsequent sections, examples of elastic scattering as well as intermolecular, intramolecular and electronic energy losses by slow electrons in dielectric and molecular solids are provided to explain how the fundamental properties of the isolated scattering phenomena are modified by the presence of neighboring targets. Scattering involving the largest energy losses are described first since they are more likely, as explained in Sec. II, to resemble their gas-phase counterpart and therefore easier to explain. Molecular solids are of particular interest in the present context, because they often retain many of the properties of their individual component atoms or molecules, making possible links with gas-phase data easier.

Linking the Gaseous and Condensed Phases of Matter
Edited by L.G. Christophorou *et al*, Plenum Press, New York, 1994

INTERACTION OF LOW-ENERGY ELECTRONS WITH ATOMS AND MOLE-CULES

The interaction of a low-energy electron with an isolated target may be described, outside the target itself, by a potential which takes into account the force between the two particles due to the dipole, quadrupole and higher moments of the target as well as induced polarization effects.[4] By analysing the various scattering cross sections as function of incident electron energy and scattering angle, it may be possible in particular cases to determine which part of the potential is dominant. For example, certain energy losses may be characterized to be the result of "dipolar scattering", which mean that the interaction with the permanent dipole of the molecule is dominant. However, if the electron penetrates significantly into the target "region" the interaction becomes more complex due to the possibility of spin-exchange and constructive interference of the electron wave within the target. This latter phenomena gives rise to electron resonances.[1] Thus, resonances occur when the scattered electron resides for a time much longer than the usual scattering time in the neighborhood of the target atom or molecule. From an atomic or

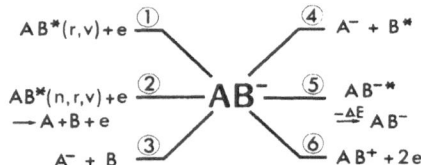

Figure 1. Decay channels of a temporary anion AB⁻

molecular orbital point of view, a resonant state may be considered as a negative ion formed by an electron which occupies temporarily an orbital of the target.[1]

Fig. 1 illustrates the possible decay channels of a diatomic transient anion AB^-. The departing electron may leave the molecule in a rotationally, vibrationally [reaction (1) in Fig. 1] or electronically (2 in Fig. 1) excited state (in the condensed state librational and translational excitation must also be considered as a possibility). If the resulting electronically excited neutral state is dissociative, ground state or excited fragments can be produced (2). If the lifetime of the resonance is, at least, of the order of a vibrational period, and if the AB^- state is dissociative in the Franck-Condon region and at least one of the possible fragments has a positive electron affinity, then the transient molecular anion may dissociate into a stable anion and a neutral fragment in the ground (3 in Fig. 1) or an excited (4) state. This process is called dissociative attachment (DA). If during its lifetime, the transient anion transfers energy to another system (e.g., to another molecule by collisional interaction), it can become stable, when the parent molecule has a positive electron affinity (5 in Fig. 1). Finally, when the transient anion is formed at energies above the ionization potential, two-electron emission is possible (6 in Fig. 1). Excluded from the reactions of Fig. 1 is the spontaneous emission of a photon by the transitory anion followed by stabilization of the electron on the molecule (i.e., $AB^{-*} \rightarrow AB^- + h\nu$). Autodetachment lifetimes of transient anions vary from about 10^{-15}s up to the millisecond time scale for large polyatomic molecules, where the excess energy of the anion may be accommodated among its many degrees of freedom, thereby delaying autodetachment.

For molecules, the amplitudes of the vibrational excitation cross sections are affected by the lifetime Δt of transitory anions. Long-lived resonances ($\Delta t > 10^{-14}$ s.) cause a significant displacement of the nuclei of the molecule when the additional electron occupies a strongly bonding or antibonding orbital. When the electron leaves the molecule, nuclear motion is initiated toward the initial internuclear distance, causing strong vibrational excitation including many overtones of the molecule. On the other hand, when the lifetime is much smaller than a typical vibrational period ($\Delta t << 10^{-14}$ s.), the nuclei are not displaced significantly and only the lower vibrational levels become excited with considerable amplitude.

We may now ask ourselves the following question: "Are the mechanisms of interaction just described still valid in the condensed phase, where the target is surrounded by neighboring atoms and/or molecules?". It is usually the case for high energy electrons, whose wavelength is short in comparison with the "diameter" of the elementary constituents of condensed matter. The short wavelength of these electrons make it possible to consider that each projectile interacts individually with the atoms and/or molecules of the condensed system. Thus, the scattered amplitudes within or outside the solid can be considered as the sum of the individually scattered waves. This concept is no longer valid at low energies, where the electron wavelength is of the order of the interatomic or intermolecular distances. In this case, the electron is scattered collectively from many targets and the scattered intensity must be derived from the sum of the interaction potentials between the electron and each of the elementary constituents of condensed matter.[5] The geometrical arrangement of condensed matter must therefore be taken into account when describing the interference of low-energy electron waves arising from multiple scattering.

Intermolecular interference is expected to play a dominant role in elastic scattering due to preservation of the short- or long-range coherence. However, when electrons lose energy into intramolecular modes, interferences between electron waves is less likely to occur due to energy localisation and loss of coherence in the scattered waves. In this case, description of the scattering process in terms of intramolecular resonant and direct potential scattering mechanisms may still be valid provided that the modifications to the isolated electron-molecule system introduced by the presence of other neighboring targets are properly taken into account.[2] In the condensed state, we expect not only the single-electron-target potential to be modified by the proximity of neighbouring atoms or molecules, but also the amplitude, lifetime, symmetry and energy of isolated transient anions as well as the number and branching ratios of their decay channels (Fig. 1). Among the "environmental effects" or factors which may contribute to these modifications we note: (i) the change in the symmetry of the scattering problem introduced by the presence of other nearby targets; (ii) the polarisation of these targets by charge of the electron and transient anion; (iii) the fixed orientation of the molecule with respect to a given frame of reference and (iv) the possible distortion and modification of the target atom or molecule by the adjacent matter. The change in symmetry may be viewed to arise from multiple scattering of the electron waves between the atoms or molecules of a solid. This condition changes the partial wave content of a resonant electron which in turn influences the lifetime of the state.[6] By acting on the isolated electron-target potential, the polarization force not only lowers the anion's energy but can also modify the resonance lifetime, by promoting electron transfer to the neighbouring targets with the introduction of an attractive potential. The lower energy tends to reduce the number of decay channels but new intermolecular energy loss and electron exchange channels appear. These changes in type and number of decay channels may affect drastically certain cross sections and the lifetime of the resonance. Lowering of the resonance energy may be viewed as a change in the electron affinity (EA) of the target molecule (even from negative to positive). In the case of well ordered solids, the orientation of the molecule is fixed so that the differential cross sections and angular distributions are greatly modified[3,7] from those of the gas phase, where they are measured from randomly oriented molecules.

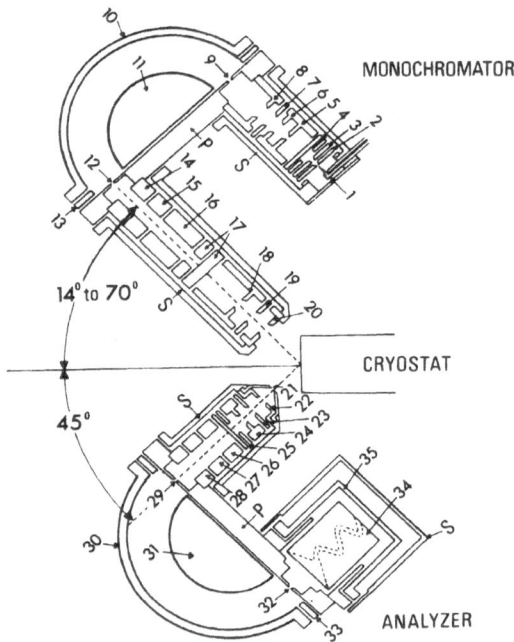

Figure 2. Schematic diagram of the hemispherical electron energy loss spectrometer. The monoenergetic electron beam produced by the monochromator strikes the target cooled by a cryostat. Electrons scattered at 45° from the plane of the target are dispersed in energy by the analyser. Electrons emitted from filament (1) are focused at the entrance (9) of the deflector region (10 and 11) by the zoom lens (6, 7 and 8). Energy dispersion is produced by applying a potential of about 1 volt between two concentric hemispheres (10, 11, 30, and 31). Imaging and energy control of the electron beam is achieved by means of electrostatic zoom lenses with cylindrical (14 to 16, 26 to 28) and aperture (6 to 8, 23 to 25) electrodes. Electrons of a specific energy are detected with a channeltron (34).

EXPERIMENTS

The interaction of low-energy electrons with dielectric and molecular solids can be investigated by allowing monoenergetic electrons to impinge on a thin multilayer film grown in an ultra-high-vacuum system by the condensation of gases or organic vapors on a clean metal substrate held at cryogenic temperatures (15-100 K).[2,3] The substrate can be either a single crystal or polycrystalline. In electron scattering experiments, it is usually possible to measure the dependence on primary electron energy of the current transmitted through,[8] or reflected from the film.[9] As a general rule, the film thickness must be larger than the total electron mean free path in order to minimize effects of the metal substrate. Evidence that a given process occurs within the bulk of a multilayer film arises from the presence of multiple scattering effects in the data.[10,11] It can also be provided by specific experiments such as matrix isolation.[12] In order to study electron scattering from a molecule adsorbed on a dielectric surface the monochromatic electron beam is incident on a target composed of a multilayer film covered by submonolayer amounts of the molecule under investigation.[13] The film serves as a spacer between the metal substrate and the molecular adsorbate. Varying the thickness of the spacer layers makes it possible to study systematically the effect of the metal surface. Experiments analysing reflected electrons are particularly powerful in order to determine losses to particular energy modes of the solid including phonons, intramolecular vibrations and electronic excitations. Only this type of experiments is reported in this article.

A high resolution electron energy loss (HREEL) spectrometer of the type used to analyse reflected electrons is shown in Fig. 2. It consists essentially of two concentric hemispherical deflectors with appropriate electron optics and a closed-cycle refrigerated cryostat.[9] Electrons leaving the monochromator are focused on the film condensed on a metal substrate secured by a press fit to the cold end of the cryostat. The angle of incidence θ_o can be varied from 14 to 70° from the film normal. Electrons reflected at $\theta_d = 45$° from the film are energy analyzed. The current transmitted at the substrate can also be recorded with this apparatus. This allows the absolute electron energy scale to be calibrated within ±0.15 eV. Energy-loss spectra are recorded by sweeping the energy of either the monochromator or the analyser. The energy dependence of a given energy loss event (i.e. the excitation function) is obtained by sweeping the energy of both deflectors with a potential difference between them corresponding to the probed energy loss. HREEL spectra are usually recorded with overall resolution ranging from 8 to 20 meV FWHM with corresponding currents lying in the 10^{-10}-10^{-9} A range.

ELECTRONIC EXCITATION

Electrons having produced electronic excitation in condensed atoms or molecules are highly incoherent and they are therefore not expected to be much influenced by coherence effects characteristic of the condensed phase. In fact, comparison of low energy HREEL spectra indicate that, in general, valence states are excited with about the same magnitude in the gas and solid phases.[14] An example is shown in Fig. 3 which displays HREEL spectra for 16-eV electrons incident on a 15-layer N_2 film.[14] Each peak can be ascribed to a vibrational progression belonging to the electronic states indicated in the figure. These results are representative of electronic excitation in molecular solids induced by slow electrons; such excitations have been found to be similar to the gas-phase only below the band gap. Above this energy, Rydberg transitions usually disappear in molecular solids as well as electron resonances associated with such states.[2,14] Valence transitions have larger widths and slightly different energies than in the gas-phase.[2,14] As expected from the strength of the exchange interaction at low energies, spin- forbidden states are prominent below 20 eV. In fact, below 10.5 eV in Fig. 3, the data are similar in amplitude and energy to those recorded in gas-phase N_2 at similar primary beam energies. Above 11 eV, in a region where an abundance of Rydberg states appears in the gas phase, only the $B^1\Pi_u$ valence state is observed in solid N_2 (Fig. 3).

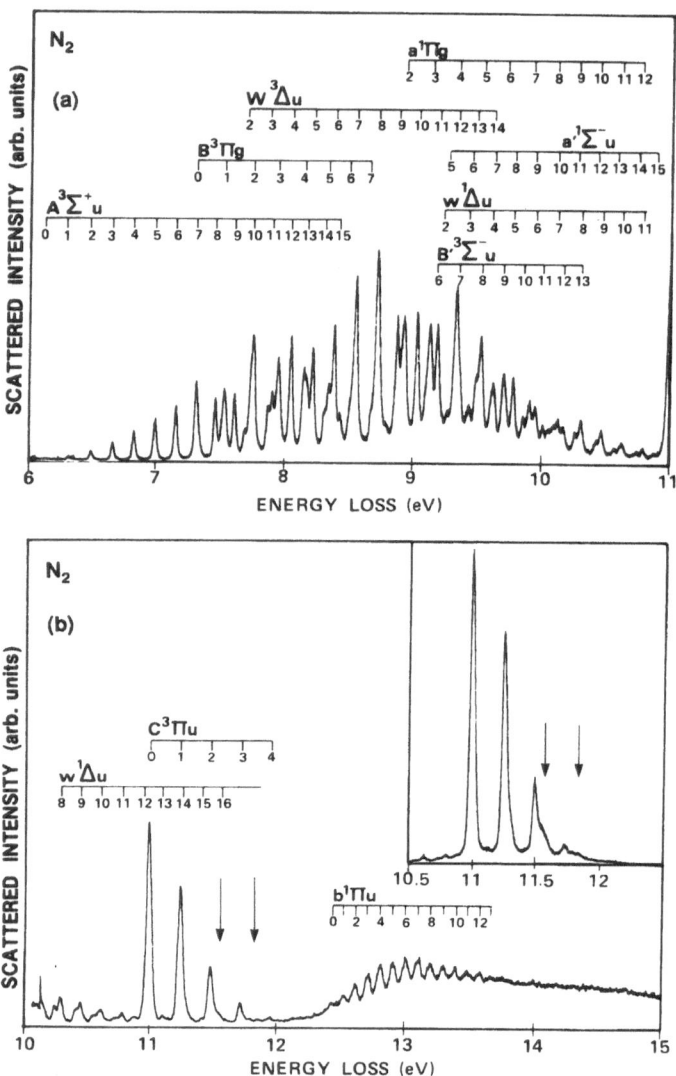

Figure 3. (a) High-resolution electron-energy-loss (HREEL spectrum in the range 6-11 eV, of a 50 Å N_2 film condensed on a Nb substrate. The incident energy was 16 eV. (b) HREEL spectrum for same target and incident energy in the range 10-15 eV.

INTRAMOLECULAR VIBRATIONAL EXCITATION

It is well known that the temporary formation of an anion can considerably enhance the vibrational excitation cross sections of gas-phase molecules. Similarly in the condensed phase, it has been shown that a temporary attachment of an electron at a particular molecular site[2] or on a molecule deposited on a surface[3] can also increase by order of magnitude the intramolecular vibrational electron energy losses. The major differences between gas and solid phase resonance scattering into the vibrational excitation channel are illustrated in this section from the behavior of the well-known $^2\Pi_g$ transient state of N_2^- reported as a function of its distance from the Pt substrate and embedded in the bulk of RGS multilayer films.

The $^2\Pi_g$ shape resonance of N_2^-, located at 2.3 eV in the gas phase,[1] involves the temporary localization of an electron in the next vacant antibonding orbital of N_2 with the ground state configuration $(\sigma_g 1s)^2(\sigma_u^* 1s)^2(\sigma_g 2s)^2(\sigma_u^* 2s)^2(\pi_u 2s)^4(\sigma_g 2p)^2(\pi_g^* 2p)$. The coupling of this transitory state with intramolecular vibrations leads to the observation of a large enhancement and of an oscillatory structure in the isolated-molecule elastic and vibrational excitation cross sections.[1] This latter structure is also observed in the excitation function for vibrations of ground state N_2 physisorbed on a metal, in a pure solid film, as well as for a N_2 in rare gas matrices.[2,3] This information allows an accurate determination of the resonance energy in various media.

The excitation function[13] for the v=1 vibrational loss of ground state N_2 recorded with a HREEL spectrometer between 0 and 4 eV[2] is shown in Fig. 4. The numbers on the right of each curve indicate the number of monolayers of the Ar spacer on which 0.1 ML of N_2 was physisorbed. These excitation functions for N_2 isolated on Ar were recorded with $\theta_o = 14°$; electrons emerging at $\theta_d = 45°$ were energy analyzed. The oscillatory structure, due to vibrational motion of N_2^-, is similar to that recorded at 90° in the gas-phase differential cross section[15] shown in the bottom of Fig. 4. Once appropriately translated by -0.62 eV all peak positions of the gas-phase measurement agree within ±0.02 eV with those of the 32-layer result. The larger overlap in the condensed-phase oscillatory structure may be attributed either to a reduction in the lifetime of the resonance or to an inhomogeneous broadening due to the molecule occupying several adsorption sites. For thicknesses smaller than 32 layers, the overall intensity of the resonance feature decreases by a factor of up to 5, while the oscillatory structure shifts to lower energy and becomes even broader. Even though, the relative separation between the peaks appears almost unperturbed, a substantial energy shift varying from 0.62 to 1.6 eV is obtained from alignment of the condensed phase spectra with the gas-phase results. Calculations[13] of the electronic polarization energy of N_2^- as a function of the thickness of the Ar film can reproduce the energy shift to an accuracy of a few percent down to a thickness of 2 layers; indicating that the lowering of the N_2^- state with decreasing distance from the metal is essentially due to the increase in induced polarization. Close to the metal surface, the classical picture of a point charge near a semi-infinite dielectric breaks down and gives an overestimate of the electronic polarization energy. Near the surface, details of the perturbed charge distribution of the anion as well as that of induced charge at the metal surface must also be taken into account. Furthermore, the symmetry of the scattering problem and molecular orientation may differ. In this case, it is preferable to compare these results to more elaborate calculations including multiple scattering of the electron waves[16] within the substrate and between the substrate and the molecule such as those performed with the coupled-angular mode method of Teilly-Billy and Gauyacq.[17] Using this method, these authors and Rous[18] were able to provide a description of the variation of the resonance energy and width of the $N_2^- \, ^2\Pi_g$ state with its distance from a metal surface to which the experimental data of Fig. 4 can be compared.

As seen in Fig. 4, the relative magnitude of each peak of the oscillatory structure is different for the isolated and condensed molecule. This is probably due to the nature of the incoming and outgoing electron wave functions. Rous et al[16] have clearly shown for O_2 on graphite that the free electron wavefunctions are appreciably modified by the presence of the substrate. The incident electron waves which interact with N_2 on Ar arise principally from two sources: those which arrive directly from vacuum and those which are reflected from the solid surface. Similarly, scattered waves can be directly emitted in vacuum or first

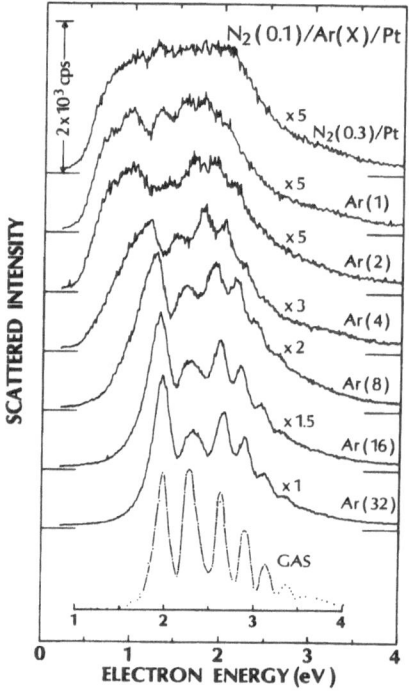

Figure 4. Excitation function for the v=1 level of N_2 condensed on an Ar film of thicknesses ranging from 0 to 32 layers. The N_2 coverage is 0.1 layer on Ar except for 0 layer of Ar. At the bottom is shown, aligned with the 32-layer result, the corresponding differential cross section measured in the gas phase.

be reflected from the surface. The multiple scattering of electron waves in the Ar solid and between the molecule and the solid changes the partial wave content of the isolated N_2^- resonance and consequently the parameters of the state (i.e., lifetime, energy, electron capture cross section and symmetry). As the N_2 molecule is placed closer to the substrate (i.e., as the number of Ar layers is reduced from 32 to 0) these parameters are further affected by the metal surface; the lineshape of the $N_2^- \, ^2\Pi_g$ excitation function is consequently modified exhibiting appreciable changes in the magnitude and width of oscillatory struc-

modified exhibiting appreciable changes in the magnitude and width of oscillatory structure. At large Ar thicknesses these modifications are probably due to changes in multiple scattering conditions in the RGS film substrate and the increased contributions from multiple scattering in the metal, but when N_2 is very close to the metal surface the appearance of new decay channels in the latter can considerably shorten the lifetime. The orientation of the N_2 molecule may be modified near the metal surface, thus also causing changes in the resonance parameters.

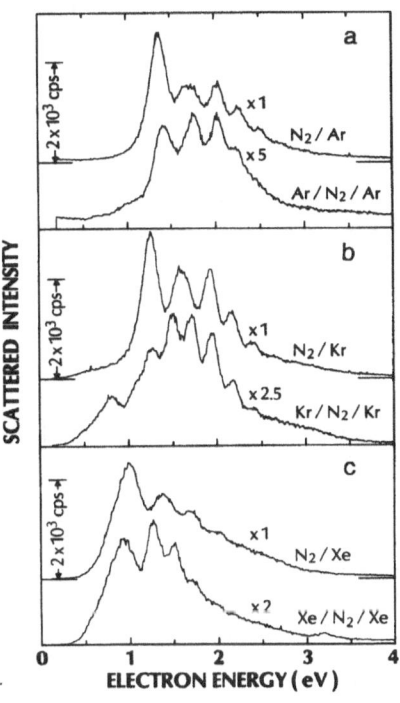

Figure 5. Excitation function for the v=1 level of N_2 deposited on the surface of a 32-layer rare gas film and with a 16-adlayer of the same rare gas is shown for Ar, Kr and Xe in (a), (b) and (c), respectively.

The effect of multiple scattering of the incoming and outgoing electron waves is even more obvious from Fig. 5 where the excitation function of the v=1 level of N_2 deposited onto and embedded inside[19] a 32-ML film of Ar, Kr and Xe are shown in (a), (b) and (c), respectively.[13] Here, it may be seen that both the energy and line shape of the $^2\Pi_g$ N_2^- state are dependent on the nature of the substrate and coverage by additional RGS layers, whereas the energy-integrated magnitude of the resonant phenomena is practically the same for all cases. However, the relative magnitude of each peak in the oscillatory structure may differ considerably. One notices that the lowest energy peak in the bulk spectra of Ar and Kr has considerably diminished compared to that recorded from the surface; it is also shifted down in the bulk by 0.27 and 0.43 eV, respectively, due to the increased electronic polarization in the overlayer. For Xe the lowest energy peak is no longer visible in the overlayer excitation function [Fig. 5(c)].

INTERATOMIC AND INTERMOLECULAR EXCITATION

In solids, it is possible to excite vibrational modes which involve the motion of two or more fundamental constituents (i.e., atoms or molecules) of the media. When these interatomic or intermolecular excitations involve oscillations of a large number of the fundamental constituents having a specific wave vestor, they are referred to as phonon

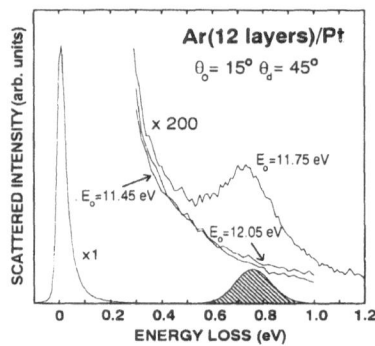

Figure 6. Electron-energy-loss spectra of a 12-layer film of Ar recorded for three incident energies E_o. The hatched curve results from the calculation of the interaction between a time-dependent charge located on a lattice site and the acoustical phonons.

modes. Although it appears that there can be no link between these excitations and the gas-phase energy loss vibrational modes which can only be intramolecular, intermolecular modes in solids can sometimes be related to frustrated rotations (i.e., librational modes) and momentum energy exchange (translational mode) of the same target in the gas phase. Furthermore, it is always possible that the mechanisms responsible for gas-phase electron scattering, be effective in producing excitations which are only found in the condensed phase (e.g., intermolecular excitations). In this section, it is shown that electron resonances found in solid Ar and amorphous H_2O films can enhance considerably the phonon and intermolecular excitations in these solids, respectively.

Fig. 6 shows HREEL[20] spectra of a 12-layer film of Ar recorded at $\theta_o = 15°$ and for three incident energies slightly below the n = 1,1' (the prime denoting the j = $\frac{1}{2}$ spin-orbit partner) bulk-exciton energies, which are located at 12.06 and 12.24 eV, respectively.[21]

The hatched curve results from the calculation of the energy-loss distribution to be discussed later. The peak near 0 eV corresponds to electrons scattered quasielastically from the film whereas the tail arises from multiple energy losses to phonon modes of the crystal. In the $E_0 = 11.75$ eV spectrum, a maximum around 0.75-eV energy loss with a FWHM of about 0.3 eV appears in the multiphonon excitations of the crystal. This feature depends strongly on the incident energy as it can be seen from the comparison with the measurements at $E_0 = 11.45$ and 12.05 eV. To show this behavior more clearly, the scattered electron intensity as a function of the incident energy (i.e., the excitation function) at the

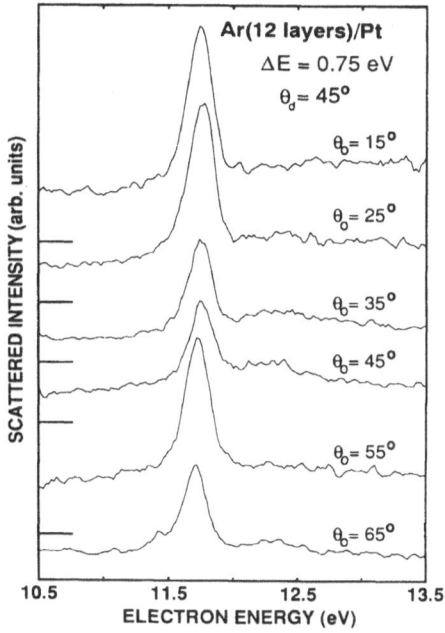

Figure 7. Scattered-electron intensity at the energy loss $\Delta E=0.75$ eV as a function of the incident electron energy recorded for several angles of incidence θ_0 and analysis at $\theta_d = 45°$ on a 12-layer film of Ar condensed on Pt.

energy loss $\Delta E = 0.75$ eV is shown in Fig. 7 for $15° \leq \theta_0 \leq 65°$ on a 12-layer film of Ar.[20] As one can see, the peaks are located at the same energy independent of the incident angle, thus ruling out the possibility that this feature could arise from diffraction effects characteristic of the scattering geometry.

A vibrational enhancement taking place over such a narrow energy range is expected if the scattering mechanism proceeds through an electron resonance (i.e., formation of a transient anion). Electron resonances in rare gases are well known from isolated electron-atom scattering.[1] In the case of Ar, the two lowest-lying core-excited resonances (i.e., one-hole, two-electron states) $(\ldots3s^2\, 3p^5\, 4s^2)\, {}^2P_{3/2}, {}^2P_{1/2}$ occur at 11.10 and 11.27 eV, respectively, with a FWHM of about 3 meV.[22] Since they appear below their parent states $(\ldots3s^2\, 3p^5\, 4s)\, {}^3P_2, {}^3P_1, {}^3P_0,$ and ${}^1P_1$ at 11.549, 11.624, 11.724, and 11.829 eV, respectively, they correspond to Feshbach resonances.[1] The present feature in Fig. 7 can be correlated to this spin-orbit-split resonant pair that would be shifted by about 0.6 eV to higher energy as well as broadened in the condensed phase. More specifically, the fact that the sharp peak in Fig. 7 lies 0.25 eV below the $n = 1$ bulk exciton, but 0.09 eV above the surface counterpart, suggests strongly that the anion formation arises mainly from the binding of the electron to the $n = 1,1'$ excitons in the bulk. Depending on the lifetime of the resonance and the magnitude of the interaction between the anion and the surrounding atoms, a lattice distortion is induced resulting in multiphonon excitations of the crystal after the decay of the resonance.

At first sight the observed shift of the resonance to higher energy from the gas to the solid may appear surprising since one would expect a net shift to lower energy simply from the electronic polarization of the medium. Actually, this shift can be qualitatively understood if we consider the following hypothetical process to calculate the anion (i.e., one-hole, two-electron state) energy in the solid E_r^S. Suppose that we extract an atom from the bulk of the crystal into the vacuum by providing the cohesive energy per atom E_C for a van der Waals solid, and that we ionize the atom by supplying the energy E_I^G to obtain a positive ion (i.e., a hole) in its ground state and an electron at rest. Then, we move the positive ion and the electron back into the bulk of the solid along with an extra electron. Finally, we allow the binding of the hole and the two electrons to form the anion state. From the conservation of total energy this process yields

$$E_n^S - V_p - B_r^S = E_r^S \; , \tag{1a}$$

where

$$E_n^S \equiv E_C + (E_I^G - V_A - B_n^S) \; . \tag{1b}$$

The expression (1b), which involves only one electron and the hole, stands for the process that corresponds to the formation of the exciton n of energy E_n^S (i.e., the parent state) in the solid. In Eqs. (1), V_p is the polarization energy of the surrounding medium by the hole (i.e., positive ion). The term V_A is the electron affinity of the crystal; it corresponds to the energy of the bottom of the lowest conduction band with respect to the vacuum level. Finally, B_n^S and B_r^S are the electron-hole and electron-exciton binding energies referred to the bottom of the lowest conduction band, respectively. Thus, one can see from Eq. (1a) that three sources may contribute to the shift of the resonance energy from the gas to the solid phase. We have the change of the transition energy from the ground to the parent state, the modification of the binding energy of the extra electron to the parent state, and the electron affinity of the crystal. Considering the mean energy E_n^S for the $n = 1,1'$ excitons of the bulk[21] of 12.15 eV, the value for the resonance energy E_r^S of 11.8 eV, and V_A of 0.2-0.3 eV above the vacuum level,[23] one deduces from Eq. (1a) a value of 0.55-0.65 eV for B_r^S. Since this compares well to the binding energy in the gas phase,[22] the present shift at higher energy is explained mainly by the blueshift of the transition energy to the parent states and the negative electron affinity of the Ar crystal.

The effect of a transient anion state on the surrounding Ar matrix can be described by considering the interaction between a time-dependent charge located on a lattice-site and the acoustical phonons treated by a quantum-mechanical description. Such a problem has been addressed in detail recently by Mills[24] in connection with the suppression of the resonance elastic peak[25] in electron scattering from a van der Waals solid. The result of the analysis leads, among other things, to an expression for the multiphonon loss feature shown in Fig. 6 whose intensity and mean energy loss depend mainly on the electron residence time τ_R and the polarizability of the surrounding atoms or molecules (i.e., dielectric constant ε of the medium). Using the expression of Mills[24] to analyse the HREEL results in solid Ar and τ_R as the only adjustable parameter, the hatched curve in Fig. 6 is obtained.

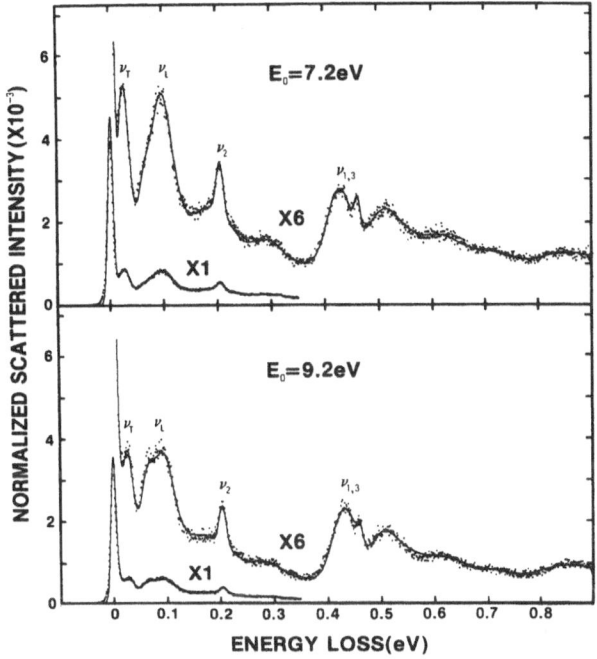

Figure 8. HREEL spectra of a 30-layer ice film recorded at $\theta_o = 14°$ and $\theta_d = 45°$ (dotted curves). The superimposed continuous curves result from multiple-scattering calculations of the electron backscattered intensities. E_o represents the incident electron energy.

From this fit, the lifetime of the core-excited Ar^- state in solid Ar has been estimated to be 2.2×10^{-14} sec.[20] Similar results have been obtained in solid Kr and Xe, indicating that in both of these solids the lowest energy core-excited resonance has a lifetime of the order of 10^{-14} sec.

HREEL spectra[26] for a 30-ML film of amorphous ice recorded at $\theta_o = 14°$ and $\theta_d = 45°$ and for incident energies of 7.2 and 9.2 eV are shown in Fig. 8. These spectra can be compared with infrared and Raman spectra of ice. The major energy-loss peaks are identified as a translational mode (ν_T) at 24 meV, librational modes (ν_L) at 62 and 95 meV, the bending mode (ν_2) at 205 meV and the unresolved stretching modes $(\nu_{1,3})$ at 425 meV.

The remaining features along with the background signal are mainly attributed to overlapping and multiple-scattering.[11] In order to extract scattering cross sections per scatterer from such spectra, HREEL spectra recorded at different incident electron energies must be

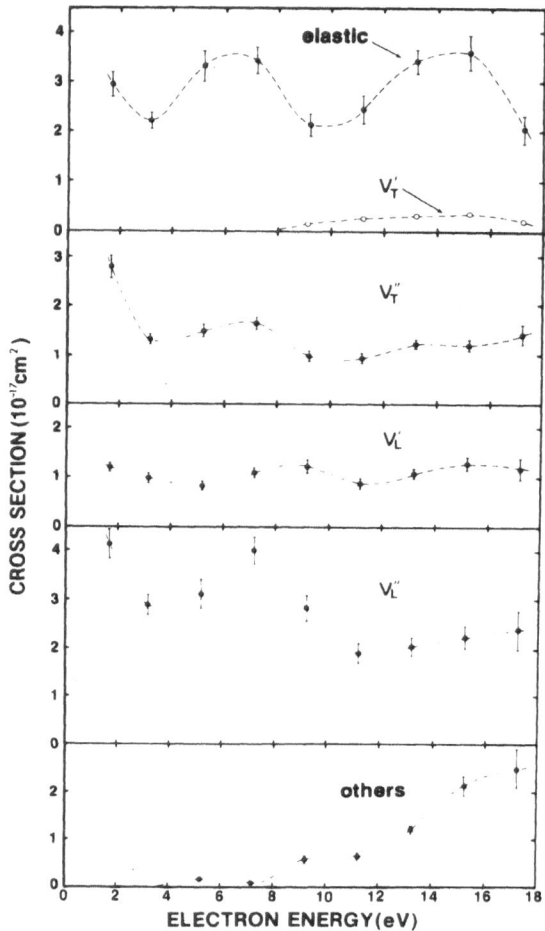

Figure 9. Energy dependence of the elastic and intermolecular vibrational electron scattering cross sections per scatterer in amorphous ice derived from multiple scattering analysis of energy loss spectra. The bottom curve in (a) represents the cross section ascribed mainly to the sum of electronic excitations.

fitted to multiple scattering calculations[10,11] based on a model which takes into account the magnitudes of the relevant cross sections. Such fits are shown by the continuous curves which pass through the experimental points in Fig. 8. In this manner, the energy dependence of the cross sections for the major energy losses can be generated, as shown in Fig. 9 for intermolecular excitations.

The absolute electron scattering cross sections for exciting the translational (v'_T and v''_T) and librational (v'_L and v'_L) modes of amorphous ice in the 2-18 eV range[11] are shown in Fig. 9. The bottom curve labeled "others" represents essentially contributions to the inelastic cross section arising from electronic excitation and ionization. The behavior of the elastic cross section is shown on top of Fig. 9. The v'_T and v'_L lattice modes are clearly characterized by a peak around 7 eV having $\simeq 2$ eV FWHM and a rise at low energy. In the curves of Fig. 9, any monotonic increase toward lower energies has been ascribed to non-resonant scattering (i.e., electron-dipole and electron-polarization interaction)[11] which is present throughout the spectra and possibly accounts for the constant portion in the magnitude of the cross sections. In the absence of a large quadrupole moment, the increasing magnitude of the cross section for vibrational excitation at very low energies in gaseous H_2O can be ascribed to the dipolar interaction.[4] Interestingly, the same behavior is found for intermolecular excitation and also possibly due to the dipolar interaction, since no such a rise at low energies is found in the overtones of these modes.

From comparison of the energies and widths of the features of Fig. 9 with those found in the gas-phase cross sections, the strong enhancement near 7 eV of the amplitude of the broad peak in the v'_L cross section correlates to a $^2B_1$ anion state. While the excitation of an intramolecular vibrational mode by a resonance process can be inferred from gas-phase mechanisms, the excitation of lattice vibrational modes by the long-lived $^2B_1$ state requires a new mechanism. Because of the overlapping of electron-molecule interaction potentials in the condensed phase, a transient anion state is also coupled with the nuclear and electronic polarization modes of the surrounding medium. When an electron is localized in the vicinity of an H_2O molecule, a force is exerted between the H_2O^- state and the permanent dipoles of the surrounding molecules which pull these latter toward the anion, initiating translational and rotational motion. Calculations with a classical harmonic oscillator model[26] indicate that, a resonance whose lifetime is greater than 10^{-14} sec, can transfer a significant amount of energy to librational motion by this mechanism, which is also effective in producing translation motion for resonance times of the order or larger than 10^{-13} sec.

ELASTIC SCATTERING

With the exception of amorphous compounds, it has not been possible to link elastic scattering in the gas-phase with that occurring at the surface and within solids. In crystalline material, elastic scattering is dominated by long-range order and thus reflects the geometrical arrangement of the constituent atoms or molecules. The structure in the energy dependence of elastically transmitted or reflected currents in molecular films can be explained by invoking multiple scattering of the electron waves between the potentials of the individual atoms or molecules, as if *intramolecular* multiple scattering was irrelevant.[27] This is somewhat surprising, since in gases electron resonances cause easily identifiable structures in the elastic cross section.[1] For example, the N_2^- ($^2\Pi_g$) anion state increases elastic scattering by a factor of 3 in gaseous N_2 around 2 eV.[1] In condensed N_2, no change can be perceived in the elastic cross section although the vibrational excitation cross sections, are increased up to two orders of magnitude by the $^2\Pi_g$ resonance state.[28]

In the case of amorphous solids, the situation is different because the electron coherence and interference is only short range. The electron energy dependence of the elastically scattered intensity from amorphous materials can be expressed via a multiple scattering theory[10,11] as an ensemble averaged scattering cross section per scatterer. This cross section is represented[29] in the single event approximation as

$$Q_s = \frac{1}{\hbar}\left(\frac{m}{2\pi\hbar^2}\right) \int_0^\pi \int_0^{2\pi} \frac{d\,Q_g(\bar{K})}{d\,\Omega}\, S(|\bar{K}|)\, \sin\theta\, d\theta\, d\phi \qquad (2)$$

where Q_g is the gas-phase total elastic cross section, $|\bar{K}| = [2(1-\cos\theta)]^{1/2}\,|\bar{k}_o|$ and θ is the angle between the incident $\bar{k}_o$ and scattered electron wave vectors. In this expression, the scattered amplitude is the coherent sum of amplitudes originating from individual sites, whereas the sum of amplitudes from multiple scattering from different sites is neglected.[29]

$S(\bar{K})$ is a structure factor related to the static correlation between sites in a condensed medium.[30] Expressions 2 have been applied[11,31] to explain qualitatively the results of elastic scattering in amorphous solid Xe and H_2O. In particular, it can reproduce the two maxima found in the energy dependence of the elastic scattering cross section of amorphous H_2O (at the top, Fig. 9) which are thus found to be related to similar structures in the magnitude of $S(\bar{K})$ for amorphous ice.[32]

CONCLUSIONS

It has been shown that recent advances in electron collision experiments make it possible to obtain reliable information on low-energy electron scattering phenomena at the surface of and within solids. Comparing these results with those previously obtained in the gas-phase, allows the information on isolated electron-target interactions to be transferred, with some modifications, to the solid phase. The scattering of a low-energy electron with an atom or a molecule and the production of phonons, intermolecular, and intramolecular excitation were shown to be sensitive to the "environment" in the condensed phase (i.e., they are dependent on the nature of the neighbors surrounding the target molecule). This may sometimes render the transfer of information to particular condensed systems more difficult; but on the other hand, this dependence makes it possible to modify the parameters of the scattering process, and thereby the energy deposited into solids, by changing the "environment" of a specific target. This control of a relatively large amount of energy offers promising possibilities for numerous applications including the manufacture of better insulators and radiosensitizing drugs for radiotherapy.

ACKNOWLEDGMENTS

The author would like to thank Dr. Michael Huels for useful suggestions and corrections and Mrs Francine Lussier for the preparation of this manuscript. This work was supported by the Medical Research Council of Canada and the Canadian Center of Excellence in Molecular and Interfacial Dynamics.

REFERENCES

1. For a review see Schulz, G.J., Resonances in electron impact on atoms, *Rev. Mod. Phys.* 45:378 and Resonances in electron impact on diatomic molecules, 423 (1973); Christophorou, L.G., *Electron-Molecule Interactions and their Applications*, vol. 1 and vol. 2, Academic Press, Orlando (1984); Allan, M., Studies of triplet states and short-lived negative ions by means of electron impact spectroscopy, *J. Electr. Spectr. Rel. Phenom.* 48:219 (1989).

2. Sanche, L., Primary interactions of low-energy electrons in condensed matter, Chap. 1 in *Excess electrons in dielectric media*, Ferradini C. and Jay-Gerin J.-P. eds. CRC Press, Boca Raton (1991).

3. Palmer R.E. and Rous, P.J., Resonance in electron-scattering by molecules on surfaces, *Rev. Mod. Phys.* 64:383 (1992); Sanche, L., Low-energy electron scattering from molecules on surfaces, *J. Phys. B.* 23:1597 (1990).

4. Takayanagi, K., Scattering of slow electrons by molecules, *Prog. Theor. Phys. Suppl. Japan* 40:216 (1967).

5. Pendry, J.B., , *Low-energy Electron Diffraction*, Academic Press, New York (1971).

6. Gerber A. and Herzenberg, A., Resonance scattering of electrons from N_2 adsorbed on a metallic surface, *Phys. Rev. B* 31:6219 (1985).

7. Davenport, J.W., Ho, W. and Schrieffer, J.R., Theory of vibrationally inelastic electron scattering from oriented molecules, *Phys. Rev. B* 17:3115 (1978).

8. Sanche, L., Transmission of 0-15 eV monoenergetic electrons through thin-film molecular solids. , *J. Chem. Phys.* 71:4860 (1979); Sanche, L., Bader, G., Caron, L., Transmission of 0-15 eV monoenergetic electrons through aliphatic and alicyclic hydrocarbons films, *J. Chem. Phys.* 76:4016 (1982).

9. Sanche L. and Michaud, M., Interaction of low-energy electrons (1-30 eV) with condensed molecules: II. Vibrational-librational excitation and shape resonances in N_2 and CO films , *Phys. Rev. B.* 30:6078 (1984).

10. Michaud M. and Sanche, L., Total inelastic cross section for slow electrons (1-20 eV) scattering in solid H_2O, *Phys. Rev. A* 36:4672 (1987).

11. Michaud M. and Sanche, L., Absolute vibrational excitation cross sections for slow electrons (1-18 eV) scattering from solid H_2O, *Phys. Rev. A.* 36:4684 (1987).

12. Sanche L. and Michaud, M., Electron energy-loss vibronic spectroscopy of matrix isolated benzene and multilayer benzene films, *Chem. Phys. Lett.* 80:184 (1981).

13. Michaud M. and Sanche, L., The $^2\Sigma_g$ shape resonance of N_2 near a metal surface and in rare gas solids, *J. Electr. Spectr. Rel. Phenom.* 51:237 (1990).

14. Marsolais, R.M., Michaud, M. and Sanche, L., Near-threshold electronic excitation in electron impact of multilayer physisorbed N_2 and CO, *Phys. Rev. A* 35: 607 (1987).

15. Wong, S.F.,Michejda J.A., and Stamatovic, A., personal communication.

16. Rous, P.J., Palmer, R.E., and Willis, R.F., Resonance electron scattering from adsorbed molecules: Angular distribution of inelastically scattered electrons and application to physisorbed O_2 on graphite, *Phys. Rev. B* 39:7552 (1989).

17. Teillet-Billy D. and Gauyack, J.-P., Angular-distribution of electrons resonantly scattered by CO molecules adsorbed on metal surface, *Nucl. Instrum. Methods* B58:393 (1991).

18. Rous, P.J., Resonance behavior of electron-scattering from nitrogen molecules adsorbed on metallic surfaces, *Surf. Sci.* 260:361 (1991).

19. When small molecules are sandwiched between rare gas layers some diffusion may occur; Sanche, L., Bass, A.D., Parenteau L., and Gortel, Z., Measuring diffusion in thin films by dissociative electron attachment: O_2 in Kr, *Phys. Rev.* B48:5440 (1993).

20 Michaud, M. and Sanche, L., Direct observation of a Feshbach-type electron resonances in solid Ar, *Phys. Rev. B* 47:4131 (1993).

21. Zimmerer, G., in *Excited-State Spectroscopy in Solids*, edited by U.M. Grassono and N. Terzi, North-Holland, Amsterdam (1987), p. 37.

22. Brunt, J.N.H., King, G.C. and Read, F.H., Resonance structure in elastic electron-scattering from helium neon and argon, *J. Phys. B* 10:1289 (1977).

23. Perluzzo, G., Bader, G., Caron, L.G., and Sanche, L., Direct determination of electron band energies by transmission interference in thin films, *Phys. Rev. Lett.* 55:545 (1985).

24. Mills, D.L., Resonant scattering of slow-electrons in molecular-solid - suppresssion of the elastic beam, *Phys. Rev. B* 45:36 (1992).

25. Fano, U., Stephens, J.A., and Inokuti, M., Absence of resonances in the elastic-scattering of electrons in molecular solids, *J. Chem. Phys.* 85:6239 (1986).

26. Michaud, M. and Sanche, L., Opening of new decay channels for core-excited resonances, *Phys. Rev. Lett.* 59:645 (1987).

27. Michaud, M., Sanche, L., Gaubert, C., and Baudoing, R., Low-energy electron reflection and transmission spectroscopy of Ar film grown on platinum, *Surf. Sci.* 205:447 (1988).

28. Sanche L. and Michaud, M., Vibrational excitation via resonances in electron scattering from N_2 multilayer films, *Chem. Phys. Lett.* 84:497 (1981).

29. Davis, T.H., Schmidt, L.D., and Minday, R.M., Kinetic theory of excess electrons in polyatomic gases, liquids, and solids, *Phys. Rev. A,* 3:1027 (1971).

30. Van Hove, L., Correlations in space and time and Born approximation scattering in systems of interacting particles, *Phys. Rev.* 95:249 (1954).

31. Bader, G., Perluzzo, G., Caron, L.G., and Sanche, L., Elastic and inelastic electron mean free path determination in solid xenon from electron transmission experiments, *Phys. Rev. B.* 26:6019 (1982).

32. Narten, A., Venkatesh, C. G., and Rice, S.A., Diffraction pattern and structure of amorphous solid water at 10 and 77 K, *J. Chem. Phys.* 64:1106 (1976).

ANION FORMATION IN LOW ENERGY ELECTRON IMPACT TO GASEOUS AND CONDENSED MOLECULES

Eugen Illenberger

Institut für Physikalische und Theoretische Chemie
Freie Universität Berlin
Takustrasse 3
D-14195 Berlin
Germany

INTRODUCTION

Low energy electrons can be very reactive in that they are effectively captured by many molecules which then undergo rapid unimolecular decompositions (see also the contribution *Electron Attachment to Molecules* in this Volume). The cross section for such processes can be very large, several orders of magnitude larger than ionization or excitation cross sections.[1,2] Electron attachment reactions to molecules in the gas phase have been studied to some extent within the last two or three decades. Analogous experiments in clusters and in condensed molecules, on the other hand, have only recently been performed.

This contribution focuses on reactions in weakly bound van der Waals clusters and condensed molecules induced by low energy (0-15 eV) electron capture. Particular emphasis will be placed on the question how the basic quantities of such processes (i.e. attachment energy, evolution of the negatively charged compound, and energy distribution of the products ultimately formed) behave on proceeding from isolated molecules under single collision conditions over molecular aggregates to molecules condensed in multilayer amounts on a metallic substrate.

During the past decade there has been an enormous increase of experimental and theoretical work concerning the properties of clusters.[3-5] Since the term "cluster" is frequently used with different meanings, we will in the context of this contribution assign a group of molecules in the gas phase which are held together by weak intermolecular forces as a molecular cluster or a molecular aggregate. In that sense, a free molecular cluster can be viewed as representing a link between molecules in the gas phase and in the condensed phase (bulk liquids or solids).

Traditionally, science has concentrated on understanding these two separate forms of matter and only recently fundamental questions such as the minimum number of atoms necessary to evolve an electric conduction band or the number of atoms or molecules required to support a bulk crystal structure have emerged.

Electron attachment and detachment processes or, more generally, electron transfer reactions play a key role in many fields of pure and applied science, in the gas phase (discharges, gaseous dielectrics, atmospheric processes, etc.) as well as in the condensed phase (electrochemistry and chemical reactions in general[6]). Within these different fields the area of non covalent van der Waals interactions between molecules and their behavior when they are subjected to excitation or ionization plays a central role in many fields of present-day natural science.

Negatively charged clusters are of particular interest since they are considered to provide models for excess electrons in liquids.[7] Apart from this, as we will demonstrate, the study of electron attachment reactions in homogeneous and heterogeneous clusters allows insight into very fundamental properties like electronic states of molecular ions, selection rules to populate them as well as intra- and intermolecular energy and electron transfer processes.

Consider a neutral molecule M (isolated or in close proximity to other molecules) interacting with free electrons. At certain specific energies, resonant electron capture can directly form a "temporary negative ion" (TNI):

$$e^- + M \quad \rightarrow \quad M^{-(*)} \tag{1}$$

This ion can principally evolve according to the following scheme:[2,8]

$$M^{-(*)} \xrightarrow{\tau_a} M^{(*)} + e^- \tag{2}$$

$$M^{-(*)} \xrightarrow{\tau_d} R + X^- \tag{3}$$

$$M^{-(*)} \xrightarrow{\tau_s} M^- + Energy \tag{4}$$

Process (2) is autodetachment recovering the neutral molecule (eventually in an excited state), (3) is dissociative electron attachment leading to stable fragments, and (4) is stabilization into the thermodynamically stable parent anion. This last process can only occur for molecules possessing a *positive electron affinity*.

It is evident that the *formation* and *decomposition* of a TNI strongly depends on whether the overall process occurs on a single and isolated molecule ("unimolecularly") or within molecules at some stage of aggregation.

In the gas phase under single collision conditions the only stabilization mechanism (4) is radiative cooling which is generally much slower than the competing processes (2) and (3), so that in a mass spectrometric experiment one usually detects fragment ions. Only in certain larger polyatomic molecules, autodetachment may be delayed to an extend that the parent anion can be observed in a mass spectrometer, i. e. the lifetime of the TNI with respect to autodetachment, τ_a, must be at least in the μs range. Such long lived parent anions are observed from molecules possessing a resonance near zero eV and when dissociation channels of the TNI are not yet accessible at that energy.[2,8] Decomposition into stable fragments (3) occurs on a time scale between $\tau_d \approx 10^{-14}$ s (direct electronic dissociation along repulsive energy surfaces) to $\tau_d \gtrsim 10^{-12}$ s (electronic and vibrational predissociation).

On proceeding to van der Waals clusters, the situation may change with respect to the *formation* of a TNI, its *energy* and its *decomposition*: (a) in addition to direct electron capture by an individual molecule, inelastic electron scattering by one molecule and capture of the slowed down electron by another molecule within one cluster (self- or autoscavenging) can considerably contribute to anion formation at higher primary electron energies,[9-11] (b) due to polarization interaction of the negative charge, the *energy* of the TNI in the cluster is shifted to lower energies, and (c) the *evolution* of the TNI can largely be affected by the cluster environment which can lead to stable anions M$^-$ by collisional stabilization, thereby evaporating the target cluster ("evaporative electron attachment") or the formation of solvated ions of the form $M_n \cdot X^-$ in the course of scattering processes within the ionized target cluster.[11-13]

The following substantial questions then arise:

1.   The electronic states populated by electron capture are controlled by certain symmetry rules.[16] Are these selection rules which hold in the single molecule-electron frame of reference violated in clusters as it is sometimes observed in the condensed phase?

2.   Does a cluster environment provide the proper medium to carry off excess energy necessary to prepare thermodynamically stable molecular anions in their relaxed configuration, not accessible under isolated conditions (collisional stabilization instead of the slow radiative stabilization (4b))?

3.   Which are the products ultimately formed? The point is then to record the *resonance profiles* of the different product ions, i.e., their formation probability as function of the initial electron energy. It must be emphasized that the energy profile of any product ion reflects the primary step of electron capture, regardless of the mechanism and complexity of its formation process. In other words, the ion yield curve of any product reflects the "electron absorption" profile of the target from which it is formed. This is true as long as the anion is formed in the primary step, i.e. inelastic scattering processes (secondary reactions) can be neglected.

4.   Many dissociative attachment reactions are known to proceed directly along repulsive potential energy surfaces and hence via the release of considerable amounts of excess translational energy.[2,8] How does the distribution of excess energy change when the repulsive TNI is formed within a cluster?

Electron impact to condensed molecules can also lead to temporary negative ions. When *desorbed* fragment anions are recorded as function of the primary electron energy, the ion yield curve can exhibit characteristic resonances similar to the gas phase as shown in a number of studies by Sanche et al.[14,15]

Although many aspects of anion formation in the condensed phase and in clusters are similar, some significant differences must be noted: Consider a TNI decomposing into R+X$^-$ due to the R-X$^-$ repulsive nature of its electronic state. In a cluster this can give rise to the observation of X$^-$, or solvated ions of the form X$^- \cdot M_n$. "Cage effects" can also stabilize the TNI leading to M$^-$ or M_n^-. In the condensed phase, on the other hand, we have to distinguish between the two possibilities

$$M^{*-} \quad \rightarrow \quad R_{ad} + X_{free}^- \hspace{6cm} (5a)$$
$$\rightarrow \quad R_{free} + X_{ad}^- \hspace{6cm} (5b)$$

where the subscript (ad) and (free) denotes the particle remaining at the surface or emitted into vacuum, respectively. Due to polarization interaction of the anion X_{ad}^- with the surrounding molecules, the energy of limit (5b) is expected lower, thus prefering the decomposition into this channel. In the present surface experiment only ions emitted into vacuum can *directly* be observed while reaction (5b) results in charging of the film. In clusters the situation is similar, however, in both cases the negative charge (localized on X^- or $X^- \cdot M_n$) is detected with the mass spectrometer. Furthermore, in the gas phase, electron attachment always occurs on a *neutral particle* since capture of more than one electron by a single target aggregate is most unlikely under the present experimental conditions.

In this contribution we present results of attachment of energy selected free electrons in the range from 0 to 15 eV to molecules in different stages of aggregation: the gas under single collision conditions, a cluster beam and a multilayer film adsorbed onto a metallic substrate.

In the first case, an effusive beam of molecules is crossed with an electron beam of defined energy and the resultant anions are recorded with a mass spectrometer. In the cluster experiment the effusive molecular beam is replaced by a supersonic jet containing a distribution of aggregates. Finally, molecules are condensed in the ultra high vacuum (UHV) in multilayer amounts on a cold metallic substrate and negative ions ejected from the multilayer film into vacuum as a result of low energy electron impact are observed.

EXPERIMENTAL

The experiments were performed with two different arrangements, one for the gas phase experiments and another for the condensed phase experiment. The "effusive beam" and "supersonic beam" techniques, have been described in detail before.[2,8,11] Electron stimulated desorption of anions is studied in an UHV apparatus which will be described below. In both equipments the electron beam is produced by a "trochoidal electron monochromator" (TEM)[17] which uses a magnetic field to collimate and guide the electron beam. This magnetic field is established by a pair of Helmholtz coils which in both set ups is mounted outside the vacuum system. The monochromator generates a beam of nearly monoenergetic electrons (FWHM = 0.1-0.2 eV) at reasonable intensities (5-50 nA) down to very low electron energies.

Figure 1 shows the experimental setup for the gas phase experiments. The electron beam is either crossed with a molecular beam effusing from a capillary directly connected with the reaction chamber or with a supersonic beam produced by adiabatic expansion of the gas seeded in Ar or He through an 80 μm nozzle. The supersonic beam contains a *distribution* of clusters. The average size can be varied to some extent by the expansion conditions (stagnation pressure, nozzle temperature).

The expansion converts the enthalpy associated with random molecular motion into directed mass flow.[2,18,19] This is illustrated in Fig. 2: at low stagnation pressure, when the mean free path of the molecules is larger than the nozzle diameter, $\lambda > d$, the molecular beam effusing from the nozzle possesses a velocity weighted Maxwell - Boltzmann distribution. When $\lambda \ll d$, we have a free jet expansion associated with a narrow velocity distribution. For a monoatomic gas, the terminal velocity can be expressed as

$$v_t = (5kT_0/M)^{1/2} \qquad\qquad (6)$$

where T_0 is the temperature of the *source*. The (translational) temperature of the molecules in the beam is determined by the width of the velocity distribution Δv. The temperature T in the beam is approximately given by the Poisson equation

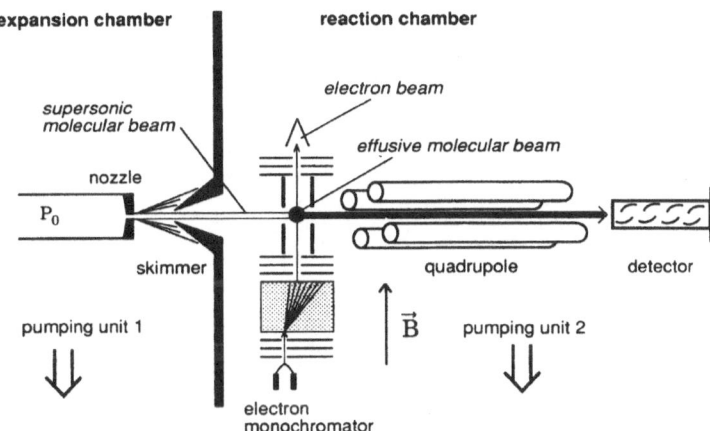

Figure 1. Schematic of the experimental setup for the study of free electron attachment to molecules and clusters. From ref.20.

$$\frac{T_0}{T} = \left(\frac{p_0}{p}\right)^{\frac{1-\kappa}{\kappa}} \qquad\qquad (7)$$

where $\kappa = c_p/c_v$ and p_0 and p is the stagnation pressure and the pressure in the beam respectively. In polyatomic molecules we have $\kappa \approx 1$ and cooling is less effective. In this case one applies the *seeded beam* technique. Here the molecules of interest are diluted in an inert carrier gas (He, Ar) which acts as a refrigerant to cool the molecules until they polymerize. Since the molecules are diluted in small amounts in the carrier gas the velocity of the beam is then still approximately given by Eqn. (6) with M now the mass of the carrier gas. If helium is used this can result in large values for the translational energy of clusters in the laboratory frame of reference. As an example, if CF_3Cl is seeded in He, the pentamer $(CF_3Cl)_5$ will move with a translational energy of more than 8 eV along the direction of the beam! In actual practice, there is always some "velocity slip" which will lower the beam velocity to some extent. To avoid discrimination in the detection of product ions due to large translational energies perpendicular to the axis of the mass spectrometer, we have chosen to place the mass spectrometer *in line* with the molecular beam (Fig. 1).

Electron stimulated desorption of anions from condensed molecules is studied with the arrangement shown in Fig. 3. The apparatus consists of a TEM, a quadrupole mass spectrometer, a manipulator which carries the metallic substrate (polycrystalline gold), and a closed cycle He refrigerator. The substrate can be heated (by resistive heating) and cooled down to 20K (by the refrigerator). All components are housed in an UHV chamber equiped with a turbomolecular pump and an ion getter pump. The working pressure is in the range of 10^{-10} mbar.

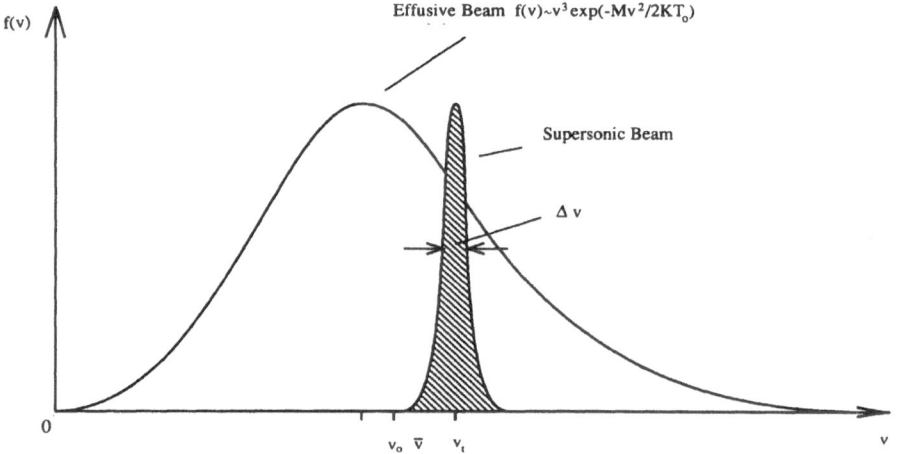

Figure 2. Velocity distribution in an effusive and a supersonic molecular beam.

The metallic substrate is mounted on the manipulator. It is *thermally connected* to the head of the refrigerator with a flexible copper cord, but *electrically isolated* by a saphir plate. The thickness of the molecular film is roughly estimated by directly exposing a defined amount of gas through a capillary to the cold surface.

The energy resolution of the electron beam and calibration of the energy scale can principally be performed by removing the metallic substrate from the electron beam and recording the SF_6^{-*} gas phase resonance. SF_6 captures electrons within a very narrow resonance near zero eV and thus gives the energy resolution of the beam and defines zero of the vacuum level.[2,8] However, since the electric potential of the metallic substrate and the ion entrance electrode of the mass spectrometer are slightly different, the energy of the electrons crossing the axis of the mass spectrometer depends on the position of the sample. Moreover, the interaction of the electron beam with the molecular film can lead to charging of the film and hence affect the energy of the electrons impinging the surface. We therefore use the onset of the electron current into the metallic substrate as zero for the vacuum level.

For the gas phase experiments a time-of-flight (TOF) analysis is applied to obtain information on the translational excess energy of the corresponding product ions. This technique has elsewhere been described in detail.[2,8] (See also the article *Electron Attachment to Molecules* in this Volume.) In brief, the electron beam is now pulsed (pulse width < 1 μs) and the flight time of the ions (generated within the short time of the electron pulse) from the reaction volume through the quadrupole mass filter to the detector is recorded. Ions

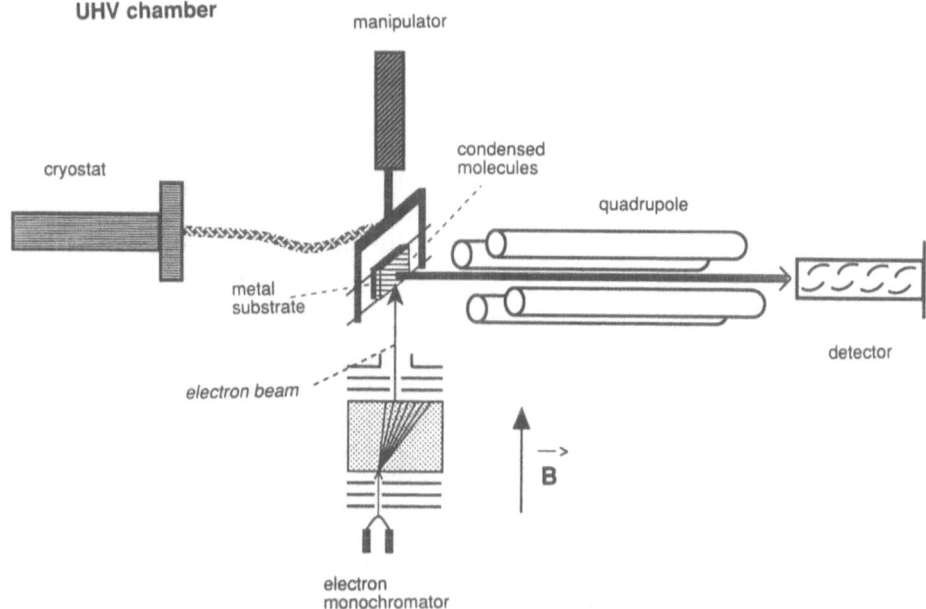

Figure 3. Schematic of the experimental setup for the study of electron induced desorption processes from condensed molecules. From ref.20.

created with low kinetic energy (e.g. thermal ions) exhibit one single peak in the TOF spectrum while fragment ions ejected from the precursor with considerable kinetic energy produce two separated peaks due to "direct" and "turn around" ions. In this latter case the ion draw out field discriminates against initial velocity components perpendicular to the quadrupole axis. This results in a separation (in time) of ions ejected parallel (direct ions) and antiparallel (turn around ions) to the flight tube axis. From the experimentally determined flight time difference, the initial translational energy of the fragment ion can be calculated.

EXPERIMENTAL RESULTS AND DISCUSSION

THE HALOGENATED METHANES CF₃I, CF₃Cl AND CF₄

Isolated CF₃I

It has been shown in a recent publication by our laboratory[20] that CF_3I possesses two resonances, at 0.0 and 3.8 eV. While the 0.0 eV resonance exclusively decomposes into I^- + CF_3, the TNI at 3.8 eV yields the ions F^-, CF_3^- and FI^-. The channel CF_3^- + I was characterized by an extraordinary amount of translational excess energy, namely 2.6 eV which is 85% of the total available excess energy. No parent molecular anion CF_3I^- could be observed under effusive gas inlet (see also the article *Electron Attachment to Molecules* in this Volume).

CF₃I Clusters

In electron attachment to clusters the following new features appear:
1. Formation of ionic products of the form M_n^- ($n \geq 1$) and $M_n \cdot I^-$ ($n \geq 1$) (M = CF_3I) with M_n^- considerably less abundant than $M_n \cdot I^-$ ($\approx$ 1:200).
2. The appearance of distinct new features in the energy profiles of I^-, $M_n \cdot I^-$ and CF_3^-.

By increasing the stagnation pressure, the intensity of the cluster products ($I^- \cdot M_n$, M_n^-) varies in the way that smaller products have a maximum at lower pressure (e.g. $M \cdot I^-$ at 0.9 bar) than larger products (e.g. $M_4 I^-$ at 1.8 bar). Although a mass spectrum is not a direct image of the neutral cluster distribution in the beam (due to fragmentation following ionization), the observed pressure dependence follows the expected behavior of an increasing average cluster size with increasing stagnation pressure.

Figure (4a) shows that the molecular anion M^- appears near zero eV. It is generated via formation of ground state CF_3I^- within the target cluster and subsequent collisional stabilization thereby evaporating the target cluster ("evaporative electron attachment") according to

$$e^- (\approx 0 \text{ eV}) + (CF_3I)_m \rightarrow CF_3I^- \cdot (CF_3I)_{m-1} \rightarrow CF_3I^- + (m-1) CF_3I$$

Of course, the stabilization energy may not in every case be sufficient to produce the maximum of (m-1) evaporated molecules.

Figure (4b) shows the ion yield profiles of I^- and $M \cdot I^-$ (the features remain the same in the homologous series $M_n \cdot I^-$). Apart from their dominant formation near zero eV, distinct new features with maxima near 4 eV and 6 eV become apparent. At zero eV all product ions must have the ground state TNI as precursor which dissociates in the target cluster. In the case of $M \cdot I^-$ the reaction can be expressed as

$$e^- (\approx 0 \text{ eV}) + (CF_3I)_m \rightarrow CF_3I^- \cdot (CF_3I)_{m-1} \rightarrow I^- \cdot CF_3I + CF_3(CF_3I)_{m-2}$$

In Fig. (4b) the intensity of I^- and $M \cdot I^-$ at zero eV is normalized to the same height (the intensity ratio at zero eV and 0.8 bar stagnation pressure is

$I^-:M \cdot I^- \approx 10:1$). Above 2 eV the scale for I^- and $M \cdot I^-$ is then multiplied by 100 and 2, respectively.

Finally, while the ion yield curves of fragments F^- and FI^- are unaffected on going from the effusive beam to clusters, the CF_3^- fragment additionally appears near 6 eV.

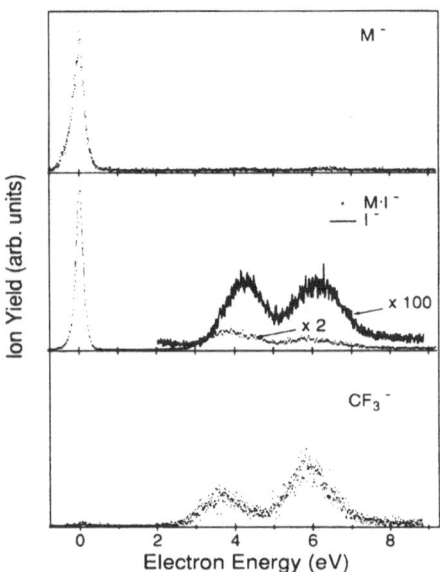

Figure 4. Ion yield curve of some products following electron attachment to CF_3I clusters at 0.8 bar.

For the products I^- and $M_n \cdot I^-$ ($n \geq 1$) we interpret the features at 4 eV and 6 eV as a result of *inelastic electron scattering* in the target cluster prior to electron localization: In the first step the electron is inelastically scattered from one molecule and the slowed down electron is captured by a second molecule *within the same cluster.* Such a mechanism was first proposed by Klots and Compton[9] in CH_3I clusters and was called *self scavenging* to emphasize the involvement of a single van der Waals cluster in contrast to the usual scavenging mechanism which is observed at higher pressures between two uncorrelated molecules (see also next paragraph). For the 4 eV feature the following alternative mechanisms leading to I^- can be formulated:

(i) direct inelasic scattering of the incident electron from one molecule:

$$e^- (\approx 4 \text{ eV}) + (CF_3I)_m \rightarrow (CF_3I^*) \cdot (CF_3I)_{m-1} + e^- (\approx 0 \text{ eV})$$

and capture of the scattered electron by a second molecule within one aggregate leading to the ejection of I^- or the ion-molecule complex $M_n \cdot I^-$.

(ii) resonant electron capture associated with a core excited resonance:

$$e^- (\approx 4 \text{ eV}) + (CF_3I)_m \rightarrow (CF_3I^{*-}) \cdot (CF_3I)_{m-1}$$

and transfer of the extra electron to a second molecule resulting in an excited neutral molecule and a ground state molecular anion:

$$(CF_3I^{*-})\ (CF_3I)_{m-1} \rightarrow (CF_3I^*) \cdot (CF_3I^-) \cdot (CF_3I)_{m-2}$$

and subsequent emission of I^- or $I^- \cdot M_n$ like in mechanism (i).

The formation of I^- and $M \cdot I^-$ near 6 eV is also considered as a result of inelastic electron scattering prior to attachment. In the neutral molecule, an optical transition at 6 eV is assigned to the first member of the Rydberg series converging to the first ionization limit (5p, 6s).[21]

Isolated CF3I Molecules Beyond Single Collision Conditions (10^{-4}-10^{-3} mbar)

The cluster results presented above can be compared with anion formation using effusive gas inlet but under higher pressure. We have recorded

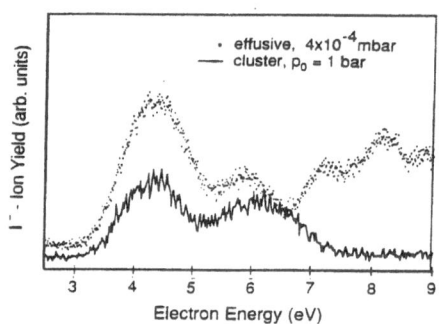

Figure 5. Comparison of the scavenger features on the channel I^- in clusters and isolated molecules beyond single collision conditions. From ref. 20.

ions at target gas pressures up to 8×10^{-4} mbar. Under these conditions, there is no indication of M^- and $M_n \cdot I^-$ formation and the ion yield curves of F^-, CF_3^- and FI^- remain virtually unaffected. However, for the I^- signal we observe the evolution of new features above 3 eV. Figure 5 compares the energy profile of I^- arising from the cluster beam at 1 bar stagnation pressure to that from the effusive beam at 4×10^{-4} mbar on an arbitrary scale for the intensities. Above 3 eV the effusive signal now increases with the *square* of the pressure as expected for a *secondary* reaction. The zero eV peak is not presented in Fig. 5, since the large amount of I^- ions causes saturation of the detector and the counting electronics. The "effusive" I^- signal above 3 eV arises from inelastic (either direct or resonant) scattering of the primary electron and capture of the slow electron by another molecule. In contrast to the process in clusters, this occurs between two *uncorrelated* monomers.

To establish this interpretation, we have added of a few % SF_6 to the CF_3I gas and recorded SF_6^-. This results in an SF_6^- energy profile which completely coincides with that of I^- (Fig. 5).

The "effusive" high pressure spectrum of I⁻ thus represents the threshold excitation spectrum (TEES) of CF₃I. The first two peaks of the TEES resemble the features of I⁻ and M·I⁻ formation from clusters (Fig. 4 and Fig. 5) with the 6 eV feature blue shifted in clusters. Interestingly, the TEES spectrum exhibits distinct further structures above 7 eV while this is not seen for I⁻ and M_n·I⁻ formation from clusters.

Molecules of the type CH_3X show many further optical transitions above 6 eV which are usually assigned to members of Rydberg series[21] and which contribute to the TEES spectrum. In clusters these higher excited states do obviously not result in the formation of stable negative ions. A similar observation was made by Klots and Compton in CH_3I clusters[9] and by our laboratory in the case of O_2 clusters.[11] In both cases only the *lower* part of the TEES spectrum was imaged in anion formation from clusters. These higher excited states are either not existent in clusters or the delocalization of the excitation destroys the ability of the rest of the cluster to capture an electron.

Anion Formation from Condensed CF₃I

Low energy (0-20 eV) electron impact to CF₃I condensed on a cold (20 K) polycrystalline Au-substrate resulted in the desorption of F⁻ and CF₃⁻. The ion I⁻ which is by far the most abundant in the gas phase (isolated molecules and clusters) could only be detected in minor amounts. Its energy profile showed a certain but reproducible time dependence in the way that the resonant

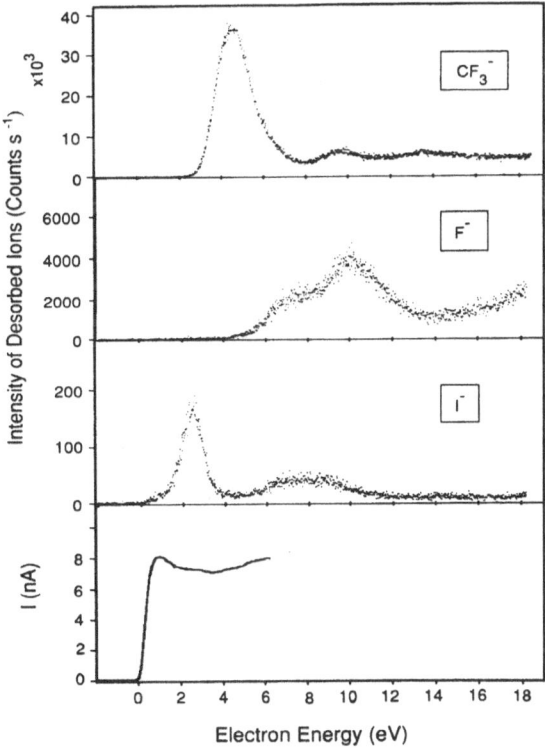

Figure 6. Electron stimulated desorption (ESD) of anions from a 8 ML film of CF₃I. The onset of the current transmitted to the metallic substrate is also shown. From ref. 20.

contribution at low energy only appeared after a number of scans from a newly deposited film. Along with that, one always observes a small shift (< 0.5 eV) in the onset of the electron current transmitted to the metallic substrate. This shift is caused by charging of the molecular film due to electron localization. As pointed out by Marsolais et al.[22] charging of the film creates a potential barrier near the vacuum/film interface which acts as a retarding potential for the impinging electrons.

Figure 6 shows the results for a 8-monolayer (ML) CF_3I film. Most remarkable is the emission of the 4-atomic fragment CF_3^- which is more abundant than the light F^- ion in multilayer films. CF_3^- appears within a resonance peaking at 4.3 eV. In contrast to that, F^- is only ejected from a structured resonance like distribution above 5 eV and as a continuously increasing signal above $\approx$ 12-13 eV. This last contribution is due to desorption induced by dipolar dissociation (DD) or ion pair formation which is also seen in the gas phase.

We interpret the formation of F^- between 5 and 10 eV as decomposition of highly excited TNIs formed directly in electron impact to the condensed molecules. Although the F^- channel in the isolated molecule does not show much contribution in that energy region, there are small but clearly visible structures around 6-8 and 10 eV.[20] A secondary process due to inelastic electron scattering in the film is unlikely: we have deposited CF_3I down to submomolayer amounts on a multilayer film of Xe as spacer between CF_3I and the metallic substrate which resulted in the same F^- feature (but weaker in intensity) between 5 eV and 12 eV as in Fig. 6(b).

The CF_3^- resonance around 4.3 eV is most likely associated with the core excited resonance which has a maximum at 3.8 eV in the gas phase. As has been shown[20] this electronic state is characterized by its significant F_3C-I repulsive nature. The two other products (F^-, FI^-) which also appear in the gas phase around 3.8 eV via comparatively slow dissociation reactions are not observed from condensed CF_3I.

It is interesting to note that the CF_3^- desorption resonance peaks at higher energy than the gas phase resonance. From the polarization interaction of the negative charge one expects the precursor molecular ion to be lower in energy with respect to the gas phase. As mentioned in the experimental section, the electron energy scale is always calibrated with respect to the onset of the transmitted current (see Fig. 6).

Recent detailed studies[23] have shown, however that for molecules with high capture cross sections the calibration of the electron energy scale is a particular delicate problem. Even by performing the very first scan, the film is subjected to considerable charging, which will affect the direction and energy of the electron beam. A calibration with respect to the onset of the electron beam may then place the energy of the resonance artificially to higher energies.

Although I^- is formed with some translational energy and with exceeding intensity in the gas phase, it appears only very weak in ESD and requires some "conditioning" of the film through electron beam exposure. It is likely that the resonant contribution near 1 eV is associated with the precursor ion in its electronic ground state. In contrast to the gas phase, however, only transitions at shorter nuclear distances within the Franck-Condon region generate sufficient repulsion for the ion I^- to leave the surface.

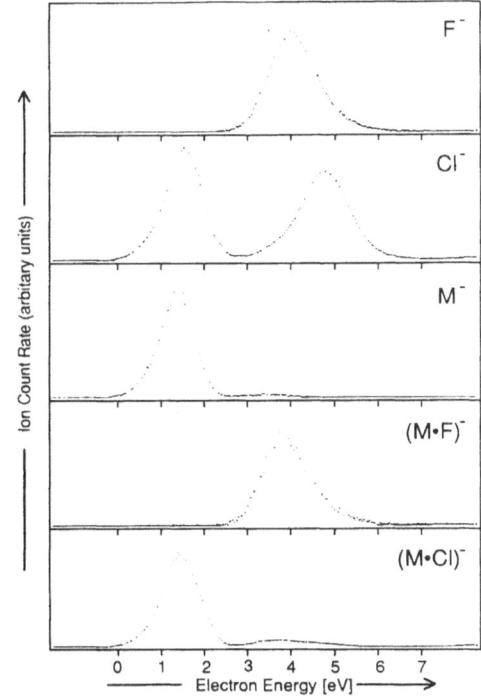

Figure 7. Some selected product ions from CF_3Cl clusters. From ref.11.

CF_3Cl: Distribution of Excess Energy in the Decomposition of Clusters

CF_3Cl possesses two resonances centered around 1.4 eV and 4-5 eV.[2,8] While the low energy state exclusively yields Cl^-, the ion near 4-5 eV decomposes into a variety of negatively charged fragments such as Cl^-, F^-, CF_3^- etc. The low energy state is interpreted as one particle resonance, with the additional electron in the first virtual MO which has a strongly localized $\sigma^*(C-Cl)$ antibonding character. TOF experiments, in fact, revealed that the temporary negative ion decomposes *directly* into $Cl^- + CF_3$ via electronic dissociation along a repulsive potential energy surface.[24]

On proceeding to clusters, we observe a variety of additional, negatively charged complexes. Figure 7 shows a few representative ion yield curves taken at 2 bar stagnation pressure. These spectra have been recorded with a nonmonochromatized electron beam (hair pin filament followed by a series of electrodes instead of the electron monochromator, Fig. 1). In spite of the low energy resolution for the present experiment ($\Delta\varepsilon \approx 0.7$ eV, fwhm), we see that all the ions are formed within a low energy resonance and/or a resonance between 4-5 eV already known from the isolated compound. The shapes and energetic positions of these ion yield curves do not depend on the stagnation pressure, at least not to an extent observable at the poor resolution.

The ions M_n^- and $M_n \cdot Cl^-$ are predominantly formed from the resonance of low energy and $M_n \cdot F^-$ is solely associated with the second resonance and closely resembles the F^- ion yield.

We will now restrict the discussion to low-energy attachment associated with the electronic ground state of CF_3Cl^-. In the isolated molecule this resonance decomposes by direct electronic dissociation exclusively into $Cl^- + CF_3$. In the aggregate, we have formation of a localized anion in a repulsive state:

$$e^-(1.4eV) + (CF_3Cl)_n \quad \rightarrow \quad (CF_3Cl^-) \cdot (CF_3Cl)_{n-1} \quad (8)$$

which evolves according to

$$(CF_3Cl^-)(CF_3Cl)_{n-1} \rightarrow \quad (CF_3Cl)_m^- + (CF_3Cl)_{n-m} \quad (9a)$$
$$\rightarrow \quad Cl^- + (CF_3) \cdot (CF_3Cl)_{n-1} \quad (9b)$$
$$\rightarrow \quad (CF_3Cl)_m \cdot Cl^- + (CF_3) \cdot (CF_3Cl)_{n-m-1} \quad (9c)$$

In Eqs. (9a-c) each neutral channel is assigned to consist of only one compound. Of course, as in the case of oxygen clusters, the reactions will generally release enough excess energy for a further evaporation of the target cluster. Channel (9a), $m=1$, leaves the parent radical anion in its relaxed configuration. This implies that the potential energy surface of M^- in its electronic ground state must possess a minimum. For the present system, a relaxation energy of more than 1.4 eV (at resonance maximum) has to be distributed in the target cluster.

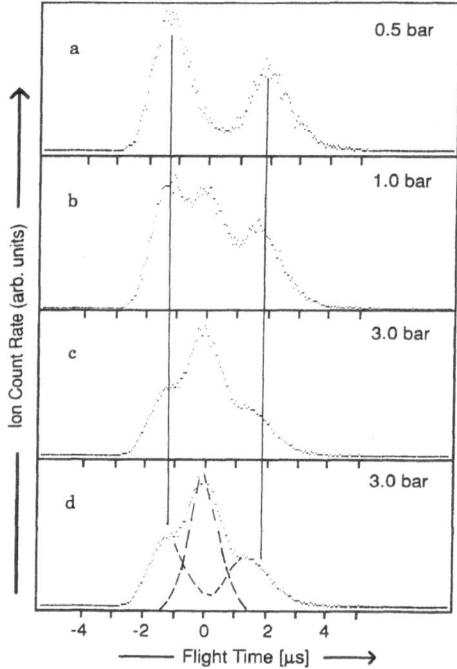

Figure 8. Evolution of the Cl^-/CF_3Cl TOF spectrum with stagnation pressure at the low energy resonance. From ref.11.

Figure 8 demonstrates how the TOF distribution evolves with increasing stagnation pressure and hence the average size of clusters in the beam.

At low stagnation pressure (0.5 bar, Fig. 8a) one observes a separated TOF doublet similar to that for the isolated system.[24] At that pressure there is no indication of cluster formation and Cl⁻ arises solely from electron capture by monomers. The quantitative evaluation of the TOF spectrum shows that virtually all of the available excess energy is transferred into kinetic energy of Cl⁻ + CF₃. The decomposition of the isolated ion is an example of the "rigid radical" limit within the impulsive model for dissociation.[8] In the present example, CF₃ behaves like a rigid radical and the decomposition of the polyatomic molecule is analogous to that of a diatomic molecule.

By increasing the stagnation pressure (1 bar) an additional feature near time zero becomes apparent. This contribution dominates the spectrum at 3 bar. Any further increase in pressure leaves the TOF distribution virtually unchanged. Figure 8d shows the result of a graphical subtraction which indicates that the TOF spectrum consists of two components, one due to Cl⁻ ions with considerable kinetic energy and an additional one yielding low-energy ions.

So far, we have demonstrated that cluster formation coincides with the appearance of low energy Cl⁻ ions. The question concerning the origin of the high-energy component (which is still present in the TOF spectra at high stagnation pressures), however, is still open. This feature may be due to Cl⁻ emission from monomers (in the beam or from background gas) or from particular clusters.

TOF experiments with helium as carrier gas have shown[25] that the high-energy component is, in fact, predominantly due to Cl⁻ emission from scattered background monomers and only a small amount is due to Cl⁻ emission from targets in the supersonic beam: If helium is used as carrier gas, the velocity of the paricles in the beam (i. e. along the axis of the flight tube) is very high and in a TOF spectrum one can easily distinguish between Cl⁻ emission from particles traveling with the seeded beam velocity and Cl⁻ emission from background molecules.

The present example demonstrates that intermolecular energy transfer in the aggregate allows the preparation of molecular anions not accessible by other techniques. To our knowledge, CF₃Cl⁻ has not been observed in the gas phase before.

Dissociative Electron Attachment to Condensed CF4

Electron capture by gaseous CF_4 results in F^- and CF_3^- formation from an exceedingly broad resonance centered near 7 eV.[2,8,24] On going to condensed molecules, F^- was the only ionic fragment found to desorb during exposure to 0-20 eV electrons.[26] The F^- was detected as a broad resonance around 7 eV, and a continuously rising signal above ≈12 eV. In Fig. 9 the ESD result is compared with F^- and CF_3^- formation from gas phase CF_4. In the gas phase the relative intensity of the two fragment ions is $F^- : CF_3^- = 1 : 0.6$. Figure 9 shows that there is a gross similarity in the resonance position of anion formation between gas phase and condensed phase. A closer inspection of the gas phase spectra reveals that the F^- curve consists of two overlapping resonances associated with ground state CF_4^- and an electronically excited state CF_4^{*-} (Fig. 9). The electronic ground state decomposes along repulsive energy surfaces via the complementary channels

$$CF_4^- \text{ (6.8 eV)} \quad \rightarrow \quad F^- + CF_3 \quad (10a)$$
$$\rightarrow \quad F + CF_3^- \quad (10b)$$

while the excited state exclusively yields F^-. This results in an F^- yield composed of two overlapping resonances as indicated in Fig. 9. It is likely that excited CF_4^{*-} correlates with the limit consisting of F^- and the CF_3^* radical in its first electronically excited state (excitation energy 6.41 eV [27,28]) finally leading to three fragments

$$CF_4^{*-} \ (7.6 \ eV) \ \rightarrow \ F^- + F + CF_2 \tag{11}$$

through predissociation of the excited precursor state. This picture is supported by an analysis of the translational energy release in the corresponding process. Both reactions (10a) and (10b) are associated with high kinetic energy. In channel (10) 33% of the available excess energy is released as translational energy and in channel (10b) it is 62% accounting for direct electronic dissociation via repulsive potential energy surfaces. In contrast to that, channel (11) yields low energetic F^- ions indicating a less direct decomposition mechanism involving electronic and vibrational predissociation.

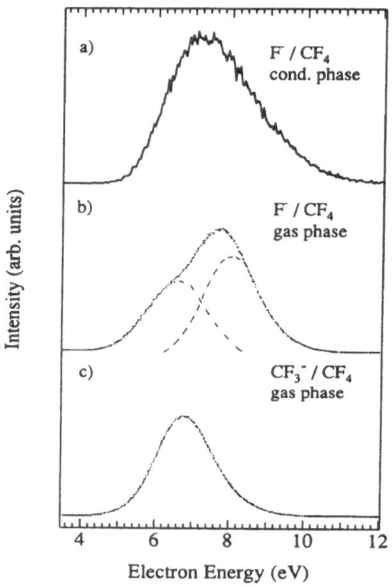

Figure 9. Comparison of DA in the gaseous and condensed phase. From ref.26.

Returning to the ESD spectrum, it shows the *same* onset as the gas phase reaction (10a) (decomposition of ground state CF_4^-) and is most likely due to decomposition of the strongly repulsive ground state molecular ion CF_4^-. This is also confirmed by an analysis of the translational energy, indicating that desorbed F^- ions also appear with remarkably high energy, (i. e. approx 60% of the value from the unimolecular decomposition in the gas phase). Although in the gas phase 62% of the total excess energy appears as translational energy in the products $CF_3^- + F$ the unfavourable mass ratio prevents CF_3^- from leaving the surface.

O₂ Molecules and Clusters

Oxygen is a particular interesting and comparatively simple system where some essential features concerning electron capture by isolated molecules and the corresponding van der Waals clusters can clearly be observed.

Figure 10 shows potential energy curves relevant in resonant electron scattering from gaseous oxygen molecules[2,8,29] which indicates that the *adiabatic* electron affinity of the oxygen molecule is positive, namely 0.440 eV.

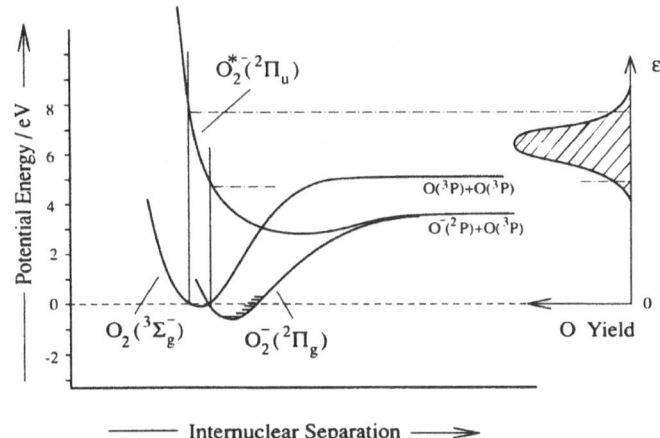

Figure 10. Potential energy curves of O₂ and O₂⁻ relevant in electron capture by gaseous oxygen molecules.

Capture of low energy electrons, i.e., a Franck-Condon transition from $O_2(^3\Sigma_g^-) + e^-$ to $\bar{O}_2(^2\Pi_g)$ forms the anion in its electronic ground state, but vibrationally excited ($v \geq 4$). Due to the absence of operative stabilization mechanisms, $\bar{O}_2^*$ thus formed has a finite lifetime with respect to autodetachment (10^{-12} to 10^{-10} s) far beyond the scale for a mass spectrometric detection. In a high pressure environment, the ion can be stabilized in a three-body process to a vibrational level ($v < 4$) where autodetachment is no longer possible (Bloch-Bradbury mechanism[30]).

A further Franck-Condon transition at higher electron energies generates the anion in a repulsive, electronically excited state, $O_2^{*-}(1\pi_u^{-1}, 1\pi_g^2)^2\Pi_u$. In this case, the incoming electron simultaneously excites an electron in the target molecule resulting in a state with one hole in the $1\pi_u$ MO (assigned as $1\pi_u^{-1}$) and two *additional* electrons in the $1\pi_g$ MO (assigned as $1\pi_g^2$). We have thus a *two particle one hole (2p-1h)* resonance as mentioned in the introduction.

Figure 11 shows the experimental result obtained at different stagnation pressures. At low pressures there is no cluster formation and only O⁻ is observable arising from a broad resonance between 4.5 and 10 eV. The continuous increase of the O⁻ signal above 17 eV is due to the (non resonant) ion formation

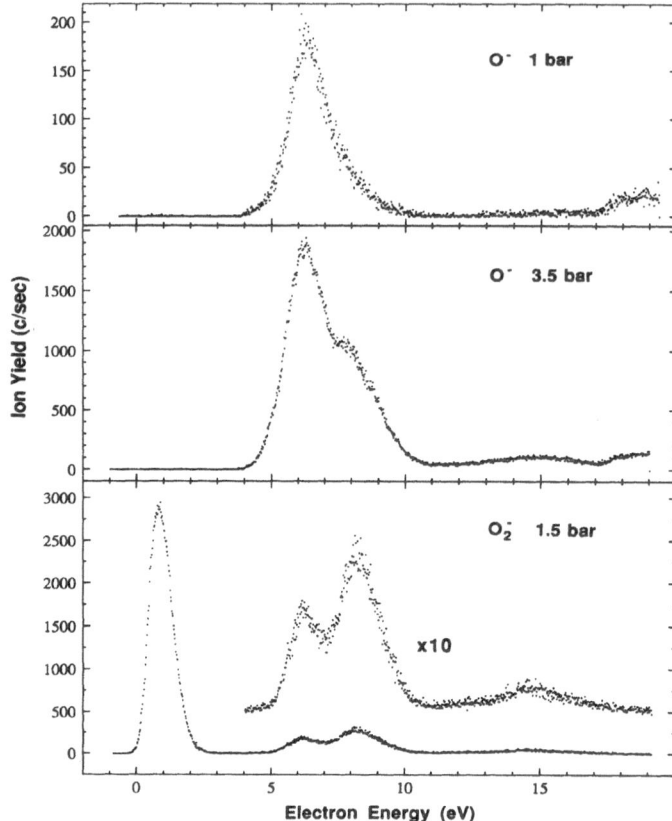

Figure 11. O⁻ and O_2^- formation following electron attachment in the molecular beam.

$$e^- + O_2 \rightarrow O_2^{**} + e^- \tag{15}$$
$$\longrightarrow O^+(^4S) + O^-(^2P)$$

which has an energetic threshold of 17.2 eV.

The shape of the resonance profile of O⁻ is explained by the *reflection principle*[31] which states that the Franck-Condon factors as a function of energy simply *reflect* the probability density of the ground vibrational state.

At higher pressures we observe the two homologous series $(O_2)_n^-$ and $(O_2)_n \cdot O^-$. This is in agreement with previous experiments on O_2 clusters by Märk, Stamatovic and coworkers.[32] In Figure 11 only the ion yield curves of O⁻ and $\bar{O}_2$ at two different stagnation pressures are recorded. Here we observe the evolution of two distinct new features at ≈ 8.3 eV and ≈ 14.5 eV. Interestingly, the new feature is even more pronounced in the $\bar{O}_2$ profile at a comparatively low stagnation pressure (1.5 bar). Formation of $(O_2)_n^-$ at very low energies has clearly been demonstrated in the previous work of Märk et al.;[32] it is an illustrative example of the "built in" many body stabilization mechanism in a cluster. As will be shown below, formation of ions of the form $(O_2)_n^-$ and $(O_2)_n \cdot O^-$ at energies above 4 eV is likely to originate from completely different mechanisms. For the explanation of the additional features at 8.3 eV and 14.5 eV In the channels $(O_2)_n \cdot O^-$ we refer to recent experiments on electron stimulated desorption (ESD) of O⁻ from O₂ multilayers performed by Sanche et al.[33] and to a study of dissociative electron attachment to singlet oxygen, O_2 ($^1\Delta_g$).[34] Analysis of the O⁻ ion yield between 4 and 15 eV in the electron stimulated desorption study can be explained by considering the negative ion states $\bar{O}_2$ lying below 15 eV. Although a variety of $\bar{O}_2$ states can be formed from O and O⁻ (24 states, e.g., from ground state O(^{3}P) and O⁻(^{2}P)!) only few of these possess the proper characteristics for dissociative electron attachment. Molecular orbital analysis of the states joining the two lowest limits indicates that only the $^2\Pi_u$, $^2\Sigma_g^+$ and $^2\Sigma_u^+$ states are relevant for dissociative attachment (they must be repulsive in the Franck-Condon region, and spin conservation requires a doublet or quartet state).

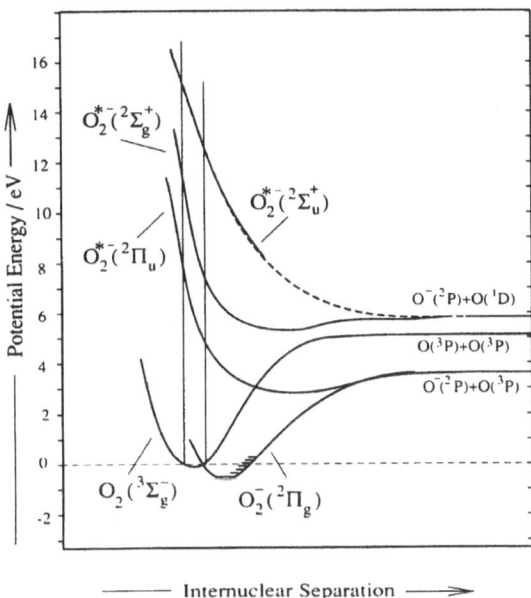

Figure 12. Potential energy diagram of O_2 and O_2^- states relevant in electron attachment to O_2 clusters (see text).

Out of these states the Σ^+ configurations cannot be formed in electron attachment to isolated O_2 from the $^3\Sigma_g^-$ ground state. Since in the single-electron-molecule frame of reference a one electron wavefunction must principally be σ^+, transitions from Σ^- to Σ^+ and vice versa are forbidden (σ^- selection rule).[16] This selection rule holds in electron attachment and auto-ionization. The electron stimulated desorption experiments, along with the theoretical analysis, lead to the conclusion that the excited negative ion states states, $^2\Sigma_g^+$ and $^2\Sigma_u^+$, are responsible for the O⁻ yield observed near 8 and 14 eV, respectively. This was considered the first experimental evidence for the violation of the σ^- selection rule in electron attachment.

In light of these findings we conclude that the structures appearing near 8.3 and 14 eV in the channels $(O_2)_n \cdot O^-$ are also due to electron attachment via the symmetry forbidden states O_2^{*-} ($3\sigma_g^{-1} 1\pi_g^2$) $^2\Sigma_g^+$ and O_2^{*-} ($2\sigma_u^{-1} 1\pi_g^2$)$^2\Sigma_u^+$. The existence of an additional state between 8 and 9 eV is further supported by our recent dissociative attachment study to O_2 ($^1\Delta_g$)[34] where population of the O_2^{*-} ($3\sigma_g^{-1} 1\pi_g^2$) $^2\Sigma_g^+$ state could be observed.

Figure 12 shows a schematic of the associated potential energy curves. From analysis of the kinetic energy release[34] it is established that the O_2^{*-} ($^2\Sigma_g^+$) state predominantly decomposes into the second dissociation limit, $O^-(^2P)$ + $O(^1D)$. We thus conclude that all odd number cluster anions $(O_2)_n \cdot O^-$ have O^- as precursor which is formed via dissociative electron attachment to the repulsive Π_u, Σ_g^+ and Σ_u^+ states.

The question now arises on the mechanism of $(O_2)_n^-$ formation at energies above 4 eV. While it is clear that at low energies these anions are formed through the electronic ground state O_2^- ($^2\Pi_g$) and subsequent collisional stabilization, it is rather unlikely to establish a reasonable mechanism to eplain for-

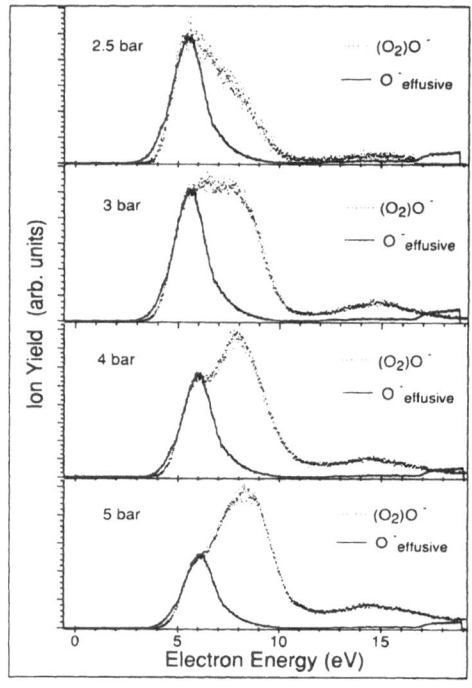

Figure 13. Energy profile of product $O^-(O_2)$ at increasing stagnation pressure showing the enhanced contribution of symmetrie forbidden transitions with increasing cluster size.

of $(O_2)_n^-$ through the initial population of the strongly repulsive Π_u, Σ_g^+ and Σ_u^+ states. In large clusters, on the other hand, cage effects may indeed prevent the dissociation of the pair $O^- + O$ ultimately leading to $(O_2)_n^-$. The experimental observation is that the relative intensities in the $(O_2)_n^-$ yield are virtually independent of stagnation pressure[35] (i. e. the average cluster size in the beam) while this is not the case for $(O_2)_n \cdot O^-$. As an example Fig. 13 shows the $(O_2) \cdot O^-$

yield at increasing stagnation pressure indicating the enhanced contribution of the symmetry forbidden transitions as the average size of the cluster increases.

In light of this we interpret $(O_2)_n^-$ formation at electron energies above 4 eV as a result of a *secondary reaction*: the incoming electron is inelastically scattered (through direct and resonant scattering, the latter involving the O_2^{*-} resonances) in the cluster and the slowed down electron is captured forming $(O_2)_n^-$. This picture is supported by the threshold excitation spectrum of oxygen which exhibits pronounced maxima at 6 and 8.3 eV.[36] A threshold excitation spectrum is obtained when the primary energy is scanned and only those electrons are recorded which have lost their total energy. Recent studies on mixed clusters by Märk et al.[10,37] and our own laboratory[35,38] have further

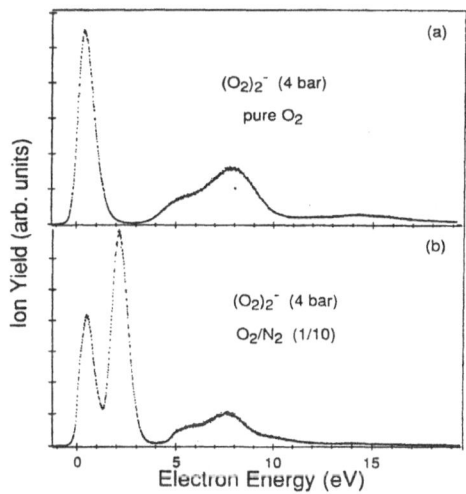

Figure 14. O_4^- formation from mixed N_2/O_2 clusters: (a) O_4^- from pure oxygen clusters; (b) O_4^- from N_2/O_2 clusters. The strong contribution at 2.3 eV is due to inalastic scattering via the $N_2^-(^2\Pi_g)$ resonance.

shown the role of inelastic scattering for anion formation in mixed clusters. It appears that these secondary processes contribute considerably to anion formation when one component of the system possesses a low energy resonance which is generally associated with a high capture cross section. As an example, Fig. 14 shows the result when a mixture of oxygen and nitrogen (mixing ratio 1:10) is expanded. The dominant feature in the O_4^- channel is now a contribution near 2.3 eV which arises from *resonant inelastic* scattering via the $N_2^-(^2\Pi_g)$ negative ion state. This clearly demonstrates the strong contribution of inelastic scattering (in the present case through a resonance of a non-attaching molecule) to negative ion formation in clusters.

In conclusion, it can be seen that negative ion formation in oxygen clusters proceeds via different processes (i) dissociative electron attachment involving the three repulsive states $^2\Pi_u$, $^2\Sigma_g^+$ and $^2\Sigma_u^+$ generating ions of the form $(O_2)_n \cdot O^-$. In single molecules, population of the Σ^+ states is symmetry forbidden, (ii) formation of $(O_2)_n^-$ at low energies (<2eV) by associative attachment with subsequent collisional stabilization thereby evaporating the target cluster ("evaporative electron attachment"), and (iii) inelastic electron scattering followed by associative attachment of the slowed down electron.

Acknowledgments

This work has been supported by the Deutsche Forschungsgemeinschaft, Stiftung Volkswagenwerk, Verband der Chemischen Industrie, and the NATO Scientific Affairs Division.

References

1. L.G. Christophorou, "Atomic and Molecular Radiation Physics", Wiley Interscience, London (1971).

2. E. Illenberger, J. Momigny, "Gaseous Molecular Ions. An Introduction to Elementary Processes Induced by Ionization", Steinkopff Verlag, Weinheim; Springer Verlag, New York (1992).

3. J. Jortner, A. Pulman and B. Pulman (Eds.), "Large Finite Systems", Reidel, Dordrecht (1987).

4. J.P. Maier (Ed.), "Ion and Cluster Ion Spectroscopy and Structure", Elsevier, Amsterdam (1989).

5. C.Y. Ng (Ed.), "Vacuum Ultraviolet Photoionization and Photodissociation of Molecules and Clusters", World Scientific, Singapore (1991).

6. R. A. Marcus, Elektronentransferreaktionen in der Chemie - Theorie und Experiment (Nobel Vortrag), *Angew. Chem. 105:1161 (1993)*.

7. H. Haberland, Solvated Electron Clusters, in "The Chemical Physics of Atomic and Molecular Clusters", G. Scoles (Ed.), North Holland, Amsterdam (1990).

8. T. Oster, A. Kühn and E. Illenberger, Gas Phase Negative Ion Chemistry, *Int. J. Mass Spectrom. Ion Proc. 89:1 (1989)*.

9. C.E. Klots and R.N. Compton, Self-Scavenging of Electrons in van der Waals Molecules of Methyl Iodide, *Chem. Phys. Lett. 73:589 (1980)*.

10. T. Rauth, M. Foltin and T.D. Märk, Autoscavenging of Electrons in Mixed van der Waals Clusters: A New Approach to the Spectroscopy of Cluster Anions, *J. Phys. Chem. 96:1528 (1992)*.

11. E. Illenberger, Electron Attachment Reactions in Molecular Clusters, *Chem. Rev. 92:1589 (1993)*.

12. T.D. Märk, Free Electron Attachment to van der Waals Clusters, *Int. J. Mass Spectrom. Ion Proc. 107:143 (1993)*.

13. C.E. Klots, Van der Waals Molecules-Linking the Gas and Liquid Phases, *Radiat. Phys. Chem. 20:51 (1980)*.

14. L. Sanche, Low Energy Electron Stimulated Desorption from Condensed Molecules, *Comments At. Mol. Phys. 26: 321 (1991)*.

15. L. Sanche, Primary Interactions of Low Energy Electrons in Condensed Matter, in "Excess Electrons in Dielectric Media", C. Ferradini and J.-P. Jay-Gerin (Eds.), CRC, Boca Raton (1991).

16. H. Sambe and D.E. Ramaker, The σ^- Selection Rule in Electron Attachment and Autoionization of Diatomic Molecules, *Chem. Phys. Lett. 139:386 (1987)*.

17. A. Stamatovic and G.J. Schulz, Characteristics of the Trochoidal Electron Monochromator, *Rev. Sci. Instrum. 41:423 (1970)*.

18. J.B. Anderson, Molecular Beams from Nozzle Sources, in "Molecular Beams and Low Density Gas Dynamics", P.P. Wegener (Ed.), Marcel Dekker, New York (1974).

19. D.R. Miller, Free Jet Sources, in "Atomic and Molecular Beam Methods", G. Scoles (Ed.), Oxford University Press, Oxford (1988).

20. T. Oster, O. Ingolfsson, M. Meinke, T. Jaffke and E. Illenberger, Anion Formation from Gaseous and Condensed CF_3I on Low Energy Electron Impact, *J. Chem. Phys. 99: 5141 (1993)*.

21. M.B. Robin, "Higher Excited States of Polyatomic Molecules", Vol I, Academic, New York (1974).

22. R.M. Marsolais, M. Dechênes and L. Sanche, Low Energy Electron Transmission Method for Measuring Charge Trapping in Dielectric Films, *Rev. Sci. Instrum. 60:2747 (1989)*.

23. M. Meinke and E. Illenberger, Electron Stimulated Desorption of Anions from Condensed and Adsorbed $CFCl_3$ Down to Very Low Electron Energies (<2 eV), *J. Chem. Phys. (submitted 1993)*.

24. E. Illenberger, Measurement of Translational Excess Energy in Dissociative Electron Attachment Processes, *Chem. Phys. Lett. 80:153 (1981)*.

25. T. Oster and E. Illenberger, TOF Measurements of DA Products from CF_3Cl and CF_4 Jets Seeded in Ar and He, *unpublished results*.

26. M. Meinke, L. Parenteau, L. Sanche and E. Illenberger, Low Energy Electron Stimulated Desorption of Anions from Condensed CF_4, *Chem. Phys. Lett. 205:213 (1993)*.

27. M. Suto and N. Washida, Emission Spectra of CF_3 Radicals. I. UV and Visible Emission Spectra of CF_3 Observed in the VUV Photolysis and the Metastable Argon Atom Reaction of CF_3H, *J. Chem. Phys. 78:1007 (1983)*.

28. M. Suto and N. Washida, Emission Spectra of CF_3 Radicals. II. Analysis of the UV Emission Spectrum of CF_3 Radicals, *J. Chem. Phys. 78:10129(1983)*.

29. D. Rapp and D.D. Briglia, Total Cross Sections for Ionization and Attachment in Gases by Electron Impact. II. Negative Ion Formation, *J. Chem. Phys. 43:1480 (1965)*.

30. F. Bloch and N. E. Bradbury, On the Mechanism of Unimolecular Electron Capture, *Phys. Rev. 48: 689 (1935)*.

31. T.F. O'Malley, Theory of Dissociative Attachment, *Phys. Rev. 150:14 (1966)*.

32. T.D. Märk, K. Leiter, W. Ritter and A. Stamatovic, Low Energy-Electron Attachment to Oxygen Clusters Produced By Nozzle Expansion, *Phys. Rev. Lett. 55:2559 (1985)*.

33. L. Sanche, Low Energy Electron Scattering from Molecules on Surfaces, *J. Phys. B: At. Mol Opt. Phys 23:1597 (1990)*.

34. T. Jaffke, M. Meinke, R. Hashemi, L.G. Christophorou and E. Illenberger, Dissociative Electron Attachment to Singlet Oxygen, *Chem. Phys. Lett. 193:62 (1992)*.

35. T. Jaffke, R. Hashemi, L.G. Christophorou and E. Illenberger, Mechanisms of Anion Formation in O_2, O_2/Ar, and O_2/Ne Clusters: The Role of Inelastic Scattering, *Z. Phys. D 25:77 (1992)*.

36. G.J. Schulz and J.T. Dowell, Excitation of Vibrational and Electronic Levels in O_2 by Electron Impact, *Phys. Rev. 128:174 (1962)*.

37. M. Foltin, V. Grill and T.D. Märk, The Observation of Unusual Resonance Channels in the Electron Attachment to Mixed Argon-Oxygen Clusters, *Chem. Phys. Lett. 118:427 (1992)*.

38. T. Jaffke, R. Hashemi, L.G. Christophorou and E. Illenberger, The Role of Inelastic Scattering by N_2 in the Formation of $(O_2)_2{}^-$ Anions in Mixed O_2/N_2 Clusters, *J. Phys. Chem. 96:10605 (1992)*.

SECTION II. IONIZATION IN DILUTE AND IN CONDENSED MATTER

IONIZATION OF ATOMS OR MOLECULES BY RADIATION AS A FUNCTION OF PHASE

Werner F. Schmidt

Hahn-Meitner-Institut Berlin/Germany

ABSTRACT

Ionization by light or high energy radiation is one of the primary processes of the interaction of radiation with matter. In this article, the fundamental differences of the ionization process in gaseous, liquid, and solid matter are discussed. The importance of recombination losses in the condensed phase and the theoretical treatment by means of Onsager´s theory are described. Examples are taken from studies on organic nonpolar liquids and solids.

INTRODUCTION

Interaction of radiation with matter leads to ionization and excitation. Both are important fundamental electronic processes which form the basis for a variety of applications. While the understanding of ionization and excitation in the gas phase and in crystals has achieved a high level of sophistication these processes are less well understood for liquids and polymers. In this chapter, we will discuss ionization processes induced by light and high energy radiation in non-polar liquids and in some organic polymers.

SINGLE PHOTON IONIZATION

The ionization of an isolated atom or molecule in vacuum consist in removing an electron from its shell and transporting it to infinity. A well defined energy, the ionization potential (IP) is required. Ionization potentials can be measured by a variety of methods. In this article we shall be mainly concerned with conductivity measurements.

In photoconductivity measurements, a low pressure vapor of the compound to be studied is illuminated with light of variable wavelength. The experimental set-up is shown schematically in Fig. 1. The current flowing through the conductivity cell is monitored as a function of wavelength. When the photon energy, $h\nu$, exceeds the ionization potential of the least bound electron, ionization occurs and charge carrier pairs (positive ion/electron) are generated. The electron leaves the positive ion with an excess kinetic energy given by,

$$\langle \varepsilon_{kin} \rangle = h\nu - IP \tag{1}$$

This excess energy is dissipated by numerous collisions with the atoms or molecules of the gas until the mean electron energy reaches $k_B T$. The electron is then said to be thermalized. Thermalized electrons and positive ions lead to an increase of the electrical current flowing through the cell. The methods relies on the fact that in a low pressure gas at moderate light

Linking the Gaseous and Condensed Phases of Matter
Edited by L.G. Christophorou *et al*, Plenum Press, New York, 1994

intensities all charge carriers can be swept out of the volume without loss due to recombination by application of a moderate electric field strength. Above a certain field strength, the current remains constant with increasing voltage. It is called saturation current. All charge carrier pairs produced per unit time in the volume are collected. Near the photo ionization threshold, highly excited states are formed which upon collisions with neutral gas molecules may autoionize leading to a diffuse onset of the photoconductivity current. By suitable extrapolation methods, the photoconductivity threshold which is equal to the ionization potential can be determined. For the determination of ionization potentials >11eV cells without windows or other methods (e.g. photoelectron spectroscopy) must be employed.

Separation of an electron/positive ion pair may be assumed to be complete when the Coulomb energy of electrostatic attraction equals $k_B T$ of the gas. In a dielectric medium of relative dielectric constant ε_r this critical distance, r_c, is given by

$$\frac{e_0^2}{4\pi\varepsilon_0\varepsilon_r r_c} = k_B T \tag{2}$$

ε_0 is the permittivity of free space, and e denotes the electronic charge. In a low pressure gas, $\varepsilon_r \approx 1$, and at room temperature, $r_c \approx 60$nm is obtained. On the other hand, the mean free path of an electron , λ_e, is given by,

$$\lambda_e = \frac{1}{n\sigma} \tag{3}$$

The number density of an ideal gas at room temperature and a pressure of 1 bar is $n=2.68$ 10^{19} cm^{-3}. Cross sections for inelastic electron scattering on molecules or atoms depend on the electron energy but seldom exceed 10^{-16} cm^2 [ref.1]. Taking this as an upper limit, for λ_e a value of 3700 nm is obtained which is much larger than r_c. In other words, in the photoionization of low pressure gases ($p\leq 1$bar) the electron becomes thermalized well outside a sphere of radius r_c and the influence of neutrals on the ionization process can be neglected.

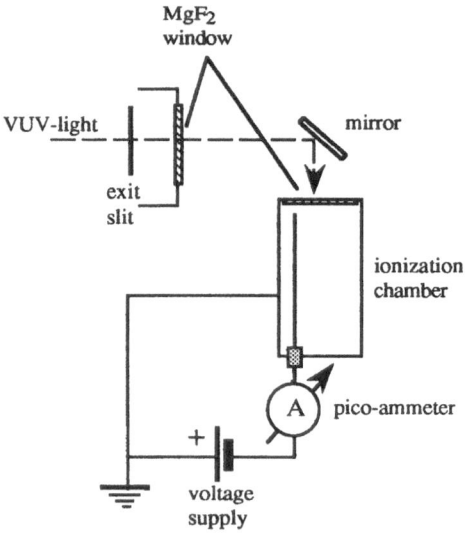

Fig. 1. Experimental set-up for measurement of the ionization potential of atoms or molecules in the gas phase by the photoconductivity method [2].

Quite a different situation is found in the condensed phase. We shall restrict ourselves to non-polar fluids and molecular or polymer solids with ε_r between 2 and 3. For $\varepsilon_r = 2$, the critical distance at room temperature is then $r_c \approx 30nm$. Due to the much higher number density, the interaction of the hot electrons from the photoionization process with the atoms or molecules of the liquid or solid takes place on a much shorter range. The distance at which the electron acquires thermal energies is now of the same magnitude as the critical distance r_c. A good portion of the initially produced electron/ion pairs exhibit separation distances $r < r_c$. For these electron/ion pairs the Coulomb attraction dominates over the thermal diffusion process which would lead to further separation. Eventually, recombination takes place. Since the electron is recombining with its parent positive ion, this process is also called geminate recombination. This process leads to much smaller photocurrents when compared to the gas phase. In addition, increase of the applied electric field strength leads to an increase of the photocurrent. An example is shown in Fig. 2. No saturation current has been obtained in molecular liquids or solids. The increase of the photocurrent with field is due to the influence of the external field on the geminate recombination process.

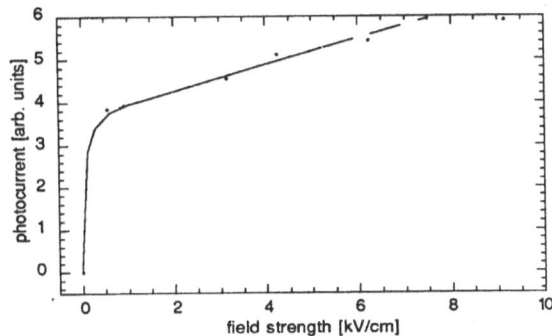

Fig. 2. Photoconductivity of 1-methylstyrene in neopentane, $h\nu > 6.2$ eV (after ref.3)

IONIZATION BY HIGH ENERGY PARTICLES OR QUANTA

High energy particles or quanta passing through matter lose energy by ionization and excitation processes. The energy deposition is spatially inhomogeneous along the track of the individual particle or quantum. Different types of radiation are characterized by their linear energy transfer (LET) which is given as the amount of energy deposited per unit length along the track . High energy electrons ($E_{el} \approx 1MeV$) and γ-quanta are characterized by low LET values while α-particles exhibit high LET values. A major portion of the ionization events produced by low LET radiation are found in spurs, which are energy loss events of the order of 50 to 100eV. Small fractions of energy are found in blobs (500eV) and short tracks (1keV). A schematic representation of the spatial distribution of energy losses is shown in Figs. 3 and 4. The ionization produced by high LET radiation can be envisaged as a dense column of positive ions along the track of the particle surrounded by a cylindrical cloud of electrons (see Fig. 5).

In low pressure gases, the same arguments as in the case of single photon ionization apply with respect to the initial separation of the electron/ion pairs for low LET radiation (In the case of high LET radiation the influence of recombination processes is a function of the gas density; we shall not discuss these effects). All charge carrier pairs exhibit an initial separation much larger than r_c and the separation of individual ionization events is also larger than r_c. In liquids, all distances are reduced approximately by the density ratio vapor/liquid which is in the range of 1/500 to 1/700 for most liquids of interest in this article.

A basic quantity, which describes the ionization process by high energy radiation in gases, is the W-values which is the energy required to produce one electron/ion pair. In radiation chemistry, out of convenience, a different quantity, the G-value, is used, which gives the number of electron/ion pairs produced per expended energy of 100 eV. Both quantities are related by $W = 100/G$. Many W-values for gases have been reported since they are of fundamental importance for the operation of ionization chambers as dosimeters [4].

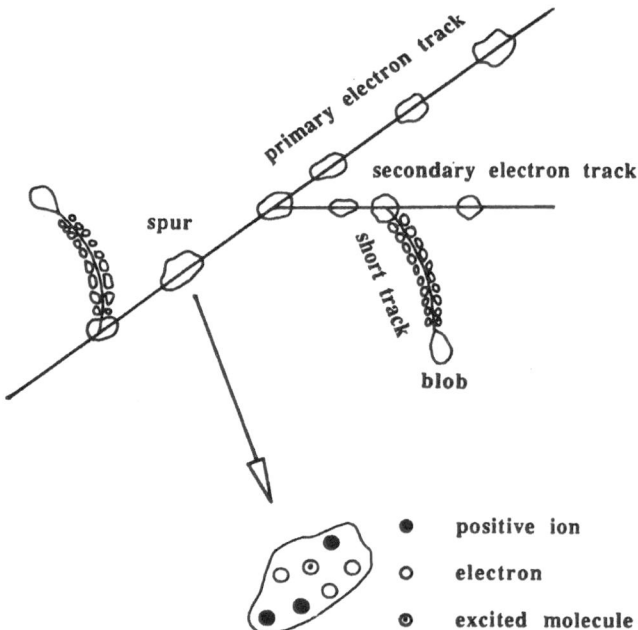

Figure 3. Spatial distribution of ionization events along the path of a low LET particle

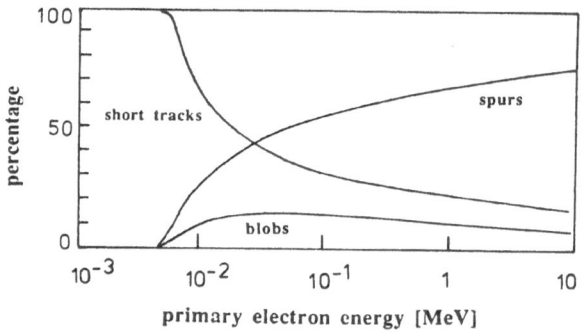

primary electron energy [MeV]

Figure 4. Distribution of spurs, blobs, and short tracks (after ref. 5)

The W-values are two to three times larger than the corresponding ionization potentials of the compounds. The excess energy goes into electronic excitations which depending on the type of gas lead to scintillation or to radiationless transitions to the ground state. Details on the ionization of gases by high energy radiation can be found in the book of Christophorou[1].

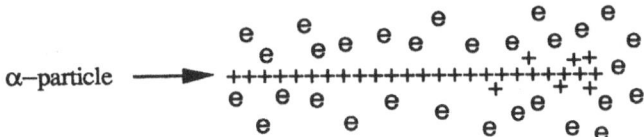

Figure 5. Schematic representation of the ionization events along the track of an α-particle

TECHNIQUE OF IONIZATION CURRENT MEASUREMENT

In this chapter we are entirely concerned with the measurement of radiation- or photo-induced ionization currents in all three phases of matter. The general experimental set-up is shown schematically in Fig. 6. The sample is contained in a parallel-plate measurement cell which is connected via an ampere-meter to the voltage source. Light or radiation is admitted into the cell. In the cases of photo ionization of solutes and of ionization by high energy γ-radiation, the ionization events are distributed homogeneously through the measurement volume. VUV-photoionization and ionization by α-particles occurs in a thin layer behind the entrance window or at the radioactive electrode, respectively, and depending on the polarity electrons or ions have to travel through the non-irradiated part of the liquid to the counter electrode. The ionization current, i_{ion}, is given by

$$i_{ion} = \sigma_{ion} \, U \frac{q}{l}$$

(4)

where σ_{ion} denotes the radiation induced conductivity, U is the applied voltage, and q and l are the cross section of the measurement electrode and the electrode distance, respectively.

Ionization currents in the gas phase increase with applied voltage up to a plateau value (saturation current, i_s) which remains constant at further increase of voltage. The saturation current is given by the condition that the charge carrier generation rate balances the rate of collection at the electrodes, i.e.

$$i_s = e_0 \, g \, V$$

(5)

where e_0 denotes the electron charge, g is the rate of generation ($cm^{-3}s^{-1}$) and V is the measurement volume. g and also i_s are proportional to the intensity of radiation (quantum flux or dose rate).

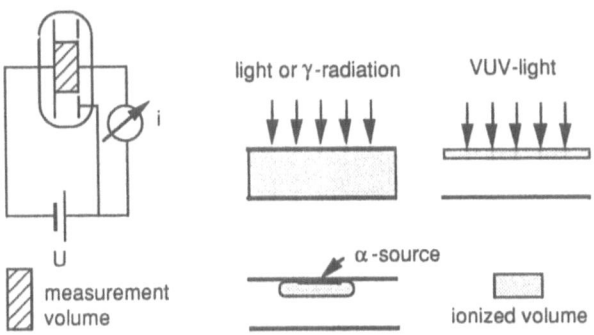

Figure 6. Schematic representation of the electric circuit for the measurement of ionization currents and of the spatial distributions of ionization events produced by different types of radiation.

In molecular liquids no saturation currents have been observed. At constant radiation intensity the current first rises fast with voltage turning over into a linear dependence of ionization current on voltage. The initial rise is due to the influence of the applied electric field on the process of volume recombination while the linear rise reflects the geminate recombination of single electron/ion pairs. In the linear region, the current is proportional to the intensity of the radiation (see ref. 6).

Ionization currents in solids exhibit a complex behavior. In some materials the ionization current is proportional to the applied voltage and it exhibits a power law dependence on the radiation intensity. Many studies have demonstrated that charge carrier trapping in defects or at impurities is a major factor which determines the overall electric response of a solid in a radiation field.

In the next section we shall present a general discussion of the process of geminate recombination in condensed systems.

ONSAGER'S THEORY OF GEMINATE RECOMBINATION

A theoretical treatment of the geminate recombination process of ions is due to Onsager[7]. He considered the motion of an oppositely charged ion pair separated by a distance r in an isotropic medium of relative dielectric constant ε_r. An external field, E, is applied, the temperature of the medium is T. The ion pair carries out diffusive motion which may be considered as changes of position by jumps. At the same time it is under the influence of their mutual Coulombic attraction and of the force exerted by the external field. This situation is depicted schematically in Fig. 7. The potential energy of the ion pair is given as

$$U(r) = -\frac{e_0^2}{4\pi\varepsilon_0\varepsilon_r} - e_0 E \cos\theta$$

(6)

The diffusion of the ion pair can be described by the Smoluchowski equation. The positive ion is assumed to rest at the origin; the probability $P(r,t)\,d^3r$ of finding the negative ion at r in a volume d^3r at time t is given by,

$$\frac{\partial P}{\partial t} = D\nabla\left[\nabla P + \frac{P}{k_B T}\nabla U\right]$$

(7)

D is the sum of the diffusion coefficients of both ions,

$$D = D_+ + D_-$$

(8)

The critical distance r_c is defined by by Eq. 2. If the initial separation of the ion pair is much greater than r_c, then the influence of diffusion is prevailing and the ion pair escapes its mutual attraction. If the initial separation is smaller than r_c, then the Coulombic attraction leads to recombination, eventually. Onsager (1938) solved Eq.7 for the steady state, when $\partial P/\partial t = 0$.

Here, we follow the treatment presented by Pai and Enck [8]. The probability, $P(r,\theta, E)$, for the escape of the ion pair was derived to be,

$$P(r,\theta,E) = \exp(-A)\exp(-B)\sum_{n=0}^{\infty}\sum_{m=0}^{\infty}\frac{A^m}{m!}\frac{B^{m+n}}{(m+n)!}$$

(9)

where

$$A = \frac{e_0^2}{4\pi\varepsilon_0\varepsilon_r k_B T r}$$

(10)

and

$$B = \left[\frac{e_0 E r}{2k_B T}\right](1 + \cos\theta)$$

(11)

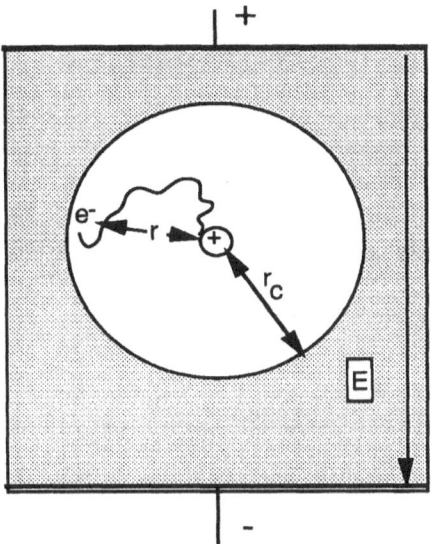

Fig. 7. Schematic representation of the random walk of an electron in the Coulomb field of its parent positive ion

If we consider many ion pairs which have an isotropic distribution function $g(r,\theta)$ given as,

$$g(r,\theta) = \frac{\delta(r - r_0)}{4\pi r_0^2}$$

(12)

where r_0 is the separation length then the fraction of escaping ion pairs $P(r_0, T, E)$ is given as

$$P(r_0, T, E) = \int P(r,\theta, E)g(r,\theta)d^3r$$

(13)

The first terms of Eq. 9 can then be written as,

$$P(r,\theta,T) = \exp\left(-\frac{r_c}{r}\right)\left\{1 + \frac{1}{2!}\left(\frac{e_0}{k_B T}\right)r_c E + \frac{1}{3!}\left(\frac{e_0}{k_B T}\right)^2 r_c\left(\frac{r_c}{2} - r\right)E^2\right.$$
$$\left. + \frac{1}{4!}\left(\frac{e_0}{k_B T}\right)^3 r_c\left(r^2 - r_c r + \frac{r_c^2}{6}\right)E^3 + ...\right\}$$

(14)

For numerical calculations of $P(r,\theta, T)$, especially at high values of E, a series derived by Mozumder [9] is very suitable.

$$P(r,\theta,E,T) = 1 - S_{-} \exp\left[-\eta - \zeta(1+z)\right] \tag{15}$$

where

$$\eta = \frac{r_c}{r} \tag{16}$$

ζ and z, stand for,

$$\zeta = \frac{e_0 E \, r}{2k_B T} \tag{17}$$

and

$$z = \cos \theta \tag{18}$$

S_{-} is a series containing only negative powers of r. It is the expansion of

$$\frac{1 - P(r,\theta,E,T)}{\exp\left[-\eta - \zeta(1+z)\right]} = S_{-} \tag{19}$$

Mozumder expanded $\exp(-\eta - \zeta(1+z))$ into a series with the required properties and after averaging of P over a random distribution of $\cos\theta$, he obtained the escape probability as,

$$\bar{P} = 1 - (2\zeta)^{-1} \sum_{k=0}^{\infty} A_k(\eta) \, A_k(2\zeta) \tag{20}$$

with the coefficients A_k given by,

$$A_k = 1 - \exp(-\eta)\left[1 + \eta + \frac{\eta^2}{2!} + \dots + \frac{\eta^k}{k!}\right] \tag{21}$$

In the limit of low electric field strength E, or for $\zeta \ll 1$, $\bar{P}$ is obtained as,

$$\bar{P} = (1 + \eta\zeta)\exp(-\eta) = e^{-r_c/r}\left[1 + \frac{e_0 E \, r_c}{2k_B T}\right] \tag{22}$$

or

$$\bar{P} = e^{-r_c/r}\left[1 + \frac{e_0^3}{8\pi\varepsilon_0\varepsilon_r k_B^2 T^2} E\right] \tag{23}$$

From Eq. 23 it follows that for $E = 0$, a finite escape probability $\bar{P}(0,T)$ exists given by,

$$\bar{P}(0,T) = e^{-r_c/r} \tag{24}$$

Besides, the escape probability increases linearly with field strength and the ratio of slope to intercept, S/I, given by,

$$S/I = \frac{e_0^3}{8\pi\varepsilon_0\varepsilon_r k_B^2 T^2} \tag{25}$$

82

depends on temperature and relative dielectric constant, only. The extent of linearity of the $\bar{P}$ vs E dependence depends on the initial separation distance. For small distances the linearity extends over a greater interval of the electric field strength than for large separation distances. Model calculations for a liquid hydrocarbon with $\varepsilon_r = 1.8$ and $T = 295$ K are shown in Fig. 8. As r increases, the $\bar{P}$ vs E dependence becomes less linear. Escape of the ion pair is enhanced by increase of the relative dielectric (see Fig. 9) constant and by increase of temperature.

The Onsager theory of ionic escape is widely used for the analysis of photo- and radiation-induced ionization in non-polar liquids. Here, an electron is trying to escape from its positive parent ion. Application of the model of Brownian motion requires the mean free path of the Brownian particle to be small compared to the characteristic length of the problem [10]. In the case of electron/ion recombination this length is the critical Onsager distance, r_c (see Eq. 2). While the mean free path of a positive ion is of atomic or molecular dimensions (an exception may be the fast positive carrier observed in C_6-cyclo-alkanes) the mean free path of the electron becomes comparable to r_c in liquids where high electron mobilities have been observed. Examples are liquid xenon, krypton, argon, methane, tetramethylsilane, and neopentane. In liquids with small electron mobility (small mean free path), the electron is frequently scattered by the molecules of the liquid and its kinetic energy can be considered as always being in equilibrium with $k_B T$ of the liquid. The motion of such slow electrons resembles that of ionic motion. In the limit of a large mean free path, the electron is allowed to make many revolutions around the positive ion before it is scattered by solvent molecules or atoms into another orbit. In such a case the electron/ion recombination is controlled by collision-induced transitions of the electron between Rydberg-like states. The motion of the electron can then be described more appropriately as diffusion in energy space [11].

If many ion pairs with different initial separation distances are considered, a distribution function of separation distances $g(r,\Theta, b)$, may be introduced. The total fraction of ion pairs escaping recombination is then given as,

$$P_{esc} = \int f(r,\Theta,E,T) \, g(r,\Theta,b) \, dv$$

(26)

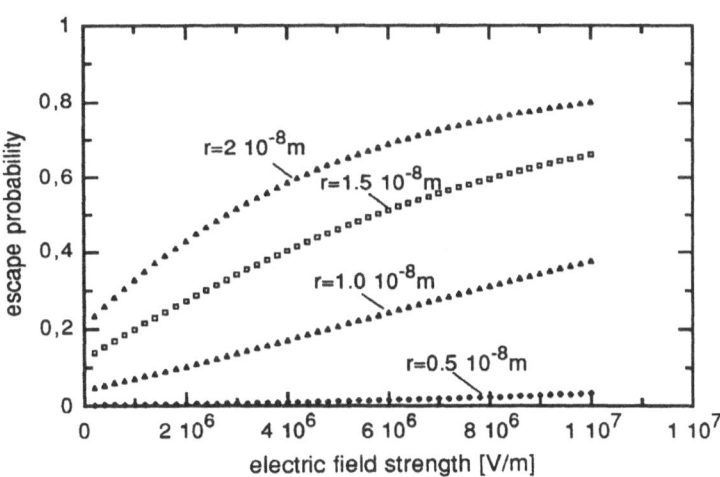

Figure 8. Calculations of the Onsager escape probability for $\varepsilon_r = 1.8$ and $T = 295$K; parameter: mean electron/ion separation distance

83

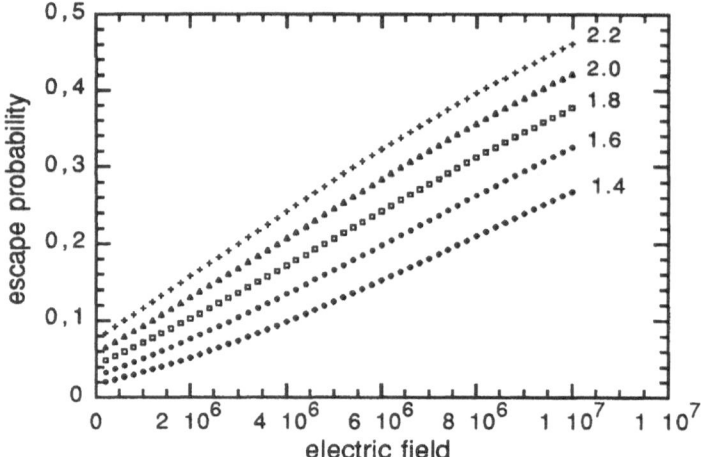

Figure 9. Calculations of the Onsager escape probability for different values of ε_r ; electron/ion separation distance $1\ 10^{-8}$m; T = 295K

In the application of the Onsager formalism to the case of low LET radiation it is implied that in a spur all ion pairs but the last recombine very fast and do not contribute to the observed ionization current.

PHOTOCONDUCTIVITY OF NONPOLAR LIQUIDS

Measurements of the photoconductivity of pure solvents yields information on the ionization energy in the liquid phase, I_{liq} , which in terms of solid state physics is called band gap. Compared to the gas phase ionization energy, I_{gas} , the liquid phase value is reduced on account of three factors: (i) upon condensation a broadening of the valence levels is observed, ΔE_{val} , which is of the order of 0.1 eV; (ii) the ionization energy is reduced by the polarization energy of the positive ion; (iii) the electron is promoted into the electronic conduction level of the liquid which is characterized by an energy V_0 . These levels are depicted schematically in Fig. 10.

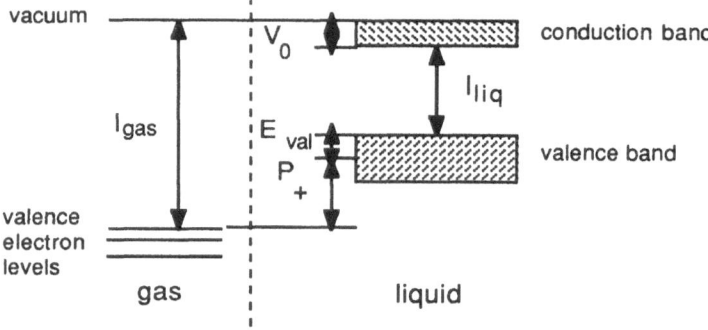

Figure 10. Schematic diagram of the electronic energy levels in the gas phase and in the liquid phase

The relationship between the various energy levels is given as,

$$I_{liq} = I_{gas} + P_+ + V_0 + \Delta E_{val} \tag{27}$$

P_+ is always negative while V_0 can be either negative of positive. Since no detailed information on ΔE_{val} is available, this contribution is usually neglected. In Table 1, values of the photoconductivity threshold of a number of non-polar liquids are listed. These energies are in the range of the vacuum ultraviolet part of the spectrum. Deuterium lamps or Synchrotron radiation were used as light sources.

Since no saturation current is obtained quantum yields for the generation of charge carrier pairs are a function of the applied electric field strength. The escape probability given by Eq. 20 has to be multiplied by the initial quantum yield η_0 which is usually taken as the value of the isolated molecule or atom. Values of η (E) for the molecular liquids listed in Table 1 are in the range of 10^{-4} to 10^{-2} for a few kV/cm (see also ref.12).

Table 1. Ionization energies, I_{liq}, of nonpolar liquids (after ref. 13)

Liquid	T[K]	I_{liq} [eV]	Liquid	T[K]	I_{liq} [eV]
Xenon	161	9.2	2,2,4-Trimethylpentane	294	8.3±0.05
Krypton	121	11.56	(Isooctane)		
Tetramethylsilane	294	8.1±0.05	2,2,4-Tetramethylpentane	295	8.2
		8.05±0.05	n-Tridecane	294	9.25±0.05
Tetramethylgermanium	294	7.6±0.05	Tetrakis(dimethylamino)ethylene	295	3.77±0.02
Tetramethyltin	294	6.9±0.1			
2,2-Dimethylpropane	294	8.85±0.05	n-Pentene-1	294	8.33±0.05
(Neopentane)		8.55±0.05	Tetramethylethylene	294	6.80±0.05
Methylbutane (Isopentane)	294	9.15± 0.05	Cyclopentane	294	8.80±0.05
n-Pentane	294	9.15±0.1	Cyclohexane	294	8.75±0.1
		8.85±0.05			8.43±0.05
2,2-Dimethylpropane	294	8.73±0.05	Hexamethyldisilane	294	6.75±0.1
(Neohexane)		8.50±0.05	Triethylsilane	294	8.25±0.1
n-Hexane	294	8.6±0.053	Bis(trimethylsilyl)ethane	295	6.85±0.05
Methylpentane	294	8.85±0.1	Polydimethysiloxane (50cSt)	296	7.29±0.05

RADIATION INDUCED CONDUCTIVITY OF NONPOLAR LIQUIDS

The radiation induced conductivity of nonpolar liquids has received extensive attention due to the possibility of using liquids in detectors and imaging devices for high energy particle and quantum radiation [6]. Instead of the quantum yield, the radiation chemical yield, G, is used which describes the number of electron/ion pairs produced by absorption of 100 eV of radiation energy. In radiation physics the W-value as the energy required for the generation of one electron/ion pairs is given. G - and W-value are connected via

$$W = \frac{100}{G} \tag{28}$$

G-values at zero applied field are determined from measurements of the ionization current as a function of field. Over a certain range of field strengths, the yield increases linearly with field (see Eq. 23). Extrapolation to $E = 0$ gives the yield of electron/ion pairs which escaped their geminate recombination. The corresponding G-value is called the free ion yield, G_{fi}. The G-value for the initial generation of pairs is usually assumed to be the same as in the gas phase. Values of G_{fi} are compiled in Table 2. It is apparent from the data that strong recombination processes in the track of an α-particle in the liquid lead to very small apparent yields at zero field.

A test of fidelity for the application of the Onsager formalism to the ionization currents in nonpolar liquids is the comparison of the experimental values of SI/I with the predictions given by Eq. 23. In Table 3 a few data are compiled which show the good agreement between the two values.

Table 2. G-values for ion-pair production in non-polar liquids

liquid	T[K]	radiation	G-value	IP/100eV
LAr	84	β - rays	tot.[a]	4.2
LKr	121	β - rays	tot.[a]	4.
LXe	161	β - rays	tot.[a]	6.1
Neopentane	*)	1.5 MeV x	fi	0.86
TMSi		1.5 MeV x	fi	0.74
TMSn		1.5 MeV x	fi	0.62
2,2,4,4-TMP		1.5 MeV x	fi	0.83
Isooctane		1.5 MeV x	fi	0.33
		α - rays	fi	0.0062
n-Hexane		1.5 MeV x	fi	0.1
		α - rays	fi	0.0058

*) all molecular liquids at t=295K
[a] total yield = initial yield, a kind of saturation current was observed

Table 3. Slope to intercept values (data from ref. 3 and ref. 6)

liquid	radiation	T [K]	ε_r	slope/intercept (10^4 cm V^{-1}) observed	theoretical value
Silicone oil (AK 10)	γ-rays	293	2.61	0.43	0.43
n-hexane	x-rays	298	1.98	0.6	0.58
lN$_2$	x-rays	77	1.43	11.0±2	11.2
1-methyl naphthalene/ neopentane	light	298	1.84	0.57	0.62

RADIATION-INDUCED IONIZATION OF SOLIDS

In this section we will be concerned with the action of low LET radiation on organic polymers. At a constant radiation intensity, generally, proportionality between ionization current, i, and applied voltage, U, is found. Values of i/U at different dose rates, D_r, (radiation intensities) show a functional dependence which can be approximated empirically by a power law expressed as,

$$\frac{i}{U} = AD_r^\Delta$$

(29)

with $0.5 \le \Delta \le 1$. Δ is a function of temperature. A is an empirical constant. Selected data are compiled in Table 4. It has been speculated that this dependence on Δ might be due to the effect of charge carrier trapping on the ionization current. It seems that an important quantity is the schubweg, x_s, of the charge carriers which is given by,

$$x_s = \mu E \tau \tag{30}$$

Here, μ denotes the carrier mobility, E is the electric field strength, and τ is the lifetime of the carrier in the mobile state. For cases where the sample thickness, d, is greater than x_s trapping is dominant while for $d \ll x_s$ the effects of geminate recombination manifest themselves in the current/field strength dependence.

Table 4. γ-Radiation induced conductivity of some polymers

Material	T[K]	Δ	A $[\Omega^{-1}cm^{-1} (rad\ s^{-1})]^{1/\Delta}$
low density	190	0.96	5.9 10^{-18}
polyethylene	253	0.86	
	300	0.84	7.4 10^{-17}
	380	0.78	8.2 10^{-16}
polystyrene	322	0.97	4 10^{-17}
	333	0.97	
	296	0.81	5.4 10^{-18}
	293	0.89	1.2 10^{-17}
teflon	311	1.0	1.2 10^{-16}
polypropylene	311	0.88	3.8 10^{-17}

The charge carrier yield induced by x-rays in poly-N-vinyl carbazole was found to exhibit the linear rise with field and a slope predicted by the Onsager theory of gemnate recombination[14] (see Fig. 11). Similar conditions were found for anthracene crystals[15]. Another polymer where an Onsager type dependence of the ionization current was found is polyethylene-terephthalate. Here, high electric field strengths and thin samples gave a dependence of the γ-ray induced current on field strength in satisfactory agreement with Onsager´s prediction[16] (see Fig. 12).

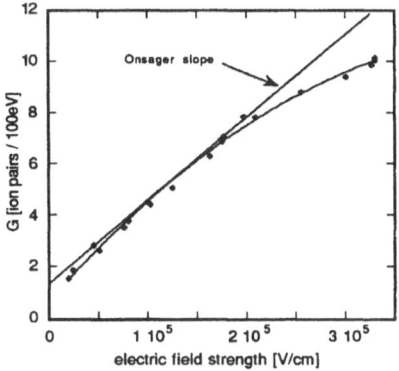

Figure 11. Electron/ion yield for x-irradiation of polyvinylcarbazole (after ref. 14]

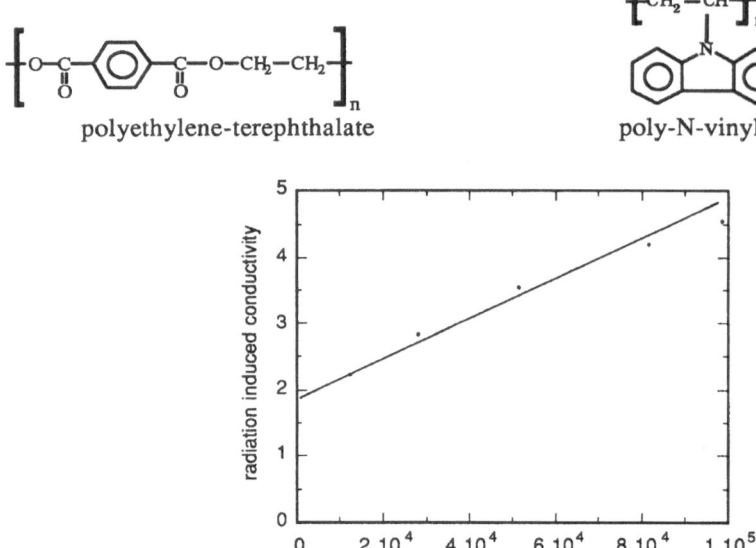

polyethylene-terephthalate   poly-N-vinyl carbazole

Figure 12. Electron/ion yield for γ-irradiation of polyethylene terephthalate; $(S/I)_{theor} = 3.5 \ 10^{-5}$ cm/V, $(S/I)_{exp} = 1.8 \ 10^{-5}$ cm/V (after ref. 16)

CONCLUSION

The ionization process in the condensed phase is governed by the close spatial correlation of electron and parent cation. Geminate recombination is the process which leads to a much smaller apparent yield of photo-induced charge carriers than is obtained in the gas phase. The influence of applied electric field and temperature on the apparent yield can be described by Onsager´s theory for liquids and organic solid where the electron can be treated as an ionic species. Advancement of the theory for liquids with high electron mobilities requires more reliable data of single photon induced and radiation induced ionization currents.

REFERENCES

1. Christophorou, L. G. "Atomic and Molecular Radiation Physics." Wiley Monographs in Chemical Physics. Birks and McGlynn ed. 1971 Wiley-Interscience. London, New York, Sydney, Toronto.

2. Koizumi, H., K. Lacmann and W. F. Schmidt, Light-induced electron emission from silicone oils, J. Phys. D: Appl. Phys. **25**, 857-861 (1992)

3. Böttcher, E. H., "Experimentelle Untersuchung der photoelektrischen Leitung reiner und mit aromatischen Molekülen dotierter organischer Flüssigkeiten", HMI-Report B406 (1984)

4. ICRU, "Physical Aspects of Irradiation", Handbook 85 (1964)

5. Mozumder, A. and J. L. Magee, Model of tracks of ionizing radiations for radical reaction mechanisms, Radiat. Res. **28**, 203-214 (1966)

6. Hummel, A. and W. F. Schmidt, Ionization of dielectric liquids by high energy radiation studied by means of electrical conductivity methods, Radiat. Res. Rev. **5**, 199-300 (1974)

7. Onsager, L., Initial recombination of ions, Phys. Rev. **54**, 554-557 (1938)

8. Pai, D. M. and R. C. Enck, Onsager mechanism of photogeneration in amorphous selenium, Phys. Rev. B. **11**(12), 5163-5174 (1975)

9. Mozumder, A., Effect of an external electric field on the yield of free ions. I General results from the Onsager theory, J. Chem. Phys. **60**(11), 4300-4304 (1974)

10. Chandrasekhar, S., Stochastic problems in physics and astronomy, Rev. Mod. Phys. **15**, 1-89 (1943)

11. Tachiya, M. and W. F. Schmidt, Escape probability of geminate electron-ion recombination in the limit of large electron mean free path, J. Chem. Phys. **90**(4), 2471-2475 (1989)

12. Böttcher, E. H. and W. F. Schmidt, Photoconductivity of nonpolar liquids induced by vacuum-ultraviolet light, J. Chem. Phys. **80**(3), 1353-1359 (1984)

13. Schmidt, W. F., "Electronic energy levels in non-polar dielectric liquids." Excess electrons in dielectric media. Ferradini and Jay-Gerrin ed., 1991, CRC Press. Boca Raton, Ann Arbor, Boston, London.

14. Hughes, R. C., Radiation-induced conductivity in polymers: poly-N-vinylcarbazole, IEEE Trans. NS. **18**(6), 281-287 (1971)

15. Hughes, R. C., Geminate recombination of x-ray excited electron-hole pairs in anthracene, J. Chem. Phys. **55**(12), 5442-5447 (1971)

16. Maeda, H., M. Kurashige and T. Nakakita, Gamma-ray-induced conduction in polyethylene-terephthalate under high electric fields, J. Appl. Phys. **50**(2), 758-64 (1979)

HIGH ENERGY IONIZATION IN LIQUIDS - THE FREE ION YIELD

Richard Holroyd
Department of Chemistry
Brookhaven National Laboratory
Upton, NY 11973

INTRODUCTION

Ionizations can be produced in liquids in many ways, for example by exposure to high energy radiation (fast electrons or alphas) or by interaction with ultraviolet photons of sufficient energy. The passage of a high energy electron results in well separated clusters, each containing a few ionizations or in the case of alpha particles a dense column of ionization. Absorption of ultraviolet light creates a single ion-electron pair. In any case only a fraction of the ion-electron pairs formed initially separate to become free; in most liquids recombination within the track or cluster dominates. The yield of ion electron pairs that separate and become free depends on several factors including the molecular structure of the solvent, the density of ionization along the track, the liquid density, the applied field and the temperature[1].

First we will consider the situation of a single isolated ion-electron pair. The electron is formed with initial kinetic energy which allows it to travel some distance away from its geminate cation. The electron may lose energy to vibrational modes but a significant fraction of the separation occurs while the electron has subvibrational (near thermal) energy. When the electron finally thermalizes or localizes it is within the Coulombic field of its parent cation and the two ions constitute a geminate pair. The free ion yield is determined by the fraction of geminate pairs which diffuse out to form free ions as against those that recombine to form excited states. The probability of escape of an isolated geminate pair, where the thermalization distance is r, is given at zero field by:

$$P^o_{esc}(r) = \exp(-r_c/r) \qquad (1)$$

where r_c is the Onsager radius, which is that distance at which the Coulombic energy drops to the thermal energy (kT); i.e., $r_c = e^2/\varepsilon kT$. Generally P_{esc} is a function of the electric field, the temperature, the dielectric constant as well as the thermalization distance, r. Thus the yield of free ions or number of ion pairs that separate per 100 eV absorbed is:

$$G_{fi} = G_{tot} \int_0^\infty D(r) \, P(r,E) dr \qquad (2)$$

Linking the Gaseous and Condensed Phases of Matter
Edited by L.G. Christophorou *et al*, Plenum Press, New York, 1994

The initial yield of ionization, G_{tot}, is not known exactly for molecular liquids. A value equal to the gas phase yield (100/W) is often assumed, where W is the number of eV required to form an ion pair. For neopentane the extrapolated yield at high field is 4.0 per 100 eV.[2] Thus the free ion yield depends on G_{tot} and P(r,E), the escape probability, which depends strongly on the thermalization distance. Note that a distribution of thermalization distances D(r) is assumed, usually Gaussian or exponential. Although irradiation by electrons of MeV energy leads to clusters of ionization, the above formalism is often used to analyze the results.

In this talk we consider some recent experimental results which show how G_{fi} depends on conditions. The results provide further insight into the factors which determine the free ion yield.

EXPERIMENTAL METHODS

The methods employed for measurements of the free ion yield differ in the way in which charge is collected. All methods require a measurement of the dose absorbed in order to calculate G_{fi}. In the steady irradiation method the current is measured with an electric field (E) applied to a sample exposed to a beam of ionizing radiation. The ion yield increases with field, as discussed below, but the results can be extrapolated to zero field to get G_{fi}^{o}, the zero field yield. When this method is used care must be taken to avoid volume recombination of ions. The efficiency of ion collection for continuous irradiation is given by[3]:

$$ Eff. = \frac{2}{1 + (1 + z^2)^{1/2}} $$

$$ \text{where } z^2 = \frac{2\ ed^2\ DR}{3\varepsilon_0\varepsilon\ E^2}\left(\frac{1}{\mu_+} + \frac{1}{\mu_-}\right) $$

Thus the efficiency decreases as d, the distance between plates, or DR, the dose rate increases. The efficiency increases as E, the field, or μ, the mobility, increase. The efficiency of ion collection is improved by working at low dose rates which means measurement of low currents (generally less than 10^{-10}A).[3]

Other methods involve collection of charge following a pulse of irradiation. Ion yields for many liquids were determined by the clearing field method.[4,5] In this technique a field large enough to avoid volume recombination is applied quickly after a short pulse in order to collect all the ions. This method gives G_{fi}^{o} directly since there is no field present during the pulse. Another method involves measuring the charge following a pulse of irradiation where a steady electric field is applied.[6] Here, as in the steady state method, the field enhances the ion yield and the observed charges are extrapolated to zero field to obtain G_{fi}^{o}. The efficiency of ion collection is given for the pulsed case by:[7]

$$ Eff = \frac{1}{U} \ln (1 + U) $$

$$ \text{where } U = e\ d\ N_0/\varepsilon_0\varepsilon\ E. $$

The efficiency will decrease as d and N_0, the initial concentration of ions, increase and will increase as E increases. A recent study[7] shows that the different methods give similar results for hexane, where $G_{fi}^{o} = 0.13 \pm .01$ ions/100 eV.

Another method of determining the ion yield is based on the electron yield from passage of a single ionizing particle. The concentration of charged species is much lower; a charge sensitive amplifier is employed. Here the efficiency is a function of the concentration of electron trapping impurities, or specifically R which is equal to τ/t_d, where τ is the electron lifetime and t_d is the drift time. The efficiency of collection is:[8]

$$Eff = 2R + 2R^2 \, [\exp(-1/R) -1]$$

The efficiency increases with R and is not very good unless R>> 1.

MOLECULAR STRUCTURE EFFECTS

Examination of the data shown in Table 1 indicates that free ion yields for liquids are dependent on molecular structure. G_{fi}^o ranges from a low value of 0.01 for ethene to values nearly a hundred-fold larger for the tetramethyl compounds. The reason for this large range of values is not entirely clear. G_{fi}^o depends on the initial separation of electron and ion, which in turn depends on the rate of energy loss of the electron and on the smoothness of the potential seen by epithermal electrons slowing down in the conduction band. Electrons must lose energy rapidly in ethene and acetylene. The thermalization distance in these liquids is quite short (<4 nm),[1] but the process responsible is as yet unidentified.

Table 1 - Free Ion Yields[a]

Liquids	Temp °K	Electrons G_{fi}^o	Alphas G_{fi} @ E=10 kV/cm
Ethene	170	0.01	
Acetylene	198	0.02	
1,3-Butadiene	269	0.04	
Benzene	296	0.05	
trans -2-Butene	293	0.08	
n-Hexane	296	0.13	0.017[d]
n-Tetradecane	296	0.12	
Carbon tetrachloride	296	0.10	
Germanium tetrachloride	296	0.13	
Tetraethylsilane	295	0.18[b]	
Tetraethylgermane	295	0.18[b]	
Tetraethylstannane	295	0.20[b]	
2,2,4-Trimethylpentane	295	0.33	0.018[e]
Tetramethylsilane	295	0.65[c]	0.029[e]
Tetramethylgermane	295	0.63[c]	0.029[f]
Tetramethylstannane	295	0.64[c]	
2,2,4,4-Tetramethylpentane	295	0.83	0.025[e]
Neopentane	296	1.10	0.036[e]

a) Data from Ref 9 unless indicated otherwise
b) Ref 14
c) Ref 15
d) Ref 16
e) Ref 17
f) Ref 18

Free ion yields are also quite low for compounds with conjugated double bonds indicating an efficient energy loss mechanism is present. Here temporary anion formation is likely to be involved.[10] Electron attachment to 1,3-butadiene and benzene has been shown to occur in liquids at high pressure[11] and in glasses at low temperature.[12,13] Such temporary anions provide a mechanism for energy loss by the electron which limits its separation distance from the positive ion.

For tetraalkyl compounds containing a central Si, Ge or Sn atom the free ion yields span quite a range.[14,15] The yields depend strongly on the nature of the alkyl group. The yields are comparable, $G_{fi}^o = 0.64\pm0.01$, for the three tetramethyl-compounds in Table 1. The ion yields are much less for the tetraethyl-compounds, $G_{fi}^o = 0.19\pm.01$, and even lower for compounds with larger side groups. For these liquids the density of the Sn-containing liquid is about twice the density of the corresponding Si-containing liquid showing that the density alone does not determine the yield. Rather the thermalization distance is dependent on the nature of the alkyl group.

The smoothness of the potential seen by the epithermal electron is important. Evidently symmetrical structures with many methyl groups provide the smoothest potential leading to the largest range. The lower yield for the tetraethyl compounds indicates the important role of methyl groups. Although the tetrachloro-compounds are symmetrical, the yields are lower because electron attachment occurs.

The product $b\epsilon$ also appears to be an important parameter. For the tetramethyl compounds discussed earlier G_{fi}^o is nearly the same and $b\epsilon$ is also the same. Recently from an analysis of free ion yield data for 52 liquids at room temperature it was shown[19] that log G_{fi}^o is linearly dependent on $(b\epsilon)^{-1}$, a dependence expected on the basis of equation 1.

Implicit in the above discussion is the assumption that thermalization precedes electron trapping. However electrons may be trapped before complete thermalization as has been suggested.[20] This provides another mechanism limiting the separation distance. It also provides a rationalization for the known correlation of thermal electron mobility with mean thermalization distance (see Ref 1) since the magnitude of the mobility is related to the probability of trapping.[10]

FIELD AND TEMPERATURE EFFECTS

The probability of separation of an electron-ion pair in an electric field as developed by Onsager[21] is of the form:

$$P(E,r) = \exp\left(-r_c/r\right) \left[1 + F(E,r)\right] \tag{3}$$

When averaged over all angles between the direction of the external field and the electron-ion axis, $F(E,r)$ becomes:[22]

$$F(E,r) = \exp(-U_o) \sum_{n=1}^{\infty} \frac{U_o^{\,n}}{(n+1)!} \sum_{j=0}^{n-1} \frac{(n-j)(r_c/r)^{(j+1)}}{(j+1)!} \tag{4}$$

where $U_o = eEr/k_BT$. At low field (small U_o):

$$F(E,r) = e^3E/2\epsilon(k_bT)^2 \text{ or } (9.68/\epsilon T^2)E \tag{5}$$

and the ratio of slope-to-intercept of plots of ion yields versus field should be $9.68/\epsilon T^2$ cm/V. Typically the values of slope-to-intercept are close to this value for liquids where G_{fi} is small. See for example the data for n-hexane in figure 1 where G_{fi} is near 0.1 per 100 eV. For higher fields and for liquids of higher free ion yield, where the dependence of yield on field becomes nonlinear at a relatively low field, the full equation (4) is required to adequately describe the data[1].

Free ion yields increase with increasing temperature. This is associated with the temperature dependence of the Onsager relation, equation (3). For single ion pairs created by photoionization the temperature dependence is given well by equation (3)[23]. For high energy radiation the mean thermalization distance, b, increases with increasing temperature because the density(d) of the liquid is decreasing. Then the product bd is considered an important parameter. This parameter is nearly constant for n-pentane and benzene over a large range of temperatures. However, bd varies considerably for 2,2,4-trimethylpentane and neopentane.[5] For such liquids made up of spherelike molecules with many methyl groups, temperature studies show that the product bd increases with temperature and goes through a maximum at a temperature just below the critical temperature. Examples of liquids exhibiting such behavior are 2,2,4-tetramethylpentane[24] and neopentane (see Fig 2). The mobility of thermal electrons also shows a maximum at nearly the same temperature for these compounds. This indicates the scattering of thermal electrons is similar to that of epithermal electrons and that the potential seen by both types of electron is relatively smooth at the maximum.

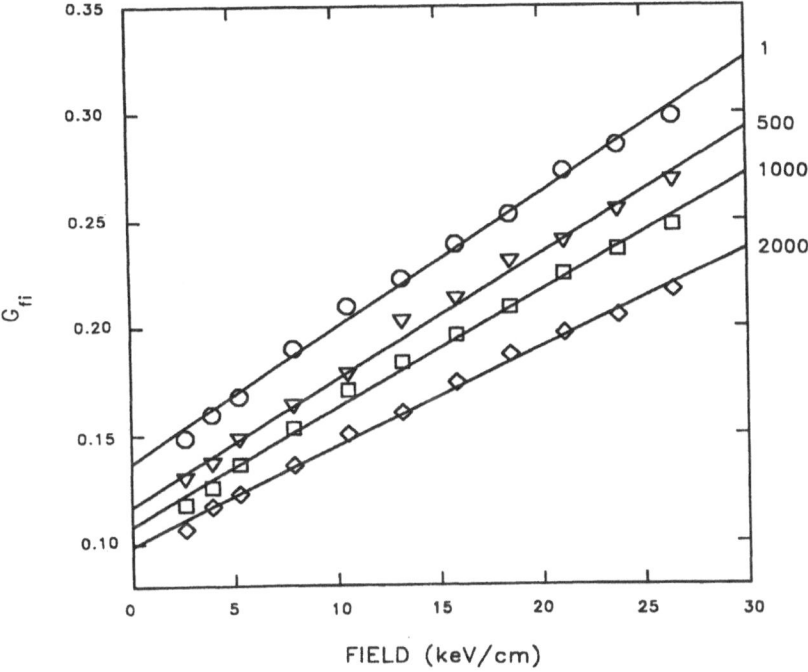

Fig. 1 Plot of G_{fi} versus field for n-hexane at indicated pressures (in bar).

It is important to realize that determinations of the mean thermalization range usually are done by assuming some distribution of ranges, D(r), typically Gaussian, and fitting the data to equation 2 for an assumed value of G_{tot}. However G_{tot} is not known for most liquids and the value of the range parameter derived changes when G_{tot} is changed. Also a Gaussian distribution may not be the best description of reality. Recently it was shown for the tetra-alkyl liquids[25] that if the form of D(r) is the same in all liquids then G_{tot} must vary from liquid to liquid, or conversely that if G_{tot} is the same then the form of D(r) changes.

TRACK EFFECTS

The discussion so far has dealt with ion yields for liquids exposed to high energy electrons (in the MeV range). Such energetic electrons are minimum-ionizing and the clusters of ionization are well separated. As the energy of the electron decreases the clusters will occur closer together because the rate of energy loss increases with decreasing energy. For a 10 keV electron the initial positions of the clusters will be, on average, 10 nm apart. Because of interactions between clusters the situation is best described as a track. The free ion yield will be less than for minimum ionizing radiation because of the enhanced possibility of ion-electron recombination in such a track.

One way to study how ion yields vary with electron energy is with X-rays since absorption of an X-ray of energy in the 2 to 30 keV range by a hydrocarbon results in a photoelectron of comparable energy. The new synchrotron radiation sources provide tunable X-rays in this range. Dosimetry is readily done with an ion chamber inserted in the X-ray beam ahead of the cell containing the liquid sample.

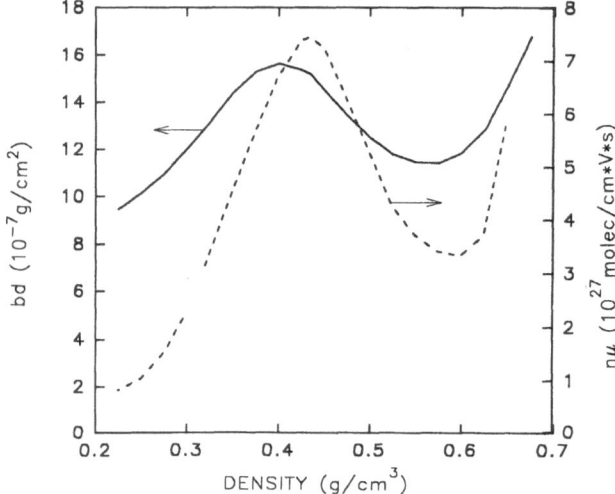

Fig. 2 Plot of bd vs density for neopentane; and plot of $n\mu_e$ vs density for neopentane (Ref 1).

The variation of the free ion yield with X-ray energy is shown in figure 3 for three hydrocarbons.[26,27] For all three the yield decreases with decreasing X-ray energy to minimum values around 2 keV. The magnitude of the effect depends on the hydrocarbon. For n-hexane G_{fi}^o decreases a factor of about two but for 2,2,4-tetramethylpentane the yield decreases a factor of about eight. Qualitatively this effect can be understood as an effect of the rate of energy loss, dE/dx, of the photo-electron which increases with decreasing X-ray energy. This results in a higher density of ionization which increases the probability of electrons combining with other ions in the tracks.

A comparison of experimental data with a theoretical predication[28] is shown in Fig 4. A computer was used to simulate diffusion and drift of ions and electrons in tracks, taking the nonhomogeneous kinetics of ion recombination into account. The theory predicts G_{fi}^o decreases with decreasing energy for both n-hexane and 2,2,4-trimethylpentane in agreement with experiment. The theory also predicts the yields increase again at very low energy.

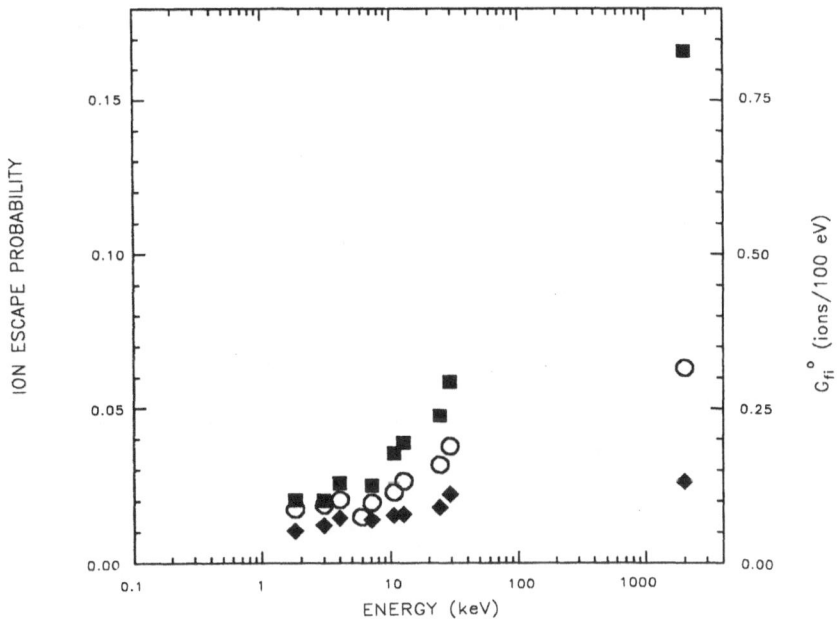

Fig. 3 Variation of zero field free ion yield with X-ray energy for ■ 2,2,4,4-tetramethylpentane, ○ 2,2,4-trimethylpentane and ♦ n-hexane.

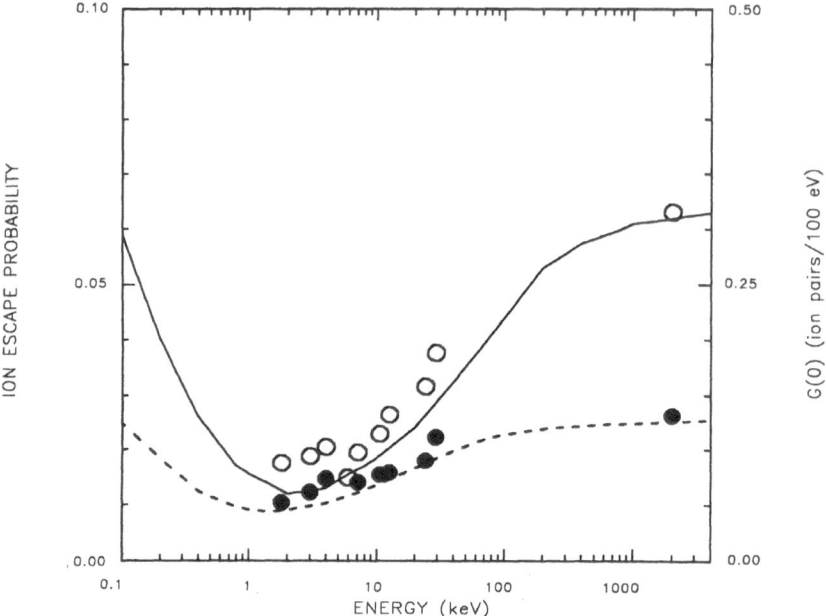

Fig 4 Dependence of G_{fi}^o on electron energy: experimental from ref 25 and 26 for ○ 2,2,4-trimethylpentane and ● n-hexane lines are theory, ref 28, G_{tot} = 5 and b = 10.6 nm for 2,2,4-trimethylpentane and 6.8 nm for n-hexane.

There are a few studies of ion yields in liquid hydrocarbons exposed to very densely ionizing particles like alpha particles.[16-18] Results for some liquids at 10 kV/cm are given in Table 1. In contrast to high energy electrons there is less of a molecular structure effect for alphas; the yields are within a factor of two for those liquids that have been studied. The field dependence of the ion yields for alpha radiation of hydrocarbons can be empirically expressed by an equation of the form:

$$G_{fi} = G_{fi}^o + \alpha E^n \qquad (6)$$

where n is 0.7 to 0.8 and α depends on the liquid. The value of the intercept, G_{fi}^o is very small, about 0.005 per 100 eV for alphas. A yield of zero is expected for the cylindrical geometry of a track in contrast to the spherical geometry of isolated ion pairs where G_{fi}^o is finite (see equation 1).

The theoretical understanding of alpha ray yields dates back to the analysis of Jaffé[29] published in 1913. A cylindrical distribution of positive and negative ions was assumed to be formed initially. The calculation of the ion yield under an applied field requires consideration of the diffusion as well as the recombination of the ions. To solve the differential equation Jaffé first ignored the recombination, solved the simplified equation, and added the recombination later. His well known equation predicts the dependence of the free ion field to be given by

$$1/G_{fi} = 1/G_{fi}^\infty + B \frac{dE}{dx} \frac{S(z)}{E} \qquad (7)$$

where S(z) is a function of E, r, and temperature (see Ref 26). Thus the yield should increase with field and decrease as dE/dx (the rate of energy loss) increases. Figure 5 shows some results for exposure of hydrocarbons to electrons (from X-rays)[26,27]. The data points are G_{fi} for an applied field of 5kV/cm. The dashed lines are best fits of equation 7 to the

X-ray data. The values of G_{fi}^{∞} used are 0.16, 0.25 and 0.42 per 100 eV for n-hexane, 2,2,4-trimethylpentane and 2,2,4,4-tetramethylpentane, respectively. The slopes, B S(z)/E, are 6.30, 4.70 and 3.58 x 10^{-8} cm for n-hexane, 2,2,4-trimethylpentane and 2,2,4,4-tetramethylpentane, respectively. The lines have been extrapolated to higher values of dE/dx. The values predicted for alphas in this way for dE/dx = 10×10^8 eV/cm, are higher (by 60% or less) than the experimental points.

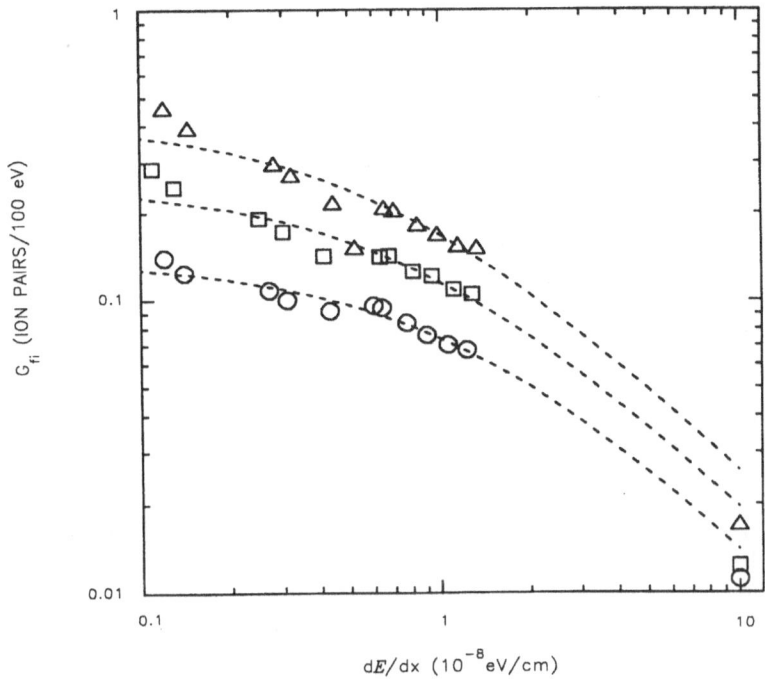

Fig. 5 Dependence of ion yields at 5 kV/cm on dE/dX for △ 2,2,4,4-tetramethylpentane, ▢ 2,2,4-trimethylpentane, and ○ n-hexane. Dashed lines eq 7-see text.

Further progress in understanding alpha tracks must take into account the detailed structure of the track. For example the electrons initially are expected to have a wider distribution than the positive ions. The difference in distribution will produce an internal electric field. Consequently, recombination will be very important. In fact Kramers' theoretical analysis[30] neglected diffusion initally and he solved the equation including recombination first and added diffusion later. For hydrocarbons the probability that electrons will escape the main track of an alpha particle is very low. Electrons which thermalize some distance from the track are more likely to contribute to the observed yield. These may be electrons on the tail of the distribution or may be ionizations caused by delta rays which are secondary electrons with enough energy to leave the track. Escape of ion pairs formed in the delta ray track is much more likely.

DENSITY EFFECTS

To determine how the free ion yield varies with liquid density, measurements are made as a function of pressure. In this way the temperature can be held constant. Applications of pressure to 3 kbar increases the density of a hydrocarbon by about 20%. Measurements of this sort are illustrated for n-hexane in figure 1. G_{fi}^{o} decreases with increasing pressure and at 3 kbar is 55% of its value at 1bar. Over this range the density increases by 20%. Thus the mean thermalization distance is expected to decrease with increasing pressure because of the increase in number of scattering centers. Analysis of the data in figure 1 indicates the product bd decreases by 10% over this pressure range.[31]

ACKNOWLEDGEMENT

This research was carried out at Brookhaven National Laboratory under contract DE-AC02-76CH00016 with the U.S. Department of Energy and supported by its Division of Chemical Sciences, Office of Basic Energy Sciences.

REFERENCES

1. G.R. Freeman, "Ionization and charge separation in irradiated materials " in "Kinetics of nonhomogeneous processes" G.R. Freeman, ed. J. Wiley and Sons NY. (1987).

2. W.F. Schmidt, Radiation-induced conductivity and ion yields in neopentane at high electric fields, Radiation Research 42:73 (1970).

3. W.F. Schmidt, R.A. Holroyd, Ion mobilities and yields in X-irradiated polydimethylsiloxane oils, Radiat. Phys. Chem. 39:349 (1992).

4. W.F. Schmidt, A.O. Allen, Ionization of liquids by radiation: measurements by a clearing field, Science 160:301 (1968).

5. W.F. Schmidt, A.O. Allen, Free ion yields in sundry irradiated liquids, J. Chem. Phys. 52:2345 (1970).

6. G. Ramanan, N. Gee, G.R. Freeman, Electron energy loss in fluids: thermalization distances in liquid and gaseous sulfur hexafluoride. Can. J. Phys. 68:925 (1990).

7. N. Gee, C. Senanayake, G. R. Freeman, Electron mobility, free ion yields, and electron thermalization distances in n-alkane liquids: effect of chain length, J. Chem. Phys. 89:3710 (1989).

8. R. Holroyd, Effects of radiation damage to TMP, TMS and liquid argon solutions, in Radiation Effects at the SSC, M.G. D. Gilchriese, ed. SSC-SR-1035 p339 (1988).

9. A.O. Allen, Yields of free ions formed in liquids by radiation, NSRDS-NBS-57 (1976).

10. J-P. Dodelet, K. Shinsaka, G.R. Freeman, Electron mobilities in liquid olefins: structure effects, J. Chem. Phys. 59:1293 (1973).

11. R.A. Holroyd, Electron reactions in nonpolar liquids - pressure effects, proceedings-this ASI.

12. T. Shida, W.H. Hamill, Molecular Ions VI. Electronic absorption and electron paramagnetic resonance spectra of molecular ions of conjugated dienes and allyl radicals, J. Am. Chem. Soc. 88:5371 (1966).

13. T Shida, S. Iwata, Electronic spectra of Ion radicals and their molecular orbital interpretation III aromatic hydrocarbons, J. Am. Chem. Soc. 95:3473 (1973).

14. 14.R. Holroyd, Free ion yields in liquids-molecular structure effects, in "Proceedings International Conference on Liquid Radiation Detectors: Their Fundamental Properties and Applications", Waseda Univ. Tokyo, Japan p24 (1992).

15. R. A. Holroyd, S. Geer, F. Ptohos, Free-ion yields for several silicon-, germanium-, and tin-containing liquids and their mixtures, Phys. Rev. B 43:9003 (1991).

16. M. Chybicki, Ionization currents induced by alpha radiation in liquid hexane and heptane, Acta Phys. Pol, 30:927 (1966).

17. R. C. Munoz, J. B. Cumming, R. A. Holroyd, Ionization of liquid hydrocarbons and tetramethylsilane by [241]Am alpha particles, J. Chem. Phys. 85:1104 (1986).

18. R. A. Holroyd, in "SSC Detector R and D at BNL", B. Yu and V. Radeka, eds, BNL 52244 p23 (1990).

19. J. -P. Jay-Gerin, T. Goulet, I. Billard, On the correlation between electron mobility, free-ion yield, and electron thermalization distance in nonpolar dielectric liquids, Can J. Chem. 71:287 (1993).

20. B.S. Yakovlev, Low energy electron localization in hydrocarbon glasses. Method of photoassisted ion pair separation, Radiat. Phys. Chem 40:37 (1992).

21. L. Onsager, Initial recombination of ions, Phys. Rev, 54:554 (1938).

22. J. Terlecki, J. Fiutak, Onsager's recombination theory applied to liquid alkanes, Int. J. Radiation Phys. and Chem. 4:469 (1972).

23. R.A. Holroyd, R.L. Russell, Solvent and temperature effects in the photoionization of tetramethyl-p-phenylenediamine, J. Phys. Chem. 78:2128 (1974).

24. T.G. Ryan, G.R. Freeman, Electron mobilities and ranges in methyl substituted pentanes through the liquid and critical regions, J. Chem. Phys., 68:5144 (1978).

25. S. Geer, R.A. Holroyd, Electron thermalization lengths and total initial ionization yields in tetra-alkyl liquids, Phys. Rev. B. 46:5043 (1992).

26. R.A. Holroyd and T.K. Sham, Ion yields in hydrocarbon liquids exposed to X-rays of 5-30 keV Energy, J. Phys. Chem. 89:2909 (1985).

27. R.A. Holroyd, T.K. Sham, B.-X. Yang and X.-H. Feng, Free Ion yields exposed to synchrotron X-rays, J. Phys. Chem. 96:7438 (1992).

28. W. M. Bartczak, A. Hummel, Computer simulation study of spatial distribution of the ions and electrons in tracks of high energy electrons and the effect on the charge recombination, J. Phys. Chem. 97:1253 (1993).

29. G. Jaffé, Zur theorie der ionisation in kolonnen, Annalen der Physik 42:303 (1913).

30. H.A. Kramers, On a modification of Jaffé's theory of column - ionization, Physica 18:665 (1952).

31. R.A. Holroyd and E. Stradowska, unpublished.

PHOTO- AND PENNING IONIZATION OF MOLECULES IN THE GAS PHASE AND IN THE LIQUID PHASE

Harald Morgner
Institut für Experimentalphysik
Universität Witten/Herdecke
Stockumer Str.10, D-58448 Witten

INTRODUCTION

Electron spectroscopy has contributed a great deal to the understanding of the properties of matter, let it be atoms, molecules or matter in the condensed phase. The common feature of these spectroscopies consists in analysing the kinetic energy of electrons that have experienced an interaction with the probe to be investigated. One way of performing the experiment is to start with a beam of electrons of known kinetic energy. If the loss of kinetic energy due to interaction with the probe is recorded one talks about electron energy loss spectroscopy (EELS). The energy loss spectrum is characteristic of the target atoms or molecules, but the detailed interpretation of such spectra is not always straightforward since the theoretical description of the electron molecule interaction is not simple. Still, many studies of this kind have been performed on gas phase molecules[1,2] and first attempts to use this experimental tool for the investigation of molecules in the liquid phase have been reported[3,4]. Another widely used technique, applicable for primary electron beams of several keV energy, is to observe electrons which originate from ions produced by the impact of the primary beam. If the ion is created by removal of an electron out of a core hole then the ion is in a highly excited state and will decay via an Auger process by emission of a second electron into a doubly charged ion. The spectroscopy of these electrons (Auger electron spectroscopy=AES) has developed into a broad field in atomic and molecular physics[5] and has become a routine tool in the analysis of solid surfaces[6]. Liquid surfaces have been investigated by this technique as well[2], but its specific contribution to the understanding of liquid surfaces is yet to be understood.

A particular fruitful type of experiment employs photons of well defined energy hν as the primary beam. Since the photon energies are kept below a few keV the Compton process is still irrelevant which implies that the photon either passes without energy loss or puts the total amount of its energy into the process of ionization of one electron. It results the simple equation:

$$E_{el} = h\nu - E_{bind} \tag{1}$$

which relates the observed quantity, i.e. the kinetic energy of the released electron E_{el}, with the binding energy E_{bind} of the electron within the atom or molecule investigated. Thus, the measured spectrum can easily be converted into a spectrum of binding energies which contains very direct information on the orbital structure of the molecule. If the molecule is transferred from the gas phase to the liquid phase the new environment shows in a shift of the binding energies of its orbitals.

If x-rays are used to photoionize the target one talks about XPS(=x-ray photoelectron spectroscopy). Kai Siegbahn introduced this technique in the fifties. The common x-ray line sources ($MgK_{\alpha} \cong 1253.6eV$ or $AlK_{\alpha} \cong 1486.6$ eV) have sufficient energy to remove electrons from core orbitals of all elements, in case of the second row elements even from the 1s-orbitals ($E_{bind} = 287eV$, 402eV, 531eV for C,N,O respectively). Indeed, core electron spectroscopy has been and still is the main object of XPS. The large differences in binding energy even between elements which are neighbors in the periodic table allow to identify the elemental composition of a probe with ease. However, if an element is not isolated but forms part of a larger entity, say a molecule, then the binding energies of its core levels are slightly, but distinctly shifted by several eV, depending on the chemical environment. This chemical - in addition to the elemental- sensitivity led Kai Siegbahn (Noble Prize 1981) to introduce the method as Electron Spectroscopy for Chemical Analysis (=ESCA).

Applications to the gas phase[7] are common and for the analysis of surfaces ESCA or XPS has become a standard tool[8]. As for liquids, the first successful attempt to make use of ESCA has been reported as early as 1973[9].

Ultraviolet photoelectron spectroscopy has traditionally used the HeI resonance line ($h\nu = 21.218$ eV) for ionization[10]. The available energy restricts the method to valence orbital ionization. However, the good energy resolution (20 meV and better) has given UPS a wide popularity in characterizing molecules in the gas phase[11], but as well in the adsorbed state on solid surfaces[12]. The operator which describes the UPS ionization process can be approximated to high accuracy by the dipole operator. This establishes a particularly simple relation between the symmetry of the molecular orbital from which an electron is removed and the observable angular distribution of the emitted electrons[13]. In case that the orbital symmetry of a molecule is known one can deduce from photoelectron angular distributions a possible molecular orientation in space, e.g. of molecules adsorbed on a regular surface[12]. The first to employ UPS for studying the surface of organic liquids were Delahay[14] and lateron Ballard[15].

Again a different type of electron spectroscopy is based on the excitation energy of metastable rare gas atoms. They carry energy by way of electronic excitation which they transfer readily to any target - gas phase or surface - upon close contact. This spectroscopy is often named after F.M. Penning who inferred the existence of this ionization process from indirect evidence in 1927 [16]. Penning ionization electron spectroscopy of atoms and molecules was pioneered by V. Cermak [17] and Hotop and Niehaus [18]. Unlike in photoionization where the whole target molecule is 'flooded' by the primary energy, in Penning ionization the available energy is localized in the excited metastable atom and it can be transfered only to those target orbitals which are in direct contact with the metastable atom. It follows that the orientation of the target molecule is of outermost importance for the ionization probability of a given orbital. Of course, in a common experimental situation one does not know the molecular orientation but would rather like to learn from spectroscopy whether the observed molecules do have a preferred orientation and how it looks like. Indeed, in this respect the spectroscopy based on metastable helium atoms He*(1s2s, ^{3}S) has been found to be very valuable. Under the name of Metastable Impact Electron Spectroscopy (= MIES) it has successfully been used even to the investigation of liquid surfaces, the first study being reported in 1986 [19].

GAS PHASE IONIZATION

As an example, we will briefly discuss ionization of the nitrogen molecule N_2 by UPS and MIES. Both processes lead to rather similar spectra, cf. fig. 1. The electronic orbital structure of N_2 is reflected in the spectra. Every band of the electron energy spectra results from the removal of an electron from one specified orbital[21]:

$$N_2(\sigma_u^2 \ \pi_u^4 \ \sigma_g^2 \ ; \ ^1\Sigma) \longrightarrow \begin{array}{l} N_2^+(\sigma_g^{-1} \ ; \ X^2\Sigma_g^+) + e^- \\ N_2^+(\pi_u^{-1} \ ; \ A^2\Pi_u \ ) + e^- \\ N_2^+(\sigma_u^{-1} \ ; \ B^2\Sigma_g^+) + e^- \end{array} \qquad (2)$$

Since the ion N_2 may be created in a vibrationally excited state each band is split into several peaks. To describe this situation, the energy balance given in eq. (1) for photoelectron spectroscopy has to be modified:

$$E_{el} = h\nu - E_{bind} - E_{vib} \qquad (3)$$

The vertical binding energy E_{bind} of the individual orbitals can be read off the v' = 0 position in the spectrum. The upper panel of fig. 1 tells us that the energy supplied by He*(2^3S) leads to the population of the same final N_2^+ states as observed in UPS. The fact that all peaks are slightly broader is caused by a degree of freedom that He*(2^3S) atoms have in addition to the electronic excitation. We talk about the kinetic energy of the atoms. If the kinetic energy of the helium is smaller after the ionization than before this shows in a slight increase in electron energy as observed in fig.1. The energy balance of the process

$$He^*(1s2s;^3S) + N_2 \longrightarrow He(1s^2) + N_2^+ + e^- \qquad (4)$$

can be written as

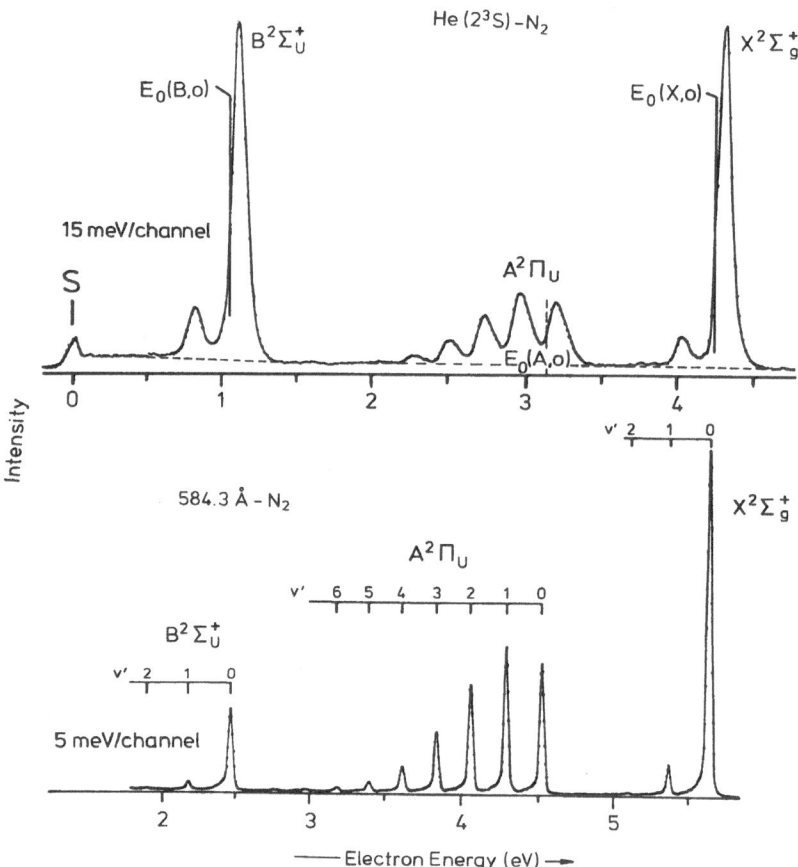

Fig.1 Ionization of N_2 by MIES using metastable helium $He^*(1s2s;^3S)$ and by HeI-UPS. The respective electron energy scales are displaced against each other in order to account for the difference between $h\nu$[HeI] = 21.218 eV and the excitation energy of the metastable atom E^*[He^*] = 19.819 eV. The figure is based on measurements by Hotop and Hübler[20].

$$E_{kin} + E^*[He^*] = E_{bind} + E_{vib} + E'_{kin} + E_{el} \tag{5}$$

which takes into account the relative kinetic energies of the nuclei before (E_{kin}) and after the process (E'_{kin}). In anology to eq.(3) we rewrite

$$E_{el} = E^*[He^*] - E_{bind} - E_{vib} + (E_{kin} - E'_{kin}) \tag{6}$$

In the above case the value of $E_{kin} - E'_{kin}$ is slightly positive. However, there are cases where this quantity takes on negative values between a few meV and even several eV. It is obvious that this has enormous effects on the electron energy spectrum obtained in MIES. Thus, it is of importance to study carefully which conditions lead to those effects if one wishes to make us of MIES as an analytical tool for the investigation of molecules, in the gas as well as in the liquid phase. Therefore, we discuss as second example the ionization of the bromine molecule Br_2. Its orbital structure is denoted as $Br_2(\sigma_g^2 \pi_u^4 \pi_g^4 ; {}^1\Sigma_g)$. In a LCAO (= Linear Combination of Atomic Orbitals) - picture the quoted molecular orbitals can be conceived as linear combinations of the 4p-orbitals of the separated bromine atoms. Photoionization shows the behaviour discussed above: the three bands observed in the UP-spectrum (fig. 2) are created by removal of electrons under the auspices of Koopmans theorem[21]:

$$h\nu + Br_2(\sigma_g^2 \pi_u^4 \pi_g^4 ; {}^1\Sigma_g) \longrightarrow \begin{array}{l} Br_2^+(\pi_g^{-1} ; X^2\Pi_g) + e^- \\ Br_2^+(\pi_u^{-1} ; A^2\Pi_u) + e^- \\ Br_2^+(\sigma_g^{-1} ; B^2\Sigma_g) + e^- \end{array} \tag{7}$$

The splitting of the X and A bands is not due to vibrational excitation of Br_2^+ but is of electronic nature. It is fine structure splitting, an effect which becomes stronger with increasing atomic weight. It is of no further relevance in the present context.

Now we start to discuss the reaction of $He^*(2^3S)$ with Br_2. Here, the metastable atom plays a twofold role. Firstly, it carries energy which can cause direct Penning ionization analoguous to eq.(4). The resulting contribution to the electron energy spectrum is essentially the same as that obtained in photoionization. We see, however, that the respective features in fig. 2 are extremely small. The reason is that most collisions between $He^*(1s2s)$ and Br_2 lead to very different behaviour. We allude here to the second role that the metastable atoms is able to play. It has a 2s valence electron like $Li(1s^2 2s)$ and accordingly undergoes chemical reactions as readily as the alkali atom. Following the same scheme as known for alkali-halogen reactions the metastable helium atom transfers its 2s-electron to the Br_2 molecule

$$He^*(1s2s) + Br_2({}^1\Sigma_g) \longrightarrow He^+({}^2S) + Br_2^-({}^2\Sigma_g) \tag{8}$$

Electron emission will then occur out of this ion pair state. In this ionization process two electrons have to be removed from $Br_2^-({}^2\Sigma_g)$, one neutralizing the helium ion and the other one being emitted (and turning up in the electron energy spectrum). The similarity to the photoionization process is gone in several respects.

First, during the time the system spends in the ion pair channel a large amount of energy is imparted to the nuclear motion, giving ($E_{kin} - E'_{kin}$) a large negative value, cf. eq.(6). Second, the final state $He + Br_2^+ + e^-$ is not restricted to those Br_2^+ levels that can be formed out of the $Br_2({}^1\Sigma_g)$ ground state by a one-elec-

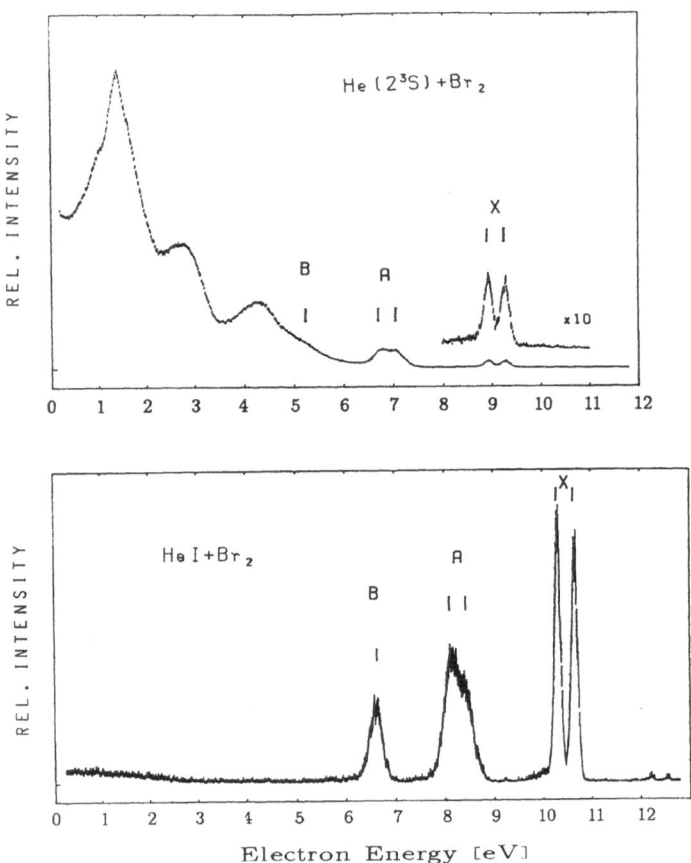

Fig. 2 HeI-UPS and He*(2³S)-MIES of Br₂. Adopted from ref. 22. Discussion is in the text.

tron process. In contrary, in principal all possible Br_2^+ states can be populated. Their number is demonstrated in fig. 3. It is obvious that UPS offers access only to a subset of molecular ion states and thus reflects more closely the electronic structure of the neutral precursor.

It is obvious that a reaction like (8) and the resulting electron energy spectrum is not apt to characterize the target molecule as is UPS. On the other hand, such a process can shed interesting light onto the details of a chemical reaction. It can be shown that the electron energy spectrum changes drastically between the initial and final stages of the 'chemical' reaction

$$He^+ + Br_2^- \longrightarrow He^+ Br^- + Br \tag{9}$$

giving an unusual chance to directly observe such a process while it is going on[23].

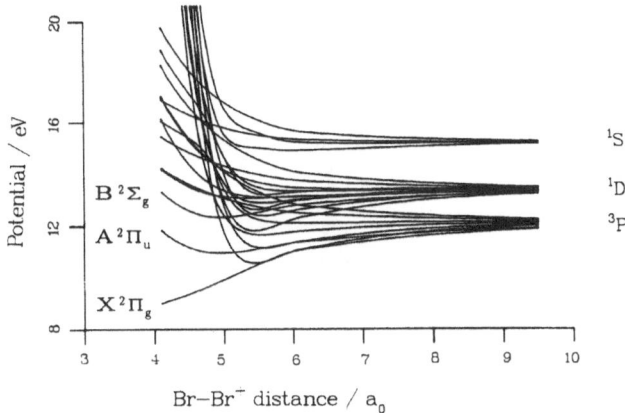

Fig. 3 Diagram of Br_2^+- potential curves. The states populated by UPS, i.e. which can be conceived as formed by a mere one electron removal, are indicated. The other states are obtained by a simple perturbation treatment and are meant to demonstrate the range of Br_2^+ potential energies rather than to predict individual Br_2^+ states. Adopted from ref. 23.

We will now address a molecule whose type of reaction with metastable helium is between N_2 and Br_2 which represent extreme cases. It is formamide (FA) which has the molecular structure

$$\overset{H}{\underset{H}{\Large\diagdown}}\overset{\oplus}{N}=C\overset{\diagup O^{\ominus}}{\underset{H}{\diagdown}}$$

and a charge distribution that results in a strong dipole moment of $\mu = 3.73$ ($\pm 2\%$) Debye. The electronic structure is described by the following orbitals which are listed from left to right according to decreasing binding energy

$$C(2s) \quad \sigma_{NC} \quad \sigma_{CO} \quad \pi_{CO} \quad \pi(n_N) \quad n_O$$

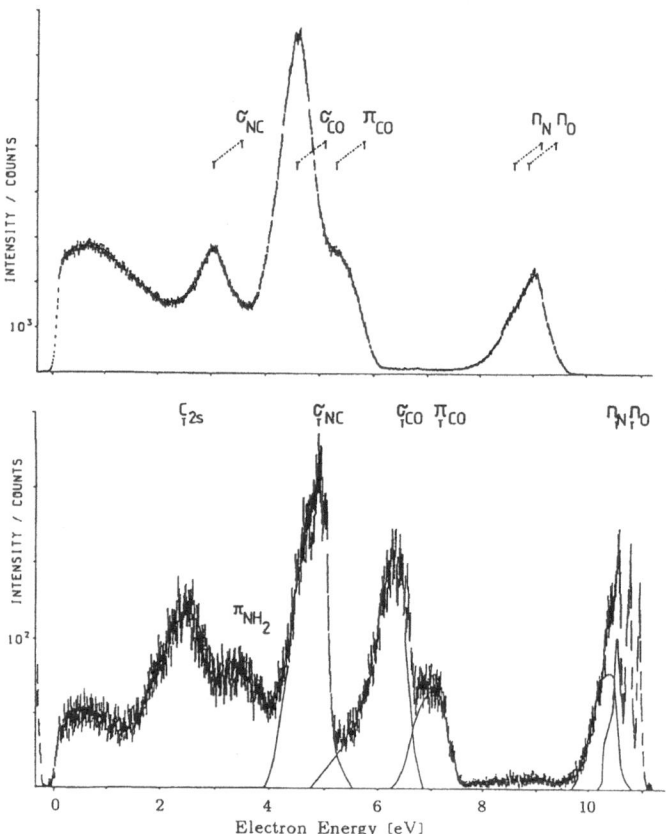

Fig. 4 Valence electron spectroscopy of formamide (FA). Adopted from ref. 24.
Upper panel: He*- MIES.   Lower panel: HeI - UPS

The HeI-photoelectron spectrum in fig. 4 shows that again most bands can be assigned to removal of an electron from one molecular orbital. There are, however, structures in the spectrum, e.g. the one denoted as π_{NH_2}, which have been analysed to indicate molecular ion states which must be described by configuration interaction of states as defined by removal of one electron accompanied - in part - by excitation of a second electron. In the following we will concentrate on the first four bands which can to good accuracy be conceived as being formed by a one-electron process. The upper panel of fig. 4 displays the He^*-MIE spectrum of formamide. The same bands can be discerned a found in UPS. On the other hand, all bands are broadened which shows well in the n_O-band: the vibrational levels cannot be distinguished any more. Further, we find an energy shift to lower electron energies which in the terms of eq.(6) corresponds to

$$\Delta E_{kin} = E_{kin} - E'_{kin} = 0.360 \text{ eV} \tag{10}$$

This shift applies uniformly to all bands observed.

The explanation of these findings is found in ref. 24. Since $He^*(2^3S)$ has the large polarizability[25] of $\alpha = 312 \text{ a}_0{}^3$ it is attracted by the static dipole of the formamide molecule. As long as He^* can be considered a neutral particle the attraction should apply equally to the positive and the negative end of the molecule. This, however, is not the case. A similar behaviour has been found between He^* and H_2O and could be analysed with the aid of ab initio calculations[26]. The negative end, i.e. the oxygen atom of the formamide molecule, exerts by far the strongest attraction to the approaching He^*. This can be rationalized by recalling that the 2s-valence electron of He^* is rather diffuse and can shield the ionic He^+ core only at large distances. At small distances the metastable accordingly acts like a positively charged particle. This results in an attractive potential of well depth $\sim$600 meV for a He^* approach towards the $C=O^\ominus$ end of FA. Compared to this situation other directions are mildly or even strongly repulsive ($^\ominus H_2N$-). Since a prerequisite for Penning ionization to occur is the close encounter between the collision partners we understand that the overwhelming number of ionization events occurs when He^* approaches the $C=O^\ominus$ group along the $C=O$ axis. That ionization out of the $\sigma_{C=O}$ orbital is most probable in Penning ionization is easily explained in this way. With respect to the other orbitals one finds that their ionization activities correlate with their electron density at the $C=O$ group of the molecule.

IONIZATION OF MOLECULES IN THE LIQUID PHASE

Formamide is an ideal candidate to discuss what happens if a molecule is brought from the gas phase into the neighborhood of other molecules in the condensed phase. First, we inspect the He^*-MIE spectrum of formamide (FA) at the surface of the pure liquid, cf. fig.5. We note that the relative weight of the $\sigma_{C=O}$ peak is strongly reduced compared to fig. 4a and that the $\pi_{C=O}$ and the $\pi(n_N)$ -bands have gained in intensity considerably. This has been interpreted in terms of a high degree of order among the molecules which form the topmost layer of liquid formamide[19]. Compared to a random orientation which should favour the ionization from the $\sigma_{C=O}$ - orbital as in the gas phase the spectrum in fig. 5 must be obtained from molecules which tend to expose the π-orbitals to the approaching metastable helium atoms and to hide at the same time their attractive

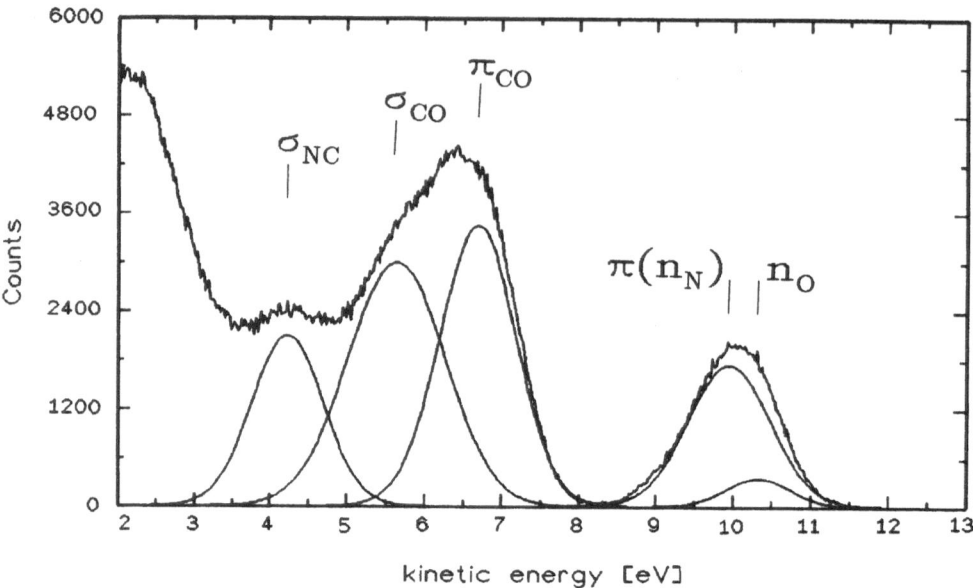

Fig. 5 He*- MIE spectrum of liquid formamide. Adopted from ref. 19.

-C=O$^\ominus$ ends. In short, the topmost molecules are oriented with their molecular plane parallel to the overall surface. This finding, based on relative band intensities, is backed by the observation that the energy shift $\Delta E_{kin} = (E_{kin} - E'_{kin})$ observed in the gas phase with respect to UPS (see discussion in the preceding section) is absent in the liquid phase[19]. Further evidence for this orientation of the topmost molecule stems from computer simulation. We have investigated the surface behaviour of a model liquid, a so called quasi-Stockmayer liquid, with parameters chosen so as to represent formamide reasonably well[27]. The calculation has been carried out on a liquid slab with periodic boundary conditions in two dimensions, thus forming two surfaces, one below and one above the slab. The thickness of the slab corresponds to about 6 molecular diameters. In fig. 6 we see how the computed density varies along the surface normal. On the left side there is the vacuum and on the right side the center of the slab. The curves refer to molecules with different orientation of their dipole moment. In the center all orientations are equally represented indicating that the slab is just thick enough to model bulk properties in its center. However, in the interface region we observe a distinct enhancement of those molecules with the dipole moment parallel to the surface. In this respect we find the computer simulation to endorse our experimental findings of preferred molecular orientation in the surface.

There is, however, a question that arises at the sight of fig. 6. If the density drops from bulk to vapour so smoothly over a distance which equals two or more molecular diameters then one might wonder which state of aggregation one should assign to a molecule in the interface between bulk and vapour. Is it in the condensed phase or has it rather to be conceived as isolated molecule? In the outskirts of the interface the density has dropped so strongly that one would hesitate to talk about condensed matter. On the other hand, the metastable helium atoms

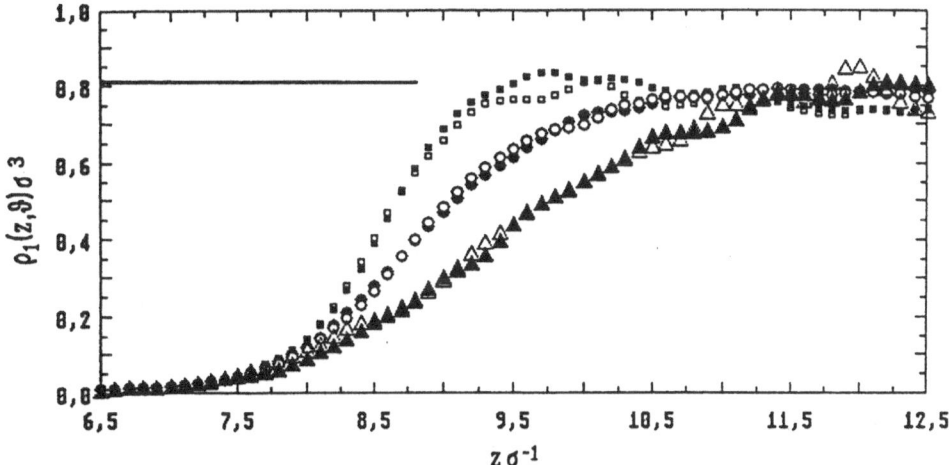

Fig. 6 Density along the surface normal from computer simulation. Quadrangles: molecules with their dipole moments oriented parallel to the surface within $10°$. Triangles: molecules with their dipole moments perpendicular to the surface within $10°$. Circles: medium orientation. The numbers are corrected for the statistical weight of the respective orientations. The filled and open symbols refer to calculations without and with taking into account long range order forces , the structure apparently not being affected. Adopted from ref. 27.

in our experiment have such a large cross section for Penning ionization of molecules ($\geqslant 10$ Å^2) that they could not reach the distance of, say, half bulk density where the degree of molecular orientation is strongest. The situation is puzzling at first glance and one could feel inclined to deny the relevance of the calculation for the experimental situation. The key to a better understanding, both of the simulation and of the real surface is offered by a graphical presentation of a typical conformation encountered in the course of the molecular dynamics calculation[28]. Fig.7 offers a view onto a snapshot of the surface of the model liquid. The molecules are represented by two spheres the molecular dipole moment lying in the common axis. The observed chain like structure is typical for all snapshots inspected. The same observation has been made before in computer simulations of bulk dipolar fluids[29]. Indeed, we find the chain like arrangement of molecules to prevail in the bulk as well as at the surface. The interesting point to note is that a chain at the surface which starts and ends in the bulk, necessarily must run more or less parallel to the overall surface. This effect causes those molecules encountered by approaching He[*] atoms to be oriented parallel. Further, both ends of the molecule, being tied up in a chain, are not accessible for a localized reaction. All this is in perfect agreement with our experimental findings. Even the gradual change of the density that could cause conceptual problems as discussed above falls into place upon inspection of fig.7. Some chains protrude from the overall surface farther than others and between chains there are regions with no molecules. If one takes the lateral average (and that over time steps) to compute the density as done for the plot of fig. 6 it is obvious that a gradual function results. It becomes obvious that the smooth gradient of laterally averaged density does

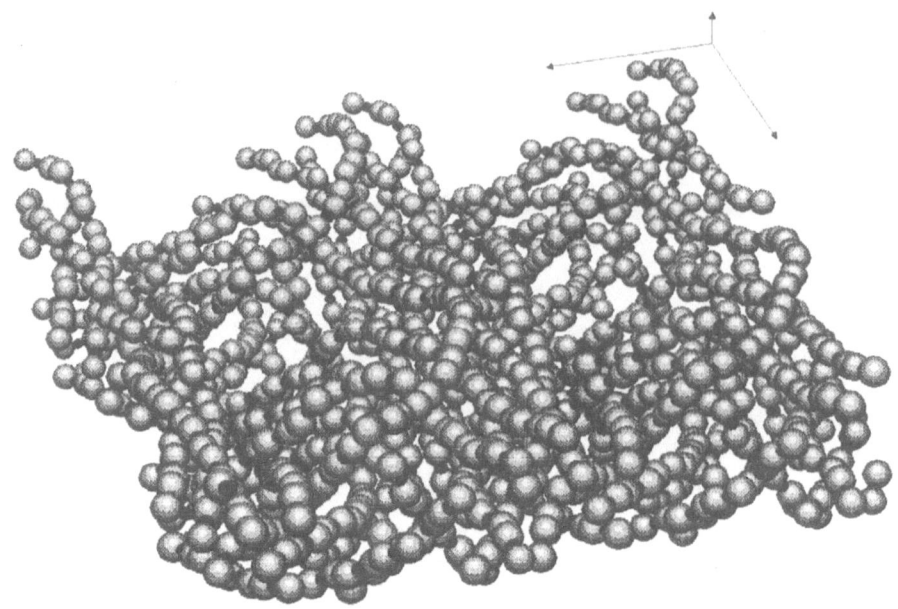

Fig.7 Graphical presentation of the ensemble of molecules used in the computer simulation[27]. Even though this is one 'snapshot' out of ca. 10^5 time steps calculated it carries features which are found to be typical[28].

not mean that there is ever any ambiguity of whether a given molecule has to be considered as being in gas phase or in condensed phase. All molecules shown in fig. 7 are unambiguously in the condensed phase in that they are closely bound to their neighbors in the chain, essentially the same situation as they experience in the bulk. We oberve, of course, evaporation of individual molecules in the simulation, but this is an extremely rare event.

To prevent misunderstandings we point out that the chains observed at a certain time step are not stable. They break and are formed again continuously, but at any given time by far the majority of molecules are tied up in chains.

An interesting experimental observation can be added which has bearing on our understanding of the liquid structure. Siegbahn et. al.[30] have evaluated the band broadening observed in electron energy spectra upon transition of formamide from gas to liquid phase. They investigated the valence orbital emission using the MgK_α - line and found that the $\sigma_{C=O}$ -band experiences a stronger broadening than other bands. They concluded that this orbital should be particularly active in forming chemical bonds in the liquid phase. Since the $\sigma_{C=O}$ - orbital extends in the molecule along the direction of the molecular dipole it certainly would be involved in hydrogen bond formation within the chain structure discussed above. Comparing with the known strength of hydrogen bonds in the solid[31] they arrived at the conclusion that the bonds leading to chain formation would more likely survive melting than the weaker bonds connecting parallel chains. This fits well with the above considerations. Since the study by Siegbahn[30] was carried out with x-ray photoelectron spectroscopy the emitted electrons had a kinetic energy of more than 1200 eV. The corresponding mean free path of the electrons can be estimated to be $\lambda_e \sim 50$ Å[32] and this value represents the observation depth of the experiment. Inspecting our MIES-data in fig. 5 we find again that the $\sigma_{C=O}$ -band is more strongly broadened than others, in particular than the oxygen lone pair orbital $n_O(\sigma)$ which would be responsible for the mentioned interchain bonding. Thus, our MIES-data support the notion of chain formation at the very surface.

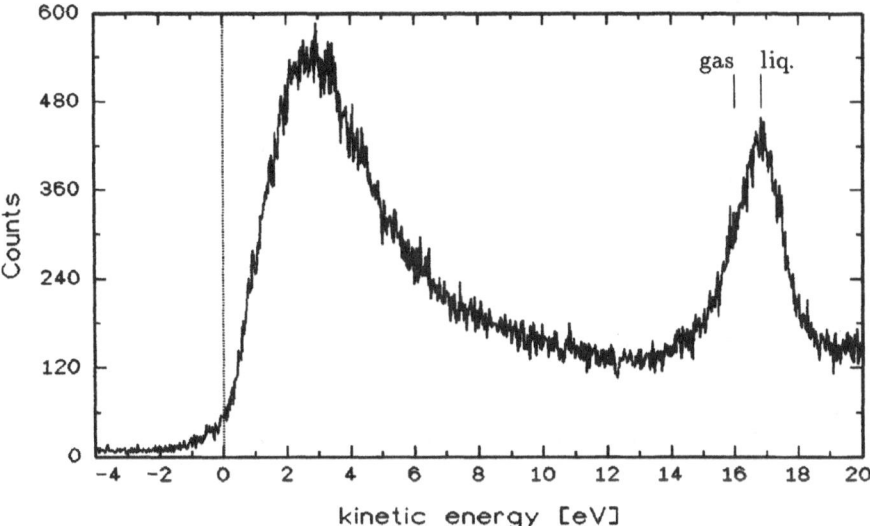

Fig. 8 Photoelectron spectrum of formamide, photon energy $h\nu = 310$ eV (Synchrotron Radiation Facility BESSY). The energy position for ionization of the C(1s) orbital in the gas phase is indicated. The same orbital, when ionized in the liquid phase, has an effective binding energy which is lower by $\sim 0.9(1)$ eV. Further explanation in text.

So far we have discussed two pieces of experimental information which are found upon the phase transition: the relative intensities in MIES and the band broadening observable in MIES as well as in photoelectron spectroscopy. There is a third quantity which can be evaluated from electron energy spectra and which may shed some further light onto our understanding of the gas/liquid interface. For illustration, we inspect the photoelectron spectrum of formamide (fig.8) taken with light of $h\nu = 310$ eV from the Berlin Synchrotron Radiation Facility BESSY. Only the C(1s)-band is shown. We observe, that the liquid peak is distinctly shifted with respect to the gas phase. This shift of the apparent binding energy upon transfer of a molecule into the condensed phase is common knowledge and has been discussed in the literature before[33,34,35]. Taking into account this phase transition shift ΔE_{PTS} we have to rewrite eq.(1) as

$$E_{el} = h\nu - E_{bind} + \Delta E_{PTS} \qquad (11)$$

Table.1 Phase transition shift ΔE_{PTS} as measured for pure formamide with different electron spectroscopies

Method	Orbital	E_{el}/eV	$\lambda_e/\text{Å}$	$\Delta E_{PTS}/eV$
MIES [a]	n_N	≈ 10	0	1.01(5)
UPS (HeI) [a]	n_N, n_O	≈ 11	≈ 6 [f]	1.06(19)
UPS (hν=40eV) [b]	n_N, n_O	≈ 31	≈ 9 [e]	1.49(15)
XPS (Mg K_α) [c]	n_N, n_O	≈ 1250	≈ 50 [e]	1.73(20)
UPS (hν=310 eV) [b]	C(1s)	≈ 16	≈ 5 [e]	0.9(1)
UPS (hν=330 eV) [b]	C(1s)	≈ 36	≈ 7 [e]	1.26(10)
UPS (hν=350 eV) [b]	C(1s)	≈ 56	≈ 8 [e]	1.26(10)
XPS (Mg K_α) [d]	C(1s)	≈ 1000	≈ 45 [e]	1.54(10)

a. H.Morgner et al. ref.35   d. H.Siegbahn et al. ref.33
b. F.Eschen et al. ref.4   e. H.Morgner ref.32
c. H.Siegbahn et al. ref.30   f. H.Morgner et al. ref.36

if we wish to maintain the definition of E_{bind}. Even though several reasons for the occurance of this energy shift could in principle be named it has quite early been argued[33] that by far the major contribution is caused by the dielectric relaxation of the neighborhood. Since a molecule at the surface has fewer neighbors than in the bulk one would expect the phase transition shift ΔE_{PTS} to be smaller at the surface than in the bulk. This has indeed been observed[35]. In table 1. we have comprised for pure formamide the ΔE_{PTS} -values as derived from MIES

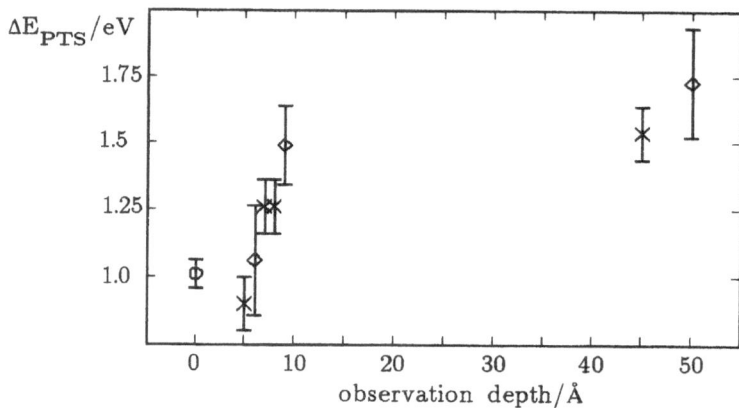

Fig. 9 The phase transition shift ΔE_{PTS} in pure formamide plotted against the respective observation depth. The data are taken from table I. Circles refer to MIES, diamonds to valence- and crosses to C(1s) core electron spectroscopy.

and photoelectron spectra. We note that the energy shift depends little on the orbital ionized but rather strongly on the kinetic energy of the released electron. This is easily understood if one recalls that the electron energy governs the observation depth by way of the energy dependent mean free path of the emitted electrons. Fig. 9 displays the phase transition shift against the observation depth of the respective experiment. We note that over a few Ångstrom ΔE_{PTS} remains constant before rising rather abruptly to the bulk value. This behaviour may reflect the observation from the computer simulations (cf. fig. 7) that the chains seem to be less closely packed at the interface than in the bulk. On one side, we emphasize again that all molecules shown in fig. 7 are unambiguously part of the condensed phase in that they are bound to their chain neighbours, on the other side, it is obvious that the environment is different at the surface in that the density of chains is lower than in the bulk. Accordingly, the electronic relaxation (or dielectric response) should be smaller in the interface as is actually observed.

One could even think of deriving a number for the surface roughness from fig. 9 and would arrive at a value of about 5 Å. However, one has to keep in mind that the exact interpretation of an observation depth based on a mean free path of the electrons depends itself on the assumed structure of the surface. Thus, the quantitative evaluation of the data in table1 and fig. 9 is a task for the future.

ACKNOWLEDGEMENTS

Thanks are due to M. Wulf for dicussions and for preparing several figures. J.Dietter has contributed the figure 7. The experimental and theoretical work has been financially supported by Deutsche Forschungsgemeinschaft (DFG), Bundesminister für Forschung und Technologie (BMFT) and Land Nordrhein-Westfalen.

REFERENCES

1. J.N.H. Brunt, G.C. King and F.H. Read, A study of resonance structure in helium using metastable excitation by electron impact with high energy resolution, *J. Phys. B* 10:433-48 (1977)
2. H. Kuppermann, W.M. Flicker and D. A. Mosher, Electronic spectroscopy of polyatomic molecules by low-energy, variable-angle electron impact, *Chem. Reviews* 79:77-90 (1979)
3. R.E. Ballard, J. Jones, D. Read, A. Inshley and M. Cranmer, Auger and electron energy loss studies on liquid surfaces, *Chem. Phys. Lett.* 147:629-31 (1988)
4. F. Eschen, M. Heyerhoff, H. Morgner and M. Wulf, unpublished results (1993) M. Heyerhoff, Diplom Thesis, University Bochum, (1993)
5. W. Mehlhorn in Atomic Inner-Shell Physics, ed. B. Craseman, Plenum, New York (1985)
6. G. Ertl and J. Küppers, "Low Energy Electrons and Surface Chemistry", VCH Verlagsgesellschaft , Weinheim (1985)
7. K. Siegbahn, C. Nordling, G. Johansson, J. Hedman, P.F. Hedin, K. Hamrin, U. Gelius, T. Bergmark, L.O. Werme, R. Manne, Y. Baer, "ESCA Applied to Free Molecules", North Holland, Amsterdam (1969)

8. D. Wagner, W.M. Riggs, L.E. Davis, "Handbook of X-ray photoelectron spectroscopy" ed. by G.E. Muilenberg, Perkin-Elmer Corp., Physical-Electronics Division, Eden Prairie (1978)

9. H. Siegbahn and K. Siegbahn, ESCA applied to liquids, *J. Electr. Spectr. Rel. Phen.* 2:319-25 (1973)

10. D.W. Turner, A.D. Baker, C. Baker and C.R. Brundle. "Molecular Photoelectron Spectroscopy, A Handbook of 584 Å Spectra", Interscience, London - New York (1970).

11. K. Kimura, S. Katsumata, Y. Achiba, T. Yamaski and S. Iwata, "Handbook of HeI Photoelectron Spectra of Fundamental Organic Molecules", Japan Scientific Societies Press, Tokyo (1981)

12. H. J. Freund and M. Neumann, Photoemission of molecular adsorbates, *Appl. Phys. A* 47:3-23 (1988)

13. R. Morgenstern, A. Niehaus and M.W. Ruf, Angular distribution of photoelectrons, *Chem. Phys. Lett.* 4:635-8 (1970)
    E.S. Chang, Angular distributions of photoelectrons with analysis on the rotational states of H_2, *J. Phys. B* 11:L69-74 (1978)

14. L. Nemec, J. M. Gaehrs, L. Chia and P. Delahay, *J. Chem. Phys.* 66:4450 (1977)

15. R.R. Ballard, J. Jones and E. Sutherland, Measuremant and calibration of the He(I) photoelectron spectra of gaseous and liquid ethanediol, propanediol and formamide, *Chem. Phys. Lett.* 112:306 (1984)

16. F.M. Penning, Über Ionisation durch metastabile Atome, *Die Naturwissenschaften* 15:818 (1927)

17. V. Cermàk, Retarding-potential measurements of the kinetic energy of electrons released in Penning ionization, *J. Phys.* 44:3781-6 (1966)

18. H. Hotop and A. Niehaus, Reactions of excited atoms and molecules with atoms and molecules, *Zeitschrift für Physik* 215:395-407 (1968)

19. W. Keller, H. Morgner and W.A. Müller, Probing the outermost layer of a free liquid surface. Electron spectroscopy of formamide under He(2^3S) impact, *Mol. Phys.* 57:623-36 (1986)

20. H. Hotop and G. Hübler, Photoelectron and Penning ionization electron spectrometry with differential retarding field analyzer, *J. Electr. Spectr. Rel. Phen.* 11:101-21 (1977)

21. T. Koopmans, *Physica* 1:104 (1933)

22. K. Beckmann, O. Leisin and H. Morgner, Excitation transfer into bound and continuum states investigated by optical and electron spectroscopy, *Mol. Phys.* 59:829-43 (1986)

23. H. Morgner and H. Seiberle, Transition state spectroscopy with electrons as studied by 3D-trajectory calculations of the reaction $He^+ + Br_2^- \longrightarrow He^+Br^- + Br$, submitted to *Can. J. Physics, (Polanyi-Special Issue)*, to appear in 1994

24. W. Keller, H. Morgner and W.A. Müller, He(2^3S) and hydrogen bonding molecules. A comparative study of He(2^3S) Penning ionization versus HeI photoionization for formamide and N-methylformamide, *Mol. Phys.* 57:637-44 (1986)

25. A. Dalgarno and A.E. Kingston, Properties of the metastable helium atoms, *Proc. Phys. Soc.* 72:1053-60 (1958)

26. B. Haug, H. Morgner and V. Staemmler, Experimental and theoretical study of Penning ionisation of H_2O by metastable helium He(2^3S), *J. Phys. B* 18:259-74 (1985)

27. A.W. Hertzner, M. Schoen and Morgner, The influence of long range electrostatic forces on static properties of a quasi-Stockmayer fluid, *Mol. Phys.* 73:1011-29 (1991)

28. A.W. Hertzner and H. Morgner, 1991, unpublished results

29. M. Matsumoto and K.E. Gubbins, Hydrogen bonding in liquid methanol, *J. Chem. Phys.* 93:1981-94 (1990)

30. H. Siegbahn, L. Asplund, P. Kelfve, K. Hamrin, L. Karlsson and K. Siegbahn, ESCA applied to liquids. II. valence and core electron spectra of formamide, *J. Electr. Spectr. Rel. Phen.* 5:1059-79 (1974)

31. J. Ladell and B. Post, The crystal structure of formamide, *Acta Crystallographica* 7:559-64 (1954)

32. H. Morgner, The investigation of liquid surfaces by electron spectroscopy, 5th *Int. Conf. Electr. Spectr.*, Kiev (1993) and *J. Electr. Spectr. Rel. Phen.* submitted (1993)

33. H. Siegbahn, L. Asplund, P. Kelfve and K. Siegbahn, ESCA applied to liquids. III*. ESCA phase shifts in pure and mixed organic solvents. *J. Electr. Spectr. Rel. Phen.* 7:411-9 (1975)

34. H. Siegbahn, Electron spectroscopy for chemical analysis of liquids and solutions, *J. Phys. Chem.* 89:897-909 (1985)

35. H. Morgner, J. Oberbrodhage, K. Richter and K. Roth, The gas-liquid phase transition shift at surfaces: experimental method and interpretation, *J. Electr. Spectr. Rel. Phen.* 57:61-77 (1991)

36. H. Morgner, J. Oberbrodhage, K. Richter and K. Roth, Surface segregation of a binary liquid mixture as studied by metastable impact electron spectroscopy, *Mol. Phys.* 73:1295-1306 (1991)

POSITRON AND POSITRONIUM ANNIHILATION IN GASES AND LIQUIDS

Alexei G. Khrapak

Institute for High Temperatures Russian Academy of Sciences
Moscow 127412, Russia

1. INTRODUCTION

A positron is the antiparticle of an electron. It has the same mass, spin and absolute value of the electric charge as an electron. Therefore, the processes the characteristic time of which is less than lifetime of a positron in a medium have a lot of common features with the corresponding electron processes. For example, in dense gases and liquids positrons collide many times with atoms before annihilation and may be considered as quasi stable particles. The interaction of slow positrons with atoms has an important distinction from the interaction of electrons with atoms: the exchange component of the interaction potential is absent in the case of positrons. This results in interesting features of the behavior of positrons injected in gases and liquids.

In a matter a positron annihilates with an electron, emitting one, two or three photons. The single-photon annihilation cross section is proportional to the electron-positron relative velocity v and therefore is very small for slow positrons. The largest cross section is for two-phonon annihilation, which in the limit of small v is given by

$$q_{2\gamma} \cong \pi r_0^2 c / v \qquad (1.1)$$

where $r_0 = e^2/mc^2$ is the electron classical radius. The annihilation cross section decreases sharply with the number of emitted photons. The three-photon annihilation cross section $q_{3\gamma} = q_{2\gamma}/372$ is very small and usually this process can be omitted. However the angular distribution of the γ-quanta produced from the three-photon annihilation may be used to obtain additional information about the momentum distribution of the annihilating particles.

The electron-positron interaction may result in production of bound states, positronium atoms (Ps), that are rather like hydrogen atoms. Ps may exist in one of two states, a triplet state $^3S_1$ (ortho-positronium, o-Ps) and a singlet state $^1S_0$ (para-positronium, p-Ps). Ps has a finite lifetime. It follows from the spin conservation that o-Ps annihilates into 3γ state, while p-Ps annihilates into 2γ. The lifetime of the p-Ps τ_p is equal to $1.2 \cdot 10^{-10}$ s; o-Ps lives 1115 times longer, $\tau_0 = 1.4 \cdot 10^{-7}$ s.

Three mechanisms are known for slow positron annihilation: (i) as a result of collision with an electron, free or bound in an atom of a medium, (ii) via formation of Ps and subsequent annihilation, or (iii) via formation of positron bound state with an atom or molecule in a matter. Usually it is possible to separate free positron annihilation from those proceeding via the previously formed Ps stage or localized state on molecules or density fluctuations. This separation is made using the substantial difference between the lifetimes for free positrons and Ps, the different dependence of the lifetimes on the density and temperature of the matter, and the different number of γ-quanta emitted after the annihilation.

Linking the Gaseous and Condensed Phases of Matter
Edited by L.G. Christophorou *et al*, Plenum Press, New York, 1994

In most experiments positrons are obtained from the β decay of radioactive nuclei, such as ^{22}Na, ^{58}Co or ^{64}Cu. After the positron emission, the radioactive nucleus is transformed into the excited nucleus with the reduced atomic number, which decays very rapidly (after a time less than 10^{-12} s.) emitting γ–quanta with a definite energy. For example, in the case of β decay of ^{22}Na excited nucleus of ^{22}Ne emits photon with energy 1.28 MeV. Positrons injected into the medium have broad spectrum of energies of about several hundreds KeV. They lose the energy rapidly through ionization and excitation of atoms or molecules. The annihilation cross section $q_{2\gamma} \sim 1/v$ (1.1) is very small at high energies and practically all positrons slowed down to the minimal excitation energy without annihilation. Therefore two or three photons accompanying the positron annihilation have an energy less than $mc^2 = 0.51$ MeV. The difference in the energies of photons, corresponding to positron emission and annihilation, enables one to measure the lifetime of positron in the system. Usually in the lifetime measurements the positron source is used to be sufficiently weak so that the presence of only a single positron in the system is most probable.

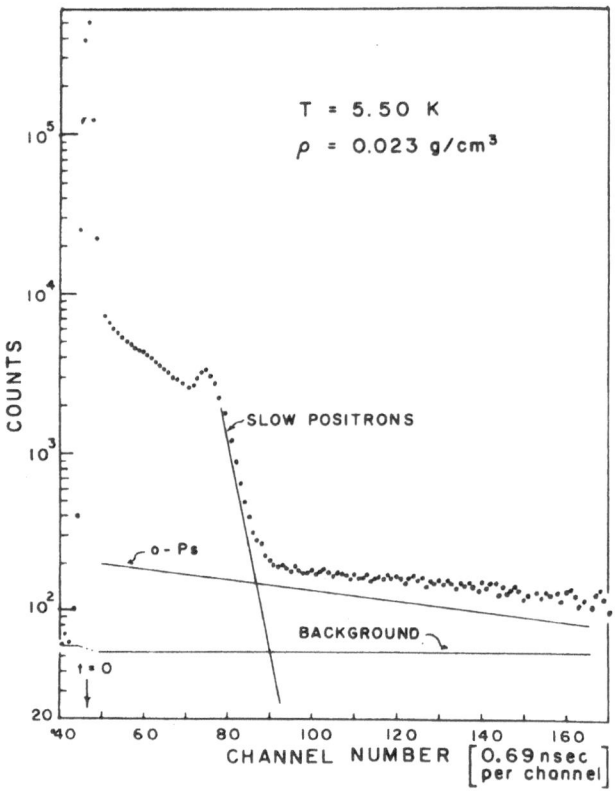

Figure 1. Time spectrum of the positron annihilation in ^{4}He at T = 5.5 K and ρ = 0.023 g cm^{-3} (from Canter *et al*, 1975).

A typical time spectrum for positron annihilation in gases is shown in Figure 1. A prompt peak is observed at the initial time which is due to positron annihilation inside the source, at the walls of the sample chamber, and due to para-positronium annihilation. The peak is followed by a rather smooth non-monotonous region, the so-called "shoulder". This shoulder is caused by the annihilation of non-thermalized positrons. The end of the shoulder corresponds to the time necessary for the positron to be thermalized. Sometimes a maximum is observed at the end of the shoulder, what is evidence that the annihilation probability rises with a decrease in positron energy.

The shoulder is followed by two components having exponential time dependencies. The short-lived component corresponds to the annihilation of thermalized positrons, while the long-lived one is due to annihilation of ortho-positronium. Analysis of the annihilation time spectra enables one to determine the annihilation rates for slow positrons, λ_1, and for slow positronium, λ_2. The rates are present in arguments of the exponents, $\exp(-\lambda_i t)$, approximating the spectrum components shown in Figure 1. A detailed analysis of the time spectra and the methods needed to use the spectra for the determination of other characteristics of the annihilation process was given in the reviews by Goldanskii (1968), Fraser (1968a), Griffith and Heyland (1978), and Charlton (1985).

At low gas densities the annihilation rate of free slow positrons λ_1 is proportional to the annihilation cross section $q_{2\gamma}$, the gas density N, the positron velocity v and the effective number of atomic electrons Z_{eff}, with which positron may annihilate

$$\lambda_1 = \pi r_0^2 c N Z_{eff}. \qquad (1.2)$$

Z_{eff} may differ from the number of atomic electrons Z . In Table 1 the experimental values of Z_{eff} are presented for a number of atoms and molecules (Griffith and Heyland, 1978; Charlton, 1985). The difference between Z_{eff} and Z is connected with two effects. The Coulomb repulsion between the positron and the atomic nucleus prevents the positron from penetrating into the inner electronic shells of the atom and thus the contribution from inner electrons to λ_1 is diminished. On the other hand, the Coulomb attraction between the positron and atomic electrons results in an effective increase in the electron concentration near the positron. For inert gases Z_{eff} is comparable with the total number of atomic electrons and decreases with a rise in N for Ar and Kr. For gaseous hydrocarbons, halogens and other molecular gases very high values of Z_{eff} are observed, which rise with an increase in N. The reason is due to an important role played by positron attachment to molecules.

Table 1. Experimental values of Z_{eff} and $^1Z_{eff}$ (from Griffin and Heyland, 1978; Charlton, 1985).

Molecule	Z	Z_{eff}	$^1Z_{eff}$
He	2	3.94	0.125
Ne	10	5.99	0.235
Ar	18	27	0.314
Kr	36	65.7	0.478
Xe	54	320	1.26
H_2	2	14.7	0.186
N_2	14	30.5	0.260
O_2	16	26	44
CO	14	38.5	0.285
CO_2	22	53	0.500
NO	15	34	
NO_2	23	78	
SF_6	70	97	0.52
CH_4	10	140	0.446
C_2H_6	18		0.625
CCl_2F_2	58	750-1500	0.57

Some positrons, injected into the medium, form Ps. The o-Ps annihilation rate λ_2 is always higher than its vacuum magnitude $\lambda_v = 1/\tau_0 = 7.0$ ns^{-1}:

$$\lambda_2 = \lambda_v + \Delta\lambda_2. \qquad (1.3)$$

The quenching rate $\Delta\lambda_2$ is due to the fact that positron during Ps-atom collision may annihilate not only with the electron bound in Ps but also with atomic electrons. It is proportional to $q_{2\gamma}$, N, v, and the effective number of electrons $^1Z_{eff}$, with which the positron may annihilate in this case

$$\Delta\lambda_2 = 4\pi r_0^2 cN^1 Z_{eff}. \tag{1.4}$$

The experimental values of $^1Z_{eff}$ are also presented in Table 1. They are many times less than Z for all atoms and molecules. It is connected mainly with the absence of the Coulomb attraction between the positronium and atomic electrons and the presence of the strong exchange repulsion.

Usually positrons injected into a medium have an energy of about 0.5 Mev. They lose it rapidly, ionizing and exciting atoms and molecules. Most positrons are slowed down to the minimal atomic excitation energy, E_1, without being annihilated. During thermalization some fraction of positrons may form Ps by capturing an electron from an atom of the medium or by picking up one of the free electrons created along the ionization trail. To form Ps by capturing an atomic electron the positron energy ε must be in so called Ore-gap, $E_1 > \varepsilon > E_0 = I - Ry/2$, where I is the ionization potential of the atom and Ry/2 is the ground state energy of Ps, equal to 6.8 eV. At higher energies, $I > \varepsilon > E_1$ positron is likely to undergo excitation of the atom in competition with the Ps formation and if $\varepsilon > I$ then the Ps kinetic energy after formation makes it likely that the Ps will break up in subsequent collision. At lower energies, $\varepsilon < E_0$, positron may still form Ps by recombination with an electron from the ionization trail. These process is described by the "spur" model (Mogensen, 1974). Ps is formed predominately within the Ore-gap (Stewart *et al*, 1990) and positrons with $\varepsilon < E_0$ lose their energy mainly in elastic collisions. The annihilation cross section is proportional to 1/v and so is very small for fast positrons. Therefore a considerable number of positrons and Ps attain thermal equilibrium with the gas.

Investigations of the positron-atom scattering were performed, for the most part, during the last two decades, since mono-energetic sources of slow positrons were invented. Now the total cross sections for scattering of positrons are known for atoms of all rare gases and a number of molecules. The cross sections for all investigated gases have a minimum at the energy about 1 eV. The presence of the Ramsauer minimum is an indication that the positron-atom (or molecule) scattering length is negative. It is well known, that the Ramsauer minimum is observed for the scattering of slow electrons also, but only in the case of atoms or molecules which have a large enough polarizability. This feature is connected to the absence of exchange repulsion in the positron-atom interaction, which is very important in the case of the electron-atom interaction.

Ps-atom scattering has been investigated not so well. Since there is no polarization interaction in this case and the van der Waals forces are weak, the exchange repulsion plays the most important role in the Ps-atom scattering. As a result, the Ps-atom scattering length is always positive and practically independent of the atomic number. Scattering of slow positrons and positronium atoms on isolated atoms and molecules will be discussed in the next Section.

The linear dependence of the annihilation rate λ_1 and quenching rate $\Delta\lambda_2$ on the atomic density is observed only in rarefied gases. With increasing of the density or decreasing of the temperature deviations from the linear dependence take place. They may be interpreted as a dependence of the effective charge on the density. For example, Z_{eff} is given by:

$$Z_{eff} \cong \int dr \rho_a(r) |\psi(r)|^2 \tag{1.5}$$

where $\rho_a(r)$ is the atomic electron density and $\psi(r)$ is the incident positron wavefunction. Only if $\psi(r)$ is a plane wave, then $Z_{eff} = Z$. As we told earlier, the differences between Z_{eff} and Z appear even in a rarefied gases, where the positron interaction with any atom may be considered independently. In a dense gas a positron may interact with two or more atoms simultaneously. Situations are possible where the positron-atom interaction itself may be still considered to be pair-like, but features of this interaction, e.g. the atom polarizability, already depend on the positions of other atoms. These and other multi-particle effects distort the incident positron wavefunction and result in a density dependence of Z_{eff}. Similar reasoning is also applicable to Ps. The density effects in moderately dense gases, which are common for electrons, positrons and Ps, will be also discussed in the next Section.

Usually the annihilation rate of Ps is larger for a higher density. However, studying the annihilation of Ps in liquid helium it was unexpectedly found that the measured lifetime of Ps was near its value in vacuum (Paul and Graham, 1957; Wackerle and Stump, 1957). It was considerably larger than the value that might be expected considering the density of the investigated liquid. The increase in the lifetime is due to creation of a cavity around the Ps, called a "bubble" (Ferrel, 1957). The formation of the bubble is possible because of a strong exchange repulsion between the electron belonging to Ps and atomic electrons. The lifetime of Ps is considerably increased in the bubble as a result of the self-trapping, since there are almost no atoms inside the bubble and as the wavefunction of Ps is exponentially small outside the bubble, the probability of annihilation with the atomic electrons is also sharply reduced. Later such bubbles were detected in various liquids. Existence of the positronium bubbles is also possible in gaseous phase. If the density of the gas is increased the bound state of Ps becomes thermodynamically more preferable sooner or later. There is a boundary sharply separating the regions of the extended and bound states. Transition through the boundary induce substantial changes in the lifetime of Ps.

There is a similarity between phenomena arising in dense media, when positronium annihilation or electron mobility are investigated. Electrons, as well as positronium atoms, are strongly repelled from helium atoms (and from several other atoms or molecules) because of the strong exchange interaction. A sharp decrease in the injected electron mobility is observed in dense gaseous helium. This effect is also related to the electron self-trapping in bubbles. Influenced by the electric field the bubble moves as a whole, feeling a hydrodynamic friction from the viscous gas or liquid. Self-trapping of Ps and electron inside bubbles is described in greater detail in the Section 3.

The main reason for the creation of positronium and electron bubbles is a strong exchange repulsion between the particles and atoms in the medium. In the case of the positron there is no exchange interaction, so the polarization attraction is of more importance. Naturally, no positron bubbles can be created. Nevertheless, positron self-trapping is possible. It may be a result of the formation of clusters, i.e. regions of higher atomic densities stabilized by the positron. In rarefied gases the annihilation rate is proportional to the atomic density and is independent of the temperature. It was observed that, starting from a certain value of the density, its slight increase results in a rapid rise of the annihilation rate. The annihilation rate increases to such a level that it seems as if gas is compressed to the liquid state in the vicinity of the positron (Roellig and Kelly, 1965; Canter and Roellig, 1970). The cluster formation results in a phase transition in microscopically small region around the positron. The clusters may exist in a restricted region adjacent to the gas-liquid coexistence curve (Khrapak and Iakubov, 1976; Hautojarvi et al, 1977). Properties of the positron clusters in dense gases is described in the Section 4.

In the last Section the kinetics of self-trapping is discussed.

2. INTERACTION OF FREE POSITRONS WITH A MATTER

The self-trapping of positron or positronium is a result of their interaction with atoms or molecules in the medium. In order to describe this interaction it is necessary to know the nature of the positron and positronium interaction with isolated particles. This information is provided by experiments on positron scattering, results of which are shortly discussed below. A detailed analysis of the experimental and theoretical works on slow positron scattering may be found in several review articles (Griffith and Heyland, 1978; Kauppila and Stein, 1982; Iakubov and Khrapak, 1982; Schrader and Svetic, 1982; Charlton, 1985).

The total cross sections for positron scattering have been measured for a number of atoms and molecules. In Figure 2 the low energy total cross sections for rare gas atoms are presented. The existence of the Ramsauer-Taünsend minimum for all rare gases is due to the absence of the exchange repulsion and domination of the polarization attraction in the positron-atom interaction. It is the main reason why the positron self-trapping is possible in dense gases in clusters only, not in bubbles.

The positrons can be self-trapped only after termalisation, when its energy is very low, about 10^{-2} eV or less. At such energy most important characteristic of the scattering is the scattering length L. The determination of the dependence of L on the atomic polarizability α and other parameters of the interaction potential can be made with help of simplest model for electron- or positron-atom interaction

$$V(r) = \{\infty, r \le R; -\alpha e^2/2r^4\varepsilon, r > R\} \tag{2.1}$$

where ε is the dielectric constant of the matter. The radius of the effective solid core R may be different for electrons and positrons. The scattering length L on model short-range potential (2.1) can be easy calculated by the "joining" of the radial part of the wave function at small distances ($\chi \sim r \cdot \sin (k/r + \Delta)$, $kr \ll 1$ where k is the wave number of electron or positron) with its asymptotic at large distances ($\chi \sim \sin (kr + \delta_0)$, $kr \gg 1$, δ_0 is the scattering phase shift with orbital quantum number $l = 0$) at the point $r = 1/k$. In the limit $k \rightarrow 0$ it gives the connection between L, R and α

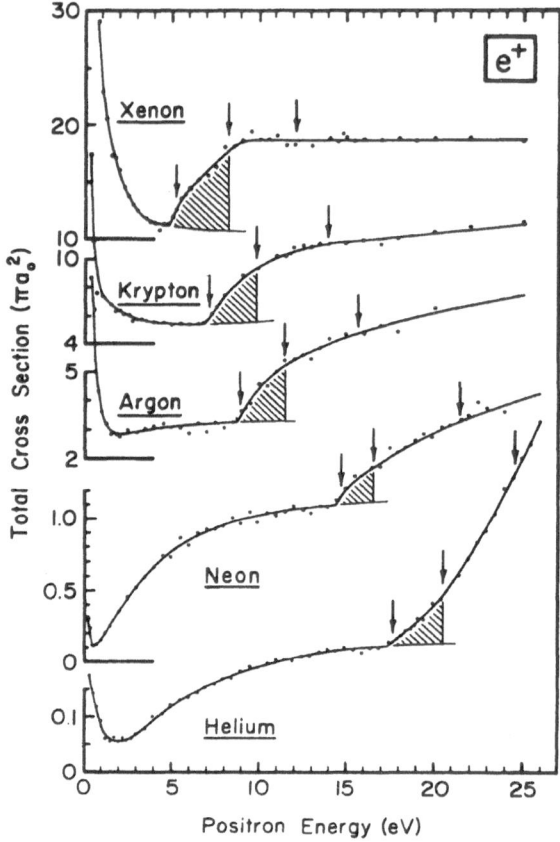

Figure 2. Total cross section curves for low-energy positron-atom scattering of the rare gases. The arrows refer to the threshold energies for positronium formation, atomic excitation and ionization in order of increasing energy (from Kauppila and Stein, 1982).

$$L = (\frac{\alpha}{a_0 \varepsilon})^{1/2} \text{ctg} \, [(\frac{\alpha}{a_0 \varepsilon})^{1/2}/R] \tag{2.2}$$

where a_0 is the Bohr radius. Scattering experiments are usually performed in low density gases where $\varepsilon \cong 1$. In the cases when the scattering length is known from the experiment, the equation (2.2) gives a possibility to define the effective radius R. These magnitudes for electrons and positrons are presented in Table 2 together with the values of the scattering length (from Gus'kov et al, 1978 for electrons and from Plenkiewicz et al, 1993 for positrons). As it was to be expected the effective radius R_p for the same atom is always less than R_e. It is due to the absence of the exchange repulsion in the case of positrons.

The scattering experiments are usually performed at not very low energy, where scattering cross section is not independent from energy. In that cases the scattering length may be determined by extrapolation with help of effective-range method (O'Malley, 1963; Iakubov and Khrapak, 1982).

In the case of positronium there is no polarization interaction, the van der Waals forces are weak and the Ps scattering with atoms or molecules may be described by only one parameter, R, which coincides with the scattering length L. There are no practically direct measurements of the elastic scattering cross section for positronium. Recently Nagashima *et al* (1992) had determined the momentum transfer cross section σ_m between Ps and helium gas atom by analysis of 2γ angular correlation spectrum of annihilation radiation, which gives information about time evolution of the o-Ps momentum distribution. They obtained $\sigma_m = 3.4 \cdot 10^{-16}$ cm^2 or L = 1 a_0. This value of the scattering length is in a good agreement with calculations of Fraser (1968b), L = 1.88 a_0, and Drachman and Houston (1970), L = 1.389 a_0. The positronium scattering length has to be positive for all atoms and molecules because of the dominant role of the exchange interaction between the electron in Ps and electrons in atoms. Its magnitude has to be close to the effective radius of the atom or molecule for electron scattering R_e which, as it can see from Table 2, is practically constant and equal about 1-2 a_0.

Table 2. Values of the atomic polarizability α, the experimental scattering length for electron L_e and for positrons L_p, and the effective radius of the solid core for electrons R_e and for positrons R_p.

Quantity	He	Ne	Ar	Kr	Xe
α, a.u.	1.39	2.67	11.1	16.8	27.3
L_e, a_0	1.16	0.45	-1.63	-3.8	-6.8
R_e, a_0	1.49	1.14	1.64	1.77	2.10
L_p, a_0	-0.47	-0.56	-4.1	-7.8	-26.7
R_p, a_0	0.60	0.86	1.35	1.54	1.77

In a dense matter the slow free positron (or electron, positronium atom) interacts simultaneously with a lots of atoms. Its wave function $\psi(r, R_1, R_N)$ is a solution of the Shrodinger equation with potential energy:

$$V(r) = \sum_{j=1}^{N} U(r-R_j) \qquad (2.3)$$

where $U(r-R_j)$ is the positron interaction energy with the j-th atom. The average potential created by the scatters is given:

$$V_0 = \left\langle \sum_{j=1}^{N} \int U(r-R_j) |\psi(r, R_1, \ldots, R_N)|^2 \, dr \right\rangle. \qquad (2.4)$$

At low densities, when potentials from different atoms are not crossed, $NL^3 \ll 1$, the low energy scattering can be described by optical model with the pseudopotential $U(r-R_j) = 2\pi\hbar^2 L \delta(r-R_j)/m$ (Landau and Lifshitz, 1987). So in the optical potential approximation

$$V_0 = 2\pi\hbar^2 LN/m. \qquad (2.5)$$

If the interaction potential is mainly repulsive, then the scattering length L is positive and foreign particle must have a kinetic energy higher than V_0 in order to penetrate in the matter. In this case the interaction with atoms or molecules results in a potential barrier preventing light particle penetration into the matter. The scattering length is positive for scattering of electrons on atoms or molecules having small polarizability α (He, Ne, H_2, N_2) and for scattering of positronium atoms on all atoms or molecules. If $V(r)$ is attractive, then L is negative, injected particle acquires an additional kinetic energy equal to V_0 and after thermalization the interaction with atoms or molecules prevents light particles emitting from the matter. The scattering length is negative for scattering of electrons on atoms and molecules with large α and for scattering of positrons on all atoms and molecules.

At higher densities several effects have to be taken into account, which lead to a non-linear density dependence of V_0. One of them consists in the screening of the electrostatic field of an electron or positron and was taken into account in (2.2) by introducing in it ε instead unity. Already this effect can result in the appearance of the minimum of V_0 and even in the change of its sign in the case of $L < 0$, as it is observed for electrons injected in dielectric liquids (Holroyd and Schmidt, 1989).

In dense gases and condensed matter strong interaction between atoms plays very important role and optical approximation does not valid. An alternative approach usually used in this case is based on the Wigner-Seitz model. This model is especially simple in the case of positronium, when its interaction with atoms can be described by a hard core of radius $L \simeq R_e$. In the Wigner-Seitz model the medium is divided into equivalent atomic spheres of radius r_s ($4\pi r_s^3 N/3 = 1$). Each sphere contains a hard core of radius L in its center. A free Ps may be in any cell with equal probabilities. Therefore Ps wave function $\psi(r) \sim r^{-1}\sin[k_0(r-L)]$ and $\psi'(r)$ have to be continuous at the cell boundaries what is possible only if $\psi'(r_s)=0$. This gives for the energy of the bottom of the conduction band

$$V_0 = \frac{\hbar^2 k_0^2}{2m}, \qquad k_0 r_s = \mathrm{tg}\,[k_0(r_s-L)]. \tag{2.6}$$

In spite of that Wigner-Seitz model is well-grounded only for systems with strong interaction between atoms like liquids or solids, it gives correct results for V_0 (2.5) in the limit of small densities and may be used for interpolation at the whole density region, from ideal gas to solid matter. Calculations of V_0 by means off the Wigner-Seitz model are in good agreement with the potential barrier measured at the injection of electrons into liquid helium (Broomall et al, 1976). It has to give not worst results for ground state energy of positronium.

In the case of positrons and electrons long-range interaction is important. The potential energy of the light particle being inside Wigner-Seitz cell can be expressed as a sum of the potential produced by the atom at the origin and of the ensemble-averaged potential energy of the interaction with the atoms lying outside the cell (Springett et al, 1968; Plenkiewicz et al, 1991). At present there are calculations of the electron V_0 for all rare-gas fluids (Plenkiewicz et al, 1989a; 1989b; 1991; Space et al, 1992; Boltjes et al, 1993) which are in a good agreement with the available experimental data (Holroyd and Schmidt, 1989). Recently Plenkiewicz et al (1993) had calculated the positron V_0 in all rare-gas fluids as a function of

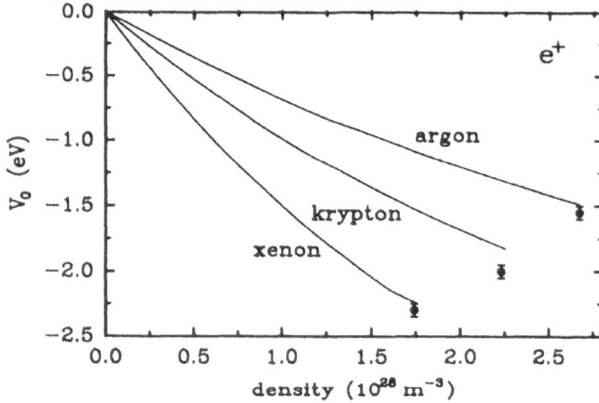

Figure 3. Conduction band energy minimum V_0 as a function of the fluid density for quasifree positrons in rare-gas liquids. Points are the experimental results obtained by Gullikson et al (1988) in crystalline samples at 4K (from Plenkiewicz et al, 1993).

density. Their results for Ar, Kr and Xe are shown in Figure 3. On the same figure, the experimental estimates of Gullikson et al (1988) in crystals, deduced from positron Bragg diffraction measurements at low temperatures are shown. There is a very good agreement between these calculations and experimental data.

Another important characteristic of the positron interaction with atoms or molecules is the effective charge Z_{eff}. The value of the Z_{eff} is determined by the overlap integral (1.5) of the atomic electrons density $\rho_a(\mathbf{r})$ and the positron wave function $\psi(\mathbf{r})$. At low densities the positron-atom interaction is pair-wise and Z_{eff} is independent from density. When the interaction is very small, the positron wave function is close to the plane wave and $Z_{eff} \cong Z$. The positron-atom interaction distorts the wave function and results in a difference between Z_{eff} and Z. Due to the prevalence of the polarization attraction between positron and atom $\psi(\mathbf{r})$ has a maximum near the atom and Z_{eff} usually exceeds Z. Wright et al (1983) found an empirical relationship between Z_{eff} and the dipole polarizability, α, of the form $Z_{eff} \sim \alpha^{1.05}$. On the contrary, $^1Z_{eff}$ is usually many times less than Z due to the prevalence of the exchange repulsion.

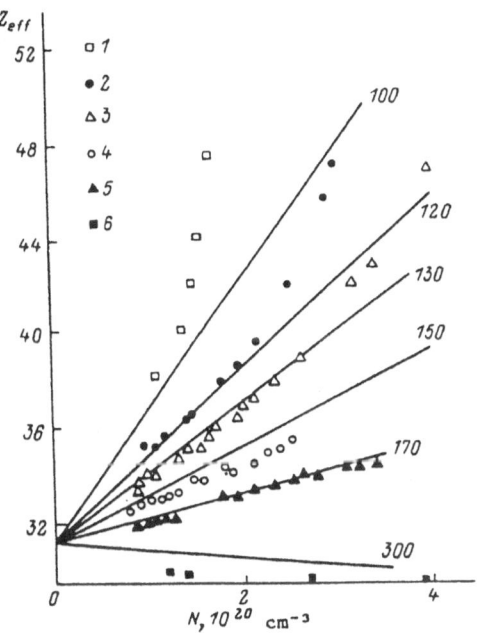

Figure 4. Plot of Z_{eff} versus the density at various temperatures: 1, 100; 2, 120; 3, 130; 4, 150; 5, 170; 6, 300 K (Kawaratani et al, 1985a). The straight lines are results of calculations (from Artem'ev et al, 1988).

The first density corrections to the positron or positronium effective charges (or nonlinear density dependence of the corresponding annihilation rates) results from the multiple scattering of positrons by atoms and from existence of the correlation in the position of the atoms (Nieminen, 1980; Iakubov and Khrapak, 1982; 1992). Artem'ev et al (1988) has calculated by Green-function method the density dependence of the Z_{eff} in dense gaseous nitrogen. Their results are shown in Figure 4 together with the experimental results of Kawaratani et al (1985a). The obvious nonlinearity in $Z_{eff}(N)$ at T = 100 K occurs because the measurements were carried out at overly high densities, at which a transition of positrons to the self-trapped state is beginning. The decrease of the Z_{eff} with density at room temperature is first of all due to the decrease of the role of the long-range attraction between molecules (in full analogy with the change of sign of the second virial coefficient at high temperatures).

At very high densities outside the clusterization region the short-range repulsion between atoms or molecules is most important. The Wigner-Seitz model may be used. The role of the interatomic attraction is negligible, thus the dependence of Z_{eff} on temperature is absent. The value of Z_{eff} in this region is less than its vacuum value and can be even less than Z, as in Ne, Ar and Xe (Iakubov and Khrapak, 1992). It is connected with the strong overlapping of the long-range polarization tails of the positron-atom interaction potential. This results in the displacement of the position of the positron wave function maximum further from the center of the atom and in decrease of the Z_{eff}. The estimation of the density dependence of Z_{eff} in the post-self-trapping region of the dense Ar carried out by Iakubov and Khrapak (1992) is in a good agreement with experiment by Tuomisaari et al (1985).

3. POSITRONIUM BUBBLES

At sufficiently small densities and high temperatures the positronium annihilation rate λ_2 is independent of temperature and increases linearly with density or has a slight non-linearity connected with the multiple scattering and interaction between atoms. Qualitative changes in the density dependence of λ_2 are observed in dense low-temperature media. The density and temperature dependence of the pick-off rate of o-Ps in dense ^4He measured by Rytsola et al (1984b) is shown in Figure 5. The pick-off rate increases with increasing temperature and density, but is many times smaller than the extrapolated low density rate.

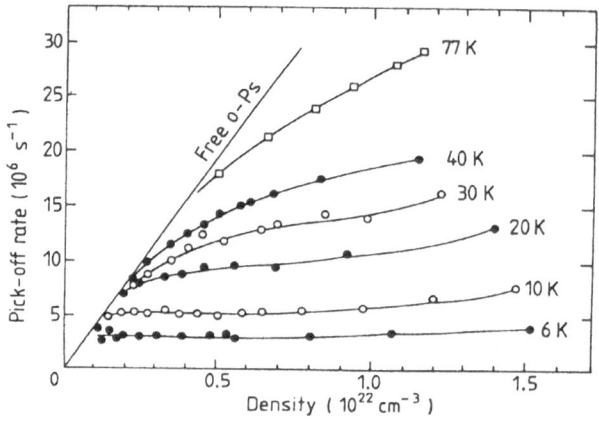

Figure 5. The pick-off rate of ortho-positronium as a function of density at various temperatures in ^4He (Rytsola et al, 1984b). The data at 77K are due to Fox et al (1977). The "free o-Ps" line corresponds to $Z_{eff} = 0.125$.

The reason for the observed anomalies is the possibility of the formation of self-trapped states of positronium. In the interaction of Ps with atoms and molecules the strong exchange repulsion dominates. So for Ps it may be energetically favorable to be in a region of lower than average density. The Ps can push the remaining helium atoms away to reduce the exchange repulsion thereby creating a bubble. The creation of the bubble results in diminishing of the annihilation rate, because the atomic density is less inside the bubble, while outside the bubble the positron wave function is exponentially small. Thus the annihilation probability is also reduced. The density of atoms "seeing" by the Ps is determined by the integral $\int N(r)|\psi(r)|^2 dr$ where $N(r)$ is the distribution of the atoms around Ps and $\psi(r)$ is the Ps wavefunction. This effective density may be substantially less than the mean gas density N.

The density and temperature dependence of the annihilation rate suggesting bubble formation has also been found in neon (Canter and Roellig, 1975; Mao and Paul, 1975), argon (Orth and Jones, 1969; Tuomisaari et al, 1986), xenon (Tseng et al, 1977; Tuomisaari et al, 1988), hydrogen (McNutt et al, 1979b), nitrogen (Kawaratani et al, 1985b), methane (McNutt et al, 1975; McNutt and Sharma, 1978), ethane (Sharma et al, 1982; Sharma, 1992), CO_2 (Wright et al, 1983; Sharma, 1992), benzene, toluene, isooctane, n-hexane and carbon disulfide (Kobayashi, 1992). In Figure 6 the annihilation rate of o-Ps λ_2 in gaseous, liquid and solid Xe is shown as typical example (Tuomisaari *et al*, 1988). The overall results are similar to those of helium, argon and other substances. At high densities the annihilation rate is clearly lower than that could be expected for the free state. This testifies about the bubble formation around the Ps atom not only in liquid Xe but in solid xenon also. The transition to the localized state is very diffuse, in contrast to He where the transition is sharp close to the critical temperature. It is connected first of all, with the fact that in the case of helium the transition takes place at the region of so small densities where density corrections to the annihilation rate of the free Ps are negligible. Practically the same situation is observed in the electron self-trapping in dense gases: very sharp transition in low temperature helium gas (Levine and Sanders, 1967) and diffuse transition in neon near the critical point (Borghesani and Santini, 1990).

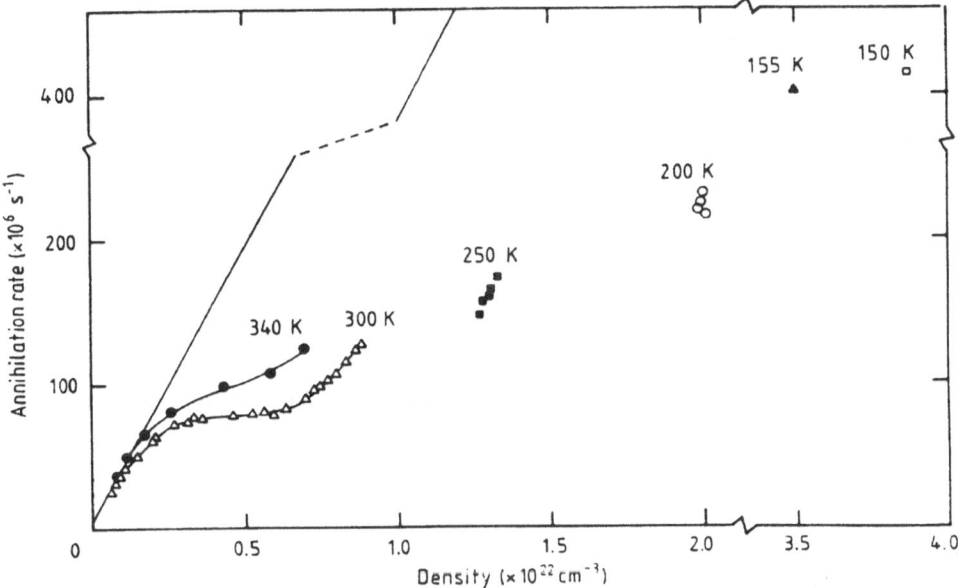

Figure 6. The annihilation rate of orthopositronium in Xe. The sample was gaseous at 340 and 300 K, liquid at 250 and 200 K, and solid at 155 and 150 K (from Tuomisaari *et al*, 1988).

The self-trapped states of light quantum particle on density fluctuations will be preferred thermodynamically if the change in the free energy of the system, ΔF, on going from the quasi-free to localized state is negative:

$$\Delta F = -\varepsilon + \Delta F_m \tag{3.1}$$

where ε is the particle binding energy in the density fluctuation associated with the localized state and ΔF_m is the energy necessary to form the fluctuation. So the existence of the bound states at the density fluctuations is only necessary condition of the self-trapping but not sufficient. Localized states will be formed if the binding energy to the fluctuation is greater than the energy necessary to create it.

The simplest model used to describe the self-trapping is the square-well bubble model. In this model the light particle is assumed to be trapped in an empty spherical cavity of radius R. The change of the free energy (3.1) may be written as:

$$\Delta F = -\varepsilon(R) + \frac{4\pi}{3} R^3 p + 4\pi R^2 \sigma \tag{3.2}$$

where second term corresponds to the work required to remove the atoms from the bubble, and the third term is the surface energy with the surface tension coefficient σ. The binding energy ε in the bubble is equal to the difference between V_0 and the energy of the zero point motion of the light particle. The equilibrium bubble radius R_0 can be estimated by minimizing ΔF with respect to the bubble radius R. In the limit of the very large well depth $(V_0 \gg \varepsilon)$ $\varepsilon \cong V_0 - \pi^2 \hbar^2/2mR^2$ and the bubble radius R_0 is determined by :

$$R_0^4 \cong \frac{\pi \hbar^2}{8m\sigma} [1 + \frac{pR_0}{2\sigma}]^{-1}. \tag{3.3}$$

For example, in ^{4}He at T = 1 K R_0 is equal 36 a_0 for electron bubbles and 30 a_0 for positronium bubbles; in liquid Ne at the triple point R_0 is equal 18 a_0 for electron bubble and 15 a_0 for positronium bubble.

The electron and positronium bubbles or voids can exist in solids also, in solid methane, for example, where Ps localization recently was discovered by Wang et al (1991). On the Figure 7 the dependence of the o-Ps lifetime on temperature in solid and liquid methane is shown. Above the triple point (T > 90.66 K) Ps is localized in the bubble. In solids the main role in ΔF_m (3.1) takes the surface energy. V_0 can be estimated in the frame of the Wigner-Seitz model discussed above. The surface tension on the boundary between solid matter and vacuum may be estimated with the help of the Parakhore formula (Hirschfelder et al, 1964)

$$\sigma = P(N_1 - N_2)^4 \tag{3.4}$$

where P is the constant, N_1 and N_2 are the densities of the being in contact phases outside and inside the void. The density inside void or bubble is always small, $N_2 \cong 0$. This allows to rewrite (3.4) in the form

$$\sigma \cong \sigma_t (N/N_t)^4 \tag{3.5}$$

where the index t corresponds to the liquid at the triple point. This formula may be applied with the equal success to liquids and solids. Similar formula was used earlier for determination of the surface energy of the positron clusters at the supercritical region (T > T_c) (Iakubov and Khrapak, 1982) and for electron bubbles near critical point (Khrapak et al, 1991), where the usual conception of the surface tension is lost.

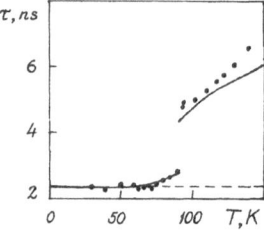

Figure 7. The variation of the o-Ps lifetime τ with temperature in condensed methane. Experimental points are from Wang et al (1991). Calculated curve is from Iakubov and Khrapak (1992).

The results of the calculation of the o-Ps life-time by Iakubov and Khrapak (1992) are shown in Figure 7. The scattering length of Ps on methane molecule was taken to be L = 2.2 a_0. It was the single adjusted parameter. The received results describe well the measured data at all temperatures including the triple point. At very low temperatures the surface tension is too high, ΔF is positive and the probability of the Ps self-trapping is low. The Ps lifetime does not depend practically on T ($\tau \cong 2.35$ ns). With the growth of the temperature the density of solid decreases slightly and in connection with (3.5) the surface tension also decreases. Above 65 K $\Delta F < 0$, the localization becomes possible and begins to influence on the lifetime: τ increases. We suppose that the jump in the value of τ at the triple point is connected only with the jump of the methane density ($\Delta N/N \cong 8\%$) which changes the value of σ and V_0. It was found that the radius of the Ps bubble increases slowly with the growth of the temperature and the binding energy decreases. At the triple point: $R = 7.2$ a_0, $\varepsilon = 0.18$ eV in solid phase and $R = 7.6$ a_0, $\varepsilon = 0.15$ eV in liquid phase.

The square well model has a specific defect. The shape of the well, and consequently the atomic density distribution inside the bubble, is postulated a priori. The model is very good for condensed matter far from the critical point. In dense gases and liquids near the critical point the distribution of atoms around the localized particle has a smooth profile. The adequate description of the properties of the self-trapped states can be obtained in this case by density functional method. In this method the atomic density distribution $N(\mathbf{r})$ and the wave function of the localized light particle $\psi(\mathbf{r})$ are determined self-consistently by minimizing the free energy of the system $\Delta F\{N, \psi\}$ (3.1) with respect to N and ψ. The binding energy of the localized particles and its wave function satisfy the Shrodinger equation:

$$-\frac{\hbar^2}{2m} \nabla^2 \psi(\mathbf{r}) + [V_0\{N(\mathbf{r})\} - V_0(N)]\psi(\mathbf{r}) = -\varepsilon\psi(\mathbf{r}) \qquad (3.6)$$

where N is the density of the homogeneous system and $V_0\{N(\mathbf{r})\}$ is the local value of the conduction band energy. A light particle can become localized if the total energy has a minimum with a non-uniform atomic density profile. As usual, the localized state is stable if the change in the total energy ΔF after localization is negative. Moreover, if the localization is possible, the ratio of the densities of free, N_f, and bound, N_b, particles or the ratio of the probabilities to find particle in the free, P_f, and bound, P_b, states is determined by:

$$\frac{N_f}{N_b} = \frac{P_f}{P_b} \sim \exp\{\frac{\Delta F}{T}\}. \qquad (3.7)$$

Calculations of the free energy of the gas or fluid is usually carried out within the local density approximation, i.e. $F\{N(\mathbf{r})\} = \int f[N(\mathbf{r})] \, d\mathbf{r}$. To construct $f(N)$ it is possible to use semi-empirical equations of state. For example, in the case of ideal gas:

$$f(\mathbf{r}) = N(\mathbf{r})\ln\frac{N(\mathbf{r})}{N} - N(\mathbf{r}) + N \qquad (3.8)$$

and from the condition $\delta\Delta F/\delta N = 0$ it is easy to obtain

$$N(\mathbf{r}) = N\exp[-\beta\tilde{V}(\mathbf{r})], \qquad \tilde{V}(\mathbf{r}) = \int V(\mathbf{r}-\mathbf{r}')|\psi(\mathbf{r}')|^2 \, d\mathbf{r}' \qquad (3.9)$$

where $\beta = 1/T$. This expression differs from the classical one by replacement of the interaction potential between light particles and host atoms $V(\mathbf{r})$ on its quantum-mechanical average $\tilde{V}(\mathbf{r})$. Now it should be noticed that because of the zero-point oscillations the average value of the kinetic energy of the trapped particle is non-zero and equal to $(\hbar^2/2m)\int |\nabla\psi(\mathbf{r})|^2 \, d\mathbf{r}$. Thus, taking into account (3.8), the change of the free energy because of the trapping of a single light particle is

$$\Delta F = \frac{\hbar^2}{2m} \int |\nabla\psi(\mathbf{r})|^2 \, d\mathbf{r} - NT \int \{\exp[-\beta\tilde{V}(\mathbf{r})] + \beta\tilde{V}(\mathbf{r}) - 1\} \, d\mathbf{r}. \qquad (3.10)$$

The search of the optimum density distribution is essential simplified for the optical model potential $V(\mathbf{r}) = 2\pi\hbar^2 L\delta(\mathbf{r})/m$ where L is the scattering length. In this case it is possible to rewrite equation (3.6) and (3.10) in dimensionless form with dependence from only one parameter $D = NL^{5/3}\lambda_\beta^{4/3}$, where $\lambda_\beta = \hbar(2mT)^{-1/2}$ is the thermal wavelength of the light particle. Thus, all characteristics of the self-trapped states in ideal gases may depend only on the same combination of external parameters. Particularly the relation between density N_k and temperature T at the line of transition from the free to bound state (where $\Delta F=0$) can be written as

$$N_k L^{5/3}\lambda_\beta^{4/3} = C. \qquad (3.11)$$

This relation shows the correct temperature dependence of the transition density, $N_k \sim T^{2/3}$, both for electron and positronium bubbles in the gaseous helium (Shikin, 1977; Iakubov and Khrapak, 1982).

The ideal gas approximation is invalid for dense matter and the interatomic interaction should be taken into account. The repulsion between atoms is most important for self-trapping into the bubble. It works against bubble formation and may result in the total disappearance of the bubbles. To account for the interaction between the atoms one can use van der Waals approximation or lattice gas model (Iakubov and Khrapak, 1978; 1982). Recently more complex numerical techniques were used for investigation of the properties of light quantum particles in classical liquids (Reese and Miller, 1993, for example). Various path-integral, Monte Carlo and molecular dynamics simulation methods were reviewed by Hernandez (1991).

The existence of Ps bubbles was directly demonstrated for the first time by Briscoe et al (1968) by measuring the angular correlation of the two photons emitted from annihilation of para-positronium in liquid He, Ne, Ar and H_2. Recently these experiments were repeated in updated version (Stewart et al, 1990). The results of these experiments are shown on Figure 8. Two components are seen clearly for all liquids except O_2. A broad component corresponds to the slow free positron annihilation and a narrow one corresponds to the annihilation of the free or localized positronium. By measuring the width of the narrow component it is possible to determine one of the components of the Ps momentum before annihilation. But the kinetic energy of Ps localized inside a bubble and its momentum depends on the size of the bubble. Taking into account the conservation of the energy and momentum of the system in the process of the annihilation of Ps localized in the bubble it is possible to obtain a relation between the bubble radius R, and the half-width $\theta_{1/2}$ of the angular distribution (Stewart et al, 1990). In the simplest model of an infinite square well potential this relation is given by:

$$R = 3.14 \cdot 10^{-2}/\theta_{1/2}$$

(3.12)

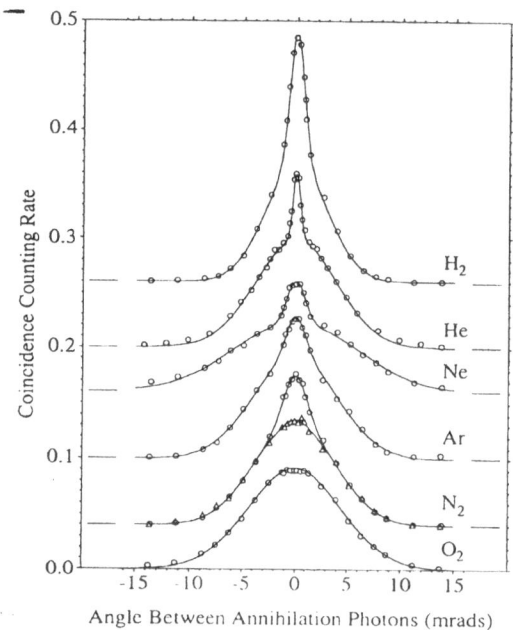

Figure 8. The angular correlation of photons from positrons annihilation in several liquids (from Stewart et al, 1990)

where R is given in units of a_0, and $\theta_{1/2}$ in radians. Handing the data presented in Figure 8 by means of (3.12) Stewart *et al* (1990) obtained R = 36.4, 24.5, 16.8, 21.7 and 14.5 a_0 for liquid He, H_2, Ne, N_2 and Ar, respectively. Varlashkin (1971) obtained R = 12.1 and 11.1 a_0 for Kr and Xe.

Stewart *et al* (1990) had conducted also a series of experiments in solid and liquid helium at high pressures up to 168 atm in the temperature range from 1.7 to 4.2 K. They did not find experimental evidence for any abrupt change in the equilibrium bubble radius in going through the liquid-solid phase transition. This is to be expected for helium. Actually the dependence of equilibrium radius R of the Ps bubble on external parameters is determined by (3.3). During the solidification the pressure is constant and the possible change of R can be connected only with the change of the surface tension σ. But the pressure on the solidification line of helium is very high (more than 25 atm for ^{4}He) and the dependence on σ practically disappears from (3.3). In others substances near the triple point pressure is very small, the dependence of R on σ is important. This results in the abrupt decrease of the bubble radius after solidification as we discussed above in connection with the measurements of the annihilation rate in solid and liquid methane.

4. POSITRON CLUSTERS

The annihilation rate of positrons in a gas at high temperatures is proportional to density and independent of temperature in conformity with (1.2). However unusual behavior of λ_1 was observed in dense gaseous ^{4}He at low temperatures (Roellig and Kelly, 1965; Canter and Roellig, 1970). They found that if T is falling there is a certain temperature T_k at which λ_1 has an abrupt rise and amounts to magnitudes specific for liquid. The same effect appears when the density is increasing along the isotherm. It is remarkable that λ_1 attains magnitudes specific for the liquid at the gas densities which are still less than the liquid density by several times.

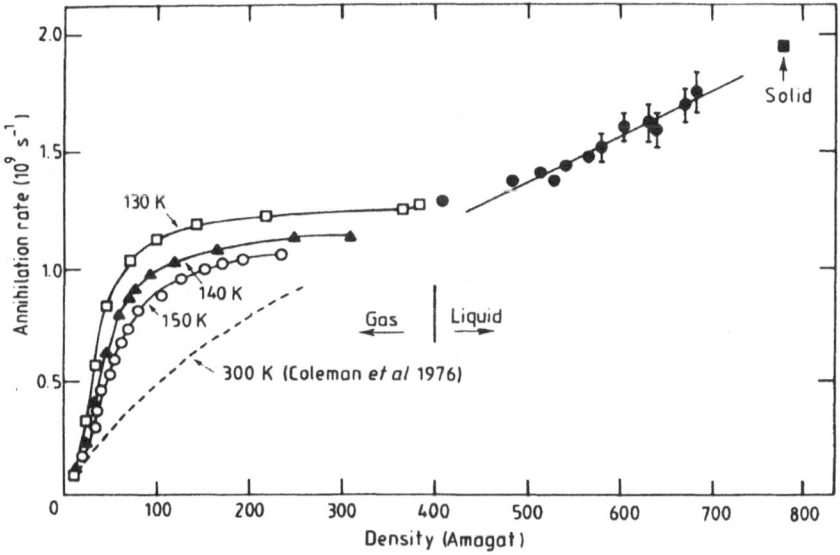

Figure 9. Positron annihilation rate in nitrogen (from Rytsola *et al*, 1984a).

Similar anomalies have been detected also at least in argon (Canter and Roellig, 1975; Albrecht and Jones, 1980; Tuomisaari *et al*, 1985), xenon (Tuomisaari *et al*, 1988), hydrogen (McNutt *et al*, 1979a), nitrogen (Rytsola *et al*, 1984), methane (McNutt *et al*, 1975), ammonia (McNutt *et al*, 1974), CO_2 and SF_6 (Heyland *et al*, 1985). In figure 9 the annihilation rate of positrons at a gaseous, liquid and solid nitrogen is shown as typical example (Rytsola *et al*, 1984a). The annihilation rate varies approximately linearly with density at the lowest densities at all temperatures. With increasing density, however, an abrupt increase λ_1 is seen at low T. The transition becomes less sharp with increasing T but at the same time it moves to a higher densities. After the transition there is a plateau region the height of which corresponds roughly to the annihilation rate in liquid. The height of the plateau is also temperature dependent. At liquid densities an approximately linear behavior is resumed and no temperature dependence is seen.

The anomalous behavior of the positron annihilation rate in dense gases at low temperatures has been shown theoretically to be due to droplet-like clustering of atoms around the positron(Khrapak and Iakubov, 1976; Stott and Zaremba, 1977; Manninen and Hautojarvi, 1978; Worrell and Miller, 1992). As we discussed above, the exchange repulsion is absent in the interaction of positrons with atoms and thus the polarization attraction prevails. The positron scattering length L is always negative as well as the conduction band energy V_0. In this case the self-trapping of positrons with creation of clusters is possible. It was found that clusters exist in restricted density and temperature region which is adjacent to the "gas-liquid" phase-coexistence curve from higher temperatures and has its own "critical" temperature T^*. Above T^* the clusters do not exist at arbitrary high densities.

Main properties of the positron cluster follow from a simple model:

(i) the positron-atom interaction is described by the optical pseudopotential (2.5);
(ii) the lattice-gas approximation is used for ΔF_m (3.1);
(iii) it is assumed that the electron is trapped in an empty spherical volume of radius R and the highest possible lattice-gas density, N_s, is present in the cluster.

Within this model the change of the free energy of the system as result of the creation of the cluster is

$$\beta\Delta F = \pi^2\lambda_\beta^2 R^{-2} + 4\pi\lambda_\beta^2 L(N_s - N) + \frac{4\pi}{3}N_s R^3 \ln\frac{N_s}{N} \qquad (4.1)$$

where $\beta = 1/T$ and λ_β is the positron thermal wavelength. The region in which a considerable fraction of positrons is trapped is bounded by the curve $N_k(T)$ determined as usual by the conditions $\Delta F = 0$ and $\partial\Delta F/\partial R = 0$. It was found (Khrapak and Iakubov, 1976) that the region of existence of the clusters is bounded both by low and high N, latter since the compressibility of matter becomes very small at high densities. There is an intercept of the boundaries at a certain temperature $T = T^*$ above which no clustering occurs. The lower boundary of the cluster formation domain is determined by

$$N_k \cong N_s\exp[-C|L|^{5/2}\lambda_\beta^2 N_s^{3/2}] = N_s e^{-G} \qquad (4.2)$$

where $C \cong 1$. So N_k is an exponential function of the temperature and depends very strong both on positron-atom interaction through L and on interatomic repulsion through N_s. Obviously that the clusters can exist if $N_k << N_s$, or if $G > 1$. It is easy to estimate that for all rare gases $G > 1$ except neon where $G << 1$ (for the estimation it is possible to take L from Table 2 and to chose $T = T_c$, $N_s = 2N_c$). Thus the self-trapping of positron is possible in all rare gases except neon, in full agreement with experiments.

The simple model discussed above can explain the main peculiarity of the observed phenomena, but the assumptions involved are rather crude. For more detail investigation of the properties of the positron clusters and conditions of their existence the density functional method is usually applied in a manner very similar to that already used for positronium bubbles in the previous section. According to estimations by Khrapak and Iakubov (1976), Manninen and Hautojarvi (1978) in ^{4}He the cluster radius varies from 20 to 40 a_0 and the number of atoms in these clusters is a few hundred. The positron binding energy in the cluster is equal about 0.1 eV. In argon and nitrogen the cluster radius varies from 10 to 25 a_0 and the number of atoms or molecules is a few tens.

The theoretical methods developed for the investigation of the positron and electron clusters in dense gases may be used with small modifications for the investigation of the positronium and electron self-trapping inside voids in solids (Khrapak, 1990). In this case the self-trapping of the light particle is possible on the density fluctuations of vacancies. The relative concentration of vacancies even on the melting line is very small (about 10^{-3}-10^{-4}) and probability of the vacancy density fluctuation may be estimated in frameworks of the ideal lattice gas model. The scattering of the light particle on a vacancy can be described by scattering length L_v, which has a sign opposite to the sign of the scattering length on the same atom, L, and has an absolute value close to L. So for electron in solid helium, neon and hydrogen as well as for positronium in all solids L_v is negative and the formation of the electron- or positronium-vacancy clusters is possible. These clusters are very similar to the positron clusters in dense gases, which properties were discussed above. Particularly the relation between the vacancy concentration N_k, the temperature T and the scattering length L_v at the line of transition to the self-trapped state is given by (4.2) in which N_S is the maximum possible concentration of vacancies equal to the density of matter N. The self-trapping of the light particle is possible only if G >> 1. It was demonstrated (Khrapak, 1990) that only for electron in neon G ~ 1. In all other cases G >> 1. Thus, according to the vacancy cluster model and in full agreement with experiments (Keshishev et al, 1970; Sakai et al, 1982; Levchenko and Mezhov-Deglin, 1990) the self-trapping of electron inside voids or bubbles is possible in solid helium and hydrogen but impossible in solid neon. The self-trapping of positronium is possible in all rare gas solids near the triple point what was observed experimentally for He (Rytsola et al, 1984b) and Xe (Tuomisaari et al, 1988) by measurements of the o-Ps annihilation rate. It seems that it is contradict to the measurements of the angular correlation of photons from p-Ps annihilation in solids (Stewart et al, 1990) where bubble creation was seen in He and was not seen in Ne and Ar. A possible reason of this contradiction is the fact that in the solid helium vacancies, on the density fluctuation of which a bubble is appear, have very high mobility as result of quantum tunnel effects (Andreev, 1976). In other solids the mobility of vacancies is very small and apparently the bubble has not enough time for formation during the lifetime of p-Ps. The lifetime of o-Ps is considerably larger and o-Ps bubbles has enough time for formation.

5. KINETICS OF SELF-TRAPPING

There were a few attempts to measure the thermalization time of the light particle injected in a dense matter and the characteristic time of its self-trapping. Onn and Silver (1969, 1971) found that electrons injected in liquid helium at 4.2 K are thermalized in a time $\sim 2 \cdot 10^{-13}$ s, instead of the approximately 10^{-10} s predicted by the electron-atom elastic collision model. This indicates that collective mechanisms must be important in the thermalization process. One of them consist of the inelastic resonant scattering of electrons on the density fluctuations (Iakubov and Khrapak, 1982; Sakai et al, 1992). An electron with low enough energy may be captured by the small bubble which has a virtual or resonant energy level. The surface of the bubble represents deep potential well with potential walls at which the electron is scattered. Between the potential walls the electron is reflected many times, making its residence time much longer than the classical time required for passing the bubble. During its residence at the fluctuation, the electron is losing energy effectively due to elastic collisions with the potential walls. Actual capture of the electron is possible only if during time of residence at the bubble the electron loses its excess kinetic energy completely. Otherwise the electron leaves the bubble loosed part of its energy. This inelastic process can be effective only when V_0 is positive, in particularly, for the thermalization of the Ps in all liquids. This can explain experimental fact of the absence of inelastic processes during the thermalization of electrons in liquid argon (Smejtek et al, 1973). Note that similar arguments were used recently by Sakai et al (1992) for estimation of the time of electron autolocalization in liquid neon.

Tawel and Canter (1986) had investigated an influence of a pulsed electric field on the positron self-trapping in gaseous helium on the saturation line. The experimental conditions were chosen in such a way that without electric field positrons were localized in clusters. When dc-electric field was applied the heating of positrons prevented their localization. When the delay time τ_d of the pulse electric field was very large, the positrons had enough time for self-trapping and the positron annihilation spectrum did not differ from one obtained without electric field. On the contrary when τ_d was very small practically all positrons remained free and the annihilation spectrum coincided with one obtained with dc electric field. Characteristic times, when the dependence of the annihilation spectrum on the delay time appears and disappears, determine the moment of the beginning and termination of the positron self-trapping. For example, at electric field 41 v/cm and at temperature 3.9 K the positron self-trapping began after about 29 ns and all positrons were localized after about 35 ns. Monte Carlo simulations by Farazdel (1986) agree well with experiment. Nevertheless a simple model similar to discussed above for the self-trapping in bubbles would be useful in this case also.

ACKNOWLEDGMENTS

The author is pleased to acknowledge helpful discussions with I.T. Iakubov and is grateful for the support of the Japan Society for the Promotion of Science and for the hospitality of the Electrical Engineering Department at the Hokkaido University during a visit.

REFERENCES

Albrecht, R.S., and Jones, G., 1980, Influence of self-nucleating clusters on the annihilation of positrons in low temperature argon gas, *Phys. Lett. A* **79**:417.

Andreev, A.F., 1976, Diffusion in quantum crystals, *Sov. Phys. Usp.* **19**:137.

Boltjes, B., de Graaf, C., and de Leeuw, S.W., 1993, Computation of the energy V_0 of an excess electron in dense helium and argon, *J. Chem. Phys.* **98**:592.

Borghesani, A.F., and Santini M., 1990, Electron mobility and localization effects in high-density Ne gas, *Phys. Rev. A* **42**:7377.

Briscoe, C.V., Choi, S.-I., and Stewart, A.T., 1968, Zero-point bubbles in liquids, *Phys. Rev. Lett.* **20**:493.

Broomall, J.R., Johnson, W.D., and Onn, D.G., 1976, Density dependence of the electron surface barrier for fluid ^{3}He and ^{4}He, *Phys. Rev. B* **14**:2819.

Canter, K.F., McNutt, J.D., and Roellig, L.O., 1975, Positron annihilation in low-temperature rare gases. I. Helium, *Phys. Rev. A* **12**:375.

Canter, K.F., and Roellig, L.O., 1970, Critical behavior of positrons in low-temperature gaseous helium, *Phys. Rev. Lett.* **25**:328.

Canter, K.F., and Roellig, L.O., 1975, Positron annihilation in low-temperature rare gases. II. Argon and neon, *Phys. Rev. A* **12**:386.

Charlton, M., 1985, Experimental studies of positrons scattering in gases, *Rep. Prog. Phys.* **48**:737.

Drachman, R.J., and Houston, S.K., 1970, Simplified model for positronium-helium scattering, *J. Phys. B* **3**:1657.

Farazdel, A., 1986, Confirmation of positron mobility edge in gaseous helium by Monte Carlo simulation, *Phys. Rev. Lett.* **57**:2664.

Ferrel, R., 1957, Long lifetime of positronium in liquid helium, *Phys. Rev.* **108**:167.

Fox, R.A., Canter, K.F., and Fishbein M., 1977, Positron and positronium decay rates in helium at high densities, *Phys. Rev. A* **15**:1340.

Fraser, P.A., 1968a, Positrons and positronium in gases, *Adv. Atom. Mol. Phys.* **4**:63.

Fraser, P.A., 1968b, The scattering and pick-off quenching of orthopositronium in helium, *J. Phys. B* **1**:1006.

Gol'danskii, V.I., 1968, Physical chemistry of the positron and positronium, *At. Energy Rev.* **6**:3.

Griffith, T.C., and Heyland, G.R., 1978, Experimental aspects of the study of the interaction of low-energy positrons with gases, *Phys. Rep. C* **39**:169.

Gullikson, E.M., Mills, A.P., and McRae, E.G., 1988, Observation of intense two-beam positron diffraction and the precise determination of the positron band gap in rare-gas crystals, *Phys. Rev. B* **37**:588.

Gus'kov, Y.K., Savvov, R.V., and Slobodyanyuk, V.A., 1978, Time-of-flight measurements of the total cross-section for elastic scattering of low-energy electrons (E=0.025-1.0 eV) by He, Ne, Ar, Kr and Xe, *Sov. Phys.- Techn. Phys.* **23**:167.

Hautojarvi, P., Rytsola, K., Tuovinen, P., Vehanen, A., and Jauho, P., 1977, Microscopic gas-liquid phase transition around the positron in helium gases, *Phys. Rev. Lett.* **38**:842.

Hernandez, J.P., 1991, Electron self-trapping in liquids and dense gases, *Rev. Mod. Phys.* **63**:675.

Heyland, G.R., Charlton, M., Griffith, T.C., and Clark, G., 1985, The temperature dependence of free positron lifetimes and positronium fraction in gaseous CO_2 and SF_6, *Chem. Phys.* **95**:157.

Hirschfelder, J.O., Curtis, C.F., and Bird, R.B., 1964, "Molecular Theory of Gases and Liquids," Wilei, New York.

Holroyd, R.A., and Schmidt, W.F., 1989, Transport of electrons in nonpolar fluids, *Annu. Rev. Phys. Chem.* **40**:439.

Iakubov, I.T., and Khrapak, A.G., 1978, On the annihilation rate of positronium in dense rare gases, *Appl. Phys.* **16**:179.

Iakubov, I.T., and Khrapak, A.G., 1982, Self-trapped states of positrons and positronium in dense gases and liquids, *Rep. Prog. Phys.* **45**:697.

Iakubov, I.T., and Khrapak, A.G., 1992, Positron and positronium localization in liquids and gases, *Materials Sci. Forum* , **105-110**:289.

Kauppila, W.E., and Stein, T.S., 1982, Positron-gas cross-section measurements, *Can. J. Phys.* **60**:471.

Kawaratani, T., Nakayama, Y., and Mizogawa, T., 1985a, Measurements of positron annihilation rate in low-temperature gaseous nitrogen, *Bull. Inst. Chem. Res., Kyoto Univ.* **63**:177.

Kawaratani, T., Nakayama, Y., and Mizogawa, T., 1985b, Density and temperature dependencies of ortho-positronium annihilation rates in low temperature gaseous N_2, *Phys. Lett. A* **108**:75

Keshishev, K.O., Mezhov-Deglin, L.P., and Shal'nikov, A.I., 1970, Carrier mobility in He^4 crystals, *Sov. Phys.-JETP Lett.* **12**:160.

Khrapak, A.G., 1990, Self-trapping of positrons and positronia in liquids and solids, *in*: "Positron and Positronium Chemistry," Y.C. Jean, ed., World Sci., Singapore.

Khrapak, A.G., and Iakubov, I.T., 1976, Positron clusters in dense gases, *Sov. Phys.-JETP Lett.* **23**:422.

Khrapak, A.G., Sakai, Y., Bottcher, E.H., and Schmidt, W.F., 1991, Stability of electron bubbles in liquid neon, *IEEE Trans. Electr. Insulation* **26**:582.

Kobayashi, Y., 1992, Pressure dependence of ortho-positronium lifetimes in molecular liquids, *Ber. Bunsenges. Phys. Chem.* **96**:1869.

Landau, L.D., and Lifshitz, E.M., 1987, "Quantum Mechanics. Non-Relativistic Theory," Pergamon Press, Oxford.

Levchenko, A.A., and Mezhov-Deglin, L.P., 1990, Motion of injected charges in solid parahydrogen, *Sov. Phys.-JETP* **71**:196.

Levine, J., Sanders, T.M., 1967, Mobility of electrons in low-temperature helium gas, *Phys. Rev.* **154**:138.

Manninen, M., and Hautojarvi, P., 1978, Clustering of atoms around the positron and positive ions in gaseous He, Ne, and Ar, *Phys. Rev. B* **17**:2129.

Mao, A.C., and Paul, D.A.L., 1975, Positron scattering and annihilation in neon gas, *Can. J. Phys.* **53**:2406.

McNutt, J.D., Kinnison, W.W., and Ray, A.D., 1974, Annihilation of positrons in gaseous and liquid NH_3-Ar and NH_3-Ne mixtures, *J. Chem. Phys.* **60**:4730.

McNutt, J.D., Sharma S.C., and Brisbon, R.D., 1979a, Positron annihilation in gaseous hydrogen-neon mixtures. I. Low-energy positrons, *Phys. Rev. A* **20**:347.

McNutt, J.D., Sharma S.C., Franklin, M.H., and Woodall, M.A., 1979b, Positron annihilation in gaseous hydrogen-neon mixtures. II. Positronium, *Phys. Rev. A* **20**:357.

McNutt, J.D., and Sharma S.C., 1978, Dependence of orthopositronium annihilation rates on density fluctuations in methane gas, *J. Chem. Phys.* **68**:130.

McNutt, J.D., Summerour, V.B., Ray, A.D., and Huang, P.H., 1975, Complex dependence of the positron annihilation rate on methane gas density and temperature, *J. Chem. Phys.* **62**:1777.

Mogensen, O.E., 1974, Spur reaction model of positronium formation, *J. Chem. Phys.* **60**:998.

Nagashima, Y., Hyodo, T., and Fujiwara, K., 1992, Energy loss of positronium in helium gas, *Materials Sci. Forum* **105-110**:1671.

Nieminen, R.M., 1980, Nonlinear density dependence of the positron decay rate in helium, *Phys. Rev. A* **21**:1347.

O'Malley, T.F., 1963, Extrapolation of electron-rare gas atom cross sections to zero energy, *Phys. Rev.* **130**:1020.

Onn, D.G., and Silver, M., 1969, Attenuation and lifetime of hot electrons injected into liquid helium, *Phys. Rev.* **183**:295.

Onn, D.G., and Silver, M., 1971, Injection and thermalization of hot electrons in solid, liquid, and gaseous helium at low temperatures, *Phys. Rev. A* **3**:1773.

Orth, P.H.R., and Jones, G., 1969, Annihilation of positrons in argon I. Experimental; II. Theoretical, *Phys. Rev.* **183**:7; **183**:16.

Paul, D.A.L., and Graham, R.L., 1957, Annihilation of positrons in liquid helium, *Phys. Rev.* **106**:16.

Plenkiewicz, B., Frongillo, Y., and Jay-Gerin, J.-P., 1993, Calculation of the ground-state energy V_0 of quasifree positrons in rare-gas fluids, *Phys. Rev. E* **47**:419.

Plenkiewicz, B., Frongillo, Y., Plenkiewicz, P., and Jay-Gerin, J.-P., 1991, Density dependence of the conduction-band energy and of the effective mass of quasifree excess electrons in fluid neon and helium, *Phys. Rev. A* **43**:7061.

Plenkiewicz, B., Plenkiewicz, P., and Jay-Gerin, J.-P., 1989a, Conduction-band energy V_0 of an excess electron in liquid krypton, *Phys. Rev. A* **39**:2070.

Plenkiewicz, B., Plenkiewicz, P., and Jay-Gerin, J.-P., 1989b, Density dependence of the conduction-band energy of excess electron in liquid argon, *Phys. Rev. A* **40**:4113.

Reese, T., and Miller, B.N., 1993, Positronium in Xe: The path-integral approach, *Phys. Rev. E* **47**:2581.

Ritsola, K., Rantapuska, K., and Hautojarvi, P., 1984a, Positron annihilation and cluster formation in nitrogen fluid, *J. Phys. B* **17**:299.

Ritsola, K., Vettenranta, J., and Hautojarvi, P., 1984b, An experimental study of positronium bubbles in helium fluids, *J. Phys. B* **17**:3359.

Roellig, L.O., and Kelly, T.M., 1965, Positron lifetimes in low-temperature helium gas, *Phys. Rev. Lett.* **15**:746.

Sakai, Y., Bottcher, E.H., and Schmidt, W.F., 1982, On the electron drift velocity in solid neon, *Z. Naturforsch. A* **37**:87.

Sakai, Y., Schmidt, W., Khrapak, A., 1992, High- and low-mobility electrons in liquid neon, *Chem. Phys.* **164**:139.

Schrader, D.M., and Svetic, R. E., 1982, The interaction of positrons with large atoms and molecules: theory, *Can. J. Phys.* **60**:517.

Sharma, S.C., 1992, Positronium localization in spontaneous density fluctuations in molecular gases, *Materials Sci. Forum* **105-110**:451.

Sharma, S.C., Eftekhari, A., and McNutt J.D., 1982, Sensitivity of orthopositronium annihilation rates to density fluctuations in ethane gas, *Phys. Rev. Lett.* **48**:953.

Shikin, V.B., 1977, Mobility of charges in liquid, solid, and dense gaseous helium, *Sov. Phys. Usp.* **20**:226.

Smejtek, P., Silver, M., Dy, K.S., and Onn, D.G., 1973, Hot electron injection into dense argon, nitrogen, and hydrogen, *J. Chem. Phys.* **59**:1374.

Space, B., Coker, D.F., Liu, Z.H., Berne, B.J., and Martyna, G., 1992, Density dependence of excess electronic ground-state energies in simple atomic fluids, *J. Chem. Phys.* **97**:2002.

Springett, B.E., Jortner, J., and Cohen, M.H., 1968, Stability criterion for the localization of an excess electron in a nonpolar fluid, *J. Chem. Phys.* **48**:2720.

Stewart, A.T., Briscoe, C.V., and Steinbacher, J.J., 1990, Positron annihilation in simple condensed gases, *Can. J. Phys.* **68**:1362.

Stott, M.J., and Zaremba, E., 1977, Positron self-trapping in ^{4}He, *Phys. Rev. Lett.* **38**:1493.

Tawel, R., and Canter, K.F., 1986, Observation of a positron mobility threshold in gaseous helium, *Phys. Rev. Lett.* **56**:2322.

Tseng, P.K., Chen, S.-H., Chuang, S.Y., and Tao, S.J., 1977, Observation of critical slowing down of ortho-positron quenching rate near gas liquid transition, *Phys. Lett. A* **60**:14.

Tuomisaari, M., Ritsola, K., and Hautojarvi, P., 1985, Localized state of positron in argon, *Phys. Lett. A* **112**:279.

Tuomisaari, M., Ritsola, K., and Hautojarvi, P., 1988, Positron annihilation in xenon, *J. Phys. B* **21**:3917.

Tuomisaari, M., Ritsola, K., Nieminen, R.M., and Hautojarvi, P., 1986, Localization of ortho-positronium in argon, *J. Phys. B* **19**:2667.

Wackerle, J., and Stump, R., 1957, Annihilation of positrons in liquid helium, *Phys. Rev.* **106**:18.

Wang, S.J., Nakanishi, H., and Jean, Y.C., 1991, Positronium localization in solid and liquid methane, *J. Phys.: Condens. Matter* **3**:2167.

Worrell, G.A., and Miller, B.N., 1992, Positron annihilation in xenon: The path-integral approach, *Phys. Rev. A* **46**:3380.

Wright, G.L., Charlton, M., Clark, G., Griffith, T.C., and Heyland, G.R., 1983, Positron lifetime parameters in H_2, CO_2 and CH_4, *J. Phys. B* **16**:4065

SELF TRAPPING OF LIGHT PARTICLES IN FLUIDS:
THE PATH INTEGRAL APPROACH

Bruce N. Miller, Jiqiang Chen, Terrence Reese, and Gregory Worrell

Department of Physics
Texas Christian University
Fort Worth, TX 76129, U.S.A.

INTRODUCTION

This paper concerns the status of some current theoretical models which are useful for representing the behavior of an excess light particle (elp) which has equilibrated in a simple fluid. For our purposes, fluid has the traditional meaning of a homogeneous substance above the liquid-vapor critical temperature, and the elp is typically an electron, positron, or positronium atom. Over a wide range of temperature, these low mass particles have a large mean thermal (or deBroglie) wavelength, and are thus able to simultaneously interact with numerous atoms in the host fluid. This opens the door for local collective behavior of the elp-fluid system on mesoscopic length scales.

Experimental measurements of the mobility of the electron,[1] the lifetime of the positron and ortho-positronium atom,[2] and the angular distribution of the decay products of para-positronium,[3] all suggest that in a region of fluid temperature and density surrounding the critical point localized, or self-trapped, states may occur. Consider, for example, the decay of ortho-positronium (o-Ps). Because the vacuum lifetime of o-Ps is about 140 ns, in addition to its partner, the positron in thermalized o-Ps has the opportunity to annihilate with electrons associated with the atoms of the host fluid. The probability for this "pick off" process depends sensitively on the density of fluid surrounding the o-Ps.[2]

In many fluids, the decay rate of o-Ps dips dramatically in the critical region, suggesting that the local density is sharply reduced from the mean density of the bulk fluid. The density reduction arises from the Fermionic repulsion of the fluid atoms by the electron in o-Ps, resulting in a net o-Ps atom repulsion and a positive scattering length. The long o-Ps wavelength joins with the large isothermal compressibility of the host fluid in the critical region to enhance the local inhomogeneity. The popular interpretation is that the region of reduced density creates a potential well for the o-Ps, which then becomes stabilized in the ground state. Because the presence of the o-Ps has induced the potential well, this localization process is referred to as self-trapping.[2]

In contrast with Ps, the atomic polarization induced by the net charge of the positron results in an attractive elp-atom potential at large separations. This attraction dominates the

Linking the Gaseous and Condensed Phases of Matter
Edited by L.G. Christophorou *et al*, Plenum Press, New York, 1994

e+-nucleus repulsion, resulting in a negative scattering length and a tendency for cluster, or droplet, formation in the vicinity of the positron. In highly polarizable fluids, such as xenon, cluster formation is initiated well below the critical density, while in other gases either a droplet may be produced in the critical region, (e.g. He) or there is little tendency to localize, and the local density is nearly that of the uniform fluid (e.g. Ne).[4] In common with Ps, the droplet results in a potential well which can stabilize the positron in a bound state, so the positron also experiences self-trapping. The experimental manifestation of droplet or cluster formation is an enhancement in the decay rate over that which would result if the fluid density remained uniform. Depending on the fluid, since the electron-atom interaction is characterized both by Fermionic repulsion at short distances and polarization induced attraction at larger distances, each type of structure may occur.[5]

The earliest theoretical models assumed that either the elp was trapped in the ground state of a spherical square well potential, or was propagating through the fluid. This approach was then generalized to a density functional theory (DFT) in which a free energy, $F[\psi, \rho]$, was constructed which depends both on the localized elp wavefunction ψ and the local fluid density, ρ.[2] Solutions for ρ and ψ were found which minimize F. Numerous realizations of DFT were employed which differed chiefly in their treatment of the inter-atomic correlations. Both the square well and the more sophisticated versions of DFT were examples of mean field theory, in which the phenomena was modeled by a continuous fluid density and a single optimized wavefunction corresponding to the ground state of the potential induced by the continuous density. These mean field theories were successful in modeling the properties of localized positrons and positronium in helium, which has a critical temperature of roughly 5K.[2] However, for heavier gases such as argon and xenon, with critical temperatures on the order of a few hundred degrees, the mean field theories were not so successful. They predicted that the existence of localized states turned on and off at specific densities on each isotherm, although the experimental data varied smoothly.[2,4,6,7]

The underlying paradigm on which DFT is based is the adiabatic model, in which a quantum mechanical elp interacts with a fluid of classical atoms. The justification for treating the atomic translational degrees of freedom classically is that, for the temperatures of interest, their mean deBroglie wavelength is less than an atomic diameter. The object of DFT is to provide an analytic approximation which mimics the adiabatic model as closely as possible. Because DFT is a mean field theory, and hence ignores fluctuations, its ability is limited. However, until recently, it was not possible to investigate the adiabatic model directly, and alternative analytic models which include the influence of fluctuations did not exist. Both of these situations changed in the last decade. As a result of the enormous improvement in computer hardware during this period, it is now possible to directly simulate the adiabatic model using either path integral Monte Carlo (PIMC)[8] or quantum molecular dynamics (QMD).[9] Each still requires prodigious blocks of cpu time so it is not possible to sample many locations of the thermodynamic state space. In addition, QMD requires regularization of the interaction potential and mode truncation of the wavefunction in Fourier space. Thus there is still a need for good analytic machinery for investigating the density and temperature dependence of the elp-fluid system.

In the following pages we will first describe the path integral representation of the equilibrated elp-fluid system and, as an application, results concerning the quantum states of positrons[10] and positronium[11] in xenon obtained by us at T. C. U.. We will show that PIMC is able to reproduce the essential features of the experimental observations from an ab initio microscopic model, and that large fluctuations in the local environment occur in the neighborhood of the elp even when it is highly localized.[12] We will then describe the analytic RISM (reference interaction site model)-polaron theory developed by Chandler et. al. to model the elp-fluid system which is based on the path integral.[13] Finally we will discuss recent work we have completed which compares the predictions of the RISM-polaron theory both for the case of a positron[14] and an electron[15] with the numerically accurate PIMC calculations.

THE DISCRETIZED PATH INTEGRAL

In the adiabatic model the atoms (or molecules) are classical objects and the elp obeys quantum mechanics. Represent the total inter-atomic interaction potential by $U(\mathbf{R})$, and the total elp-atom interaction by $W(\mathbf{r}, \mathbf{R})$, where $\mathbf{R}$ represents the set of atomic (or molecular) positions $\{R_j\}$ and $\mathbf{r}$ denotes the position operator of the elp (for positronium, the center of mass). Then the Hamiltonian of the quantum particle is

$$H = -(\hbar^2/2m)\Delta + W(\mathbf{r}, \mathbf{R}) \tag{1}$$

and the statistical average of the quantum operator $O(\mathbf{r}, \mathbf{R})$ is given by

$$\langle O(\mathbf{r}, \mathbf{R})\rangle = (1/Z) \int d\mathbf{R} \exp(-\beta U) \int d\mathbf{r} \langle \mathbf{r}|O(\mathbf{r}, \mathbf{R}) \exp(-\beta H)|\mathbf{r}\rangle , \tag{2}$$

where Z is the configurational partition function, $\int d\mathbf{r}$ represents the usual trace over the quantum states of H in the position representation and, as usual, $\beta=1/kT$.[16] Typically U and W are assumed to be pairwise additive,

$$U(\mathbf{R}) = \Sigma_{1 \leq i < j \leq N} u(R_i - R_j) , \qquad W(\mathbf{r}, \mathbf{R}) = \Sigma_{1 \leq j \leq N} w(\mathbf{r} - R_j) , \tag{3}$$

e.g. u may be a Lennard-Jones (12, 6) potential while the form of w will depend on whether the elp is an electron, positron, or positronium atom.[2,5]

For the purposes of computing the statistical average, Feynman proved, by inserting p-1 intermediate positions in (2), that a single quantum particle can be represented by a classical harmonic chain consisting of p pseudo-particles.[17] In this representation,

$$\int d\mathbf{r} \langle \mathbf{r}|O(\mathbf{r}, \mathbf{R}) \exp(-\beta H)| \mathbf{r} \rangle \rightarrow \int d\underline{\mathbf{r}} O(\mathbf{r}_1, \mathbf{R}) \exp[-\beta \Phi(\underline{\mathbf{r}}, \mathbf{R})] , \tag{4}$$

$$\Phi(\underline{\mathbf{r}}, \mathbf{R}) \equiv \Sigma_{1 < \alpha < p} [(2mp/\beta^2 \hbar^2) |\mathbf{r}_{\alpha+1} - \mathbf{r}_\alpha|^2 + (1/p) W(\mathbf{r}_\alpha, \mathbf{R})] , \tag{5}$$

where $\underline{\mathbf{r}}$ represents the set of positions $\{\mathbf{r}_\alpha\}$ of the pseudo-particles and $\mathbf{r}_{p+1}=\mathbf{r}_1=\mathbf{r}$ (the chain is closed). Thus each pseudo-particle (pp) interacts with the fluid atoms via the reduced potential W/p , and with its nearest neighbors via a harmonic potential with effective temperature dependent force constant $4mp/\beta^2\hbar^2$. The equivalence is exact in the limit $p\rightarrow\infty$. The problem of computing the equilibrium properties of the system is consequently reduced to sampling the scalar (classical) distribution $\exp[-\beta(U+\Phi)]$ in the configuration space of atomic and pp coordinates $(\mathbf{R}, \underline{\mathbf{r}})$. Specific algorithms have been developed to accomplish this task and go under the name of path-integral Monte Carlo (PIMC).[8,18,19,20]

Two structural properties of fundamental importance are the rms separation between a pair of pp's separated by s harmonic bonds, R_s, and the elp-atom radial distribution function, g(r):

$$R_s^2 = (1/p) \Sigma_\alpha \langle |\mathbf{r}_{\alpha+s} - \mathbf{r}_\alpha|^2 \rangle , \qquad \rho g(r) = (1/p) \Sigma_{\alpha,j} \langle \delta(\mathbf{r}_\alpha - R_j - r)\rangle . \tag{6}$$

Here the angle brackets imply the complete statistical ensemble average, and ρ is the mean

fluid density. In the absence of the elp-atom interaction (i.e. if W=0), g is unity and $p^2R_s^2=$
$3\Lambda^2$ s(p - s), where $\Lambda=\sqrt{(2\pi\beta\hbar^2/m)}$ is the thermal wavelength of the elp. Clumping of the
chain is indicated by strong reductions in R_s at the midpoint of the chain (s=p/2) below the
free particle value ($\Lambda\sqrt{(3/4)}$) and is attributed to the dominance of localized elp states, while
deviations of g from unity describe changes of the average fluid environment in the
neighborhood of the elp induced by its presence.

The average annihilation rate of the positron is proportional to the overlap between the
positron wavefunction and the atomic electrons. It is given by [21]

$$\langle\lambda\rangle = \rho\int d\mathbf{r}\, f(r)\, g(r) \tag{7}$$

in appropriate units. For the pick-off decay rate of ortho-positronium (7) must be altered to
account for the penetration of the positron into the electron cloud of a fluid atom.[11] We note
in passing that the momentum distribution of the annihilating photons can also be expressed
in terms of a path integral, but here the complete off-diagonal density matrix is required and
the chain is no longer closed.[21]

APPLICATION TO E⁺ AND PS IN XENON

Our first application of PIMC was to the e+-Xenon system.[10] Xenon was chosen
because experimental measurements of the annihilation rate exhibited an anomalously large,
nonlinear, increase with density on two isotherms (300K and 340K) above the critical
temperature (289K).[22] A Lennard-Jones (12, 6) potential was chosen for u and an effective
e+-atom interaction potential w(r - R) was constructed by smoothing the sharp cut-off in the
potential Nakanishi and Schrader produced from fits to e+-Xe scattering data.[21] In units
where the Lennard-Jones length parameter $\sigma=1$, simulations were carried out at $\rho=0.01$,
0.05, 0.15, and 0.5 (in these units the critical density is about 0.35) at T=340K. They
showed that the positron was highly localized (small R_S) in a large cluster (strongly peaked
g), or droplet, consisting of about 35 Xenon atoms, at the two intermediate densities and, to
within a constant factor, reproduced the shape of the experimental plot of $\langle\lambda\rangle$ versus ρ. It
was surprising that the positron was less localized at the highest density considered, and that
the strongest localization occurred well below the critical density. This may be attributed to
the formation of long percolation channels of reduced potential in the dense fluid within
which the elp can propagate.

In contrast with e+, analytic potentials have not been developed for the Ps-Xe
interaction potential, and Ps-Xe scattering data is not available. Since the Fermionic repulsion
of the electron in Ps dominates the interaction, we simply used a hard sphere potential with a
radius of $a_0+\sigma/2$, corresponding to the geometric separation of the two atoms.[11] Simulations
were carried out on two isotherms (T=300K, and T=340K) at scaled densities $\rho=0.017$,
0.088, 0.17 and 0.35 (the critical density). The infinitely repulsive interaction caused a very
low acceptance rate in the simulations and forced us to develop a new sampling
algorithm.[11,20]

Consistent with the experimental data, the model predictions here were much less
dramatic than for e+. As determined from studies of R_S and g, the degree of localization and
the size of the cavity carved out by the positronium atom gradually increased with density on
each isotherm. The effect was slightly stronger at the lower temperature. While the
experimental plots of the pick-off decay rate versus density show a characteristically flat
plateau in the critical region, the predicted reduction in slope was smaller.[24]

In addition to determining the average annihilation rates at a specified density and
temperature, PIMC was used to simultaneously generate the complete decay rate equilibrium

distribution.[12] This is not possible with the earlier models since they are mean field theories and, for the first time, provides the opportunity to answer questions concerning the constancy of the elp's environment. This information is quantified by the relative dispersion, $\zeta_\lambda = \sigma_\lambda / \langle \lambda \rangle$, where σ_λ^2 is the variance of the decay rate calculated for each distribution. For o-Ps in Xenon at 340K, $0.83 > \zeta_\lambda > 0.44$ for $0.017 < \rho < 0.35$.[11] For e+ in Xenon at T= 340 K, $1.2 > \zeta_\lambda > 0.18$ for $0.01 < \rho < 0.5$.[10] Thus the simulations clearly demonstrate that, in each case, the local environment of the elp fluctuates strongly. These large fluctutations can only be reconciled with the sharp values of λ reported by experimentalists if the fluctuations vary rapidly on the timescale of a positron lifetime. Crude estimates suggest that this criteria is met in the case of the pick-off decay of o-Ps, but is only marginally satisfied for positrons.

RISM-POLARON THEORY

Great efforts have been put forth for developing an analytical theory to treat the localization of an excess light particle in a disordered medium, such as a fluid. However, most existing theories such as DFT employ some type of mean field theory, and do not consider the fluid density fluctuations, which are believed to have a crucial influence on the quantum states of light particles such as excess electrons, positrons and positronium atoms. The RISM (reference interaction site model) -polaron theory developed by Chandler and coworkers provides a new analytic approach for treating these systems.[13] In this theory, the light particle is represented by the path integral, and therefore the quantum fluctuations are naturally included. Although the theory inherits a weak mean field approximation, the fluid density fluctuations are partially included as well. The RISM-polaron theory has been successfully applied to a few systems, such as an electron solvated in a molten salt,[25] a hydrated electron,[26] and a thermalized positronium.[27]

The main equation in the theory[13]

$$\rho h(r) = \int dr' \int dr'' \, \omega(|r - r'|) \, c(|r' - r''|) \, \chi(|r''|) \tag{8}$$

provides the connection between the particle-fluid direct correlation function $c(r)$ and the particle-fluid pair correlation function $h(r)$. In equation (8), the solvent density-density correlation function $\chi(|r-r'|) = \langle \delta \rho(r) \, \delta \rho(r') \rangle$ is an input to the theory and represents the structural influence of the fluid on the state of the light particle. $\omega(r)$ is the zero frequency component of the polymer probability density function $\omega(r,\tau)$

$$\omega(r) = (\beta \hbar)^{-1} \int_0^{\beta \hbar} d\tau \, \omega(r, \tau) \tag{9a}$$

where

$$\omega(r, r'; i, j) \equiv \omega(|r - r'|; i\text{-}j) \rightarrow \omega(|r - r'|; \tau - \tau') \quad (p \rightarrow \infty) \tag{9b}$$

represents the probability density for finding polymer sites i and j near r and r', respectively, in the limit of a continuous chain, and is approximated within the context of Feynman's polaron theory.[16]

To obtain a self-consistent solution for Eq. (8), we need a closure between $h(r)$ and $g(r)$. The closure most appropriate to the RISM-polaron theory has not been determined, and could be system-dependent, as in the case of the equilibrium theory of fluids. The HNC (Hypernetted chain) closure[28]

$$c(r) = \exp\{-\beta w(r) + [h(r) - c(r)]\} - [h(r) - c(r)] - 1, \tag{10}$$

which is widely used for fluids with long range, attractive, inter-atomic potentials, was employed as a closure for the system of an electron solvated in a molten salt, where the particle-solvent interaction consists of the coulomb potential. For both e^+ and e^-, the elp-atom potentials are characterized by a strongly repulsive core and a long range attractive tail, so it would seem that the best choice at present for these systems is the HNC closure, which we adopted.

To complete the calculations we also need $\chi(r)$ for fluid xenon as an input. We have selected the solutions of the reference-hypernetted-chain (RHNC) equation of simple classical fluids due to Lado et. al. since the difference between them and the results of simulation studies have been shown to be comparable to the uncertainties in the simulation data.[29]

Our calculations, and their comparison with PIMC simulation results, show that the RISM-polaron theory is generally applicable to systems such as an excess electron or a positron in xenon.[14,15] Although the theory underestimates the localization of the light particle, indicated by polymer compression in the path integral representation, it predicts a good particle-fluid radial distribution function, particularly at densities above the liquid-vapor critical value. Thus, from (7), we see that it should be especially useful for computing the positron annihilation rate.

CONCLUSIONS

Coupled with modern computer hardware, PIMC has provided us with a powerful instrument for investigating light particles in thermal equilibrium. It has clearly demonstrated that the source of the anomalous, nonlinear positron annihilation rate in xenon at low to moderate density is the formation of large elp-atom clusters, and the mechanism for enhanced electron or positron mobility above the critical density is the formation of extended "percolation" channels of reduced potential energy in which the elp can propagate. With it, we have learned that the correlation between the degree of localization and the local deformation of the fluid is much weaker than the predictions of DFT would suggest, and that local fluctuations are much larger.

Although PIMC is numerically accurate, it is also a very time-consuming method, and it is all but impossible to use it to cover the complete range of the parameters needed for describing the elp-fluid system. On the other hand, since its predictions are qualitatively similar to those of PIMC calculations, the RISM-polaron theory has demonstrated its ability for describing an excess electron and a positron in xenon. It has the advantage of allowing computations which cover a much larger range of parameters using present hardware. Moreover, in contrast with DFT, it includes the effects of fluctuations, and its solutions are not discontinuous functions of the thermodynamic parameters. For the case of a purely repulsive potential, RISM-polaron theory works well over a wide density range.[27] However, our computations have shown that it fails to reproduce the clustering revealed by PIMC below the critical density for a strongly polarizable fluid, such as xenon.[14,15] It is hoped that our investigations will provide motivation for its improvement.

Acknowledgements

The authors are grateful for the support of the Welch Foundation, the Pittsburgh Supercomputing Center, and the Research Foundation of Texas Christian University.

REFERENCES

1. N. Gee and G. R. Freeman, Electron transport in dense gases: Limitations on Ioffe-Regel and Mott criteria, *Can. J. Chem.* **64**, 1810 (1986).
2. I. T. Iakubov and A. G. Khrapak, Self-trapped states of positrons and positronium in dense gases and liquids, *Prog. Phys.* **45**, 697 (1982).
3. A. T. Stewart, C. V. Briscoe, and J. J. Steinbacher, Positron annihilation in simple condensed gases, *Can. J. Phys.* **68**, 1362 (1990).
4. M. Tuomisaari, K. Rytsola, and P. Hautojarvi, *Positron and positronium localization in gases, in Positron Annihilation Studies of Fluids*, ed. S. Sharma (World Scientific, Singapore, 1988), p. 444 .
5. J. Hernandez, Electron self-trapping in liquids and dense gases, *Rev. Mod. Phys.* **63**, 675, (1991).
6. B. N. Miller and T. L. Reese, Self-trapping of a light particle in a dense fluid: A mesoscopic model, *Phys. Rev. A* **39**, 4735 (1989)
7. T. Reese and B. N. Miller, Self-trapping of a light particle in a dense fluid: Application of scaled density-functional theory to the decay of orthopositronium, *Phys. Rev. A* **42**, 6068 (1990).
8. D. F. Coker, B. J. Berne and D. Thirumalai, Path integral Monte Carlo studies of the behavior of excess electrons in simple fluids, *J. Chem. Phys.* **86**, 5689 (1987).
9. R. Kalia, P. Vashishta, and S. W. deLeeuw, Quantum molecular dynamics study of electron transport in an external filed, *J. Chem. Phys.* **90**, 6802 (1989).
10. G. A. Worrell and Bruce N. Miller, Positron annihilation in xenon: The path integral approach, *Phys. Rev. A* **46**, 3380 (1992).
11. T. Reese and B. N. Miller, Positronium in xenon: The path integral approach, *Phys. Rev. E,* **47**, 2581 (1993).
12. B. N. Miller, T. Reese, and G. Worrell, Positron lifetime distributions in fluids, *Phys. Rev. E* **47**, 4083 (1993).
13. D. Chandler, Y. Singh, and D. M. Richardson, Excess electrons in simple fluids. I. General equilibrium theory for classical hard sphere aolvents, *J. Chem. Phys.* **81**, 1075 (1984); A. L. Nichols III, D. Chandler, Y. Singh and D. M. Richardson, Excess electrons in simple fluids. II. Numerical results for the hard aphere solvent, ibid **81**, 5109 (1984).
14. J. Chen and B. N. Miller, Positron localization in xenon: RISM-polaron theory, *Phys. Rev. E*, in press.
15. J. Chen and B. N. Miller, Excess electrons in xenon: RISM-polaron theory, submitted to J. Chem. Phys.
16. R. P. Feynman, *Statistical Mechanics* (Benjamin; Reading, Mass.; 1979).
17. R. P. Feynman and A. R. Hibbs, *Quantum Mechanics and Path Integrals* (McGraw-Hill, New York, 1965).
18. E. L. Pollock and D. M. Ceperley, Simulation of quantum many-body systems by Path-integral methods, *Phys. Rev. B* **30**, 2555 (1984); D. M. Ceperley and E. L. Pollock, Path-integral computation of the properties of Liquid ^{4}He, Phys. Rev. Lett. **56**, 351 (1986).
19. M. Sprik, M. L. Klein and D. Chandler, Staging: A sampling technique for the Monte Carlo evaluation of path interals, *Phys. Rev. B* **31**, 4234 (1985); Computer

simulation of a quantun particle in a quenched disordered system: Direct observation of Lifshitz traps, *Phys. Rev. B* **32**, 545 (1984).

20. T. Reese and B. N. Miller, The modified staging algorithm: a convergent technique for valuating path integrals, submitted to J. Comp. Phys.

21. B. N. Miller and Y. Fan, Localization in fluids: A comparision of competing theories and their application to positron annihilation, *Phys. Rev. A* **42**, 2228 (1990).

22. M. Tuomisaari and K. Rytsola, Positron annihilation in xenon. ref. 4, p. 77.

23. H. Nakanishi and D. Schrader, Polarization potentials for positron- and electron-atom systems, *Phys. Rev. A* **34**, 1810 (1986).

24. M. Tuomisaari, K. Rytsola, and P. Hautojarvi, Localized state of positron in argon, *Phys. Lett.* **112A**, 279 (1985); M. Tuomisaari, K. Rytsola, R. M. Nieminen, and P. Hautojarvi, Localization of ortho-positronium in argon, *J. Phys. B* **19**, 2667 (1986).

25. G. Malescio and M. Parinello, Polaron theory of electron solvated in molten salts, *Phys. Rev. A* **35**, 897 (1987).

26. D. Laria, D. Wu and D. Chandler, Reference interaction site model polaron theory of the hydrated electron, *J. Chem. Phys.* **95**, 4444 (1991).

27. J. Chen and B. N. Miller, Positronium in xenon: RISM-polaron theory. unpublished.

28. J. P. Hansen and I. R. McDonald, *Theory of Simple Liquids* (Academic Press, London, 1986).

29. F. Lado, S. M. Foiles and N. W. Ashcroft, Solutions of the reference-hypernetted-chain equation with minimized free energy, *Phys. Rev. A* **28**, 2374 (1983).

SECTION III. ELEMENTARY PROCESSES INDUCED IN CLUSTERS BY ELECTRONS AND PHOTONS

CLUSTERS: AN INTRODUCTION

Eugen Illenberger

Institut für Physikalische und Theoretische Chemie
Freie Universität Berlin
Takustrasse 3, D-14195 Berlin, Germany

Physicists and chemists are used to label quite different things with the term "cluster". Therefore, to avoid confusion, we will in the context of this Volume assign a group (or an accumulation) of atoms or molecules in the gas phase, held together by weak intermolecular forces, as a cluster. In that sense, clusters represent *the* link between the gaseous and condensed phase. It is well known since the work of van der Waals[1] that the existence of the condensed phase stems from the forces between molecules. Therefore, these forces are commonly referred to as van der Waals forces and, consequently, an accumulation of atoms or molecules is labelled as a van der Waals cluster. Although at the time of van der Waals the origins of such weak and short range intermolecular forces were not understood, the fundamental relation between the *macroscopic properties of matter* and the *forces between the constituent molecules* was already evident.

In the past, science has focused on describing and understanding the properties of the two separate forms of matter, gaseous molecules and bulk liquids or solids. However, within the last 15 years or so there has been an enormous increase in experimental and theoretical work concerning the properties of clusters. New experimental techniques became available allowing the preparation of clusters in the gas phase under fairly controlled conditions. This provides ways and means to explore the "no man's land" between the gaseous and condensed phase. References 2-6 represent just a somehow arbitrary selection of some of the many books, reviews and articles which have appeared. In the meantime, chapters on clusters can already be found in textbooks.[6]

Clusters exhibit unique physical and chemical phenomena, e.g.,

- size dependence of structural, electronic, thermodynamic or chemical properties,
- evolution of condensed matter properties or macroscopic surface characteristics with increasing cluster size,
- large surface/volume ratio.

The large surface to volume ratio can be rationalized from elementary considerations. For the simple case of spherical noble gas clusters,[6] the number of atoms at the surface (N_s) to the total number of atoms (N) can be approximated by the equation $N_s/N = 4\,N^{-1/3}$. For a cluster containing 100 atoms we have 88% of the atoms at the surface, this drops to 8% in a 10,000 atom cluster. In a macroscopic ensemble ($N \approx 10^{23}$), on the other hand, the ratio becomes $N_s/N \approx 10^{-7}$.

In relation to these properties fundamental questions such as the following arise:

Linking the Gaseous and Condensed Phases of Matter
Edited by L.G. Christophorou *et al*, Plenum Press, New York, 1994

- How many metal atoms are necessary to evolve an electric conduction band?
- What is the minimum of NH_3 molecules to bind an excess electron?
- How small is the number of atoms or molecules to support the bulk crystal structure?

These and many other questions have extensively been discussed in the literature within the last decade. In the case of metal atoms like Hg, e.g., the transition from van der Waals binding over covalent to metallic binding as the cluster size increases has beeen followed.[7,8]

Nowadays, free clusters are often generated in supersonic molecular beams making the isolated cluster amenable to microscopic studies. The clusters are usually formed upon adiabatic expansion of the gas under high pressure (several bars) into vacuum. The expansion converts the random motion into directed mass flow. This results in a very low (translational) temperature (sometimes less than 1 K) in the molecular jet leading to cluster formation.

Most of the articles in this chapter deal with electron and photon induced processes in *neutral clusters*. Experiments with neutral clusters suffer from two shortcomings which should be mentioned here:

1. A cluster beam formed by supersonic expansion always contains a *distribution* of clusters of different size. The average size can be controlled to some extent by certain experimental parameters (pressure in the source, temperature, size and shape of the nozzle, etc.). However, by measuring a product from a reaction in a cluster, one cannot *a priori* conclude on the *exact* size of the initial cluster: since the atoms or molecules are held together by weak van der Waals forces, a reaction in the cluster can easily be accompanied by a concomitant evaporation of cluster constituents.

2. Problems in obtaining *structural* information for neutral clusters.

The first shortcoming can principally be resolved by subjecting the neutral cluster beam to a scattering process with a rare gas beam as invented by Buck et al.[9] Due to momentum transfer the clusters in the beam are dispersed according to their size, i.e., light clusters scatter into large angles and *vice versa*. This technique, however, is restricted to small clusters. It has been used for size selected experiments in clusters like[6] $(C_2H_4)_n$, $(C_6H_6)_n$ and $(SF_6)_n$ up to $n \approx 6$.

The access to *structural* information on clusters remains a problem. Although high level *ab initio* calculations yield reliable information on small elemental clusters like Na_n, Li_n, etc.[11], these methods are generally not adequate to treat van der Waals clusters of organic molecules. This stems from the generally weaker intermolecular forces (as compared to the metal clusters) and hence small energy differences between the different structures. This, in turn, may indicate that many cluster experiments are just faced with a mixture of (energetically nearly degenerate) isomers.

The articles of this chapter can only highlight some aspects in the area of present day cluster research. Yet they do give some flavour of the fascinating problems and questions which have emerged in this relatively new field.

T.D. Märk introduces the basic processes when electrons collide with clusters, i.e., the mechanisms of how cluster ions are generated and how they decay. This concerns the modes in which the excess energy is stored and the time scales for the associated decay mechansims. Particularly interesting are the ionization and electron attachment experiments on fullerenes (although these molecules are not clusters according to our definition introduced above). M.A. Johnson tracks the exciting problem of excess electron localization in charged clusters of the form $(H_2O)_n^-$ and the exchange of energy between the excess electron and the vibrational levels of the water cluster. Due to the excess charge, it is in this case possible to separate the clusters according to their size and perform photoabsorption and photodetachment experiments on *size selected cluster anions*. J.C. Miller uses multiphoton ionization techniques on NO-rare gas van der Waals clusters to obtain information on electronic and vibrational properties, the dissociation dynamics and "magic numbers" of the corresponding ions. E. Rühl describes how core level excitation with synchrotron radiation can be used to bridge the gap between the isolated gas phase molelcule and the condensed phase. Remarkable progress has been achieved by extending EXAFS studies and studies on the fragmentation of multiply charged cluster cations from noble gas clusters to benzene and methanol clusters. C. Chauvet studies the photoreaction of NO_2 in the NO_2-C_2H_4 van der Waals complex by means of pump and probe experiments and compares the reaction pathways with the isolated NO_2 molecule. Using molecular dynamics simulation, S.R. Berry examines phase transitions in clusters. It appears that even small clusters containing as few as seven atoms exhibit distinct solid-like and liquid-like forms. In a second contribution the relation between the contours of potential energy landscapes and the dynamics of the system is examined. Finally, in an article appearing in chapter 1, E. Illenberger describes how the mechanisms and dynamics of anion formation by electron capture change when proceeding from isolated molecules over clusters to condensed molecules.

References

1.   J.D. van der Waals, Doctoral Dissertation, Leiden (1873).

2.   J. Jortner, A. Pullman and B. Pullman (Eds.), "Large Finite Systems", Reidel, Dordrecht (1987).

3.   G. Benedek, T.P. Martin and G. Paccioni (Eds.), "Elemental and Molecular Clusters", Springer, Berlin (1988).

4.   Reactions in and with Clusters, Special Issue, *Ber. Bunsenges. Phys. Chem.* 96, No 9 (1992).

5.   J. Jortner, Cluster Size Effects, *Z. Phys. D* 24:247 (1992).

6.   H. Haberland, Cluster, in "Vielteilchensysteme", Bergmann-Schäfer, Lehrbuch der Experimentalphysik, Band 5, de Gruyter, Berlin (1992).

7.   H. Haberland, H. Kornmeier, H. Langosch, M. Oschwald and G. Tanner, Experimental Study of the Transition from van der Waals over Covalent to Metallic Bonding in Mercury Clusters, *J. Chem. Soc. Faraday Trans.* 86:2437 (1990).

8.   K. Rademann, Photoionization Mass Spectrometry and Valence Photoelectron-Photoion Coincidence Spectroscopy of Isolated Clusters in a Molecular Beam, *Ber. Bunsenges. Phys. Chem.* 93:653 (1989).

9.   U. Buck and H. Meyer, Scattering Analysis of Cluster Beams: Formation and Fragmentation of Small Ar_n Clusters, *Phys. Rev. Lett.* 52:109 (1984).

10.   F. Huisken and M. Stemmler, Vibrational Predissociation of Size-Selected SF_6 Clusters, *Chem. Phys.* 132:351 (1989).

11.   V. Bonacic-Koutecky, P. Fantucci and J. Koutecky, Quantum Chemistry of Small Clusters of Elements of Groups Ia, Ib, and IIa: Fundamental Concepts, Predictions and Interpretation of Experiments, *Chem. Rev.* 91:1035 (1991).

MECHANISMS AND KINETICS OF ELECTRON IMPACT IONIZATION OF ATOMS, MOLECULES AND CLUSTERS

Tilmann D. Märk

Institut für Ionenphysik
Leopold Franzens Universität
A 6020 Innsbruck, Austria

INTRODUCTION

Whereas inelastic interaction of free electrons with gas phase atoms or molecules results in changes of the electronic (ionization) and/or nuclear motion (vibrational and rotational excitation), respectively, interaction of electrons with atomic or molecular clusters (van der Waals cluster) may comprise multiple inelastic electron collisions and may involve subsequent intra- and intermolecular reactions within the cluster. These sequence of events leads to the production of excited cluster ions (different in composition and stoichiometry from the neutral precursors) involving the deposition of excess energy into various degrees of freedom. Energy and charge exchange between different sites in the cluster may in turn cause additional spontaneous reactions not present in gas phase atoms and molecules. These reactions on time scales ranging from a few vibrational oscillations up to the millisecond time regime will strongly contribute to the fragmentation pattern observed.

After a short introduction concerning ionization mechanisms and processes of atoms and molecules[1-3], the different production mechanisms of cluster cations[4,5] (and anions[6]) including recently discovered (non-statistical) energy storage modes and concomitant decay mechanisms of ionized van der Waals (vdW)[7-9] and fullerene clusters[10,11] will be discussed with particular emphasis on the transition from few-body to many-body impact dynamics. This will be followed by a short review on the experimental and theoretical aspects of the kinetics of electron impact ionization[2,12] including a discussion of the change in ionization cross section as a function of cluster size thus elucidating the transition from isolated-particle behavior to condensed-phase property.

IONIZATION MECHANISMS AND PROCESSES

Ionization of Atoms and Molecules

Electrons accelerated through a potential of several tens of volts have a de Broglie wavelength of ~ 0.1 nm. This wavelength and the molecular dimensions are similar and

Linking the Gaseous and Condensed Phases of Matter
Edited by L.G. Christophorou *et al*, Plenum Press, New York, 1994

mutual quantum effects occur, i.e. a bound electron may be promoted from a lower to a higher orbital (excitation) or-if the electron energy exceeds a critical value (the ionization energy IE or appearance energy AE) - an electron may be ejected from the target, thus producing a positive ion (cation). Conversely, direct attachment of the incident electron to an atomic target to give a stable anion is less likely. The reason for this is that the translational energy of the attaching electron and the binding energy (electron affinity EA) must be accommodated in the emerging anion. Usually, this anionic excess energy leads to either fragmentation (dissociative attachment to a molecular target) of the anion or detachment of the electron.

As the electron energy is increased, the variety and abundance of the cations produced from a specific molecular target will increase, because the electron ionization process may proceed via different reaction channels, each of which gives rise to characteristic ionized and neutral products. These include the following types of ions[13]: parent ion, fragment ion, multiply charged ion, excited ion, metastable ion, rearrangement ion, and ion pair. For the simple case of a diatomic molecule AB these reaction channels are (including excitation and dissociation):

$$
\begin{aligned}
AB + e \rightarrow AB^+ + e_s + e_e & \qquad \text{(single ionization),} & (1) \\
AB^{2+} + e_s + 2e_e & \qquad \text{(double ionization),} & (2) \\
AB^{z+} + e_s + z.e_e & \qquad \text{(multiple ionization)} & (3) \\
AB^{K+} + e_s + e_e & \qquad \text{(K-shell ionization)} & (4) \\
AB^{**} + e_s \rightarrow AB^+ + e_s + e_e & \qquad \text{(autoionization),} & (5) \\
A^+ + B + e_s + e_e & \qquad \text{(dissociative ionization),} & (6) \\
A^+ + B^- + e_s & \qquad \text{(ion pair formation),} & (7) \\
AB^* + e_s & \qquad \text{(excitation)} & (8) \\
AB^{-*} & \qquad \text{(attachment)} & (9) \\
A^- + B & \qquad \text{(dissociative attachment)} & (10) \\
A + B + e & \qquad \text{(dissociation into neutrals)} & (11) \\
A^* + B + e & \qquad \text{(dissociation into excited neutrals)} & (12)
\end{aligned}
$$

where e_s is a "scattered" electron and e_e an "ejected" electron. Other products may be obtained, especially for complex atomic targets such as polyatomic molecules or clusters.

Most of the ionization reactions summarized above (e.g., process (1) through (4), (6), and (7)) can be classified as direct ionization events, in which the ejected and the scattered electron leave the ion within 10^{-16} s. Conversely, there exist alternative ionization channels (competing with direct ionization) in which the electrons are ejected one after the other. For instance autoionization (e.g., process (5)) can be described as a two-step reaction: First, a neutral molecule (or atom) is raised to a superexcited state, which can exist for some time. Then, radiationless transition into the continuum occurs. For molecules, the upper autoionization rate is limited by the characteristic energy-storage mode frequency. In addition, if predissociation (into two neutrals) is faster than autoionization, the latter will not occur at an appreciable rate. Moreover, autoionization is a resonance process and this will complicate the respective ionization cross-section dependence (i) at lower electron energy (e.g. deviation from the threshold law[13]), but also (ii) at higher energies (e.g., see the partial ionization cross-section function for the production of Ar^+ shown in Fig. 1). The top curve in Fig. 1 shows the variation of the cross-section function obtained by summing over all possible ionization mechanisms. The two curves in the middle illustrate the variation of the strengths of the 3d and 4p autoionization processes (e.g., $Ar + e \rightarrow Ar^*(3s3p^63d) + e \rightarrow Ar^+ (3s^23p^5) + 2e$). The bottom curve shows the behavior of the direct ionization mechanisms. Quite similarly, multiply charged ions can be formed in a two-step autoionization process. First, a singly charged ion is produced by the ejection of an electron from an inner shell (inner-shell ionization process (4)). This internally ionized atom (or molecule) may then be transformed into a multiply charged ion by a series of radiationless transitions (Auger effect).

Electron impact ionization of an atom involves the transition between electronic states, whereas in the case of molecules also vibrational and rotational excitation has to be considered. The energy transferred into vibrational or rotational excitation is, however, small compared to the energy of electronic transitions. The changes in vibrational levels during ionization can be described for diatomic molecules with help of the Franck-Condon principle, whereas rotational excitation depends on the validity of the electric dipole selection rule[2,13]. The Franck-Condon principle may be summarized as follows: No (or only negligible) changes occur in the nuclear separation and velocity of relative nuclear motion during the ionization event. This is due to the large ratio of nuclear to electronic mass and the short electron interaction time. Hence, the arrival point on the final potential energy curve lies directly above the starting point on the initial potential energy curve (vertical

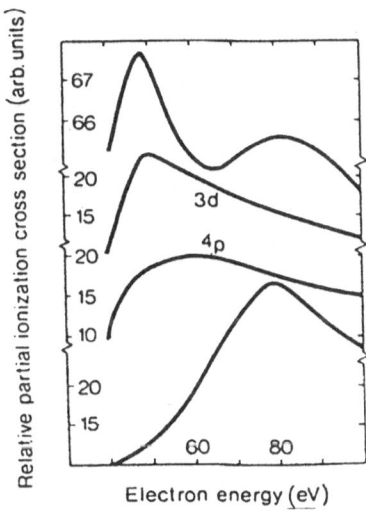

Figure 1. Partial ionization cross section curves for Ar^+ production via electron impact ionization of Ar after Crowe et al.[14]. Top curve: sum of all channels, middle curves: production via the 3d and 4p autoionization channels, bottom curve: production via direct ionization.

transition). Depending on the relative shape of the potential energy curves in a specific system, different reaction channels are possible, i.e. the final level lies (i) within the region of discrete (bound) vibrational states, (ii) not only within this region, but includes part of the continuum and (iii) entirely within the continuum of a repulsive state and all transitions lead to dissociatove ionization. If the corresponding potential energy curves of the neutral and ionized diatomic (or pseudo-diatomic) molecules are known, the resulting fragmentation ratios between atomic and diatomic ion may be derived from calculated normalized vibrational Franck Condon overlap integrals[1,2,13]. Fig. 2 shows as an example calculated and measured results for the H^+/H_2^+ ratio for H_2 as a function of electron energy from threshold up to 25 eV (i.e. including therefore only contributions from the $^2\Sigma_g^+$ state of H_2^+).

The electron impact ionization of large polyatomic compounds also proceeds without nuclear displacement, however, the few two-dimensional potential energy curves have to be

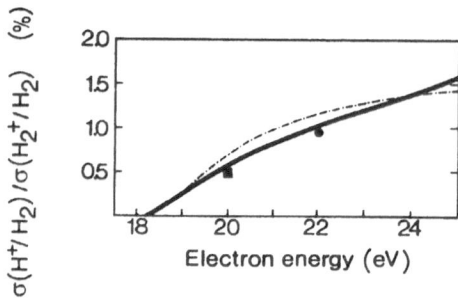

Figure 2. Partial electron impact ionization cross section ratio $\sigma(H^+/H_2)$ / $\sigma(H_2^+/H_2)$ versus electron energy. Experimental results designated • (Hipple, see Bleakney et.al[15]), ■ (Adamczyk et al.[16]), — (Crowe and Mc Conkey[17]) and theoretical results designated ----- (Browning and Fryar[18]).

replaced by a multitude of n-dimensional potential energy hypersurfaces. Moreover, subsequent dissociation is not occurring directly, but only via (delayed) unimolecular decomposition reactions after the excess energy has been transferred and distributed to the many degrees of freedom. In order to describe this ionization and fragmentation process it is possible to use statistical methods, i.e. the Quasi Equilibrium Theory (QET) developed in 1952 by Rosenstock et al.[19] or its equivalent the RRKM theory[1,20], i.e. dissociation is described as the motion along a reaction coordinate via a critical "activated" complex if enough internal energy has accumulated in this coordinate to overcome the energy barrier (activation energy E_0) of the activated complex. If the internal energy is E, the dissociation rate k(E) is given as

$$k(E) = \frac{s}{h} \cdot \frac{w(E - E_o)}{\rho(E)} = \nu \left(\frac{E - E_o}{E} \right)^{N-1} \tag{13}$$

with h Planck constant, s symmetry factor (number of identical reaction paths), w (E-E_0) number of vibrational and rotational states of the activated complex with energy $\leq E - E_o$, $\rho(E)$ density of states of the excited ion with energy E, ν frequency factor (ratio of the products of the vibrational frequencies of the excited ion and the activated complex) and N number of degree of freedoms. In the framework of QET ionization cross section ratios may be derived by (i) determining the rate of dissociation, k(E), for each fragmentation channel as a function of the internal energy E, (ii) constructing the breakdown graph from these individual dissociation rates and (iii) convoluting this breakdown graph with the total internal energy distribution (which can be obtained from photoelectron spectroscopy, photoionization studies or from the second derivative of the total electron impact ionization cross section function). The whole procedure hinges on the assumption of a linear electron ionization threshold law and needs detailed knowledge of the polyatomic compound in terms of vibrational frequencies, activation energies and structure of the transition state. Fig. 3 shows as an example calculated[21] and recently measured[22] normalized partial cross section as a function of electron energy for C_3H_8. Because of the statistic nature of the decomposition of the excited primary (parent) ion breakdown graphs, normalized fragmentation patterns and also relative partial cross sections will depend on the time elapsed since the primary ionization event. Lifshitz and coworkers[23] have used recently a trapped ion mass spectrometry technique to investigate this effect in detail for C_6H_6. Moreover, delayed (metastable) dissociation reactions are also known to occur for small molecular ions, though to a much lesser degree and via different energy storage and disposal mechanisms, i.e. forbidden electronic predissociation and tunneling through a (rotational) barrier[2,13].

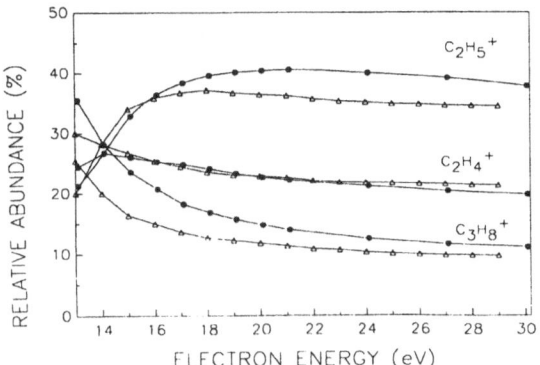

Figure 3. Some normalized partial electron impact ionization cross section functions for C_3H_8 after Vestal[21] (theoretical results) designated by full dots and after Grill et al.[22] (experimental results) designated by open triangles.

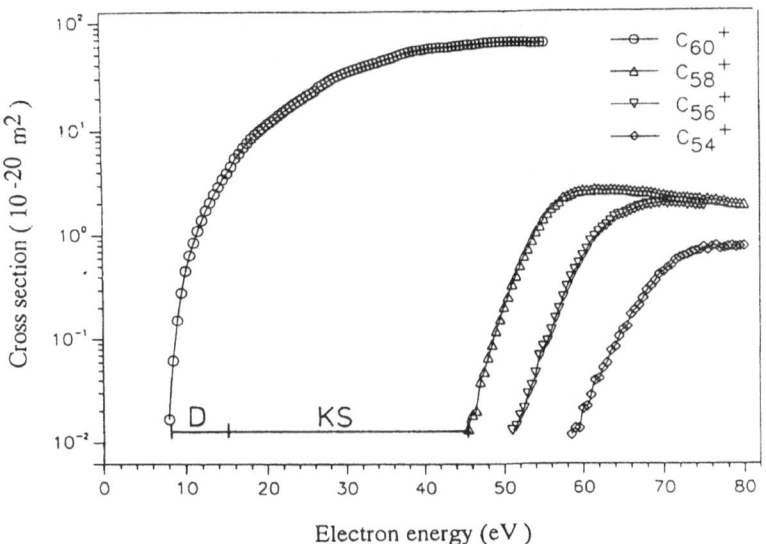

Figure 4. Partial ionization cross sections as a function of electron energy for the reactions $C_{60} + e \rightarrow C_{60}^+$, C_{58}^+, C_{56}^+ and C_{54}^+, respectively, after Foltin et al.[11] and Lezius et al.[27] (D .. Dissociation energy, KS ...Kinetic shift, see text).

Another important consequence of the statistical nature of the unimolecular decay of polyatomic ions is the occurrence of a kinetic shift[24] of the AE of fragment ions, i.e. the observed appearance energy is larger than the activation energy E_0 and depends on the time elapsed since the primary ionization event. Although the kinetic shift is negligible for many molecular ions[1] and also vdW cluster ions[4-6], for certain aromatic-hydrocarbons[1] it reaches values of up to appr. 2 eV (e.g. see results on $C_5H_5^+$ formation from benzene[25]). In contrast, Foltin et al.[11] have recently reported a huge kinetic shift (KS) of more then 34 eV for the dissociation of C_{60}^+ into C_{58}^+ (see Fig. 4). This huge kinetic shift is a direct consequence of the unique structure (truncated icosahedron) of C_{60}, where for a large number of degrees of freedom exists a high degree of symmetry with a rather large activation energy. Using two different methods (i.e. the finite heat bath model of Klots[26] (EEM) and the RRKM expression) Foltin et al. calculated the corresponding decay rates and the breakdown graph for C_{60}^+. From a comparison of the results of these calculations with the experimental breakdown graphs a dissociation energy (D) of (7.1 ± 0.4)eV could be deduced.

Ionization of van der Waals Clusters

Inelastic interaction between an electron and a vdW cluster may proceed through a variety of different ionization channels (similar to the ones discussed above), i.e. including simple one-step ionization reactions (direct ionization)

$$Ar_m + e \rightarrow Ar_m^{+*} + 2e \tag{14}$$

and also two-step processes where the initial electron interaction leads to an intermediate state followed by an autoionization reaction

$$Ar_m + e \rightarrow Ar_m^{**} + e \rightarrow Ar_m^{+*} + 2e \tag{15}$$

The primary ionization event will be followed by further intra-and intermolecular reactions, i.e. isomerization, solvation, and other ion-molecule reactions. As an example, we shall briefly consider the sequence of events for the overall reaction

$$Ar_2 + e \rightarrow Ar_2^+ + 2e \tag{16a}$$
$$\rightarrow Ar^+ + Ar + 2e \tag{16b}$$

Close to the adiabatic ionization threshold of Ar_2 electron impact can only lead to the production of a molecular[28] Rydberg state Ar_2^* due to the poor Franck Condon overlap between Ar_2 and Ar_2^+ in their ground states (see transition 2 in Fig 5). A subsequent associative ionization process, i.e. coupling of the Rydberg electron and the nuclear motion, leads to the production of Ar_2^+ in its stable ground state. If the density of Rydberg states close to the adiabatic ionization threshold is high, little vibrational energy will be deposited into the dimer ion via this two-step (autoionization) ionization process. At higher electron energies (above the adiabatic ionization threshold) this threshold mechanism will be replaced by that of direct ionization (see vertical transition 1 in Fig. 5). In this case vibrational (and/or electronic) excess energy will be deposited into the ion. Depending on the storage mode, the final outcome is either (i) a vibrational excited dimer ion or (ii) in the case of a repulsive state the production of a fragment ion Ar^+ via a dissociative ionization process. Moreover, as the stabilizing entity in a rare gas cluster ion[29-31] appears to be Ar_2^+ or Ar_3^+, similar ionization events will take place in larger clusters, resulting in turn in extensive heating of the cluster ion via vibrational excitation transfer from the intramolecular ion mode to the weak bound van der Waals modes (see below), via the reaction sequence:

$$\text{Ar}_m + e \rightarrow \text{Ar}_2{}^+(v') \cdot \text{Ar}_{m-2} + 2e \qquad\qquad (17a)$$

$$\downarrow$$

$$\text{Ar}_2{}^+(v'') \cdot \text{Ar}_{m-2}(v) \text{ with } v'' < v' \qquad\qquad (17b)$$

Besides ionization, inelastic electron interaction may also result in the formation of electronically excited states, the lowest states being the $^3P_0$ and $^3P_2$ metastable and the $^3P_1$ and $^1P_1$ resonant states (see Fig. 5). Extensive information regarding exciton trapping in liquids and solids indicate that two of these excited states (excitons) become subsequenty trapped[32] by self-localization into the bound excimer state $^1\Sigma_u{}^+$ and $^3\Sigma_u{}^+$ (see Fig. 5). An ultrafast vibrational energy flow from these vibrational excimer states $\text{Ar}_2{}^*$ (v) into the cluster leads also to extensive heating of the cluster entity[33]. Moreover, radiative decay to the ground state may lead to additional energy disposal (see below).

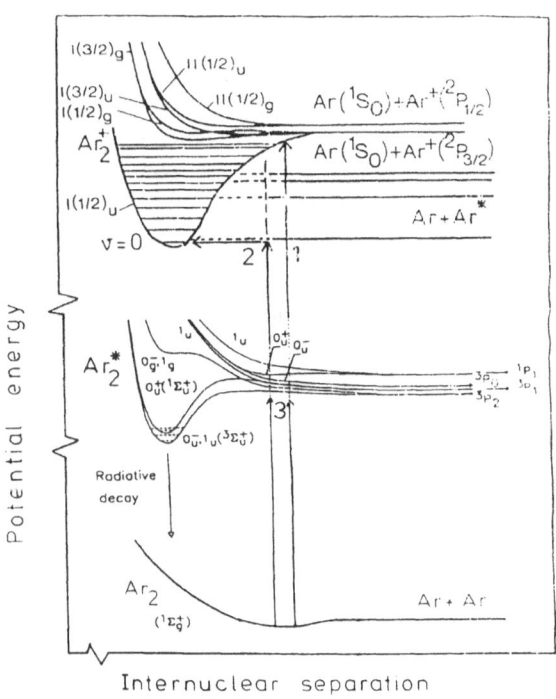

Figure 5. Schematic view of the Ar_2, $\text{Ar}_2{}^*$ and $\text{Ar}_2{}^+$ potential energy curves and possible electron excitation channels. 1: direct ionization, 2: two-step autoionization and 3: electronic excitation.

Energy degradation and relaxation and possible ion-molecule reactions following the primary (single collision) ionization event as outlined above leads to various types of predissociation processes. These spontaneous decay reactions and their properties[7-9] will be discussed in the next section of this chapter. Furthermore, besides these single collision events, electron interaction with a cluster may lead to multiple collisional cascades via interaction of the secondary electrons with cluster constituents. This yields on the one hand multiply charged cluster ions [37-39] consisting of several singly charged localized states within the cluster or on the other hand - as proved quite recently and described below-excited cluster ions consisting of a singly charged ion and an electronically excited state at different sites within the cluster[40-44].

Unimolecular reactions of cluster ion

The common unimolecular (metastable) decay reaction in the metastable time regime for cluster ions is monomer evaporation[4,45,26]. Electronic predissociation and barrier penetration have been named as possible mechanisms for small cluster ions[46-48]. A typical example (see Fig. 5) is the metastable decay of Ar_2^+ $(II(1/2)_u)$ via Ar_2^+ $(I(1/2)_g)$ into $Ar^+(^2P_{3/2})$ + Ar with a lifetime of appr. 90 μs[46,47]. A particular variant of barrier penetration is the tunneling through a centrifugal barrier, which has been proposed to account for the slow decay of small Ar cluster ions[46,48]. Conversely, vibrational predissociation is thought to be the dominant metastable dissociation mechanism for large (n > 10-20) cluster ions[49,50]. If a cluster ion is complex enough, the random motion of an activated ion on its potential hypersurface will be complicated enough to increase its lifetime into the metastable time regime. As in the case of polyatomic molecules this process has to be treated theoretically in the framework of statistical theories (RRKM, QET). The primary excitation process is assumed to have no influence on the values of k, that is, k is (i) independent of the ionization mode and (ii) slow relative to the rate of redistribution of the initial vibrational and electronic excitation energy over all degress of freedom[1,20]. A metastable dissociation reaction of an excited cluster ion P_n^{+*} via

$$P_n^{+*} \xrightarrow{k} P_{n-1}^+ + P\ (+T) \tag{18}$$

may be characterized by the rate k (which is the inverse of the mean lifetime) and the kinetic energy T released in the decay channel. Reaction (18) constitutes the simplest case of ions being excited to a specific energy E and decaying via one reaction channel. Cluster ions produced by electron impact ionization of a neutral cluster beam normally comprise, however, a broad range of energies due to the broad range of energies deposited into the ions by the primary ionization process (see above). Parent ions with different energies will have different decay rates and fragment ions produced will receive different internal energies. Moreover, in some cases, P_n^{+*} may decay by competing reactions and the produced daughter ion P_{n-1}^+ may not be stable and decay again by further decomposition reactions[51]. This situation makes analysis of experimental data very difficult and usually only averaged or relative values for the rates (and kinetic energy release) may be obtained leading to an apparent time dependence of the rate k (see Fig.6). Considering the kinetics of monomer evaporation within the frame of QET and using the concept of an evaporate ensemble, Klots[26] predicted the time dependence of an evaporating parent population, P_n^{+*}, isolated and normalized to unity at a time t_o, to be given by a non-exponential decay function

$$[P_n^{+*}](t) = 1 - \frac{C}{(\gamma')^2}\ \ln\left(\frac{t}{t_o + (t-t_o)\exp(-(\gamma')^2/C)}\right) \tag{19}$$

with C the heat capacity of the cluster (in units of the Boltzmann constant k_B) and γ' the modified Gspann parameter γ, defined by

$$\gamma' = \gamma/(1-(\gamma/C_n)^2/12)\ \ \text{with}\ \ \gamma = \frac{E_{v,n}}{k_B \cdot \sqrt{TT^*}} \tag{20}$$

and containing the energy of evaporation $E_{v,n}$ and a geometric means of the before-and-after temperatures T and T^*, respectively. The Gspann parameter is very nearly independent of

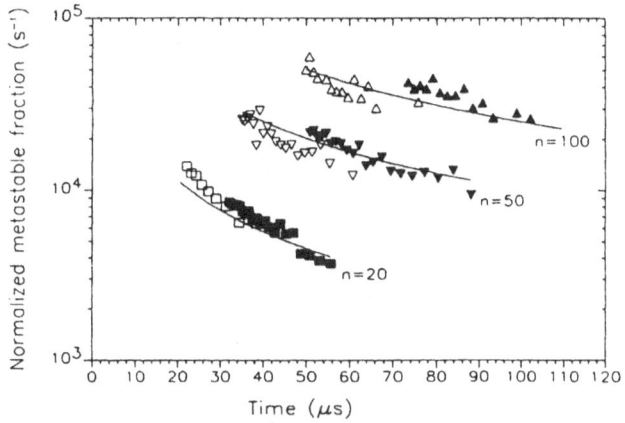

Figure 6. Comparison of measured and predicted (solid curve) apparent decay rates (metastable fractions divided by the time of the metastable windows) versus time since ionization after Ji et al.[52] for three different cluster ion sizes n = 20,50 and 100 and two different rare gases (open symbols: argon, filled symbols: krypton), respectively.

the size and composition of the cluster and on a typical laboratory time scale of tens of μs equal[26] to about 21.5. The single unknown parameter in the Klots formula is the heat capacity. Choosing plausible values for C there is very good agreement (at least for larger clusters) with the experimental findings obtained recently[52] using a double focussing mass spectrometer system in order to measure the time dependence of metastable fractions (ratio between the fragment and the parent ion currents), see Fig. 6.

Moreover, equ. (19) may be used to predict the dependence of the rate k on cluster size n. Again, there is good agreement in the general trend between existing experimental data[53] and the predict curve[26] for atomic cluster ions. It is interesting to note that for certain cluster sizes the experimental values deviate from the predicted curve beyond quoted error bars. In most cases these anomalous small or large rates (metastable fractions) coincide in a mirror-like fashion with enhanced or depleted ion abundances in the ordinary mass spectrum (Fig. 7). The reason for these anomalies are additional structural stability ("magic numbers")

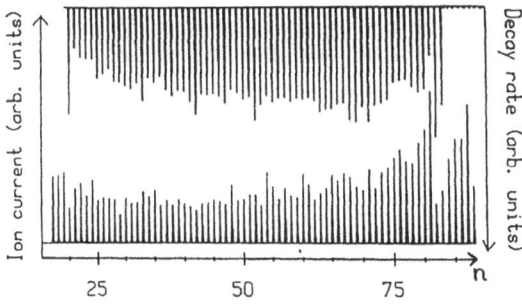

Figure 7. Section of mass spectrum for argon cluster ions and apparent metastable decay rates as a function of cluster size after Scheier and Märk.[53]

not included in the continuum based model of Klots (see also RRKM calculations reported by Stace and coworkers[50]). According to calculations and experimental evidence especially stable atomic clusters are formed at n = 13,55,147,309 etc, their corresponding structure being icosahedral shell closures[54]. Engelking proposed recently a modified RRK-QET method to determine the binding energies of monomers to the cluster ion from previously measured k and T values[55]. Similarly, Klots[26] derived recently a correction factor to account for these observed deviations thus allowing to calculate the ratio of monomer kinetic energies $E_{v,n+1}/E_{v,n}$, respectively (see also for negative cluster ions Ref. 56).

Considering also thermochemical aspects Klots[26] predicted that due to the influence of the surface energy sequential evaporations of monomers should be a more likely cooling process for excited cluster ions than single step splitting off of larger fragments. Such sequential evaporations have been shown in molecular dynamics calculations to occur after ionization of neutral vdW in the ps time regime. Moreover, multiple evaporation (i.e. the loss of more than one monomer) in the metastable time regime (μs) have been observed by numerous authors. However, only recently it became possible to determine the true nature of this process. Using both field free regions of a double focussing sector field mass spectrometer as independent observation windows it was possible to demonstrate recently that certain ions P_n^{+*} (with P = Ar[51] and N_2[57]) decay by sequential series (and not single fissioning), i.e.

$$P_n^{+*} \rightarrow P_{n-1}^{+*} \rightarrow P_{n-2}^{+*} ... P^+ \tag{21}$$

evaporating a single monomer in each of these successive decay steps. It turns out that individual apparent decay rates of these cluster ions are depending on time and on the parent ion due to the fact that each ion may exist (in an ion ensemble probed) in a variety of energy states.

Excimer-induced fragmentation of rare gas cluster ions

Recently a rather unusual metastable fragmentation channel has been discovered in argon cluster ions[7,40]. In contrast to the well known case of single monomer evaporation due to vibrational predissociation, in this alternative decay reaction (see also earlier results on a similar phenomenon for nitrogen cluster ions, which has been attributed to the radiationless vibrational relaxation in the metastable time regime of one N_2 (v=1) within the cluster ion)[57-60] the number of ejected Ar monomers rises from 2 for Ar_4^+ up to 10 for Ar_{30}^+. After studying the dependence of the metastable fractions on (i) the electron energy (Fig. 8), (ii) parent cluster size and the number of ejected monomers (Fig. 9), and (iii) the time interval between ion formation and dissociation (Fig. 10), it has been concluded that a metastable, electronically excited excimer Ar_2^* ($^3\Sigma_u^+$) localized inside the cluster ion is responsible for the observed unusual decay pattern, i.e. the radiative decay of this excimer leads to repulsion of Ar atoms in the ($^1\Sigma_g$) ground state of Ar_2 (see Fig. 5) and to subsequent disintegration of the cluster. In order to further investigate this phenomenon studies have recently been extended to the metastable fragmentation of neon cluster ions[41], in particular to the reaction

$$Ne_4^{+*} \rightarrow Ne_2^+ + product(s) \tag{22}$$

Figure 8 shows the dependence of the fragment ion current resulting from the metastable decay reaction (22) on the electron energy. For comparison, the electron energy dependence of the Ne_4^+ parent ion current is also shown. Linear extrapolation in the onset regions shows, that the appearance energy (AE) of the metastable decay is 37.2 ± 0.8 eV, which is

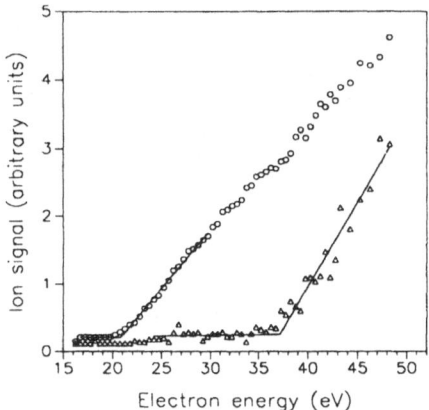

Figure 8. Ion current as a function of electron energy for the parent ion Ne_4^+ (designated o) and for the fragment ion Ne_2^+ produced via the metastable decay $Ne_4^+ \rightarrow Ne_2^+$ + products (designated Δ) after Foltin and Märk[41].

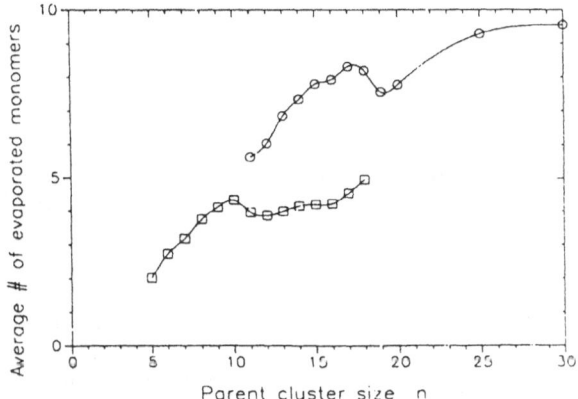

Figure 9. Average number of ejected monomers as a function of cluster size n for the fragmentation of Ar_n^{+*} ions for the two different decay channels induced by excimer decay after Märk et al.[7].

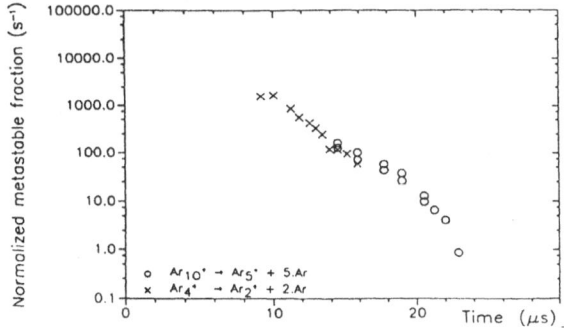

Figure 10. Normalized metastable fraction versus time since ionization for the metastable decay $Ar_{10}^+ \rightarrow Ar_5^+$ and $Ar_4^+ \rightarrow Ar_2^+$, respectively, after Märk et al.[7].

approximately 16.4 eV more than the AE of the parent ion Ne_4^+. The first electronically excited state of the Ne^+ ion is the $(2s2p^6)\ ^2S_{1/2}$ state, 26.9 eV above the ground ionic state. The lowest excited states of the neutral Ne are the $(2p^53s)\ ^3P_2,\ ^3P_0$ metastable states and the $^3P_1,\ ^1P_1$ resonant states lying 16.62 to 16.85 eV above the ground state. Taking into consideration the measured AE of the metastable decay (22), we have to conclude therefore that electronically excited states of neutral Ne are involved in the fragmentation process, populated most likely by either the scattered or ejected electron after a successful ionization process inside the same cluster.

Fluorescence studies in liquid and solid neon indicate[32] that after electron or photon impact excitation and subsequent multi-step fast nonradiative transitions, the excimer dimer Ne_2^* in the 0_u^- and $1_u(^3\Sigma_u^+)$ or $0_u^+(^1\Sigma_u^+)$ bonding state is formed (see also Fig. 1). The second excited 0_u^+ state (correlated with the $^3P_1$ atomic level) is radiatively coupled to the $0_g^+\ (^1\Sigma_g^+)$ ground electronic state and therefore is short living (lifetime $t \approx 3$ns). The first excited, nearly degenerate 0_u^- and $1_u\ (^3\Sigma_u^+)$ states, are metastable, but its radiative lifetime is considerably reduced with respect to the corresponding atomic state $^3P_2$ and lies in the microsecond region. From the above mentioned AE measurements it is possible to propose, that the metastable $Ne_2^*(^3\Sigma_u^+)$ excimer is also present in the Ne_4^{+*} ions. It's radiative decay followed by repulsion of Ne atoms in the ground $Ne_2(^1\Sigma_g^+)$ state leads to fragmentation of the cluster ion in times of the order of microseconds after ionization and excitation. In argon cluster ions this unusual decay reaction exhibits a single exponential decrease of the metastable fraction with increasing time interval between formation and fragmentation of those ions (see Fig. 10). This is evidence for a definitive, unique fragmentation rate. The corresponding fragmentation lifetime deduced[40] was appr. 1.5 µs, agreeing well with the radiative lifetime of the $Ar_2^*\ (^3\Sigma_u^+)$ excimer in solid and liquid argon, respectively[32] and in argon clusters[61]. Contrary to the argon case, in neon the metastable decay (22) was not only observed in the first field free region but also in the second field free region, i.e. appr. 17 µs after the ionization. This indicates a much longer radiative lifetime for $Ne_2^*\ (^3\Sigma_u^+)$. This longer lifetime is consistent with the fluorescence measurements, which indicate radiative lifetimes between 5 µs in solids and 11 µs in the gas phase, respectively[62].

In conclusion, this work gives clear evidence that the unusual metastable fragmentation of rare gas cluster ions is initiated by radiative decay of an excimer dimer being present on the surface (trapped surface exciton) or in the bulk (trapped bulk exciton) of the cluster ion. Recently, Hertel and coworkers[63] reported the direct observation of such an excitation process during photoionization of argon clusters using synchroton radiation and threshold photoelectron photoion coincidence TOF analysis. They observed two distinct maxima of the metastable TPEPICO spectrum interpreting the second peak as being due to the excitation of the n = 2 exciton, just 1.5 eV above the n = 1 state. This mode selective excitation of the excimer in the cluster ion (see also the photofluorescence measurements of Bondarenko et al.[64] and of Möller and coworkers[65]) constitutes a beautiful example for the violation of vibrational energy equipartitioning in a large finite system due to the existence of an "isolated" electronic state[66]. The nature of this state, in particular the interaction between the Ar_2^+ and Ar_2^* within the cluster, is still open to question. Using a simple ion/induced-dipole interaction model molecular dynamics calculations carried out recently in our laboratory give a first hint as to the dynamics of the fragmentation mechanisms occurring on a femto-to-picosecond time scale and involving caging of the two outgoing excimer related atoms. These simulations also elucidate the nature of the bimodal distribution shown in Fig. 9, which not only gives information on the structural stability of the decaying parent, but also contains information on the dynamics of the fragmentation mechanism.

Excimer mediated electron attachment to mixed clusters: Autoscavenging

Another interesting question arising from the discovery of the excimer induced metastable fragmentation process described above, concerns the mechanism of a one-electron impact process leading to the ionization of the cluster and simultaneously to the formation of the excimer in the cluster. To address this question, we carried out recently an experiment on the electron attachment to argon and xenon clusters doped with O_2[42,43] and SF_6[44] molecules, respectively, and we observed a distinct resonance in the electron attachment efficiency curve at the energy of the first excited state (s) of the cluster atoms, i.e. Ar and Xe, respectively (see Fig. 11). This resonance results from a multiple-scattering of the incoming electron within the cluster which involves an excitation of cluster atoms and subsequent trapping of the zero-energy scattered electron in the zero-energy resonance of O_2 or SF_6 thus leading to the production of stable (mixed) cluster anions. We termed this process as autoscavenging and concluded that this mechanism can be utilized in detection of excited states in clusters. Similar observations[67] have been recently reported in the case of Ar/N_2 clusters by Illenberger and coworkers[68].

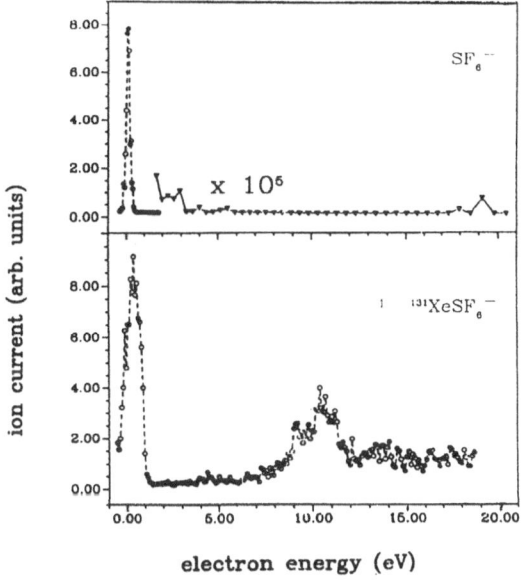

Figure 11. Ion yield of SF_6^- for electron attachment to a SF_6 cluster beam (upper part) showing only ion production via the zero energy nuclear excited Feshbach resonance mechanism and ion yield of $XeSF_6^-$ for electron attachment to mixed Xe/SF_6 clusters exhibiting additional resonance structures above an electron energy of about 8 eV corresponding to the autoscavinging process desbribed in the text after Rauth et al.[44].

Metastable decay initiated by isomerization

Although there exist many studies of neat or mixed clusters of atoms and/or small molecules (see above), much less is known about the metastable fragmentation of ionized clusters of polyatomic molecules. Because the energetics and kinetics of the prompt and metastable fragmentation of the ionized single molecule have been thoroughly studied and are well understood[1-3], simple hydrocarbons such as ethane or propane are especially suited for dissociation studies of clusters consisting of polyatomic molecules. For instance, it is well known that the parent ions $C_3H_8^+$ produced by electron impact ionization of propane gas are formed in excited states with internal excitation energies of up to 6 eV. Under non-collisional conditions this excess energy is subsequently released via unimolecular fragmentation reactions giving rise to a variety of fragment ions (see above).

In contrast, the dominant metastable dissociation process observed for stoichiometric and non-stoichiometric cluster ions of propane is single monomer evaporation[34-36]. None of the metastable decay reactions which occur for monomer ions (e.g. $C_3H_8^+ \rightarrow C_3H_7^+$ etc.[69]) occur in the cluster environment, apparently due to the fast energy transfer from intramolecular modes to intermolecular vibrations of cluster constituents. Moreover, besides the usual monomer evaporation, however, an additional metastable decay reaction

$$C_3H_7^+ (C_3H_8)_n \rightarrow C_3H_7^+ (C_3H_8)_{n-p} + p(C_3H_8). \qquad (23)$$

involving the loss of up to 4 monomers from the $C_3H_7^+(C_3H_8)_n$ ion has been observed (see Fig. 12). Taking into account the various features investigated (temporal evolution, energetics and fragmentation pattern) of this novel metastable decay process, it was concluded that this decay reaction is driven by an isomerization reaction within the cluster of n-propyl ions to isopropyl ions in the μs time regime. This reaction sequence constitutes again a new mechanism for metastable dissociation of cluster ions (i.e. isolated state driven reactions), thereby adding a new variety to the metastable decay mechanisms of this kind discussed above. The mass spectrum of propane clusters shows, besides stoichiometric parent ions, groups of ions which correspond to non-stoichiometric fragment ions known from the fragmentation patterns of the monomer molecular ion. The abundance of these ions are, however, different from the abundances of fragment ions originating from the monomer. In particular, the relative abundance of the $C_3H_7^+ (C_3H_8)_n$ is higher than that of other clustered fragment ions (e.g., clustered $C_2H_5^+$, $C_2H_4^+$, CH_3^+ ions), a situation which is completely different from the monomer molecule mass spectrum.

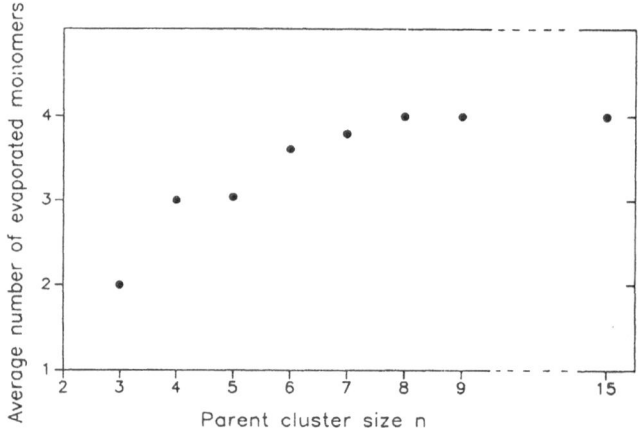

Figure 12. Average number of ejected monomers as a function of cluster size n for the fragmentation of $(C_3H_8)_n$ $C_3H_7^+$ ions induced by isomerization of the propyl ion after Foltin et al.[35].

The metastable current resulting from the decay process (23) decreases approximately exponentially with the increase of the time interval between ionization and fragmentation indicating a non-statistical decay from a well defined single state. The corresponding mean lifetime deduced from this exponential dependence is approximately 15 μs. On the other hand, the process of single (statistical) monomer evaporation shows a non-exponential decrease in accordance with the evaporative ensemble mechanism outlined by Klots[26]. The amount of energy needed to dissociate 4 monomer propane molecules from the cluster ions $C_3H_7^+(C_3H_8)_{n\geq8}$, may be estimated from the enthalpy of vaporization of liquid propane (19.4 kJ mol^{-1} at the boiling point[70]). This estimation leads to a lower limit for the energy needed to evaporate the four monomers of 0.78 eV. It is improbable that such an amount of energy could be stored for a relatively long time in the cluster ion as electronic excitation energy or as vibrational energy in specific vibrational modes of the central ion. The propane molecular ion is a prime example of the successful application of the quasiequilibrium theory of mass spectra, and the rapid (and complete) degradation of electronic excitation energy into vibrational energy of the molecular propane ion has been proven in many experiments and supported by theory[69]. In trying to explain this new metastable decay reaction several points need to be taken into account: (i) The process concerns $C_3H_7^+$ ions clustered by propane molecules. $C_3H_7^+$ can exist in several isomeric forms, i.e. n-propyl, isopropyl (sec-propyl), corner-protonated cyclopropane (c-propyl)[71]. (ii) An analogous process with the release of several monomer molecules was not detected in ionized ethane cluster decomposition (isomeric structures of $C_2H_5^+$ ions do not occur). (iii) On the other hand, an analogous process was observed in ionized butane clusters for $C_4H_7^+(C_4H_{10})_n$ ions, involving metastable decomposition with the ejection of 2-4 and 4-7 monomers (note that $C_4H_7^+$ can exist in several isomeric forms)[34,36]. This leads to the suggestion that the metastable decomposition process (23) in propane results from energy release due to the isomerization of a more energetic form to a more stable form of the propyl ion in the cluster. The energies of the propyl isomers are well known[70]: The energy difference between the n-propyl ion (ΔH_f = 209 kcal/mol) and the isopropyl ion (ΔH_f = 190 kcal/mol) is 0.82 eV; the corner-protonated cyclopropane ion has a ΔH_f = 199 kcal/mol and thus lies 0.39 eV above the energy of the isopropyl ion, the most stable of all of the isomers. The energy difference between the n-propyl and isopropyl isomer thus accounts well for the energy necessary to cause the evaporation of 4 propane molecules from the $C_3H_7^+$ $(C_3H_8)_{n\geq8}$ cluster ions. The fact, that less than 4 molecules are evaporated for smaller clusters than n = 8 can be explained, if one takes into account that the binding energy of a monomer molecule to the central ion is expected to increase with decreasing size of the cluster.

The n-propyl ion may be formed by electron ionization and subsequent fragmentation of the propane molecular ion in the cluster. Alternatively, the propyl ion can be formed by hydride-ion transfer reactions of various propane fragment ions within the cluster. The following processes are known to occur with large rate constants in the gaseous phase[72]:

$$C_2H_5^+ + C_3H_8 \rightarrow C_2H_6 + C_3H_7^+ \qquad (24)$$
$$C_2H_4^+ + C_3H_8 \rightarrow C_2H_5 + C_3H_7^+ \qquad (25)$$
$$C_2H_3^+ + C_3H_8 \rightarrow C_2H_4 + C_3H_7^+ \qquad (26)$$
$$CH_3^+ + C_3H_8 \rightarrow CH_4 + C_3H_7^+ \qquad (27)$$
$$C_3H_5^+ + C_3H_8 \rightarrow products \qquad (28)$$

Reactions (24) - (27) are exoergic both for the formation of isopropyl and n-propyl ions; reaction (28) is exoergic for the formation of the isopropyl ion, but endoergic (by 10.7 kcal/mol), if the n-propyl ion is formed. Formation of these propyl ions by hydride-ion transfer reactions within the cluster presumably accounts for the increased relative intensities of clustered $C_3H_7^+$ ions in the mass spectra of ionized clusters in comparison with the abundances of other fragment ions (removed by the reaction (24) - (27)). Also the participation of these highly fragmented ions (involving higher appearance energies) may at least qualitatively explain the tailing of the ionization efficiency curves for clustered propyl ions as shown in Fig. 13.

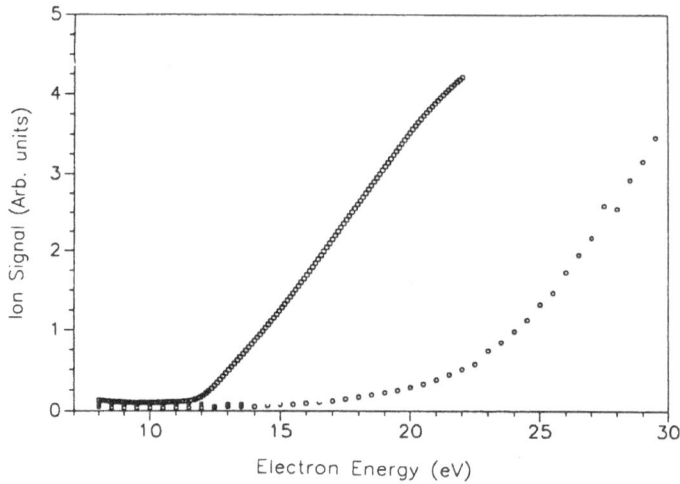

Figure 13. Ion current versus electron energy for $C_3H_7^+$ (produced via electron impact ionization of propane gas) designated □ and $(C_3H_8)_9\, C_3H_7^+$ (produced via electron impact ionization of propane clusters) designated o.

IONIZATION CROSS SECTIONS

Cross Sections of Atoms and Molecules

Despite numerous studies in this century about electron impact ionization, quantitative knowledge in terms of absolute (and accurate) cross section functions is still not satisfactory[2]. In particular, the measurement and/or calculation of partial ionization cross sections for the production of a specific ionic species are not yet as detailed or accurate as would be necessary for the many areas of applications. Even in the case of such simple target systems as the rare gases (see for example Fig. 14) up to recently large differences existed

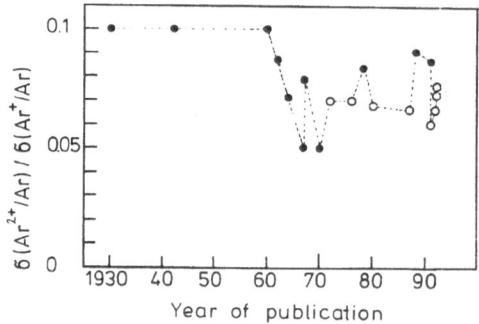

Figure 14. Measured partial ionization cross section ratios $\sigma(Ar^{2+}/Ar)$ / $\sigma(Ar^+/Ar)$ at 100 eV electron energy versus year of publication after Märk[12]. Data designated by open circles are measured with improved and controlled experimental conditions (see text).

among the various experimental (and theoretical) studies[12,73]. Nevertheless, there has been recently considerable progress in this field, experimentally as well as theoretically, i.e. the development of (i) improved experimental methods and (ii) new theoretical concepts. Because of the huge amount of potential neutral targets (more than 100 atoms, more than 50 000 molecular compounds as known in mass spectrometry, and legends of newly synthesized clusters), it is virtual impossible to determine experimentally ionization cross sections for all of the existing compounds. Therefore, theoretical treatments and concepts (analytical expressions) are necessary to bridge this gap between demand and supply. According to Inokuti et al.[74] these expressions should be based on sound physical and mathematical reasoning, rather than mere expediency.

In a usual experimental setup a parallel and monoenergetic beam of electrons will cross a target medium at rest. Assuming single collision conditions (i.e. $N_t.L. \sigma \ll 1$, with N_t target density in particles/m^3, L collision interaction length in m and σ cross section in m^2) the total positive ion current i_t produced along the interaction volume is given by

$$i_t = i_e . N_t . L . \sigma_t \tag{29}$$

with i_e electron current and σ_t the total ionization cross section. If the ions produced are analyzed with respect to their mass to charge ratio m/z.e with a mass spectrometer, the respective individual ion current produced is

$$i_i = i_e . N_t . L . z . \sigma_p \tag{30}$$

with σ_p the partial ionization cross section, that is, the cross dection for the production of a specific ion irrespective of its charge z.e. Therefore, the total ionization cross section of an atomic system is related to the partial cross sections via the weighted sum

$$\sigma_t = \Sigma z . \sigma_p \tag{31}$$

Experimental determination of σ_p or σ_t therefore involves the accurate measurement of i_e N_t, L and i_i or i_t, respectively. As will become apparent (and has been discussed in detail earlier, see Märk and Dunn[2]) the most difficult parameter to be measured accurately in the case of absolute total cross sections is the target density N_t, whereas for partial cross sections the measurement of the individual ion currents i_i is the crucial point.

One of the earliest and widliest used experimental methods to determine total ionization cross sections is the condenser plate method of Tate and Smith[75]. This is the method which has been later used very successfully by Rapp et al.[76] to produce their benchmark total ionization cross section functions for the rare gases and several small molecules. In this method a magnetically collimated electron beam is directed through a target gas of known density N_t. All ions which are produced in a well defined region are completely removed and collected. The main limitation of this method is the absolute measurement of the gas density, a difficult matter for many gases[77]. Djuric et al.[78] have recently overcome some of the difficulties by using a capacitance manometer to determine the gas pressure in their parallel plate ionization chamber. De Heer and Inokuti[79], in their definitive 1985 review on total electron impact ionization cross sections, discussed and summarized experiments and results up to this year, including also the Lozier tube[80], the summation method[81], gas filled counters[82] and crossed beam methods[83]. Excellent data on total (and partial, see below) ionization cross sections have been obtained recently by Freund and coworkers[84] using a modified crossed-beam method where an electron beam is crossed with a fast neutral beam, prepared by charge transfer neutralization of a mass selected ion beam. This approach was first used for atomic ionization cross-section measurements by Peterson and coworkers[85]; it has since been used by Ziegler et al.[86] and has been refined by

Harrison and coworkers[87]. Extensions by Freund and coworkers[88-97] have made it a powerful method (see also Becker and coworkers[98]).

The first mass spectrometric determination of partial cross section functions date back to the 1930's. Some of these studies were repeated later, however, up to recently large differences existed in both, magnitude and shape, of partial ionization cross section functions (e.g. see Fig. 14). As pointed out by many workers this was due to large discrimination effects occurring at the ion source exit and mass spectrometer slits[77,99]. Moreover, another common problem is the absolute calibration. Closely related to this is the fact that discrimination may occur at the ion detector.[100] There exist, however, several recent experimental studies (see recent reviews[2,12,73,84,101,102]) using new and sophisticated approaches in order to overcome those difficulties, including (i) improved crossed beam investigations, (ii) crossed fast (after charge transfer neutralization) atom beam techniques, (iii) improved metastable ion detection, (iv) trapped ion mass spectrometry, (v) improved extraction and transmission techniques (such as the use of cycloidal mass spectrometry, large acceptance sector field mass spectrometer, field free diffusive extraction and the penetrating-field extraction and deflection method). Some of the recent studies come very close to meet the main condition for measuring accurate cross section functions, i.e. a constant and/or complete ion source-mass spectrometer collection efficiency independent (i) of the mass to charge ratio of the ion under study, (ii) of the incident electron energy and (iii) of the initial kinetic ion energy. The main technical difficulty in measuring accurate partial cross sections is caused by the initial kinetic energy of ions resulting from the dissociative ionization process. Very recent developments (see also the fast beam technique as applied by Freund, Becker and coworkers[84,98,101]) of a pulsed electron beam/time of flight apparatus[103-106] and computer based correction techniques[22,107] have for the first time provided satisfactory results also for kinetic fragment ions (see Fig. 15). For more detail see the aforementioned reviews.

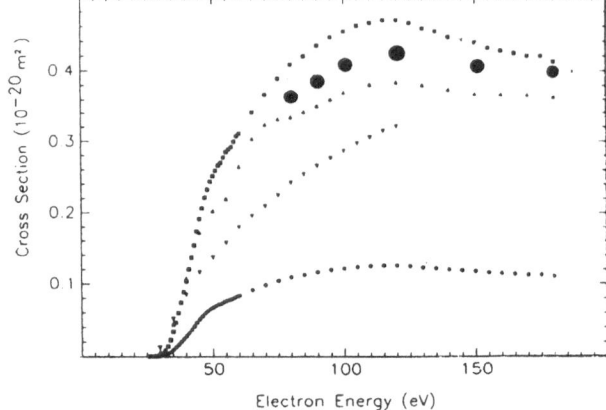

Figure 15. Absolute partial ionization cross section versus electron energy for $CF_4 + e \rightarrow CF^+$ after Märk[12]. The two uppermost data sets are from experimental determinations[103,107] taking explicity into account possible discrimination effects due to kinetic excess energy of the fragment ions CF^+.

The theoretical treatment of the kinetics of electron impact ionization (i.e. in the exit channel a full three body problem) has received a great deal of attention. Quantum mechanical (approximation) calculations are difficult, few and some of them not as accurate as necessary[108]. Therefore, other methods have been developed with the goal to obtain reasonably accurate cross sections. Three different approaches have been used, i.e. (i) empirical and semiempirical formulae, (ii) classical collision theories, and (iii) semiclassical collision theories. Theoretical methods have been reviewed several times[2,12,99,102,108,109], in particular the accuracy and reliability of the most widely used formulae. The use of classical mechanics to describe electron impact ionization has been pioneered by Thomson[110]. The original approach has been modified by several authors by adopting different initial conditions in the classical description of the problem. According to Rudge[108], none of these formulae represent a substantial improvement over the early Thomson theory. In order to obtain a logarithmic decrease in the ionization functions (as predicted correctly by the Born-Bethe approximation, see Bethe[111]) Gryzinski[112] reconsidered the problem in 1965 by assuming that the atomic electron has a continuous velocity distribution. Others have suggested means of improving the Thomson theory further by incorporating certain features of the quantal treatment into the approximation, e.g. exchange effects. Although these formulae[109] are a significant improvement, they and also other empirical or classical formulations do not give the correct magnitude and behavior of the ionization functions for certain, rather simple, atoms such as neon, nitrogen, and fluorine[113,114].

Based on a comparison between these classical formulae and the Born Bethe formula[111] (which is only accurate at higher energies due to the use of the Born approximation),

$$\sigma = \sum_n 4\pi \, a_o^2 \, \xi_n \, \frac{M_{nl}}{u} \, \ln(4c_{nl}u) \tag{32}$$

(with a_o Bohr radius, ξ_n number of equivalent electrons in the n-th subshell, $u = E/E_{in}$ (E_{in} ionization energy of an electron in the nth subshell), M_{nl} dipole matrix element and c_{nl} collisional parameter) Deutsch and Märk[114] recently suggested to replace the Bohr radius in the classical formulation by the radius of the corresponding electron subshell, r_{nl}. This step is in line (i) with a result of Bethe's calculation that the ionization cross section of an atomic electron with quantum numbers (n,l) is approximately proportional to the mean square radius $<r^2>_{nl}$ of the electron shell (n,l) and (ii) with the experimental observation of a correlation between the maximum of the atomic cross section and the sum of the mean square radii of all outer electrons[115]. Following this suggestion, this approach was successfully applied[114,116,117] to a large number of ground state and excited state atoms using the DM formula

$$\sigma = \sum_{n,l} \pi r_{nl}^2 \, \xi_{nl} \, g_{nl} \cdot f(u) \tag{33a}$$

with

$$f(u) = \frac{1}{u}\left(\frac{u-1}{u+1}\right)^{3/2}\left\{1 + \frac{2}{3}\left(1 - \frac{1}{2u}\right)\ln\left(2.7 + (u-1)^{1/2}\right)\right\} \tag{33b}$$

the energy dependent term from the classical Gryzinski approach and r_{nl}^2 the mean square radius of the nl shell, and g_{nl} weighting factors (following the original approach of Bethe[111], who calculated these "Ionisierungsfaktoren" as a function of the quantum numbers n and l using hydrogenic wave functions). Margreiter et al.[116-118] determined via a fitting procedure these generalized weighting factors g_{nl} using accurate experimental data of the rare gases and uranium as test cases. It was demonstrated that formula (33) leads in general to an improved agreement with the experimental results not only in the case of ground state atoms[114,117], but also in case of excited state atoms[116]. Recently, these authors derived relationships for the product $g_{nl} \cdot E_{in}$ yielding with its use even better agreement with the experimental data[117], allowing also the calculation of inner-shell ionization cross sections[119] and outer-shell ionization cross sections[120]. In a number of recent papers it has been demonstrated that the DM formulation may be applied successfully not only to calculate ionization cross sections of atoms, but also to the calculation of cross sections for molecules[118] (see also previous determination using the additivity rule[113,115] and the Jain-Khare[122] approach), for radicals[121] and even for cluster targets[117] (see below). In contrast to the situation for total (single) ionization cross sections just discussed above, almost no theoretical treatments are available for the determination of absolute partial cross sections (for details see Ref. 102) except those concepts based on Franck Condon overlap integrals or RRKM/QET calculations discussed above (see also Fig. 2 and 3).

Cross Sections for Clusters

In recent years, there has been a growing interest in a new category of molecules, i.e. the clusters[4-6,123-125]. Neutral atomic and/or molecular clusters (bound by weak forces such as dispersion forces) are produced in free jet nozzle expansions, and most of these experiments on van der Waals clusters use electron impact ionization in combination with mass spectrometry for the detection of these species[4]. However, very little quantitative information is known yet in terms of ionization cross sections. This is mainly due to the fact that it is not possible to produce beams of neutral clusters of known density and defined size. Therefore it is not surprising that no absolute total or partial ionization cross sections for a specific cluster size (except for dimers, see Fig. 16) have been reported up to date.

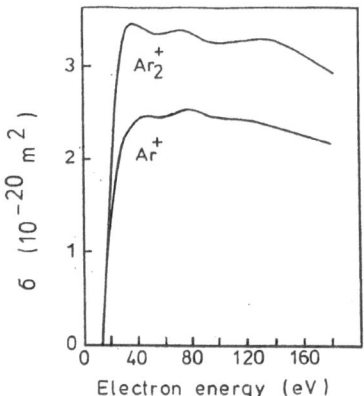

Figure 16. Partial electron impact ionization cross section for the reactions $Ar + e \rightarrow Ar^+ + 2e$ and $Ar_2 + e \rightarrow Ar_2^+ + 2e$, respectively, as a function of electron energy after Märk[4].

Conversely, total ionization cross sections have been measured for cluster distributions of H_2 and CO_2[126,127]. Fig. 17 shows as an example the total ionization cross section functions divided by the averaged number of cluster constituents ("effective" cross section) for various CO_2 cluster distributions. The position of the maximum of the cross section shifts to higher electron energies with larger averaged cluster size (in part due to the loss of energy as the electron passes through the cluster; see also below) and the magnitude of the effective cross section decreases for larger sizes. Also shown in Fig. 17 are theoretical estimates from Deutsch et al.[117] using a modified DM approach (see above).

According to theoretical considerations[128,129] supported by the limited number of experiments, the total ionization cross section of a cluster with size m may be envisioned to be the product of three terms times the total cross section of the monomer, $\sigma_{t,1}$, i.e.,

$$\sigma_{t,m} = \sigma_{t,1} \; T_1 . T_2 . T_3 \tag{34}$$

The first term T_1 describes the probability that an electron strikes the cluster. This probability depends for a vdW-cluster on the geometric cross-sectional area of the cluster and is therefore proportional to $m^{2/3}$ for a spherical (or cubic) cluster of m constituents if the density of the cluster is assumed to be independent of cluster size. The next term, T_2, corresponds to the probability that an impacting electron can cause an ionizing event within

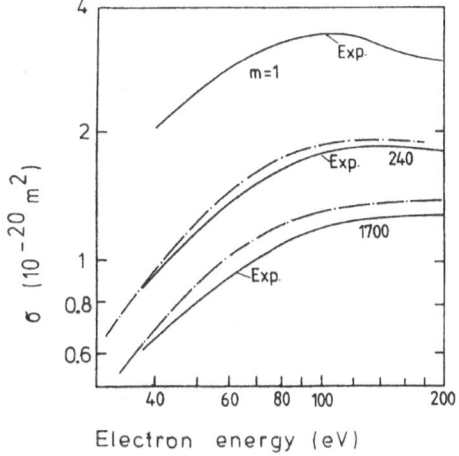

Figure 17. Effective total electron impact ionization cross section functions for CO_2 cluster distributions with an average number of 240 and 1700, respectively, after Hagena and Henkes[126] (designated Exp.). Also shown for comparison the measured monomer cross section (m=1) and calculated values using a modified DM approach reported by Deutsch et al.[117]

the cluster. This term can be expressed as $1\text{-}e^{-d/\lambda}$ where d is the mean distance traveled inside the cluster by the incident electron and λ is the mean free path for ionization within the cluster. The mean free path depends only on the energy of the incident electron, whereas d depends on the radial dimension of the cluster which is proportional to $m^{1/3}$. Thus, for a small cluster (d $\ll \lambda$), $1\text{-}e^{-d/\lambda}$ is proportional to $m^{1/3}$ and for a large cluster (d $\gg \lambda$), the term is unity. The last term, T_3, considers the probability that the ensuing secondary electron escapes the cluster. This probability is given by $1\text{-}e^{-D/\lambda}$ where D is the mean distance that a secondary electron can travel in the cluster and still have sufficient energy to escape the cluster. In this case, for a small cluster (d $\ll$ D), the probability is unity, whereas for a large cluster the probability is proportional to $m^{-1/3}$. Therefore, the total ionization probability for a small cluster (d $\ll \lambda$ or D) is directly proportional to the cluster size m. For a larger cluster (d $\gg \lambda$ or D) the probability becomes proportional to $m^{2/3}$. The dimension D is expected to be smaller than λ, so the expected criterion for this proportionality is d $\gg$ D. If the cluster is very large (d $\gg \lambda$ and D), the proportionality should fall to $m^{1/3}$. Bottiglioni et al.[128] point out that electrons of about 80 eV incident energy can be captured only by clusters exceeding 10^8 molecules (i.e. see also the dependence of the ratio for the total anion to cation yield for large H_2 and N_2 clusters on the cluster size as shown in Fig. 18). From the dependence of the total cross section on cluster size Henkes and coworkers determined the ionization shell thickness D to be about three molecular layers for CO_2 clusters[131] and five and one-half for the H_2 clusters[127]. Moreover, it is important to note that the production (and its probability as a function of electron energy and cluster size m) of multiply charged clusters (e.g. see relative effective partial cross sections measured by Gspann and Körting[130] for H_2 and N_2 cluster distributions) is also influencing the shape of the total effective cross section curve in such a way that for larger cluster sizes and larger electron energies the relative probability of charge production is increased[130,131].

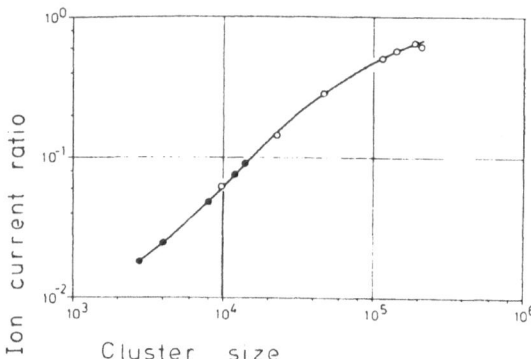

Figure 18. Ratio of anion to cation yield for nitrogen (full dots) and hydrogen (open circles) clusters as a function of cluster size after Gspann and Körting[130].

A few relative partial ionization cross section functions have been reported (the most extensive study concerns Ar clusters[38]; others are summarized in Ref.4). For the reasons given above no absolute values are available. Owing to this lack of data, the additivity rule[113,119] (despite being only valid for total cross sections) has been used occasionally to calibrate cluster ion signals detected by mass spectrometry, i.e. assuming for instance that the dimer ion signal is twice the monomer ion signal and so forth. This procedure, however, is at variance with the results presented above (Fig. 16). It neglects possible fragmentation of the neutral cluster under study and possible cascading from larger neutral clusters present in the distributions. Using a modified additivity rule and taking into account dissociative channels[99], it is possible to deduce at least for rare gas dimers absolute partial ionization cross sections (e.g. Fig. 16). Moreover, in a few cases accurate partial ionization cross section ratios have been determined for small clusters using either spectroscopic methods[132] or the method of producing size selected neutral clusters by momentum transfer via scattering[133]. As expected from the mass spectrometry of ordinary molecules[13] considerable fragmentation is occurring for some of the clusters studied[99,134,135].

ACKNOWLEDGEMENT

This paper is based on work supported by the Österreichischer Fonds zur Förderung der Wissenschaftlichen Forschung, Wien. It is a pleasure to thank all of my coworkers who have contributed to our research results described within this review.

REFERENCES

1. K. Levsen. "Fundamental Aspects of Organic Mass Spectrometry", Verlag Chemie, Weinheim (1978).
2. T.D. Märk and G.H. Dunn. "Electron Impact Ionization", Springer, Wien (1985).
3. E. Illenberger and J. Momigny. "Gaseous Molecular Ions", Steinkopff Verlag, Darmstadt (1992).
4. T.D. Märk, Cluster ions: production, detection and stability, Int. J. Mass Spectrom. Ion Proc. 79:1 (1987).
5. T.D. Märk, Metastable decay of rare gas cluster ions: mechanism and kinetics, Z. Phys. D12: 263 (1989).
6. T.D. Märk, Free electron attachment to van der Waals clusters, Int. J. Mass Spectrom. Ion Proc. 107 : 143 (1991).
7. T.D. Märk, M. Foltin, V. Grill, T. Rauth and G. Walder, Excimer and isomer mediated reactions in ionized van der Waals clusters, Ber. Bunsenges. Phys. Chem. 96 : 1125 (1992).
8. T.D. Märk, M. Foltin and P. Scheier, Spontaneous decay of ionized atomic clusters, in "Clustering Phenomena in Atoms and Nuclei", M. Brenner, L. Lönnroth, F.B. Malik eds., Springer, Heidelberg (1992).
9. T.D. Märk, Fission versus metastable decay series of rare gas cluster ions, in: "Nuclear Physics Concepts in Atomic Cluster Physics", H.O. Lutz, R. Schmidt, R. Drezler eds., Springer, Berlin (1992).
10. M. Lezius, P. Scheier and T.D. Märk, Free electron attachment to C_{60} and C_{70}, Chem. Phys. Lett. 203 : 232 (1993).
11. M. Foltin, M. Lezius, P. Scheier and T.D. Märk, On the unimolecular fragmentation of C_{60}^+ fullerene ions: the comparison of measured and calculated breakdown patterns, J. Chem. Phys. 98 : 9624 (1993).
12. T.D. Märk, Ionization by electron impact, Plasma Phys. Contr. Fusion. 34:2083 (1992)
13. T.D. Märk, Ionization of molecules by electron impact, in "Electron Molecule Interactions and their Applications", L.G. Christophorou, ed., Academic Press, New York (1984).
14. A. Crowe, J.A. Preston and J.W. Mc Conkey, Ionization of argon by electron impact. J. Chem. Phys. 57:1620 (1972).
15. W. Bleakney, E.U. Condon and L.G. Smith, Ionization and dissociation of molecules by electron impact, J. Phys. Chem. 41:197 (1937).
16. B. Adamczyk, A.J.H. Boerboom, B.L. Schram and J. Kistemaker, Partial ionization cross sections of He, Ne, H_2 and CH_4 for electrons from 20 to 500 eV, J. Chem. Phys. 44:4640 (1966).
17. A. Crowe and J.W. Mc Conkey, Dissoziative ionization by electron impact I. Protons from H_2, J. Phys. B6:2088 (1973).
18. R. Browning and J. Fryar, Dissoziative photoionization of H_2 and D_2 through the $1s\sigma_g$ ionic state, J. Phys. B6:364 (1973).
19. H.M. Rosenstock, M.B. Wallenstein, A.L. Wahrhaftig and H. Eyring, Absolute rate theory for isolated systems and the mass spectra of polyatomic molecules, Proc. Natl. Acad. Sci. 38:667 (1952).
20. W. Forst. "Theory of Unimolecular Reactions". Academic Press, New York (1973).
21. M.L. Vestal, Theoretical studies on the unimolecular reactions of polyatomic molecule ions.I. Propane, J. Chem. Phys. 43:1356 (1965).
22. V. Grill, G. Walder, D. Margreiter, T. Rauth, H.U. Poll, P. Scheier and T.D. Märk, Absolute partial and total electron impact ionization cross sections for C_3H_8 from threshold up to 950 eV, Z. Phys. D25:217 (1993).
23. C. Lifshitz and S. Gefen, Time dependent mass spectra and breakdown graphs. I. 1,5- hexadiyne, Int. J. Mass Spectrom. Ion Proc. 35:31 (1980).
24. W.A. Chupka, Effect of unimolecular decay kinetics on the interpretation of appearance potentials, J. Chem. Phys. 30:191 (1959).
25. S.M. Gordon and N.W. Reid, An investigation of the kinetic shift in mass spectrometry, Int. J. Mass Spectrom. Ion Phys. 18:379 (1975).
26. C.E. Klots, Evaporative cooling, J. Chem. Phys. 83:5854 (1985); Evaporation from small particles, J. Phys. Chem. 92:5864 (1988); Evaporative corrections to high-pressure ionic clustering data, Int. J. Mass Spectrom. Ion Proc. 100:457 (1990); Kinetic methods for quantifying magic, Z. Phys. D21:335 (1991).
27. M. Lezius, P. Scheier, M. Foltin, B. Dünser, T. Rauth, V.M. Akimov, W. Krätschmer and T.D. Märk, Interaction of free electrons with C_{60}: ionization and attachment reactions, Int.J. Mass Spectrom. Ion Proc. in print (1993).

28. P. Dehmer, Photoionization of Ar_2 at high resolution, J. Chem. Phys. 76:1263 (1982).

29. T.D. Märk, Production and reactions of cluster ions studied with molecular beam ionization techniques, Europhys. Conf. Abstr. 6D:29 (1982).

30. H. Haberland, A model for the processes happening in a rare-gas cluster after ionization, Surf. Science. 156:305 (1985).

31. H.U. Böhmer and S.D. Peyerimhoff, Stability and structure of singly-charged argon clusters Ar_n^+, n = 3-27. A Monte-Carlo simulation, Z. Phys. D11:239 (1989).

32. E. Morikawa, R. Reininger, P. Gürtler, V. Sailer and P. Laporte, Argon, krypton and xenon excimer luminescence: from the dilute gas to the condensed phase, J. Chem. Phys. 91:1469 (1989).

33. D. Scharf, J. Jortner and U. Landman, Excited-state dynamics of rare-gas clusters, J. Chem. Phys. 88:4273 (1988).

34. M. Foltin, V. Grill, T. Rauth and T.D. Märk, Isomer-induced metastable decay of $(C_4H_{10})_n\ C_4H_7^+$ cluster ions, Int. J. Mass Spectrom. Ion Proc. 110:R7 (1991).

35. M. Foltin, V. Grill, T. Rauth, Z. Herman and T.D. Märk, Slow metastable decay process of $(C_3H_7^+)$ $(C_3H_8)_n$ cluster ions induced by isomerization of the propyl ion, Phys. Rev. Lett. 68:2019 (1992).

36. M. Foltin, V. Grill, T. Rauth and T.D. Märk, Metastable fragmentation of $(C_4H_{10})_n\ C_4H_7^+$ cluster ions induced by delayed isomerization reactions of $C_4H_7^+$, J. Chem. Phys. 96:5213 (1992).

37. P. Scheier and T.D. Märk, Triply charged argon clusters: production and stability (appearance energy and appearance size), Chem. Phys. Lett. 136:423 (1987).

38. M. Lezius and T.D. Märk, Direct experimental evidence for the Coulomb explosion of doubly charged argon cluster ions Ar_n^{2+}, Chem. Phys. Lett. 155:496 (1989).

39. O. Echt and T.D. Märk, Multiply charged clusters, in "Clusters of Atoms and Molecules. Vol.II". H. Haberland, e.d., Springer, Heidelberg (1994).

40. M. Foltin, G. Walder, A.W. Castleman Jr. and T.D. Märk, Magic metastable decay of singly charged argon clusters, J. Chem. Phys. 94:810 (1991).

41. M. Foltin and T.D. Märk, Metastable fragmentation of Ne_4^+ cluster ion initiated by excimer decay, Chem. Phys. Lett. 180:317 (1991).

42. M. Foltin, V. Grill and T.D. Märk, The observation of unusual resonance channels in the electron attachment to mixed argon-oxygen clusters, Chem. Phys. Lett. 188:427 (1992).

43. M. Foltin, T. Rauth and T.D. Märk, SF_6^- production via excimer-mediated electron attachment to mixed rare gas/SF_6 clusters, Int. J. Mass Spectrom. Ion Proc. 116:273 (1992).

44. T. Rauth, M. Foltin and T.D. Märk, Autoscavenging of electrons in mixed van der Waals clusters: a new approach to the spectroscopy of cluster anions, J. Phys. Chem. 96:1528 (1992).

45. K. Stephan and T.D. Märk, Metastable Ar_3^+ cluster ions: temperature dependence of the decay rate, Chem. Phys. Lett. 90:51 (1982).

46. K. Stephan, A. Stamatovic and T.D. Märk, Unimolecular and collision-induced dissociation of Ar_2^+ produced by electron ionization of Ar_2, Phys. Rev. A28:3105 (1983); P. Scheier, A. Stamatovic and T.D. Märk, Dynamics and kinetics of the metastable decay series: $Ar_3^{+*} \rightarrow Ar_2^{+*} \rightarrow Ar^+$, J. Chem. Phys. 89:295 (1989).

47. K. Norwood, J.H. Guo and C.Y. Ng, A photoion-photoelectron coincidence study of Ar_n (n= 2-4), J. Chem. Phys. 90:2995 (1989).

48. E.E. Ferguson, C.R. Albertoni, R. Kuhn, Z.Y. Chen, R.G. Keesee, and A.W. Castleman Jr., The role of rotational tunneling in the metastable decay of rare gas cluster ions, J. Chem. Phys. 88:6335 (1988).

49. T.D. Märk, P. Scheier, K. Leiter, W. Ritter, K. Stephan and A. Stamatovic, Unimolecular decay of metastable Ar cluster ions. Evolution of magic numbers in Ar cluster mass spectra, Int. J. Mass Spectrom. Ion Proc. 74:281 (1986).

50. P.G. Lethbridge and A.J. Stace, Reactivity-structure correlations in ion clusters: A study of the unimolecular fragmentation patterns of argon ion clusters, Ar_n^+, for n in the range 30-200, J. Chem. Phys. 89:4062 (1988).

51. P. Scheier and T.D. Märk, Observation of sequential decay series in metastable Ar clusters: $Ar_n^{+*} \rightarrow Ar_{n-1}^{+*} \rightarrow Ar_{n-2}^+$, Phys. Rev. Lett. 59:1813 (1987).

52. Y. Ji, M. Foltin, C.H. Liao and T.D. Märk, Size and temporal dependence of the metastable decay probabilities of ionized rare gas clusters, J. Chem. Phys. 96:3624 (1992).

53. P. Scheier and T.D. Märk, Metastable decay of singly charged argon cluster ions Ar_n^{+*}, Int. J. Mass Spectrom. Ion Proc. 102:19 (1990).

54. O. Echt, A. Reyes-Flotte, M. Knapp, K. Sattler and E. Recknagel, Magic numbers in mass spectra of Xe, $C_2F_4Cl_2$ and SF_6 clusters, Ber. Bunsenges. Phys. Chem. 86:860 (1982).

55. P.C. Engelking, Determination of cluster binding energy from evaporative lifetime and average kinetic energy release: application to $(CO_2)_n^+$ and Ar_n^+ clusters, J. Chem. Phys. 87:936 (1987).

178

56. M. Lezius, T. Rauth, V. Grill, M. Foltin and T.D. Märk, Production and properties of CO_2 cluster anions, Z. Phys. D24:289 (1992).

57. P. Scheier and T.D. Märk, Quantized sequential decay series of metastable N_2 cluster ions $(N_2)_n^{+*} \rightarrow (N_2)_{n-1}^{+*} \rightarrow .. N_2^+$, Chem. Phys. Lett. 148:393 (1988).

58. T.F. Magnera, D.E. David and J. Michl, "Magic" losses of one to five N_2 molecules from metastable N_{2n}^+ clusters, Chem. Phys. Lett. 123:327 (1986).

59. T. Leisner, O. Echt, O. Kandler, X.J. Yan and E. Recknagel, Quantum effects in the decomposition of nitrogen cluster ions, Chem. Phys. Lett. 148:386 (1988).

60. G. Walder, C. Winkler and T.D. Märk, Transfer of vibrational energy in nitrogen clusters doped with O_2: new evidence for the origin of quantum effects in the metastable decay series of $(N_2)_n^{+*}$, Chem. Phys. Lett. 157:224 (1989).

61. T. Möller, private communication, 1990.

62. B. Schneider and J.S. Cohen, Ground and excited states of Ne_2 and Ne_2^*. II. Spectroscopic properties and radiative lifetimes, J. Chem. Phys. 61:3240 (1974); R. Gaethke, P. Gürtler, R. Kink, E. Roick and G. Zimmerer, Vibrational relaxation and hot luminescence of Ne_2^* centers in solid neon, Phys.stat. sol. b124:335 (1984).

63. H. Steger, J. de Vries, W. Kamke and I.V. Hertel, Metastable decay of argon clusters after photoionization at high excess energies, Z. Phys. D21:85 (1991).

64. E.A. Bondarenko, E.T. Verkhovtseva, Y.S. Doronin and A.M. Ratner, New emission continua of rare-gas custers in the VUV region, Chem. Phys. Lett. 182:637 (1991).

65. R. Müller, M. Joppien, S. Wolf and T. Möller, Fluoreszenz von ionischen Argon Clustern: Strahlender Zerfall von Zwei-Zentren Anregungen, HASYLAB Jahresbericht, Hamburg (1992).

66. C. Lifshitz, Intramolecular energy redistribution in polyatomic ions, J. Phys. Chem. 87:2304 (1983).

67. T. Jaffke, R. Hashemi, L.G. Christophorou and E. Illenberger, Mechanisms of anion formation in O_2, O_2/Ar and O_2/Ne clusters; the role of inelastic electron scattering, Z. Phys. D25:77 (1992).

68. R. Hashemi, T. Jaffke, L.G. Christophorou and E. Illenberger, The role of inelastic electron scattering by N_2 in the formation of $(O_2)_n^-$ anions in mixed O_2/N_2 clusters, J. Phys. Chem. 96:10605 (1992).

69. C. Brunnee and H. Voshage. "Massenspektrometrie". Thiemig, München (1964).

70. D.R. Lide, ed., "Handbook of Chemistry and Physics". CRC Press, Boston (1990).

71. D.J. Mc Adoo, F.W. Mc Lafferty, P.F. Bente, Ion cyclotron resonance spectroscopy in structure determination. II. Propyl ions, J. Am. Chem. Soc. 94:2027 (1972).

72. W. Lindinger, private communication, 1992.

73. T.D. Märk, Total and partial electron impact ionization and attachment cross sections of atoms, molecules and clusters (quasi-liquid): a review of experimental and theoretical methods and data for radiotherapy, IAEA-TECDOC. 506: 179 (1989).

74. M. Inokuti, M. Kimura and M.A. Dillon, Analytic representation of cross section data, Adv. Atomic Mol. Opt. Phys. 32: in print (1993).

75. J.T. Tate and P.T. Smith, The efficiencies of ionization and ionization potentials of various gases under electron impact, Phys. Rev. 39:270 (1932).

76. D. Rapp and P. Englander-Golden, Total cross sections for ionization and attachment in gases by electron impact. I. Positive ionization, J. Chem. Phys. 43:1964 (1965).

77. L.J. Kieffer and G.H. Dunn, Electron impact ionization cross-section data for atoms, atomic ions and diatomic molecules: I. Experimental data, Rev.Mod. Phys. 38:1 (1966).

78. N.L. Djuric, I.M. Cadez and M.V. Kurepa, H_2O and D_2O total ionization cross-sections by electron impact, Int. J. Mass Spectrom. Ion Proc. 83:R7 (1988); Total electron impact ionization cross sections of methanol, ethanol and n-propanol molecules, Fizika.21:339(1989); Electron impact total ionization cross-sections for methane, ethane and propane,Int. J. Mass Spectrom. Ion Proc. 108:R1(1991).

79. F.J. de Heer and M. Inokuti, Total ionization cross sections, in "Electron Impact Ionization". T.D. Märk G.H. Dunn, eds., Springer, Wien (1985).

80. W.W. Lozier, The heat of dissociation of CO and the electron affinity of O,Phys. Rev. 46:268 (1934).

81. T.D. Märk and F. Egger, Ionization of phosphine and deuterated phosphine by electron impact from threshold to 180 eV, J. Chem. Phys. 67:2629 (1977).

82. F.F. Rieke and W. Prepejchal, Ionization cross sections of gaseous atoms and molecules for high-energy electrons and positrons, Phys. Rev. A6: 1507 (1972).

83. H. Funk, Über die Ionisierung von Alkaliatomen durch langsame Elektronen, Ann. Physik. 4:149 (1930).

84. R.S. Freund, Electron impact ionization cross sections for atoms, radicals and metastables, in "Swarm Studies and Inelastic Electron Molecule Collisions", L.C. Pitchford, B.V. Mc Koy, A. Chutjian, S. Trajmar, eds, Springer, New York (1987).

85. C.J. Cook and J.R. Peterson, Direct and dissociative ionization cross sections for electrons in N_2, Phys. Rev. Lett. 9:164 (1976).

86. D.L. Ziegler, J.H. Newman, K.A. Smith and R.F. Stebbings, Double ionization of atomic oxygen by electron impact, Planet. Space Sci. 30:451 (1982).

87. E. Brook, M.F.A. Harrison and A.C.H. Smith, Measurements of the electron impact ionization cross sections of He, C, O and N atoms, J. Phys. B11:3155 (1978); R.G. Montague, M.F.A. Harrison and A.C.H. Smith, A measurement of the cross section for ionization of helium by electron impact using a fast crossed beam technique, J. Phys. B17:3295 (1984).

88. P.B. Armentrout, S.M. Torr, A. Dosi and R.S. Freund, Electron impact ionization cross section of metastable $N_2(A\Sigma_u^+)$, J. Chem. Phys. 75:2786 (1981).

89. F.A. Baiocchi, R.C. Wetzel and R.S. Freund, Electron-impact ionization and dissociative ionization of the CD_3 and CD_2 free radicals, Phys. Rev. Lett. 55:771 (1984).

90. T.R. Hayes, R.C. Wetzel and R.S. Freund, Absolute electron-impact-ionization cross-section measurements of the halogen atoms, Phys. Rev. A35:578 (1987).

91. R.C. Wetzel, F.A. Baiocchi, T.R. Hayes and R.S. Freund, Absolute cross sections for electron-impact ionization of the rare-gas atoms by the fast-neutral-beam method, Phys. Rev. A35:559 (1987).

92. T.R. Hayes, R.C. Wetzel, F.A. Baiocchi and R.S. Freund, Absolute cross sections for electron-impact ionization and dissociative ionization of the SiF free radical, J. Chem. Phys. 88:823 (1988).

93. T.R. Hayes, R.J. Shul, F.A. Baiocchi, R.C. Wetzel and R.S. Freund, Electron-impact ionization cross sections of the SiF_3 free radical, J. Chem. Phys. 89:4035 (1988).

94. R.J. Shul, T.R. Hayes, R.C. Wetzel, F.A. Baiocchi and R.S. Freund, Electron impact ionization cross sections of SiF_2, J. Chem. Phys. 89:4042 (1988).

95. R.J. Shul, R.C. Wetzel and R.S. Freund, Electron-impact-ionization cross sections of Ga and In atoms, Phys. Rev. 39:5586 (1988).

96. R.J. Shul, R.S. Freund and R.C. Wetzel, Electron-impact-ionization cross sections of GaCl, GeCl and SnCl, Phys. Rev. 41:5856 (1990).

97. R.S. Freund, R.C. Wetzel, R.J. Shul and T.R. Hayes, Cross section measurements for electron-impact ionization of atoms, Phys. Rev. A41:3575 (1990).

98. V. Tarnovsky and K. Becker, The electron-impact ionization of Ar and Kr revisited: a critical analysis of double-to-single ionization cross section ratio measurements using the fast-atom-beam technique, Z. Phys. D22:603 (1992); Absolute partial cross sections for the parent ionization of the $CF_x(x=1-3)$ free radicals by electron impact, J. Chem. Phys. 98:7868(1993).

99. T.D. Märk, Mass Spectrometric determination of partial ionization cross sections, Beitr. Plasmaphysik. 22:257 (1982).

100. J.A. Syage, Electron impact cross sections for multiple ionization of Ar: detector gain effects revealed, J. Phys. B24:L527 (1991).

101. K.H. Becker, Electron impact induced dissociative excitation and ionization of halogen containing molecules, in "Book of Invited Papers", XVIIIth ICPEAC, Aarhus (1993).

102. T.D. Märk, Y. Hatano, F. Linder and M. Hayashi, Electron collision cross sections, IAEA-TECDOC: in print (1994).

103. C. Ma, C.R. Sporleder and R.A. Bonham, A pulsed electron beam time of flight apparatus for measuring absolute electron impact ionization and dissociative ionization cross sections, Rev. Sci. Instrum. 62:909 (1991); Absolute partial and total electron-impact-ionization cross sections for CF_4 from threshold up to 500 eV, Phys. Rev. A44:2921(1991); M.R. Bruce, C. Ma and R.A. Bonham, Positive ion pair production by electron impact dissociative ionization of CF_4, Chem. Phys. Lett. 190:285(1992); C. Ma, M.R. Bruce and R.A. Bonham, Erratum, Phys. Rev. A45:6932 (1992); M.R. Bruce and R.A. Bonham, Problems in the measurement of the Ar^{2+}/Ar^+ partial ionization cross-section ratio by use of the pulsed-electron-beam time-of-flight method, Z. Phys. D 24:149 (1992).

104. G. Spekowius and B. Brehm, Cross sections and thresholds for electron impact triple and quadrupole ionization processes of CO, N_2 and O_2, Chem. Phys. Lett. 187:442 (1991).

105. E. Krishnakumar and S.K. Srivastava, Ionization cross sections of rare-gas atoms by electron impact, J. Phys. B21:1055 (1988); Cross sections for the production of N_2^+, $N^+ + N_2^{2+}$ and N^{2+} by electron impact on N_2, B23:1893 (1990); Cross sections for electron ionization of O_2, Int. J. Mass Spectrom. Ion Proc. 113:1 (1992); E. Krishnakumar, A pulsed crossed beam approach for measurement of electron impact partial ionization cross sections: results on $CO_2 + e \rightarrow CO_2^+ + 2e$, Int.J. Mass Spectrom. Ion Proc. 97:283 (1990).

106. M.B. Shah, D.S. Elliott and H.B. Gilbody, Pulsed crossed-beam study of the ionization of atomic hydrogen by electron impact, J. Phys. B20:3501 (1987); M.B. Shah, D.S. Elliott, P. Mc Callion and H.B. Gilbody, Single and double ionization of helium by electron impact, J. Phys. B21:2751 (1988);

P. Mc Callion, M.B. Shah and H.B. Gilbody, Multiple ionization of magnesium by electron impact J. Phys. B25:1051 (1992); A crossed beam study of the multiple ionization of argon by electron impact, J. Phys. B25:1061 (1992).

107. H.U. Poll, C. Winkler, D. Margreiter, V. Grill and T.D. Märk, Discrimination effects for ions with high initial kinetic energy in a Nier-type ion source and partial and total electron ionization cross-sections of CF_4, Int. J. Mass Spectrom. Ion Proc. 112:1(1992).

108. M.R.H. Rudge, Theory of the ionization of atoms by electron impact, Revs. Mod. Phys. 40:564 (1968).

109. S.M. Younger and T.D. Märk, Semi-empirical and semi-classical approximations for electron ionization, in "Electron Impact Ionization", T.D. Märk, G.H. Dunn, eds., Springer, Wien (1985).

110. A. Burgess and I.C. Percival, Classical theory of atomic scattering, Adv. Atomic Molecular Phys. 4:109 (1968).

111. H. Bethe, Zur Theorie des Durchgangs schneller Korpuskularstrahlen durch Materie, Ann. Phys. 5:325 (1930).

112. M. Gryzinski, Two particle collisions. I. General relations for collisions in the laboratory system, Phys. Rev. A135:305 (1965).

113. H. Deutsch, P. Scheier and T.D. Märk, Calculation of electron impact ionization cross-sections. The fluorine anomaly, Int. J. Mass Spectrom. Ion Proc. 74:81 (1986).

114. H. Deutsch and T.D. Märk, Calculation of absolute electron impact ionization cross-section functions for single ionization of He, Ne, Ar, Kr, Xe, N and F, Int. J. Mass Spectrom. Ion Proc. 79:R1 (1987).

115. J.W. Otvos and D.P. Stevenson, Cross-sections of molecules for ionization by electrons, J. Am. Chem. Soc. 78:546 (1956); J.B. Mann, Ionization cross sections of the elements calculated from mean-square radii of atomic orbitals, J. Chem. Phys. 46:1646 (1967).

116. D. Margreiter, H. Deutsch and T.D. Märk, Absolute electron impact cross sections for single ionization of metastable atoms of He, He, Ne, Ar, Kr, Xe and Rn, Contr. Plasma Phys. 30:487 (1990).

117. H. Deutsch, D. Margreiter and T.D. Märk, Determination of ionization cross sections for atoms, excited atoms, molecules and clusters by means on an improved semiclassical formula, in "Proc. Pentagonale Workshop on Elementary Processes in Clusters, Lasers and Plasmas", T.D. Märk, R. Schrittwieser, eds., Studia Innsbruck (1991).

118. D. Margreiter, H. Deutsch, M. Schmidt and T.D. Märk, Electron ionization cross sections of molecules. Part II. Theoretical determination of total (counting) ionization cross sections of molecules: a new approach, Int. J. Mass Spectrom. Ion Proc. 100:157 (1990).

119. H. Deutsch, D. Margreiter and T.D. Märk, A semi-empirical approach to the calculation of absolute inner-shell electron impact ionization cross sections, Z. Phys. D. in print (1993).

120. H. Deutsch and T.D. Märk, Calculation of absolute outer-shell electron impact ionization cross sections, Contr. Plasma Physics. in print (1993).

121. H. Deutsch, C. Cornelissen, L. Cespiva, V. Bonacic-Koutecky, D. Margreiter and T.D. Märk, Total electron impact ionization cross sections of free molecular radicals: the failure of the additivity rule revisited, Int. J.Mass Spectrom. Ion Proc. in print (1993).

122. D.K. Jain and S.P. Khare, Ionizing collisions of electrons with CO_2, CO, H_2O, CH_4 and NH_3, J. Phys. B9:1429 (1976); Ionizing collisions of electrons with C_2H_4 and C_2H_6, Ind. J. Pure Appl. Phys. 14:201 (1976); S.P. Khare and A. Kumar, Mean energy expended per ion pair by electrons in molecular nitrogen, J. Phys. B10: 2239 (1977); Mean energy expended per ion pair by electrons in molecular oxygen, B11:2403 (1978).

123. T.D. Märk and A.W. Castleman Jr., Experimental studies on cluster ions, Adv. Atomic Molec. Physics. 20:65 (1985).

124. G. Benedek, T.P. Martin and G. Pacchioni. "Elemental and Molecular Clusters", Springer, Berlin (1988).

125. E.R. Bernstein. "Atomic and Molecular Clusters," Elsevier, Amsterdam (1990).

126. O.F. Hagena and W. Henkes, Die Bestimmung des effektiven Ionisierungsquerschnitts in kondensierten Molekularstrahlen, Z. Naturforschg. 20a:1344 (1965).

127. W. Henkes and F. Mikosch, The effective cross section for ionization by electrons of molecules in hydrogen clusters, Int. J. Mass Spectrom. Ion Phys. 13:151 (1974).

128. F. Bottiglioni, J. Coutant and M. Fois, Ionization cross sections for H_2, N_2 and CO_2 clusters by electron impact, Phys. Rev. A6:1830 (1972).

129. R.G. Keesee, A.W. Castleman Jr. and T.D. Märk, Electron-cluster interaction, in "Swarm Studies and Inelastic Electron Molecule Collisions", L.C. Pitchford, B.V. Mc Koy, A. Chutjian, S. Trajmar, eds,. Springer, New York (1987).

130. J. Gspann and K. Körting, Cluster beams of hydrogen and nitrogen analyzed by time-of-flight mass spectrometry, J. Chem. Phys. 59:4726 (1973).

131. H. Falter, O.F. Hagena, W. Henkes and H. Wedel, Einfluß der Elektronenenergie auf das

Massenspektrum von Clustern in kondensierten Molekularstrahlen, Int. J. Mass Spectrom. Ion Phys. 4:145 (1970).

132. T.E. Gough and R.E. Miller, Infrared laser and mass spectrometric analysis of cluster beams: dimer fragmentation due to electron impact, Chem. Phys. Lett. 87:280 (1982); J. Geraedts, S. Stolte and J. Reuss, Vibrational predissociation of SF_6 dimers and trimers, Z. Phys. A304:167 (1982).

133. U. Buck, Properties of neutral clusters from scattering experiments, J. Phys. Chem. 92:1023 (1988).

134. H. Helm, K. Stephan and T.D. Märk, Electron-impact ionization of Ar_2, ArKr, Kr_2, KrXe and Xe_2, Phys. Rev. A19:2154 (1979).

135. U. Buck and H. Meyer, Electron bombardment fragmentation of Ar van der Waals clusters by scattering analysis, J. Chem. Phys. 84:4854 (1986).

PHOTOFRAGMENTATION AS A PROBE OF ELECTRON THERMALIZATION IN SIZE-SELECTED CLUSTER ANIONS

David J. Lavrich, Paul J. Campagnola
and Mark A. Johnson

Department of Chemistry
Sterling Chemistry Laboratory
Yale University
225 Prospect St.
New Haven, CT 06511

INTRODUCTION

The localization of electrons in liquids has fascinated chemists for over one hundred years beginning with the early work on the metal ammonia solutions and later with electrons in rare gases and a variety of solvents. These systems represent a class of liquids where an excess electron is not associated with a specific molecular framework, but is physically stabilized by the medium. Early efforts by Ogg[1] captured the essence of the localization phenomenon using simple considerations of the trade-off between solvation energy, which tends to localize the excess charge, and the zero point energy of the localized system, which tends to favor delocalization. These ideas were extended and clarified by Jortner[2] in the 1960's, and more recent calculations have been carried out on the molecular level using quantum path integral methods by several groups.[3-8]. While it should be noted that the cavity interpretation is not embraced by all workers in this area, the qualitative cavity idea is basically born out by recent theory, and the electron in water has a spatial extent of about 3 Å[9]. Since the excess electron is localized by interaction with the medium, the extent of the medium would appear to influence the degree of localization. It has been of particular interest to establish if the cavity or "internal" states are indeed stable in small particles where surface binding can provide an alternative means of trapping the charge. Landman has recently considered electron binding onto a dielectric sphere, for example, and reported analytical results for the delocalization of a 1s type orbital as the radius of the sphere is reduced. Landman continued this work to simulate the behavior of "molecular" water clusters, where the delocalization of the charge is evident as the cluster size is decreased down to the dimer, where the electron is barely bound by the combined dipoles of the two water monomers. Thus, the evolution of localization in the water system would appear to begin at small size as dipole-bound states, evolve into surface trapped states, and then finally emerge as the bulk or internal states at yet larger size.[10]

In an early experimental attempt to quantify the types of binding operative in the $(H_2O)_n^-$ clusters, Bowen measured the electron binding energies, or more correctly the Vertical Detachment Energies (VDE) of these clusters in the range $n=2-69$.[11,12] The almost exact $n^{-1/3}$ scaling of the VDE values led those authors to the surprising conclusion that all clusters down to $n=6$ can be regarded as "internal states." This in turn led us to attempt to measure the absorption spectra of the $(H_2O)_n^-$ clusters to determine if indeed the characteristic spectrum[13] of the hydrated electron, e_{aq}^-, was present in the clusters. The spectra of $n=18$ and 30 were similar in shape and magnitude to e_{aq}^-, but significantly shifted to the red, with $n=18$ farther red-shifted than $n=30$, and about 0.75 eV lower in energy relative to e_{aq}^-. At that time, the connection to the localization issue was unclear; however, it would appear that for this pseudo-one electron system, the spectral moment methods[14-16] used to define the spatial extent of the bulk electron can be applied to the cluster absorption spectra to quantify the degree of localization. In the first section of this paper, we discuss the application of moment theory to the $(H_2O)_n^-$ spectra to establish the connection between the red shifts in the spectra and the localization of the excess electron.

While the localization concepts apply to rigid or locally frozen water clusters which host the electron, real clusters are often thermally excited, and we are led to consider the dynamics of how the motion of the water molecules affects the binding of the electron. We present the results of two experiments which probe the exchange of energy between the excess electron and vibrational levels of the water cluster. In one experiment, the electron is pumped *via* the characteristic electronic absorption band, and the degradation of the electronic energy into vibrational energy is followed by subsequent evaporation of water monomers from the cluster. In a second experiment, we explore the complementary case in which a cold $(H_2O)_n^-$ cluster is vibrationally excited by condensation heating and then loses the electron *via* thermionic emission. Both experiments reveal a close coupling between the electronic and vibrational degrees of freedom.

APPLICATION OF SPECTRAL MOMENTS TO CLUSTER ACTION SPECTRA

We wish to interpret the position of the absorption bands of $(H_2O)_n^-$ clusters in terms of delocalization of the electron. We begin describing the cluster ions by a one-electron approximation where the interaction with the remaining electrons is assumed to be adiabatic. The size of the electron can be expressed as the standard deviation in position $\langle\Delta R^2\rangle$. Simple manipulations:

$$\langle\Delta R^2\rangle = \Sigma \, \langle 0|R|n\rangle\langle n|R|0\rangle \qquad\qquad [1]$$

$$\langle\Delta R^2\rangle = \Sigma \, \frac{\mu^2}{e^2} \qquad\qquad [2]$$

indicate that $\langle\Delta R^2\rangle$ can be related to the dipole transition moment, μ, and hence to the integrated absorption cross section:[17]

$$\sigma_0 = \frac{\pi\omega\mu^2}{3\varepsilon_0 hc^2} \qquad cm^2 cm^{-1} \qquad\qquad [3]$$

The size can then be expressed as:

$$\langle\Delta R^2\rangle = \Sigma \, \frac{\mu^2}{e^2} = \frac{3\varepsilon_0 hc^2}{e^2\pi} \Sigma \, \frac{\sigma_0}{\omega} \qquad\qquad [4]$$

We will shortly show that in the case of $(H_2O)_n^-$, the final states are either fragmentation or electron continua, so the continuum form of Eq. [4]:

$$\langle\Delta R^2\rangle = \frac{3\varepsilon_0 hc}{2\pi^2 e^2} \int \frac{d\omega\, \sigma(\omega)}{\omega} \qquad\qquad [5]$$

is required for the cluster systems. This expression is equivalent to that obtained by Tuttle[14] in his application of spectral moment theory to the absorption band of solvated electrons in the bulk. This is a specific example of a "spectral moment"[15,16] generally defined as:

$$S(k) = \int d\omega\, \omega^k \langle f(\omega)\rangle \qquad\qquad [6]$$

where ω is the angular frequency and $\langle f(\omega)\rangle$ is the spectral density of oscillator strength, directly proportional to the absorption cross section. Obviously $S(0)$ represents the familiar Thomas-Kuhn-Reiche sum rule where the integrated oscillator strength for a one electron system is unity. The -1th moment can be related to the standard deviation in position of the excess electron (in atomic units):

$$\langle\Delta R^2\rangle = \frac{3c}{4\pi^2} \int \frac{d\omega\, \sigma(\omega)}{\omega} \qquad\qquad [7]$$

In our application of Eq. [7], we will interpret red-shifting of the $(H_2O)_n^-$ electronic absorption spectrum as a manifestation of increasing electron delocalization in the ground electronic state. At first glance, this notion might appear counter-intuitive since a red shift in absorption is often regarded as due to enhanced stabilization of the excited state relative to the ground state. Such considerations are usually invoked, for example, to explain the red shift of charge-transfer electronic transitions in the presence of solvent. There is no ambiguity in interpretation, however, since both statements can be simultaneously true. Consider the case of a particle in an infinite box in which we take the dipole spectrum and then open the box into a square well by lowering the potential so that only one bound state is left in the original box. The square well spectrum will red shift as excitation is now occurring to continuum final states, but the ground state wavefunction also spreads out as it tunnels into the now-finite potential. In other words, we cannot change the excited states without changing the ground state.

EXPERIMENTAL PROCEDURES

The ion source and mass spectrometer have been described in detail previously[18] and only a brief account will be given here. A high energy cw electron beam (1 keV) ionizes a pulsed supersonic expansion of water and argon (298 K), forming hydrated electron clusters *via* attachment of slow secondary electrons on to neutral clusters. After drifting for approximately 15 cm, the ions are injected into a mass spectrometer and separated by standard time of flight techniques. The branching ratio between photodetachment and photofragmentation is important in our analysis of kinetic shifts, and was determined by comparison of parent ion photodepletion relative to the integrated $(H_2O)_n^-$ daughter ion intensities.

The application of spectral moment methods requires absolute absorption cross sections. Obviously, we can only measure absolute photodestruction cross sections, and the connection between these values and the true absorption cross sections must be carefully constructed. Absolute photodestruction cross sections are measured at several discrete photon energies *via* depletion of the parent ionic cluster ensemble:

$$\frac{C}{C_0} = \exp^{-\sigma F} \qquad [8]$$

where σ is the photodestruction cross section, F is the laser fluence, and C is the ion signal. As these clusters decay by both photofragmentation and photodetachment, total ion destruction must be used to measure the absorption cross section. The "absorption spectra" are a map of these measurements. Laser configurations used were Nd:YAG harmonics, Nd:YAG pumped dye, and difference frequency mixing Nd:YAG fundamental and the dye laser in $LiIO_3$. To ensure accurate cross section measurements, the laser fluence must be extremely homogeneous such that all the ions in the ensemble experience identical intensity. To this end, the Nd:YAG fundamental was converted to a Gaussian profile in the far field

(~20 meters) with subsequent generation of the harmonics. The dye beams were further smoothed by spatial filtering. The resulting beams were then expanded and the central region of the Airy pattern was selected with a circular aperture having an area much larger than the size of the ion packet at the laser interaction region. The spatial homogeniety of the laser was established by rastering a pinhole across the beam. The resulting profiles were constant to within 10% over the area of the beam.

In order to use Eq. [8] for an absolute cross section, however, it must also be established that all the ions in the mass selected packet experience the same laser fluence. This in turn requires uniform laser overlap with all of the ions. There is a potential for

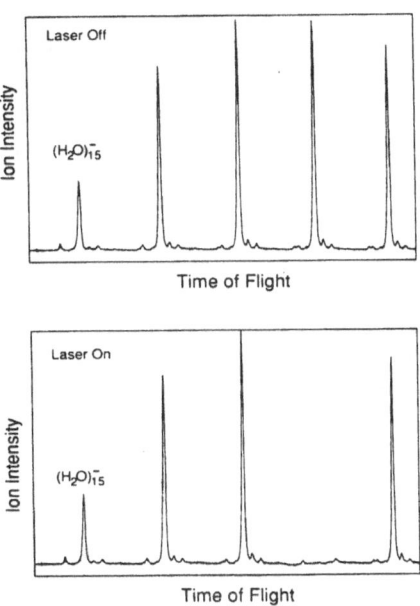

Figure 1. Time of flight mass spectra of the $(H_2O)_n^-$ cluster distribution a) before, and b) after selective laser interaction with $n=18$.

error here in our experiment since we intersect the laser transverse to the axis of the ion beam at the transient focus. We establish the overlap by counter-propagating a "standard" 1064 nm beam through identical apertures as those used to define the scanned laser, while the standard beam is sufficiently intense (about 50 mJ/pulse) to deplete the ion ensemble. The observed depletion is displayed in figure 1, where the lower trace results from selective excitation of $n=18$; >97% depletion of this peak indicates that essentially all of the ions are contained within the cylindrical volume of the laser beam. To avoid multiphoton effects, laser fluences were adjusted such that 10-15% of the parent ions were depleted for the cross section measurements.

USING BREAKDOWN CURVES TO AVOID KINETIC SHIFTS IN THE SPECTRA

To accurately measure an absorption cross section through photodestruction, all the ions which absorb a photon must experience either electron loss or at least one monomer evaporation prior to entering the second mass spectrometer (10 µs). This constraint is particularly important in the case of the $(H_2O)_n^-$ clusters where the monomer binding energy (0.42 eV)[19] is on the same order of magnitude as the low energy tails of the photodestruction spectra. We must be sure that "invisible" photoabsorption is not occurring due to slow decomposition of the excited ensemble. Electron loss will dominate the photodecomposition when the excitation energy is far above the vertical electron detachment energy, in which case direct photodetachment leads to neutralization of the cluster on a very short time scale (10^{-14} s). Direct detachment appears to be the dominant decay mechanism for the smaller clusters ($n<7$) where fragmentation has not been observed at any excitation wavelength. The larger clusters present a more difficult circumstance, however, since direct detachment, monomer evaporation, and electron evaporation all compete as decomposition pathways, as we will discuss in detail later in this paper.

Fragmentation is the dominant decay process[20] at the absorption threshold for $n>15$, and monomer evaporation kinetics should control the dissociation time scale. Note that because the initial ensemble has significant internal energy prior to laser interaction, excitation with photon energies near the monomer binding energy *does not imply* that the dissociation is occurring with an internal energy near the dissociation asymptote. We have investigated the internal energy content of the $(H_2O)_n^-$ clusters in a previous publication,[21] and observed that more than 0.7 eV of internal energy can be removed from our ensembles by adjusting the expansion conditions, so that at least two bond energies of internal excitation are in the clusters *prior* to photolysis.

We have previously reported[21] the observation of metastable decay of the $(H_2O)_n^-$ clusters by monomer evaporation, and have succeeded in controlling the rate *via* the average internal energy content of the clusters. For an absolute cross section measurement, however, we must know not only about the *mean* decay behavior but that of the entire ensemble; in other words, we are concerned with the fraction of the clusters which have insufficient energy to decay even when energized by photoexcitation. We illustrate here a method based on the photon energy dependence of the photofragment distribution or "breakdown curve" to qualitatively determine the photon energy range in which all clusters in the ensemble will experience at least one evaporation in the detection time window.

The method will be demonstrated on $(H_2O)_{25}^-$. The breakdown curve[22,23] is defined by:

$$f_n(E) = \frac{I_n(E)}{\Sigma I_n(E)} \qquad\qquad [9]$$

where $f_n(E)$ is the fraction of daughter ions corresponding to evaporation of n monomers so that $\Sigma f_n(E)=1$. The photofragmentation distribution in $(H_2O)_n^-$ clusters typically consists of two or three daughter ions. The problem is essentially the f_0 or "loss of 0" channel in which the cluster absorbs a photon but does not experience an evaporation on the timescale of the experiment. The experimental breakdown curve for $n=25$ is shown is figure 2. In general the $f_n(E)$ curves are peaked around $E=nD_0$, the minimum energy required to evaporate n water molecules, and the f_n peaks are therefore separated by the monomer binding energy (0.42 eV). Since the f_2 and f_3 curves are well developed, each of these channels were fit to a Gaussian function with the parameters of full width at half maximum, the band maximum, and central energy. The width and height parameters for the f_2 and f_3

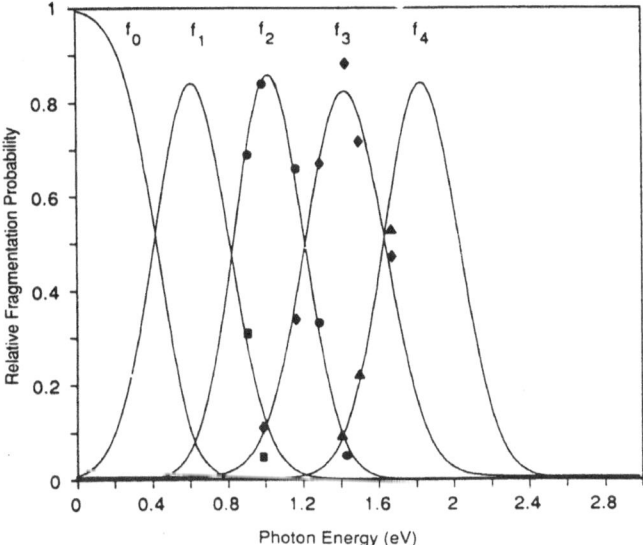

Figure 2. Experimental photofragmentation breakdown curves for $(H_2O)_{25}^-$; $f_2 = \bullet$, $f_3 = \blacklozenge$, $f_4 = \blacktriangle$. Solid lines present gaussian fits to f_2 and f_3 which are used to estimate f_0 and hence the range of photon energies where the absorption spectra are affected by a kinetic shift.

functions were remarkably similar with mean energies separated by 0.41 eV, close to the monomer binding energy of 0.42 eV. Simple modelling of the sequential evaporation processes anticipates the similarity of the $f_n(E)$ functions for small n, and based on these simulations we suggest that average parameter values from $n=2$ and 3 can be used to extrapolate the f_1 and f_0 functions. Note that these functions are completely dependent on the character of the internal energy distribution in the $(H_2O)_n^-$ ensemble and therefore can only be used extrapolate the behavior of the ions created at the laser interaction region *in this particular experiment.*

Examination of the breakdown curves in figure 2 indicates that excitation with more than about 0.6 eV photon (or any statistically distributed) energy will result in >98% decomposition of the $(H_2O)_{25}^-$ ensemble of ions created in the source. In a previous communication[24], we showed the photodestruction spectrum for $(H_2O)_{30}^-$ showed a dramatic fall off at 0.75 eV. Analysis of the $(H_2O)_{30}^-$ breakdown curve allows us to conclude this drop reflects the absorption cross section rather than a kinetic artifact. We emphasize that because the water binding energy is rather large compared to the onset of absorption, this photon energy is close to the limit where photodestruction is an accurate measurement of the absorption cross section.

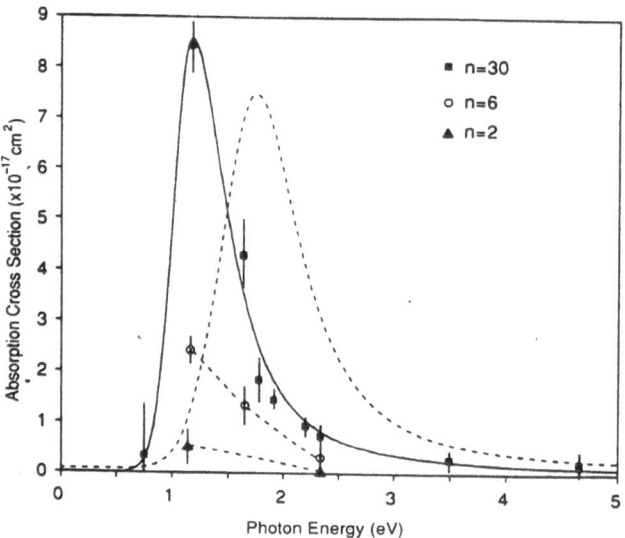

Figure 3. Absorption spectra of $(H_2O)_n^-$, (■) $n=30$, (○) $n=6$, (▲) $n=2$. Dotted line presents the spectrum the bulk hydrated electron.

RESULTS AND DISCUSSION

The absorption spectra[24] for $n=30$ is shown with preliminary data for $n=2$ and 6 along with that of the bulk hydrated electron.[13] The data for $n=7$ and 18 are omitted here for clarity. The $n=18$ and 30 spectra are developed and are clearly red-shifted relative to the hydrated electron.

Since the data for $n=2$ and 6 are obviously sparse, we must examine the possibility that significant features exist higher in energy than the points at 2.33 eV. This possibility is unlikely for the small species since photoelectron spectroscopy has established that the observed points are far above the vertical electron detachment energies (.02 and .4 eV for $n=2$ and 6, respectively) and absorption undoubtedly results directly to the photodetachment continuum. Note that no photofragments have been observed for these smaller parents at the wavelengths displayed in figure 3. The fact that excitation in the visible results in a decreasing detachment cross section with increasing energy is consistent with excitation far above the electron detachment threshold, and there are no low lying electronically excited states of H_2O to support autodetaching resonances in the uv below about 6 eV. For the case of excitation far above threshold, we generally expect[25] the cross section to fall off as $1/v^{3.5}$, which accurately describes the high energy tails of the $n=18$ and $n=30$ clusters for which we have a significant data base. In essence, this fall-off can be attributed to the size of the ground state orbital relative to the De Broglie wavelength of the photoejected electron, and therefore the shape of the photodetachment cross section already reveals clues to the spatial extent of the excess electron. We can use this expected functional dependence of the photodetachment cross section to estimate the integrated oscillator strength contained in the spectrum for $h\upsilon > 1.16$ eV. The balance of the oscillator strength below 1.165 eV can then be obtained by the difference from unity. This analysis indicates that 0.97 and 0.84 oscillator strength lies to the red of 1.165 eV for $n=2$ and 6, respectively, indicating that the trend in increasing red-shift found from $n=30$ to 18 becomes very pronounced at the smaller sizes. We have tested this conclusion with a preliminary study which found significant photoabsorption of $n=2$, 6, and 7 at 10.6 microns (CO_2 laser).

The measured oscillator strengths for $n=18$ and 30, where the overall bands lie within the range of the measurements, are 0.6 and 0.65 $\pm$ 0.2, respectively. Since the quasi-one electron treatment requires the observed oscillator strength in the lowest electronic band to be unity, we must either have an error in the measurements or the one-electron approximation. The primary sources of error in the integrated oscillator strengths are the uncertainty about whether we have indeed located the band maximum and the uncertainty in each cross section measurement ($\pm 20\%$). Since all the f values are below unity, we doubt the error results from the cross section error and favor the former explanation. In our first report,[24] we noted that the cluster f values were actually quite close to the $f=0.75$[13] value observed for the bulk hydrated electron, and therefore assumed we had recovered most of the oscillator strength. Upon closer inspection, however, it appears that the bulk value is reduced from unity by almost exactly the index of refraction, η, of water (i.e. $\eta f \approx 1$). This reduction results from the screening of the electric field in the medium, an effect which should not contribute significantly in the case of the clusters since they are much smaller than the wavelength of light and well within the "skin depth" of the medium. We are therefore concerned about the "missing" oscillator strength of about 30%, and we must regard our f values as lower bounds on the true values. We can nonetheless use our measured spectra to obtain lower bounds on the spatial extent of the excess electron in its ground state orbital by assuming that all of the remaining or "missing" oscillator strength lies just below the lowest photon energy covered in our experiment, 0.75 eV. The results of our analysis are shown in figure 4, where the estimated size of the electron (lower bound) is plotted along with the physical size of the cluster. If we assume the clusters are spherical, the radius scales as $r=r_0 n^{1/3}$, where r_0 is the radius of a water monomer (about 1.5 Å). Sketching a smooth curve through the experimental data for the electron reaveal that the physical size of the cluster is the same as the electron orbital around $n=15$. Field detachment studies[26] of the weakly bound dimer anion $(H_2O)_2^-$ indicate its electron radius is on the order of 20 Å, far off scale on the plot in figure 4. Hence it seems clusters

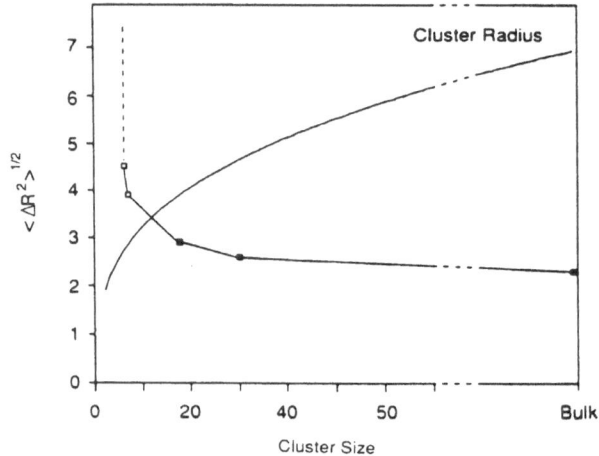

Figure 4. Effective radius of the excess electron on $(H_2O)_n^-$ clusters extracted from absorption spectra, along with the value for the bulk hydrated electron. An estimate of the average radius of the clusters (using bulk density and spherical geometry) is included in the smooth curve.

smaller than $n=15$ cannot contain the electron in a well defined cavity. This is in qualitative agreement with calculations by Landman and Jortner[5,10] where they find the reorganization energy of the cluster necessary to form a cavity is greater than the binding energy in this range. Interestingly, it is also in this range where Landman predicts interior states should begin to be stable. The size of the electron extracted by this procedure for $(H_2O)_{30}^-$ is 2.6 Å, which is not significantly larger than the bulk value of 2.3 Å,[9] suggesting that even in a relatively small cluster the charge is almost fully localized by the medium.

THERMIONIC EMISSION FROM $(H_2O)_n^-$ FOLLOWING PHOTOEXCITATION AND CONDENSATION HEATING

Until now, we have focussed on the localization of the electron in the ground electronic state of $(H_2O)_n^-$, and discussed how the localization depends on the number of water molecules present. It is clear, however, that the state of motion of these water molecules should also be related to the efficacy of a given number of water molecules to localize the excess electron. Consider, for example, the case of the water dimer binding an extra electron where the binding energy is derived from the electron interacting with the combined dipoles of each water monomer. If we then vibrationally excite the water dimer along the soft bending or "hinging" coordinate, this net dipole will decrease and consequently so will the electron binding energy. In this case we expect the degree of localization to be highly dependent on this normal coordinate, yielding a strong dependence of the electronic wavefunction on the relative position of the two water molecules. Such a situation would appear ideal for Born-Oppenheimer breakdown in which energy is transferred between the excess electron and the soft modes of the cluster core. This process can be regarded as "thermalization" of electronic energy into vibrational excitation and *vice versa*. With this in mind, we now explore the photophysics and reaction chemistry of $(H_2O)_n^-$ to look for signatures of such thermalization effects.

One of the manifestations of thermalization is, of course, the formation of the clusters in the first place,[27] where an impinging free electron is captured onto a neutral water cluster and relaxed into the ground state. Rossky and co-workers[4] have discussed such non-adiabatic processes at length from a purely quantum mechanical point of view. Let us now consider a kind of microscopic reverse where we photoexcite the excess electron back into the continuum, and ask whether the excited electronic state can relax back to the ground state before the electron can escape. If the $(H_2O)_n^-$ cluster is size selected prior to laser interaction, this relaxation can be monitored by the subsequent evaporation of water monomers while the electron ejection channel is trivial to detect by direct amplification of the free electron signal. Therefore, we have investigated the branching ratio for the processes:

$$(H_2O)_n^-$$

$$\downarrow h\upsilon$$

[10]

$$[(H_2O)_n^-]^*$$

$$\swarrow \qquad\qquad \searrow$$

$$(H_2O)_n + e^- \qquad\qquad (H_2O)_m^- + (n\text{-}m)\, H_2O$$

as a function of cluster size as well as photon energy. In general, fragmentation dominates near the red side of the absorption spectrum but gradually gives way to electron detachment as the laser is scanned toward higher energy through the band. A typical data set for the energy dependence of the branching is presented in figure 5, which displays the absorption

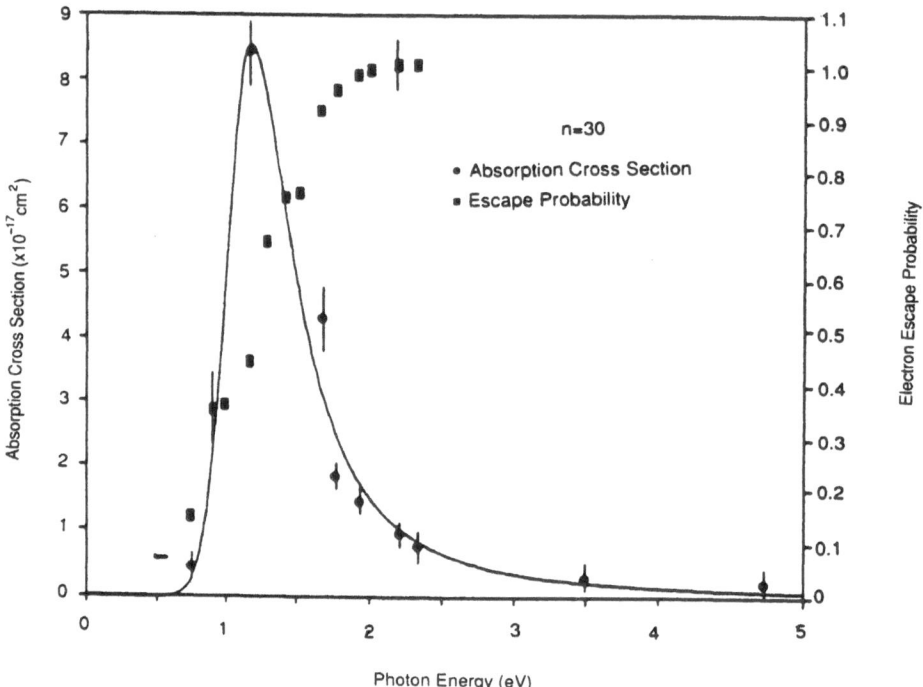

Photon Energy (eV)

Figure 5. Plot of the branching ratio for electron escape (right side vertical axis ■) vs photon energy. The absorption band (●) is also reproduced.

spectrum for $n=30$ on the same energy scale as the electron ejection (i.e. escape) probability. We have previously shown that the maxima of the absorption spectra are quite close to the Vertical Detachment Energies (VDE) of the clusters.[24] Inspection of figure 5 indicates that the excitation must be significantly above the VDE before they are ejected with unit probability. Very far above threshold, we expect the excitation to occur directly to the continuum. One of signatures of direct detachment in which the electron is ejected quickly from the cluster is that the angular distribution of the electron is asymmetric with respect to the electric vector of linearly polarized laser light. Indeed, the angular distribution of photoelectrons at 2.33 eV excitation is strongly skewed toward preferential ejection along the electric vector with an asymmetry parameter, β, of +0.92.[28] When the electron is recaptured at lower energy, however, it is possible that some of the ejected electrons are slowed by interaction with the water cluster prior to ejection. This is plausible since the photoelectrons are captured with increasing probability as the laser is scanned down toward the photoelectric threshold where the nascent electrons possess very little kinetic energy.

We have carried out a systematic study to determine how the electron ejection probability depends on cluster size as well as photon energy. This amounts to a three dimensional data base in which the data presented in figure 5 are extended along a depth axis *vs* cluster size. A projection of this data is displayed in figure 6, which shows the size

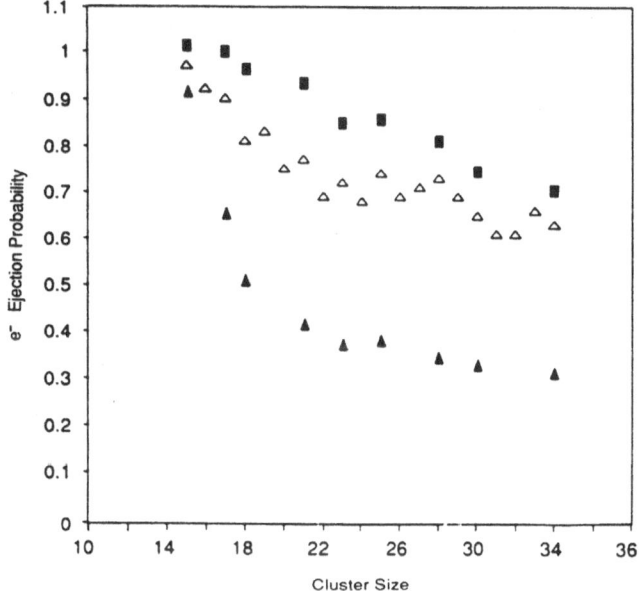

Figure 6. Electron escape probability as a function of cluster size for four different energies (■) 1.43 eV, (▵) 1.29 eV, (▲) 0.99 eV.

The cause for the increasing electron ejection propensity at the smaller sizes is not immediately obvious, since the absorption spectra and the photoelectron spectra are slowly dependence of the electron escape probability as a function of cluster size at three different energies throughout the band. For $n>22$, the escape probability gradually decreases with increasing cluster size such that the slopes for all the wavelengths are nearly identical. There is a rather rapid increase toward smaller clusters, however, beginning in the vicinity of $n=20$ and becomeing more pronounced as the excitation energy is decreased.

varying over the range $13<n<25$ with a size dependence more akin to slow variation in escape probability at $n>22$. One intriguing possibility is that the intrinsic capture of the electrons is, in fact, given by extrapolation from the higher clusters, but the rapid increase at smaller cluster size arises from the re-ejection of a captured and thermalized electron. Such a suggestion is plausible since the vertical binding energy is about 1 eV around $n=15$, and decreases with decreasing cluster size. If the *adiabatic* electron binding energy begins to approach the evaporation energy of a water monomer from the cluster, then the two channels, electron and molecular water evaporation, will compete to cool the hot anionic cluster. This competition has already been suggested to account for the observation of a size-dependent autodetachment around $n=15$,[28] and this type of thermionic emission should generally occur in negative ion systems whenelectrons are more weakly bound than monomers.[29]

We have therefore investigated the propensity of the $(H_2O)_n^-$ clusters to thermally eject the electron. To ensure that the available energy is indeed thermal, however, it was obviously not desirable to excite the clusters using their electronic absorption bands. While vibrational excitation *via* infrared absorption should be possible, there is an alternative method to warm the clusters by simply condensing a species on the outside of the cluster which releases a large condensation enthalpy. Water itself is a good candidate, with a condensation energy of 0.42 eV, but alcohols are also strongly bonded to water. The condensation experiments were carried out by exposing the clusters to a background gas of either D_2O or ethanol in the source after the cluster ions are formed but before they were injected into the mass spectrometer. This performance of the experiment is illustrated in figure 7, showing the control case in which $CO_2^- \cdot (H_2O)_n$ clusters (upper trace) are exposed to ethanol (lower trace). New peaks appear in the exposed clusters throughout the size range $10<n<18$, which correspond to incorporation of one ethanol molecule:

$$CO_2^- \cdot (H_2O)_n + CH_3CH_2OH \rightarrow CO_2^- \cdot (H_2O)_{n-1} \cdot CH_3CH_2OH + H_2O \qquad [11]$$

We have carried out the control experiment with hydrated CO_2^- clusters since the excess electron is very stable in these ion-solvent complexes and cannot be thermally ejected. Figure 8 presents the results from the analogous experiment using $(H_2O)_n^-$ parent ions, and unlike the $CO_2^- \cdot (H_2O)_n$ case, only the water clusters above $n=15$ incorporate either an ethanol molecule (upper trace) or deuterium labelled water (lower trace). Since the results on the $CO_2^- \cdot (H_2O)_n$ experiment indicate that all clusters are intrinsically about the same in incorporating the hetero-molecule, we conclude that the smaller $(H_2O)_n^-$ species are destroyed upon condensation by the associative detachment reaction:

$$M + (H_2O)_n^- \rightarrow M \cdot (H_2O)_n + e^- \qquad M = D_2O, CH_3CH_2OH \qquad [12]$$

Since neither D_2O nor CH_3CH_2OH molecules bind an electron, it is sure that the collision with these species only heats the $(H_2O)_n$ clusters without introducing charge transfer. Moreover, this heating is initially injected into the low frequency, inter-molecular modes

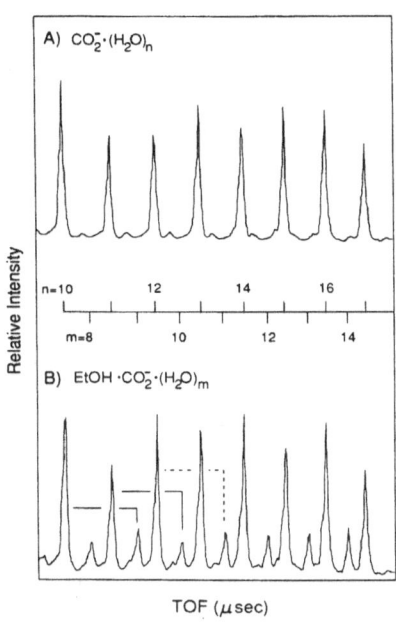

Figure 7. Parent mass spectra of a) $CO_2^- \cdot (H_2O)_n$ clusters and b) cluster distribution after exposing the clusters in the upper trace to ethanol. Brackets indicate the products of the solvent exchange reaction.

of the $(H_2O)_n^-$ clusters, the lowest (most degenerate) form of internal excitation. Thus, the fact that the electron is ejected from such a collision is substantial evidence that the electron ejection is indeed thermal; the excess electron is completely coupled to the low frequency modes of the cluster and interchanges energy with them.

We can quantify this effect by using the propensity to incorporate the condensing molecule as a measure of the branching ratio between evaporation of water molecules and ejection of the electron:

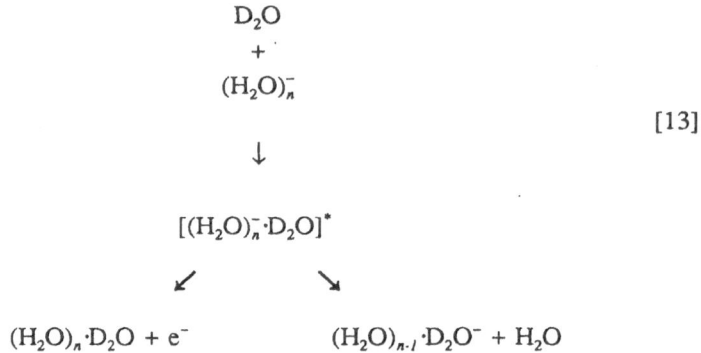

$$D_2O$$
$$+$$
$$(H_2O)_n^-$$

[13]

$$\downarrow$$

$$[(H_2O)_n^- \cdot D_2O]^*$$

$$\swarrow \qquad \searrow$$

$$(H_2O)_n \cdot D_2O + e^- \qquad\qquad (H_2O)_{n-1} \cdot D_2O^- + H_2O$$

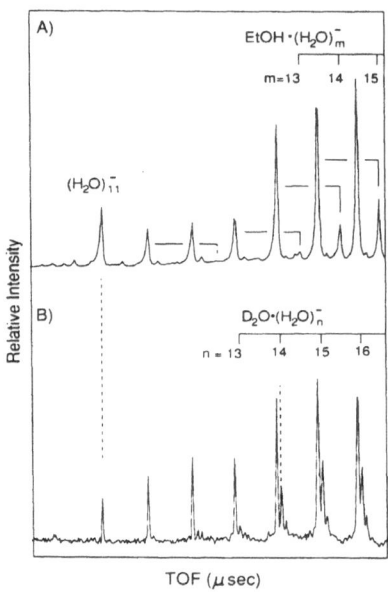

Figure 8. Cluster ion distributions obtained after exposure of $(H_2O)_n^-$ clusters to a) ethanol and b) D$_2$O. Only the larger clusters incorporate the hetero-solvent molecules.

This branching ratio is displayed in figure 9, along with the data from photoexcited $(H_2O)_n^-{}^*$ in figure 6. The condensation branching ratio for D_2O and ethanol are very similar, each indicating a very rapid increase in electron ejection at $n=14$. The data displaying the behavior of the photoexcited species are clearly converging toward the condensation heating results as the photon energy is lowered. It appears that $n=14$ represents a cross-over in the evaporation kinetics where the electron becomes so weakly bound that it is lost with a higher rate than monomer evaporation even though the latter is preferred by the channel

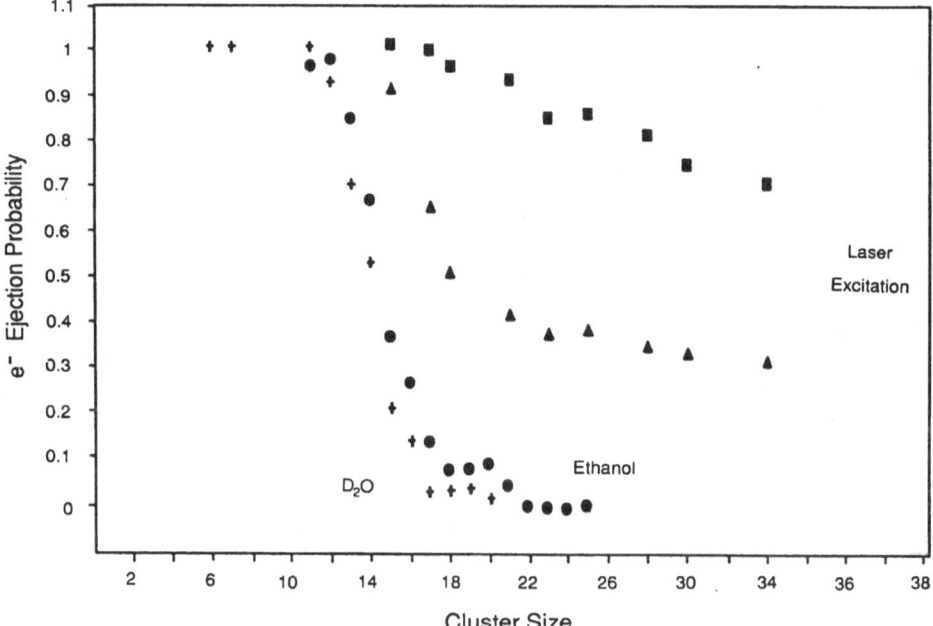

Figure 9. Electron ejection probability as a function of cluster size for both laser excitation, ($\blacksquare$)=1.43 eV, ($\blacktriangle$)=0.99 eV, (reproduced from figure 6) and heating by condensation,($\bullet$)=CH_3CH_2OH, (+)=D_2O.

degeneracy (i.e. 14/1). We speculate that excitation of the electronic spectrum at 0.5 eV would result in essentially the same electron capture propensity as given by the condensation results. In summary, the excess electron is efficiently coupled to the motion of the water molecules and energy exchange between the electron and the cluster molecules is so efficient that the electron is ejected as simply one of $n+1$ particles competing for evaporation from the cluster.

GENERAL REMARKS ABOUT $(H_2O)_n^-$ AND e_{aq}^-

We are beginning to accumulate sufficient data on the $(H_2O)_n^-$ systems to make a few remarks on the approach to e_{aq}^- with increasing size. Earlier field detachment work on $(H_2O)_2^-$ by Haberland[26] revealed that the excess electron was quite delocalized (r=20 Å). This result is quite consistent with Bowen's observation from the $(H_2O)_2^-$ photoelectron spectrum that the intra-molecular water stretches are essentially unperturbed by the charge.[11,12] In this work, we only find the high energy tail of the photodetachment spectrum above 1 eV which we infer is another manifestation of the extreme electron delocalization. After a gap in the parent distribution from n=3 to 5 (thought to occur due to electron capture kinetics), the n=6 and 7 species also display weak electron binding. The photoabsorption cross section is still decreasing above 1 eV and photodetachment remains the dominant photochemical channel. Bowen has argued that these small species might possess two forms, one which binds the electron internally and another with the electron bound to the surface, with the internal state population comprising about 80% of the ensemble. However, the bound for the electron "radius" estimated from the absorption spectra (Fig. 4) is clearly larger than the spatial extent of the cluster. Thus, it would appear that the extrapolation of the internal state down to these small species is not valid since the electron cannot be localized "within" the clusters. On the other hand, the increased absorption cross section at 1 eV does indicate significant localization (i. e. blue shift) relative to the dimer anion, and it is important to emphasize that this analysis provides no insight into the topology of the ground state orbital in which it resides.

After another gap in parent intensity from n=8 to 10, the $(H_2O)_n^-$ clusters from n=11 to 18 represent an intermediate regime. Electron evaporation and monomer evaporation compete around n=15, complicating the interpretation of photoexcitation experiments. This effect implies that we are blind to the existence of locally excited electronic states for clusters smaller than n=15 since electron evaporation is the lowest energy decomposition pathway. Thus, direct detachment and internal conversion both lead to electron ejection. Photofragmentation is observed for n≥15, indicating that an electronically excited state is sufficiently long-lived to internally convert its energy to the ground state with subsequent evaporation. The decrease in cross section on the red edge of the absorption is now observed for the first time at n=18, with an onset that closely parallels the shape of the photoelectron spectrum. The photofragmentation yield is correlated to production of low energy photoelectrons. One possibility for this excited state is that it is indeed derived from the electron continuum but the photoexcited electron loses its energy by electron-phonon scattering (i.e. is partially thermalized) before it can leave the cluster. Our lower bound on the size of the electron at n=18 is smaller that the cluster itself, consistent with more significant interaction between the excess electron and the water framework. The angular distribution of the photoejected electrons at high energy indicates that the electron is ejected along the electric vector of the laser with an anisotropy parameter β≈+1,[30] consistent with a mostly s→p type of excitation expected for the bulk electron.[4,7] $(H_2O)_n^-$ in this size range was found to undergo electron scavenging reaction chemistry[31] similar to that of the bulk hydrated electron.

ACKNOWLEDGEMENTS

We gratefully acknowledge support from the National Science Foundation under CHE-9207894.

REFERENCES

1. R.A. Ogg, Jr., Physical interaction of electrons with liquid dielectric media. The properties of metal ammonia solutions, *Phys. Rev.* 69:668 (1946).
2. J. Jortner, S.A. Rice, and E.G. Wilson, *in*: "Metal Ammonia Solutions," G. Lepoutre and M.J. Sienko, eds., W.A. Benjamin Inc., New York (1964).
3. M. Marxhi, M. Sprik, and M. L. Klein, Solvation of electrons, atoms, and ions in liquid ammonia, *Farad. Disc. Chem. Soc.* 85:373 (1988).
4. P. J. Rossky and J. Schnitker, The hydrated electron: Quantum simulation of structure, spectroscopy, and dynamics, *J. Chem. Phys.* 92:4277 (1988).
5. R. N. Barnett, U. Landman, C.L. Cleveland, and J. Jortner, Electron localization in water clusters. II. Surface and internal states, *J. Chem. Phys.* 88:4429 (1988).
6. R. N. Barnette, U. Landman, G. Makov, A. Nitzan, Theoretical studies of the spectroscopy of excess electrons in water cluster, *J. Chem. Phys.* 93:6226 (1990).
7. A. Wallqvist, G. Martyna, B. J. Berne, Behavior of the hydrated electron at different temperatures: Structure and absorption spectrum, *J. Phys. Chem.* 92:1721 (1988).
8. N. R. Kestner, *in*: "Radiation Chemistry", Farhataziz and M. A. J. Rodgers, eds., VCH Publishers, Inc., New York (1987).
9. D-F Feng and L. Kevan, Theoretical models for solvated electrons, *Chem. Rev.* 80:1 (1980).
10. R.N. Barnett, U. Landman, C.L. Cleveland, and J. Jortner, Size dependence of the energetics of electron attachment to large water clusters, *Chem. Phys. Lett.* 145:382 (1988).
11. J.V. Coe, G.H. Lee, J.G. Eaton, S.T. Arnold. H.W. Sarkas, K.H. Bowen, C. Ludewigt, H. Haberland, and D.R. Worsnop, Photoelectron spectroscopy of hydrated electron cluster anions, $(H_2O)^-_{n=2-69}$, *J. Chem. Phys.* 92:3980 (1990).
12. G.H. Lee, S.T. Arnold, J.G. Eaton, H.W. Sarkas, K.H. Bowen, C. Ludewigt, and H. Haberland, Negative ion photoelectron spectroscopy of solvated cluster anions, $(H_2O)^-_n$ and $(NH_3)^-_n$, *Z. Phys. D.* 20:9 (1991).
13. F.-Y Jou and G.R. Freeman, Temperature and isotope effects on the shape of the optical absorption spectrum of solvated electrons in water, *J. Phys. Chem.* 83:2383 (1979).
14. S. Golden and T.R. Tuttle, Jr., Nature of solvated electron absorption spectra, *J. Chem. Soc. Faraday Trans. II* 75:474 (1979).
15. J.O. Hirschfelder, W.B. Brown, and S.T. Epstein, Adv. Quant. Chem. 1:255 (1964).
16. U. Fano and J.W. Cooper, Spectral distribution of atomic oscillator strength, *Rev. Mod. Phys.* 40:441 (1968).
17. R.C. Hilborn, Einstein coefficients, cross sections, f values, dipole moments, and all that, *Am. J. Phys.* 50:982 (1982).

18. L.A. Posey, M.J. DeLuca, and M.A. Johnson, Demonstration of a pulsed photoelectron spectrometer on mass-selected negative ions: O^-, O_2^-, and O_4^-, *Chem. Phys. Lett.* 131:170 (1986).

19. C. E. Klots, Evaporative Cooling, *J. Chem. Phys.* 83:5854 (1985).

20. L.A. Posey, P.J. Campagnola, M.A. Johnson, J.G. Eaton, G.H. Lee and K.H. Bowen, On the origin of the competition between photofragmentation and photodetachment in hydrated electron clusters, $(H_2O)_n^-$, *J. Chem. Phys.* 91:6536 (1989).

21. P. J. Campagnola, L. A. Posey, and M. A. Johnson, Controlling the internal energy content of size-selected cluster ions: An experimental comparison of the metastable decay rate and photofragmentation methods of quantifying the internal excitation of $(H_2O)_n^-$, *J. Chem. Phys.* 95:7998 (1991).

22. W.A. Chupka and J. Berkowitz, Photoionization of ethane, propane, and n-butane with mass analysis, *J. Chem. Phys.* 47:2921 (1967).

23. M.L. Vestal, Theoretical studies on the unimolecular reactions of polyatomic molecule ions. I. Propane, *J. Chem. Phys.* 43:1356 (1965).

24. P.J. Campagnola, D.J. Lavrich, M.J. DeLuca, and M.A. Johnson, Photodestruction spectra of the anionic water clusters, $(H_2O)_n$, $n = 18$ and 30: Absorption to the red of e_{aq}^-, *J. Chem. Phys.* 94:5240 (1991).

25. P. K. Kabir and E. E. Saltpeter, Radiative corrections to the ground-state energy of the helium atom, *Phys. Rev.* 108:1256 (1957).

26. H. Haberland, C. Ludewigt, H. G. Schindler, and D.R. Worsnop, Experimental observation of the negatively charged water dimer and other small $(H_2O)_n^-$ clusters, *J. Chem. Phys. 81:3742 (1984)*.

27. M. Knapp, O. Echt, D. Kreisle, and E. Recknagel, Trapping of low energy electrons at preexisting, cold water clusters, *J. Chem. Phys.* 85:636 (1986).

28. M. Knapp, O. Echt, D. Kreisle, and E. Recknagel, Electron attachment to water clusters under collision-free conditions, *J. Phys. Chem.* 91:2601 (1987).

29. C. E. Klots, Quasiequilibrium rate constants for thermionic emission from small particles, *Chem. Phys. Lett.* 186:73 (1991).

30. P. J. Campagnola, L. A. Posey, and M. A. Johnson, The angular distribution of photoelectrons ejected from the hydrated electron cluster $(H_2O)_{18}^-$

31. L.A. Posey, M.J. DeLuca, P.J. Campagnola, and M.A. Johnson, Reactions of hydrated electron clusters $(H_2O)_n^-$: scavenging the excess electron, *J. Phys. Chem.* 93:1178 (1989).

MULTIPHOTON IONIZATION STUDIES OF
VAN DER WAALS MOLECULES AND CLUSTERS[1]

John C. Miller

Chemical Physics Section
Health Sciences Research Division
Oak Ridge National Laboratory
Post Office Box 2008
Oak Ridge, TN 37831-6125

INTRODUCTION

Multiphoton ionization (MPI) following irradiation of atoms in an intense laser field was first observed by Veronov and Delone in 1965. They tightly focused a ruby laser into xenon gas and observed seven-photon, nonresonant ionization.[1] This work paved the way for what is now a major field of atomic and molecular physics. The first MPI studies of molecules appeared in 1972 and 1973.[2,3] Spurred on by the development and availability of widely-tunable, high-power lasers, studies of both atoms and molecules increased dramatically. Tunability allows the exploitation of resonances which both enhance the MPI cross section and allow spectroscopic studies.

Further advances followed the use of effusive and supersonic atomic and molecular beams and the development of mass-resolved ion detection and energy-resolved photoelectron spectroscopy. Comprehensive reviews of these latter two topics have recently appeared.[4,5] Although the first use of supersonic expansions for MPI in 1975 centered on the ability to rotationally cool the resonance spectra of nitric oxide,[6] the later combination of supersonic jets and mass-resolved MPI detection indicated the presence of weakly- bound clusters of atoms and molecules. In the 1980s the use of both MPI and laser-induced fluorescence for the study of such van der Waals (vdW) molecules has increased exponentially.

In the present paper, we will present results of studies which combine high peak power MPI with supersonic jets and mass-resolved detection to examine atomic and molecular clusters. Examples which will be discussed include clusters of rare gases, nitric oxide molecules, and mixed clusters involving vdW species such as ArNO. Binary molecular clusters of the form $(NO)_m(Y)_n$ where Y represents another small molecule such as CO_2, H_2O, CH_4, etc. have also been extensively studied.

---

[1] Research sponsored by the Office of Health and Environmental Research, U. S. Department of Energy under contract DE-AC05-94OR21400 with Martin Marietta Energy Systems, Inc.

This review will discuss three different types of experiments. The first is resonant MPI studies of NO· rare gas complexes with a nanosecond dye laser.[7] The emphasis here is on learning about the electronic and vibrational properties of small, weakly-bound van der Waals molecules. In the second part, the use of picosecond resonant MPI will be demonstrated where the dynamics of the dissociation of such molecules can be probed.[8] Finally, non-resonant MPI with a picosecond laser will be shown to be an excellent tool for the study of extended cluster distributions such as $(NO)_m$ or $(NO)_m(Y)_n$.[9] The discussion of these last results will center around so-called "magic numbers" which are clusters of a particular size (m,n) which are more stable than other clusters with different numbers of molecules. Evidence for several unique structures of nitric-oxide-containing species will be presented.

All of the weakly-bound species discussed here contain one or more nitric oxide molecules. There are two reasons for this. First, NO is admirably suited for the MPI technique. Its low ionization potential (9.264 eV) and well-studied manifold of states allows for very sensitive detection of nitric-oxide-containing species using visible and near uv lasers. Second, nitric oxide is one of the most ubiquitous molecules in nature and is important to mankind in roles ranging from being an atmospheric pollutant to acting as a neurotransmitter in the human body.

APPARATUS

The experimental arrangements for the study of MPI of clusters have evolved thorough several versions. Figure 1 is a schematic showing its present form. Briefly, the isentropic core of a pulsed supersonic expansion is selected by a skimmer and intersected at 90° with a focused laser beam. The mass-to-charge ratios of the ions which are formed are analyzed using a linear time-of-flight mass spectrometer. The current version of the apparatus differs from those used previously in that a skimmer and differential pumping have been added to the mass spectrometer region and the mass resolution has been improved.

The pulsed valve is commercially available (R. M. Jordan Co.). A 0.5-mm diam nozzle aperture and backing pressures of several atmospheres were typically used in these experiments. The nozzle-to-ionization-region distance is adjustable from 10-17 cm, whereas the 1 or 2 mm diameter skimmer (Beam Dynamics) is fixed at 7.5 cm from the interaction region. Gas pulse durations range from 70-100 μs depending on the carrier gas.

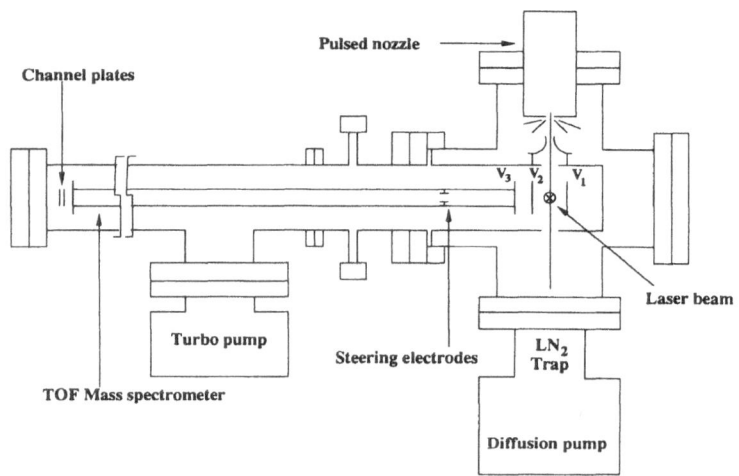

Fig. 1. Experimental apparatus for the multiphoton ionization of molecular clusters.

The laser beam is focused with a 75 mm focal length lens into the jet in the region between the extraction plates of a time-of-flight (TOF) mass spectrometer. The spectrometer is of the design introduced by Wiley and McLaren[34] and has a mass resolution (m/Δm) of about 300 in the range of 1-2500 amu. A dual channel-plate electron multiplier is used to detect the photoions. The field-free region of the flight tube is approximately 0.75 m long and incorporates steering electrodes to counter the cluster kinetic energy perpendicular to the TOF axis. The nozzle, TOF assembly, and focusing lens are mounted within an 8 in. six-way cross and pumped to 10^{-7} Torr with a 6 in. liquid-N_2 trapped, diffusion pump. Additionally, the TOF region is differentially pumped with a 200 l/s turbo-pump.

For the nanosecond experiment an excimer-pumped dye laser system (Lambda Physik) was used. Typically 1-5 mJ of dye laser energy was available. For picosecond experiments, the laser system consists of an Nd:YAG laser (Quantel YG571C) and an H_2 Raman cell (Quanta Ray). The Nd:YAG laser delivers 1200 mJ in a 10 ns pulse (Q-switched operation) or 75 mJ in a 20 ps pulse (mode-locked operation) at the 1064 nm fundamental wavelength. The second, third, or fourth harmonic (532, 355, or 266 nm) of the 1064 nm output is available in either mode of operation, and additional wavelengths are obtained by Raman shifting these harmonics in an H_2 Raman cell. The peak power as a function of wavelength is calculated to be approximately 214, 71, and 28 TW/cm^2 for the 532, 355, and 266 nm outputs of the Nd:YAG laser in the picosecond mode. These calculations assume a Gaussian, diffraction-limited beam and perfect focusing optics. The actual peak powers are unknown but are probably about one order of magnitude less.

The laser is operated at a 10 Hz repetition rate, and the opening of the nozzle aperture is triggered by a signal from the laser. The laser pulse is incident on the early edge of the gas jet in order to optimize cluster detection. After each laser pulse a distribution of masses is detected by the TOF mass spectrometer, and the mass spectrum is recorded with a digital oscilloscope (Tektronix 11402). The signal-to-noise ratio was improved by averaging spectra from about 4000 consecutive laser shots.

The temperature of the neutral clusters is not measured in the present experiments. In previous studies, rotational temperatures of 1-5K for nitric oxide have been measured under similar expansion conditions.[10] The neutral cluster temperatures are assumed to be similar in dilute expansion mixtures but may be considerably higher when extensive clustering occurs.

RESONANT MPI OF NITRIC OXIDE -RARE GAS MOLECULES

Although nitric oxide dimers produced in supersonic expansions were observed as early as 1967,[11] ArNO was not observed until 1973 when Novick et al.[12] detected this species in a molecular beam electric deflection experiment. The first spectroscopic studies of rare gas-NO clusters were presented in 1981 by Langridge-Smith et al.[13,14] who used laser-induced fluorescence to probe NO/rare gas expansions near the $A^2\Sigma^+$ state of NO. For ArNO, the excitation spectrum obtained while monitoring NO fluorescence yielded a broad structureless feature blue shifted about 350 cm^{-1} from the $A \leftarrow X$ transition of the uncomplexed NO. This feature was attributed to excitation to the repulsive part of the ArNO potential well, leading to dissociation and subsequent fluorescence. Similar but weaker features were seen for the Ne and He analogs.

The first mass-resolved multiphoton ionization studies were applied to the NO/rare gas molecules by Sato et al.[15] They observed a structured spectrum of ArNO near the $C^2\Pi \leftarrow X^2\Pi$ transition of NO. The MPI photoelectron spectra also yielded information about the ground state of the ArNO$^+$ ion. Subsequent work from the author's laboratory[7] extended these MPI studies to higher vibrational levels of the $C^2\Pi$ state as well as to the $A^2\Sigma$, the $D^2\Sigma^+$, and $B^2\Pi$ states of ArNO. The $C^2\Pi$ state spectra were obtained and compared for NeNO, ArNO, KrNO, XeNO, and CH$_4$NO. Bound-free spectra similar to those observed by laser-induced fluorescence[13,14] were recorded near the $v = 0$ levels of the A-X transition of NO. In addition, some structured spectra were obtained which probed the bound-bound region. Most recently , in an elegant two-color MPI study, Sato et al,[16] have photodissociated ArNO with one laser via a bound-free absorption and probed the rotational distribution of the nascent NO by $(1+1)$ MPI using a second tunable laser. From this work they were able to obtain upper limits to the ground state dissociation energy of the van der Waals molecules. A very recent threshold photoelectron study has provided details of the NO^+Ar ion.[17]

Finally, using microwave and radio frequency spectroscopy, Mills et al.[17] have analyzed the rotational spectrum to yield the ground state geometry. The ArNO molecule is nearly T shaped with an averaged van der Waals bond distance of 3.71 Å and bond angle of 84.845°.

Direct spectroscopic investigations for ArNO have thus yielded substantial understanding of the ground states of both the neutral and the ion as well as of excited states of the neutral.

As an example of the spectra obtained by resonant MPI, Fig. 2 shows the spectrum of ArNO near the $C^2\Pi \leftarrow X^2\Pi$ transition of uncomplexed nitric oxide. The origin of the van der Waals molecule's spectrum is shifted by 335 cm[-1] which is a typical "matrix shift" and indicates that the excited state is more strongly bound than the ground states. The spectrum shows a vibrational progression of about 58 cm[-1] which is assigned to the stretching mode of Ar-NO. The Franck-Condon envelope peaks for $v = 1$, consistent with a different internuclear distance for the excited state. The smaller peaks probably represent the bending mode.

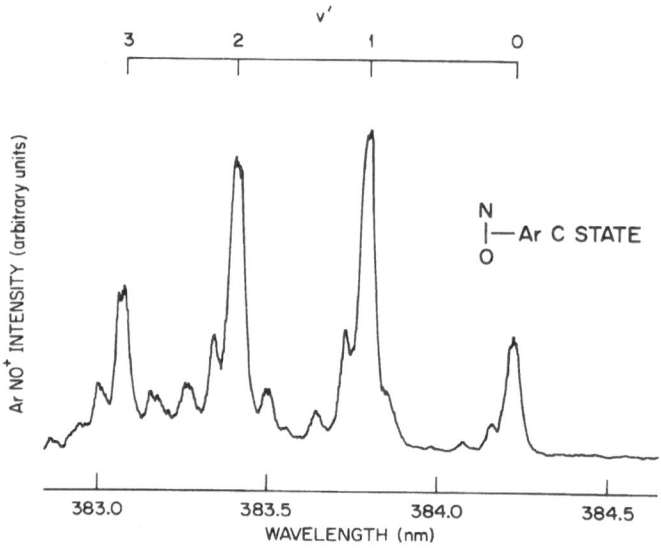

Fig. 2. Multiphoton ionization spectrum of NOAr associated with the C←X transition of NO.

Similar spectra were obtained for NeNO, KrNO, and XeNO and the results are summarized in Table 1. Most of the spectroscopic parameters scale with the strength of the van der Waals interaction, which is seen to correlate with the polarizability, α, of the rare gas partner. Table 2 collects all of the known spectroscopic data for both neutral and ionic ArNO. A rather complete picture of these weakly-bound nitric oxide rare gas species has emerged.

TABLE I. SUMMARY OF SPECTROSCOPIC CONSTANTS FOR THE KNOWN STATES OF ArNO AND ArNO+

	T_0 (cm^{-1})	ω_0 (cm^{-1})	D_0 (cm^{-1})	Ref.
ArNO X $^2\Pi_{1/2}$ (v=0)	0	-	~ 87	7,16
ArNO A $^2\Sigma^+$ (v=0)	44,201	-	54-101	7
ArNO B $^2\Pi_{1/2}$ (v=9)	54,138[a]	30	130	7
ArNO C $^2\Pi_{1/2}$ (v=0)	52,038[a]	58	422	7,13
ArNO D $^2\Sigma^+$ (v=0)	52,450[a]	63	931	7
ArNO+ X $^1\Sigma^+$ (v=0)	73,869	94	951	17

[a]First observed band assumed to be the origin

TABLE II. SUMMARY OF C STATE DATA FOR NO-RG SPECIES

	T_0 (cm^{-1})	ΔT_0 (cm^{-1})	ω'_0 (cm^{-1})	K^a (10^{-2}dyn/A)	a^b (A^{3}0)
NO	52 373	. . .	2395	2522	. . .
NeNO	52 287	-86	34	0.82	0.3946
ArNO	52 038	-335	58	3.39	1.642
KrNO[c]	51 978	-395	53	3.63	2.480
XeNO[c]	51 711	-662	59	5.01	4.044
CH$_4$NO	51 605	-768	96	5.66	2.699

[a]Force constant for vdW bond (diatomic approximation).
[b]Polarizability of vdW partner.
[c]Data in table for higher energy series.

RESONANT MPI OF DISSOCIATING MOLECULES

Similar MPI studies on other important vdW molecules have been impeded by the presence of dissociative neutral resonances or of ion fragmentation. For instance, in none of the previously cited MPI studies of ArNO was the nitric oxide dimer ion, $(NO)_2^+$, observed, although the neutral dimers were known to exist in the molecular beam. Similarly, previous studies of iodine/rare gas expansions failed to detect the ArI$_2^+$ ion.[19] In both of these cases efficient dissociative pathways prevented observation of the parent ion. Use of the present picosecond laser can ameliorate these problems in two ways. First, the high peak power allows nonresonant MPI to be observed, thus avoiding dissociative resonances of the neutral cluster. Furthermore, when such resonances are encountered, the short pulse (10-20 ps) and high ionization rate allow efficient competition with bond breaking. These points will be illustrated by the results of picosecond MPI of nitric oxide clusters at a number of fixed wavelengths.

Nitric oxide and $(NO)_n$ van der Waals molecules have been the subjects of numerous spectroscopic studies, not only due to their intrinsic interest but because NO plays a prominent role in the chemistry of the upper atmosphere. Although NO dimers and rare gas-NO clusters are readily formed in a free jet expansion, nanosecond MPI experiments with these species, such as those described in the last section, have failed to detect the $(NO)_2^+$ parent ion. This failure to detect the dimer ion may be attributed to the presence of dissociative states in the $(NO)_2$ and $(NO)_2^+$ manifolds. The only known excited state of the neutral dimer dissociates very rapidly,[20] and the dimer ion is readily photodissociated by visible light.[21] The electronic states of NO and $(NO)_n$ are shown schematically in Fig. 3 along with the multiphoton steps for the wavelengths used in the present study.

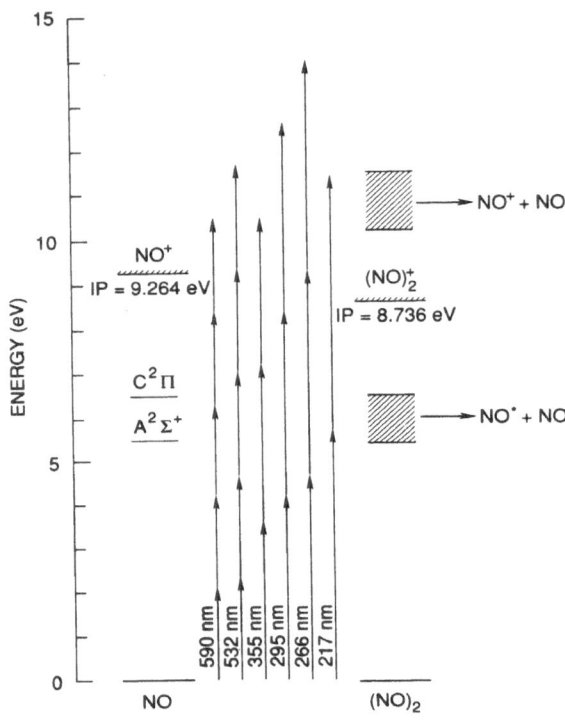

Fig. 3. Energy level diagram for NO and $(NO)_2$

We do not observe the dimer ion (or larger polymers) when we use visible light (532 nm or 585-605 nm). However, when either nanosecond or picosecond UV light (355 nm, 266 nm, or frequency-doubled 585-605 nm) is used, $(NO)_n^+ (n = 1-5)$ ions are readily observed. These results are consistent with rapid dissociation of the ion following further excitation with visible light. When UV photons are used to ionize, this dissociating resonance is avoided. Fig. 4a and 4b show TOF spectra following MPI with 217 nm light (the second anti-Stokes Raman-shifted wavelength from the 266 nm output). In this case the two-photon ionization is resonant with the dissociative excited state in the neutral manifold. Only NO^+ ions are observed with nanosecond pulses, while $(NO)_n^+ (n = 1-4)$

ions are observed with picosecond pulses. Clearly, the higher peak power available in the picosecond pulses enables MPI to compete effectively with the fast neutral dissociation channel. In principle, the lifetime of the excited dimer state might be determined from careful laser power dependence studies of the dimer ion intensity at 217.8 nm. A straightforward rate equation analysis for such a four-level system leads to a functional form of the power dependence from which the lifetime could be extracted if the excitation and ionization cross-sections were known. In fact, these quantities are not known for the dimer and, furthermore, power dependence studies are difficult to perform accurately in this wavelength region, so no such analysis is warranted at the present time. Further details of this work may be found in Reference 8.

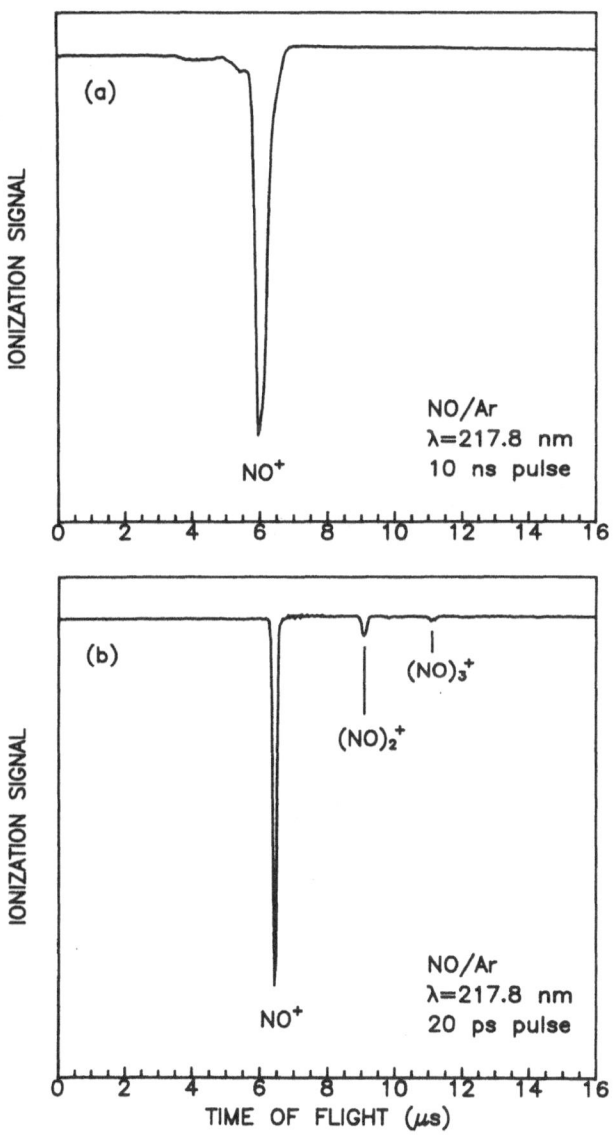

Fig. 4. Masss spectra following a) 10 ns and b) 20 ps MPI at 217.8 nm

NONRESONANT MPI OF CLUSTERS

The success of the picosecond MPI studies of $(NO)_n$ and $NOAr_n$ clusters described above, particularly those using the 266 nm fourth harmonic of the Nd:YAG laser, led to a systematic study of $(NO)_m(Y)_n$ clusters. For clusters containing nitric oxide, the nonresonant two-photon ionization at 266 nm avoids dissociative resonances and ionizes just above the NO ionization potential, thus leaving little excess energy in the cluster ion. The high peak power of the laser applied to such a low order multiphoton ionization leads to very high detection sensitivity. This is essential as large clusters are present at very low concentrations in the cluster beam.

A major motivation of these studies is to investigate the properties of solvated nitric oxide molecules and ions. This is important for understanding the role of atmospheric or cellular nitric oxide where it is probably best represented as a solvated species such as $NO(H_2O)_n$. A second emphasis is the observation of certain "magic" sized clusters which indicate especially stable or strongly-bound species due to electronic or geometric factors.

In this section, we will first describe non-resonant MPI studies of $(Xe)_n$ to illustrate the concepts of magic numbers. The results on several nitric oxide containing clusters will be presented which show several other types of "magic."

Xenon Clusters

Since the first observation of magic numbers in rare gas cluster distributions, researchers have been intrigued with the possible reasons for the extra stability of certain sized clusters.[22] A model has evolved which involves assigning the magic numbers to especially symmetric structures of the ionized cluster and recognizing that extensive fragmentation occurs during the ionization process.[23,24] Magic numbers have been identified in cluster distributions of many atomic and molecular species including the now famous Buckyball or C_{60}.

Fig. 5 shows the distribution of Xe_n^+ clusters observed on the present apparatus following three-photon ionization. Magic numbers, marked with asterisks, are observed for clusters with n = 13, 16, 19, 23, 25, 29 and 55. The most prominent of these correspond

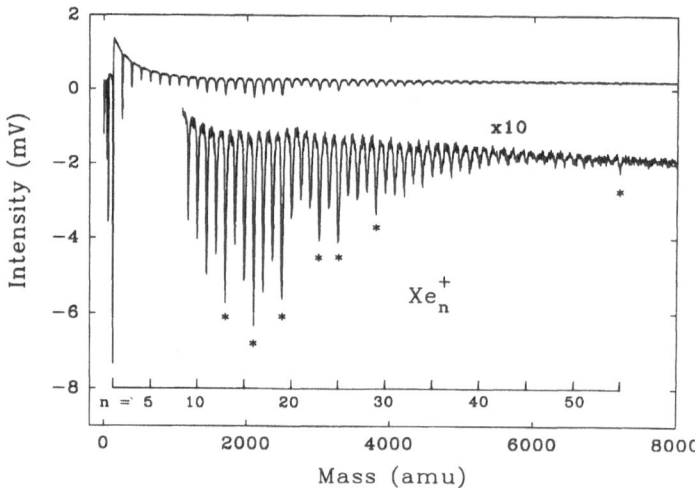

Fig. 5. The $(Xe)_n^+$ cluster distribution following three photon ionization with 266 nm light of a neat xenon expansion. The asterisks indicate magic numbers.

to highly symmetric icosahedral structures. Calculations[25] indicate that Xe_{13}^+ and Xe_{19}^+ are especially stable when arranged as an icosahedron and double icosahedron respectively about a central Xe^+ ion. Xe_{54}^+ arises after completing another solvation sphere. A similar spectrum, obtained with nanosecond laser pulses, has been published previously.[26]

Nitric Oxide-Argon Clusters

Small clusters composed of nitric oxide and various rare gas partners have been studied in our labs previously, as outlined in earlier sections. Recently we have been able to synthesize very large clusters of the form $(NO)_m(Ar)_n$.[9] The ranges of the indices m (1-8)

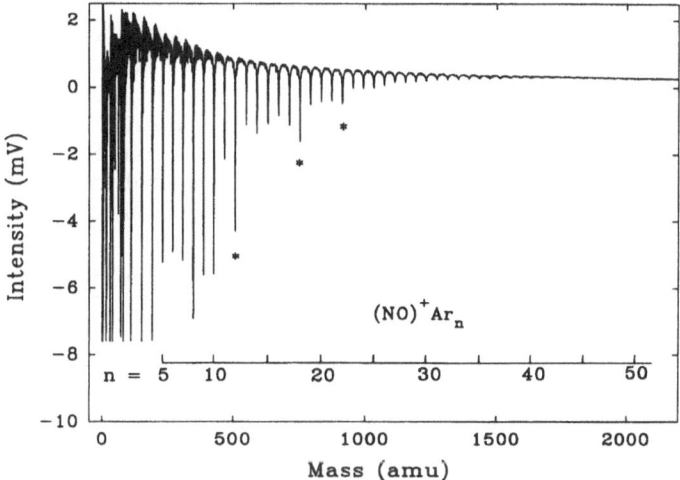

Fig. 6. The $(NO)^+Ar_n$ cluster distribution following four photon ionization with 532 nm light of a 1 % NO/Ar expansion. The asterisks indicate magic numbers.

and n(1-54) depend on the relative concentrations and expansion conditions.

Fig. 6 shows a spectrum of $NO^+Ar_n(n \leq 54)$ following 532 nm ionization. As discussed previously, all cluster ions with m > 1 are dissociated by visible light. Since the ionization potential of NO is lower than that of Ar, one expects structures where the charge is localized on the NO^+ core, and the argon atoms arrange themselves in solvent spheres. The magic numbers for n = 12, 18, 22, and 54 indicate that icosahedral structures are especially stable, similar to the xenon clusters of Fig. 5. Using 266 nm photons, the solvated nitric oxide dimer $[(NO)_2^+Ar_n]$ magic numbers are only observed for n = 17 and 21. Apparently, the dimer is too big to be stabilized within a single icosahedron. A double icosahedron, however, has two central sites and can accommodate the dimer.[9]

Neat Nitric-Oxide Clusters

In contrast to the $(NO)_m(Ar)_n$ clusters formed in a 1% NO/Ar mixture described previously, when a 5% NO/Ar mixture was expanded in a supersonic jet and two-photon ionized, the mass spectrum of Fig. 7 was obtained. Pure NO clusters up to about $(NO)_{40}^+$ are readily observed. The overall envelope of the distribution is seen to be a monotonically decreasing function of m. Superimposed on this envelope, however, is a very striking odd-even intensity alternation, with the odd-m clusters being more intense. This alternation is evident only for m > 8. For m > 19, *only* the odd-sized clusters are observed. Weaker intermediate peaks in the low x region are due to $(NO)_m^+N_2O$ species from natural nitrous oxide contaminant in the gas. If no purification is done, this series can be seen out to m = 40 and also exhibits odd-even alternation. Similar odd-even effects are seen in the negative ion cluster spectrum produced by Rydberg electron transfer.[27]

Because NO is a free radical having an unpaired electron, even-m neutral clusters have all of their electrons paired leading to stronger bonding than for the odd-m neutral clusters. For the ions, however, the argument reverses. Odd-m positive or negative ions have all of their electrons paired and should be more strongly bound and hence more stable.

Relating relative bond strengths and stabilities to the intensity distributions in Fig. 7 is not trivial, however, because dynamical factors also play a role. For instance, the observed positive ion abundances reflect the original neutral distribution, the m-dependent photoionization cross section (related to the x-dependent ionization potential) and m-dependent cluster evaporation rates. Likewise, the negative ion distribution reflects electron attachment cross sections, cluster evaporation rates, and possibly electron autodetachment processes.

Similar odd-even alternations have been observed in cluster distributions of one-electron atoms such as the group IA[28] or IB[29] metals. The present work represents the first example of such odd-even effects for covalent molecular clusters.

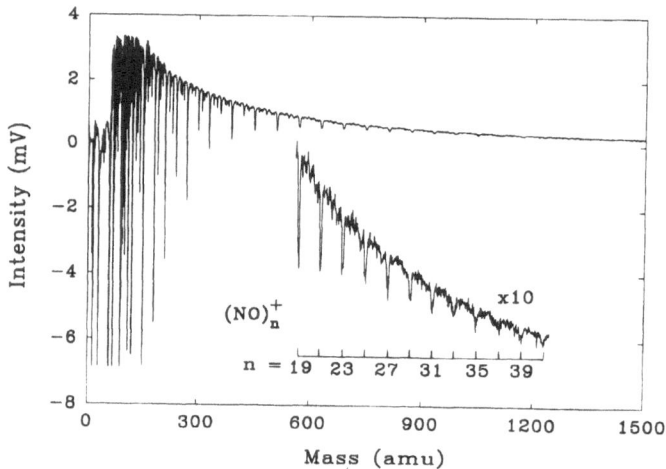

Fig. 7. The $(NO)_n^+$ cluster distribution following two photon ionization with 266 nm light of a 5% NO/Ar expansion.

212

Nitric Oxide-Methane Clusters

Fig. 8 shows the cluster distribution obtained after ionizing an expansion of nitric oxide, methane and argon (5%, 5%, 90%). This distribution is quite different from that observed for any other binary molecular cluster studied by the author or reported in the literature. From inspection of Fig. 8, it is immediately obvious that only some of the possible $(NO)_m^+(CH_4)_n$ clusters are observed. First, for the bulk of the distribution, only mixed clusters containing odd numbers of NO molecules are observed. This is similar to the odd-even alternation observed for pure $(NO)_m^+$ clusters, but much more extreme for low m than previously observed. [$(NO)_2^+(CH_4)$ and $(NO)_4^+(CH_4)$ can be weakly observed.] The second striking feature is that for each cluster with m nitric oxide molecules only one peak is strongly observed corresponding to a set number, n, of CH_4 molecules. In each case $n = \frac{m-1}{2}$ or rearranged $m = 2n+1$. For the particular concentration of Fig. 8, the $n = \frac{m}{2}$ peak is also observed weakly.

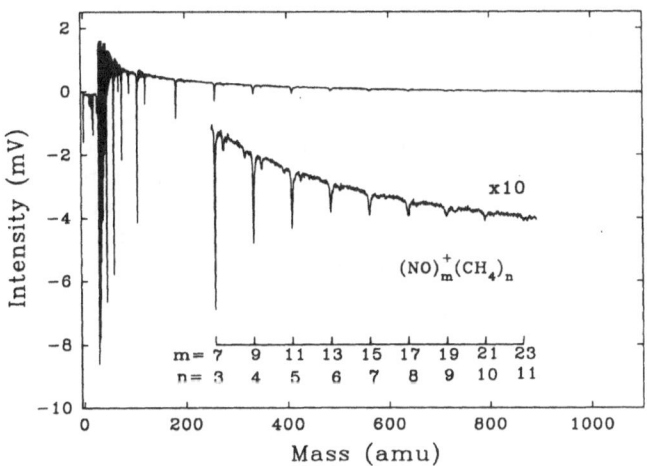

Fig. 8. The $(NO)_m^+$ $(CH_4)_n$ cluster distribution following two photon ionization with 266 nm light of an NO/CH$_4$/Ar expansion.

The implication of the fixed relationship between m and n is that a very stable configuration of mixed NO/CH$_4$ must exist. Furthermore, the polymer, which can be observed out to $(NO)_{23}(CH_4)_{11}$, appears to be infinitely repeating. Such a stoichiometric relationship within heteroclusters is not observed in any other mixed clusters. Furthermore, no similar examples for relatively weakly-bound molecular clusters are known to the author. Similar stoichiometric hetero-clusters have been observed for mixed metal/carbon atom clusters, such as the metallo-carbohedrenes (met-cars).[30] Also, recently some novel molybdenum carbide species of formula $MO_nC_{4n}(n = 1-4)$ have been reported.

Clearly, more experimental studies as well as some theoretical guidance are required before speculating about possible structures. More details may be found elsewhere.[32]

Other Binary Clusters

We have also prepared binary clusters of nitric oxide and several other partners.[9] These include N_2O, CO_2, CS_2, H_2O, SF_6, and SO_2. In each case extensive clustering is observed and some magic peaks are seen in the mass spectra. Space limitations preclude discussing these results here.

CONCLUSIONS

Clearly, the coupling of multiphoton ionization, mass spectrometry and supersonic nozzles produces a powerful tool for the study of atomic and molecular clusters. Resonant MPI is most useful for understanding the electronic and rovibronic spectroscopy of small van der Waals molecules. Its application is much more difficult for large clusters however. Non-resonant MPI, on the other hand, seems ideal for the study of cluster distributions but, of course, provides little spectroscopic information.

We suggest that laser mass spectrometry using high-power lasers may constitute a near "universal" detector of atomic and molecular species. Unlike resonant MPI, no knowledge of intermediate states is required. Although more complex and more expensive than the traditional electron impact source, MPI has some unique advantages (and disadvantages) which are discussed below.

A major factor involved in the MPI technique is that atoms or molecules with a high ionization potential (IP) are discriminated against relative to those with a low IP as, typically, an additional photon is required for ionization. An nth order MPI process usually has a cross-section several orders of magnitude larger than that for n + 1 MPI. This need not always be a disadvantage, however, For instance, common expansion gases such as He, Ar, or N_2 all have high IPs and thus would appear much less prominently (if at all) in the MPI mass spectrum. The spectrum of the dopant and its clusters would then be less congested. Furthermore, space-charge problems are also reduced if the most abundant species are not ionized. Since clusters typically have a reduced IP relative to monomers, a judicious choice of the wavelength for MPI could allow discrimination against the nonomer while all of the cluster ions would be efficiently ionized. For instance, the E_i of $(NO)_2$ is 8.736 eV compared to 9.264 eV for uncomplexed NO. The choice of a laser wavelength between 284.8 and 267.7 nm would allow two-photon detection of all $(NO)_n$ clusters while three photons would be required to ionize NO.

Compared to resonant MPI, the non-resonant analog gives a more "complete" representation of the populations of species in the beam. (Of course, the ionization selectivity is also lost.) Furthermore, since precise tuning is not required, the higher power and better beam quality of the Nd:YAG harmonics may be used. The option of using different order MPI schemes by choosing different harmonics can be used to avoid resonance, as simply demonstrated in the present study. Finally, many of the adjunct techniques developed for resonant MPI can also be used in the non-resonant experiment. These might include, for instance, photoelectron or ion kinetic energy analysis, two-color techniques, or polarization dependencies.

ACKNOWLEDGMENTS

The contributions of many collaborators - W. C. Cheng, D. B. Smith, S. Desai, C. S. Feigerle and H. S. Carman - are gratefully acknowledged as are numerous discussions with C. E. Klots and R. N. Compton. The use of a NATO International Collaboration Grant (0474/87) during the course of this work is also acknowledged.

All research was sponsored by the Office of Health and Environmental Research, U.S. Department of Energy under contract DE-AC05-84OR12400.

REFERENCES

1. G. S. Veronov and N. B. Delone, Ionization of the xenon atom by the electric field of ruby laser emission, *JETP Lett.* 1:66 (1965).
2. W. C. Lineburger and T. A. Patterson, Two photon photodetachment spectroscopy: The C_2^- $^2\Sigma$ states, *Chem. Phys. Lett.* 13: 40 (1972).
3. C. B. Collins, B. W. Johnson, D. Popescu, G. Musa, M. E. Pascu, and I. Popescu, Multiphoton ionization of molecular cesium with a tunable dye laser, *Phys. Rev. A* 8:2197 (1973).
4. "Lasers and Mass Spectrometry," D. M. Lubman, ed., Oxford University Press, 1990.
5. R. N. Compton and J. C. Miller, Multiphoton ionization photoelectron spectroscopy: MPI-PES, *in* "Laser Applications in Physical Chemistry," Marcel Dekker, Inc., New York (1989), pp. 221-306.
6. D. Zakheim and P. N. Johnson, Two- and three-photon resonances in the four-photon ionization spectrum of nitric oxide at low temperatures, *J. Chem. Phys.* 68:3644 (1978).
7. J. C. Miller and W. C. Cheng, Multiphoton ionzation of NO-rare gas van der Waals species, *J. Phys. Chem.* 89:647 (1985); J. C. Miller, Multiphoton spectroscopy of X-NO (X=Kr, Xe, CH_4) van der Waals molecules, *J. Chem. Phys.* 86:3166 (1987); The $A^2\Sigma^+$ state of ArNO van der Waals molecules probed by $1+1$ muiltiphoton ionization spectroscopy, *J. Chem. Phys.* 90:4031 (1989).
8. D. B. Smith and J. C. Miller, Picosecond multiphoton ionization of molecular clusters, *J. Chem. Phys.* 90:5203 (1989); Nonresonant and resonant multiphoton ionization of $(NO)_n$ and Ar_nCO clusters with picosecond laser pulses, *J. Chem. Soc. Faraday Trans. II* 86: 2441 (1990).
9. S. R. Desai, C. S. Feigerle and J. C. Miller, Magic numbers in $(NO)_m^+ Ar_n$ heteroclusters produced by two-photon ionization in a supersonic expansion, *J. Chem. Phys.* 97:1793 (1992); Multiphoton ionization and dissociation of $(NO)_m(Ar)_n$ clusters, *Z. Phys. D.* 26:220 (1993); Magic cluster ion distributions $(NO)_3^+(N_2O)_n$ and $(NO)_3^+(CO_2)_n$, *Z. Phys. D.* 26: S183 (1993).
10. J. C. Miller and R. N. Compton, Multiphoton ionization of ultracold nitric oxide, *J. Chem. Phys.* 84:675 (1986); J. C. Miller, Ultra sensitive, isotopically-selective detection of nitric oxide in a supersonic beam, *Anal. Chem.* 58:1702 (1986); C. S. Feigerle and J. C. Miller, Multiphoton ionization of vibrationally hot nitric oxide produced in a pulsed supersonic glow discharge, *J. Chem. Phys.* 90:2900 (1989).
11. T. A. Milne and F. T. Greene, Mass spectrometric detection of dimers of nitric oxide and other polyatomic molecules, *J. Chem. Phys.* 47:3668 (1967).
12. S. E. Novick, P. B. Davies, T. R. Dyke, and W. Klemperer, Polarity of van der Waals molecules, *J. Am. Chem. Soc.* 95:8547 (1973).
13. P. R. R. Langridge-Smith, E. Carrasquillo M., and D. H. Levy, The direct photodissociation of the van der Waals molecule NO · Ar, *J. Chem. Phys.* 74:6513 (1981).
14. E. Carrasquillo M., P. R. R. Langridge-Smith, and D. H. Levy, The direct photodissociation of van der Waals molecules *in*: Laser Spectroscopy V, ed., A. R. W. McKellar, T. Oka, and B. P. Stoicheff ,Springer, New York (1981), p. 333.
15. K. Sato, Y. Achiba, and K. Kimura, The Ar · NO van der Waals complex studied by resonant multiphoton ionization spectroscopy involving photoion and photoelectron measurements, *J. Chem. Phys.* 81:57 (1984).
16. K. Sato, Y. Achiba, H. Nakamura, and K. Kimura, Anomolous rotational-state distribution of NO A state in UV photodissociation of rare gas-NO van der Waals complexes. Rotational rainbow effect, *J. Chem. Phys.* 85:1418 (1986).
17. M. Takahashi, Two-color $(2+1')$ multiphoton ionization threshold photoelectron study of the Ar · NO van der Waals complexes: Observation of intermolecular vibrational progressions of the $ArNO^+$ cation, *J. Chem. Phys.* 96:2594 (1992).
18. P. D. A. Mills, C. M. Western, and B. J. Howard, Rotational spectra of rare gas-nitric oxide van der Waals molecules. 2. The structure and spectrum of argon-nitric oxide, *J. Phys. Chem.* 90: 4961 (1986).
19. J. C. Miller, Multiphoton ionization of jet-cooled iodine, *J. Phys. Chem.* 91:2589 (1987).
20. J. Billingsly and A. B. Callear, Investigation of the 2050Å system of the nitric oxide dimer, *Trans. Faraday Soc.* 67:589 (1971).
21. M. F. Jarrold, A. J. Illies and M. T. Bowers, Investigation of the dynamics and energy disposal in the photodissociation of small ion clusters using a high-energy ion beam crossed with a laser beam: Photodissociation of $(NO)_2^+$ in the 488-660 nm range, *J. Chem. Phys.* 79:6086 (1983).
22. O. Echt, K. Settler and E. Recknagel, Magic numbers for sphere packings: Experimental verification in free xexon clusters in *Phys. Rev. Lett.* 47:1121 (1981).

23. H. Haberland, A model for the processes happenings in a rare-gas cluster after ionization, *Surf. Sci.* 156:305 (1985).

24. C. E. Klots, Kinetic methods of quantifying magic; *Z. Phys. D.* 20:1001 (1991); Systematics of evaporation, *Z. Phys. D.* 20:105 (1991).

25. H. U. Böhmer and S. D. Peyerimhoff, Stability and structure of singly-charged xenon-argon clusters $[Xe_1, Ar_{n-1}]^+$, n = 3-27, a Monte-Carlo simulation, *Z. Phys. D.* 8:91 (1988).

26. O. Echt, M. C. Cook and A. W. Castleman, Multiphoton ionization studies of xenon clusters, *J. Chem. Phys. Lett.* 135:229 (1987).

27. H. S. Carman, Jr., Low energy electron attachment to clusters of nitric oxide, *J. Chem. Phys.* (in press)

28. K. Hansen, J. T. Khoury, M. L. Homer, F. E. Livingston and R. L. Whetten, Picosecond resonant two-photon ionization of cold sodium clusters, *Z. Phys. D.* 26:187 (1993).

29. D. E. Powers, S. G. Hansen, M. E. Guesic, D. L. Michalopolous and R. E. Smalley, Supersonic copper clusters, *J. Chem. Phys.* 78:2866 (1983).

30. S. F. Cartier, Z. Y. Chen, G. S. Walder, C. R. Sleppy and A. W. Castleman, Jr., Production of metallo-carbohedrenes in the solid state, *Science* 260:195 (1993).

31. C. Jin, R. E. Hauffler, R. L. Hettich, R. N. Compton, A. A. Puretzky and A. V. Dem'yanenko, Novel molybdenum carb ide clusters: Mo_nC_{4n} (n = 1-4), *Science*, in press (1993).

32. S. R. Desai, C. S. Feigerle and J. C. Miller, Laser ionization of binary molecular clusters containing nitric oxide, to be published

CORE LEVEL EXCITATION IN FREE CLUSTERS:
NEXAFS, EXAFS AND COULOMB EXPLOSION

J. Geiger, S. Rabe, C. Heinzel, H. Baumgärtel and E. Rühl

Institut für Physikalische und Theoretische Chemie
Freie Universität Berlin
Takustr. 3
D-14195 Berlin, Germany

INTRODUCTION

Photoionization of clusters in the regime of valence electron excitation has provided in the past information on ionization potentials, thermochemical and spectroscopic properties as a function of cluster size.[1-3]

More recently, core level excitation with soft X-rays has been investigated in order to bridge the gap between spectroscopic and photochemical properties of the isolated gas phase species and the condensed phase.[4-11] First systems under investigation have been rare gas clusters,[4-10] and aggregates of simple molecules,[11] which show characteristic changes of spectroscopic properties as a function of their size. Specifically, argon clusters show in the Ar 2p excitation regime the evolution of atomic Rydberg states to surface and bulk exciton states with increasing cluster size.[4,5] Surface and bulk exciton states are clearly distinguished because of their blue-shift with respect to the atomic Rydberg states. The regime of core ionization continua is dominated by broad oscillations, which have been interpreted in terms of extended X-ray absorption fine structure (EXAFS).[5-7] The EXAFS analysis allows the determination of the number of nearest neighbors as well as the Ar-Ar distance. Considerably different spectral behavior is found for total fluorescence yield spectra in the Ar 2p regime.[8] Dispersed fluorescence spectra indicate that the "*third continuum*" of argon is observed, which is assigned to the radiative decay of doubly charged argon clusters before charge separation ("*Coulomb explosion*") occurs. *Coulomb explosion* is the dominant decay process of small multiply charged clusters, where both symmetric and asymmetric charge separation processes are observed.[9,10]

We present in this paper recent progress in the field of core excitation of molecular clusters: Benzene and methanol are investigated in the regime of C 1s excitation (280-400 eV). The aim of this study is (i) the investigation of characteristic changes in the near-edge (NEXAFS) spectra as a function of cluster size; (ii) to find evidence for EXAFS in molecular clusters in order to determine local structures within clusters; (iii) to investigate fragmentation of multiply charged clusters, which are formed efficiently in the regime of core level excitation.

EXPERIMENTAL

The experimental setup has been described earlier in detail.[5,8,9,10] Briefly, it consists of a supersonic jet expansion for cluster preparation, devices for detection of cations (total cation yields and a time of flight mass spectrometer), electrons, and fluorescence, as well as synchrotron radiation as a tunable source of soft X-rays in the energy range between 280-400

Linking the Gaseous and Condensed Phases of Matter
Edited by L.G. Christophorou *et al*, Plenum Press, New York, 1994

eV. The HE-TGM-2 monochromator at the storage ring facility BESSY (Berlin) is utilized for the measurements at the C 1s edge. The experiments were performed typically with an energy resolving power of (E/ΔE) of 300-500. Coincidences between charged particles (electrons and cations) are measured with a multihit time-to-digital converter (MIPSYS; FLY-TDC). Clusters of benzene and methanol are produced by seeding the liquid samples with rare gases (Ar and He). The samples are of commercial quality (Merck, purity: 99.9%) which are used without further purification. No impurities are observed by photoionization mass spectrometry.

RESULTS AND DISCUSSION

NEXAFS in Molecular Clusters

Benzene Clusters. Figure 1a shows photoion yield curves of the benzene dimer (Bz_2^+) in the regime of the C 1s excitation recorded at different stagnation conditions, as well as photoion yields for larger singly and doubly charged benzene cluster cations. The top spectrum shows the Bz^+/Bz_2^+ charge separation channel, which is obtained from photoion photoion coincidence (PIPICO) experiments. The spectral features in the 280-320 eV energy regime are well-known.[12] As in the case of other molecular species one observes all features in the gas and condensed phase,[13] although there are differences in relative intensities. The lowest energy feature is the C 1s→π* (e_{2u}) transitions at 285.2 eV. We find no shift of this resonance as a function of the cluster size. Three other maxima occur at higher energy: 288.9 eV (π* (b_{2g})), 293.5 eV (σ*(e_{1u})), and 300.2 eV (σ*($e_{2g}+a_{2g}$)).[13] It should be noted that the feature occurring at 293.5 eV has been assigned to a doubly excited state decaying by autoionization.[14] The three photoion yield curves of Bz_2^+ show increased intensity in the C 1s continuum as the cluster size is increased (see Fig. 1a). The threshold of dimer formation is observed at about p_0=0.6 bar. The corresponding dimer spectrum consists mainly of a strong π* resonance and weak intensity at higher photon energy. A similar shape is observed for Bz^+ (m/z=78) under effusive jet conditions (not shown) indicating that the dimer cation is

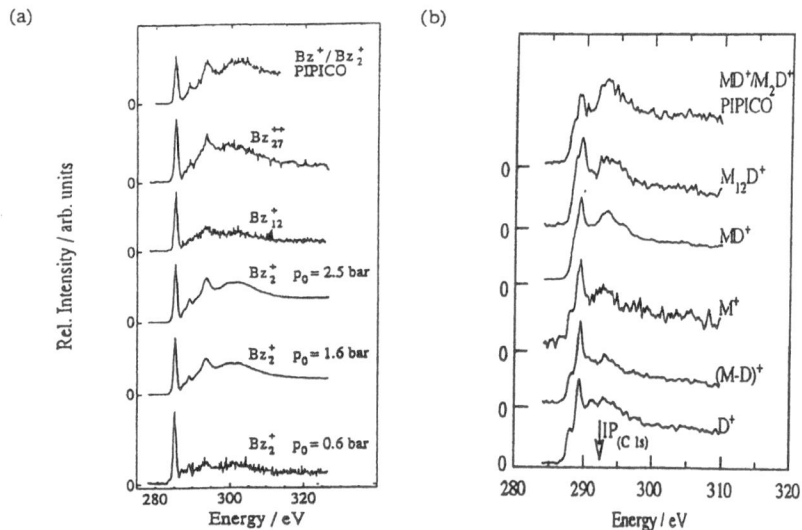

Figure 1. (a) Photoion yield curves of the benzene dimer (Bz_2^+) at different expansion conditions (seed gas: He), Bz_{12}^+, Bz_{27}^{++} and the yield of the Bz^+/Bz_2^+ PIPICO channel. **(b):** Photoion yield curves of methanol-d_4 (M), its fragments, methanol-d_4 cluster fragments and the yield of the MD^+/M_2D^+ PIPICO channel.

formed as a result of single ionization, if no larger clusters are present in the beam. The photoion yields at p_0=1.6 bar and 2.5 bar are quite similar in shape but there is significantly greater intensity in the C 1s continuum as a result of multiparticle fragmentation of C 1s ionized large benzene clusters. The critical size for stable dications has been reported recently to be Bz_{23}^{++}.[15] The photoion yield of Bz_{27}^{++} is remarkably different from that of Bz_{12}^{+} which is close in time of flight and both spectra are recorded at p_0=3.6 bar. The singly charged fragment shows small contributions from dissociative double ionization (as the dimer if no larger clusters are present at p_0=0.6 bar), whereas the stable dication shows a spectrum of similar shape as the dimer cation yields and the PIPICO yield of the Bz^{+}/Bz_2^{+} coincidences (top spectrum in Figure 1a).

These findings indicate that photoion yields of mass selected molecular clusters change their shape as a function of the neutral cluster size. Small cluster fragments, such as dimers, are formed efficiently via dissociative double ionization.

Methanol Clusters. The C 1s excitation regime of gaseous methanol (M) has been investigated earlier.[16,17] Three intense resonances govern this spectral region. The lowest energy feature at 287.9 eV corresponds to the lowest Rydberg state (C 1s→3sa' transition). The maximum at 289.4 eV has been originally assigned to the 3p Rydberg excitation,[16] whereas this feature has been assigned later to the C 1s→$\pi^*$$_{CH_3}$ transition.[17] A broad maximum at 292.4 eV corresponds to the C 1s→$\sigma^*$$_{CO}$ excitation. This energy is close to the C 1s ionization energy (292.42 eV).[18] The spectra shown in Figure 1b indicate that these spectral features have different intensities in the ion yield curves of mass selected species. The molecular fragments (D^+, CD_3O^+, CD_3OD^+) show clearly the lowest Rydberg state as a shoulder, if no clusters are present in the beam. The intensity of the $\sigma^*$$_{CO}$ resonance is relatively weak compared to the $\pi^*$$_{CH_3}$ resonance. In contrast, the lowest Rydberg state is not observed for cluster fragments (($CD_3OD)_nD^+$) as well as those mass channels which are also produced via cluster fragmentation. One typical example is the cation CD_3O^+ which changes its shape considerably when clusters are present in the beam (not shown). This finding is characteristic for Rydberg excitations in molecular clusters, which are known to be washed out as the cluster size is increased.[19] The dominant feature (C 1s→$\pi^*$$_{CH_3}$) is not shifted in energy in the cluster spectra as it is typical for valence transitions. However, the intensity of this feature changes relative to the continuum resonance (C 1s→$\sigma^*$$_{CO}$) as well as the intensity in the C 1s continuum if clusters are present in the beam. The top spectrum in Figure 1b shows the PIPICO yield of the $CD_3OD_2^+/(CD_3OD)_2D^+$ (MD^+/M_2D^+) cation pair, indicating that charge separation is comparably weak below the C 1s ionization threshold. Therefore, high PIPICO yield is found in the C 1s continuum.

Changes in the shape of photoion yield curves of molecular and cluster fragments reflect different fragmentation processes, as already discussed for benzene clusters.

EXAFS in Molecular Clusters

Photoion yield curves of molecular clusters have been investigated with respect to possible intermolecular EXAFS structure. Clusters containing small molecules with light elements do not show distinct evidence for intermolecular EXAFS.[11] One obvious reason for this finding is, that EXAFS is mostly sensitive for backscattering from nearest neighbors rather than from more distant shells. However, evidence for weak intermolecular EXAFS has been found in the case of benzene. Intramolecular EXAFS of the isolated molecule has been investigated earlier in the gas and condensed phase.[20,21]

Figure 2a shows the cation yield of benzene clusters (Bz_2^+-Bz_8^+) in an extended energy range. Weak oscillations of the cross section are observed in the C 1s continuum, which suggests evidence for *intermolecular* EXAFS in benzene clusters. The upper part of Figure 2b shows that the experimental and theoretical EXAFS signal (from F_{eff} calculations)[22] are in good agreement for an *intermolecular* C-C-distance of 3.45 Å (see lower part of Figure 2b). Six nearest neighbors are obtained from the EXAFS analysis, which is consistent with a cluster structure containing T-shaped dimer units. This structure agrees with earlier experimental and theoretical work for benzene dimers as well as solid benzene.[23] One expects from the intermolecular potential for a T-shaped geometry that the C-C distance should be 3.75 Å, which is in agreement with the value that is deduced from the EXAFS analysis. Our preliminary analysis is different from that reported earlier for solid benzene,[21] since we have analyzed the low k range, where essentially no *intramolecular* EXAFS is expected to occur.

This is shown in Figure 2a, where the experimental spectrum is compared to the calculated *intramolecular* EXAFS (r_{C-C}= 1.41 Å) and the above mentioned *intermolecular* EXAFS. The results indicate that the regime close to the C 1s edge is more sensitive to *intermolecular* EXAFS, whereas at higher energy the *intramolecular* contribution dominates. The quality of the EXAFS analysis is somewhat limited for *intermolecular* EXAFS because of the overlap with the *intramolecular* part and damping of the EXAFS-signal with increasing excitation energy.

The results indicate that structural information of free molecular and atomic clusters can be obtained from EXAFS experiments.[6,7]

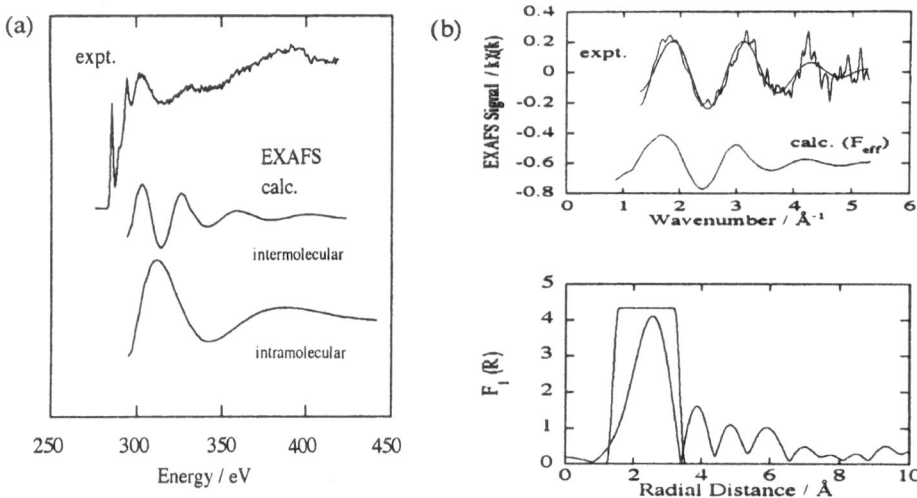

Figure 2. (a) Experimental photoion yield curve of benzene clusters and the calculated EXAFS signal from F_{eff} calculations for intermolecular and intramolecular EXAFS in benzene clusters. (b) upper part: experimental EXAFS signal, back transformed EXAFS signal, and the result from F_{eff} calculations, lower part: Fourier transform of the k^1 weighted EXAFS signal.

COULOMB EXPLOSION

Benzene Clusters. Charge separation ("*Coulomb explosion*") in small doubly charged benzene (Bz) clusters is investigated with the photoelectron photoion photoion coincidence (PEPIPICO) technique. The threshold for stable dications which do not decay via charge separation is Bz_{23}^{++}.[15] Smaller clusters undergo rapid fission where singly charged fragments are formed via symmetric or asymmetric charge separation. This process is accompanied with evaporation of neutrals before or after charge separation. Correlated pairs of cations are the result of this decay mechanism, which are experimentally measured with coincidence spectroscopies (PIPICO or PEPIPICO). A typical PEPIPICO spectrum of benzene clusters is shown in Figure 3a. The spectrum shows that channels of symmetric charge separation are Bz^+/Bz^+ and Bz_2^+/Bz_2^+. Symmetric charge separations of larger species are not found (Bz_n^+/Bz_n^+, with n>2). The regime of asymmetric charge separation is governed by the intense Bz^+/Bz_2^+ cation pair. Series of asymmetric charge separations (Bz^+/Bz_n^+ and Bz_2^+/Bz_n^+, with n>2) are found with weak intensity, which is different from results on argon clusters.[9,10] This difference is rationalized by the considerably lower critical size of stable doubly charged benzene clusters as well as efficient evaporation of neutrals. The shape of the Bz^+/Bz_2^+-PEPIPICO signal is box-like, indicating that the momenta of both fragment ions are not directed into opposite direction, as it is typical for Coulomb explosion. Similar peak shapes have been observed for charge separation in large argon clusters.[10] This behavior is due to the considerable charge separation distance (CSD) within benzene clusters. The CSD is estimated from the width of the Bz^+/Bz_2^+-PEPIPICO signal yielding 6±0.8 Å, corresponding to 2.4±0.3 eV kinetic energy release for a two body charge separation of a doubly charged benzene trimer. This result is in agreement with the van der Waals distance of benzene, where 6.31 Å are obtained from Lennard-Jones potentials.[24]

Methanol Clusters. Figure 3b shows the PEPIPICO spectrum of methanol-d_1 ($CH_3OD \equiv M$) clusters at 300 eV excitation energy. The cation pairs indicate that predominantly deuterated fragments are formed in the charge separation process. It is therefore concluded that the hydrogen centers at the methyl group have minor importance for the charge separation process of doubly charged methanol clusters. Similar findings are reported for singly charged acetaldehyde clusters.[25] Two different charge separation mechanisms may explain the coincidences between deuterated cluster fragments: (i) double ionization of one molecule within the cluster with subsequent D^+-transfer, and more likely (ii) secondary ionization of a neighboring molecule via the Auger electron followed by charge separation and secondary stabilization of both singly charged fragments via proton transfer.

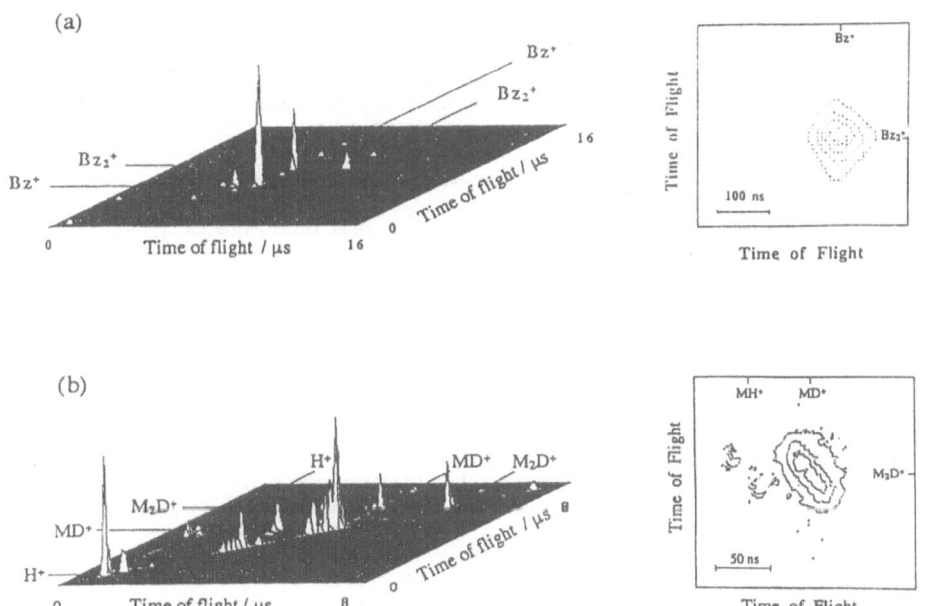

Figure 3. (a) PEPIPICO spectrum of benzene (Bz) clusters at 300 eV excitation energy (top); projection of the Bz^+/Bz_2^+ coincidence; (b) PEPIPICO spectrum of methanol-d_1 (M) clusters at 300 eV excitation energy (left); projection of the MD^+/M_2D^+ coincidence (right).

The slope of the MD^+/M_2D^+ signal is -1.5 which is compatible with the following decay mechanism:

$$
\begin{align}
M_7^{++} &\rightarrow M_3^+ + M_4^+ \tag{1a} \\
M_3^+ &\rightarrow MD^+ + M + (M\text{-}D) \tag{1b} \\
M_4^+ &\rightarrow M_2D^+ + M + (M\text{-}D) \tag{1c}
\end{align}
$$

Considerably different slopes are observed for other charge separation channels of clusters (not shown) indicating that a variety of different fragmentation routes are active. Figure 3b shows also evidence for a "twisted" PEPIPICO signal, where the center of the peak has a different slope than the base, as it has been observed for argon clusters.[10] The signal shape depends on the expansion conditions, i.e. the neutral cluster size, suggesting that different cluster sizes contribute to different signal slopes via evaporation of neutrals.

ACKNOWLEDGEMENTS

We would like to thank A.P. Hitchcock for helpful discussions. Financial support by the Bundesministerium für Forschung und Technologie is gratefully acknowledged (grant No. 05 5KEFXB5-TP3).

REFERENCES

1. C.Y. Ng, Molecular beam photoionization studies of molecules and clusters, *Adv. Chem. Phys.* 52:263 (1983).

2. "Structure and Dynamics of Weakly Bound Molecular Complexes", NATO-ASI Series C-212, A. Weber, ed., Reidel, Dordrecht (1987).

3. "Reactions in and with Clusters", *Ber. Bunsenges. Phys. Chem.* 96:1091 (1992).

4. E. Rühl, H.W. Jochims, C. Schmale, E. Biller, A.P. Hitchcock, and H. Baumgärtel, Core level excitation in argon clusters, *Chem. Phys. Lett.* 178:558 (1991).

5. E. Rühl, C. Heinzel, A.P. Hitchcock, and H. Baumgärtel, Ar 2p spectroscopy of free argon clusters, *J. Chem. Phys.* 98:2653 (1993).

6. E. Rühl, C. Heinzel, A.P. Hitchcock, H. Baumgärtel, H. Schmelz, C. Reynaud, W. Drube, and R. Frahm, K-Shell spectroscopy of Ar clusters, *J. Chem. Phys.* 98:6820 (1993).

7. E. Rühl, C. Heinzel, H. Baumgärtel, W. Drube, and A.P. Hitchcock, Ar 2p and Ar 1s spectroscopy of free argon clusters, *Jpn. J. Appl. Phys.* 32, Suppl. 2:791 (1993).

8. E. Rühl, C. Heinzel and H.W. Jochims, Fluorescence of Ar(2p)-excited argon clusters, *Chem. Phys. Lett.* 211:403 (1993).

9. E. Rühl, C. Schmale, H.W. Jochims, E. Biller, M. Simon and H. Baumgärtel, Charge separation in core excited argon clusters, *J. Chem. Phys.* 95:6544 (1991).

10. E. Rühl, C. Heinzel, H. Baumgärtel, M. Lavollée, and P. Morin, Fragmentation of doubly charged argon clusters, *Z. Physik D*, submitted (1993).

11. E. Rühl, Core-level excitation in molecular clusters, *Ber. Bunsenges. Phys. Chem.* 96:1172 (1992).

12. A.P. Hitchcock, Bibliography of atomic and molecular inner-shell excitation studies, *J. El. Spectrosc. Relat. Phenom.* 25:245 (1982) and update, Nov. 13 (1991).

13. J.A. Horsley, J. Stöhr, A.P. Hitchcock, D.C. Newbury, A.L. Johnson, and F. Sette, Resonances in the K-shell excitation spectra of benzene and pyridine: gas phase, solid, and chemisorbed states, *J. Chem. Phys.* 83:6099 (1985).

14. M.N. Piancastelli, T.A. Ferrett, D.W. Lindle, L.J. Medhurst, P. A. Heimann, S.H. Liu, and D.A. Shirley, Resonant processes above the carbon 1s ionization threshold in benzene and ethylene, *J. Chem. Phys.* 90:3004 (1989).

15. N.G. Gotts, P.G. Lethbridge, and A.J. Stace, Observations of Coulomb explosion in doubly charged atomic and molecular clusters, *J. Chem. Phys.* 96:408 (1992).

16. G.R. Wight and C.E. Brion, K-shell excitation spectra of CH_4, NH_3, H_2O, CH_3OH, CH_3OCH_3, and CH_3NH_2 by 2.5 keV electron impact, *J. El. Spectrosc. Relat. Phenom.* 4:25 (1974).

17. I. Ishii and A.P. Hitchcock, The oscillator strengths for C-1s and O-1s excitations of some saturated and unsaturated organic alcohols, acids and esters, *J. El. Spectrosc. Relat. Phenom.* 46:55 (1988).

18. W.L. Jolly, K.D. Bomben, and C.J. Eyermann, Core-electron binding energies for gaseous atoms and molecules, *At. Data Nucl. Data Tables* 31:433 (1984).

19. E. Rühl, B. Brutschy, and H. Baumgärtel, Autoionization resonances in homogeneous benzene clusters, *Chem. Phys. Lett.* 157:379 (1989).

20. A.P. Hitchcock and I. Ishii, Carbon and oxygen K-shell EXAFS of gases studied by electron energy loss spectroscopy, *J. Phys. (Paris)* 47, C-8:199 (1986).

21. G. Comelli and J. Stöhr, SEXAFS studies of C-C bond lengths in condensed and chemisorbed molecules, *Surf. Sci.* 200:35 (1988).

22. J. Mustre, Y. Yacoby, E.A. Stern, and J.J. Rehr, Analysis of experimental extended-x-ray-absorption fine structure (EXAFS) data using calculated curved-wave, multiple scattering EXAFS spectra, *Phys. Rev. B* 42:10843 (1990).

23. G. Karlström, P. Linse, A. Wallqvist and B. Jönsson, Intermolecular potential for the H_2O-C_6H_6 and the C_6H_6-C_6H_6 systems calculated in an ab initio SCF CI approximation, *J. Am. Chem. Soc.* 105:3777 (1983); E.G. Cox, Crystal structure of benzene, *Rev. Mod. Phys.* 30:159 (1958).

24. J.O. Hirschfelder, C.F. Curtiss, and R.B. Bird, "Molecular Theory of Gases and Liquids", Wiley, New York (1964).

25. B. Brutschy, P. Bisling, E. Rühl, and H. Baumgärtel, Photoionization mass spectrometry of molecular clusters using synchrotron radiation, *Z. Physik D* 5:217 (1987).

REACTION IN THE NO_2-C_2H_4 VAN DER WAALS COMPLEX

J.C. Loison, C. Dedonder-Lardeux, C. Jouvet and D. Solgadi

Laboratoire de photophysique moléculaire du CNRS
Bât 213 - Université de Paris Sud.
91405 Orsay Cedex - France

ABSTRACT

The photodissociation of NO_2 within NO_2-C_2H_4 complexes or small clusters excited at 355 nm leads to the formation of vibrationally and rotationally cold vinoxy radicals ($CH_2CHO\cdot$). On the other hand, the second reaction product, the H atom has been probed through laser induced fluorescence of the Lyman α. The analysis of the Doppler profile indicates that the H atom leaves with a high kinetic energy. This is in agreement with the calculated late barrier in the reaction path.

INTRODUCTION

Reactions in van der Waals complexes have become a very useful and powerful technique to study chemical reactions in details.

In this approach, recently reviewed by Shin et al[1], the reacting molecules frozen within a van der Waals complex are excited by a photon, which promotes the system on the reactive surface. The reaction proceeds, and the products are detected by laser induced fluorescence. The geometry of the reacting molecules is thus fixed before the excitation and therefore the entrance channel geometry is restricted as well as the symmetry of the entrance surface.

The reaction we choose to study in the complex is the reaction of ground state oxygen $O(^3P)$ with ethylene (C_2H_4). This reaction has been fairly extensively studied theoretically[2-4] and experimentally[6-17], and is important in combustion processes.

Recently a crossed beam experiment of Schmoltner et al[6] has shown that the open channels at low collisional energy (less than 6 Kcal/mol) are:

$$O(^3P)+C_2H_4 \rightarrow CH_3 + CHO\cdot(\text{formyl radical}) \qquad (1a)$$

$$H\cdot + CH_2CHO\cdot(\text{vinoxy radical}) \qquad (1b)$$

$$H_2 + CH_2CO \quad (\text{ketene}) \qquad (1c)$$

Linking the Gaseous and Condensed Phases of Matter
Edited by L.G. Christophorou *et al*, Plenum Press, New York, 1994

REACTION WITHIN THE NO_2---C_2H_4 COMPLEX

We have undertaken the study of the O^3P + C_2H_4 reaction through the reactivity of the NO_2--C_2H_4 complexes:

The NO_2 molecule was chosen as a precursor of atomic oxygen since laser induced photodissociation of NO_2 is easy and has been well studied[18-20].

Excitation of NO_2, within the complex, above its dissociation threshold is expected to give $NO...O(^3P)...C_2H_4$ and therefore will allow to study the O + C_2H_4 reaction within a restricted geometry entrance channel. This

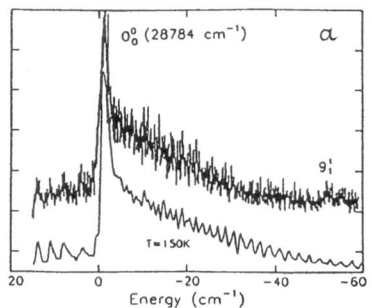

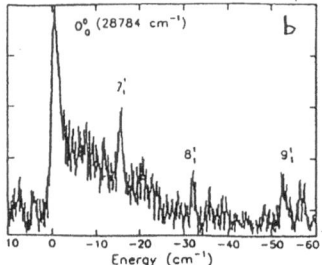

Figure 1. (a) Rotational contour of the Vinoxy radical obtained after the excitation of the NO_2-C_2H_4 complex at 355 nm. (b) Excitation at 266 nm

restricted geometry is determined by the geometry of the ground state complex. Moreover the entrance channel symmetry of the oxygen atom attack is determined by this geometry and by the symmetry of the dissociative state of the NO_2 molecule.

RESULTS

On figure 1a and 1b are shown the laser induced fluorescence spectra of the vinoxy radical produced in photolyzing the complex at 355 and 266 nm.

As previously shown[21,22], the vinoxy produced at 355 nm is very cold: the rotational contour can be simulated with a Boltzmann distribution at 150K (i.e. E_{rot} = 150 cm^{-1} ≅ .4 kcal), and the vibrational energy content is around 250 ± cm^{-1} ≅ .7 kcal, whereas the available energy for the reaction is 19 kcal.

At 266nm, NO_2 still dissociates to NO X $^2\Pi$ and O^{3}P, the threshold for formation of O^{1}D lying at 244 nm. Assuming that NO_2 dissociates in the same way in the complex as in the free molecule, the mean kinetic energy of the oxygen atom is 3 kcal (1.2 kcal at 355 nm) so that the available energy is now 21 kcal.

The vibrational energy distribution in the vinoxy radical is clearly different from that at 355 nm: in particular the 7_1 and 8_1 band appear clearly on this spectrum. A *contrario*, the rotational contour of the 0-0 transition does not change with the dissociation energy: at 266 nm as well as at 355 nm, it can be simulated with a Boltzmann distribution at 150K.

These results should be compared with the ones obtained in a bulk experiment in which the NO_2 is photodissociated at 355nm and the leaving O^{3}P atom undergoes a collision with the C_2H_4 molecule. In this case, as shown by the figure 2 the rotational and vibrational energy within the vinoxy product is very high (more than 2000cm^{-1}).

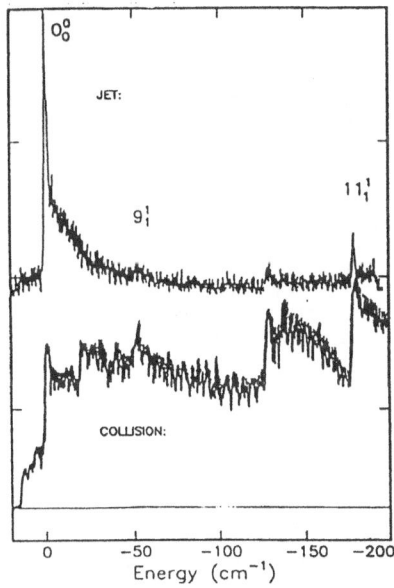

Figure 2 : Upper trace: L.I.F. of vinoxy issued from the reaction in the NO_2-C_2H_4 complex. The rotational and vibrational energy is very low. Lower trace : L.I.F. of vinoxy issued from the reaction in the bulk. The rotational and vibrational energy is very large

The H atom kinetic energy has been obtained through the Doppler profile of the Liman α VUV laser induced fluorescence. In the profile presented in figure 3, the kinetic energy of the H atom is more than 7 kcal. Since the H atom is very fast and since the signal was obtained in a fairly dense region of the supersonic jet the H atoms undergo few collisions before detection.

Extrapolation at 0 collision obtained by varying the delay between the pump and probe laser have shown that the kinetic energy of the H atom obtained in the reaction lies between 14 and 20 kcal.

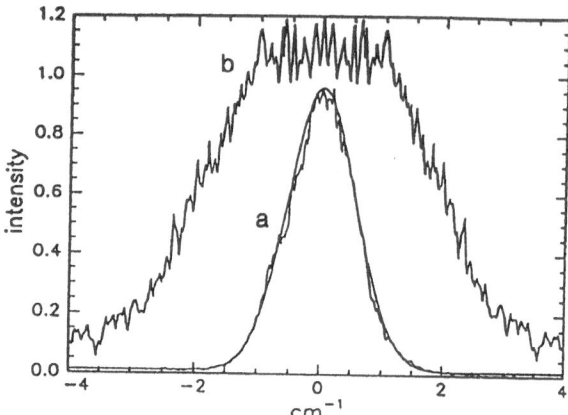

Figure 3 : Ha Doppler profile of the H atom. (a) experimental resolution. (b) H issued from the reaction

DISCUSSION

The small energy content found in the vinoxy radical after the photoinduced reaction in the complex could be understood in terms of an impulsive dissociation at the late transition state:

Indeed, theoretical calculations[2-5] predict that the formation of vinoxy proceeds through a triplet surface which presents a late barrier in the exit channel. The late transition state has a structure (concerning the CCO squeleton) similar to that of the vinoxy product, the leaving H atom being already at 1.7Å from the nearest carbon. We have assumed that this late barrier is higher by 3 Kcal than calculated (this is still within the estimated uncertainty of the calculation), therefore that there is no internal energy at the transition state. when exciting at 55nm. We can then assume an impulsive rupture of a CH bond at the transition state. With the impulsive model we can calculate the average energy in rotation as $E_{rot} \simeq$ 120 cm^{-1} and in vibration as $E_{vib} \simeq$ 200 cm^{-1}, in agreement with the experimental observations.

This model will also explain the energy content of the vinoxy radical obtained when the $NO_2-C_2H_4$ complexes are excited at 266 nm. From experimental works done on the photodissociation of NO_2 we can estimate that the mean kinetic energy of the oxygen atom is increased by $\simeq$ 2 kcal when NO_2 is excited at 266 nm as compared to the excitation at 355 nm. Assuming a statistical distribution of the energy at the transition state, these 2 kcal will be found mainly in vibration at the late transition state, and remain in vibrational energy of the vinoxy radical when the H atom leaves, as observed experimentally.

But this model does not explain why the vinoxy radical is produced with an important internal energy in thermal collisions. Our results for the reaction in complex are in agreement with the crossed beam experiment of Kleinermanns and Luntz[15]. However the kinetic energy release measured by Schmoltner et al[6] (also in a crossed beam experiment) is less important than what is expected if the vinoxy is cold.

Different hypothesis based on theoritical calculation can be tested to interpret the contradictory results obtained on this system, in our experiments as well as in the crossed beam experiments :

1) the mechanism of formation is the same in the $O^3P + C_2H_4$ collision and in the $NO_2-C_2H_4$ photoinduced reaction : the vinoxy product obtained in the excitation of complexes or clusters is produced hot by the chemical reaction and is cooled by evaporation effect.

2) There are two competing mechanisms, and the selection of the entrance channel realised in the reaction in complexes favors channel which leads to the cold product.

These hypothesis will now be discussed in view of the previous experiments (thermal collisions, crossed beam experiments and complex experiment) [6,15,22] and of theoretical calculations.

Reaction and cluster.

In our previous papers[21-22] we already discussed hypothesis 1a. The vinoxy would be produced rotationally and vibrationally hot in the reaction within a cluster, and cooled through the evaporation of molecules. The process would then be:

$$(NO_2-C_2H_4)_n + h\nu \rightarrow NO + (CH_2CHO^* - C_2H_4 - NO_2)_{n-1} \qquad reaction$$
$$\rightarrow CH_2CHO + C_2H_4 + NO_2 \qquad evaporation \; and \; cooling$$

In the simplest approximation, where each evaporating molecule takes away the same amount of energy, one expects that the cooling is more efficient in large clusters than in small ones. Thus, the rotational energy in the vinoxy product, which is the most easily relaxed in a collisional process, should depend on the backing pressure because the concentration and the size of the clusters depends strongly upon the backing pressure in the jet. But we have shown previously that the rotational contour obtained for backing pressures varying by a factor 3 does not change, (within the experimental error ± 50K) whereas the cluster concentration changes drastically.

Therefore either the rotational relaxation induced by the evaporation process is unefficient, and the observed energy in the vinoxy molecule is very close to that obtained in the one to one complex or the observed signal comes mainly from the one to one complex.

It should be noticed that the "evaporative cooling" process does not explain the result obtained when clusters are excited at 266nm. Furthermore one should have seen some cold H atom, leading to a different Doppler profile than what observed.

Two different mechanisms in competition : selection of the entrance channel in the excitation of the clusters.

We will try to reconcile the different experimental results: our experiment in gas phase and the crossed beam experiment of Schmoltner et al[6] in which the vinoxy product is internally excited on one hand, our experiment in complexes and the crossed beam experiment of Kleinermanns and Luntz[15] where the vinoxy product is cold on the other hand.

The explanation is based on the following and experimental results :
-In the triplet surface the vinoxy is obtained after a late barrier
-In the singulet surface there is a reactive path which leads to the vinoxy product without late barrier. Such a reactive path can be found through the formation of the vinyl alcohol[23].

-The LIF detection in jets favours the observation of the cold vinoxy product because fewer levels are populated, as compared to hot product. We can estimate that the detectivity is a factor 1000 higher for cold vinoxy.

Under these assumptions we prpose that two channels are leading to the vinoxy product:
- A direct one on the triplet surface, which leads to cold vinoxy product.
- An indirect one, which is induced by an intersystem crossing to the singlet surface. Since there is no barrier in the exit channel[23] the energy distribution in this case will be statistical.

These two channels are in competition and the branching ratio depends on the impact parameter and on the collisional energy. The direct route is obtained if vibrations which are not along the reaction coordinate are not excited in the collisional process.

If some energy is lost in exciting these vibrations (the CC stretch, CCO bend for example) there will not be enough energy to cross over the late barrier on the triplet surface. The system being stuck behind the barrier may undergo an intersystem crossing to the singlet surface, which leads to HCO and hot vinoxy.

The direct mechanism would give an important translational energy as observed for the H atom, but its yield may be low. Whereas the indirect mechanism would lead to a large spread of translational energies. The kinetic energy measured in the experiment of Schmoltner et al[6] may then be the sum of these two contributions: as the sensitivity is smaller at high energy, a double distribution will not appear if the yield for the direct channel is less than 10%.

In our gas phase experiment, the two channels should be open. The observation of hot vinoxy indicates that the yield for the direct mechanism is much lower (1%) than for the indirect one. Moreover, secondary collisions thermalize part of the oxygen atoms, and this will favour the indirect process.

In single collision regime, Kleinermanns and Luntz[15] detected only cold vinoxy, because the laser induced fluorescence measurements, as stated above, highly favor the detection of cold products. When the pressure is higher, they observe the appearence of hot bands; this may result from two effects: the decrease of the kinetic energy of the oxygen atoms by collisions, which will decrease the yield of the direct mechanism, and/or the vibrational relaxation of hot vinoxy radicals.

In the photoinduced reaction in complexes, the two mechanisms are effective since we have enough energy to cross over the late triplet barrier, but once again LIF favours the detection of the direct channel. Besides in these experiments, the oxygen atom attacks the ethylene in a well defined direction, unlike in collision experiments; this limited impact parameter may also increasse the yield of the direct mechanism.

CONCLUSION

Photoinduced reaction in the NO_2-C_2H_4 complex produces cold vinoxy radicals, whereas thermal gas phase collisions of O^3P with ethylene produce vibrationally and rotationally excited vinoxy. This can be understood if we assume that two different reaction mechanims are in competition: a direct mechanism through the triplet surface yielding cold products, and an indirect channel following a singlet reaction path, producing hot radicals.

REFERENCES

1. Shin S.K., Chen Y., Nickolaisen S., Sharpe S.W., Beaudet R.A., and Wittig C., Photoinitiated reactions in weakly-bonded complexes, *Advances in Photochemistry* (1991), *and references therein.*

2. Yamaguchi K., Yabushita S., Fueno T., Kato S., and Morokuma K., Geometry optimization of the ring-opebed oxirane diradical : mechanism of the addition reaction of the triplet oxygen atom to olefins, *Chem. Phys. Lett.* 70:27 (1980).

3. Fueno T., Takahara Y., Yamaguchi K., Approximately projected UHF Moller-Plesset calculations of the potential energy profiles for the reaction of the triplet oxygen with ethylene, *Chem. Phys. Lett.* 167:291 (1990).

4. Dupuis M., Wendoloski J.J., Takada T., and Lester W.A., Theoretical study of electrophilic addition: $O(^3P)+C_2H_4$, *J. Chem. Phys.* 76:481 (1982).

5. Bigot B., Sevin A., Devaquet A., Ab Initio SCF calculations on the photochemical behavior of the three-membered rings.3.Oxirane : ring opening, *J.A.C.S.* 101:1095 (1979).

6. Schmoltner A.M., Chu P.M., Brudzynski R.J., and Lee Y.T., Crossed molecular beam study of the reaction $O(^3P)+C_2H_4$, *J. Chem. Phys.* 91:6926 (1989).

7. Cvetanovic R.J. and Singleton D.L., Reaction of oxygen atoms with ethylene. *Rev. Chem. Int.* 5:183 (1984).

8. Pruss F.J., Slagle I.R., and Gutman D., Determination of Branching Ratios for the Reaction of Oxygen Atoms with Ethylene, *J. Phys. Chem.* 78:663 (1974).

9. Endo Y., Tsuchiya S., Yamada C., Hirota E., and Koda S., Microwave kinetic spectroscopy of reaction intermediates: O + Ethylene reaction at low pressure, *J. Chem. Phys.* 85:4446 (1986).

10.a. Koda S., Endo Y., Hirota E., and Tsuchiya S., Deuterium isotope effect on the branching ratio in $O(^3P)$ + Ethylene reaction, *J. Phys. Chem.* 91:5840 (1987).

10.b. Koda S., Endo Y., Hirota E., and Tsuchiya S., Branching Ratios in $O(^3P)$ Reactions of Terminal Olefins Studied by Kinetic Microwave Absorption Spectroscopy *J. Phys. Chem.* 95:1241 (1991).

11. Klemm R.B., Sutherland J.W., Wickramaaratchi M.A., and Yarwood G., Flash Photolysis-Shock Tube Kinetic Study of the Reaction of $O(^3P)$ with Ethylene : $1052 \text{ K} \leq T \leq 2284 \text{ K}$, *J. Phys. Chem.* 94:3354 (1990).

12. Mahmud K., Marshall P., and Fontijn A., A High-Temperature Photochemistry Kinetics Study of the Reaction of $O(^3P)$ Atoms with Ethylene from 290 to 1510 K *J. Phys. Chem.* 91:1568 (1987).

13. Clemo A.R., Davidson F.E., Duncan G.L., Grice R., Translational energy threshold functions for oxygen atom reactions, *Chem. Phys. Lett.* 84:509 (1981).

14. Klemm R.B., Nesbitt F.L., Skolnik E.G., Lee J.H., and Smalley J.F., Direct Rate Constant Measurements for the Reaction of Ground-State Atomic Oxygen with Ethylene, 244-1052 K, *J. Phys. Chem.* 91:1574 (1987).

15. Kleinermanns K., and Luntz A.C., Laser-Induced Fluorescence of CH_2CHO Produced in the Crossed Molecular Beam Reactions of $O(^3P)$ with Olefins, *J. Phys. Chem.* 85:1966 (1981).

16. Sridharan U.C., and Kaufman K., Primary Products of the $O + C_2H_4$ Reaction *Chem. Phys. Lett.* 102:45 (1983).

17. Hunziker H.E., Kneppe H., and Wendt H.R., Photochemical Modulation Spectroscopy of Oxygen Atom Reactions with Olefins, *J. Photochem.* 17:377 (1981).

18. Zacharias H., Geilhaupt M., Meier K., and Welge K.H., Laser photofragment spectroscopy of the NO_2 dissociation at 337 nm. A nonstatistical decay process, *J. Chem. Phys.* 74:218 (1981).

19. Busch G.E., and Wilson K.R., Triatomic Photofragment Spectra. I. Energy Partitioning in NO_2 Photodissociation*, *J. Chem. Phys.* 56:3626 (1972).

20. Mons M., Dynamique de la Photodissociation de NO_2. Etude par ionisation multiphotonique résonnante de la répartition d'énergie et des anisotropies dans les fragments. Ph.D Thesis, Orsay (1988).

21. Loison J.C., Dedonder-Lardeux C., Jouvet C., and Solgadi D., Photoinduced Chemical Reaction in the $No_2-C_2H_4$ van der Waals Complex, *J. Phys. Chem.* 95:9192 (1991).

22. Loison J.C., Dedonder-Lardeux C., Jouvet C. and Solgadi D. Comparison between the Photo Induced Chemical Reaction in the $NO_2-C_2H_4$ van der Waals Complex and the $O + C_2H_4$ Gas Phase Reaction, *Ber. Bunsenges. Phys. Chem.* 96:1142 (1992).

23. Smith B.J., Guyen M.T., Bouma W.J., and Radom L., Unimolecular Rearrangements Connecting Hydroxyethylidene (CH_3-C-OH), Acetaldehyde ($CH_3-CH = O$), and Vinyl Alcohol ($CH_2 = CH-OH$), *J.A.C.S.* 113:6452 (1991).

PHASE TRANSITIONS IN CLUSTERS:
A BRIDGE TO CONDENSED MATTER

R. Stephen Berry

Department of Chemistry and the James Franck Institute
The University of Chicago
Chicago, Illinois 60637-1403, U. S. A.

PHASES OF CLUSTERS

Clusters of atoms or molecules, consisting of as few as three and as many as many thousands of particles, exhibit some properties of individual molecules, some properties of bulk matter and some properties characteristic specifically of small finite systems, that is, of clusters themselves. One of the unique sets of properties of clusters is their behavior with respect to phase equilibrium and phase changes--or at least to the analogues of the phases of bulk matter as they manifest themselves in small systems. Yet the unique phase behavior of clusters gives us new insights into the nature of phase equilibrium, phase transitions and metastability of the phases of bulk matter. This Chapter is a review of the status of this subject, which still has many major open questions.

The phases or phase-like forms that clusters may exhibit include solid, soft solid, surface-melted with a solid core, homogeneous liquid, amorphous or glassy, and coexisting crystalline and amorphous forms. There may also be a phase-like form consisting of a solid surface around a liquid core. The most studied of these by far are the solid and homogeneous liquid forms. The first evidence for distinguishable solid and liquid clusters came from simulations[1-9]. The structures of solid-like clusters of van der Waals, argon-like solids had been explored by Hoare and Pal by then and subsequently [10-15], so much was known of the structure and dynamics of simple solid-like clusters.

The surprising result of the simulations of the 1970's was the appearance of unambiguous, distinguishable solid and liquid forms of argon-like clusters containing as few as *seven* atoms. This seemed at first glance to contradict the dogma that sharp phase changes are characteristic of very large systems. More specifically, it raised the question of what characteristics are responsible for the existence of distinguishable phases of such small systems, what is the nature of the equilibrium between the phases, and how do the transitions occur between them.

The phase-like forms that clusters may exhibit are observable because certain necessary conditions are satisfied by the complex set of forces that hold the cluster together. It is easiest to describe those conditions if the cluster can be described accurately by the adiabatic or Born-Oppenheimer approximation, that is, if the energy and quantum states--or the classical motion--of the component particles can be well described in terms of a single potential energy function $V(r_1, r_2, \ldots, r_N)$ that depends only on the 6N-6 independent coordinates of the N particles after the center-of-mass motion and rotational motion have been transformed away. This restriction is not necessary; only the 3 center-of-mass coordinates might be transformed away so that the rotation of the cluster is left in the description; or the system might suffer exchange of electronic and vibrational energy on a time scale rapid enough to affect the quantum states and the spectrum, in which case the Born-Oppenheimer approximation would be inapplicable and the treatment would require more than one electronic state and more than one potential surface. These would complicate the discussion without providing extra insight, at least without clarifying the basic phenomenon. Hence we proceed, invoking the concept of a single potential surface that governs the atomic or molecular motion.

On the multidimensional surface of $V(r_1, r_2, \ldots, r_N)$, there are minima corresponding to the locally stable structures of the cluster. For all but the simplest clusters, there are several equivalent minima corresponding to each of the stable geometries of the system. These correspond to all the geometrically indistinguishable, permutationally different structures. The equilateral triangular structure of the homonuclear three-particle cluster and the tetrahedral structure of the four-particle cluster are represented by only one minimum (if they are indeed stable structures for the forces in question) because the permutations of the identical nuclei can all be achieved by rigid-body rotations of the cluster. All other locally stable structures of homonuclear clusters, however symmetrical, correspond to two or more minima on the surface of $V(r_1, r_2, \ldots, r_N)$ because there are permutations of the vertices of all other figures that cannot be reached by rigid-body rotations alone.

There are, in general, several kinds of minima on the surface of $V(r_1, r_2, \ldots, r_N)$, corresponding to all the different, locally stable geometric structures the cluster may have. The number of geometrically different minima increases roughly exponentially with N, the number of particles comprising the cluster. The number of permutationally different, geometrically equivalent locally stable structures (and therefore minima) increases roughly factorially. Hence the total number of minima on $V(r_1, r_2, \ldots, r_N)$ increases roughly as N!

exp(aN) where a is of order unity. The six-particle cluster held together by pairwise Lennard-Jones interactions, a fairly good model for Ar_6, has two geometrically distinct kinds of minima, the regular octahedron of which there are 30 geometrically equivalent permutational isomers, and a distorted octahedron, of which there are 12 accessible from each of the 30 octahedra, 360 in all. The seven-particle cluster held together by the same Lennard-Jones forces has four kinds of locally stable geometric structures, the lowest in energy a pentagonal bipyramid, and a total number of potential minima close to 10,000 including those of all the permutational isomers of the four geometric structures. The 8-particle cluster held together by the same kind of forces has eight kinds of locally stable structures and over 40,000 permutational isomers of just the lowest-energy structure among these eight. Hence the potential surfaces of even very simple clusters are rather complex, with many kinds of hills, valleys and mountain passes. This point is pursued in more detail in the accompanying paper.

A stable phase-like form is certainly not equivalent to a stable structure. A stable solid consists of a set of atoms or molecules vibrating about well defined sites, which are sites on a regular lattice if the solid is crystalline and sites in an amorphous array if the solid is a glass or amorphous material. A solid exhibits almost no diffusion, rotates like a rigid body and is rather (or extremely) incompressible. The extent to which the particles make excursions from their equilibrium positions is frequently measured by the "Lindemann criterion," the fractional root-mean-square deviation of the nearest-neighbor distance, usually symbolized as δ. It is of course given by $\langle (R - R_{eq})^2 \rangle^{1/2}/R_{eq}$. For typical solids and solid-like clusters, δ is typically in the range 0.05-0.1.

A liquid consists of a set of atoms or molecules that are relatively free to deform in response to changing forces, such as the change in gravitational forces when the liquid is transferred between vessels of different shapes, or to move in response to a finger's push on its surface. A liquid is characterized by its *dynamical* behavior, not by its static characteristics; by its compliance, not by its disorder. From a microscopic standpoint, a liquid is a form of condensed matter that moves among the catchment basins on its multidimensional potential surface at an average rate not very different from its rate of vibration within a well. "Not very different" may mean within two or three powers of 10, in contrast to solids, whose rates of interwell passage may be 10 to 50 powers of 10 slower than vibrations within a well.

Most of the modes of motion of a liquid are vibrational motions in roughly the same range of frequencies as the modes of solids. Liquids, however, have some modes that have very low frequencies and large amplitudes of motion, and are, in general, extremely anharmonic. These are the modes that carry the system from one basin to another, so that the potential surface governing their motion is very unlike a simple paraboloid. Rather, it will, in general, include two or more wells that are paraboloidal near their minima, but the relevant region will also include the saddle between the wells and therefore portions of potential surface where the curvature is negative, the Hessian has at least one negative eigenvalue, at least one of the "frequencies" one would guess from a local second derivative is negative and the corresponding local "force constant" is imaginary. In general, the motions of atoms or molecules in a liquid are more anharmonic than those of the same particles in a solid, and the interwell motions are among the most anharmonic of these.

Solid phases of clusters may be regular, either crystal-like, based on lattice structures, or polyhedral; they may be amorphous, with little or no short-range order; or they may be "soft," passing among a small number of catchment basins in what Sugano[16] has called the "fluctuating state" and would be called "fluxional" in the context of inorganic and organic chemistry[17-19]. An example is the cluster of six atoms bound by pairwise Lennard-Jones potentials mentioned previously; in its solid-like form, it is octahedral, undergoing only small-amplitude vibrations around that octahedron. In its liquid-like form, the cluster moves among all the 360 catchment basins so that all the atoms can take any of the places in the structure. In the soft-solid or fluctuating state, the cluster can pass from one octahedral structure to any of the 12 distorted structures that can be reached directly from that octahedron, but cannot pass over to any other octahedron or any of the 12 distorted structures around any of the octahedra.

Clusters of more than about 50 particles may also exhibit phase-like forms in which part of the cluster is solid and part, liquid-like. If the liquid-like form wets the solid, then such "phases" are surface-melted states, liquid surrounding solid[6, 20-24]--or, in principle, especially if the solid is less dense than the liquid, clusters with liquid cores surrounded by solid shells[23, 24]. There is evidence from simulation and perhaps even from experiments[25, 26] that clusters may exhibit surface-melted "phases" but there is, at present, no evidence for any core-melted forms of clusters. These forms contrast sharply with the forms taken on by clusters whose liquid forms do not wet the corresponding solids, for example alkali halide clusters[27-29]. Instead of surface-melted forms, these exhibit what we have called "nonwetting" forms, in which one side of the cluster is regular and solid-like, while the other side is amorphous and liquid-like.

THE BASIS FOR PHASE EQUILIBRIUM

While the first evidence for distinct phase-like forms of clusters came from simulations, the basis for this behavior appeared soon after[30, 31], in the form of a quantum statistical argument supplemented by computations for a model system. The argument applies equally for quantum and classical systems, and goes as follows. Identify the solid-like "phase" of a cluster with a near-rigid, conventional, structured, molecule-like form, and the liquid-like "phase" with a nonrigid form, akin to a very floppy molecule. Assume that if a form is stable, then it has well-defined states, quantum or classical; granting this, we can make estimates of the densities of states of the various phases from models. For example solid-like phases can be mimicked by an Einstein model in which all modes of internal motion are harmonic and have the same fundamental frequency, or by a Debye model in which the motions are also assumed harmonic but the states are distributed over a range of frequencies with a density of fundamental modes that increases linearly with energy up to a maximum frequency, or by a still more complex model that incorporates anharmonicity. The simplest that incorporates the conditions necessary to replicate the early simulations is an Einstein model with a size-dependent single frequency. Liquid-like clusters can be described in a very oversimplified way by the Gartenhaus-Schwartz model[32], which treats all the internal modes of motion as harmonic oscillators with a common fundamental frequency; in contrast to the Einstein model for clusters, there are no distinct rotational modes. More realistic models can of course be used instead, particularly models that incorporate the very low-frequency or "soft" modes characteristic of liquids, but the Gartenhaus-Schwartz model is already adequate to rationalize the early simulations.

Models can be selected or constructed as needed for other phase-like forms. For example, simulations show that an appropriate model for the surface-melted cluster is a solid core, a surface containing a specified number of vacancies and a number of particles that float around the cluster just outside the surface layer. The model is characterized by the number of "floaters," which move relatively freely in their own spherical shell, while the atoms in the normal outer shell vibrate around fixed sites with frequencies lower than those of the other atoms in the solid core[22-24].

From the models, one estimates the densities of states of solid-like and liquid-like forms (or whatever other phase-like forms one is describing). From the densities of states, one constructs the partition functions of the clusters in each phase of interest, and from them, the free energies, $F_{solid}(T)$ and $F_{liquid}(T)$. For solid-liquid phase equilibrium, the next level of the line of reasoning is simply that at low energies, the solid-like form has a much higher density of states than the liquid-like form; in fact if the liquid-like form requires any promotion energy, such as the energy to create some minimum density of defects, then the density of states of the liquid-like form is zero for some interval of energy bounded by the ground state energies of the solid and the liquid. This means that the partition function of the solid cluster is larger, and the free energy of the solid cluster is lower, than that of the liquid counterpart, so that the solid is stable relative to the liquid. However as the energy increases, and the densities of states of both forms increase, the density of states of the liquid increases considerably faster than that of the solid. If Einstein and Gartenhouse-Schwartz models are used, the density of states of the liquid form of an N-particle cluster increases faster than that of the N-particle solid cluster. Hence, in an isothermal system or a canonical ensemble, as the temperature is raised, the number of liquid-like states accessible to the cluster grows faster than the number of solid-like states, and, as a result, the entropic contribution to the partition functions and the free energies of the two phases tends to favor the liquid form more and more as the temperature increases.

In more precise terms, we can think of the free energy (Helmholtz for systems of constant volume, Gibbs for systems of constant pressure) as a function of not only N and T but of a parameter γ reflecting the degree of nonrigidity of the cluster. Various quantities can serve as γ; one quantum mechanical measure is the ratio of the lowest-energy transition with one quantum of angular momentum added to the cluster, to the lowest-energy transition from the ground state which involves no change in angular momentum: $\gamma = \Delta E(\Delta J=+1)/\Delta E(\Delta J=0)$.

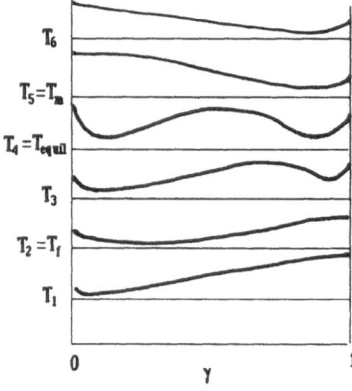

Figure 1. The schematic behavior of the free energy (Helmholtz if T and V are constant, Gibbs if T and p are constant) as a function of the nonrigidity parameter γ for successive higher temperatures. At all temperatures below the freezing limit T_f, only a solid-like form is stable; at all temperatures above the melting limit T_m, only a liquid-like form is stable. Between T_f and T_m, the solid-like and liquid-like forms can exist in dynamic equilibrium. Near $\gamma=0$, the system is solid-like; near $\gamma=1$, the system is liquid-like and nonrigid.

A microscopic definition of γ is a normalized density of defects. This was used in a very specific form by Stillinger and Weber[33] and then generalized[34-36]. We suppose that $0 \le \gamma \le 1$, with 0 the rigid, solid-like limit and 1, the nonrigid, liquid-like limit. This can always be done by judicious definition of the scale for γ. The important point here is that γ is definable and that free energies can be expressed as functions $F(N,T, \gamma)$ and $G(N,T, \gamma)$. With this as a basis, we can now ask how F or G behaves, for fixed N and various values of T, as a function of γ. To illustrate this, we refer to Figure 1.

At low T, $F(N,T, \gamma)$ increases monotonically from some value of γ at or near 0 to $\gamma=1$. As T increases, the energy term of F increases--linearly with T if the specific heat C_v is constant, faster if C_v increases with T--but the entropy term TS grows faster for both solid and liquid clusters, so the free energy decreases with increasing T. It decreases faster for the liquid cluster than for the solid cluster because the density of states of the liquid cluster grows faster with energy than that of the solid. This means that as T increases, the monotonic curve from $\gamma=0$ (or near it) to $\gamma=1$ droops near the high end; it droops until, at some temperature T_f, the curve of $F(N,T, \gamma)$ vs. γ develops a flat spot, i.e. a point at which the slope of this curve is zero, *in the vicinity of* $\gamma=1$. At temperatures above T_f, the free energy has two minima, implying two locally stable forms of the cluster--provided these forms persist long enough to develop and exhibit equilibrium-like properties that we associate with specific phases. Whether this is the case or not is a question that can only be answered by examining the dynamical behavior of individual systems; simulations of clusters of various sizes indicate that some clusters do exhibit well-defined, phase-like forms but others pass too rapidly between the minima to exhibit clear, solid-like or liquid-like behavior, and instead would look like slush in an experiment. Continuing, we ask what happens to the free energy as the temperature is increased further. The density of states of the liquid-like form continues to increase with increasing energy at a rate greater than that of the density of states of the solid. Hence the free energy of the liquid-like minimum drops faster than that of the solid-like minimum as T increases and the curve of $F(N,T, \gamma)$ vs. γ tips further toward the minimum near the liquid limit of $\gamma=1$. A temperature T_m is eventually reached at which the solid-like minimum of $F(N,T, \gamma)$ disappears and becomes a point of inflection. This is the upper limit of temperature for the stable existence of the solid-like form, the melting limit. At temperatures above T_m, $F(N,T, \gamma)$ has only a single minimum as a function of γ. Thus we come to the conclusion that a canonical ensemble of clusters of a fixed number N of particles has a sharply bounded range of temperature, between T_f and T_m, within which solid and liquid clusters coexist.

Within this coexistence band, the total concentration of N-particle clusters is divided, at any instant, into liquid and solid clusters in the ratio given by the traditional chemical equilibrium constant $K \equiv [\text{liquid}]/[\text{solid}] = \exp[-(F_{liq} - F_{sol})/kT]$. At temperatures below T_f, K is zero because no locally stable liquid exists at those temperatures; likewise at temperatures above T_m. K is infinite because no solid exists there. Between these limits, K increases smoothly from some finite small value to some finite large value, presumably (but not necessarily) passing through the point at which $K = 1$ and the free energies of the solid and liquid are equal. Thus K has two discontinuities, one at T_f and one at T_m, is constant and zero for $T < T_f$, constant and infinite for $T > T_m$, and increases continuously in the interval between the discontinuities[30, 31, 37-41].

If it is valid to invoke ergodicity for atomic clusters, then the time history of any single cluster must display a mean distribution of times spent in each phase equal to the ensemble average distribution at any instant. In other words, following any individual cluster under conditions of constant temperature should reveal the value of K just as well as the ensemble average does. However nothing in the theory thus far indicates how long one must follow the system in order to see this average behavior. It appears from simulations that these times are nanoseconds at very least, and may be much longer for many systems.

A variety of simulation studies have borne out that clusters of some but not all sizes, and with a variety of effective potentials, exhibit coexistence of solid and liquid forms[35-37, 42-50]. Laboratory evidence supports the existence of liquid and solid clusters and transformations between them, but not yet of their coexistence[51-60]. There is not, as yet, any evidence from simulation or experiment either for or against the sharp discontinuities at the limits of the coexistence band.

THE IMPLICATIONS FOR BULK MATTER; PHASE DIAGRAMS FOR CLUSTERS

Next we turn to the implications that this behavior has for bulk matter. First, one might suspect that T_f and T_m converge to a common value as N grows very large. This is not the case in most situations. Specifically, if one uses a defect-based model for the softening that leads to melting[61], even a generalized version that assumes no specific form for the defects but only that the free energy can be expanded in a power series in the average density of defects, then one arrives at this conclusion. If the defects repel each other or decrease the density of vibrational states, then T_f and T_m converge as N grows, but if, as is usual, the defects attract each other or the defects lower the vibrational frequencies of the cluster and increase the density of vibrational states, then T_f and T_m remain apart as N grows large[34-36].

To see how solid-liquid phase equilibrium develops as N increases, it is convenient to use, instead of K, the quantity $D \equiv (K - 1)/(K + 1) = [(\text{liquid}) - (\text{solid})] / (\text{total})$, a quantity that varies from -1 when the system is all solid to $+1$ when the system is all liquid. This quantity, which we call a "distribution," is a function of temperature T, pressure p and cluster size N: $D = D(T,p,N)$. Consider first a small cluster. At a fixed pressure, as T increases from a low value, D begins and remains at −1 until T reaches T_f. At this point, D jumps discontinuously to its equilibrium value for $T=T_f + \varepsilon$. Then, as T increases, D is expected to follow a smooth S-curve until $T=T_m$, at which point D becomes +1, discontinuously. This is illustrated by the most gradual of the curves in Fig. 2, with its two discontinuities. Because K can be written as $\exp(−N\Delta\mu/kT)$; $\Delta\mu$ is the difference in the mean chemical potentials of the liquid and solid cluster. In the region of a phase change, $\Delta\mu$ changes sign. If N is of order 10-100, then K and D change slowly and smoothly within the coexistence range, K from a number less than 1 to a number greater than 1, and D, from a negative number greater than −1 to a positive number less than 1, over a measurable range of T. As N becomes larger, the exponential swings more sharply so that for a moderately large cluster, D swings continuously from a value very close to −1 to a value very close to +1 within a narrow range

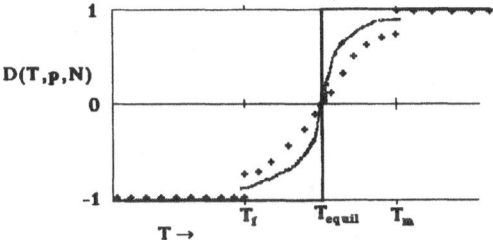

Figure 2. The behavior of the distribution $D \equiv (K − 1)/(K + 1) = [(\text{liquid}) − (\text{solid})] / (\text{total})$ as a function of temperature for small, medium-sized and very large clusters. The crosses represent the behavior of a small cluster; the shaded curve, the behavior of a medium-size cluster and the heavy, solid lines, the behavior of a very large system and of bulk matter.

of T. This is shown by the intermediate curve of Fig. 2. If N is very large, e.g. 10^{10} or greater, then D swings from a value indistinguishable from −1 to a value indistinguishable from +1 within a temperature range too small to be measured, in the vicinity of the temperature T_{equil} where $\Delta\mu=0$, as illustrated by the steepest curve of Fig. 2. In this limit, we have the apparent paradox that the distribution D passes through a discontinuity too small to be observed at T_f, swings continuously but so sharply as to be indistinguishable from a discontinuous change from −1 to +1 and then passes through a second unmeasurably small discontinuity at T_m. This, then, is a first-order melting-freezing transition, interpreted as the limit of the freezing-melting phase change of a succession of ever-larger clusters.

The interpretation just given of how first-order phase transitions evolve from phase changes of clusters is, in many ways, only an amplification of what had been presented by Hill[62-64]. However it teaches us several things:

a) The phase changes of clusters are *not* first-order or second-order transitions; they are different, and lie outside that traditional classification. However the transitions we have examined are related to first-order transitions. They are the succession of transitions that become the first-order melting/freezing transition at the temperature T_{equil}, in the limit of infinite N;

b) The local stability that is responsible for coexistence of solid and liquid clusters is the same phenomenon as the local stability responsible for the metastability of undercooled (supercooled) liquids and of, rare as they are, superheated solids of bulk matter.

c) Because of the sharp limits T_f and T_m, the spinodal curves have temperature limits on each branch, a lower limit T_f on the liquid branch and an upper limit T_m on the solid branch. These limits depend on pressure, of course.

Next we summarize briefly the result that constant-pressure, constant-temperature simulations of clusters may be done just as constant-temperature simulations can. In fact constant-pressure molecular dynamics simulations have been carried out for some time[65] and only later was the algorithm developed that is now commonly used for constant-temperature molecular dynamics simulations[66-68]. By simulating clusters under conditions of constant temperature and pressure, and evaluating the distributions of energy and volume from the results, one can see the sharp peak of the solid-like cluster at low temperatures and all pressures, the broad peak of the liquid-like cluster at high temperatures and also at all pressures, and the bimodal distribution in energy and volume associated with coexistence of solid and liquid clusters at intermediate temperatures. In some earlier work, bimodality was not found for the distributions of mean temperature for Lennard-Jones clusters of 33 particles[43], which was taken to suggest that the coexistence range of temperature was too narrow to make coexistence observable for clusters of this size or larger. However the results for clusters of 55 and 147 particles show that coexistence of solid and liquid clusters is quite normal for clusters of these sizes[23, 24], and the theoretical basis for the finite range of coexisting solid and liquid clusters indicates that except in unusual circumstances, the finite interval between the limiting freezing and melting temperatures should persist as N goes to infinity[34, 36]. From these results, one can construct curves analogous to the coexistence curves of bulk phase diagrams, that is, curves along which the free energies and mean chemical potentials of solid-like and liquid-like clusters are equal. These include curves such as pressure vs. (mean) volume, pressure vs. temperature, density vs. temperature, and heat capacity vs. temperature. However one must construct such a set of curves for clusters of every size, and as yet there are only very crude ideas about the scaling of such properties[34,

$^{36, 69, 70}$. Moreover these curves do not reflect the full richness of the phase equilibria of finite systems; in particular, they do not reflect the coexistence ranges and how the equilibrium composition varies through these regions. We describe shortly how it is possible now to construct phase diagrams for the melting and freezing of small finite systems.

One point that needs further exploration is the sensitivity of the results of constant-temperature, constant-pressure simulations to the definition of the volume of the cluster. Three definitions were suggested and only the crudest one of these has been implemented. That definition takes the volume of the smallest enveloping sphere as the volume of the cluster. The other two are the method of the convex hull$^{71, 72}$ and the so-called α-shape method, which allows for concavity^{73}. Neither of the latter methods has been used in simulations of clusters; both fairly cry out to be used, in order to test the robustness of the constant-pressure simulations.

Now we turn to the construction of the phase diagram for a finite system. To begin, instead of displaying only the curves of coexistence in planes of various variables, we introduce the distribution D discussed above, the difference between the fraction of liquid and the fraction of solid, in addition to the usual thermodynamic variables. It is easiest to work with pressure and temperature as the other two variables. Hence we construct the diagrams in a (p,T,D) space, where p and T can range from zero to infinity and $-1 \leq D \leq 1$. We put pressure on the ordinate, T on the abscissa in the plane of the paper, and D on their mutual perpendicular coming out of the plane. The structure of the diagram, shown in Figs. 3a and 3b for very large and moderately small systems, respectively, is easiest seen by following isobaric paths of increasing temperature. In Fig. 3a, at a low pressure and low temperature, D is -1, corresponding to all solid clusters; D remains at -1 or just above it--indetectably far, in fact--until T almost reaches T_{equil}, at which the free energies of solid and liquid clusters are equal and D=0. Then D swings very sharply, like the heavy solid curve of Fig. 2, from almost -1 to almost +1 within a very tiny interval about T_{equil}, so sharply that the continuous change of D looks discontinuous. Then D continues at almost +1 and strictly becomes +1 only at temperatures above T_f. The same qualitative behavior occurs at each higher pressure, but with each increase in pressure, the shift of D occurs at a higher temperature. Thus the phase diagram for this system in (p,T,D) space is two planar regions separated by a curved step.

The phase diagram of a smaller system, displayed in (p,T,D) space is also easy to construct from the information in Fig. 2. Along an isobar, starting at a low temperature, we find D=-1 up to T_f, where D jumps discontinuously to a value above -1 and presumably (but not yet proven to be) less than 0. D then climbs with increasing T, presumably through D=0, increasing smoothly until $T=T_m$, where D again jumps discontinuously, here to +1. At each higher pressure, this entire sequence is repeated but at a higher set of temperatures. Consequently the phase diagram for the small system has the form of two planar regions separated by curved step-like link consisting of discontinuous, sharp steps away from each plane, connected by an s-shaped curved surface. The fraction of the total step that is included in the discontinuities diminishes as the size of the cluster increases, but the steepness of the continuous, s-curved portion also increases with N, so that the gradual step with jumps becomes one single step, in the large-N limit.

SURFACE MELTING OF CLUSTERS

Clusters of more than about 50 particles may exhibit a form in which their surfaces have characteristics of liquids and their interiors, the properties of solids. In particular, the phemonenon occurs if the solid is wet by its own melt. This was suspected by Briant and Burton[6] and Hoshino[74] and shown for a 55-atom Lennard-Jones cluster by simulations of Nauchitel and Pertsin[20]. The subject was pursued subsequently[21, 22, 75] to show that the melted surface of a Lennard-Jones or metal cluster does indeed exhibit the characteristics normally associated with a liquid: large mean square displacements and diffusion coefficients, large Lindemann parameters (root-mean-square relative fluctuations in nearest-neighbor distances), soft modes of vibration and broad pair distribution functions, all while the atoms of the inner layers of the cluster remain solid-like. The most striking aspect of this surface melting shows itself in animations but not in the numerical diagnostics or in snapshots; most of the surface atoms undergo large-amplitude, highly anharmonic motions so that at any instant, a snapshot will indeed show an amorphous-looking structure for the outermost layer of the cluster. However most of the surface particles, in fact all but a few that have popped out of the surface and float in the space just outside the surface layer, undergo highly organized collective oscillations around a well defined polyhedral structure. Only the "floaters" carry most of the free motion and give the surface most of its liquid-like character. At any instant the surface seems amorphous but averaged over perhaps a few picoseconds, the surface-melted cluster looks more like a polyhedron than like a swarm around an ordered core.

As recently as 1988, it was not clear what order to assign to the surface-melting transition[76]. It is now clear that at least the kind of surface melting that occurs in small clusters is a finite-system precursor to a first-order transition, with a nonzero (but small) latent heat and a region of coexistence for finite systems that becomes the regions of metastable supercooling and superheating in the bulk limit. In fact simulations and analytic statistical-mechanical theory both show that clusters may exhibit coexistence of solid with a surface-melted form, of that surface-melted form with liquid clusters, and even of these three phase-like forms together, provided the parameters governing the shape of the multidimensional potential surface of the cluster are suitable[23, 24]. This condition seems to be well satisfied by clusters of identical Lennard-Jones particles.

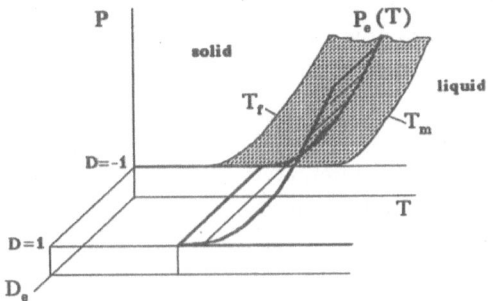

VERY LARGE SYSTEM

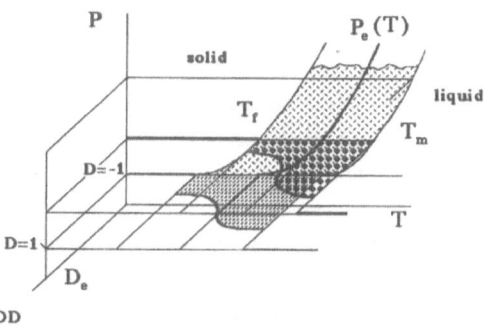

MEDIUM–SIZE CLUSTER

Figure 3. Phase diagrams for a large cluster (a), and a smaller cluster (b), in (p,T,D) space. The former shows the sharp step in D that allows the use of a simple, planar diagram with a single coexistence curve for solid-liquid equilibrium of bulk matter. The latter requires that the range of values of D be contained in the diagram, so a three-dimensional picture is needed.

The behavior of systems whose melts do not wet the corresponding solids is very different. Alkali halide crystals, for example, are not uniformly wet by their melts[77]. This has the consequence that instead of exhibiting a form in which their surfaces are melted, alkali halide clusters exhibit a form in which one side of the cluster is solid and the other side is liquid[27, 28]. This phenomenon appears in both instantaneous snapshots and in the quenched structures obtained by tracking the path from an instantaneous configuration down to the locally stable structure around which the cluster was oscillating in that instantaneous configuration. The indications now are that it is likely to be very difficult to find such forms experimentally because if the cluster is cooled fast but not instantaneously, it finds its way to a rocksalt-like structure of low energy because of the highly focusing nature of the alkali halide potential surface. On the other hand if the cluster is quenched very fast, it will almost certainly be trapped in a totally amorphous structure simply because the number of such structures is so much larger than the number of partly-liquid structures.

CONCLUSION

The solid, soft solid and liquid phases of clusters share many of the properties of phases of bulk matter. Even glassy phases of clusters, which were not included in this discussion(see [29, 78-83]) probably occur. Because we can study clusters in so much more detail than we can bulk matter, and because we can often find the limiting behavior for the limit of large N, we get new insights into the behavior of bulk matter. For example we learn much about the first-order transition away from the critical region, something much less understood than the nature of the critical region. Because of the small differences in free energy of different phase-like forms of clusters, we can expect to see coexistences of various "phases" of clusters under conditions in which such coexistence is quite impossible in bulk matter.

We have not discussed molecular clusters here, for want of space. They show phase-like behavior that is somewhat less dramatically different from the behavior of bulk matter simply because the short-range anisotropic forces connected with their shapes are strong enough to determine the properties of clusters and bulk matter primarily on the basis of near neighbor relationships. Hence the structures of molecular van der Waals clusters are, for the most part, very much like the structures of the corresponding solids[53, 55, 84-86]. Moreover molecular clusters may exhibit phase changes, as bulk phases may, involving the rotational disordering and even nearly-free rotation of molecules in their lattice sites. The extent to which they may exhibit coexistence over finite ranges of temperature and pressure, or display glassy forms, is still quite unexplored. Their complexity has only made them an even richer topic than atomic clusters, and one therefore that takes longer to explore.

ACKNOWLEDGMENTS

The author would like to express his thanks to the students and research associates who have contributed much of the work from The University of Chicago which has been cited here, notably Julius Jellinek, Gregory Natanson, François Amar, Thomas Beck, Heidi Davis, Hai-Ping Cheng, John Rose, David Wales and Ralph Kunz. The research carried out at The University of Chicago was supported by the National Science Foundation.

REFERENCES

1. D.J. McGinty, Molecular dynamics studies of the properties of small clusters of argon atoms, *J. Chem. Phys.* 58:4733 (1973).
2. R.M. Cotterill, W. Damgaard Kristensen, J.W. Martin, L.B. Peterson and E.J. Jensen, *Comput. Phys. Comm.* 5:28 (1973).
3. R.M.J. Cotterill, W. D. Kristensen and E. J. Jensen, *Phil. Mag.* 30:245 (1974).
4. R.M.J. Cotterill, *Phil. Mag.* 32:1283 (1975).
5. W. Damgaard Kristensen, E.J. Jensen and R.M.J. Cotterill, Thermodynamics of small clusters of atoms: A molecular dynamics simulation, *J. Chem Phys.* 60:4161 (1974).
6. C.L. Briant and J.J. Burton, Molecular dynamics study of the structure and thermodynamic properties of argon microclusters, *J. Chem. Phys.* 63:2045 (1975).
7. R. D. Etters and J. B. Kaelberer, Thermodynamic properties of small aggregates of rare-gas atoms, *Phys. Rev. A.* 11:1068 (1975).
8. R. D. Etters and J. B. Kaelberer, On the character of the melting transition in small atomic aggregates, *J. Chem. Phys.* 66:5112 (1977).
9. J.B. Kaelberer and R.D. Etters, Phase transitions in small clusters of atoms, *J. Chem. Phys.* 66:3233 (1977).
10. M.R. Hoare and P. Pal, Physical Cluster Mechanics: Statics and Energy Surfaces for Monatomic Systems, *Adv. Phys.* 20:161 (1971).
11. M.R. Hoare and P. Pal, *J. Cryst. Growth.* 17:77 (1972).
12. M.R. Hoare and P. Pal, *Nature.* 236:75 (1972).
13. M.R. Hoare and P. Pal, *Nature.* 230:5 (1972).
14. M.R. Hoare and P. Pal, Physical cluster mechanics: statistical thermodynamics and nucleation theory for monatomic systems, *Adv. Phys.* 24:645 (1975).
15. M.R. Hoare, Structure and Dynamics of Simple Microclusters, *Adv. Chem. Phys.* 40:49 (1979).
16. S. Sugano. "Microcluster Physics," Springer–Verlag, Berlin (1991)
17. W. v. E. Doering and W. R. Roth, *Angew. Chem.* 75:27 (1963).
18. E.L. Muetterties, *Inorg. Chem.* 4:769 (1965).
19. Z. Slanina. "Contemporary Theory of Chemical Isomerism," D. Reidel, Dordrecht (1986)
20. V.V. Nauchitel and A.J. Pertsin, A Monte Carlo study of the structure and thermodynamic behaviour of small Lennard-Jones clusters, *Mol. Phys.* 40:1341 (1980).

21. R. Averbach, ed., "Surface Melting and Surface Diffusion on Clusters," 206, Proc. Symposium on Clusters and Cluster-Assembled Materials, 1991, Materials Research Society, City (Year).

22. H.-P. Cheng and R.S. Berry, Surface melting of clusters and implications for bulk matter, *Phys. Rev. A.* 45:7969 (1992).

23. R.E. Kunz and R.S. Berry, Coexistence of multiple phases in finite systems, *(submitted to Phys. Rev. Lett.)*. (1993).

24. R.E. Kunz and R.S. Berry, Multiple phase coexistence in finite systems, *(submitted to Phys. Rev. E)*. (1993).

25. B. Baguenard, M. Pellarin, J. Lermé, J. L. Vialle and M. Broyer, Competition between atomic shell and electronic shell structures in aluminum clusters, *(submitted to J. Chem. Phys.)*. (1993).

26. J. Lermé, C. Bordas, M. Pellarin, B. Baguenard, J. L. Vialle and M. Broyer, Influence of surface softness on supershell structure of metal clusters: Application to gallium, *(submitted to Phys. Rev. B)*. (1993).

27. J. Luo, U. Landman and J. Jortner, Isomerization and Melting of Small Alkali-Halide Clusters, *in* : "Physics and Chemistry of Small Clusters," P. Jena, B. K. Rao and S. N. Khanna. ed., Plenum Press, New York, N. Y. (1987).

28. J. P. Rose and R. S. Berry, Freezing, Melting, Nonwetting and Coexistence in $(KCl)_{32}$, *J. Chem. Phys.* 98:3246 (1993).

29. J. Rose and R.S. Berry, $(KCl)_{32}$ and the possibilities for glassy clusters, *J. Chem. Phys.* 98:3262 (1993).

30. R.S. Berry, J. Jellinek and G. Natanson, Unequal freezing and melting temperatures for clusters, *Chem. Phys. Lett.* 107:227 (1984).

31. R.S. Berry, J. Jellinek and G. Natanson, Melting of clusters and melting, *Phys. Rev. A.* 30:919 (1984).

32. S. Gartenhaus and C. Schwartz, Center-of-mass motion in many-particle systems, *Phys. Rev.* 108:482 (1957).

33. F.H. Stillinger and T.A. Weber, Inherent pair correlation in simple liquids, *J. Chem. Phys.* 80:4434 (1984).

34. R.S. Berry and D. J. Wales, Freezing, melting, spinodals and clusters, *Phys. Rev. Lett.* 63:1156 (1989).

35. D.J. Wales and R. S. Berry, Melting and freezing of small argon clusters, *J. Chem. Phys.* 92:4283 (1990).

36. D.J. Wales and R. S. Berry, Freezing, melting, spinodals and clusters, *J. Chem. Phys.* 92:4473 (1990).

37. R.S. Berry, T. L. Beck, H. L. Davis and J. Jellinek, Solid-Liquid Phase Behavior in Microclusters, *in* : "Evolution of Size Effects in Chemical Dynamics, Part 2," I. Prigogine and S. A. Rice. ed., John Wiley and Sons, New York (1988).

38. R.S. Berry, Clusters, Melting, Freezing and Phase Transitions, *J. Chem. Soc. Faraday Trans.* 86:2343 (1990).

39. R.S. Berry, Structure and dynamics of clusters: Phase equilibrium and phase change, *in* :

"The Chemical Physics of Atomic and Molecular Clusters," G. Scoles. ed., North-Holland, Amsterdam (1990).

40. R.S. Berry, Structure and dynamics of clusters: An introduction, *in* : "The Chemical Physics of Atomic and Molecular Clusters," G. Scoles. ed., North-Holland, Amsterdam (1990).

41. R.S. Berry, Melting and Freezing of Clusters, *in* : H. Haberland. ed., (1992).

42. J. Jellinek, T. L. Beck and R. S. Berry, Solid-liquid phase changes in simulated isoenergetic Ar_{13}, *J. Chem. Phys.* 84:2783 (1986).

43. T.L. Beck, J. Jellinek and R. S. Berry, Rare gas clusters: Solids, liquids, slush and magic numbers, *J. Chem. Phys.* 87:545 (1987).

44. T.L. Beck and R. S. Berry, The interplay of structure and dynamics in the melting of small clusters, *J. Chem. Phys.* 88:3910 (1988).

45. F. Amar and R.S. Berry, The onset of nonrigid dynamics and the melting transition in Ar_7, *J. Chem. Phys.* 85:5943 (1986).

46. E. Blaisten-Barojas and D. Levesque, Molecular-dynamics simulation of silicon clusters, *Phys. Rev. B.* 34:3910 (1986).

47. E. Blaisten-Barojas and D. Levesque, A Molecular Dynamics Study of Silicon Clusters, *in* : "Physics and Chemistry of Small Clusters," P. Jena, B. K. Rao and S. N. Khanna. ed., Plenum Press, New York (1987).

48. J.D. Honeycutt and H. C. Andersen, Molecular Dynamics Study of Melting and Freezing of Small Lennard-Jones Clusters, *J. Phys. Chem.* 91:4950 (1987).

49. S. Sawada, Dynamics of Transition Metal Clusters, *in* : "Microclusters," S. Sugano, Y. Nishina and S. Ohnishi. ed., Springer–Verlag, Berlin (1987).

50. S. Sawada and S. Sugano, Structural fluctuations of Au_{55} and Au_{147}: Substrate effect, *Z. Phys. D.* 24:377 (1992).

51. E.J. Valente and L. S. Bartell, Electron diffraction studies of supersonic jets. VI. Microdrops of benzene, *J. Chem. Phys.* 80:1451 (1984).

52. E.J. Valente and L. S. Bartell, Electron diffraction studies of supersonic jets. VII. Liquid and plastic crystalline carbon tetrachloride, *J. Chem. Phys.* 80:1458 (1984).

53. L.S. Bartell, Diffraction Studies of Clusters Generated in Supersonic Flow, *Chem. Rev.* 86:492 (1986).

54. L.S. Bartell, L. R. Sharkey and X. Shi, Electron Diffraction and Monte Carlo Studies of Liquids. 3. Supercooled Benzene, *J. Am. Chem. Soc.* 110:7006 (1988).

55. L.S. Bartell, L. Harsanyi and E. J. Valente, Phases and Phase Changes of Molecular Clusters Generated in Supersonic Flow, *J. Phys. Chem.* 93:6201 (1989).

56. S. W. Rick, D. L. Leitner, J. D. Doll, D. L. Freeman and D. D. Frantz, The quantum mechanics of clusters: the low-temperature equilibrium and dynamical behavior of rare-gas systems, *J. Chem. Phys.* 95:6658 (1991).

57. J. Bösiger and S. Leutwyler, Surface-Melting Transitions and Phase Coexistence in Argon Solvent Clusters, *Phys. Rev. Lett.* 59:1895 (1987).

58. J. Bösiger, R. Knochenmuss and S. Leutwyler, Wetting-nonwetting transitions in Argon solvent clusters, *Phys. Rev. Lett.* 62:3058 (1989).

59. R. Knochenmuss and S. Leutwyler, Selective spectroscopy of rigid and fluxional carbazole-argon clusters, *J. Chem. Phys.* 92:4686 (1990).

60. T. Troxler and S. Leutwyler, Electronic spectra of naphthalene-Ar_n solvent clusters (n=1-30), *J. Chem. Phys.* 95:4010 (1991).

61. F. H. Stillinger and T. A. Weber, Point defects in bcc crystals: Structures, transition kinetics, and melting implications, *J. Chem. Phys.* 81:5095 (1984).

62. T. L. Hill, On First-Order Phase Transitions in Canonical and Grand Ensembles, *J. Chem. Phys.* 23:812 (1955).

63. T. L. Hill. "The Thermodynamics of Small Systems, Part 1," W. A. Benjamin, New York (1963)

64. T. L. Hill. "The Thermodynamics of Small Systems, Part 2," W. A. Benjamin, New York (1964)

65. H.C. Andersen, Molecular dynamics simulations at constant pressure and/or temperature, *J. Chem. Phys.* 72:2384 (1980).

66. S. Nosé, *Mol. Phys.* 52:255 (1984).

67. S. Nosé, A unified formulation of the constant temperature molecular dynamics methods, *J. Chem. Phys.* 81:511 (1984).

68. S. Nosé, Constant Temperature Molecular Dynamics Methods, *Prog. Theor. Phys.* Supplement 103:1 (1991).

69. V. Privman and M. E. Fisher, Finite-size effects at first-order transitions, *J. Stat. Phys.* 33:385 (1983).

70. M. E. Fisher and V. Privman, First-Order Transitions in Spherical Models: Finite-Size Scaling, *Commun. Math. Phys.* 103:527 (1986).

71. S. Saxena, P. C. P. Bhatt and V. C. Prasad, *IEEE Trans. Comput.* 39:400 (1990).

72. R. J. Zauhar and R. S. Morgan, *J. Comput. Chem.* 11:603 (1990).

73. H. Edelsbrunner, *in* : "Algorithms in Combinatorial Geometry," W. Brauer, G. Rozenberg and A. Saloma. ed., Springer-Verlag, New York (1987).

74. K. Hoshino and S. Shimamura, A simple model for the melting of fine particles, *Phil. Mag. A.* 40:137 (1979).

75. Z.B. Güvenç and J. Jellinek, Surface Melting in Ni_{55}, *Z. Phys. D.* 26:304 (1993).

76. K. Strandburg, Two-dimensional melting, *Revs. Mod. Phys.* 60:161 (1988).

77. G. Grange, R. Landers and B. Mutaftschiev, Contact Angle and Surface Morphology of KCl Crystal-Melt Interface Studied by the "Bubble Method", *Surf. Sci.* 54:445 (1976).

78. A. Rahman, M. J. Mandell and J. P. McTague, Molecular dynamics study of an amorphous Lennard–Jones system at low temperature, *J. Chem. Phys.* 64:1564 (1976).

79. M. Amini and R. W. Hockney, Computer Simulation of Melting and Glass Formation in a Potassium Chloride Microcrystal, *J. Non-Cryst. Sol.* 31:447 (1979).

80. M. Amini, D. Fincham and R. W. Hockney, A molecular dynamics study of the melting of alkali halide crystals, *J. Phys. C.* 12:4707 (1979).

81. V. Buch, Identification of two distinct structural and dynamical domains in an amorphous water cluster, *J. Chem. Phys.* 93:2631 (1990).

82. H. Miyagawa, Y. Hiwatari and S. Itoh, Molecular-Dynamics Study for the Glass Transition in LiI, *Prog. Theoret. Phys.* Supplement No. 103:47 (1991).

83. H.-P. Cheng and U. Landman, Controlled Deposition, Soft Landing, and Glass Formation in Nanocluster-Surface Collisions, *Science*. 260:1304 (1993).

84. L. S. Bartell, F. J. Dulles and B. Chuko, Structure and Dynamics of Molecular Clusters. Diagnostic Criteria in Monte Carlo Computations, *J. Phys. Chem.* 95:6481 (1991).

85. L. S. Bartell and S. Xu, Molecular Dynamics Examination of an Anomalous Phase of TeF_6, *J. Phys. Chem.* 95:8939 (1991).

86. L. S. Bartell and J. Chen, Structure and Dynamics of Molecular Clusters. 2. Melting and Freezing of CCl_4 Clusters, *J. Phys. Chem.* 96:8801 (1992).

EXPLORING POTENTIAL SURFACE LANDSCAPES AND HOW THEY GOVERN DYNAMICS

R. Stephen Berry

Department of Chemistry and The James Franck Institute
The University of Chicago
Chicago, Illinois 60637, U. S. A.

THE MULTIDIMENSIONAL POTENTIAL SURFACE

The subject of multidimensional potential surfaces and the dynamics on those surfaces was just reviewed by this writer[1], and the closely related topic of the analytic representation of potential surfaces, largely of small systems, had been reviewed shortly before by Schatz[2]. Consequently we will here very tersely review some of the fundamentals, briefly survey aspects of the subject treated in those reviews and discuss some aspects not covered in those reviews.

The description of any system of two or more atoms usually begins with consideration of how to deal with the combination of relatively fast electronic motions and the relatively slow nuclear motions. Traditionally this involves invoking the approximate separability of electronic and nuclear motions, which justifies the concept of an effective potential energy for the nuclear motion, and that is the starting point for this discussion. However, in light of the concern about cases in which this separation is inapplicable, we begin with the caveat that here we shall not concern ourselves with situations in which energy flows between electronic and nuclear degrees of freedom so as to change any quantum numbers. (This is the condition for the validity of the *adiabatic* approximation, which was long ago justified by Ehrenfest's Theorem.) Rather, we will assume that the electronic degrees of freedom always equilibrate so fast, with respect to nuclear motion that the energy of the electrons--kinetic and potential-- plus the instantaneous, configuration-dependent potential energy of the nuclei, constitute an effective potential energy. The limitations of this assumption were recently discussed by Sutcliffe[3]. The effective potential energy is a function of the nuclear coordinates only; it, together with initial conditions, governs the motion of the nuclei, whether classical or quantum-mechanical. We can think of this potential as defining a surface.

Linking the Gaseous and Condensed Phases of Matter
Edited by L.G. Christophorou *et al*, Plenum Press, New York, 1994

We suppose that the polyatomic system we are describing, be it molecule or cluster, consists of N atoms, and that these are assigned positions relative to the center of mass or to some other convenient choice such as that of Radau coordinates. Thus three coordinates fix the position of the system in the laboratory frame. We also assume that the system is not rotating (or that the coordinate system rotates with the molecule or cluster) so that three (angular) coordinates are also fixed. This leaves 3N-6 independent coordinates on which the effective potential depends. Hence we can think of the effective potential as a dependent variable which defines a surface in a space of 3N-5 dimensions, 3N-6 of which are those of the independent variables. For a three-atom system, 3N-6=3 so the surface is the dependent variable in a 4-dimensional space. We shall be concerned here with clusters not only of three but of 6, 7 and even many more atoms, as many as 55 and 64, so we shall be dealing with spaces of roughly 12 to about 185 dimensions, and the surfaces in these spaces.

All problems concerning potential surfaces, even within the limited context we have chosen, begin with the issue of constructing the potential surface itself. This is an industry of moderate size and considerable sophistication which does not need further review here[4-6]. This subject is far from closed. Most approaches to it are based either of two approaches. One involves the construction of effective interaction potentials such as sums of pair interactions, or such pair potentials supplemented with three-body potentials or even four-body contributions and sometimes with mean-field effects such as one-body, density-dependent terms. The other approach requires finding the eigenvalues of the electronic Schrödinger equation at a sufficient number of configurations, that is, at a sufficient number of points on the surface, to permit us to construct an adequate approximation to the surface from those points.

Some of the difficult, open issues regarding construction of multidimensional potential surfaces are these: What is the minimum number of points for which the potential surface can be constructed to yield results of a given desired accuracy? How should these points be distributed? Presumably they should be taken rather densely in the vicinity of stationary points; how can we locate the regions of stationary points from some very economical procedure, just to get the procedure started? If we can fit each region of the surface near a stationary point to an analytical representation, how do we join the representations from regions of different stationary points?

If one pursues the line of finding relatively simple, effective potentials, one is on firm ground with the rare gas atoms and with simple alkali halides, possibly even with other ionic solids such as alkaline earth oxides. For the rare gases, the Lennard-Jones distance-dependent pair potential[7] has been an excellent approximation to the accurate potential surfaces for these systems[8-11]. Likewise, the Born-Mayer potential, consting of long-range Coulombic interactions and exponential short-range repulsions[12], is a good approximation to more accurate, more complex forms of potential energies of interacting polarizable ions[13-15]. However the use of such potentials to describe covalently bound insulators and semiconductors is hazardous, and metals are even less amenable to such descriptions--despite the use of such methods for modeling metals, even by this writer. "Effective medium" and "embedded atom" potentials, two popular, closely related approximate potentials that combine explicit pair interactions with mean-field many-body effects, can be used with caution to carry out indicative calculations to help guide intuition, but should only be used in that spirit.

EXPLORING THE SURFACE

The topography of a multidimensional potential surface is complex, so complex that for any system of more that eight or ten particles we are unlikely to know all the information we might obtain, even if we had a reliable analytic representation of the surface. However there are useful things to know, even for quite complex surfaces: the most important minima, the saddles that link these minima with other, higher minima, the density of locally stable states as a function of energy, and the microcanonical entropies of the important basins of the surface--in effect, the logarithms of the areas of many-dimensional lakes at all levels of their surfaces. Contemporary techniques enable us to find all but the last of these. Finding minima is routinized to the level of a textbook subject[16]; finding saddles is not as thoroughly explored but satisfactory methods are available now[17-34]. The development of this subject is included in the review mentioned previously. One of the most interesting open problems in this field now is how to find reliable estimates of the microcanonical entropy.

Another kind of exploration of potential surfaces is that carried out by the clusters themselves. This can be traced by comparing isomerization rates from simulations with those predicted by RRKM theory, for example[35], and the results imply that RRKM theory is a good predictor of such isomerizations. A more subtle kind of investigation explores the extent of chaotic and ergodic behavior of the cluster, as a function of energy and of time. This is a relatively new topic which has been reviewed very recently[36] so we will give only a brief summary of the findings. Small clusters are only slightly chaotic and ergodic but become more so as their energy increases. However when very small clusters, of 3, 4 or 5 particles, become energized enough to pass over saddles on their potentials, they lose some chaotic and ergodic character because they spend long intervals in saddle regions where their kinetic energies are low and their trajectories seem rather regular. Larger clusters do not show such drops because of two reasons: their many other degrees of freedom hide the characteristics of the "soft" modes, and the saddles become sharper, offering less phase and configuration space than do the saddles of small clusters. Ergodicity seems to develop with definite separations of time scales for different degrees of ergodicity. It is possible to associate the curvature of the potential surface with a local contribution to the extent of chaotic behavior, as measured by the Kolmogorov entropy.

COMPLEX SURFACES, GLASSES AND "PROTEIN FOLDING"

One fairly complex system that has been studied in some detail is the $(KCl)_{32}$ cluster. It was used as a vehicle to develop methods of estimating densities of locally stable configurational states from statistical samples of minima, and of categorizing those locally stable states[37, 38]. This system has as its lowest-energy structure a 4x4x4 rocksalt crystal. It has other low-energy structures, all essentially rocksalt-like, some in slabs or sticks, some with a defect or two or three. There are many more locally stable structures that are rocksalt-like on one side and amorphous on the other, structures that emerge from quenching hot clusters that are solid on one side and liquid on the other, a situation possible because alkali halide melts do not uniformly wet their corresponding solids. At still higher energies are many, many more entirely amorphous structures. If a simulated molten $(KCl)_{32}$ cluster is quenched "instantaneously" in the sense of computer time steps, then the number of available amorphous structures is greater than the number of rocksalt structures by a factor of roughly 10^{12} to 10^{14} to 1, depending on the precise energy from which the cluster is quenched. On this basis one might expect the $(KCl)_{32}$ cluster to be a good glass former.

To the contrary, if one simulates a quench at a very fast but finite rate, say 10^{12} K/s (about 50 vibrational periods) or even 5×10^{12} K/s, the cluster finds its way to a rocksalt-like structure--perhaps one of the defective structures but rocksalt nonetheless. In other words the potassium chloride cluster wins against odds of order 10^{12}:1 to find a special, low-energy structure. This is even more amazing, statistically, than the folding of a protein or other polymer of the same number of atoms. If one carries out the cooling at a rate of 10^{13} K/s or higher, or if the long-range Coulomb forces are replaced by shielded Coulomb forces, the cluster can be trapped in an amorphous structure. This leaves us with a final challenge: what makes a potential surface "focusing"? Is it perhaps better to answer this question before we try to find out why proteins fold into physiologically active shapes as they move on their potential surfaces? Might we learn more about protein folding and other systems with many complicating constraints such as maintenance of primary skeletal structure by first examining simple systems which have no such constraints? Perhaps we should begin by learning how the topography of a potential surface "guides" a system down to a particular structure, and then go on to apply what we learn to more complex molecules.

ACKNOWLEDGMENTS

The author would like to express his thanks to his many students and associates whose work is cited here and also to those on whose previous work this was built. The research described here that was done in our group at The University of Chicago was supported by the National Science Foundation.

REFERENCES

1. R. S. Berry, Potential Surfaces and Dynamics: What Clusters Tell Us, *Chem. Revs.* (in press):(1993).

2. G. C. Schatz, The analytic representation of electronic potential energy surfaces, *Revs. Mod. Phys.* 61:669 (1989).

3. B. Sutcliffe, The coupling of nuclear and electronic motions in molecules, *J. Chem. Soc. Faraday Trans.* 89:2321 (1993).

4. J. N. Murrell, S. Carter, S. C. Farantos, P. Huxley and A. J. C. Varandas. "Molecular Potential Energy Functions," Wiley, New York (1984)

5. D. M. Hirst. "Potential Energy Surfaces," Taylor and Francis, London (1985)

6. T. H. Dunning Jr. and L. B. Harding, *in* : "Theory of Chemical Reaction Dynamics," M. Baer. ed., CRC Press, Boca Raton, Florida (1985).

7. J. E. Lennard-Jones, *Proc. Roy. Soc. A.* 106:463 (1924).

8. R. A. Aziz and H. H. Chen, *J. Chem. Phys.* 67:5719 (1977).

9. R. A. Aziz, *Molec. Phys.* 38:177 (1979).

10. R. A. Aziz and M. J. Slaman, The argon and krypton interatomic potentials revisited, *Mol. Phys.* 58:679 (1986).

11. R. A. Aziz, A highly accurate interatomic potential for argon, *J. Chem. Phys.* 99:4518 (1993).

12. M. P. Tosi and F. G. Fumi, *J. Phys. Chem. Solids.* 25:45 (1964).

13. D. O. Welch, O. W. Lazareth, G. J. Dienes and R. D. Hatcher, Alkali halide molecules: configurations and characteristics of dimers and trimers, *J. Chem. Phys.* 64:835 (1976).

14. D. O. Welch, O. W. Lazareth, G. J. Dienes and R. D. Hatcher, Clusters of alkali halide molecules, *J. Chem. Phys.* 68:2159 (1978).

15. R. S. Berry, Optical Spectra of the Alkali Halide Molecules, *in* : "Alkali Halide Vapors," P. Davidovits and D. L. McFadden. ed., Academic Press, New York (1979).

16. W. H. Press, B. P. Flannery, S. A. Teukolsky and W. T. Vetterling. "Numerical Recipes," Cambridge University Press, Cambridge (1986)

17. J. W. McIver Jr. and A. Komornicki, Structure of Transition States in Organic Reactions. General Theory and an Application to the Cyclobutene-Butadiene Isomerization Using a Semiempirical Molecular Orbital Method, *J. Am. Chem. Soc.* 94:2625 (1972).

18. A. Komornicki and J. W. McIver Jr., *J. Am. Chem. Soc.* 95:4512 (1973).

19. A. Komornicki and J. W. McIver Jr., Structure of Transition States. III. A MINDO/2 Study of the Cyclization of 1,3,5-Hexatriene to 1,3-Cyclohexadiene, *J. Am. Chem. Soc.* 96:5798 (1974).

20. J. W. McIver, The Structure of Transition States: Are They Symmetric?, *Acc. Chem. Res.* 7:72 (1974).

21. J. Pancik, Calculation of the Least Energy Path on the Energy Hypersurface, *Coll. Czech. Chem. Comm.* 40:1112 (1975).

22. H. Poppinger, On the Calculation of Transition States, *Chem. Phys. Lett.* 35:550 (1975).

23. T. A. Halgren and W. N. Lipscomb, The Synchronous Transit Method for Determining Reaction Pathways and Locating Molecular Transition States, *Chem. Phys. Lett.* 49:225 (1977).

24. P. G. Mezey, M. R. Peterson and I. G. Csizmadia, Transition state determination by the X-method, *Can. J. Chem.* 55:2941 (1977).

25. K. Müller and L. D. Brown, Location of Saddle Points and Minimum Energy Paths by a Constrained Simplex Optimization Procedure, *Theor. Chim. Acta (Berl.).* 53:75 (1979).

26. W. H. Miller, N. C. Handy and J. E. Adams, Reaction path Hamiltonian for polyatomic molecules, *J. Chem. Phys.* 72:99 (1980).

27. C. J. Cerjan and W. H. Miller, On finding transition states, *J. Chem. Phys.* 75:2800 (1981).

28. D. O'Neal, H. Taylor and J. Simons, Potential Surface Walking and Reaction Paths for $C_{2v}Be + H_2 <- BeH_2 -> Be + 2H$ ($^1A_1$), *J. Phys. Chem.* 88:1510 (1984).

29. J. Simons, P. Jørgenson, H. Taylor and J. Ozment, Walking on Potential Energy Surfaces, *J. Phys. Chem.* 87:2745 (1983).

30. J. Baker, An Algorithm for the Location of Transition States, *J. Comp. Chem.* 7:385 (1986).

31. R. S. Berry, H. L. Davis and T. L. Beck, Finding saddles on multidimensional potential surfaces, *Chem. Phys. Lett.* 147:13 (1988).

32. R. S. Berry, How We and Molecules Explore Molecular Landscapes, *in* : "Mode Selective Chemistry," J. Jortner, A. Pullman and B. Pullman. ed., Kluwer Academic Publishers, Amsterdam (1991).

33. D. J. Wales, Transition States for Ar_{55}, *Chem. Phys. Lett.* 166:419 (1990).

34. D. J. Wales, Locating Stationary Points for Clusters in Cartesian Coordinates, *J. Chem. Soc. Faraday Trans.* 89:1305 (1993).

35. J. Rose and R. S. Berry, Towards elucidating the interplay of structure and dynamics in clusters: Small KCl clusters as models, *J. Chem. Phys.* 96:517 (1992).

36. R. S. Berry, "Atomic Clusters: Laboratories for Studying Chaos and Ergodicity," *in* Proc. Adriatico Research Conference "Mesoscopic Systems and Chaos, a Novel Approach,", Trieste, 1993, H. Cerdiera and G. Casati, ed., World Scientific, Singapore (1994).

37. J. Rose and R. S. Berry, $(KCl)_{32}$ and the possibilities for glassy clusters, *J. Chem. Phys.* 98:3262 (1993).

38. J. P. Rose and R. S. Berry, Freezing, Melting, Nonwetting and Coexistence in $(KCl)_{32}$, *J. Chem. Phys.* 98:3246 (1993).

SECTION IV. ELECTRON MOTION IN GASES AND LIQUIDS

DENSITY AND FIELD DEPENDENCE OF EXCESS ELECTRON MOBILITY IN HIGH-DENSITY NOBLE GASES

A.F.Borghesani and M.Santini

Dipartimento di Fisica "G.Galilei", Università di Padova
and Consorzio Interuniversitario Fisica della Materia
Via F.Marzolo 8, I-35131 Padova, Italy

INTRODUCTION

The transport properties of excess electrons in high-density non-polar gases may give substantial information on several subjects that are of interest in the physics of condensed matter. In particular, the mobility of extraelectrons can be used as a probe to study the electron states in a disordered medium and the relationship among the electron-atom interactions and the properties of the fluid.

In gases at room temperature and at atmospheric pressure the interatomic distance is of the order of ten atomic diameters. An extra electron injected into the gas interacts normally with one atom at a time. In this situation are also satisfied the assumptions that the presence of neighboring atoms has no influence on the scattering process and that the interaction time between the electron and the gas atom in a collision is negligible in comparison to the mean free time of electrons between collisions. The collision properties are thus described in terms of single collisions between electrons and isolated atoms and the gas-phase scattering cross section for momentum transfer is used to calculate the mobility of the excess electrons according to the usual Boltzmann formalism [1].

On the opposite side of the density range are liquids. In a liquid below its normal boiling point the interatomic distance is of the order of the atomic diameter and the interaction ranges of thermal electrons are such that an excess electron injected into a liquid simultaneously interacts with much more than one atom. The electron mean free path is now comparable to its thermal wavelength and multiple scattering effects are to be taken into account. According to the fact that electrons move much faster than the massive atoms of the fluid, the Born-Oppenheimer approximation is valid and the liquid is replaced by an instantaneous lattice obtained by taking a snapshot of the fluid. Multiple scattering effects in the high density environment give origin to a conduction band and electron transport is described by means of the deformation potential theory borrowed from solid state physics. Within this theory the source of scattering is the local modulation of the excess electron energy at the bottom of the conduction band due to the intrinsic density fluctuations of the disordered medium [2,3].

In between these two extreme cases high-density gases give the researchers the unique opportunity to study how gradually the single scattering picture valid in the dilute gas limit goes over into the multiple scattering picture of the high-density liquid because the interatomic distance can be varied relatively at will by adjusting the gas temperature and pressure. Moreover, high-density gases are good representatives of disordered materials and the presence of a critical point in an experimentally ac-

Linking the Gaseous and Condensed Phases of Matter
Edited by L.G. Christophorou *et al*, Plenum Press, New York, 1994

cessible (Pressure, Temperature) region allows the investigation of the role of the increasing disorder related to the intrinsic density fluctuations of the medium.

The choice of investigating the excess electron mobility in noble gases is due to several features of these particular kind of gases. First of all, their excitation energies are quite large, so that the momentum transfer cross section is uniquely determined at the low energies involved in drift mobility experiments. Normally, these energies are comparable to or a little bit larger than $(3/2)k_BT$ since the electric fields used are not very large. As a consequence, the collisions between electrons and atoms are isotropic to a very good approximation[1].

Moreover, in the noble gas family both repulsive as well as attractive electron-atom interactions are present. In He and Ne the short-range repulsive exchange forces dominate the electron-atom interaction and the scattering length is positive. On the contrary, in Ar, Kr, and Xe the long-range part of the interaction potential due to polarization forces is predominant and, as a consequence, these gases have negative scattering lengths and their cross sections show a very deep Ramsauer minimum[4] as shown in Figure 7.

One more advantage to use noble gases is of technological relevance. Namely, noble gases are relatively easy to purify and the impurity content, mainly O_2 molecules, can be readily reduced in the part per billion range. At this impurity level the attachment of the excess electrons to O_2 molecular impurities and their scattering off other non attaching impurities has a completely negligible influence on electron transport.

In the classical theory of electron motion in gases under the influence of uniform electric fields it is assumed that the zero-field density-normalized mobility $(\mu_0 N)$ is unaffected by the gas density and depends on the field strength only through the ratio E/N, where E is the electric field strength and N is the gas number density[4,5]. This is the result of the solution of the Boltzmann transport equation with the above mentioned assumptions for the dilute gas. Under these assumptions the classical prediction for the zero-field density-normalized mobility, i.e., that of thermal electrons, is then given by

$$(\mu_0 N)_c = \frac{4e}{3(2\pi m)^{1/2}(k_B T)^{5/2}} \int_0^\infty d\epsilon \, \frac{\epsilon}{\sigma_m(\epsilon)} \exp\left(-\epsilon/k_B T\right) \tag{1}$$

where $\sigma_m(\epsilon)$ is the energy-dependent electron-atom gas-phase momentum transfer scattering cross section. For a given cross section, at fixed temperature, $(\mu_0 N)_c$ is density independent. However, this classical prediction was found to be in disagreement with the experimental results as anomalous density effects were revealed in several compressed gases.

In Figure 1 we show the zero-field density-normalized mobility $(\mu_0 N)$ as a function of the number density N for He at $T = 77.4\,K$[6] and Ar at room temperature[7]. These are the typical examples of the two kinds of behavior experimentally observed in noble gases. For a comprehensive review of the experimental data we refer to literature[8]. It suffices here to recall that Helium and Neon, among others, show a negative effect, i.e., $(\mu_0 N)$ decreases with increasing density up to moderately large densities and for temperatures not too low. For larger densities (and quite low temperatures), the electron mobility in these gases shows a precipituous drop of several orders of magnitude. The high-mobility branch is generally understood in terms of quasi-free electron states, while the low-mobility branch is usually associated with self-trapped electron states (pseudobubbles). In He and Ne the interaction with excess electrons is dominated by short-range repulsive exchange forces and these gases are characterized by a positive scattering length[9].

On the other hand, gases such as Ar and the higher atomic weight noble gases show a positive density effect, i.e., $(\mu_0 N)$ increases with increasing density, even if it has to be recalled that a small, though evident, negative effect is shown by Ar vapor at T=90 K and for pressures up to approximately 0.1 MPa[10]. Negative scattering length gases, even at the largest densities, do not show evidence for localized states, at all.

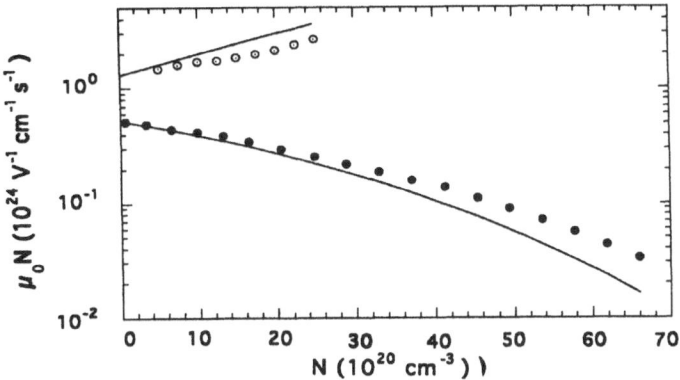

Figure 1. Zero-field density-normalized electron mobility $\mu_0 N$ as a function of the gas density N in Ar at room temperature[7] (open points) and in He[6] at $T=77.6\ K$ (closed points). The solid lines are the prediction of the multiple scattering theories of Braglia and Dallacasa[11] for Ar and of O'Malley[12] for He.

EXPERIMENTAL TECHNIQUE

During the years several different experimental techniques have been devised to determine the mobility and diffusion coefficients of the excess electrons. For a complete review of the experimental methods we refer to literature[8]. We want here to recall briefly only the so called *Pulsed-Townsend-Photoinjection* technique, since it is one of the techniques mainly used in order to study the high-density gases at low temperature. In fact, the simplicity of the apparatus design allows the construction of experimental cells that can withstand very large pressures and can be cooled to very low temperatures. One further advantage of this method is the straightforward analysis of the resulting transient wave forms to determine the electron drift velocity.

In Figure 2 we schematically show the experimental apparatus we have been using in the last few years[13]. A parallel-plate capacitor contained in the high-pressure cell delimits the drift space. A swarm of electrons is produced by irradiating the gold-coated cathode with the VUV light pulse of a laser or of a Xe flashlamp, whose duration is typically negligible compared to the drift time. The amount of charge extracted in our apparatus ranges between 4 and 400 fC, depending on the gas density and on the applied electric fields. The electrons drift towards the anode, inducing there a charge.

The anode is connected to the detection circuit. If the time constant of the anode circuit, RC, is smaller than the transit time of the swarm, the displacement current, in absence of electron attachment, generates a constant potential difference across the resistor R that is then detected by the preamplifier. As all the electrons reach the anode the current drops to zero and the same does the potential difference across R.

On the contrary, in the limit of RC much larger than the drift time, the current signal is integrated by the capacitor and the potential difference across the circuit is a linearly rising voltage wave form that can be accurately fitted to a straight line so as to find the start and end points of the swarm motion in the drift gap. The choice between integrating or not the current signal is dictated by the signal-to-noise ratio that can be achieved in the actual experimental configuration. For instance, in our setup we are forced to integrate the current signal because the low lamp intensity and the low fields used yield a small amount of photoextracted electrons. The resulting voltage signal is recorded by a high-speed digital transient analyzer and eventually fed to a personal computer for quick off-line data processing. This method allows to gather and analyze a very large number of wave forms yielding data of very good statistical accuracy.

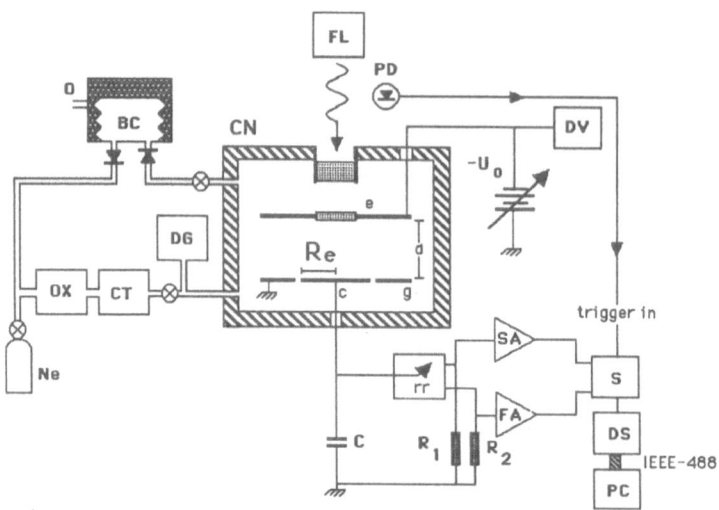

Figure 2. Schematics of the apparatus for pulsed photoinjection swarm experiment [14] FL: Xe flashlamp, OX: Oxisorb cartridge, CT:LN$_2$ active charcoal trap, CN: High pressure cell, e:emitter, c: collector, BC: bellow circulator, DV: Digital Voltmeter,PD: Triggering photodiode, C: integrating capacitor, rr: switching reed relay, FA,SA: Fast and Slow Amplifiers, S: Selector, DS: Digital Scope, PC: Personal Computer.

A gas handling system including an Oxisorb cartridge and a LN$_2$-cooled active-charcoal trap within a recirculation loop is used to remove impurities, mainly O$_2$, present in the gas. The effect of the impurities is twofold. In the first place, O$_2$ scavenges the excess electrons as the result of the well-known resonant attachment process [8]. As a consequence, the electron signal becomes faint and its form is no longer a straight line since the current decreases exponentially with time. This fact brings about an increased complexity of the waveform to be analyzed [15]. In Figure 3 we show two typical signals obtained in absence or in presence of electron attachment.

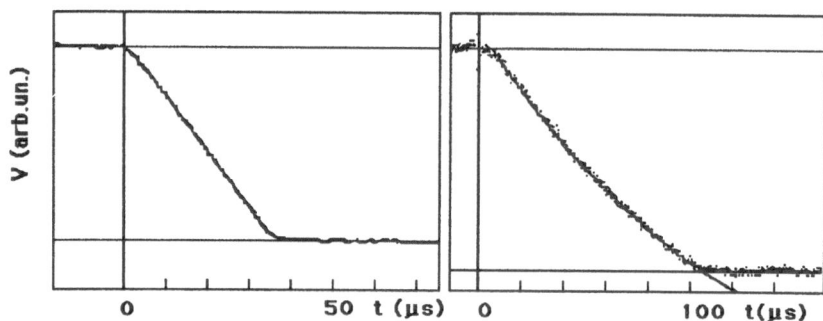

Figure 3. Signal obtained in absence (left) or in presence (right) of electron attachment to O$_2$ impurities [15]. Note that also in presence of attachment is possible to accurately fit the signal to an analytical wave form yielding both electron drift time and lifetime.

In the second place, non-attaching impurities, such as N$_2$, can be highly effective in thermalizing the excess electrons even at fairly large electric fields, especially if the host gas has a small cross section. This fact is particularly important in Ne, where the cross section at thermal energies is only a few tenths of Å^2. In order to illustrate this point, we show in Figure 4 the drift velocity of excess electrons in Neon gas as a function of the reduced field E/N at room temperature and at fairly low density [13], $N = 1.9 \times 10^{20}$ cm^{-3}. The open points were obtained after a coarse purification stage, while the closed points were obtained after recirculating the gas through the purifiers for many hours. It has to be noted, however, that, though the amount of O_2 in both cases was so low as to give no detectable distortion of the signal wave form, nonetheless only the data taken after a prolonged purification agree with the prediction of the classical theory (solid line) using the momentum transfer cross section of O'Malley and Crompton[16].

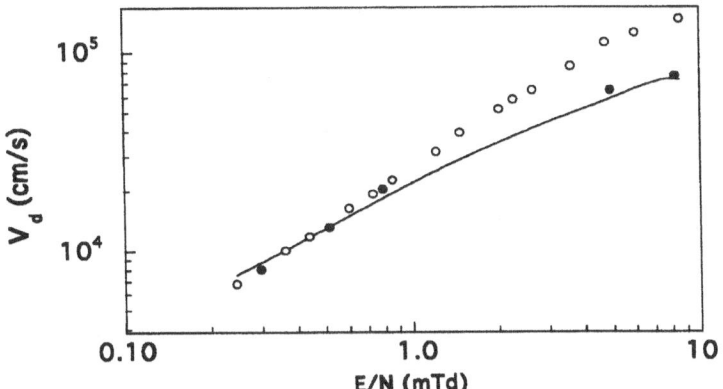

Figure 4. Electron drift velocity in Neon gas at room temperature and at a density [13] $N=1.9 \times 10^{20}$ cm^{-3}. Solid line: classical prediction. Open dots: coarse purification, closed dots: prolonged purification.

EXPERIMENTAL RESULTS AND THEORETICAL DESCRIPTION

The strikingly different behavior of the mobility of excess electrons in small- and high-polarizability noble gases is shown in Figures 5 and 6. In Figure 5 we show the behavior of the density-normalized mobility μN of excess electrons in Neon gas as a function of the reduced field E/N for several gas densities [14] at $T = 46.5\,K$.

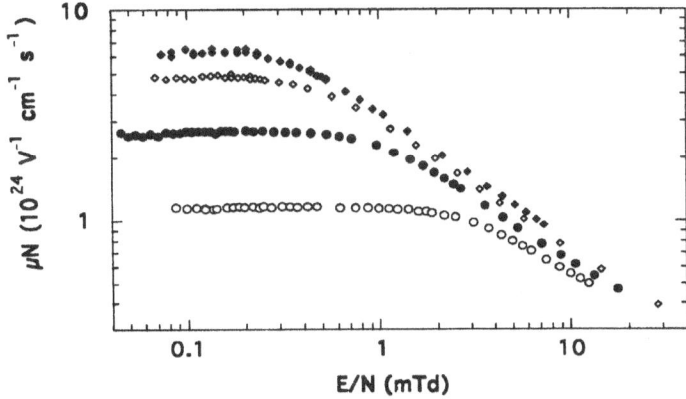

Figure 5. μN in Ne at $T=46.5\,K$ as a function of E/N [14]. From top: $N=(18.9, 35.2, 56.5, 80.2) \times 10^{20}$ cm^{-3}.

For each gas density μN shows a low-field region where it is constant, independent of the field. This is the region of thermal electrons, i .e., a region where the electric field is so small that it is not able to increase the mean electron energy above its thermal value. For larger fields μN decreases with increasing field and is roughly proportional to $(E/N)^{-1/2}$. This field dependence is truly obeyed by a hard-sphere gas[5] and the experimentally observed behavior of μN can be explained by noting that the more energetic electrons at large electric fields can approach more closely the atoms and thus they are predominantly scattered off the repulsive part of the interaction potential. Furthermore, it has to be observed that at high fields the mobility curves converge all to the low-density curve and that the largest deviations from the low-density mobility occur at thermal energies. All these features are also characteristic of Helium gas[6,17].

In Figure 6 we report μN as a function of E/N for several densities in Argon gas[18] at $T = 162.7K$.

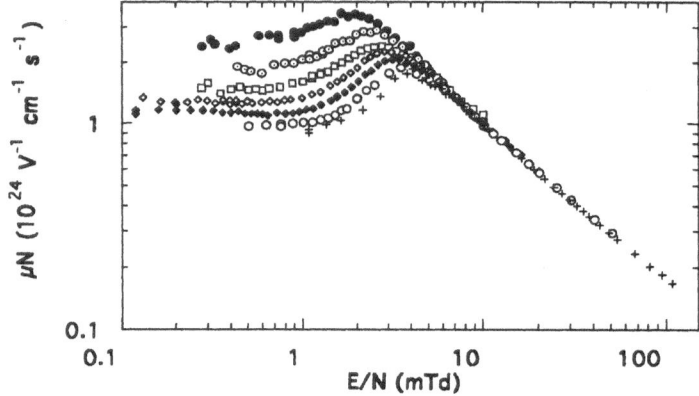

Figure 6. Density-normalized mobility μN of Argon as a function of the reduced field E/N at $T=162.7K$[18]. The densities are from top $N=(60.8,50.0,39.8,30.2,19.9,7.91,3.70)\times10^{20}$ cm^{-3}.

The main difference with the He and Ne case is the presence of a mobility maximum for reduced fields E/N in the range $(2-4)$ mTd $(1$ mTd $= 10^{-20}$ V cm$^2)$ depending on the gas density. This maximum is related to the Ramsauer minimum in the scattering cross section[4] as shown in Figure 7.

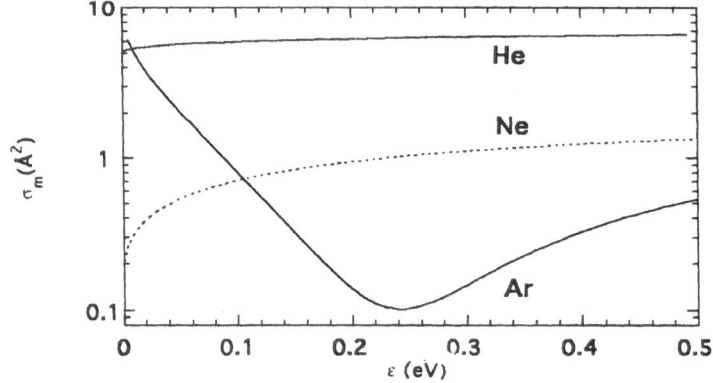

Figure 7. Momentum transfer scattering cross sections of He[19], Ne[16], and Ar[20].

However, in analogy with He and Ne, also in Ar there is a low-field region where μN is independent of E/N and there is also a high-field region where the deviations from the low-density mobility are greatly reduced and the μN data for all densities converge to a single curve. A similar behavior is also shown by Xe[21] and Kr[22].

From plots similar to those in Figures 5 and 6 the zero-field density-normalized mobility $\mu_0 N$ is determined and plotted as a function of the gas density N, as in Figure 1, in order to allow comparison with the classical, single-scattering theory, Eq. (1), which predicts a density-independent $\mu_0 N$.

A number of theories has been developed in the recent past to account for the experimental observations [11,12,23,24]. (For a review see Ref. [8].) These theories are based on the observation that at the densities and temperatures involved in the experiments the electron thermal wavelength is not much smaller than the electron mean free path, as required by the classical theory, but it is comparable to it. Therefore, the electron wave packet interacts simultaneously with several scattering centers and multiple-scattering effects come into play. As a result $\mu_0 N$ at low field strength is no longer density independent.

These multiple scattering theories were developed in the limit of vanishingly small electric fields ($E/N \to 0$). They are based on a complex shift of the electron kinetic energy in a dense disordered medium [25,26,27] and assume that the cross sections do not vary very much with energy. This approximation is fairly good for He because of its small atomic polarizability, but surely not for Ne and Ar, as can be seen in Figure 7.

Legler Theory

Legler [23] for the negative density effect in He, where the cross section is nearly energy independent, assumed that the electron kinetic energy, ϵ', in the dense medium is shifted to the value $\epsilon' = \epsilon - \Delta$ with respect to the kinetic energy ϵ in vacuo, where

$$\Delta = \frac{2\pi\hbar^2}{m} N a \qquad (2)$$

is the real part of the complex energy shift, and a is the electron-atom scattering length. The electron mobility is then calculated by evaluating the mean time between collisions at the mean velocity corresponding to the shifted kinetic energy. (For the positive density effect, the derivation is not as straightforward). Though predicting the correct direction for the positive as well as negative density effects, the Legler model had limited success because it overestimated the magnitudes for both the effects at low density, because the prediction for the positive effect curved in the opposite direction from the experimental data, and because it did not account for the negative effect observed in Ar at 90 K up to 0.1 MPa [10].

Atrazhev and Iakubov theory

Atrazhev and Iakubov [24] presented two different theories to account for the two density effects. For the negative one, they proposed a quantum interference model based on multiple scattering where the collision frequency $\nu(\epsilon) = N\sigma_m(\epsilon)\sqrt{(2\epsilon/m)}$ in the dense gas is enhanced by a factor $(1 + A\lambda_T/\ell)$, where A is a constant of order unity, $\ell = (N\sigma_m)^{-1}$ is the electron mean free path, and λ_T is the thermal electron wavelength. This model predicts a linear decrease of $\mu_0 N$ with N and agrees fairly well with the He data at small N and at fairly high temperatures.

For the positive effect Atrazhev and Iakubov proposed a model where the large polarizability of the atoms of gases like Ar and CH_4 allows electron-induced dipole-dipole interaction among the gas atoms leading to an attenuation of the electron-atom interaction. They were able to obtain a fairly good agreement with the experimental $\mu_0 N$ data in Ar [7] and CH_4 [28] but at the cost of introducing an adjustable parameter.

O'Malley theory

Also O'Malley [12] proposed two theories for the two density effects. The positive effect is again explained in terms of a kinetic energy shift in the electron-atom collision frequency. The model agrees semiquantitatively with the Ar data [7] at room temperature improving on the curvature of the Legler model, and predicts the small negative effects in Ar at 90 K, but it gives origin to a controversy about the choice of the sign of the energy shift [11].

The negative effect is explained by O'Malley in terms of a modification of the electron energy distribution in order to account for electron trapping in spatially localized states with negative kinetic energies as a consequence of the uncertainty principle. O'Malley[12] obtained a correction to the low-density mobility in the form

$$\mu_0 N = (\mu_0 N)_c \exp\left(-\bar{\Gamma}/k_B T\right) \tag{3}$$

where $\bar{\Gamma}$ is the average value of the imaginary part of the complex shift of the electric kinetic energy given by

$$\Gamma = 2\left(\frac{\hbar^2}{2m}\right)^{1/2} N \sigma_T \sqrt{\epsilon} \tag{4}$$

Here σ_T is the total scattering cross section. If σ_T does not depend very much on the energy ϵ, as in the case of He, $\bar{\Gamma}$ is given by

$$\bar{\Gamma} = k_B T \left[g_0 (1 + g_0^2/4)^{1/2} + g_0^2/2\right] \tag{5}$$

where $g_0 = 2N\langle\sigma_T\rangle\hbar/\sqrt{2mk_B T} \approx 2\lambda_T/(2\pi\ell)$ where λ_T is the thermal wavelength and $\ell = 1/(N\langle\sigma_T\rangle)\,(\approx 1/N\langle\sigma_m\rangle$ at thermal energies) is the classical mean free path. For small densities the O'Malley result agrees very well with the result of Athrazhev and Iakubov. The O'Malley model predicted well the mobility decrease with increasing N up to moderate N values in a number of gases (He[6] at $T = 293\,K$ and $T = 77.6K$, CO_2 at room temperature[29], H_2 at $T = 77.6\,K$[6]). However, in He at $T = 77.6\,K$[6] at larger N and in He at lower temperatures ($T = 4.2\,K$[30], $T = 7.3\,K$ and $T = 18.1\,K$[31]), except at the lowest densities, the agreement is quite poor. In a more recent work[32] O'Malley put his theory on firmer grounds introducing also a mobility edge, to be deduced from the Ioffe-Regel criterion[33] for localization $\lambda_T/(2\pi\ell) \approx 1$, but did not improve the agreement with the experimental data.

Braglia and Dallacasa theory

Braglia and Dallacasa[11] proposed two theories as well for both density effects based on the low-energy electron-atom scattering t-matrix formalism. Multiple scattering effects give origin to a shift of the electron kinetic energy related to the t-matrix. For negative scattering length gases (positive density effect), for which the Fermi energy shift $\Delta = 2\pi(\hbar^2/m)Na$ is negative, they arrived at the formula

$$\mu_0 N = \frac{4eN}{3m[\pi(k_B T)^5]^{1/2}} \int_0^\infty d\epsilon \frac{\epsilon^{3/2}}{\nu(\epsilon - \Delta)} e^{-(\epsilon/k_B T)} \tag{6}$$

where $\nu(\epsilon) = \sqrt{(2\epsilon/m)}N\sigma_m(\epsilon)S(0)$ is the collision frequency and $S(0)$ is the long-wavelength limit of the static structure factor introduced to take into account the multiple scattering effects arising from correlations among scatterers due to intrinsic density fluctuations[34]. This model improved a little over the O'Malley model the agreement with the Bartels data in Ar[7], but deviates substantially from the experimental values at moderately larger densities and is not able to predict the observed behavior of $\mu_0 N$ as a function of N in gaseous methane[28].

For positive scattering length gases (negative density effect and $\Delta > 0$) Braglia and Dallacasa proposed the following formula

$$\mu_0 N = \frac{4eN}{3m[\pi(k_B T)^5]^{1/2}} \int_{\epsilon_0}^\infty d\epsilon \frac{\epsilon^{3/2}}{\nu(\epsilon + \Delta)} e^{-(\epsilon/k_B T)} \left(1 - \frac{d\Delta}{d\epsilon}\right) \tag{7}$$

where the cutoff energy ϵ_0 is the root of $(1 - d\Delta/d\epsilon)$ and takes into proper account localized states due to the presence of a mobility edge. It has to be recalled that Eq. (7) has been derived by assuming a nearly constant scattering cross section. For small densities it gives the same linear N dependence as the theories of Athrazev and Iakubov and of O'Malley, if $d\Delta/d\epsilon = (d/d\epsilon)N\,\mathrm{Re}\,t(\epsilon) \approx \lambda_T/(2\pi\ell)$, where $t(\epsilon)$ is the t-matrix. However, this last equation is still matter of controversy [12]. This model is in better agreement with the experiments in He than O'Malley over a wider T and N range if a correction to the gas density according to the Ioffe-Regel criterion is introduced. In absence of this correction Eq. (7), agrees fairly well with the results of Legler.

The fairly good agreement of nearly all of these theories for the negative density effect in He, especially at low densities, can be related to the fact that their predictions give a correction to the classical mobility proportional to $\lambda_T/(2\pi\ell)$ to first order in N, as experimentally observed in He [17] over a very wide temperature range $(2.7 < T < 300)\,K$. However, these theories for the negative density effect are not able to accurately describe the density dependence of the electron mobility in He at high densities because these are transport theories for quasifree propagating electron states, while the charge transport mechanism in He at very high densities is completely different, involving trapped electrons in density fluctuations (pseudobubbles), and must be treated in a completely different way [17,30,31,35]. Moreover, all multiple scattering theories for both density effects do not account for the density dependence of the electron mobility in the transition region from the low-density gas to the liquid.

The heuristic Padua model

As already pointed out [8] and as shown here again in this brief review, none of the proposed theories are able to account for all of the observed phenomena, nor they are able to describe the observed field dependence of the mobility at fixed density. Even neglecting their disagreement with the experimental data, however, it is not satisfactory the fact that the positive and negative density effects do not emerge from a single theoretical approach, rather than from two different approaches. In fact, we believe that the description of the well-defined physical process of the scattering of (epi-)thermal electrons off noble-gas atoms at high density should be unique and should not depend on the sign of the scattering length of the host atoms. It would be also desirable that a simple kinetic approach, like the Boltzmann formalism, with suitable modifications to account for multiple scattering effects, could be retained to allow a simple physical intuition of the problem, as was done by Cohen and Lekner [36]. After all, a similar attempt has been recently carried out with relative success to explain the electron mobility in non-polar liquids [37], where, up to now, the most successful theories are intrinsically multiple-scattering theories involving electron-phonon scattering [2,38].

The first step towards this goal involves an accurate check of the theories for the negative density effect shown by gases with positive scattering length. As previously explained, they have been derived by assuming a nearly energy-independent electron-atom scattering cross section (in fact, they were thought for He), and predict corrections to the classical mobility proportional to $\lambda_T/(2\pi\ell)$ to first order in N, as a result of an increased collision frequency [24], or of a modification of the electron energy distribution function [12], or of introducing a mobility edge [11].

None of them uses here the kinetic energy shift, which is introduced when describing the negative scattering length gases. For Neon multiple scattering theories predict a very small negative density effect. In fact, Ne, as a result of the good balance between the attractive and repulsive part of the electron-atom interaction has a positive, though small, scattering length [16] ($a \approx 0.1\,\text{Å}$ to be compared to $a \approx 0.6\,\text{Å}$ for He [19]), and at thermal energies its scattering cross section is approximately one order of magnitude smaller than that of He. We carried out experimental measurements in supercritical Neon gas in a very wide temperature range $(48 < T < 300\,K)$ up to moderately large densities [13,39,40] ($N < 45 \times 10^{20}\,\text{cm}^{-3}$), in supercritical Ne close to the critical point [14] ($45 < T < 48\,K$, $T_c = 44.4\,K$) up to very large densities ($N_{max} \approx 180 \times 10^{20}$ cm^{-3}), and in Neon vapor at the liquid-vapor coexistence [13] ($25.7 < T < 40.4\,K$). We observed indeed a negative density effect at low- up to medium- densities ($N \leq 95 \times 10^{20}$ cm^{-3}), but the magnitude of this effect was much larger than predicted. Moreover, this N dependence of $\mu_0 N$ becomes stronger as the temperature is lowered. In Figures 8 and 9 we show some experimental results along with the predictions of the theories. (We note incidentally that we also found evidence of localized states at very high densities [14,41,42] ($N \geq 95 \times 10^{20}$ cm^{-3}) as expected in gases with fairly strong repulsive interactions with excess electrons [43]. For densities $N \leq 95 \times 10^{20}$ cm^{-3}, however, no localized states are present [14], and we have only quasifree propagating electron states.)

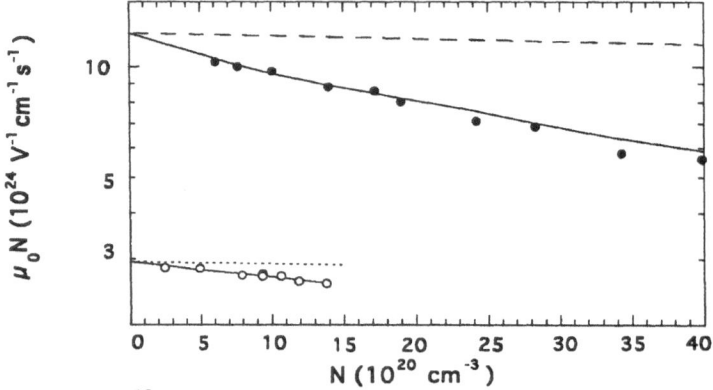

Figure 8. $\mu_0 N$ in Ne gas [13] as a function of N at $T=47.9\,K$ (closed points) and $T=293\,K$ (open points). Dashed lines: predictions of O'Malley theory [12]. Solid lines: predictions of the Padua model [13].

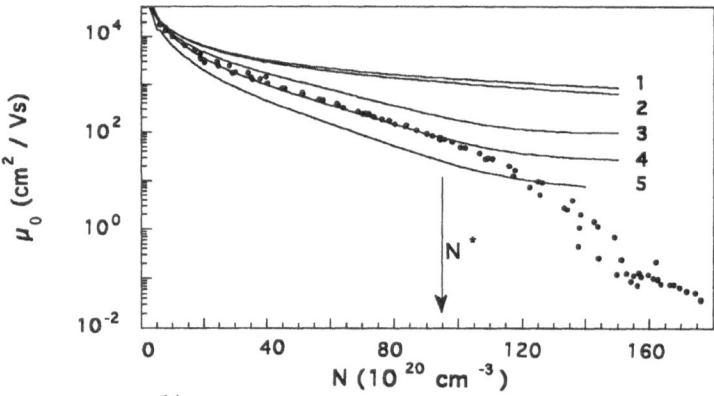

Figure 9. $\mu_0 N$ in Neon gas [14] at $45.0<T<48.4\,K$. 1-classical prediction $(\mu_0 N)_c$, 2-O'Malley theory, 3-$(\mu_0 N)_c/S(0)$, 4- Padua model (valid up to $N \approx N^*=95\times10^{20}$ cm^{-3}.), 5- Braglia and Dallacasa model.

The most relevant difference among He and Ne is the energy dependence of the scattering cross sections, especially at low energies (see Fig.7). For He, $\sigma_m(0) \approx 4.97\,\text{Å}^2$ and $\sigma_m(\epsilon)$ varies by 11 % in the range 0-40 meV. In Ne $\sigma_m(0) \approx 0.16\,\text{Å}^2$ and in the same energy range it increases by a factor 3, approximately. The different behavior of $\mu_0 N$ has therefore to be related to the different energy dependence of the cross sections. The results of multiple scattering theories have been obtained for constant cross sections, and therefore neglect a second possible multiple scattering effect due to both the energy dependence of σ_m and to the Fermi shift of the electron energy.

According to the simplified Springett, Jortner, and Cohen (SJC) model[44] developed for monoatomic fluids and taking into account the result for conduction electrons in metals[45], the energy of quasifree electron states takes the form

$$\epsilon' = V_0 + \frac{\hbar^2 k^2}{2m^*} = V_0 + \epsilon \tag{8}$$

where k is the wavevector, m^* is the effective electron mass (which we set $m^* = m$, the free electron mass), and V_0 is the ground state energy. At low densities V_0 is well approximated by the Fermi energy shift $V_0 \approx (2\pi\hbar^2/m)Na$, which can be positive or negative according to the sign of the scattering length. At higher densities V_0 can be written as $V_0 = E_k + U_p$, where E_k is the zero-point kinetic energy which essentially arises from excluding the electrons from the hard-core volume of the atoms, and U_p is a potential energy contribution arising from the polarization part of the interaction potential. E_k is usually calculated by means of the Wigner-Seitz (WS) method[46,47], as also confirmed experimentally[48].

In experiments aiming at the determination of the resonant electron attachment to O_2 impurities in High density He[49,50], where $U_p \ll E_k$ because of the small atomic polarizability of He and V_0 is essentially kinetic energy, it appeared evident that the capture process depends on the total shifted kinetic energy $\epsilon + E_k$. Moreover, the presence of a peak in the electron attachment frequency at a well-defined gas density N_R, for which the energy shift equals the resonant attachment energy $(2\pi\hbar^2/m)N_Ra \approx E_R \approx 91$ meV, suggested that the electron energy distribution was shifted to higher energies by the density dependent E_k.

By neglecting density fluctuations that are important in determining the density of states[51] only close to the critical point, we thus modified the classical formula for the mobility as follows. The density of states was assumed to start at E_k and for energies larger than E_k it was assumed to be that of free electrons.

$$\begin{cases} g(\epsilon) = 0 & \text{for } \epsilon \leq E_k \\ g(\epsilon) \propto (\epsilon - E_k)^{1/2} & \text{for } \epsilon \geq E_k \end{cases} \tag{9}$$

The energy distribution function (for $E/N = 0$) is then given by

$$\begin{cases} f(\epsilon) = 0 & \text{for } \epsilon \leq E_k \\ f(\epsilon) \propto (\epsilon - E_k)^{1/2} e^{-(\epsilon/k_B T)} & \text{for } \epsilon \geq E_k \end{cases} \tag{10}$$

Upon normalization, the classical energy distribution function is modified into the shifted classical energy distribution function given by

$$f(\epsilon) = \frac{2}{\left[\pi(k_B T)^5\right]^{1/2}} (\epsilon - E_k)^{1/2} e^{-(\epsilon - E_k)/k_B T} \tag{11}$$

and the Eq. (1), the classical prediction for the mobility, must be accordingly modified so as to yield

$$(\mu_0 N) = \frac{4e}{3[(2\pi m)(k_B T)^5]^{1/2}} \int_{E_k}^{\infty} d\epsilon \, \frac{(\epsilon - E_k)}{\sigma_m(\epsilon)} e^{-(\epsilon - E_k)/k_B T} \tag{12}$$

The lower integration limit set to E_k means that there are no states, neglecting fluctuations, below the energy shift. Of the factor $(\epsilon - E_k)$ in the integrand numerator, a part $(\epsilon - E_k)^{1/2}$ comes from the electron density of states. The remaining $(\epsilon - E_k)^{1/2}$

factor is related to the group velocity of an electron immersed in the kinetic potential
$E_k : v_g = [2(\epsilon - E_k)/m]^{1/2}$. The scattering cross section, however, has to be evaluated
at the total kinetic energy, comprising the zero-point contribution E_k. Upon changing
integration variables we get in fact

$$(\mu_0 N) = \frac{4e}{3[(2\pi m)(k_B T)^5]^{1/2}} \int_0^\infty d\epsilon \frac{\epsilon}{\sigma_m(\epsilon + E_k)} e^{-\epsilon/k_B T} \tag{13}$$

Therefore, a density dependence is introduced here through the energy dependence
of σ_m and the density dependent kinetic energy shift E_k.

The second (and weaker) density effect, that of a reduction of the classical prediction for the mobility proportional to $\lambda_T/(2\pi \ell)$, has been then introduced in the
form of the O'Malley factor $\exp -(\Gamma/k_B T)$. Thus we finally yield [13]

$$(\mu_0 N) = \frac{4e}{3[(2\pi m)(k_B T)^5]^{1/2}} \int_0^\infty d\epsilon \frac{\epsilon}{\sigma_m(\epsilon + E_k)} e^{-\left[(\epsilon + \Gamma(\epsilon))/k_B T\right]} \tag{14}$$

The values of the kinetic energy shift E_k have been calculated by means of the (WS)
eigenvalue equation

$$\tan[k_0(r_s - \tilde{a})] - k_0 r_s = 0 \tag{15}$$

k_0 is related to E_k by $E_k = \hbar^2 k_0^2/2m$. $r_s = (3/4\pi N)^{1/3}$ is the radius of the equivalent WS sphere and $\tilde{a}$ is the positive Hartree-Fock scattering length of the hard-core
repulsive Hartree-Fock pseudopotential in the SJC model [44] and has the meaning of
an equivalent atomic radius. It has to be recalled that the WS model can be properly
exploited if there is no overlap of the atomic potential at the WS radius. This approximation is good enough in our case, since even at the highest densities the experimental
data of μN at fairly large reduced electric fields E/N are in quite good agreement
with the prediction of the classical theory obtained using the gas-phase scattering
cross section of the isolated electron-atom system, as shown in Figure 10, and this
means that there is no overlap of the polarization tails of the atomic potentials.

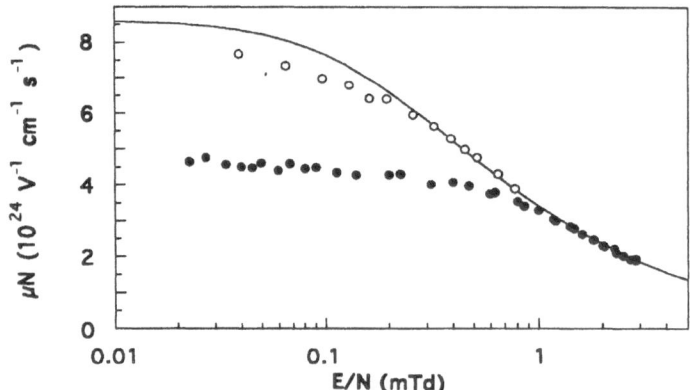

Figure 10. μN in Ne [13] at $T=77.4K$. Solid curve: classical calculation. Open dots: $N=7.7\times10^{20}$ cm^{-3},
closed dots: $N=44.3\times10^{20}$ cm^{-3}.

As radius of the hard-core scatterers to be used in Eq. (15), we used a radius deduced from the cross section for low energy electron scattering $\tilde{a} = \sqrt{(\sigma_T/4\pi)}$, with the total scattering cross section evaluated at the energy $E_k = \hbar^2 k_0^2/2m$. The WS eigenvalue equation reads then

$$\tan\left[k_0(r_s - \sqrt{\sigma_T(E_k)/4\pi}\,)\right] = k_0 r_s \tag{16}$$

Since σ_T depends on the energy very strongly and E_k is a function of k_0, Eq. (16) must be iteratively solved. In Figure 11 we show the E_k for Ne calculated as explained previously.

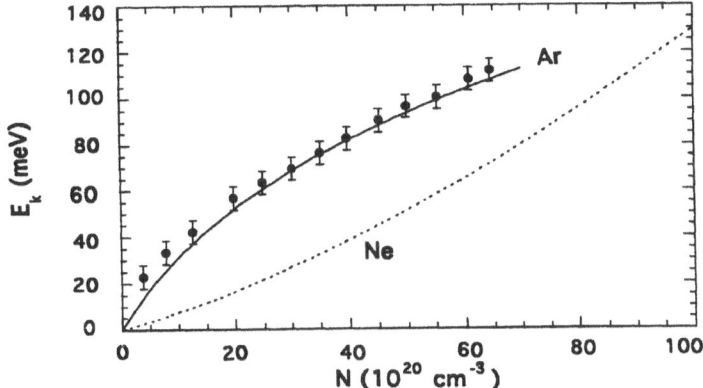

Figure 11. Kinetic energy shift E_k calculated according to Eq. (16) in Ne (dotted line), in Ar (solid line), and experimentally determined from $\mu_0 N$ data in Ar[18] at $T=162.7\,K$.

In Figure 8 the predictions of this heuristic model, Eq. (14), supplemented by Eq. (16) are shown along with the predictions of the O'Malley theoy of multiple scattering. As can be seen, the heuristic model agrees in a nearly perfect way with the experimental data and it has to be emphasized the fact that no adjustable parameters are introduced.

In Figure 9 are reported the results obtained in Ne at temperatures close to the critical point[14] ($45 \le T \le 48.4\,K$). In the conditions of this experiment, the gas compressibility is very large and $S(0) \gg 1$, so that correlations among scatterers due to intrinsic density fluctuations cannot be neglected. According to the Lekner model[34], this effect can be described by an enhancement of the cross section, which at low energies is proportional to $S(0) : \sigma_m \to \sigma_m S(0)$. Therefore, Eq. (14) must be divided by $S(0)$. As can be seen in Figure 9, this model agrees nearly perfectly with the experimental data up to $N \approx 95 \times 10^{20}$ cm^{-3}, where localization sets in. In this Figure we can also observe the failure of the other multiple scattering theories.

The agreement of this heuristic model with the experimental data has been obtained by taking into account three multiple scattering effects: 1- a quantum shift of the electron kinetic energy, which forces the evaluation of the scattering properties (i.e.,σ_m) at the shifted kinetic energy $\epsilon + E_k$; 2- an enhancement of the scattering rate due to quantum interference when the electron thermal wavelength λ_T becomes comparable to the mean free path ℓ. (This effect is related to the "weak localization" introduced to explain the Hall-mobility of electrons in liquid Xe[52] and the light localization in disordered dielectric materials[53]); 3- correlations among scatterers due to intrinsic density fluctuations described by the static structure factor. In He, far away from the critical point, the second effect is predominant because σ_m is large and nearly constant. In Ne, the second effect is weak because σ_m is quite small and the first effect is very large due to the strong energy dependence of σ_m. Close to the critical point, moreover, also the third effect has a non negligible influence.

The good results obtained in Ne are therefore mainly due to the introduction of a positive density-dependent shift in the kinetic energy of the electrons, as done by the present multiple scattering theories for the negative scattering length gases. In Ne we shifted the electron kinetic energy by the amount $E_k > 0$, calculated according to the WS model, i.e., by the kinetic contribution to the total energy shift V_0. However, there could remain a doubt whether the electron energy should be shifted by the total energy shift V_0, and not only by its kinetic energy contribution E_k. In fact, in both He and Ne, owing to their small atomic polarizability, U_p is small and both $V_0 > 0$ and $E_k > 0$. (E_k is always positive because it is a kinetic energy). In order to solve this problem, we decided to carry out new electron mobility measurements in Ar gas[18], where V_0 is surely negative[54,55].

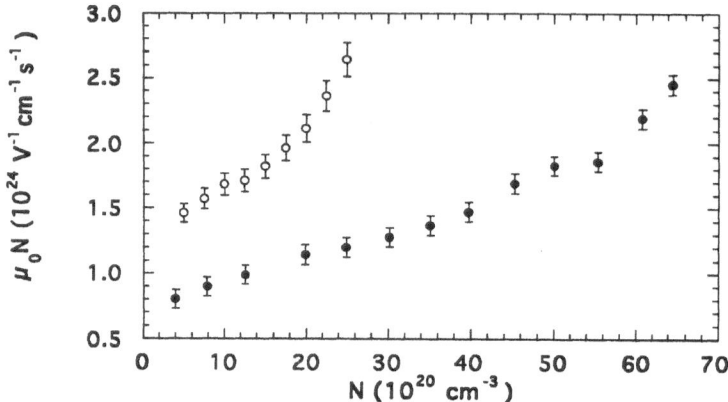

Figure 12. $\mu_0 N$ of Ar at room temperature (open points)[7] and at $T=162.7\,K$ (closed points)[18]. These latter data were used to determine the E_k values reported in Fig. 11.

Furthermore, these measurements give the opportunity to check if the quantum interference effect, which becomes important as the electron thermal wavelength gets comparable to the mean free path, is effective only in He and Ne, but not in Ar. In fact, the present multiple scattering theories assume this effect to be the only responsible of the observed behavior of the positive scattering length gases (while our Ne measurements have shown that several effects are simultaneously present), but they do not take it into account for the description of the behavior of the negative scattering length gases.

Some of our results obtained at $T = 162.7\,K$[18] ($\approx 12\,K$ above $T_c = 150.9\,K$) are reported in Figure 6. In Figure 12 we report $\mu_0 N$ as a function of the gas density. The data of Bartels[7] at room temperature are also reported once more for comparison. Our data show also a similar density dependence as the data of Huang and Freeman that were taken at the liquid-vapor coexistence[56]. The most interesting features of the experimental data are the following. For each density N up to the largest ($N_{max} \approx 65 \times 10^{20}\,\text{cm}^{-3}$) μN turns out to be field independent for $E/N \le 1$ mTd (1Td $= 10^{-17}\,\text{V cm}^2$). All μN curves show very pronounced maxima in the range ($2 < E/N < 4$) mTd, whose magnitudes and positions depend on N. The mobility maximum can be observed from the dilute gas, and must be obviously related to the Ramsauer-Townsend minimum of the scattering cross section. It is important to note that the position of the mobility maxima $(E/N)_{max}$ depends on N, namely it decreases with increasing N. This fact has been already observed as a shift of the mobility maximum to lower characteristic energies as density is increased[57]. Third and final important feature is that for $E/N \ge (8-10)\,\text{mTd}$ μN does not depend any more on N but it is only function of E/N. (This behavior is similar to that shown by Ne in Figures 5 and 10.)

In Figure 13 we plot this $(E/N)_{max}$ as a function of N. From the presence of the mobility maximum up to the largest density and from the converging of all curves to

272

a single one for large E/N, we draw the conclusion that the gas-phase scattering cross section can be still used even the largest densities. In other words, there is no overlap of the polarization potential tails of the atoms, even though at $N \approx 65 \times 10^2 0\,\mathrm{cm}^{-3}$ the WS radius $r_s \approx 3.3\,\text{Å}$ approximately equals the Ar hard-core radius used in the Percus-Yevick equation for liquid Ar[58].

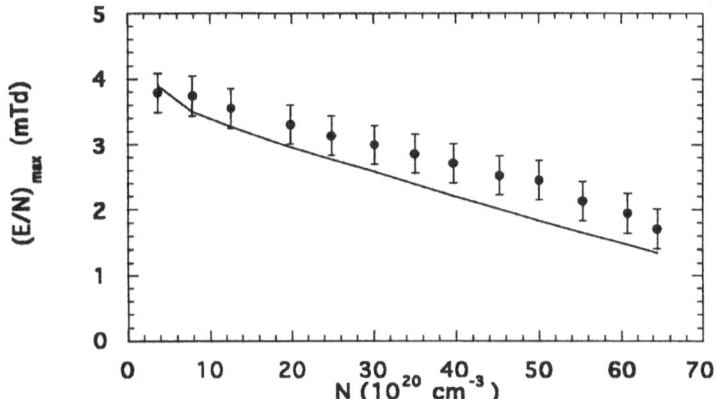

Figure 13. Position of the mobility maximum $(E/N)_{max}$ as a function of N in Ar[18] at $T=162.7 K$. The solid line is calculated according the Padua model using the scattering cross section of Weyhreter *et al.*[20].

At the mobility maximum the mean electron energy $\bar{\epsilon}$ should be approximately equal to the energy of the Ramsauer-Townsend minimum ϵ_{rt}, $\bar{\epsilon} \approx \epsilon_{rt} \approx 230\,\mathrm{meV}$. Since we have found that $(E/N)_{max}$ decreases with increasing N (Fig. 13), it appears that electrons need a smaller energy gain from the field at increasing density in order to reach the value ϵ_{rt}. This means that the missing energy necessary to reach ϵ_{rt} is supplied by a density-dependent quantum shift and that this contribution must be positive and increase with density. Therefore, this shift cannot be V_0, since we know that it is negative and that in the gas-density range it decreases[54].

Also the behavior of $\mu_0 N$ can be qualitatively understood by observing that for $E/N \to 0$ no energy is gained from the field and the mean electron energy lies below ϵ_{rt} in a region where σ_m decreases rapidly with the energy (Fig. 7). With the aid of Eq. (14) $\mu_0 N$ can be considered as a suitable weighted average of the inverse of σ_m. To a first approximation, this average value can be calculated by evaluating $1/\sigma_m$ at the mean electron energy, $\bar{\epsilon}$. If $\bar{\epsilon}$ had not any density-dependent contribution in addition to thermal one, $1/\sigma_m(\bar{\epsilon})$ would not vary with N. On the contrary, if $\bar{\epsilon}$ increases with N at constant T because of the quantum shift, the average inverse cross section and, hence, $\mu_0 N$ increase because σ_m decreases strongly with the energy. In this picture $\mu_0 N$ increases with N as a combined result of both a positive quantum contribution to the electron energy and of the decrease of σ_m with energy. This is the same explanation adopted for Ne.

In order to build a model for this positive density effect taking into account the results for Ne and He, which is also able to explain the field dependence of the mobility, we start by assuming that the electron kinetic energy in the medium is shifted only by E_k, the kinetic contribution to V_0 : $\epsilon' = \epsilon + E_k = (\hbar^2 k^2/2m) + E_k$. We also observe that it is the group velocity

$$v = \frac{\hbar k}{m} = [\frac{2}{m}(\epsilon' - E_k)]^{1/2} \tag{17}$$

that contributes to the equipartition value arising from the gas temperature[5].

We further note that as the electron energy is increased, for instance by means of the electric field, all multiple scattering effects weakens, as clearly indicated by the fact that for large E/N μN in both Ne and Ar does not depend on N and can be calculated by the classical formula using the gas-phase scattering cross section. Therefore, all multiple scattering corrections to the classical formula have to vanish as energy is increased. This fact is far more important in Ar at $T = 162.7\,K$ than in Ne at $T = 45\,K$. In fact, at $T = 45\,K$ the electron thermal wavelength is $\lambda_T \approx 110\,\text{Å}$, while at $T = 162.7\,K$ $\lambda_T \approx 58\,\text{Å}$, to be compared to a mean atomic separation $\bar{d} = N^{-1/3} = 8.5\,\text{Å}$ for $N = 4.0 \times 10^{20}\,\text{cm}^{-3}$ or $\bar{d} = 3.3\,\text{Å}$ for $N = 65 \times 10^{20}\,\text{cm}^{-3}$. At energies close to the Ramsauer minimum the electron wavelength $\lambda = h/(2m\epsilon)^{1/2}$ is $\lambda \approx 25\,\text{Å}$ and at 1 eV $\lambda \approx 12\,\text{Å}$. So, as energy is increased the size of the electron wave packet decreases and the effects of multiple scattering are reduced.

Then, we use the classical formula for μN but not restricted only to thermal electrons[5]

$$\mu N = -\frac{2}{3}\left(\frac{e}{m}\right)^{1/2} \int_0^\infty d\epsilon \, \frac{\epsilon}{\sigma_m(\epsilon)} \left(\frac{dg(\epsilon)}{d\epsilon}\right) \tag{18}$$

where $g(\epsilon)$ is the Davydov-Pidduck energy distribution function[4,5,36]

$$g(\epsilon) = A \exp\left\{ -\int_0^\epsilon \frac{dz}{k_B T + \frac{1}{6}\frac{M}{m}\left(\frac{eE}{N}\right)^2 \frac{1}{z\sigma^2_m(z)}} \right\} \tag{19}$$

where M is the Ar atomic mass, and A is given by the normalization condition $\int_0^\infty \sqrt{\epsilon} g(\epsilon) d\epsilon = 1$. The quantum interference effect has been treated by means of the linearized theory of Atrazhev and Iakubov[24] because at thermal energies $\lambda_T/\ell = \lambda_T N \sigma_m < 1$. This theory predicts an enhancement of the scattering cross section by the factor $(1 + N\lambda\sigma_m/\pi)$:

$$\sigma'_m(\epsilon) = \sigma_m(\epsilon)\left[1 + N\lambda\sigma_m(\epsilon)/\pi\right] \tag{20}$$

with $\lambda = h/\sqrt{(2m\epsilon)}$. As expected, this correction factor tends to unity for large ϵ. (On the contrary, this quantum interference correction according to O'Malley[12], $\exp[-\Gamma(\epsilon)/k_B T]$ with Γ given by Eq. (4), has not the expected limiting behavior at large energies).

The multiple scattering effect due to correlations among scatterers is treated according to the Lekner theory[34] that also predicts an enhancement of the scattering cross section by a factor

$$\mathcal{F}(k) = \frac{1}{4k^4} \int_0^{2k} dq\, q^3 S(q) \tag{21}$$

where k is the transfer wave vector such that $\epsilon = (\hbar k)^2/2m$ and $S(q)$ is the static structure factor given by[59,60,61]

$$S(q) = \frac{S(0) + (qL)^2}{1 + (qL)^2} \tag{22}$$

$L^2 = 0.1 l^2[S(0) - 1]$, $l \approx 10\,\text{Å}$ is the so called " short-range correlation length", $S(0) = N k_B T \chi_T$ is the long wavelength limit of $S(q)$ and χ_T is the isothermal compressibility of the gas. By substituting $k = (2m\epsilon/\hbar^2)^{1/2}$ we get

$$\mathcal{F}(\epsilon) = 1 + [S(0) - 1]\left\{ \frac{(8mL^2\epsilon/\hbar^2) - \ln[1 + (8mL^2\epsilon/\hbar^2)]}{8(2mL^2\epsilon/\hbar^2)^2} \right\} \tag{23}$$

The energy dependence of $\mathcal{F}(\epsilon)$ for $S(0) = 10$ is shown in Figure 14. For $\epsilon \to 0$ $\mathcal{F}(\epsilon) \to S(0)$ and this justifies the use of $S(0)$ in Ne at $T \approx 45K$ where $k_B T = 3.9\,\text{meV}$ is very small. Again, for large ϵ, $\mathcal{F}(\epsilon) \to 1$, i.e., no multiple scattering correction as expected.

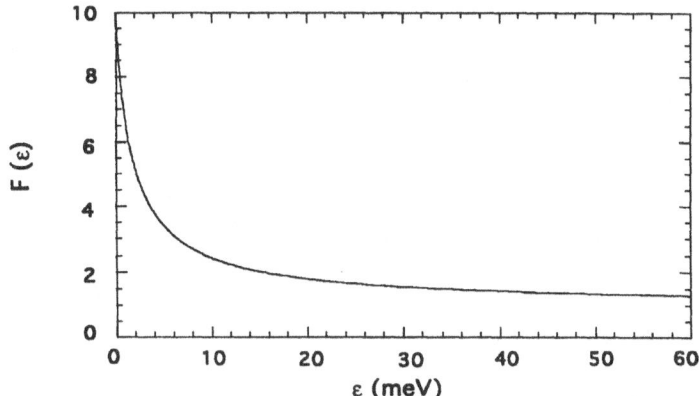

Figure 14. Lekner's cross section enhancement factor that account for correlations among scatterers due to density fluctuations [18], calculated for $S(0)=10$.

We finally recall that, as in Eq. (13), the scattering cross section has to be evaluated at the shifted energy $\epsilon' = \epsilon + E_k$. Summarizing, we obtain an effective cross section

$$\sigma_m{}^{\ast}(\epsilon) = \mathcal{F}(\epsilon)\sigma_m(\epsilon)\left[1 + \frac{N\lambda\sigma_m(\epsilon)\mathcal{F}(\epsilon)}{\pi}\right] \qquad (24)$$

which has to be inserted in Eq. (18) and there it has to be evaluated at the shifted energy $\epsilon + E_k$.

For Ar we have followed a different approach than in Ne, where we first calculated E_k from the WS equation, and then calculated $\mu_0 N$. This time we left E_k free and adjusted it in order that the $\mu_0 N$ values calculated from Eq. (18) for $E/N = 0$ (supplemented by Eqns. (20), (23), and (24)) agree within 1 % with the experimental values shown in Figure 12. In Figure 11 we report the E_k determined in this way and the values determined by the WS equation, Eq. (16), using the gas-phase total scattering cross section of Ar. We see that the experimental E_k is positive as expected, and that it agrees very well with the WS calculation of the kinetic energy shift. With the experimentally determined E_k we calculated μN as a function of E/N using Eqns. (18), (23), and (24). One such curve is reported in Figure 15 for a density $N = 30.2 \times 10^{20}\,\text{cm}^{-3}$, using two different cross sections [20,62]. Owing to the fact that the discrepancy among the cross sections can amount up to 50% especially close the Ramsauer minimum, we see that the results shown in Fig. 15 are extremely good. From these calculations we can also determine the position of the mobility maximum, $(E/N)_{max}$. The experimental $(E/N)_{max}$ values are compared to those calculated by this model in Figure 13. Again, the agreement is very good. The model shows therefore a good degree of internal consistency in the sense that the value of E_k calculated from the low-field μN data is a good value to be used also at high fields, it is independent of E/N over two to three orders of magnitude in E/N, and depends only the gas density. Further measurements in Ar gas at different T, above and below T_c, yet to be published [63], confirm the predictions of this model. In particular, it is able to describe the negative density effect observed at $T \approx 143\,K$.

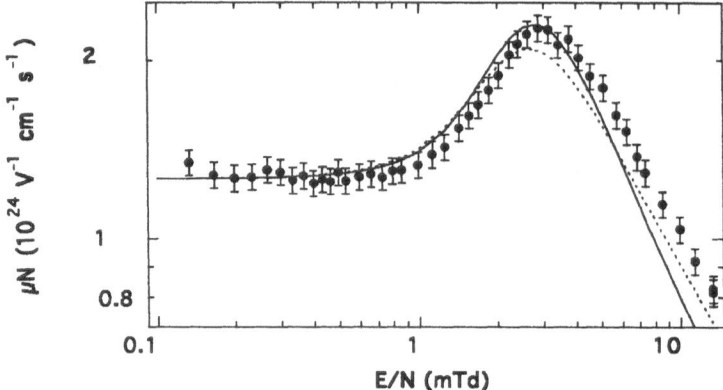

Figure 15. μN as a function of E/N in Ar gas at $N=30.2\times10^{20}$ cm^{-3} and $T=162.7\,K$ [18]. Solid line: Eq.(18) with the σ_m of Weihreter *et al.* [20]; dotted line: Eq.(18) with the σ_m of Haddad-O'Malley [62].

This extension of the model for the prediction of the density-normalized mobilities of thermal electrons to non-zero electric fields derived from the Ar measurements can be shown to work also in Neon. In Figure 16 we report the result of this model for Ne at $N = 18.7 \times 10^{20}$ cm^{-3} and $T = 46.5\,K$. Again, the agreement is very good.

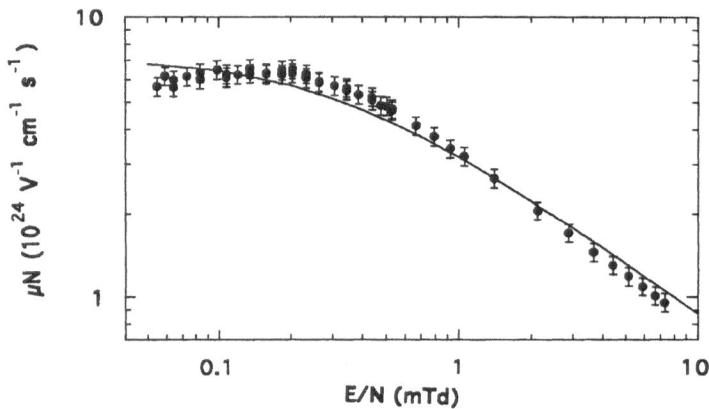

Figure 16. μN as a function of E/N in Ne gas [14] at $N=18.7\times10^{20}$ cm^{-3} and $T=46.5\,K$. Solid line: Eq. (18).

CONCLUSIONS

The classical Boltzmann equation for the calculation of the electron mobility based on a single scattering picture derived from the dilute gas conditions can still be used in high-density gases if corrections due to multiple scattering are suitably introduced.

To summarize, multiple scattering effects give origin to a shift of the electron energy, but only the kinetic part of this contribution, E_k, quite accurately calculated by means of the WS model, enters in the calculation of the scattering properties (i.e., the scattering cross sections). This contribution is of course positive and appears to be valid for both cases of positive as well as negative scattering length gases. They

They produce an enhancement of the collision rate depending on the ratio $\lambda(\epsilon)/\ell(\epsilon)$, where $\lambda = h/\sqrt{2m\epsilon}$ is the electron de Broglie wavelength and ℓ is the mean free path. There is also an enhancement of the cross section due to correlations among scatterers accounted for by the static structure factor of the gas. All these multiple scattering effects are simultaneously present IN all gases. However, their relative weight depends on several conditions.

Correlations among scatterers are effective at N and T close to the critical point. If σ_m is large and nearly constant (like in He) the most important effect is due to the increase of the scattering rate and is governed by the ratio λ_T/ℓ. If σ_m is small and strongly energy dependent (like in Ne) the effect of the energy shift in σ_m is overwhelming. In Ar we have a large and strongly energy dependent cross section, so that both the energy shift and the increase of the scattering rate are large and must be accounted for simultaneously.

Some problems, however, are left open. For instance, there is evidence [63] that at densities larger than those described here and at temperatures very close to the critical point ($T \approx 152.7K$) the model can still work if the WS model for the calculation of E_k is abandoned. Moreover, it has been found [63,64] that the electron mobilities for Ar gas just above T_c and for liquid Ar just below T_c at nearly the same number density are practically indistinguishable showing the same electric field dependence. This fact lends some credence to the efforts aimed at obtaining a kinetic description of the electron mobility in the liquid.

As a conclusion, we believe that the measurements in Ne and Ar helped clarify and unify the physical picture of the low-energy electrons scattering processes in dense non-polar gases. It would be, however, desirable that a general theory could be produced where all these multiple scattering effects are incorporated in a formal way and not only heuristically as we have done.

We also believe that if some of the still open problems will be solved a common kinetic description of the electron mobility in non-polar gases and liquids should be possible.

REFERENCES

1. G.R.Freeman and D.A.Armstrong, Electron and Ion Mobilities, in: *Advances of Molecular Physics*, vol. **20**:267, D.Bates and B.Bederson Eds., Academic Press, Orlando (1985).
2. S.Basak and M.H.Cohen, Deformation-Potential Theory for the Mobility of Excess Electrons in Liquid Argon, *Phys.Rev.* B **20**:3404 (1979).
3. R.A.Holroyd and W.F.Schmidt, Transport of Electrons in Nonpolar Fluids, *Annu.Rev.Phys.Chem.* **40**:439 (1989).
4. L.G.Huxley and R.W.Crompton, *The Diffusion And Drift Of Electrons In Gases*, Wiley, New York (1974).
5. G.H.Wannier, *Statistical Physics*, Dover Publications, New York (1987).
6. A.K.Bartels, Density Dependence of Electron Drift Velocities in Helium and Hydrogen at 77.6 K, *Appl.Phys.* **8**:59 (1975).
7. A.K.Bartels, Density Dependence of the Electron Drift Velocity in Argon, *Phys.Lett.* **44** A:403 (1973).
8. S.R.Hunter and L.G.Christophorou, Interphase Physics: Linking Knowledge on Electron-Molecule Interactions in Gases to Knowledge on Such Processes in Condensed Matter, in: *Electron-Molecule Interactions And Their Applications*, vol.**2**:221, and Electron Motion in Low- and High-Pressure Gases, *ibid.* vol.**2**:89, L.G.Christophorou and K.Siomos Eds., Academic Press, Orlando (1984), and L.G.Christophorou and S.R.Hunter, Electrons in Dense Gases, in *Swarms Of Ions And Electrons In Gases*, W.Lindinger, T.D.Maerk, and F.Howorka Eds., Springer Verlag, Wien (1984).
9. T.F.O'Malley, Extrapolation of Electron-Rare Gas Atom Cross Section to Zero Energy, *Phys.Rev.* **130**:1020 (1963).
10. A.G.Robertson, Drift Velocities of Low Energy Electrons in Argon at 293 and

90 K, *Aust.J.Phys.* **30**:39 (1977).

11. G.L.Braglia and V.Dallacasa, Theory of Electron Mobility in Dense Gases, *Phys.Rev.* **A 26**:902 (1982).

12. T.F.O'Malley, Multiple Scattering Effect on Electron Mobilities in Dense Gases, *J.Phys.* **B 13**:1491 (1980).

13. A.F.Borghesani, L.Bruschi, M.Santini, and G.Torzo, Electron Mobility in Neon at High Densities, *Phys.Rev.*, **A 37**:4828 (1988).

14. A.F.Borghesani and M.Santini, Electron Mobility and Localization Effects in High-Density Neon Gas,*Phys.Rev.* **A 42**:7377 (1990).

15. A.F.Borghesani and M.Santini, Electron Swarm Experiments in Fluids–Signal Waveform Analysis, *Meas.Sci.Technol.* **1**:939 (1990).

16. T.F.O'Malley and R.W.Crompton, Electron-Neon Scattering Length and S-Wave Phaseshifts from Drift Velocities, *J.Phys.* **B 13**:3451 (1980).

17. K.W.Schwarz, Electron Localization in Dense Helium Gas: New Experimental Results, *Phys Rev.* **B 21**:5125 (1980).

18. A.F.Borghesani, M.Santini and P.Lamp, Excess Electron Mobility in High-Density Argon Gas, *Phys.Rev.* **A 46**:7902 (1992).

19. T.F.O'Malley, P.G.Burke, and K.A.Berrington, R-Matrix Calculation of Low-Energy e-He Scattering, *J.Phys.* **B 12**:953 (1979).

20. M.Weyhreter, B.Barzick, A.Mann, and F.Linder, Measurements of Differential Cross Sections for e-Ar, Kr, Xe Scattering at E=0.05-2 eV, *Z.Physik* **D 7**:333 (1988).

21. S.S.-S.Huang and G.R.Freeman, Electron Mobilities in Gaseous, Critical, and Liquid Xenon: Density, Electric Field, and Temperature Effects: Quasilocalization, *J.Chem.Phys.* **68**:1355 (1978).

22. F.M.Jacobsen, N.Gee, and G.R.Freeman, Electron Mobility in Krypton Dense Gas as Functions of Electric Field Strength, Temperature, and Density, *J.Chem.-Phys.* **91**:6943 (1989).

23. W.Legler, Multiple Scattering and Drift of Electrons in Gases of High Density, *Phys.Lett.* **A 31**:129 (1970).

24. V.M.Atrazhev and I.T.Iakubov, The Electron Drift Velocity in Dense Gases, *J.Phys.* **D 10**:2155 (1977) and A.G.Khrapak and I.T.Iakubov, Electrons and Positrons in Denses Gases, *Sov.Phys.Usp.* **22**:703 (1979).

25. E.Fermi, Sopra lo Spostamento per Pressione delle Righe Elevate delle Serie Spettrali, *Il Nuovo Cimento* **11**:157 (1934).

26. M.Lax, Multiple Scattering of Waves, *Rev.Mod Phys.* **23**:287 (1951).

27. L.L.Foldy, The Multiple Scattering of Waves, *Phys.Rev.* **67**:107 (1945).

28. H.Lehning, Pressure Dependence of the Electron Drift Velocity in Methane, *Phys.Lett.* **29 A**:719 (1969).

29. H.Lehning, Resonance Capture of Very Slow Electrons in CO_2, *Phys.Lett.* **28 A**:103 (1968).

30. K.W.Schwarz, Anomalous Electron Mobilities in Dense Helium Gas, *Phys.Rev.-Lett.* **41**:239 (1978).

31. H.R.Harrison, L.M.Sander, and B.E.Springett, Electron Mobility and Localization in Dense He^4 Gas,*J.Phys.* **B 6**:908 (1973).

32. T.F.O'Malley, Electron Distribution and Mobility in a Dense Gas of Repulsive Scatterers, *J.Phys.* **B 22**:3701 (1989).

33. A.F.Ioffe and A.R.Regel, *Prog.Semicond.* **4**:237 (1960).

34. J.Lekner and A.R.Bishop, Electron Mobility in Simple Fluids Near the Critical Point, *Philos.Mag.* **27**:297 (1973) and J.Lekner, Scattering of Waves by an Ensemble of Fluctuating Potentials, *Philos.Mag.* **18**:1281 (1968).

35. J.L.Levine and T.M.Sanders, Jr., Anomalous Electron Mobility and Complex Negative Ion Formation in Low-Temperature Helium Vapor, *Phys.Rev.Lett.* **8**:159 (1962).

36. M.H.Cohen and J.Lekner, Theory of Hot Electrons in Gases, Liquids, and Solids, *Phys.Rev.* **158**:305 (1967).

37. K.Kaneko, Y.Usami, K.Kitahara, Gas Kinetic Approach for Electron Mobility in Dense Media *J.Chem.Phys.* **89**:6420 (1988) and Errata Corrige on ' Gas Kinetic

Approach for Electron Mobility in Dense Media' *J.Chem.Phys.* **90**:6810 (1989).

38. Y.Naveh and B.Laikhtman, Mobility of Excess Electrons in Fluid Argon: Single- and Double-Phonon Theory, *Phys Rev.* **B 47-I**:3566 (1993).

39. L.Bruschi, M.Santini, and G.Torzo, Electron Mobility in High Density Neon Gas, *Phys.Lett.* **102 A**:102 (1984).

40. A.Borghesani, L.Bruschi, M.Santini, and G.Torzo, Density Dependence of the Electronic Mobility in High Density Neon Gas, *Phys Lett.* **108 A**:255 (1985).

41. A.F.Borghesani and M.Santini, Electron Localization Effects and Resonant Attachment to O_2 Impurities in Highly Compressed Neon Gas, in *Gaseous Dielectrics VI*, L.G.Christophorou and I.Sauers Eds., Plenum, New York (1991), p.27.

42. A.F.Borghesani and M.Santini, Electron Localization-Delocalization Transition in High-Density Neon Gas, *Phys.Rev.* **A 45**:8803 (1992).

43. J.P.Hernandez, Electron Self-Trapping in Liquids and Dense Gases, *Rev.Mod.-Phys.* **63**:675 (1991).

44. B.E.Springett, J.Jortner, and M.H.Cohen, Stability Criterion for the Localization of an Excess Electron in a Nonpolar Fluid, *J.Chem.Phys.* **48**:2720 (1968).

45. J.Bardeen, An Improved Calculation of the Energies of Metallic Li and Na, *J.Chem.Phys.* **46**:367 (1938).

46. E.Wigner and F.Seitz, On the Costitution of Metallic Sodium, *Phys.Rev.* **43**:804 (1933).

47. C.Kittel, *Introduction To Solid State Physics*, Wiley, New York (1986).

48. J.R.Broomall, W.D.Johnson, and D.J.Onn, Density Dependence of the Electron Surface Barrier for Fluid ^{3}He and ^{4}He, *Phys.Rev.* **B 14**:2819 (1976).

49. A.K.Bartels, Density Dependence of the Electron Attachment Frequency in Dense Helium-Oxygen Mixtures, *Phys.Lett.* **45 A**:491 (1973).

50. L.Bruschi, M.Santini, and G.Torzo, Resonant Electron Attachment to Oxygen Molecules in Dense Helium GAs, *J.Phys.* **B 17**:1137 (1984).

51. T.P.Eggarter and M.H.Cohen, Mobility of Excess Electrons in Gaseous He: A Semiclassical Approach, *Phys.Rev.Lett.* **27**:129 (1971).

52. G.Ascarelli, Hall Mobility of Electrons in Liquid Xenon, *J.Phys.Condens.Matter* **4**:6055 (1992).

53. S.John, Localization of Light *Comment.Condens.Matter Phys.* **14**:193 (1988).

54. R.Reininger, U.Asaf, I.T.Steinberger, and S.Basak, Relationship between the Energy V_0 of the Quasi-free-Electron and its Mobility in Fluid Argon, Krypton, and Xenon, *Phys.Rev.* **B 28**:4426 (1983).

55. A.F.Borghesani, G.Carugno, and M.Santini, Experimental Determination of the Conduction Band of Excess Electrons in Liquid Ar, *IEEE Trans. Electr. Insul.* **EI-26**:615 (1991).

56. S.S.-S.Huang and G.R.Freeman, Electron Transport in Gaseous and Liquid Argon: Effects of Density and Temperature, *Phys.Rev.* **A 24**:714 (1981).

57. L.Christophorou and D.L.McCorkle, Experimental Evidence for the Existence of a Ramsauer-Townsend Minimum in Liquid CH_4 and Liquid Ar (Kr and Xe), *Chem.Phys.Lett.* **42**:533 (1976).

58. J.Lekner, Motion of Electrons in Liquid Argon, *Phys.Rev.* **158**:130 (1967).

59. H.E.Stanley, *Introduction To Phase Transitions And Critical Phenomena*, Oxford University Press, Oxford (1971).

60. M.Fixman, Density Correlations, Critical Opalescence, and the Free Energy of Nonuniform Fluids, *J.Chem.Phys.* **33**:1357 (1960).

61. G.E.Thomas and P.W.Schmidt, X-Ray Study of Critical Opalescence in Argon, *J.Chem.Phys* **39**:2506 (1963).

62. G.N.Haddad and T.F.O'Malley, Scattering Cross Sections in Argon from Electron Transport Parameters, *Aust.J.Phys.* **35**:35 (1982).

63. A.F.Borghesani, M.Santini, and P.Lamp, to be published.

64. R.Eibl, P.Lamp, and G.Buschhorn, Measurement of Electron Mobility in Liquid and Critical Argon at Low Electric-Field Strengths, *Phys.Rev.* **B 42**:4356 (1990).

EXCESS ELECTRON LOCALIZATION
IN HIGH-DENSITY NEON GAS

A.F.Borghesani and M.Santini

Dipartimento di Fisica "G.Galilei", Università di Padova
and Consorzio Interuniversitario Fisica della Materia
Via F.Marzolo 8, I-35131 Padova, Italy

INTRODUCTION

Excess electron transport in high-density non polar gases is not restricted to quasifree electron propagating states characterized by very large value of the drift mobility. In some gases, such as He and H_2, in specific conditions of temperature and density, excess electrons transport proceeds via localized states, characterized by very small values of the drift mobility. In gases such as Ar, on the contrary, even at the highest densities, it appears that only extended states of high mobility are available for excess electron transport. It is instructive to look at once at the density dependence of the mobility μ_0 of thermal excess electrons in He[1], Ne[2], and Ar[3]. (Fig.1).

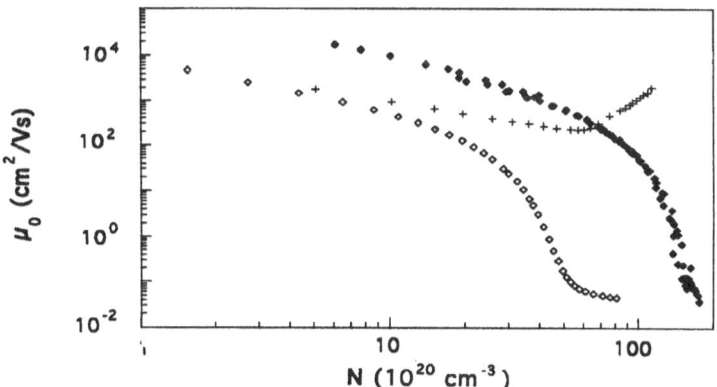

Figure 1. Zero-field electron mobility μ_0 as a function of the gas density N in Ar at $T=152.7\,K$ (crosses)[3], in He at $T=26.1\,K$ (open points)[1] and in Ne at $T=(45.0-48.4)\,K$ (closed points)[2].

In Ar μ_0 decreases a little in the low density region up to $N \approx 60 \times 10^{20}\ \mathrm{cm}^{-3}$ so as to nearly satisfy the classical prediction $\mu_0 N \approx \mathrm{const.}$[4], and then increases again with N(a behavior similar to that in CH_4[5]). By contrast, in both He and Ne the electron mobility of thermal electrons drops by several orders of magnitude as density is raised. These observations are interpreted as indicating that in these gases electrons undergo a transition from a delocalized state to a localized one as the gas density increases. The localized state is assumed to be formed by an electron in a hollow cavity [6] and hydrodynamics governs the motion of this complex.

Evidence for localized excess electron states in non polar liquids [7,8,9,10] and high-density gases was found long ago [11,12,13,14]. A short review appeared some years ago [15] and a more specific one has been recently published [16]. The problem at hand is the determination of the states available to a light quantum particle in a medium whose structure can be locally modified by the interaction with the particle itself or in a medium that is statically disordered [17]. Self-trapping of excess electrons has to be expected in lower-than-average density region if the electron-atom short-range repulsion due to exchange forces is so strong that a local dilatation of the fluid leads to a state of lower free energy than the quasi-free state [18]. A stable configuration of the system can be achieved if the average gas density is high enough that the energy of the delocalized electron state is larger than the increased kinetic energy of the electron due to localization plus the contribution due to the pressure-volume work necessary to create the cavity. As a consequence, self-trapped excess electron states are expected for gases whose scattering length is positive, as in He, Ne, H_2, as the result of the predominant influence of the short-range repulsive exchange forces with respect to the contribution of polarization interaction. In the case of the heavier rare gases, Ar, Kr, and Xe, the attractive contribution due to polarization effects overwhelms the repulsive forces and leads to negative scattering lengths for these atoms in the gas phase [19,20] and no localization effects are present, neither in the gas-phase (Figure 1) nor in the liquid-phase [21,22].

Neon lies at the borderline between He and the heavier noble gases, since, as a result of a good balance between repulsive and attractive interaction with electrons, it has a positive, though small scattering length [23]. If electron localization takes place in Neon gas, this should happen at fairly large densities and at quite low temperatures. Localization in liquid Neon has been observed, indeed [9,10,24], as expected on the basis of the calculations of the stability of the localized state [19,25]. So, we have carried out drift mobility measurements of excess electrons in high-density Neon gas to search for possible excess electrons localization effects.

EXPERIMENTAL RESULTS

In Figure 2 we report the data of the drift mobility of thermal excess electrons injected into Ne gas at temperatures close to the critical one [2] ($45.0 < T < 48.4\ K$, $T_c = 44.38\ K$ [26]) as a function of the gas number density N. These data are the same as in Figure 1, but plotted so as to put more emphasis on the localization-delocalization transition. In order to be sure that the low-mobility branch of the electron mobility curve does indeed correspond to electron transport and not to transport of impurity ions, we also report in this Figure the zero-field drift mobility of negative O_2^- ions which are formed as a consequence of the well-known resonant electron attachment process [27]. Even if in our experiment the impurity concentration was very low, as witnessed by the very low attachment rates [28] (0.5-10 kHz, electron mean lifetime in the range 100 μs - 2 ms) corresponding to impurity concentrations in the ppb range, our technique [29,30] enabled us to directly detect the ions and measure their mobility [31]. As can bee clearly seen, even when the electron mobility has reached its lowest value, nonetheless the ionic mobility is nearly one order of magnitude smaller than that of electrons. By contrast, in the liquid Ne measurements the nature of the charge carrier was inferred from the temperature behavior of the observed mobility [9].

The electron mobility reported in Figure 2 shows a drop of nearly 6 orders of magnitude in the density range $(5 < N < 180) \times 10^{20}$ cm^{-3}. In the same density range the mean interatomic distance between the Ne atoms, $\bar{d} = N^{-1/3}$, changes from 12.6 Å to 3.8 Å. The first portion of the curve, up to a density $N^* \approx 95 \times 10^{20}$ cm^{-3}, shows a mobility drop of approximately 2.5 orders of magnitude and it is believed that in this density region there only exist delocalized states. The observed behavior of their mobility has to be attributed to multiple scattering corrections to the quasi-free electron classical mobility as a result of the simultaneous interaction of the very broad electron wavepacket with many atoms at a time [2,32]. In fact, the electron thermal wavelength at $T \approx 45K$ is very large, $\lambda_T = h/\sqrt{2\pi m k_B T} \approx 111$ Å $\gg \bar{d}$.

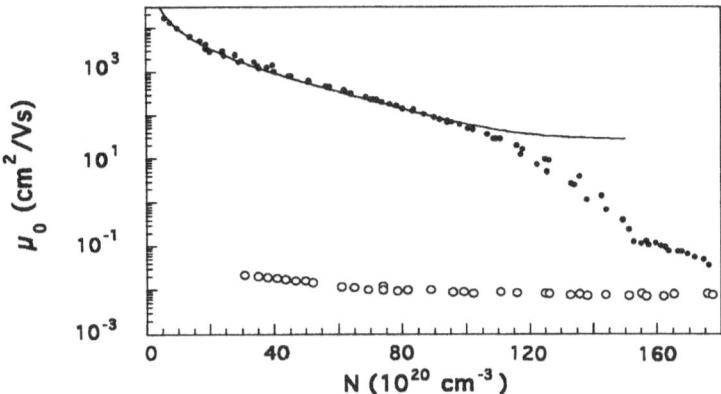

Figure 2. Zero-field mobility μ_0 of excess electrons (closed points) and of O_2^- ions [31] (open points) as a function of the gas density N in Ne at $T=(45.0-48.4) K$ [2]. The solid line is the result of the modified multiple scattering theory [2,32].

The second portion of the experimental curve, for $N > 95 \times 10^{20}$ cm^{-3}, is characterized by a further mobility drop of approximately 3.5 orders of magnitude. This mobility reduction has been attributed to the presence and gradual increase of localized states, namely (pseudo) bubbles, that are in coexistence with quasifree propagating states. The higher the density, the larger the proportion of localized states with respect to the extended ones and the smaller the average mobility. It is worth noting that this second part of the curve shows a different curvature as a function of N than the first portion and that the curvature change is located at N^*, approximately.

This interpretation of the experimental curve has been suggested by the behavior of the drift mobility as a function of the electric field [2]. Such a behavior is depicted in Figure 3 for densities below and above N^*. For $N < N^*$ the density-normalized mobility μN is constant for low reduced electric fields (typical behavior of thermal electrons) and for larger fields it starts decreasing with E/N, being roughly proportional to $(E/N)^{-1/2}$. At larger fields μN is no longer dependent on the density and it turns out to depend solely on E/N, as predicted classically. This behavior was brought into evidence at lower densities in a previous work [32].

By contrast, when $N \geq N^*$, μN is constant at first, then it increases with E/N showing a maximum. Finally, it decreases again and meets the classical high-E/N dependence. A similar supralinear dependence of the electron drift velocity was observed also in He [11,33]. This change of behavior takes place in Ne between 90 and 100 $\times 10^{20}$ cm^{-3}, as shown in greater detail in Figure 4. The qualitative explanation is that for $N > N^*$ and at low electric fields there is coexistence of quasifree and localized electron states, in the sense that electrons are free for a time τ_f and localized for a time τ_b, on average. The average mobility is given by the weighted sum $\mu = (\tau_f \mu_f + \tau_b \mu_b)/(\tau_f + \tau_b)$, where μ_f and μ_b are the mobilities of quasifree and localized states, respectively. The electric field may inhibit the initial trapping in a density fluctuation, or enhance the electron escape probability from the localized state, or change the distributions of extended and localized states [34]. In any case, the ratio τ_f/τ_b increases with the field, and at densities close to N^*, where the energy of the localized state is believed to be small, only a small value of the electric field E should be required in order to produce a detectable change in τ_f/τ_b. At larger densities, the localized state is more stable and larger fields are needed to produce this change. We have shown in Figure 4 by means of arrows the values of $(E/N)^*$ where μN starts increasing over its thermal value instead of decreasing as expected for quasifree electrons.

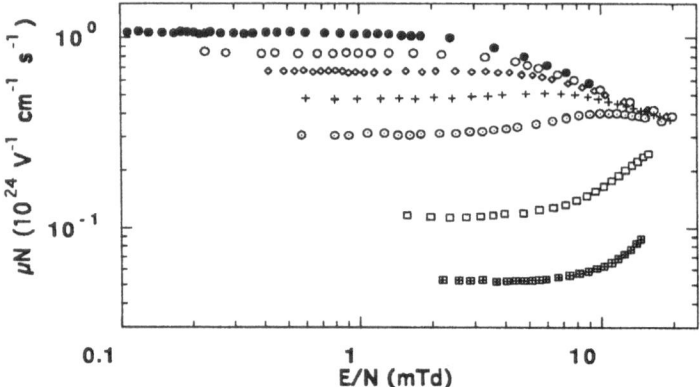

Figure 3. Density-normalized mobility μN of excess electrons as a function of the reduced electric field E/N (1 mTd=10^{-20} Vcm^2) in Ne at T=46.5 K for several densities[2]. From top: N= (83.4, 90.2, 95.6, 102.1, 110.8, 126.3, 135.9)$\times 10^{20}$ cm^{-3}.

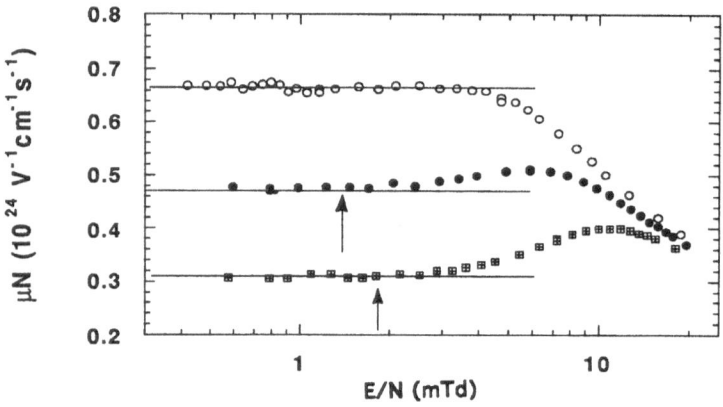

Figure 4. Density-normalized mobility μN of excess electrons as a function of the reduced electric field E/N in Ne at T=46.5 K for N= (95.6,102.1,110.8) $\times 10^{20}$ cm^{-3} (from top)[2].

It can be observed that $(E/N)^*$ increases with density. In Figure 5 we report the values E^* of the threshold electric field as a function of N to demonstrate that the electric field necessary to destroy the localized states or to inhibit their formation does indeed increase with N. The determination of E^* is a rough one and E^* can be obtained only with a large uncertainty. Nonetheless, it gives an estimate of the threshold density above which stable localized states are present.

284

DISCUSSION OF THEORETICAL MODELS

A number of models have been devised in order to account for the experimental results in liquid and gaseous He, and in gaseous Ne. In the specific case of gaseous Neon, three models have been used to interpret the experimental data, namely, the *bubble* model[2], the *percolation* model[35], and the *mesoscopic* model[16,36,37]. In the following we shall describe them briefly. However, a description of the mobility of the quasifree extended states is preliminarily required.

The density dependence of the mobility of quasifree excess electrons

In the density range below N^* the density dependence of the zero-field mobility of excess electrons in a high-density non polar gas can be described quite accurately by a means of a heuristic model[32,38] that introduces some multiple scattering corrections in the classical formula for the mobility[4] and that is also described in another paper of this volume.

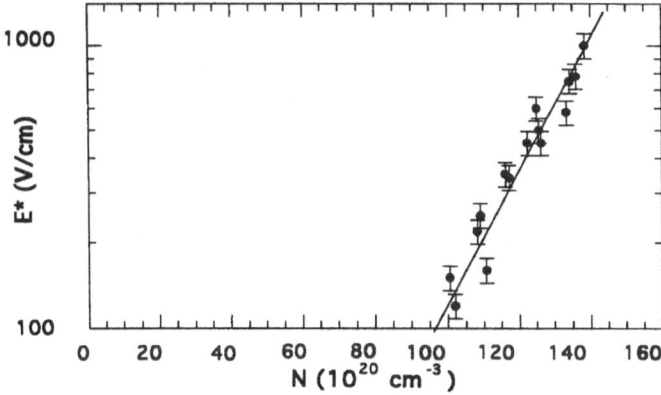

Figure 5. Threshold electric field E^* as a function of N. The solid line is only a guideline.

Owing to the large extension of the electron wavepacket, larger than the interatomic spacing, multiple scattering effects modify the behavior of the electron mobility as a function of the gas density with respect to its behavior in the dilute gas. First of all, the electron ground-state energy is shifted by an amount V_0[39,40,41] whose kinetic part E_k enters in the scattering process of the delocalized electrons. E_k represents very naturally the bottom of the "conduction band", in the sense that no quasifree propagating states exist for energies lower than E_k. In the second place, there is a reduction of the mean free path, or, conversely, an increase of the cross section, because of quantum interference effects that become important as the electron mean free path becomes comparable to its thermal wavelength[33,42,43,44]. Finally, correlations among scatterers also give an increase of the cross section accounted for, at low energies, by the long wavelength limit of the static structure factor[45] $S(0) = N k_B T \chi_T$, where χ_T is the isothermal compressibility of the gas. The kinetic energy shift depends on the scatterers density and can be obtained explicitly by replacing the fluid structure with an ordered array of scatterers, neglecting the overlap of the electron-atom potential tails, and matching the electron wave function to its asymptotic expression on the surface of a Wigner-Seitz sphere of volume $(4/3)\pi r_s^3 = 1/N$ centered on each atom[46]. One thus gets

$$E_k = \frac{\hbar^2 k_0^2}{2m} \tag{1}$$

where m is the electron mass and the wavevector k_0 is determined by the eigenvalue equation

$$\tan\left\{k_0\left[r_s - \tilde{a}(k_0)\right]\right\} - k_0 r_s = 0 \tag{2}$$

The atomic radius $\tilde{a} = \sqrt{\sigma_T/4\pi}$ is related to the total scattering cross section σ_T.

Including all these multiple scattering effects, the mobility of thermal electrons is then given by [35,36]

$$\mu = \frac{4}{3}\frac{e}{\hbar}\frac{\lambda_T \ell^*}{S(0)}\exp\left[-(\pi^{-1/2}\lambda_T/\ell^*)\right] \tag{3}$$

where λ_T is the electron thermal wavelength and

$$\ell^* = \frac{1}{N}\left(\frac{1}{k_B T}\right)^2 \int_0^\infty d\epsilon \frac{\epsilon}{\sigma_m(\epsilon + E_k)}\exp\left[-(\epsilon/k_B T)\right] \tag{4}$$

In Eq.(4) σ_m is the gas-phase momentum transfer scattering cross section that has to be evaluated at the shifted kinetic energy $\epsilon + E_k$. The prediction of this model is represented by the solid line in Fig. 2. The agreement with the experimental data is very good, indeed, up to the density N^*.

The *bubble* model

In the dense liquid the most probable state can be adequately described as an empty cavity of radius ≈ 20 Å with large potential depth and small ground-state energy. In the lower density gas the electron-atom interaction is weaker and it should be possible that some atoms occupy the region of low potential for the electron giving a smoother density profile of the cavity. Furthermore, the distribution of states with non negligible occupation probability in thermal equilibrium may not be so well defined as in the dense, low temperature liquid. This means that in the gas the transition from the delocalized state to the localized one should not be as sharp as in the liquid. Bearing in mind these observations, we now describe the bubble model [19].

Suppose electrons to be subjected to a spherically symmetrical potential with $V = V_i$ for $r < R$ and $V = V_0$ for $r \geq R$, with the condition $V_i < V_0$, where R is the bubble radius. The lowest-lying bound state (s-wave state) with energy ϵ_1 and wavefunction $u(r)$ must satisfy the radial Schroedinger equation for $f(r) = ru(r)$

$$\begin{cases} f'' + k_i^2 f = 0 & \text{for } r \leq R \\ f'' - k_o^2 f = 0 & \text{for } r > R \end{cases} \tag{5}$$

where $f'' = d^2 f/dr^2$,

$$k_i^2 = \frac{2m}{\hbar^2}(\epsilon_1 - V_i) \tag{6}$$

and

$$k_o^2 = \frac{2m}{\hbar^2}(V_0 - \epsilon_1) \tag{7}$$

Imposing the boundary conditions to the radial wavefunctions at $r = R$, as usual, we obtain the eigenvalue equation

$$\tan X = \frac{-X}{(H^2 - X^2)^{1/2}} \tag{8}$$

with

$$X = k_i R \tag{9}$$

and

$$H^2 = \left(\frac{2m}{\hbar^2}\right)(V_0 - V_i)R^2 \tag{10}$$

Let X_1 be a solution of Eq.(8), then the energy ϵ_1 of the s-state is

$$\epsilon_1 = \left(\frac{\hbar^2}{2mR^2}\right)X_1{}^2(R) + V_i \tag{11}$$

In order to have bound states, from Eq.(10) we get the condition

$$\left(\frac{2m}{\hbar^2}\right)(V_0 - V_i)R^2 \geq \frac{\pi^2}{4} \tag{12}$$

The empty bubble

Suppose that the cavity of radius R is empty. Let T be the absolute temperature, N the host gas number density, and $P = P(N,T)$ the gas pressure [47]. The potential for $r > R$ is the density dependent self energy $V_0(N)$, which we assume to be given by the Wigner-Seitz potential. Inside the bubble we must account for for the polarization energy V_p. According to Miyakawa and Dexter [25], we assume V_p constant within the hollow cavity

$$V_p = -\frac{\mathcal{E}_p N}{R} \tag{13}$$

where $\mathcal{E}_p = (1/2)\alpha(e/4\pi\epsilon_0)^2$, with the Ne atomic polarizability $\alpha = 4.362 \times 10^{-41}$ F m^2. At fixed density N we have

$$V_i = -\frac{\mathcal{E}_p N}{R} \quad \text{and} \quad V_0 = V_0(N) \tag{14}$$

Inserting Eq. (14) into Eq. (12) we obtain the miminum radius R_0, below which there are no bound states, as

$$R_0 = \frac{-\mathcal{E}_p N + \sqrt{(\mathcal{E}_p N)^2 + (h^2/8m)V_0}}{2V_0} \tag{15}$$

For $R \geq R_0$ and from Eqns. (8), (9) and (10) the total energy ϵ_T is found from Eq.(11) by adding the work $\mathcal{E}_v$ done at constant T against P to create the hollow spherical cavity

$$\epsilon_T = \left(\frac{\hbar^2}{2mR^2}\right)X_1{}^2(R) - \frac{\mathcal{E}_p N}{R} + \mathcal{E}_v \tag{16}$$

where $\mathcal{E}_v = (4/3)\pi PR^3$. If ϵ_T as a function of R shows a minimum ϵ_M for $R = R_M$, the localized state is stable. If this is not the case $\epsilon_T(R)$ and R decrease until, for $R = R_0$ the electron leaves the bubble. The stability found for ϵ_M is not absolute. Only if $(V_0 - \epsilon_M) \gg k_B T$ the bubble state is surely stable. As an example, for Ne at $T = 46.5\,K$ and using the Wigner-Seitz model for $V_0(N)$, the empty bubble is stable only for $N > 137 \times 10^{20}$ cm^{-3}. Moreover, assuming that the ratio of the number of bubble states n_b to the number of extended states n_f is $n_b/n_f = \exp[-(\epsilon_M - V_0)/k_B T]$, we obtain $n_b = n_f$ (i.e., $\epsilon_M = V_0$) only for $N > 175 \times 10^{20}$ cm^{-3}. This means that empty bubbles are very unlikely to form in Neon gas. This is due to the fact that the mechanical work PV is very large because of the large pressure ($P = 3.41\,\text{MPa}$ at $T = 46.5\,K$ and $N = 137 \times 10^{20}$ cm^{-3})

and this contribution inhibits the formation of stable empty bubbles.

<u>The partially empty *bubble*</u>

As guessed before, in a gas some atoms could be allowed to enter the region of low potential for the electron. Let therefore assume the the bubble is partially filled with density N_1 and let us define the filling fraction $F = N_1/N$, where N is the mean gas density. The electron ground state energy is thus $V_0(N)$ outside the bubble and $V_1 \equiv V_0(N_1)$ inside the bubble. The polarization energy must be corrected so as to be zero if the cavity were completely filled

$$V_p = -\frac{\mathcal{E}_p N}{R}(1 - F) \tag{17}$$

The electron potential inside the bubble is now

$$V_i = V_1 + V_p = V_0(N_1) - \frac{\mathcal{E}_p N}{R}(1 - F) \tag{18}$$

The new minimum radius for the existence of a solution of the Schroedinger equation is now

$$R_0 = \frac{1}{2(V_0 - V_1)}\left\{-\mathcal{E}_p N(1 - F) + \sqrt{\left[\mathcal{E}_p N(1 - F)\right]^2 + \frac{h^2}{8m}(V_0 - V_1)}\right\} \tag{19}$$

The new mechanical work is given by [2]

$$\mathcal{E}_v = \left(\frac{4\pi}{3}\right)PR^3\left\{(1 - F) - \frac{FN}{P}\left[BN(1 - F) + 0.5CN^2(1 - F^2) - A\ln F\right]\right\} \tag{20}$$

where A, B, and C are the coefficients of the cubic equation, $P = AN + BN^2 + CN^3$, which interpolates very well the experimental equation of state at our temperatures and densities.

Once again, we seek for a mimimum ϵ_M of the total electron energy as a function of $R > R_0$. In Figure 6 we show $(\epsilon_M - V_0)/k_B T$ as a function of the filling fraction F at $T = 46.5\,K$ for several densities.

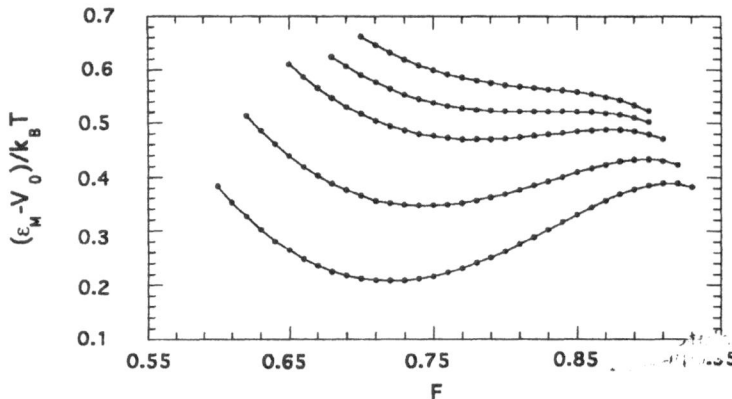

Figure 6. The dependence of $(\epsilon_M - V_0)/k_B T$ as a function of the filling factor F for Ne gas at $T=46.5\,K$ for the densities (from top) [2] $N= (95.0,\ 95.5,\ 96.0,\ 97.0,\ 98.0)\times 10^{20}\,cm^{-3}$.

The result of the model is that the bubble at lower densities gets more and more filled in order to minimize its total energy. It may happen that there exists no minimum for ϵ_T as a function of R. Even though bubbles may be generated by thermal fluctuations, they are not stable.

As N is increased, ϵ_T shows a minimum ϵ_B for given values of the bubble radius R_B and filling fraction F_B. The minimum gets more and more pronounced as N increases and the bubbles turn out to be stable against fluctuations of their radius or filling fraction. From Figure 6 we see that weakly stable partially filled bubbles can be present for $N \geq N^* \approx 95 \times 10^{20}$ cm^{-3}. In Figure 7 we show the values of the filling fraction F_B, of the bubble radius R_B, and of the difference $(\epsilon_B - V_0)/k_B T$ as a function of the gas density for $T = 46.5\,K$.

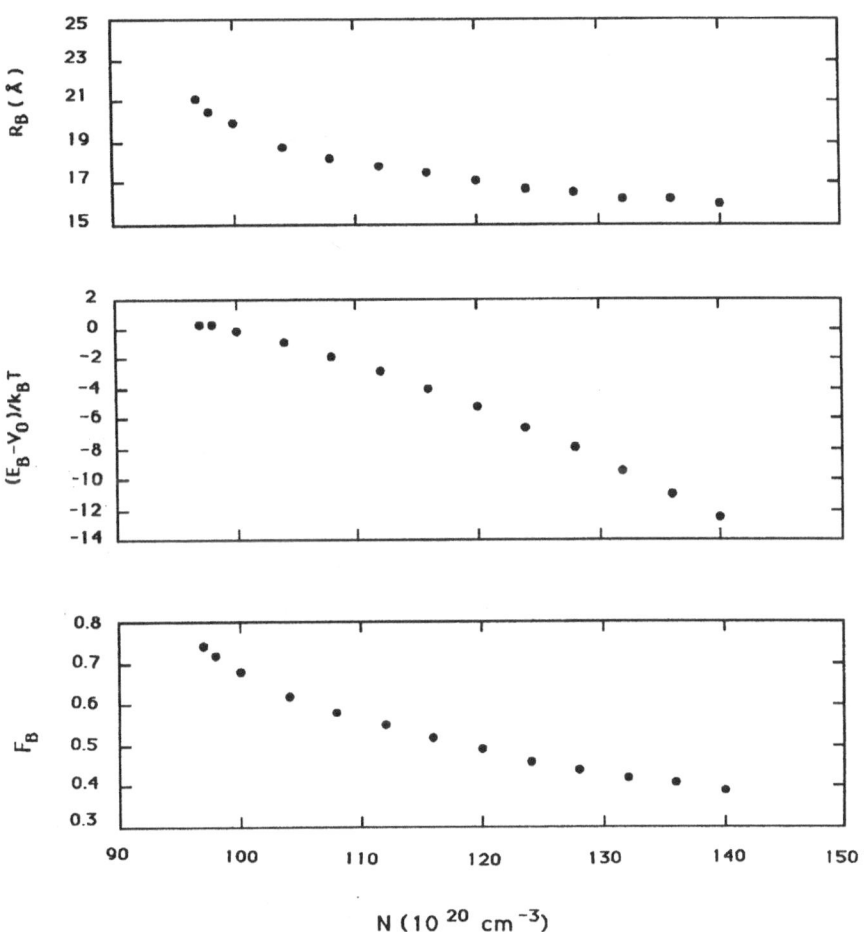

Figure 7. Density dependence of the radius R_B, of $(\epsilon_B - V_0)/k_B T$ and of the filling fraction F_B for stable bubbles [2].

From this Figure we can note that as the density increases the radius of the stable bubbles and their filling fraction decrease and that for $N \geq N^*$ the bubbles tend to become absoⁱ ᵗely stable, as indicated from the fact that $(\epsilon_B - V_0)/k_B T$ becomes large and negative. These results depend, of course, on the choice of the kinetic potential V_0.

The coexistence of localized and extended states can be described by the ratio n_b/n_f. Letting $\Delta_B = (\epsilon_B - V_0)/k_B T$ and $n = n_f + n_b$ be the total number of electrons, we have $n_f/n_b = \exp(\Delta_B)$ and $n_f/n = \exp(\Delta_B)/[1 + \exp(\Delta_B)]$. The average mobility is then calculated as the weighted sum

$$\mu = \left[\mu_b \left(\frac{n_b}{n} \right) + \mu_f \left(\frac{n_f}{n} \right) \right] = \frac{\mu_b + \mu_f e^{\Delta_B}}{1 + e^{\Delta_B}} \tag{21}$$

Assuming that the mobility of the localized states can be given by the Stokes formula $\mu_b = e/(6\pi\eta R)$ with $R = R_B$ (Fig.7). $\eta \approx 21.0 \times 10^{-6}$ Pa·s is the gas viscosity [48], and exploiting Eqns.(3) and (4) for μ_f, we calculated μ. The results are shown in Figure 8.

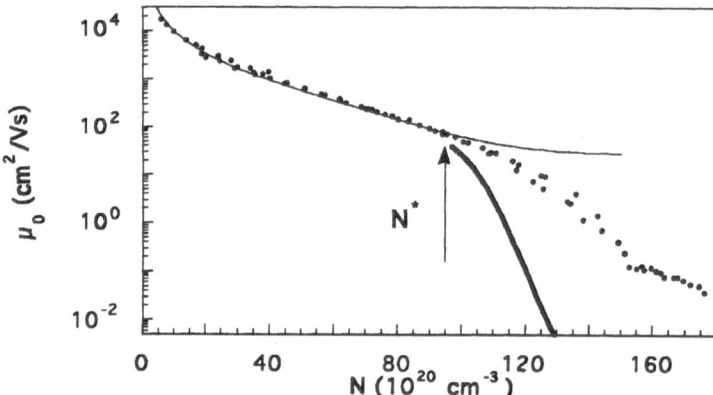

Figure 8. Results of the bubble model (thick solid line) compared to the experimental mobility [2]. The thin solid line is the prediction for the mobility of the extended states calculated according to Eqns. (3) and (4).

As can be observed, the agreement with the experiments is reasonable in spite of the crudeness of the model. The density at which localization starts to be experimentally detected is fairly well reproduced, indeed. However, the model fails to accurately reproduce the mobility values. We believe that the failure of this model, beyond the choice of an unrealistic density profile of the bubble, is due to the fact that it focuses only on the two most-probable states of the system, i.e., the localized and the extended ones, while it clearly appears that the experimental observations are the result of averaging over an equilibrium distribution of the states of the system and that the most-probable states do not possess the ensemble-averaged mobility. Moreover, the dynamics of the localization process is not known very precisely, yet [49]. It seems plausible from experiments of electron thermalization [50] that electrons are first trapped in a suitable density fluctuation (the so called *incipient bubble* [51]) and then proceed to thermalize into that region producing a stable bubble by adjusting the radius and filling density so as to minimize their free energy. (A similar picture is also proposed for liquid Ne [24].) However, in absence of a detailed knowledge of the mean lifetime of quasifree electrons and of the mean stabilization time the behavior of the mobility in the coexistence region is still an open problem.

The percolation model

The previous approach failed to explain the experimental observations in detail because it did not take into account, among other things, the configurational entropies of the system needed to calculate the density of states. The calculation of the density of states, by contrast, is the central goal of the *percolation* model developed by Eggarter and Cohen [52,53,54]. Once the density of states is known, the properties of the excess electrons can be calculated as averages over a distribution of thermally occupied states. The Eggarter and Cohen (EC) model is a semiclassical one in the sense that it takes into account the local zero-point kinetic energy of the electron in the definition of an effective potential surface and the mobility edge is simply defined as the minimum energy allowing a connected wave function to span over the whole gas sample [55].

An excess electron interacts with a collection of hard-core scatterers and electron-electron interactions are neglected owing to the low electron concentration. The medium is structurally disordered. To convert this hard-core problems into an equivalent one, where electrons move in a smooth effective potential, the material is divided into boxes of side L. The distribution of the scatterers is sampled by the electron within a box and there the average potential acting on the electron is given by the local density-dependent Wigner-Seitz (WS) energy $E_k \equiv V(N)$. The hard-core approximation is quite good for He, and not so bad for Ne, so we accept that no negative polarization contribution are considered in the local potential. The gas density and, hence, the potential fluctuate from box to box. The potential fluctuations are therefore related to the intrinsic density fluctuations. The length L sets the scale of the autocorrelation function of the potential and is the most important quantity in the EC model. This way to treat disordered material is not restricted to gas, but is also used elsewhere, as, for instance, in the semiconductor physics [56].

Within each box the local density of states n_i is assumed to be that of a free electron with energy ϵ above the local WS energy $V_i \equiv V(N_i)$, where N_i is the local number density of the gas in the i-th box,

$$n_i(\epsilon) = \frac{L^3}{2\pi^2}\left(\frac{2m}{\hbar^2}\right)^{3/2}(\epsilon - V_i)^{1/2} \qquad \text{for } \epsilon > V_i \qquad (22)$$

The global density of states is obtained by summing over all cells the single contribution weighted with a distribution function of the local density values, which is assumed to be gaussian. The effective potential is also distributed according to a gaussian function about $V = V(N)$, where N is the average gas density, with variance

$$\sigma_V{}^2 = N L^{-3} S(0) \left(\frac{\partial V}{\partial N}\right)^2 \qquad (23)$$

In Eq.(23) the long wavelength limit of the static structure factor, $S(0)$, has been introduced to account for the fluctuations-induced correlations among the gas atoms. The ensemble-averaged density of states is then given by

$$n(\epsilon) = 2(2\pi)^{1/2}\left(\frac{2m}{\hbar^2}\right)^{3/2}\sigma_V{}^{-1}\int d\tilde{V}\,(\epsilon - \tilde{V})^{1/2}\exp\left\{-[(\tilde{V}-V)^2/2\sigma_V{}^2]\right\} \qquad (24)$$

Upon a change of variables, $z = (\epsilon - \tilde{V})/\sigma_V$ and $x = (\epsilon - v)/\sigma_V$ Eq.(24) can be rewritten as

$$n(\epsilon) = (2\pi^2)^{-1}\left(\frac{2m}{\hbar^2}\right)^{3/2}\sigma_V{}^{1/2}\mathcal{F}(x) \qquad (25)$$

with

$$\mathcal{F}(x) = (2\pi)^{-1/2}\int_0^\infty dz\, z^{1/2}\exp\left[-(x-z)^2/2\right] \qquad (26)$$

This averaging over all cells gives origin to a density of states with a low-energy tail. The states deeply located in this tail ($\epsilon \ll V$) come from those cells where the gas density N and also V are very small. These states are trapped because the surrounding cells with larger N and V act as a nearly impenetrable barrier to the electron.

The electron transport properties are calculated in this model according to the percolation model [56]. Electrons are assumed to be able to percolate through the cells if their energy is large enough and have therefore high mobility. Electrons with energies below the percolation threshold are localized and have very low mobility, allowing for atomic motion. The percolation energy E_c assumes thus the role of the mobility edge and is associated naturally with the bottom of the conduction band [57]. To calculate the value of E_c the space is divided into regions accessible to the electron, if the local WS energy V is smaller than the electron energy ϵ, and in regions forbidden to the electron, if $V > \epsilon$. The fraction of space available to electrons is defined as

$$\phi(\epsilon) = \int_{-\infty}^{x} dw \, \exp\left(-w^2/2\right) = 0.5\left[1 + \mathrm{erf}\left(\frac{x}{\sqrt{2}}\right)\right] \tag{27}$$

and is the probability that a point chosen at random is in an allowed region. Replacing the original system of a gas divided into boxes with a cubic lattice of side L we have $\phi(E_c) = 0.30$. Different equivalent lattices give different values of $\phi(E_c)$ but the EC model is rather insensitive to the precise choice of $\phi(E_c)$. From Eq.(27) we obtain

$$E_c = V - 0.52\sigma_V(E_c) \tag{28}$$

and from this equation we see that the mobility edge is strongly dependent on the variance of the potential, and, hence, on the sampling length L. In Figure 9 we show E_c calculated using the classical mean free path as sampling length L along with the WS energy.

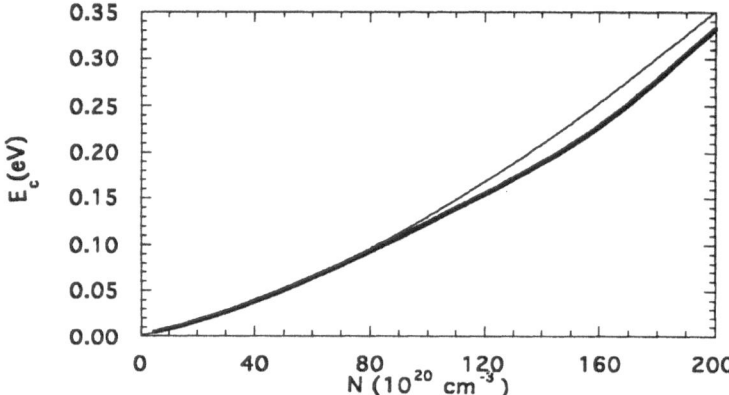

Figure 9. The percolation threshold energy E_c (thick solid line) and the Wigner-Seitz energy (thin solid line) in Neon as a function of the gas density [35]. For $N \leq 80 \times 10^{20}$ cm^{-3} the two curves are indistinguishable.

The mobility is calculated as an ensemble average over the states of the system

$$\mu = \frac{\int d\epsilon \, \mu(\epsilon) n(\epsilon) \exp\left[-(\epsilon/k_B T)\right]}{\int d\epsilon \, n(\epsilon) \exp\left[-(\epsilon/k_B T)\right]} \tag{29}$$

For the energy dependent mobility of the extended states the Drude result [58] $\mu_f(\epsilon) = (2/3)(e/m)\tau(\epsilon)$ is used, where $\tau(\epsilon)$ is the mean collision time related to the mean free path ℓ and v velocity by [59]

$$\tau(\epsilon) = \frac{\ell(\epsilon)}{v} \tag{30}$$

This result applies strictly to hard spheres but is is a very good approximation for He and a quite good one also for Ne.

The mean free path $\ell(\epsilon)$ must account for the two possible electron scattering mechanisms. The first one is the usual electron-single atom scattering described by the momentum transfer scattering cross section $\sigma_m(\epsilon)$ yielding a mean free path

$$\ell_c = \left(N \sigma_m(\epsilon) S(0) \right)^{-1} \tag{31}$$

In Eq. (31) the energy ϵ includes the energy shift $V \equiv E_k$ and is not the much smaller thermally activated contribution as in Eq.(4). Multiple scattering effects due to correlations among scatterers have also been taken into account by means of $S(0)$ according to the Lekner model[45].

The second mechanism is scattering off prohibited regions yielding a free path

$$\ell_p = L\phi(\epsilon) \left[1 - \phi(\epsilon) \right]^{-1} \tag{32}$$

The overall mean free path is thus

$$\ell(\epsilon) = \frac{\ell_p \ell_c}{\ell_p + \ell_c} = \frac{L\phi(\epsilon)}{1 - \phi(\epsilon) + LNS(0)\sigma_m(\epsilon)\phi(\epsilon)} \tag{33}$$

In Eq.(30) $(1/v)$ must be replaced by the conditional average of the inverse group velocity $v = [2(\epsilon - V)/m]^{1/2}$ over the allowed regions. The group velocity has to be used because electrons are immersed in the kinetic potential V. Finally, we obtain

$$\tau(\epsilon) = \frac{\ell(\epsilon)}{\phi(\epsilon)} \left(\frac{2m}{\sigma_V} \right)^{1/2} \left[\frac{\mathrm{d}\mathcal{F}(x)}{\mathrm{d}x} \right] \tag{34}$$

To account for the fact that an electron in an allowed region may not propagate if this region does not span over the whole gas, Eq.(34) has to be multiplied by the conditional probability $p(\epsilon)$ that a point is in an allowed and unbounded region.

$$\begin{cases} 0 & \text{if } \epsilon \le E_c \\ 1 - \exp\left\{ -25 \left[\phi(\epsilon) - \phi(E_c) \right] \right\} & \text{if } \epsilon > E_c \end{cases} \tag{35}$$

Since the gas atoms are not fixed in space, the localized states are given the semihydrodynamic mobility of a rigid sphere of radius R[60]

$$\mu_b = \frac{e}{6\pi\eta R} \left[1 + \frac{9\eta R}{4NR(2\pi M k_B T)^{1/2}} \right] \tag{36}$$

where η is the gas viscosity and is the mass of the gas atoms. Finally, we obtain for the energy-dependent mobility to insert in Eq.(29)

$$\mu(\epsilon) = \left[1 - p(\epsilon) \right] \mu_b + p(\epsilon)\mu_f(\epsilon) \tag{37}$$

Originally EC used an uncertainty principle related sampling length $L = ch/\Delta p \equiv L_u$ with $\Delta p = [\langle p^2 \rangle / 3]^{1/2} = [(2m\epsilon)/3]^{1/2}$. c is an adjustable parameter of order unity. In Figure 10 we report the result of the EC model (lower solid line) compared to the experimental results. The EC model correctly shows a high-mobility region at low density and a low-mobility region at high density, and the two branches are joined smoothly. At low densities the agreement with the data is fairly good, better than in the original model developed for He[52], because we have introduced the multiple scattering correction due to correlations among scatterers through $S(0)$, which was not introduced in the original papers. Moreover, at low density $E_c \approx V \equiv E_k$ and the EC equations turn into Eqns.(3) and (4). Nonetheless, the transition between extended and localized states appears at too small densities, suggesting that fluctuations are overestimated and that too small a sampling length L has been chosen. Unfortunately, even increasing the adjustable parameter c the overall agreement cannot be improved very much[35].

This choice of the De Broglie wavelength of the electron as the density sampling length and the semiclassical counting of the states have given origin to strong criticism [33,51,61]. In the fluctuating density problem the potential strength depends on the sampling volume. If the potential fluctuations amplitude in the sampling box is too small and if also the cell side is small, bound states might not even exist. Therefore, the use of the density of states of a free particle in the box is not adequate and the states in the tail of the distribution, i.e., the localized ones, might be over-estimated. It has been suggested [51] that the correct sampling length should be the classical mean free path $L = \ell_c' = 1/N\sigma_m$, as a consequence of the fact that the electron is affected by the gas density only upon interaction. Moreover, it is suggested [51] that within a cell the intrinsic density fluctuations have to be neglected because the electron adiabatically averages over them. The physical picture is that within a cell the electron is delocalized for $\epsilon > V$ and that the global properties of a wavefunction spanning over several scatterers should not depend strongly on minor rearrangements of their positions [33]. This decoupling of the electron from intrinsic fluctuations is valid when calculating the density of states, but it does not hold true when calculating the electron mobility because the environment of the cells, hence $S(0)$, surely influences the transport. In fact, when electrons propagate from cell to cell correlations among cells do affect the electron motion.

Since in He the thermal electron wavelength at $T = 4.2\,K$ and the mean free path at $N \approx 10^{20}\,cm^{-3}$ are nearly equal, the validity of the previous suggestion cannot be ascertained. By contrast, in Ne they are very different and we can check accurately if the use of the classical free path as sampling length is correct. The upper solid line and the dotted line in Figure 10 have been calculated according to the EC but using a length related to the classical mean free path ℓ_c' as the sampling length L. For the dotted line we used $L = \ell_c'$ and for the upper solid line we set $L = 1.05\,\ell_c'$.

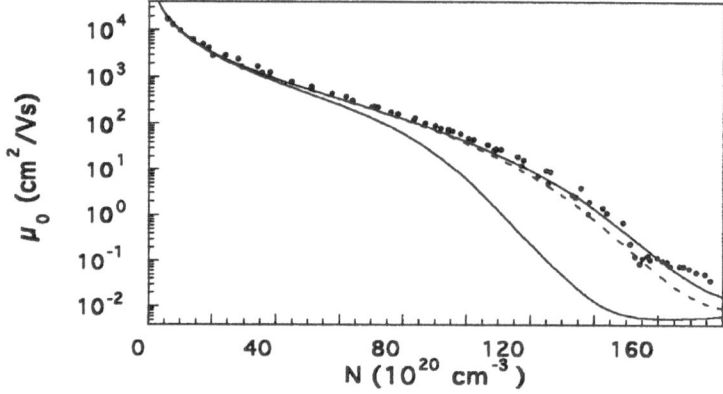

Figure 10. Comparison of the predictions of the percolation model with the experimental data [35]. Lower solid line: original EC model with $L=L_u$, $c=1$. Dotted line and upper solid line modified EC model with $L=\ell_c'$, and $L=1.05\ell_c'$.

We see from Figure 10 that the model is quite sensitive to the choice of L. In Ne at $T = 46.5\,K$ the classical mean free path is much larger than the electron de Broglie wavelength and L calculated according to the uncertainty principle is always smaller than ℓ_c' even at the largest densities. In Figure 11 we show the uncertainty principle related sampling length L_u and the classical mean free path ℓ_c' for $N = 150 \times 10^{20}\,cm^{-3}$ as a function of the electron energy. As mentioned, ℓ_c' is always larger than L_u. As a consequence the variance of the potential fluctuations is smaller using ℓ_c' rather than L_u, the transition region of the density of states from the high-energy extended states to the low-energy localized one is narrower, and the states in the tail are not overestimated[35]. With this choice $L = \ell_c'$ the EC model has been improved very much, as can be seen in Figure 10. Moreover, if one assumes that the sampling length should be only of the order of the classical mean free path, a further improvement can be obtained by taking $L = 1.05\ell_c'$. However, further increasing the sampling length does not produce any further improvement[35].

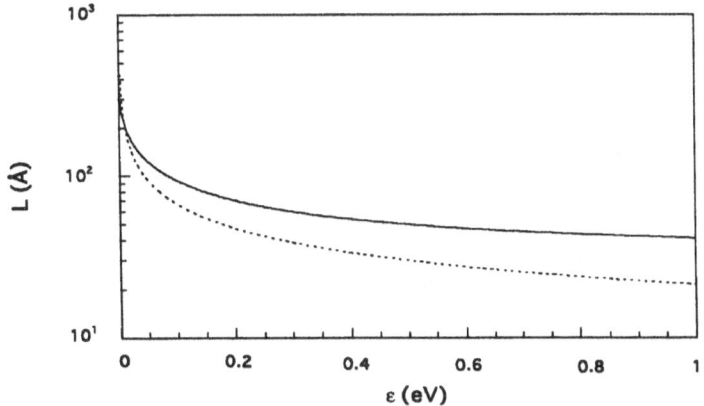

Figure 11. Sampling length for the EC model as a function of the electron energy. Solid line: classical mean free path at $N=150\times10^{20}$ cm^{-3}. Dotted line: uncertainty principle related sampling length L_u.

In Figure 12 we show the results of the EC model, modified so as to take into account multiple scattering effects as we have done here, for the He case[33]. Here, the thermal electron wavelength is comparable to the classical mean free path and the use of either of them as sampling length gives essentially the same result.

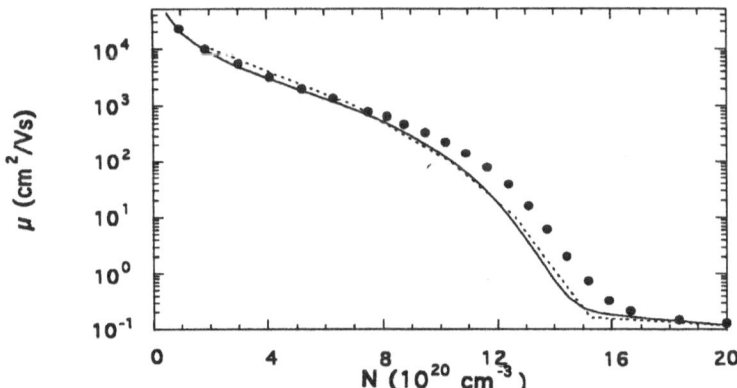

Figure 12. EC model results for He at $T=4.2$ K. Closed point: experimental data[33]. Dotted line: sampling length $L=L_u$. Solid line: $L=1.1\ell_c'$.

It is interesting to check now the arguments leading to the suggestions[33,51] that the sampling length should be $L = \ell_c'$ when calculating the density of states, and not $L = \ell_c = \ell_c'/S(0)$ as expected on the basis of the calculation of the mobility. In Figure 13 the prediction of the EC model for Ne are shown along with the experimental data. The dotted line is the prediction obtained using $L = 1.05\,\ell_c'$ as sampling length. The solid line is obtained, by contrast, using the classical mean free path reduced by correlations among scatterers as sampling length, $L = \ell_c'/S(0)$. By so doing, the fluctuations are given too much importance and the abundance of localized states is severely overestimated. As a consequence, the localization transition occurs at too small densities and is much sharper than experimentally observed. Thus, the assumption of neglecting fluctuations within the sampling cells in the calculation of the density of states seems appropriate.

As a concluding remark to this section we note that an adequate treatment of the mobility of the extended states according to multiple scattering theories must be adopted in order to achieve agreement with the experimental data at low density where localized states are of little importance. Furthermore, we stress the fact that the relationship between classical percolation threshold and quantum delocalization[55], that between Anderson-localized and self-trapped states, and even the dynamical aspects of localization are still open problems.

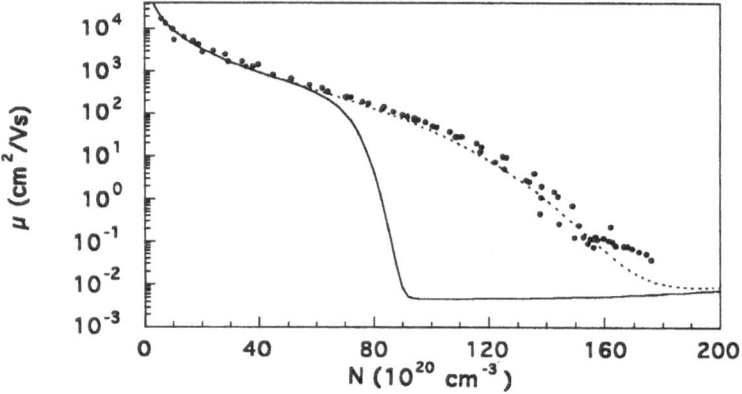

Figure 13. EC model results for Ne at $T=46.5\,K$. Points: experimental data [2,35]. Dotted line: sampling length $L=1.05\ell_c'$. Solid line: $L=\ell_c'/S(0)=\ell_c$.

The mesoscopic model

In this last section we give a brief outline of the refined formalism recently developed in order to give a better description of electron localization in Ne[16,36,37].

Coupling of a particle to its environment may give origin to a non-perturbative modification of the environment itself so as to form a new state completely different from the state in the decoupled system of particle plus environment. It also may happen that a particle in a fluid can self-consistently deform the fluid around itself so as to become self-trapped. In the previously mentioned bubble model, given the electron-fluid potential, the bubble is described as an electron trapped inside a not realistic spherical square well and the energy of formation of the bubble is minimized with respect to bubble radius and filling fraction. In the mesoscopic model[16,37] a model local-density functional approach coupled to Wigner-Seitz boundary conditions is used to obtain self-consistently the free energy and a more realistic density profile of the bubble by solving the Schroedinger equation for the ground-state pseudo-wavefunction of the electron.

The gas atoms are assumed to interact with each other by means of a Lennard-Jones pair potential[62]. The electron-atom interaction is described by means of a pseudopotential

$$V_{ps}(\mathbf{r}) = -\frac{2\pi\hbar^2}{m}\frac{\tan\left[\delta_0(k)\right]}{k}\delta(\mathbf{r}) \tag{38}$$

where δ_0 is the s-wave phaseshift and $\delta(\mathbf{r})$ is the Dirac delta. In order to treat the interaction with the many atoms of a fluid, it is assumed that the atomic potentials are non-overlapping and the electron-fluid potential is then given simply by summing V_{ps} over all atoms. By assuming that the fluid is on average ordered (we recall also that the Born-Oppenheimer approximation is valid), so that the potential is on average translationally invariant, the ground state energy of the electron in the fluid is given by the WS eigenvalue equation, Eq.(2).

Because of the presence of the electron potential, the density varies in space $N \to N(r)$. If the electron-fluid interaction is such that $N(r)$ varies slowly over distances large compared to the mean interatomic spacing, the electron potential energy can be calculated from the local density and the pseudopotential becomes then function of the WS energy, and hence of the local density

$$V_{ps}[N(\mathbf{r})] = -\frac{2\pi\hbar^2}{m}\frac{\tan\delta_0(k_0)}{k_0}N(\mathbf{r}) \tag{39}$$

with k_0 related to N by means of Eq.(2).

The energy E of the self-trapped electron is described by the usual quantum mechanical expectation value of the Hamiltonian

$$\mathcal{H} = -\frac{\hbar^2}{2m}\nabla^2 + V_{ps}[N(r)] - V_{ps}[N] \tag{40}$$

where N is the average fluid density. The zero of energy is chosen to be that of the quasifree electron. The fluid distortion energy is

$$\int d\mathbf{r} \left[N(\mathbf{r})g(\mathbf{r}) - P(\mathbf{r})\right] \tag{41}$$

where $g(\mathbf{r})$ and $P(\mathbf{r})$ are the local chemical potential and local gas pressure, respectively.

Assuming that the self-trapped state is spherically symmetric (a good approximation, since we are searching for the lowest lying level), the following one-dimensional radial s-wave equation is obtained from the Schroedinger equation

$$\left\{-\frac{\hbar^2}{2m}\frac{d^2}{dr^2} + \left(V_{ps}[N(r)] - V_{ps}[N]\right)\right\}r\Psi(r) = Er\Psi(r) \tag{42}$$

where E is the energy eigenvalue. The free energy of bubble formation is then given by

$$\Delta F = E + \int \left\{N(\mathbf{r})[g(\mathbf{r}) - g] - [P(\mathbf{r}) - P]\right\}d\mathbf{r} \tag{43}$$

where $g(\mathbf{r})$, $N(\mathbf{r})$, and $P(\mathbf{r})$ are the local chemical potential, density, and pressure respectively, and g, N, and P are their average values. Upon extremizing Eqns.(42) and (43) under the local-density approximation, one obtains the two coupled equations for the local density and the electron wave function

$$N(r)\frac{\delta g[N(r)]}{\delta N(r)} - \frac{\delta P[N(r)]}{\delta N(r)} = 0 \tag{44}$$

and

$$\{g[N(r)] - g\} + \frac{\delta V_{ps}[N(r)]}{\delta N(r)}\frac{|\Psi(r)|^2}{\langle\Psi|\Psi\rangle} = 0 \tag{45}$$

where δ means functional differentiation. Equations (42) and (45) must be solved self-consistently. This goal has been accomplished for Ne[37] and the results for the mobility are shown in Figure 14. The agreement of the mesoscopic model predictions, indicated by the thick solid line, with the experimental data is better than that of the simple partially filled bubble. This model predicts correctly the threshold density N^* above which localized states begin to be stable, as indicated by the solid arrow in the Figure. Moreover, this model agrees with the data where the experimental data are reduced to 1/2 of the quasifree electron mobility (hollow arrow in the Figure). Since the bubble states are much less mobile than the quasifree ones $\mu_b \ll \mu_f$, then $\mu = (\mu_f n_f + \mu_b n_b)/(n_f + n_b) \approx \mu_f[n_f/(n_f + n_b)]$. When $n_f = n_b$ then $\mu = \mu_f/2$. This means that the model correctly predicts the density at which $n_f = n_b$, signalled by the hollow arrow. The disagreement at still larger densities is attributed to the ignorance of the mobility of the bubble state[63].

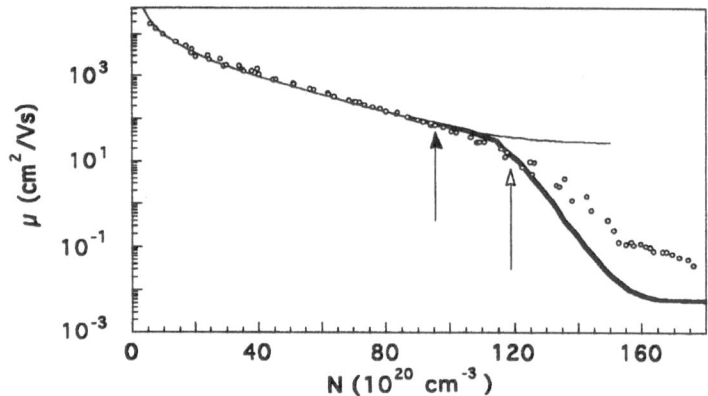

Figure 14. Results of the mesoscopic model for Ne at $T=46.5\,K$. Points: experimental data [2,35]. Thin solid line: quasifree electron mobility calculated according Eq.(3) and (4). Thick solid line: Mesoscopic model calculations [37]. The meaning of the arrows is explained in the text.

CONCLUSIONS

Self-trapping of excess electrons in high-density Neon has confirmed the predictions that localized electron states may exist in disordered media, whose interaction with excess electrons is repulsive enough, and that are at the same time compliant enough in order that the distortions of the material itself do not cost too much energy. Fairly simple models, such as the bubble model, are able to qualitatively describe the basic physics involved in localization and allow well-defined predictions about some interesting quantities such as, for instance, the temperature dependence of the localization-delocalization transition density.

More refined models, such as those aiming at the determination of the density of states of excess electrons in a non-polar medium, are in far better agreement with the experiments, probably because they implicitly account for the configurational entropy contributions.

We believe that further experiments in the search for localization at higher temperatures in He, or in the search for the existence of electron localization in mixture of positive- and negative scattering-length gases, such as He/Ar mixtures, should allow to reach a deeper understanding of the localization process.

REFERENCES

1. A.F.Borghesani and M.Santini, to be published.
2. A.F.Borghesani and M.Santini, Electron Mobility and Localization Effects in High-Density Neon Gas, *Phys.Rev.* A 42:7377 (1990).

3. A.F.Borghesani, M.Santini, and P.Lamp, to be published.

4. L.G.Huxley and R.W.Crompton, *The Diffusion And Drift Of Electrons In Gases*, Wiley, New York (1974) .

5. N.E.Cipollini, R.A.Holroyd, and M.Nishikawa, Zero-Field Mobility of Excess Electrons in Dense Methane, *J.Chem.Phys.* **67**:4636 (1977).

6. B.E.Springett, M.H.Cohen, and J.Jortner, Properties of an Excess Electron in Liquid Helium: The Effect of Pressure on the Properties of the Negative Ion, *Phys.Rev.* **159**:183 (1967).

7. L.Meyer and F.Reif, Mobilities of He Ions in Liquid Helium, *Phys.Rev.* **110**:279 (1958).

8. G.Careri, F.Scaramuzzi, and J.O.Thompson, Heat Flush and Mobility of Electric Charges in Liquid Helium, *Il Nuovo Cimento* **13**:186 (1959).

9. L.Bruschi, G.Mazzi, and M.Santini, Localized Electrons in Liquid Neon, *Phys.-Rev.Lett.* **28**:1504 (1972).

10. R.J.Loveland, P.G.LeComber, and E.W.Spear, Experimental Evidence for Electronic Bubble States in Liquid Ne, *Phys.Lett.* **39 A**:225 (1972).

11. J.L.Levine and T.M.Sanders, Jr., Anomalous Electron Mobility and Complex Negative Ion Formation in Low-Temperature Helium Vapor, *Phys.Rev.Lett.* **8**:159 (1962) and Mobility of Electrons in Low-Temperature Helium Gas, *Phys.Rev.* **154**:138 (1967).

12. H.R.Harrison and B.E.Springett, Electron Mobility Variation in dense Hydrogen Gas,*Chem.Phys.Lett.* **10**:418 (1971) and Electron Mobility Variation in Dense ^{4}He Gas, *Phys.Lett.* **35 A**:73 (1971).

13. H.R.Harrison, L.M.Sander, and B.E.Springett, Electron Mobility and Localization in Dense He^4 Gas,*J.Phys.* **B 6**:908 (1973).

14. J.A.Jahnke, M.Silver, and J.P.Hernandez, Mobility of Excess Electrons and O_2^- Formation in Dense Fluid Helium, *Phys.Rev.* **B 12**:3420 (1975).

15. G.R.Freeman and D.A.Armstrong, Electron and Ion Mobilities, in: *Advances of Molecular Physics*, vol. **20**:267, D.Bates and B.Bederson Eds., Academic Press, Orlando (1985).

16. J.P.Hernandez, Electron Self-Trapping in Liquids and Dense Gases, *Rev.Mod.-Phys.* **63**:675 (1991).

17. P.W.Anderson, Absence of Diffusion in Certain Random Lattices, *Phys.Rev.* **109**:1492 (1958).

18. J.Jortner, N.R.Kestner, S.A.Rice, and M.H.Cohen, Study of the Properties of an Excess Electron in Liquid Helium. I. The Nature of the Electron-Atom Interactions, *J.Chem.Phys.* **43**:2614 (1965).

19. B.E.Springett, J.Jortner, and M.H.Cohen, Stability Criterion for the Localization of an Excess Electron in a Nonpolar Fluid, *J.Chem.Phys.* **48**:2720 (1968).

20. T.F.O'Malley, Extrapolation of Electron-Rare Gas Atom Cross Section to Zero Energy, *Phys.Rev.* **130**:1020 (1963).

21. H.Schnyders, S.A.Rice, and L.Meyer, Electron Mobilities in Liquid Argon, *Phys.Rev.Lett.* **15**:187 (1965).

22. B.Halpern, J.Lekner, S.A.Rice, and R.Gomer, Drift Velocity and Energy of Electrons in Liquid Argon, *Phys.Rev.* **156**:351 (1967).

23. T.F.O'Malley and R.W.Crompton, Electron-Neon Scattering Length and S-Wave Phaseshifts from Drift Velocities, *J.Phys.* **B 13**:3451 (1980).

24. Y.Sakai, W.F.Schmidt, and A.Khrapak, High- and Low-Mobility Electrons in Liquid Neon, *Chem.Phys.* **164**:139 (1992).

25. T.Miyakawa and D.L.Dexter, Stability of Electronic Bubbles in Liquid Neon and Hydrogen, *Phys.Rev.* **184**:166 (1969).

26. V.J.Johnson, *A Compendium of the Properties of Materials at Low Temperature (Phase 1) Part I. Properties of Fluids*, National Technical Information Service, U.S.Department of Commerce (1960).

27. S.R.Hunter and L.G.Christophorou, Interphase Physics: Linking Knowledge on Electron-Molecule Interactions in Gases to Knowledge on Such Processes

in Condensed Matter, in: *Electron-Molecule Interactions And Their Applications*, vol.2:221, L.G.Christophorou and K.Siomos Eds., Academic Press, Orlando (1984).

28. A.F.Borghesani and M.Santini, Electron Localization Effects and Resonant Attachment to O_2 Impurities in Highly Compressed Neon Gas, in *Gaseous Dielectrics VI*, L.G.Christophorou and I.Sauers Eds., Plenum, New York (1991), p.27.

29. A.F.Borghesani, L.Bruschi, M.Santini, and G.Torzo, Simple Photoelectronic Source for Swarm Experiments in High-Density Gases, *Rev.Sci.Instrum.* **57**:2234 (1986).

30. A.F.Borghesani and M.Santini, Electron Swarm Experiments in Fluids–Signal Waveform Analysis, *Meas.Sci.Technol.* **1**:939 (1990).

31. A.F.Borghesani, D.Neri, and M.Santini, Low-Temperature O_2^- Mobility in High-Density Neon Gas, *Phys.Rev.* **E 48**:1379 (1993) and D.Neri, Mobilità di Ioni Ossigeno in Gas Neon Denso, *Tesi di Laurea*, University of Padua (1991), unpublished.

32. A.F.Borghesani, L.Bruschi, M.Santini, and G.Torzo, Electron Mobility in Neon at High Densities, *Phys.Rev.* **A 37**:4828 (1988).

33. K.W.Schwarz, Electron Localization in Dense Helium Gas: New Experimental Results, *Phys Rev.* **B 21**:5125 (1980).

34. J.P.Hernandez, Electron Drift in Gaseous He: Density, Temperature, and Field Dependences, *Phys.Rev.* **A 5**:635 (1972).

35. A.F.Borghesani and M.Santini, Electron Localization-Delocalization Transition in High-Density Neon Gas, *Phys.Rev.* **A 45**:8803 (1992).

36. J.P.Hernandez and L.W.Martin, Analysis of Excess Electron States in Neon Gas, *Phys.Rev.* **A 43**:4568 (1991).

37. L.W.Martin, Self-Trapping of Electrons in Fluid Neon, Ph.D. Thesis, Univesity of North Carolina (1991), unpublished.

38. A.F.Borghesani, M.Santini and P.Lamp, Excess Electron Mobility in High-Density Argon Gas, *Phys.Rev.* **A 46**:7902 (1992).

39. E.Fermi, Sopra lo Spostamento per Pressione delle Righe Elevate delle Serie Spettrali, *Il Nuovo Cimento* **11**:157 (1934).

40. M.Lax, Multiple Scattering of Waves, *Rev.Mod Phys.* **23**:287 (1951).

41. L.L.Foldy, The Multiple Scattering of Waves, *Phys.Rev.* **67**:107 (1945).

42. G.L.Braglia and V.Dallacasa, Theory of Electron Mobility in Dense Gases, *Phys.Rev.* **A 26**:902 (1982).

43. T.F.O'Malley, Multiple Scattering Effect on Electron Mobilities in Dense Gases, *J.Phys.* **B 13**:1491 (1980).

44. V.M.Atrazhev and I.T.Iakubov, The Electron Drift Velocity in Dense Gases, *J.Phys.* **D 10**:2155 (1977).

45. J.Lekner, Scattering of Waves by an Ensemble of Fluctuationg Potentials, *Philos.Mag.* **18**:1281 (1968).

46. E.Wigner and F.Seitz, On the Costitution of Metallic Sodium, *Phys.Rev.* **43**:804 (1933).

47. R.D.McCarty and R.B.Stewart, Thermodynamic Properties of Neon from 25 to 300 K between 0.1 and 200 Atmospheres, Natl. Bur. Stand. (U.S.) Report No.8726, U.S.GPO, Washington (1965).

48. V.A.Rabinovich, A.A.Vasserman, V.I.Nedostup, and L.S.Veksler, *Termophysical Properties Of Neon, Argon, And Xenon*, Hemisphere, New York (1988).

49. K.W.Schwarz and B.Prasad, Dynamics of Electron Localization in Dense Helium Gas, *Phys.Rev.Lett.* **36**:878 (1976).

50. D.G.Onn and M.Silver, Attenuation and Lifetime of Hot Electrons Injected into Liquid Helium, *Phys.Rev.* **183**:295 (1969) and Injection and Thermalization of Hot Electrons in Solid, Liquid, and Gaseous Helium at Low Temperatures, *Phys.Rev.* **A 3**:1773 (1971).

51. J.P.Hernandez, Self-Trapped States of an Electron in a Structurally Disordered System, *Phys.Rev.* **A 7**:1755 (1973) and Self-Trapped States of Electrons in Dense Fluids, *Phys.Rev.* **B 11**:1289 (1975).

52. T.P.Eggarter and M.H.Cohen, Simple Model for Density of States and Mobility of an Electron in a Gas of Hard-Core Scatterers, *Phys.Rev.Lett.* **25**:807 (1970).

53. T.P.Eggarter and M.H.Cohen, Mobility of Excess Electrons in Gaseous He: A Semiclassical Approach, *Phys.Rev.Lett.* **27**:129 (1971).

54. T.P.Eggarter, Semiclassical Theory of Electron Transport Properties in a Disordered Material, *Phys.Rev.* **A 5**:2496 (1972).

55. S.H.Simon, V.Dobrosavljievic, and R.M.Stratt, Semiclassical Percolation Approach to Electronic States in Simple Fluids, *Phys.Rev.* **A 42**:6278 (1990).

56. J.M.Ziman, *Models Of Disorder*, Cambridge University Press, Cambridge (1979).

57. R.Zallen and H.Scher, Percolation on a Continuum and the Localization-Delocalization Transition in Amorphous Semiconductors, *Phys.Rev.* **B 4**:4471 (1971).

58. N.W.Ashcroft and N.D.Mermin, *Solid State Physics*, Saunders College, Philadelphia (1976).

59. G.H.Wannier, *Statistical Physics*, Dover, New York (1966).

60. E.Cunningham, *Proc.Royal Soc.London*, Ser. **A 83**:357 (1910) and A.M.Tyndall, *The Mobility Of Positive Ions In Gases*, Cambridge University Press, Cambridge (1938).

61. J.P.Hernandez and J.M.Ziman, Electron Localizations by Density Fluctuations, *J.Phys.C* **6**:L251 (1973) and T.P.Eggarter and M.H.Cohen, Comment on 'Electron Localization by Density fluctuations', *J.Phys.* **C 7**:L103 (1974).

62. B.N.Miller and T.Reese, Self-Trapping of a Light Particle in a Dense Fluid: a Mesoscopic Model, *Phys.Rev.* **A 39**:4735 (1989).

63. J.P.Hernandez, private communication.

BOLTZMANN EQUATION FOR SLOW ELECTRON TRANSPORT IN GASES AND LIQUIDS

Yosuke Sakai

Department of Electrical Engineering, Hokkaido University,
Sapporo 060 Japan

INTRODUCTION

The transport properties of slow electrons in liquefied rare gases have not necessarily been understood well, although these liquids represent the most simple dense matters and their concentration can be easily changed in a relatively wide range depending on their temperature. Measurements until now have shown the electron drift velocity over a wide range of field strengths in Ar, Kr, and Xe liquids (e.g.Miller et al.,1968). However, in their characteristics, still a several points remain not explained well, such as, the marked saturation in the drift velocity vs field curves in the high field region (Miller et al.,1968;Yoshino et al.,1976), a maximum in the zero field electron mobility at a liquid temperature just below the critical point (Jahnke et al.,1971), behavior of the drift velocity in liquid mixtures (Sakai et al.,1993), and so on. These points have been tried to explain introducing the effective mass of an electron, a deformation potential model on the bottom of the conduction band and the structure of a matter (Nishikawa 1985), an inelastic scattering model for electrons with scattering centers produced by a polarized modulation associated with changes in translational states of pairs, triplets, etc., of atoms during collision encounters (Sakai et al.,1985; Nakamura et al.,1986), and so on.

On the other hand, in gaseous phase, as far as data of the various collision cross sections are known, the electron transport properties can be analyzed using a Boltzmann equation method in a wide range of E/N values (e.g. Sawada et al.,1989). Here, E is the field strength and N is the particle concentration of a matter.

In this article, first, a Boltzmann equation for slow electrons in external fields in simple (rare gas) liquids is presented, and the differences of the equation from the gases, especially with respect to scattering terms, are discussed. Using this Boltzmann equation the transport properties for slow electrons are analyzed by taking account of an inelastic scattering mode for hot electrons. The presence of this inelastic mode is discussed on the basis of the inelastic light scattering modes which have been predicted and observed in rare gas liquids (MaTague,1969;Fleury,1973). The electron transport properties in the liquids are compared with those in gases. It is shown that the difficulty in accounting for the saturation behavior of the drift velocity in the high field region can be solved essentially in the framework of a Boltzmann equation modified by Cohen and Lekner (1967) by newly introducing the inelastic scattering.

SCATTERING POTENTIAL

An electron at a distance R from an isolated atom of polarizability α induces on it a dipole of strength $\alpha e/R^2$, which in turn attracts the electron with a force of magnitude $2\alpha e^2/R^5$ (see figure 1). The interaction potential is therefore $-\alpha e^2/R^4$. Actually, the electron feels the averaged local potential as $-\alpha e^2 f(R)/R^4$ due to a multi-interaction. $f(R)$ is the factor which shows the spatial distribution of the atoms related to the liquid structure. Physically it is clear that the polarization force does not increase without limit as R tends to zero, and it must therefore be modified. This is modified, where the effects of both exchange and correlation are approximated at all distances of the following potential (Lekner 1967), as

$$U_\alpha(R) = -0.5\alpha \frac{e^2 f(R)}{(R^2 + R_\alpha^2)^2}. \tag{1}$$

Here, R_α is the adjusting parameter so as to obtain an experimental value of the scattering length (Frost and Phelps 1964). At small R this potential represents mainly exchange, since the polarization part goes to zero. It is finding the local field acting on the atom, which consists of the direct field and the sum of all other fields due to dipoles induced on neighboring atoms. In the case of external fields, the induced dipoles are parallel in regions containing many atoms.

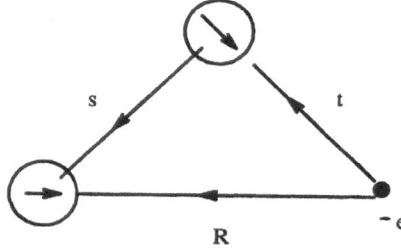

Figure 1. Local field due to dipolarization. The dipolarized field at R which influences the electron is the sum of the electron induced dipole and dipole induced dipole at any point of t.

Consequently, the single-atom potential seen by an electron in the liquid U_1 is the Coulomb field of the nucleus screened by the Hartree potential U_H, plus the screened exchange and polarization potential U_α, as $U_1 = U_H + U_\alpha$. But, as the result that the potential is superimposed together, the electron feels no force in a uniform potential U_0 region. As for the effective potential U_{eff} it is necessary to be subtracted out the U_0 from U_1, as

$$U_{eff} = U_1 - U_0. \tag{2}$$

From U_{eff} the effective scattering cross section $\sigma_{eff}(\epsilon, \theta)$ could be obtained in principle in the framework of a single scattering model (Lekner 1967).

STRUCTURE OF LIQUID

The statistical description of the structure and dynamics of liquids, in comparison with the solids and gases, is relatively incomplete. The fundamental difficulty is that the liquid state lacks any idealized abstraction such as the ideal gas or perfect solid, which can form a basis for theoretical refinement. Unlike the solid or the gas where either configuration or kinetic processes dominate the description, in liquid state physics we are confronted with the full and general statistical problem in which there are both dynamical and configurational contributions to the total energy.

In simple liquids considered now, mutual configuration of liquid particles moves continuously, so the position r and momentum p of each atom should be considered to represent the configuration perfectly. When, if a particle with the momentum p_1 at a time t_1 stays at the position r_1, the probability of the particle which is found with p_2 at t_2 and r_2 may be written as $f(r_1, p_1, t_1; r_2, p_2, t_2)$, the pair-correlation function $G(r = r_1 - r_2, \tau = t_1 - t_2)$ is defined (Engelstaff 1967) and is written as

$$G(r, \tau) = \int \int f(r_1, p_1, t_1, r_2, p_2, t_2) dp_1 dp_2. \tag{3}$$

Then, the spectral function $S(K, \omega)$ is expected to be the Fourier transform of $g(r) = G(r, \tau) - \rho$, over r and τ, where ρ is the average concentration,

$$S(K, \omega) = \frac{1}{2\pi} \int \int \exp\{j(K, r - \omega\tau)\} g(r) dr d\tau. \tag{4}$$

If equations (3) and (4) are averaged over in τ and ω respectively, the pair correlation function $g(r)$ and structure factor $S(K)$ are obtained. As an example, $g(r)$ for liquid Ar is shown in figure 2 (Engelstaff 1967). This indicates that the liquid particle feels strongly the structure around itself. Figure 3 shows $S(K)$ for liquid Ar (Lekner 1967) along with the liquid Kr and Xe(Ashcroft,1967). This $S(K)$ indicates that the scattering of an electron with a small wave vector change (momentum change) is rare, but that the structure near $K = 2A^{-1}$ appears corresponding to the point B ($r = 4A$) in $g(r)$.

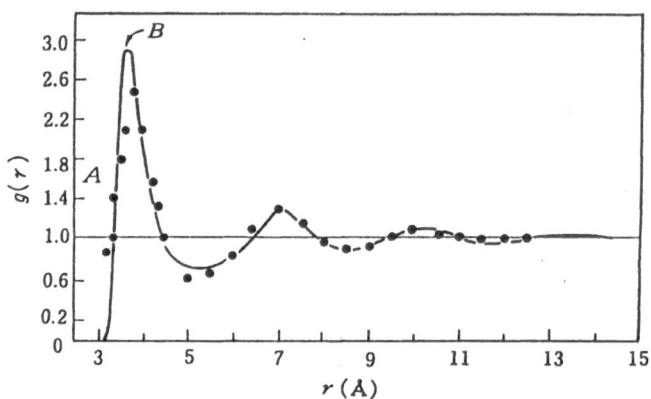

Figure 2. Pair correlation function $g(r)$ of liquid Ar at 84K. Closed circle; neutron diffraction measurement, solid line; calculation from the interatomic potential. A region; repulsive potential, B region; minimum point of the interatomic potential.

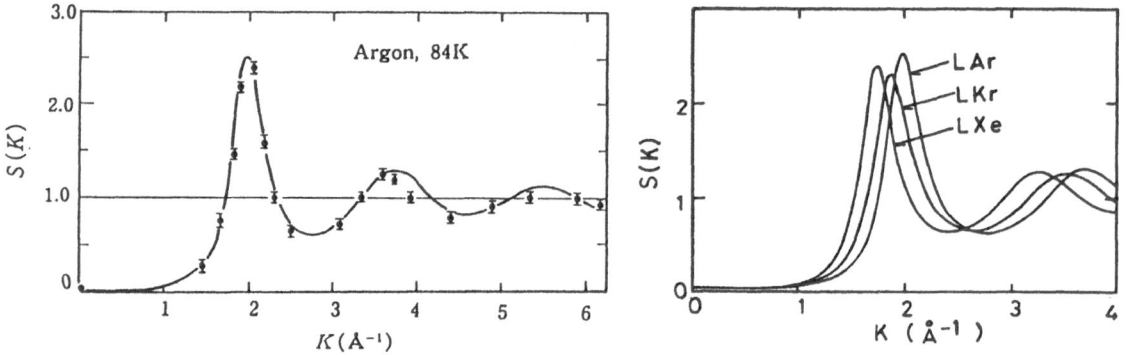

Figure 3. Structure factor $S(K)$ of liquid Ar, liquid Kr and liquid Xe. Values of $S(0)$ are 0.048 for liquid Ar at 84K, 0.051 for liquid Kr at 116K and 0.054 for liquid Xe at 163K (Atrazhev and Iakubov,1981). Closed circle; neutron diffraction measurement, solid line; calculation from the interatomic potential (Lekner,1967).

$S(\boldsymbol{K}, \omega)$ has the following relations (Cohen and lekner,1967). The first is the detailed balance condition, as

$$S(-\boldsymbol{K}, -\omega) = \exp\left(-\frac{\omega}{kT}\right) S(\boldsymbol{K}, \omega), \tag{5}$$

where k is the Boltzmann constant and T is the liquid temperature. Secondly,

$$\int S(\boldsymbol{K}, \omega)d\omega = S(\boldsymbol{K}), \tag{6}$$

$$\int \omega S(\boldsymbol{K}, \omega)d\omega = \frac{K^2}{2M}, \tag{7}$$

$$\int \omega^2 S(\boldsymbol{K}, \omega)d\omega \simeq 2kT\langle\omega\rangle. \tag{8}$$

Here $K = |\boldsymbol{K}|$ and M is the mass of a liquid particle.

SCATTERING CROSS SECTION IN LIQUID

Van Hove(1954) gave a differential scattering cross section $\sigma(\epsilon, \theta)$ for a slow electron with an incident energy of ϵ, which scatters to the direction θ with the momentum change $\boldsymbol{K}$ and energy change ω, as

$$\frac{d\sigma(\epsilon, \theta)}{d\omega d\epsilon} = \sigma_{eff}(\epsilon, \theta)S(\boldsymbol{K}, \omega) \tag{9}$$

Here Ω is the solid angle for a scatter to the direction of θ. $\sigma_{eff}(\epsilon, \theta)$ is derived in principle from U_{eff}. This model may be valid especially when L (the mean distance of the liquid particles) $\leq \lambda_d$ (de Broglie wave lengthof an electron) $\leq \Lambda$ (effective electron scattering mean free path). Usually for the slow electrons in the liquids this relation is satisfied.

If this cross section is adopted in a Boltzmann equation analysis, it is noted that two kinds of the collision cross sections for energy transfer $\sigma_0(\epsilon)$ and momentum transfer $\sigma_1(\epsilon)$ are defined, as

$$\sigma_0(\epsilon) = 2\pi \int_0^\pi \sin\theta\sigma_{eff}(\epsilon, \theta)(1 - \cos\theta)d\theta, \tag{10}$$

$$\sigma_1(\epsilon) = 2\pi \int_0^\pi \sin\theta\sigma_{eff}(\epsilon, \theta)(1 - \cos\theta)S(K_0)d\theta, \tag{11}$$

Here, K_0 is the momentum change in a perfectly elastic collision, i.e. $\omega = 0$, and is $\sqrt{8m\epsilon}\sin(\theta/2)$, where m is the mass of an electron.

The cross sections $\sigma_0(\epsilon), \sigma_1(\epsilon), \sigma_{in}(\epsilon)$ for the inelastic collision and Q_{io} for ionization adopted in this study (Sakai et al.,1985; Nakamura et al.,1986) are shown in figures 4 (a)-(c) along with the momentum transfer cross section $Q_m(\epsilon)$ in gases (Laborie 1968) for Ar, Kr and Xe respectively. Values of $\sigma_0(\epsilon)$ in a low energy region is adjusted so that the present drift velocity agrees with measurement values (Miller et al.,1968). In figure 4(a) experimentally estimated values of $\sigma_1(\epsilon)$ (Shibamura et al.,1979) are shown to agree with the present values.

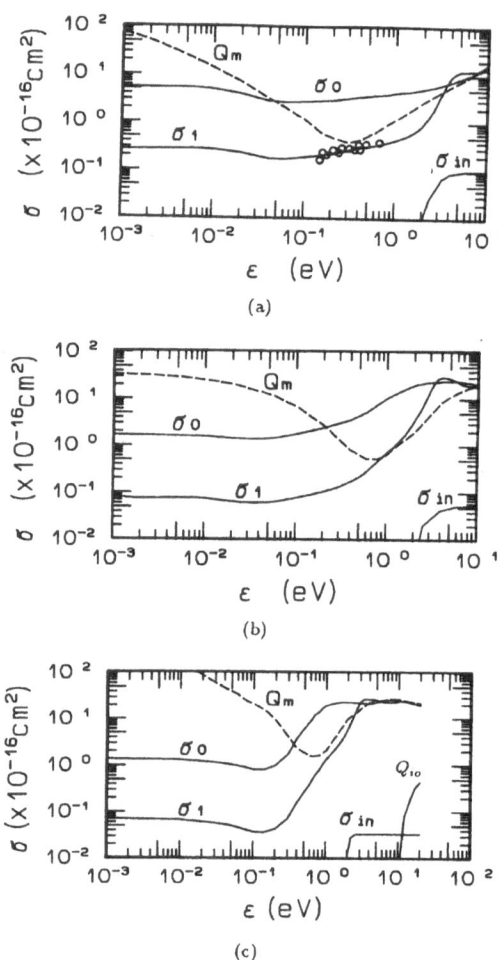

Figure 4. Electron collision cross sections $Q_m, \sigma_o, \sigma_1, \sigma_{in}$, and Q_{io}. (a) liquid Ar, (b) liquid Kr, and (c) liquid Xe. Q_m;momentum scattering cross section in gaseous phase, Q_{io} is the effective ionization cross section in liquid Xe. o;measured values of Shibamura et al(1979).

BOLTZMANN EQUATION METHOD

In Condensed Phase

At a time of t, the number of electrons at the position between r and $r + dr$, with the momentum between p and $p + dp$ is represented as $f(r, p, t)drdp$, then a Boltzmann equation is given as

$$\frac{\partial f(r, p, t)}{\partial t} + p \cdot \nabla_r f(r, p, t) + eE \cdot \nabla_p f(r, p, t) = \left\{ \frac{\partial f(r, p, t)}{\partial t} \right\}_{coll}. \qquad (12)$$

By symmetry, the steady-state momentum distribution function $f(p)(= \int \int f(r, p, t dr dt)$ of electrons moving in a uniform electric field is a function of $p_2(= \epsilon)$ and of the angle ξ between the field E and momentum P. We expand in Legendre polynomials, as

$$f(p) = f_0(\epsilon)P_0(\cos \xi) + f_1(\epsilon)P_1(\cos \xi) + f_2(\epsilon)P_2(\cos \xi) + \cdots. \qquad (13)$$

In the steady-state (t-independent) homogeneous (r-independent) electron flow, the first and second terms in the left-hand side of equation (12) become zero. Taking an approximation of neglecting $f_2, f_3 \cdots$, the third term becomes as

$$eE \frac{\partial f(p)}{\partial p} dp = eE(\frac{p}{m})d^3p \left\{ P_0 \frac{df_1/d\epsilon + f_1/\epsilon}{3} + P_1(\cos \xi)(df_0/d\epsilon + \cdots) \right.$$
$$\left. + P_2(\cos \xi) \frac{(2df_1/d\epsilon - f_1/\epsilon)}{3} \right\} + \cdots. \qquad (14)$$

This loss is compensated by the net gain in the number of electrons entering and leaving the element d^3p because of collisions. Cohen and Lekner(1967) gave the collision term as follows. The number scattered from d^3p into $d^3p' = p'^2dp'd\Omega'$ in unit time is

$$Nd\Omega' \int d\omega \left(\frac{p'}{m} \right) \sigma \left(\frac{\epsilon - \omega}{2}, \theta \right) S(K, \omega)f(p)d^3p. \qquad (15)$$

Here N is the concentration of liquid particles, p'/m is the electron velocity after a collision $(\sqrt{2\epsilon'}/m)$. The momentum transfer K and energy transfer ω are given by $K = p - p'$, $\omega = \epsilon - \epsilon'$. Similarly, the number of electrons scattered into d^3p from d^3p' is

$$Nd\Omega \int d\omega \left(\frac{p}{m} \right) \sigma \left(\frac{\epsilon - \omega}{2}, \theta \right) S(-K, -\omega)f(p')d^3p'. \qquad (16)$$

Here, $(\epsilon - \omega)/2$ is the average of the initial and final energies. Considering the momentum and energy conservation during a collision,

$$d^3p' = \frac{p'd\Omega'}{pd\Omega}d^3p. \qquad (17)$$

Using equations (15) - (17), the net number of electrons scattered into d^3p per unit time is, integrating over $d\Omega'$,

$$N \int d\Omega' \int d\omega \left(\frac{p'}{m}\right) \sigma \left(\frac{\epsilon - \omega}{2}, \theta\right) \{S(-\mathbf{K}, -\omega)f(\mathbf{p}') - S(\mathbf{K}, \omega)f(\mathbf{p})\} d^3p. \qquad (18)$$

Combining equation (14) and (18) with equation (5), equation (12) becomes as

$$e\mathbf{E}\frac{\partial f(\mathbf{p})}{\partial \mathbf{p}} = N \int d\Omega' \int d\omega \left(\frac{p'}{m}\right) \sigma \left(\frac{\epsilon - \omega}{2}, \theta\right) S(\mathbf{K}, \omega) \left\{\exp\left(-\frac{\omega}{kT}\right) f(\mathbf{p}') - f(\mathbf{p})\right\}. \qquad (19)$$

The left-hand side in equation (19) has already been given in equation (14). To evaluate the right-hand side, one should use equation (13) and an expansion of $S(\mathbf{K}, \omega)$ as

$$S(\mathbf{K}, \omega) = S(\mathbf{K}_0, \omega) - \frac{\omega}{4\epsilon} \mathbf{K}_0 \cdot \frac{\partial}{\partial \mathbf{K}_0} S(\mathbf{K}_0, \omega). \qquad (20)$$

Then, the equation belonging to $P_0(\cos\theta)$ in equation (19) is ,

$$\frac{eE}{3}\sqrt{\frac{2}{m}}\frac{\partial}{\partial \epsilon}\{\epsilon f_1(\epsilon)\} =$$

$$\sqrt{\frac{2}{m}}\frac{2m}{M}\frac{\partial}{\partial \epsilon}\left[\epsilon^2\left\{f_0(\epsilon) + kT\frac{\partial f_0(\epsilon)}{\partial \epsilon}\right\} N \int_0^\pi d\Omega'(1 - \cos\theta)\sigma_{eff}(\epsilon, \theta)\right] + X_{in}, \qquad (21)$$

where X $_{in}$ indicates the inelastic term for electron scattering and is written as,

$$X_{in} = \frac{\partial}{\partial \epsilon}\int_\epsilon^{\epsilon + \epsilon_{in}} \epsilon' f_0(\epsilon') N\sigma_{in}(\epsilon') d\epsilon'. \qquad (22)$$

The equation belonging to $P_1(\cos\theta)$ is,

$$\epsilon E\frac{\partial f_0(\epsilon)}{\partial \epsilon} = -N f_1(\epsilon)\int_0^\pi d\Omega'(1 - \cos\theta)\sigma_{eff}(\epsilon, \theta)S(K_0), \qquad (23)$$

f_1 in equation (23) is put into equation (21), then

$$-\frac{e^2 E^2}{3}\frac{\partial}{\partial \epsilon}\left\{\frac{\epsilon}{N\sigma_1(\epsilon)}\frac{\partial f_0(\epsilon)}{\partial \epsilon}\right\} =$$

$$\frac{2m}{M}\frac{\partial}{\partial \epsilon}\left[\epsilon^2 N\sigma_0(\epsilon)\left\{f_0(\epsilon) + kT\frac{\partial f_0(\epsilon)}{\partial \epsilon}\right\}\right] + X_{in}. \qquad (24)$$

$f_0(\epsilon)$ is normalized as $1 = \int \sqrt{\epsilon} f_0(\epsilon) d\epsilon$. Here, σ_0 and σ_1 are defined in equations (10) and (11). It may be valid to take into account the inelastic process for electron scattering, since the inelastic mode for light scattering due to van der Waals dimers, trimers, etc. in the liquid have been observed (McTague 1969).

It is clear that only the satisfactory way to obtain $f_0(\epsilon)$ in equation (24) is by a numerical technique. The Boltzmann equation for electrons in gases has been solved by a method known as "backward prolongation" using a digital computer (Sawada et al.,1989), and the same method is adopted here.

The validity of the solution is confirmed by calculating the value of the left-hand side of the following energy-balance equation,

$$\frac{\phi + \epsilon_{in} R_{in}}{eEW} = 1, \tag{25}$$

where,

$$\phi = \frac{\sqrt{2m}}{M} \int_0^\infty \sigma_0(\epsilon) \left\{ f_0(\epsilon) + kT \frac{\partial f_0(\epsilon)}{\partial \epsilon} \right\} d\epsilon,$$

and checking if the value agrees with unity. By choosing an appropriate value of the energy difference between the nearest mesh points and the maximum energy from which the computation starts towards zero energy, it is found to be possible to balance equation (25) to within 1 part in 10^3.

In Gas Phase

Naturally, even if the concentration decreases to gaseous phase, this Boltzmann equation is valid in placing $S(K) \to 1$, and $\sigma_0(\epsilon)$ and $\sigma_1(\epsilon) \to \sigma_m(\epsilon)$. Concerning linking zone between gasous and liquid states, the electron transport properties are given in principle using this theory if electrons are quasi-free, but unfortunately both values of $S(K)$ and $\sigma(\epsilon)$ are hardly available now.

ELECTRON TRANSPORT PROPERTIES

Drift Velocity

Figure 5 shows the electron drift velocity W and mean energy $\bar{\epsilon}$ obtained in the present study comparing with previous theoretical values of Cohen and Lekner(1967) and available experimental values of Miller et al(1968) and Yoshino et al(1976). This figure clearly shows that the traditional theory of Cohen and Lekner(1967) could not explain the saturation in the drift velocity in high fields. This may be because they did not consider any inelastic modes for electron scattering, although the electrons are hot. The present result shows that it is possible to find excellent agreement between the theoretical and experimental values by assigning the inelastic process effectively. In liquid Xe for $E/N \geq 10^{-17}$ Vcm^{-2} W is shown to increase again due to the strong cooling of the electrons by appearance of ionization collisions (see figure 7).

In gaseous phase the shoulders in W vs E/N curves for $5 \times 10^{-20} \leq E/N \leq 10^{-18}$ Vcm^{-2} are seen as in figure 6, which correspond to the energies where the Ramusawer-Townsend minimum in Q_m appears.

Mean Energy

The mean electron energy $\bar{\epsilon}$ is defined as $\bar{\epsilon} = \int \epsilon^{3/2} f_0(\epsilon) d\epsilon$, and if $f_0(\epsilon)$ is Maxwellian, $3eD/2\mu = \bar{\epsilon}$. In figure 5 $\bar{\epsilon}$ and $3eD/2\mu$ are shown along with experimental values of $3eD/2\mu$ (Sibamura et al.,1979). The theoretical values of $3eD/2\mu$ agree quite well with the experimental ones in the E/N values in which the measurement was carried out. For $E/N \leq 5 \times 10^{-22}$ Vcm^{-2}, $\epsilon \simeq 3eD/2\mu$, since $f_0(\epsilon)$ is almost Maxwellian (see figure 8). With increasing E/N $\bar{\epsilon}$ becomes smaller than $3eD/2\mu$ values, since $f_0(\epsilon)$ shifts from Maxwellian as shown in figure 8. In high $E/N (\geq 10^{-21}$ Vcm$^{-2})$ the electrons could not be heated up efficiently due to the increase in the energy loss by colliding frequently with the inelastic modes.

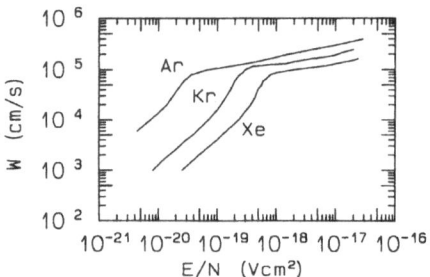

Figure 5. $W, \bar{\varepsilon}$ and $3eD/2\mu$ of electrons vs E/N in liquid Ar, liquid Kr and liquid Xe. simbols in W;measured values (Miller et al.,1968; Yoshino et al.,1976) and simbols in $3eD/2\mu$;measured values (Shibamura et al., 1979; Kubota et al.,1982).

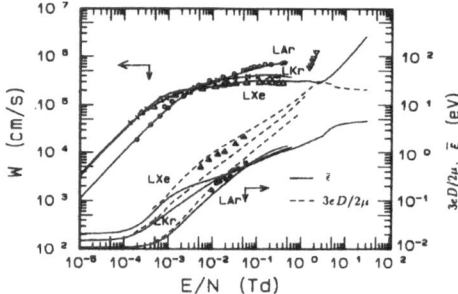

Figure 6. W in Ar, Kr, and Xe gases at room temperature (Laborie,1968).

Effective Ionization Coefficient

Derenzo et al(1972) measured the ionization coefficients $\alpha(\text{cm}^{-1})$ in liquid Xe for 8×10^{-17} Vcm^{-2} (2.8Td) $\leq E/N \leq 14 \times 10^{-17}$ Vcm^{-2}(14Td) as shown figure 7 along with the gaseous phase. The solid line is the present values calculated using the effective Q_{io} with the onset energy of 9.22 eV (Asaf and Steinberger,1974) in figure 4(c). If the present cross section is adopted, a good agreement with the experimental values is obtained. In this high E/N range the electron energy is suppressed due to the cooling by the inelastic scattering with the electronic excited level. As the result, W starts again to increase sharply but the electron energy almost saturates as seen in figure 5.

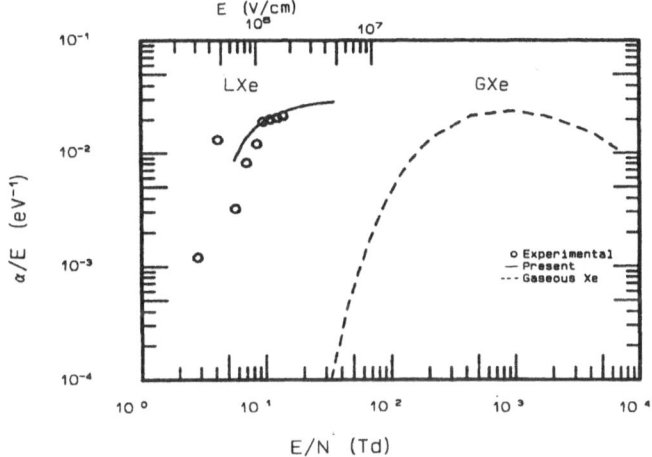

Figure 7. Ionization coefficient to field strength α/E in liquid Xe. Experimental values are from Derenzo et al.(1974).

Energy Distribution

The electron energy distribution $f_0(\epsilon)$ and $\sqrt{\epsilon} f_0(\epsilon)$ in liquid Ar, as an example, are shown in figure 8(a) and (b) respectively. In a low E/N region the energy distribution is shown to be Maxwellian. Note in figure 8(b) that with increasing E/N the tail of the distribution becomes short because of the increase in the collision frequency with the inelastic mode as well as the sharp increase in σ_0 around an energy of 1 eV as seen in figure 4. This behavior could suggest to correspond that the most electrons obtain the energy corresponding to the top of the conduction band.

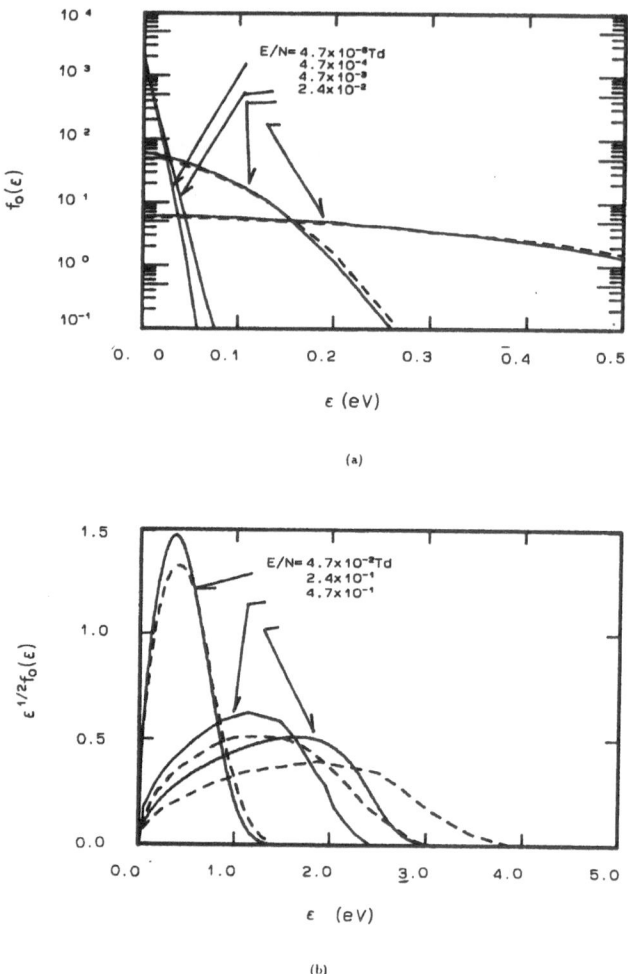

Figure 8. Electron energy distribution in liquid Ar at 84K. (a) $f_0(\epsilon)$ and (b) $\sqrt{\epsilon}f_0(\epsilon)$
———;the inelastic process is considered, ---; the inelastic process is not considered.

Effect of Molecular Impurity

If a few percents of CH_4 molecules are added in liquid Ar, the saturation drift velocity in
high fields shows nearly one order of magnitude higher values than those in liquid Ar as
shown in figure 9. The onset E/N for this jump is clearly shown, which corresponds to
the electron energy reaching the main vibrational energies (0.16 and 1.75 eV) of a CH_4
(Nakamura et al.,1987). The present analytical values agree well with experimental
values of Yoshino et al(1976).

Comparing the W with $\bar{\epsilon}$, one can understand quite well that the jump in W at high
E/N appears due to the cooling of the electrons colliding with the inelastic modes of
CH_4.

314

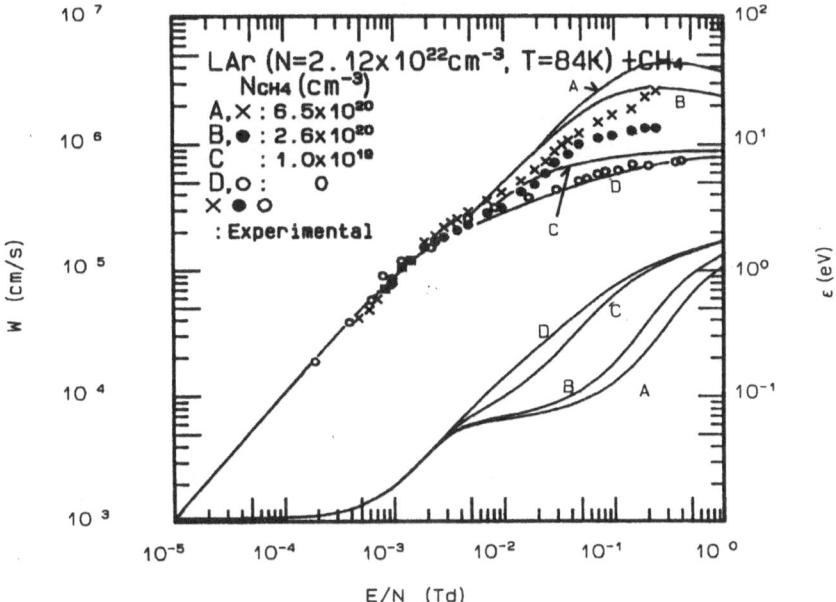

Figure 9. Influence of CH_4 solute on W and $\bar{\epsilon}$ of electrons. symboles; measured values (Yoshino et al.,1976).

The electron energy distribution is also influenced by the molecular additives as shown in figure 10. At high E/N, the electron energy distribution shifts its peak to a low energy part significantly and approaches to Maxwellian.

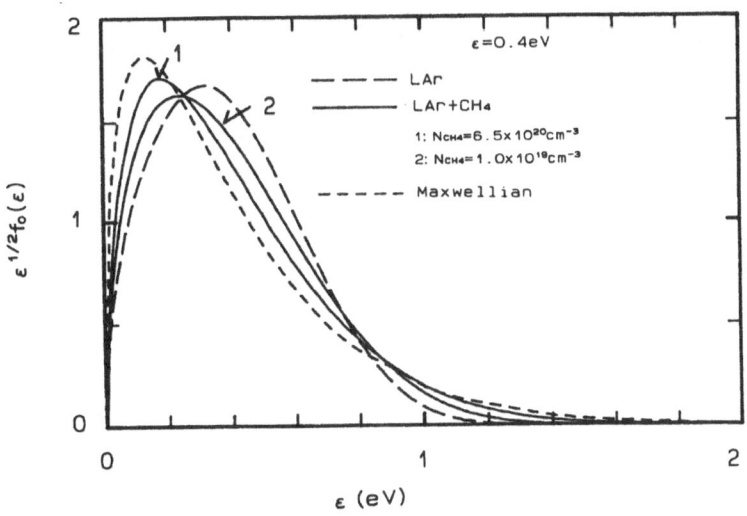

Figure 10. Influence of CH_4 additives on $f_0(\epsilon)$.

Comparison of the Present Modelling with Light Scattering

Using the present analysis, the electron transport properties both in liquids and gases in a wide E/N range could be understood properly. The cross sections used in this analysis are discussed on the light scattering observations so far (Fleury et al.,1973; McTague et al.,1969), and further evidences for the appropriateness of the present inelastic scattering model in the liquids is listed as follows.

(1) Even in rare gas liquids the inelastic light scattering modes have been observed(Fleury et al.,1973;McTague et al.,1969).

(2) It is seen from figure 4 that the ratio of σ_{in} to the maximum of σ_0 is 0.007 for Ar, 0.002 for Kr, and 0.001 for Xe. These values agree very well with the ratio of the total integrated intensity of the inelastic light scattering spectra to the Rayleigh-Brillouin spectrum, i.e. elastic light scattering mode; 0.1 for Ar (McTague et al.,1969) and 0.001 for Xe (Fleury et al.,1973).

(3) The inelastic light scattering spectra have been observed to be continuous and to decay exponentially with the frequency shift ω, as $I = I_o \exp(-\omega/\Delta)$. Here, Δ is 22 cm^{-1}(3 meV) for Ar, 17 cm^{-1}(2meV) for Kr, and 14 cm^{-1} (1.7 meV) for Xe (Fleury et al.,1973). The present onset energy ϵ_{in} for $\sigma_{in}(\epsilon)$ decreases with a similar order of Δ from Ar to Xe; 60 meV for Ar, 30meV for Kr, and 20meV for Xe. The difference between the ϵ in and Δ may come from difference of their definitions.

(4) The saturation of the drift velocity of hot electrons can be properly explained in the framework of the Cohen-Lekner's treatment, if the electron inelastic process determined on the basis of the light scattering observations is taken into consideration.

Linking of Gaseous and Liquid Phase

In gaseous phase the behavior of slow electrons is understood sufficiently well using the Boltzmann equation method, since values of the various cross sections for electron-atom collisions are obtained. However in a condensed phase an electron interacts collectively with atoms, as the result two different scattering cross section σ_0 for incoherent scattering and σ_1 for coherent scattering appear.

Actually in rare gas liquids and solids the dispersion relation is not hyperboric ($\Delta\epsilon \sim \Delta k^2$) any more (Klein,1976). This characteristics may reflect the change in the effective mass of an electron, then the loss energy of an electron per collision could change as well.

In the intermediate zone between gases and liquids, though the electron drift velocities in Ar have been measured(Huang et al.,1981), it is more complicated to explain their behavior since the structure of them become less clear owing to the increase in thermal agitation motion of the particles.

ACKNOWLEDGEMENT

The author greatly appreciates the fruitful discussion and cooperations to his coworkers; Prof.H.Tagashira, Dr.S.Nakamura, Mrs. Sukegawa, H.Kojima, T.Hayashi, T.Ando, K.Kimura in authors laboratory, Dr.W.F.Schmidt in Hahn-Meitner Institute Berlin and Prof.A.Khrapak in Institute for High Temperatures in Moscow.

REFERENCES

Ashcroft,N.S., 1967, The hard core dimensions of the noble gas atoms, Physica 35:148.

Asaf,U., and Steinberger,I.T., 1974, Photoconductivity and electron transport parameters in liquid and solid xenon, Phys.Rev.B 10: 4464.

Atrazhev,A.M., and Iakubov,I.T.,1981, Hot electrons in non-polar liquids,J.Phys.C 14: 5139.

Cohen,M.H., and Lekner,J., 1967, Theory of hot electrons in gases, liquids, and solids, Phys.Rev. 158: 305.

Derenzo,S.E., Mast,T.S., Zakla,H., and Muller,R.A., 1974, Electron avalanche in liquid xenon, Phys.Rev.A 9: 2582.

Engelstaff,P.A., 1967, "An Introduction to the Liquid State", Academic Press, London & New York.

Fleury,P.A., Worlock,J.M., and Carter,H.L., 1973, Molecular dynamics by light scattering in the condensed phase of Ar, Kr, and Xe, Phys.Rev.Lett. 30: 591.

Frost,L.S.,and Phelps,A.V.,1964,Momentum-transfer cross sections for slow electrons in He, Ar, Kr, and Xe from transport coefficients, Phys.Rev.A 136:1538.

Huang,S.S.-S., and Freeman,G.R., 1981, Electron transport in gaseous and liquid argon: Effect of density and temperature, Phys.Rev.A 24: 714.

Jahnke,J.A., Meyer,L., and S.A.Rice, 1971, Zero-field mobility of excess electrons in fluid argon, Phys.Rev.A 3: 734.

Klein, M.L., and Venables, J.A., 1976, "Rare Gas Solids", Academic Press, London & New York.

Kubota, et al., 1982, Hot electron relaxation in solid and liquid argon, krypton and xenon, J.Phys.Soc.Japan 51: 3274.

Laborie,P., Rocard,J.M., and Rees,J.A., 1968, Electronic Cross-sections and Macroscopic Coefficients, Dunod, Paris.

Lekner,J., 1967, Motion of electrons in liquid argon, Phys.Rev. 158: 130.

McTague,J.P., Fleury,P.A., and DuPre,D.B., 1969, Intermolecular light scattering in liquids, Phys.Rev. 188: 303.

Miller,L.S., Howe,S., and Spear,W.E.,1968, Charge transport in solid and Liquid Ar, Kr and Xe, Phys.Rev. 166: 871.

Nakamura,S., Sakai,Y., and Tagashira,H.,1986, Effective momentum transfer cross section for excess electrons in liquid argon,Chem.Phys.Lett. 130:551.

Nakamura,S., Sakai,Y., and Tagashira,H.,1987, Effect of molecular impurities on the high field electron conduction in liquid Ar, Kr, and Xe - analyses by Cohen Lekner theory-, TIEE Japan A 107,543.

Nishikawa,M.,1985, Electron mobility in fluid argon: applicability of a deformation potential theory, Chem.Phys.Lett. 114: 271.

Sakai,Y., Ando,T., Kimura,K., Tagashira,H., and Nakamura;S., 1993, Excess electrons in N_2/Ar liquid mixtures, Nuclear Ins. & Method A327:92.

Sakai,Y., Nakamura,S., and Tagashira,H.,1985, Drift velocity of hot electrons in liquid Ar, Kr, and Xe, IEEE Tr.Elc.Ins. EI-20: 133.

Sawada,S., Sakai,Y., and Tagashira,H., 1989, Boltzmann equation analyses of electron swarm parameters in Hg/Ar gas mixtures:effect of metastable Hg and Ar atoms, J.Phys.D 22: 282.

Shibamura,E., Takahashi,T., Kubota,S., and Doke,T., 1979, Rate of diffusion coefficient to mobility for electrons in liquid argon, Phys.Rev.A 20: 2547.

Van Hove,L., 1954,Correlations in space and time and Born approximation scattering in systems of interacting particles, Phys.Rev. 95: 249.

Yoshino,K., Sowada,U., and Schmidt,W.F.,1976, Effect of molecular solutes on the electron drift velocity in liquid Ar, Kr, and Xe, Phys.Rev.A 14: 438.

ELECTRON SCATTERING IN DENSE GASES AND LIQUIDS
AND RELATED PHENOMENA

Igor T. Iakubov

Institute for High Temperatures
Russian Academy of Sciences
Moscow, 127412 Russia

INTRODUCTION

In diluted gases interaction of excess electron with atoms and molecules can be described as a sequence of independent acts of pair scattering. The scattering effect is determined entirely by the cross-section of atom or moleculo which knows nothing about the existence of other scatterers. With the density increase this conception losses gradually its validity. Finally in liquid electron interacts at the same time with a number atoms, which are correlated strongly. Moreover electron may loss its high mobility and be captured by density fluctuations. Underline, we consider slow thermal electrons which are in the most degree sensitive to density effects. The goal of theory is to describe not only the electron state in liquid but the gradual changes in its state with the density growth beginning from gas. The peculiarities of electron scattering are clearly displayed in such important observable characteristics as electron mobility in electric field and energy barrier. Hence theory must obtain density dependencies of energy spectrum and mobility.

Linking the Gaseous and Condensed Phases of Matter
Edited by L.G. Christophorou *et al*, Plenum Press, New York, 1994

We shall not discuss here a number of interesting although delicate effects and consider two groups of liquids. Electron being injected into liquid of the first group does not loss its high mobility. It is the group of well polarized liquids. On the contrary in liquids with low polarizabilities electron may be self-trapped.

Below in the first section we shall see how differently peculiarities of interaction display themself in moderately dense gases, how they are strengthening with density growth and finally manifest themself so vividly in the phenomena of enhanced mobility and self-trapping. The physics of self-trapping became much clearer now and below in the second section attention is given to highly polarized liquids. The measurements of bottom of the conductivity band in the all density range is the achievement of the last ten years. It turns out that this dependence has the minimum, the position of which correlates to the position of the sharp maximum in the density dependence of mobility.

The consistent description of electron-atom interaction cannot begin with introducing of explicit appearance of interaction potential. Indeed the strong polarization tail characterizes long range component of the interaction. But the short range one cannot be represented explicitly as a potential. All necessary and adequate information is contained in phases of scattering. For slow electron the zero phase is the most essential. Hence the starting data for the theory of electron state in liquids are: scattering length on solitary atom, its polarizability, binary correlation function of liquid. It is enough to obtain the scattering phase in medium and to trace its evolution with density growth.

The given consideration is based on the O'Malley theory of slow electron scattering,[1] on the pseudo-potential conception,[2] and on the Cohen and Lekner kinetic theory.[3]

The overlap between potentials of adjacent atoms causes the polarization tails cut-off and electron-atom interaction in liquid becomes short-range one. The important consequences steam from it. The scattering length passes through the zero with density growth and alters its sign, it becomes positive.[4] Below the analytic expressions are obtained. It is possible to follow the observable quantities on going from gas densities to liquid ones. The calculated values of conduction zone bottom and mobility are satisfactory compared to measured data. Some another problems are discussed.

ELECTRON MOBILITY IN MODERATELY DENSE GASES

The first publication by Lowke (1963) was followed up to the end of the 1970s by the accumulation of a very large amount of experimental evidence of deviation from ideal-gas behavior of electrons at densities N of the order of 10^{20}-10^{21} cm^{-3}. The density effect varies from one gas to another. The reduced mobility μ/μ_0 in gases such as He, N_2, H_2 and also many hydrocarbons falls as N increases while in other gases (Ar, Xe, CH_4) it increases.

In the frames of gas-kinetic theory interaction between an electron and medium may be represented as a sequence of independent collisions. An electron undergoes

$$\nu(\varepsilon) = Nq(\varepsilon)(2\varepsilon/m)^{1/2} \tag{1}$$

collisions in unit time, where ε is the energy of electron, N is the particle density, $q(\varepsilon)$ is the cross section of electron scattering on atom (molecule). Electron mobility is given by $\nu(\varepsilon)$ averaged over the electron energy distribution

$$\mu_0 = (e/m) \nu(\varepsilon)^{-1} = (2/3)e(2/\pi Tm)^{1/2}(qN)^{-1} \tag{2}$$

Electron Scattering on Solitary Atom

Interaction with solitary atom is determined entirely by the set of scattering phases δ_1, which are related to $q(\varepsilon)$

$$q(\varepsilon) = 4\pi k^{-2} \sum_1 (21 +1) \sin^2\delta_1 \tag{3}$$

Here k is the electron wave number, $\varepsilon = \hbar^2 k^2/2m$, $\hbar l$ is the angular moment. Phases are small for slow electron and δ_0 is important only,

$$q = 4\pi k^{-2}\delta_0 \tag{4}$$

Consider electron scattering on short-range potential – atom with low polarizability. It is convenient to wright the Schrodinger equation for function χ

$$d^2\chi/dr^2 + [k^2 - 2m\hbar^{-2}V(r)]\chi = 0 \qquad (5)$$

where $\chi(r) = r\Psi(r)$, Ψ is the electron wave function. Outside of the range action of $V(r)$ the solution is given by

$$\chi(r) \sim \sin(kr + \delta_o) \qquad (6)$$

For slow electron

$$\chi(r) \sim (r - L_o) \qquad (7)$$

where

$$\delta_o = - L_o k \qquad (8)$$

Here L_o is the scattering length. Its value is of order of potential action range. However L_o gives no information on the electron-atom potential structure.

If L_o is known we need not look for the real potential and may use any pseudo-potential. The Fermi one

$$V(r) = 2\pi\hbar^2 m^{-1} L_o \delta(r) \qquad (9)$$

It is the simplest and it is correct in the Born approximation.
The expressions obtained are correct if δ_o is small. So

$$L_o \ll \lambda , \qquad NL_o^3 \ll 1 \qquad (10)$$

where $\lambda = \hbar(2m\epsilon)^{-1/2}$ is the electron wave length.
Above given analysis may be applied to particles with low polarizabilities such as He, H_2, Ps. Consider now the opposite case of high polarizability. At large distances electron-atom interaction is the polarization one

$$V(r) = - e^2\alpha/2r^4 \qquad (11)$$

where α is the polarizability of atom. The electron-atomic potential at short distances is unknown. An indirect information about it is given by scattering length. Now low energy scattering is determined by two parameters [1]

$$\delta_0 = - L_0 k - \pi\alpha k^2/(3a_0) \qquad (12)$$

The large polarizability and negative sign of scattering length are related. Consider it. The Schrodinger equation (5) for the case of zero energy is reduced to

$$\chi'' - (2m/\hbar^2)\, V(r)\chi = 0 \qquad (13)$$

At large distances from atom, where expression (11) for V(r) is valid, the solution of the equation has the form

$$\chi(r) \sim r\, sin\, [(\alpha/a_0)^{1/2} r^{-1} - \Delta] \qquad (14)$$

where Δ is the phase shift due to unknown short-range part of potential. Relation between Δ and L_0 follows from (7) and (14)

$$L_0 = (\alpha/a_0)^{1/2} [tg(\Delta)]^{-1} \qquad (15)$$

where the value of $tg\Delta$ is taken in the first quadrant.

The phase shift Δ gives full necessary information about short-range part of interaction. However it is useful to imagine that the short-range repulsion of range a exists,

$$a = \Delta^{-1}(\alpha/a_0)^{1/2}, \qquad \Delta = \pi + arctg(L_0^{-1}\sqrt{\alpha/a_0}) \qquad (16)$$

Here we follow the ideas of pseudo-potential theory. Pseudo-potential does not reflect as a rule the features of real potential, but it gives correct value of scattering phase and smooths wave function. Such is the pseudo-wave-function (14). The space area of pseudo-description must be small. For our problem the following inequality is required

$$a^3 \ll \bar{r}^3, \qquad \bar{r} = (3/4\pi N)^{1/3} \qquad (17)$$

At that case phase shift Δ is conserved in dense media.

Table 1. Parameters of interaction.

	α, a_0^3	L_0, a_0	a, a_0	σ, a_0
argon	11.1	− 1.63 − 1,40	1,63 1,68	6,50
xenon	27,1	− 5,9	2,17	7,69

Density Corrections to Mobility

Consider electron in the medium of short-range scatterers (9). Electron potential energy $U(r)$ may be written as a sum of interaction potentials of individual atoms

$$U(r) = 2\pi\hbar^2 m^{-1} L_O \sum_j \delta(\mathbf{r} - \mathbf{R}_j)$$

where R_j are the atomic radius-vector.

There are two other combinations containing interaction parameter – $N\lambda^2 L_O$ and $N\lambda L_O^2$. The second equals to the ratio of the electron wave length to the mean free path $l = (NL_O^2)^{-1}$. It describes the interference of subsequently scattered electron waves. The relevant correction [6] to the gas-kinetic frequency of collisions is given by

$$\nu_m(\varepsilon, N) = \nu(\varepsilon) [1 + 2\lambda(\varepsilon)Nq] \qquad (18)$$

The condition $N\lambda q \ll 1$ coexists with $NL_O^3 \ll 1$, because $\lambda \gg L_O$.

After the averaging over the Maxwell distribution we get the thermalized mobility for low-energy electrons

$$\mu/\mu_O = 1 - \pi^{1/2} Nq\lambda_T \qquad (19)$$

where $\lambda_T = \hbar(2mT)^{-1/2}$. Eq.(19) describes well the corrections to μ/μ_O in weak electric fields, Figure 1.

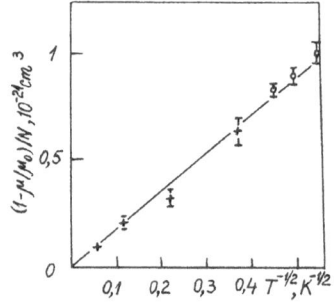 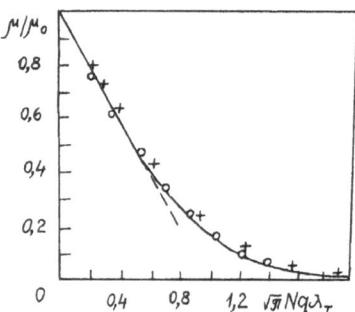

Figure 1. Temperature dependence of density correction to μ.[6] Points – experimental data,[7] line is calculated from (19).

Figure 2. μ/μ_O in He and H_2.[6] Points – experimental data,[8,9] T = 77,6 K. Line is calculated from (19).

324

Eq. (19) is to be considered as the first term of the expansion μ/μ_O with respect to $Nq\lambda_T$, this cease apply for $Nq\lambda_T > 1$. However the problem does not have another parameters. So we conclude that μ/μ_O is the universal function in the general case of non-linear behavior of $Nq\lambda_T$ for all gases with low polarizabilities. At Figure 2 measured data for He and H_2 are shown. The points lie on a single curve, which is described by (19) for small $Nq\lambda_T$.

The first of aforementioned parameters - $N\lambda^2 L_O$ gives the the boundary of energy spectrum of injected electrons

$$V_O = 2\pi\hbar^2 L_O Nm^{-1} \tag{20}$$

If $V_O > 0$ it gives energy barrier for electron injection to the matter. If $V_O < 0$ it indicates the position of the bottom of conduction zone in the matter which is lowered in comparison with vacuum. Measurements indicate that an increase in the density in some gases (Ar, CH_4) increased reduced mobility, not lowered it. This effect lacks full quantitative description up to now although qualitative explanation exists.[4,6] Dense medium of highly polarizable particles screens the long-range field. The interactions of electron with atoms is attenuated. It increases mobility.

Polarization potential in the medium takes now the form

$$V_m = - f(\alpha e^2/2r^4), \qquad f = [1 + (8\pi/3)\alpha N]^{-1} \tag{21}$$

The factor f is the reciprocal to dielectric function. We introduce in (15) the screened polarizability $\alpha_m = f\alpha$. Looking for the small correction to L_O we obtain

$$L = L_O[1 + (4\pi/3) \alpha N (\alpha/aL_O a_O)] \tag{22}$$

The absolute magnitude of L decreases as N increases (remind $L_O < 0$), the scattering becomes weaker. Figure 3 compares the calculated and experimental data for argon.

It is remarkable that the tendency of mobility variation is conserved with the farther compression of a gas. Highly polarizable media become more transparent for electron because the long tails of scatterer potentials are overlapped and resultive potential field U(r) becomes much more smooth.

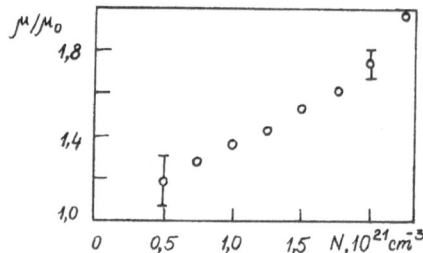

Figure 3. μ/μ_O in Ar.[6] Points – experimental data,[10] curve is calculated from (22). T = 300 K.

This phenomenon will be considered in the next section.

In gases with low polarizabilities the farther compression leads to the deep drop in mobility. At Figure 4 precursor of this phenomenon is displayed. The observed deviations from the universal dependence μ/μ_O on $Nq\lambda_T$ are to be considered as due to the new physical factor. It is the self-trapping of electrons in dense medium. The discussion of this prominent phenomenon is out the scope of the given review, see e.g.[11]

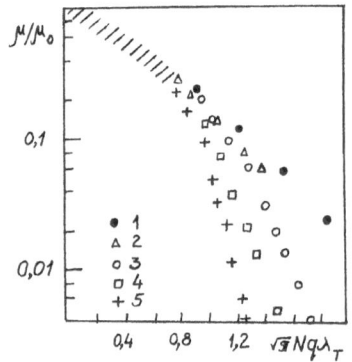

Figure 4. Measured μ/μ_O from various sources[6] 1 – H_2 77,6 K, 2 – He 77,6 and 52,8 K, 3 – He 20,3 K, 4 – He 13,8 K, 5 – He 4,2 K. The hatched region is the range of moderate densities.

ELECTRON IN HIGHLY POLARIZABLE FLUIDS

Electron Scattering in Liquids

In liquids with high polarizability electrons may be

considered as almost free. The mobilities in Ar,[12] Kr,[13] Xe,[14] and some hydrocarbons[15,16] increase with density growth and transit through the sharp maxima near $N_* = 10^{22} cm^{-3}$. The further investigations[17,20] gave information about the density dependence of the bottom of conduction zone. It has a wide minimum near N_*. The correlation between μ_{max} and $(V_o)_{min}$ is the feature of all high mobility liquids.[21]

Explanation to the enhanced electron mobility has been given in.[3,4] Account for inter-atomic correlation results in introduction to (2) structure factor of liquid $S(0)$,

$$\mu = (2/3) e (2/\pi T m_{ef})^{1/2} [4\pi L^2 NS(0)]^{-1} \qquad (23)$$

Near triple points $S(0) \ll 1$ and free path of electron in liquid $l = [4\pi L^2 NS(0)]^{-1}$ becomes much larger than electron wavelength λ and inter-atomic distance $\bar{r}$. If $S(0)$ is not so small the inequality $\lambda \ll l$ has to be guaranteed by the smallness of L. The scattering length is negative in a gas due to the long-range polarization attraction. In fluid, where $\bar{r}$ is small, the long potential tails are cut off. The electron-liquid potential becomes short-range and L becomes positive. So with compression scattering length passes through the zero. It occurs at the density N_*. This kinetic theory has been successfully developed by a number of investigators.

Another approach was developed in[22,23] It is assumed that electrons are scattered by density fluctuations. Measured $V_o(N)$ were used to calculate deformation potential and farther the dependence $\mu(N)$. Some success was achieved, but the basis of the theory seems to be not fully clear.

The discovering of the minimum of $V_o(N)$ gave a rise to a number of researches. The explicit expressions for electron-atom potentials were assumed which contain a number of adopted parameters. The $V_o(N)$ values for liquid were computed using these potentials. However it leads to important contradictions, see discussion in.[24]

Below we follow the paper.[15] It would not be adequate to begin the study of electron in a liquid with consideration of scattering on solitary atom. Every atom finds itself in a close environment of its neighbors. Strong inter-atomic correlation leads to validity of the cell model. Every atom may be considered as deposited at the center of the Wigner-Zeits cell of radius $\bar{r} = (3/4\pi N)^{1/3}$ Inside of the each cell potential is given by the sum of the potential of central atom $V(r)$ and of summary polarization potential of environment

$$V_{cell}(r) = V(r) + U(r), \qquad U(r) = \sum_j V(r-R_j), \qquad r<\bar{r} \qquad (24)$$

where R_j are the coordinates of environment atoms. Eq.(24) describes the muffin-tin structure of resultant interaction.

Let us consider a scattering of slow electron by single cell. We shall find for the beginning the exact expression for scattering length on the central atom potential which is cut off in the cell boundary $\bar{r}$. The potential $V_c(r)$ coincides with central atom potential $V(r)$ inside of a cell but vanishes outside. The wave function $\chi_c(r)$ for cut off potential satisfies the equation similar to (13)

$$\chi_c'' + (2m/\hbar^2)V_c(r)\chi_c = 0 \qquad (25)$$

Combination of Eqs.(13) and (25) and following integration lead to well known formula[26]

$$(\chi\chi_c'-\chi'\chi_c)\big|_{r=\infty} = (2m/\hbar^2) \int_0^\infty [V_c(r)-V(r)]\chi\chi_c dr \qquad (26)$$

The l.h.s. of (26) contains the asymptotes of the wave functions $\chi(r)$ (7) and $\chi_c(r) \sim (r-L_c)$ where L_c is the scattering length for cut off potential which is wanted. So, the left part of (26) is proportional to the difference (L_c-L_0). The integrand in r.h.s. of (26) differs from zero outside of the cell only, where $V_c(r) - V(r) = \alpha e^2/2r^4$. The wave function $\chi(r)$ has the form (14) and χ_c is the wave function of free motion $\chi_c = \text{const } (r-L_c)$. After integration one obtains

$$L_c= \{\bar{r}^{-1}+ (\alpha/a_0)^{-1/2}tg[(\alpha/a_0)^{1/2}(a^{-1}- \bar{r}^{-1})]\}^{-1} \qquad (27)$$

where tg value is taken in the second quadrant if its argument exceeds $\pi/2$ and transits to the first one in the opposite case. It is the exact expression for scattering length on the potential V_c. For a dilute gas (in the limit $\bar{r} \to \infty$) L_c coincides with L_0 and it is negative. With a decreasing in $\bar{r}$ (increasing in density) L_c changes its sign.

Besides the centered atom electron interacts with environment (24). This interaction is polarization one, but a screening by fluid has to be taken into account. It is given by the factor $f(r)$. It is the Lorentz local-field screening function.[4,27] Numerical calculations[24] give

$$f = [1+(8\pi/3)\alpha N]^{-1}, \quad r > \bar{r}, \qquad f=1, \quad r<\bar{r} \qquad (28)$$

The mean potential of environment may be written in a form

$$U(r) = \sum_j f(r)V(r-R_j) = - 3\alpha e^2 (2\bar{r}^3)^{-1} f \int_{\bar{r}}^{\infty} (R^2 - r^2)^{-2} R^2 g(R) dR \qquad (29)$$

where $g(R)$ is the pair correlation function of a fluid. Expending the integrand into a series one obtains

$$U(r) = - 3\alpha e^2 (2\bar{r}^3)^{-1} f \int_{\bar{r}}^{\infty} g(R) R^{-2} [1+2(r/R)^2+3(r/R)^4] dR =$$

$$= u_0 + u_2 + u_4 \qquad (30)$$

u_0 gives the shift of a conduction band bottom due to electron-environment interaction, potentials u_2 and u_4 give the additional contributions to scattering.

For estimates, one can use the approximate $g(R)$ of liquid: $g(R) = 0$, if $R \le \sigma$ and $g(R) = 1$, if $R > \sigma$. Here σ is the distance of the closest approach, the parameter of the Lennard-Jones inter atomic potential. At that case

$$U(r) = - \alpha e^2 (\sigma \bar{r}^3)^{-1} f [3/2 + (r/\sigma)^2 + 0.9 (r/\sigma)^4] \qquad (31)$$

In a gas $\sigma < \bar{r}$, one may take $g(R) = 0$, if $R \le \bar{r}$ and $g(R) = 1$, if $R > \bar{r}$, and use (31) with the replacement $\sigma \to \bar{r}$.

For the accurate calculations the realistic correlation functions may be used in (30). We arrive at

$$U(r) = - \alpha e^2 (\sigma \bar{r}^3)^{-1} f[3/2 I_0 + I_2 (r/\sigma)^2 + I_4 (r/\sigma)^4] \qquad (32)$$

$$I_n = (n+1) \int g(x) x^{-2(n+1)} dx, \qquad n = 0,2,4 \qquad (33)$$

I_n take close values for Ar, Kr and Xe: $I_0 \approx 1.2$, $I_2 \approx 1.1$, $I_4 \approx 1,2$ if $\bar{r} < 0.9 \sigma$.

Scattering on $(u_2 + u_4)$ is given by the first Born approximation. We adopt the well known expression for a change in scattering length ΔL as a result of change in potential[26]

$$\Delta L = (2m/\hbar^2) \int_a^{\bar{r}} \chi^2(r)(u_2 + u_4)dr \qquad (34)$$

At $r \leq \bar{r}$ $\chi(r)$ takes the form (14). The normalization constant is determined by the condition of the continuity with
the wave function of scattered electron at $r = \bar{r}$. It gives

$$\chi(r) = (1-L_c/\bar{r})[sin(\sqrt{\alpha/a_o}\,\bar{r}^{-1}-\Delta)]^{-1}rsin(\sqrt{\alpha/a_o}\,r^{-1}-\Delta) \qquad (35)$$

Approximate evaluation of integral in (34) gives

$$\Delta L = - [\alpha/(2a_o\sigma)](\bar{r}/\sigma)^2(1-L_c/\bar{r})^2 f[I_2 + (3/4)I_4(\bar{r}/\sigma)^2)] \qquad (36)$$

The Born approximation validity for slow electrons is guaranteed by the absence of bonded states in potential well ($u_2 +$
u_4). It is more convincing to calculate the second Born correction[26] It turns to be small if $(4/9)(\bar{r}/\sigma)^2(\alpha/a_o\sigma) \ll 1$.
    The total scattering length

$$L = L_c + \Delta L \qquad (37)$$

Both contributions to L are small and hence are additive.
Their values may be comparable. The expressions are suitable
for fluids where the cell model is valid. They cannot be re-
commended for gaseous states where the close order is weake-
ned. However (37) has a correct limit at low densities, $L(\bar{r} \rightarrow$
$\infty) = L_o$. It is proper for some another Wigner-Seitz formulae.

Mobility and Conduction Band Bottom

    Calculated values of L for are given in Figure 5. The
"experimental" L are extracted from measured mobilities making use of (23). The calculated L(N) are in a good agreement
with experimental ones. It will be recalled that the theory
contains no adjusted parameters. It is very important to hit
in point of L-zero. It is noted above, that the scattering
length for cut off potential L_c (27) passes through the zero.
But it proceeds at too small densities. Here the ΔL value
plays in (37) the important role shifting zero-L-point to domain of larger densities. Underline howovor that the pure

theory can never pretend to high numerical accuracy. So the very good hitting in N_* has to be considered as fortunate.

The comparison of calculated electron mobilities with experimental ones is more pretentious for theory because mobility changes by the order of magnitude, Figure 5. The maxima of $\mu(N)$ dependencies are satisfactory described by theory, although we shifted calculated $L(N)$ for Xe to the dotted position, Figure 5b.

The position of μ_{max} is given by the density N_* where scattering length turns to be zero. The magnitude of μ_{max} is defined by fluctuations of scattering length δL due to density fluctuations δN.[28] The quantity $L^2 S(0)$ in expression (23) is replaced by $L^2 S(0) + \overline{\delta L^2}$, where $\overline{\delta L^2}$ is the mean value of squared fluctuation of L. Using (37) choose another way of δL^2 determination than taken in.[28] Let us calculate

$$\overline{\delta L^2} = (dL/dN_*)^2 \, \overline{\delta N^2} \tag{38}$$

where $\overline{\delta N^2}$ is the mean squared particle density fluctuation in liquid and (dL/dN_*) is the derivative taken at N_*. Differentiating $L(N)$ (37) we keep the major terms and account that $L_c(N_*) = - \Delta L(N_*)$. As a result we have

$$\overline{\delta L^2} \simeq L_c^2 S(0)(9\Re)^{-1} [\alpha(a_o \bar{r} L_c)^{-1}(1-L_c/\bar{r})^2 + 2]^2 \tag{39}$$

where L_c and $\bar{r}$ are taken at N_*. The quantity $\Re$ is a number of atoms inside of the volume Ω, where the density fluctuation occurs. It comes from the expression $\overline{\delta N^2} = \Re^{-1} N^2 S(0)$.

We can not determine $\Re$ value exactly. Here the undeterminacy exists which is always proper for such a tasks. However we can estimate it. Minimal volume Ω has to contain sufficient number of atoms to form the environment potential $u_2(r)$. The magnitude of it is formed by two coordinate spheres of atom in a liquid (the first and the second maxima of $g(R)$). The estimated in this manner $\Re$ equal to 30 for Ar and 50 for Xe. However it turns to be not enough to obtain the measured μ_{max}. It is necessary to increase these $\Re$-values in 5 times to obtain in Ar, Kr, and Xe μ_{max} which are different by the order of magnitude but close to measured values.

Consider now the conduction zone bottom V_o. It consists of two components:

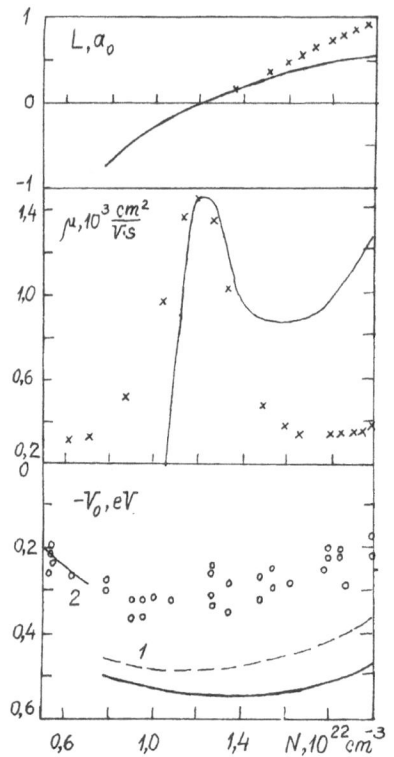
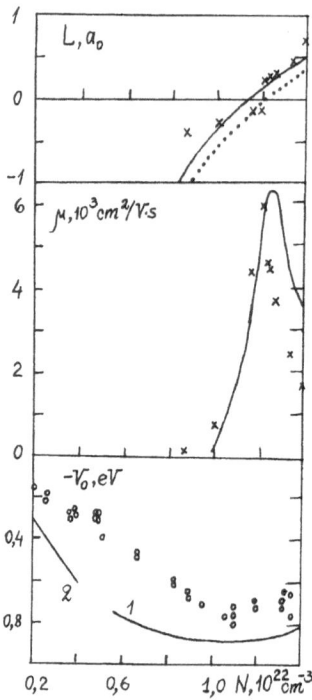

Figure 5. Measured and calculated $L(N)$, $\mu(N)$ and $V_O(N)$.[25] Points - experimental data from.[12-21] Ar - a (solid - $L_O = -1,63\ a_O$, dashed - $L_O = -1,4\ a_O$), Xe - b.

$$V_O = u_O + T_O \tag{40}$$

u_O is a result of mean polarization of environment (30). It leads to the lowering of ground state energy of electrons, $u_O < 0$. T_O is a result of electron scattering by liquid potential and its sign is defined by scattering length. T_O may be calculated in optical or modified optical approximations

$$T_O = T_{opt}\{1 - 3L\bar{r}^{-3} \int_O^\infty R[g(R)-1]dR\}, \quad T_{opt} = 3e^2 La_O/(2\bar{r}^3) \tag{41}$$

Modified optical approximation takes into account the correlation between atoms in a dense fluid. So we have

$$V_O = -3I_O\alpha e^2[2\sigma(\bar{r}^3+2\alpha)]^{-1} + 3La_O e^2\{2\bar{r}^3[1+0.2(3L\sigma^2)\bar{r}^{-3}]\} \tag{42}$$

Intogral in (41) for liquids gives factor 0,2, for a gas 0,5.

If $L > 0$, T_O is interpreted as energy of zero-oscillations of electron . The Wigner-Seitz approach[2] considers liquid as a array of hard scatterers of radius L. Then

$$T_O = \hbar^2 q_O^2/2m, \qquad tg[q_O(\bar{r} - L)] = q_O\bar{r} \qquad (43)$$

In liquids the Wigner-Seitz and modified optical approaches give close results. For a gas optical approximation is valid,

$$V_O = - 3\alpha e^2(2\bar{r}^4)^{-1} + 3e^2 L_O a_O(2\bar{r}^3)^{-1} \qquad (44)$$

The theory describes the minimum of $V_O(N)$, Figure 5. Curves 1 - Eq.(42) and 2 - Eq.(44) may be connected by a smooth line.
At the minimum of $V_O(N)$, where $L(N_*) = 0$, scattering part of V_O vanishes. So, $(V_O)_{min}$ contains u_O only

$$(V_O)_{min} = - (3I_O/2)(e^2/\sigma)(\alpha/\bar{r}_*^3)[1 + 2\alpha/\bar{r}_*^3]^{-1} \qquad (45)$$

The relation between measured $(V_O)_{min}$, σ, N_* and α is demonstrated by Figure 6 for different liquids.[30]

Effective Mass and Cross Section

Energy of almost free electron $\varepsilon = \hbar^2 k^2/2m_{ef}$. Difference between m_{ef} and m indicates the level of electron freedom. Experimental information about m_{ef} is indirect. It is obtain-

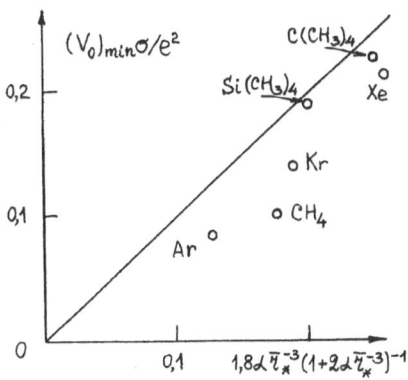

Figure 6. Relationship between $(V_O)_{min}$ and parameters of interaction.[25] Points - from.[30]

ed from analysis of spectra of rare solids and liquids. The thing is that electron, bonded on highly excited atomic state, is very similar to free electron which we study here. Bond energy contains effective mass. Hence it influences the value of radiated quantum. "Observed" $m_{ef} \approx (0,3 - 0,6)m$.[31]

At finite energies the knowledge of s-scattering phase δ_0 is not sufficient. Calculation of m_{ef} requires the p-scattering phase δ_1. In[32-33] explicit expressions for electron-liquid potential were chosen and values of δ_1 and m_{ef} were calculated. As in previous sections another way is taken below which is based on the pseudo-potential ideas.

It is known that pseudo-potential for for every phase has to be constructed independently.[5] The radius of pseudo-potential core for s-scattering a was determined from the measured value of scattering length. For determination of radius of pseudo-potential core for p-scattering a_1 some measured characteristic of this scattering on solitary atom has to be used. Or it may be well calculated characteristic of it. As such a basic value we chose zero of δ_1 – the wave-number k^*: $\delta_1(k^*) = 0$. For argon atomm $k^* \approx 0,315\ a_0^{-1}$.[34] Its knowledge permits to find a_1 and further to calculate $\delta_1(k)$ in fluid.

Consider the scattering on solitary atom. The scattering phases on polarization tail δ_1^p and on core δ_1^c are additive

$$\delta_1 = \delta_1^p + \delta_1^c \tag{46}$$

The usage of the simplest expressions for them

$$\delta_1^p = \pi\alpha k^2/(15a_0), \qquad \delta_1^c = -(ka_1)^3/3 \tag{47}$$

results in the core radius for Ar $a_1 = 2,85\ a_0^{-1}$.

The value may be corrected if to introduce to the expressions (47) the next members of expansions over the powers of ka_1. In the Born formula for δ_1^p we replace wave function of free motion by the pseudo-wave function $\chi_1(r)$ which vanishes at $r = a_1$,

$$\delta_1^p = \alpha k^2 a_0^{-1} \int_{ka_1}^{\infty} x^{-2}[x^{-2}\sin(x + \Delta) - x^{-1}\cos(x - \Delta)]^2 dx \tag{48}$$

whero $\Delta = ka_1 - arctg(ka_1)$. At the limit of small ka_1 we have

$$\delta^p_1 \cong \alpha k^2 a_0^{-1}[\pi/15 - ka_1/9 - (ka_1)^2/6] \tag{49}$$

The theory of low-energy scattering gives δ^c_1 in the form

$$\delta^c_1 \cong - (1/3)(ka_1)^3[1 - (3/5)(ka_1)^2] \tag{50}$$

It results in $a_1 = 2{,}09\ a_0$. Calculated with (48)-(50) $\delta_1(k)$ are well compared to the data,[34] points at the **Figure 7.**

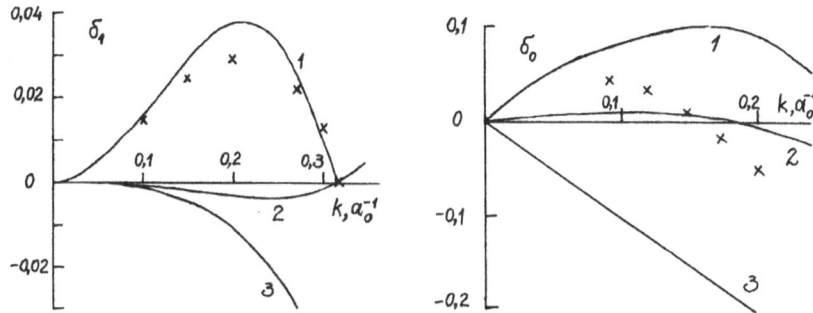

Figure 7. Scattering phases in Ar – δ_1 (a) and δ_0 (b): 1 – on solitary atom, 2 – in fluid at $N = 10^{22} cm^{-3}$, 3 – at $2 \cdot 10^{22} cm^{-3}$

In fluid

$$\delta_1 = \delta^p_1 + \delta^c_1 + \delta^n_1 \tag{51}$$

where δ^n_1 is the scattering phase on the Wigner–Seitz cell environment (31). For small $k\bar{r}$

$$\delta^n_1 \cong [\alpha k^3 \bar{r}/(3a_0)](\bar{r}/\sigma)^3 f \{ (2/21)[1 - (7/45)(k\bar{r})^2] +$$

$$+ (1/15)(\bar{r}/\sigma)^2[1 - (9/55)(k\bar{r})^2]\} \tag{52}$$

$$\delta^p_1 \cong \alpha k^3 \bar{r}(9a_0)^{-1}[1 - 9a_1/(5\bar{r}) + (a_1/\bar{r})^3 - (k\bar{r})^2/15] \tag{53}$$

Calculation results are given by Figure 7a. At triple point, where polarization potential is cut off strongly, scattering on the core dominates and $\delta_1 < 0$. At fluid of smaller density components of δ_1 almost cancelled and δ_1 becomes small. At lower densities δ_1 takes positive sign as in a gas.

Now we may calculate the effective mass. Wave function of electron near the bottom of conductivity zone is given by

$$\Psi_0 \cong exp(ikr)\, u_0, \qquad u_0 = Asin[q_0(r - L)]/(q_0 r) \qquad (54)$$

where q_0 is given by (43), A is the normalization constant

$$A^2 = (4/3)(q_0\bar{r})^3\{2q_0(\bar{r} - L) - sin[2q_0(\bar{r} - L)]\} \qquad (55)$$

The Bardin perturbation theory results in the expression for ε which contains effective mass in the following form

$$(m_{ef}/m)^{-1} = A^2(q_0\bar{r})^{-2}sin^2[q_0(\bar{r} - L)][dln\chi_1/dln\bar{r} - 1] \qquad (56)$$

where χ_1 is the wave function of electron in the cell with angular moment $\hbar l = 1$ and the minimal energy,

$$dln\chi_1/dln\bar{r} = \{(q_0\bar{r})^2 tg(q_0\bar{r}+\delta_1)\,[tg(q_0\bar{r}+\delta_1)-q_0\bar{r}]^{-1}\}^{-1} \qquad (57)$$

It is clearly seen that values of m_{ef}/m are sensitive to δ_1.

Two different dependencies of $L(N)$ were used to obtain m_{ef}/m, Figure 8. $L(N)$ were taken from Figure 5a – calculated $L(N)$ (for curve 1, $L_0 = - 1,63\ a_0$) and "experimental" one (curve 2). Both dependencies of $m_{ef}(N)/m$ behave sensible. The functions increase with the density lowering and turn to be equal to 1 at N_*, where $L = 0$. At this density electron is the most free and m_{ef} has to be very close to m. Apropos it is not clear how the curve 3 from[33] could arrive at unit with density lowering.

The effective cross section at small energies may be determined if two scattering phases are known – δ_0 and δ_1, $q \cong 4\pi k^{-2}(\delta_0^2 + 3\delta_1^2)$.

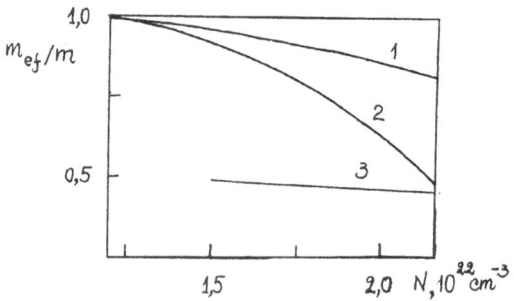

Figure 8. m_{ef}/m for fluid Ar: 1 and 2 – the given calculations, 3 – from[33]

$\delta_o(\epsilon)$ was found in the same way which was used for $\delta_1(\epsilon)$. For scattering on solitary atom the expression (12) was corrected with the account for the presence of pseudo-potential core

$$\delta_o = - L_o k - \pi \alpha k^2/(3a_o) + (4\alpha k^3/3a_o)(1-C-ln2ak), \qquad (58)$$

where $C = 0,577$, $ka \ll 1$. For scattering in fluid

$$\delta_o = -Lk - (\alpha k^3/3a_o) \{1-4(a/\bar{r})ln(\bar{r}/a)+(10a/3\bar{r})-6(a/\bar{r})^2+$$

$$+2(a/\bar{r})^3-(1/3)(a/\bar{r})^4+(1/7)(\bar{r}/\sigma)^3 f [1+(7/20)(\bar{r}/\sigma)^2]\} \qquad (59)$$

Calculated δ_o are given at Figure 7b. Comparison with the data[34] indicates that we may hope to qualitative results only. This level of accuracy turns to be enough here. We see that the theory describes qualitatively correct the evolution of δ_o with the variations in density. As to the dependence of δ_o on energy it is linear in liquid. Note also that the first phase is small. So for thermal electrons one may undoubtedly use in fluid the simple formula for cross section, $q = 4\pi L^2$. Cross section does not depend on energy. Even for comparatively heated electrons this conclusion conserves the validity.

REFERENCES

1. T.F.O'Malley, Extrapolation of electron-rare gas atom cross sections to zero energy, *Phys. Rev.* 163:1020(1963).
2. B.E.Springett, J.Jortner. and M.H.Cohen, Excess electron in nonpolar fluid, *J.Chem. Phys.* 48:2720(1968).
3. M.H.Cohen and J.Lekner, Theory of hot electrons in gases, liquids, *and* solids, *Phys. Rev.* 158:305(1967).
4. J.Lekner, Motion of electrons in liquid argon, *Phys. Rev.* 158:130(1967).
5. K.Huang. "Statistical Mechanics", Wiley, New York (1963).
6. V.M.Atrazhev and I.T.Iakubov, Electron mobility in liquids and dense gases, *High Temperatures* 18:966(1981).
7. K.W.Schwarz, Charge-carrier mobilities liquid helium, *Phys. Rev.* A6:2510(1972).
8. A.Bartels, Pressure dependence of electron drift velocity in H at 77,8 K, *Phys. Rev. Lett.* 28:213(1972).
9. A.Bartels, Density dependence of electron drift velocities in He and H at 77,6 K, *Appl. Phys.* 8:59(1975).
10. A.Bartels, Density dependence of electron drift velocity in argon, *Phys. Lett.* 44:403(1973).
11. A.G.Khrapak and I.T.Iakubov. "Electrons in Dense Gases and Plasmas." Nauka, Moscow (1981).
12. J.A.Jahnke, L.Meyer, and S.Rice, Zero-field mobility of excess electron in fluid argon, *Phys. Rev.* A3:734(1971).
13. F.M.Jacobsen, N.Gee, and G.R.Freeman, Electron mobility in liquid krypton, *Phys. Rev.* A34:2329(1986).

14. S.S.-S.Huang and G.R.Freeman, Electron mobilities in ga-
    seous, liquid and solid Xe, *J.Chem. Phys.* 68:1355(1978).
15. R.A.Holroyd and N.E.Cipollini, Correspondence of conduc-
    tion band minima and mobility maxima in dielectric li-
    quids, *J.Chem. Phys.* 69:501(1978).
16. D.L.McCorcle, L.G.Christophorou, D.V.Maaxey, and J.G.Car-
    ter, Ramsauer-Townsend minimum in cross sections of
    methane, propane, butane, and neopentane, *J.Phys.* B11:
    3067(1978).
17. W.Tauchert and W.F.Schmidt, Energy of quasi-free electron
    in liquid Ar, Kr, Xe, *Z.Naturforsch.* 30a:1085(1975).
18. W.Zdrojewski von, J.G.Rabe, and W.F.Schmidt, Photoelect-
    ric determination of V - values in solid rare gases,
    Z.Naturforsch. 35a:672(1980).
19. A.O.Allen and W.F.Schmidt, Determination of energy level
    of electrons in liquid argon over a range of densities,
    Z.Naturforsch. 37a:316(1982).
20. R.Reininger, U.Asaf, and I.T.Steinberger, Density depen-
    dence of quasi-free electron state in fluid Xe and Kr,
    Chem. Phys. Lett. 90:287(1982).
21. R.Reininger, U.Asaf, I.T.Steinberger, and S.Basak, Rela-
    tionship between energy of quasi-free-electron and its
    mobility, *Phys. Rev.* 288:2426(1983).
22. S.Basak and M.H.Cohen, Deformation-potential theory for
    mobility of excess electron in liquid argon, *Phys. Rev.*
    B20:3404(1979).
23. G.Ascarelly, Calculation of mobility of electron injected
    in liquid argon, *Phys. Rev.* B33:5825(1986).
24. B.Bolties, C.de Graaf and S.W.de Leeuw, Computation of
    energy V of excess electron in dense helium and argon,
    J.Chem. Phys. 98:275(1993). B.Bolties, PhD Thesis, 1992.
25. V.M.Atrazhev and I.T.Iakubov, Thermal electrons in li-
    quids with high polarizabilities, submitted to *J.Phys.*
26. L.D.Landau and E.M.Lifshits. "Quantum Mechanics," Perga-
    mon, Oxford (1965).
27. J. Baird, Continuum dielectric model for electron in non-
    polar fluids, *Phys. Rev.* A32:1235(1985).
28. J.Lekner, Mobility maximum in rare-gas liquids, *Phys.
    Lett.* A27:341(1968).
29. L.L.Tankersley, Energy barrier for electron penetration
    into helium, *J.Low-Temp. Phys.* 11:451(1973).
30. K.Nakagawa, K.Itoh, and M.Nishikawa, Effect of molecular
    structure on density dependence of electron mobility and
    conduction band energy, *IEEE Trans. Elect. Insul.* 23:-
    509(1988)
31. Laporte P., Subtil J.L., R.Reininger, Saule V., Berns-
    torff S., and I.T.Steinberger, Evolution of intermediate
    exitons in fluid Ar and Kr, *Phys. Rev.* B35:62701987).
32. A.A.Jahnke, N.A.W.Holzwart, and S.A.Rice, Comments on
    theory of electron mobility in simple fluids, *Phys. Rev.*
    A5:463(1972).
33. B.Plenkievicz, P.Plenkievicz, and J.-P.Jay-Gerin, Calcu-
    lation of effective mass of electron in rare-gas liqu-
    ids, *in* "10th Intern. Conf. Conduction and Breakdown in
    Dielectric Liquds," Eds P.Atten and R.Tobazeon, CNRS,
    Grenoble (1990).
34. H.Nakanishi and D.M.Schrader, Simple but accurate calcu-
    lations on scattering of electrons and positrons from Ne
    and Ar, *Phys. Rev.* A34:1823(1986).

MULTIPLE SCATTERING OF ELECTRONS IN POLAR GASES - EVIDENCE FOR SHORT LIVING DIPOLE-BOUND ELECTRON STATES IN CH$_3$CN

Th. Klahn, P. Krebs, and U. Lang

Institut für Physikalische Chemie und Elektrochemie
der Universität Karlsruhe, Kaiserstraße 12, D-76128 Karlsruhe
Federal Republic of Germany

Slow electrons drifting under the influence of an external electric field E in polar gases undergo strong scattering due to the long-range anisotropic electron/electric dipole interaction.[1] For strongly polar gases with a dipole moment D $\approx$ 2 Debye the number density n_{IR} at which the de Broglie wavelength just exceeds the mean free path of the electron is low, i.e. in the order of some 10^{19} cm^{-3} at T=300 K. Therefore, multiple scattering has to be considered already at densities n $\ll n_{IR}$. In addition, it cannot be excluded that the electron is temporarily or permanently captured by the dipole molecules. The probability of such a process increases with n. Therefore, the electron mobility in polar gases shows a large dependence on n and T at these low densities.

In this paper we analyse electron mobilities in moderately dense polar gases as NH$_3$, CH$_3$OH, and CH$_3$CN according to Polischuks[2] theoretical calculations on quantum multiple scattering contributions to the mobility of the electrons. It will be demonstrated, at least qualitatively, that due to the nonideality of these polar gases correlation of molecule positions cannot be neglected.

THEORETICAL RESULTS OF POLISCHUK

The mobility of electrons in a disordered medium of anisotropic scatterers is given by[2]

$$\mu = -\frac{4\pi}{3}\frac{e}{m}\int_0^\infty dE(\partial f(E)/\partial E)(2mE)^{3/2}(\Gamma^m(E))^{-1} \tag{1}$$

where E = $(\hbar k)^2/2m$ is the electron kinetic energy. The density of states f(E) has the form corresponding to a free electron gas:

$$f(E) = (m/2\pi k_B T)^{3/2}\exp(-E/k_B T). \tag{2}$$

The quantity $2\Gamma^m(E)$ stands for the transport collision frequency. In the binary collision approximation (Lorentz limit) it is given by

$$\Gamma_0^m(E) = n\sigma_m^{av}(E)\hbar k/2m \tag{3}$$

According to Altshuler[3] the momentum transfer cross section $\sigma_m^{av}(E)$ for dipole molecules in the point-dipole approximation takes the following form within the Born approximation

$$\sigma_m^{av}(E) = (4\pi/3)(De/\hbar)^2 m/E \equiv A\, m/2E \tag{4}$$

where D is the dipole moment of the scatterers.

It was shown by Polischuk that taking into account the interference effects in multiple scattering the collision frequency is increased in comparison with $2\Gamma_0^m(E)$

$$\Gamma^m(E) = \Gamma_0^m(E)\left[1 + (0.5\pi - 0.6)\hbar\sigma_m^{av}(E)\,n/k\right]. \tag{5}$$

Finally the following deviation from the Lorentz limit μ_L is obtained[2,4]

$$\mu/\mu_L = 1 - \frac{\sqrt{\pi}}{8}(0.5\pi - 0.6)\lambda_T\langle\sigma_m^{av}(E)\rangle n \tag{6}$$

with $\displaystyle \langle\sigma_m^{av}(E)\rangle = 4\pi\int_0^\infty dE\sigma_m^{av}(E)f(E)\sqrt{2E/m^3} = A\,m/k_BT$

which is valid, however, only for $\lambda_T/L \ll 1$, where $\lambda_T = \hbar/(2mk_BT)^{1/2}$ and $L = 1/\langle\sigma_m^{av}(E)\rangle n$ are the de Broglie wavelength and the mean free path of thermal electrons, respectively. The theoretical calculations of Polischuk could be carried out only within the Born approximation (D < 2.5 Debye). Since the experimental values of $\sigma_m^{av}(E)$ are sometimes about twice as large as the Born ones (Eq. (4)) Polischuk[2] proposed to use in Eq. (6) the averaged cross section $\langle\sigma_m^{av}(E)\rangle^{exp}$ obtained from measurements at very low density:

$$\langle\sigma_m^{av}(E)\rangle^{exp} = \frac{8e}{3}(2/\pi m k_BT)^{1/2}\left[(\mu n)_0\right]^{-1} \tag{7}$$

where $(\mu n)_0$ is the experimental density-normalized mobility in the limit of $n \to 0$. Therefore, μ/μ_L in Eq. 6 has to be replaced by $\mu n/(\mu n)_0$.

ELECTRONS IN AMMONIA

In the past we have measured electron mobilities in sub- and supercritical NH_3 (D=1.47 Debye) as a function of n and T to study the transition from the quasi free to the localized electron state.[5,6] We have analysed the experimental results in context with Polischuks theoretical calculations, using also data of Christophorou et al..[7] An example is given in Fig. 1a which demonstrates an almost quantitative agreement between the experimental results and theory (Eq. (6)). It is generally noted that although the theoretical description is only valid for gas densities where $\lambda_T/L \ll 1$ (i.e. for $n \ll [\lambda_T\langle\sigma_m^{av}(E)\rangle]^{-1}$) agreement between theory and experiment is observed - although more qualitatively - in a much larger density range (see Fig. 1a). This was also implied by Polischuk when he used our electron mobility data in H_2O[8] to prove his theoretical description (see Fig. 3 in Ref. 2). To analyse this behaviour in more detail we used a more exact form of Eq. (6) obtained from Eqs. (1) to (5). To extend the calculation formally to higher densities we have introduced (as a precaution) a mobility edge E_C defined by the Ioffe-Regel rule[9] $\lambda(E_C)/L(E_C)=1$, i.e. we used for the integration of Eq. (1)

$$E \geq E_C \equiv \left[\hbar(8\pi/3)(Dem/\hbar)^2 n\right]^{2/3} \tag{8}$$

We obtained finally[10]

$$\frac{\mu}{\mu_L} = 0.5 \int_{\varepsilon_C}^{\infty} d\varepsilon \exp(-\varepsilon)\varepsilon^2 \left[1+\xi n/\varepsilon^{3/2}\right]^{-1} \tag{9}$$

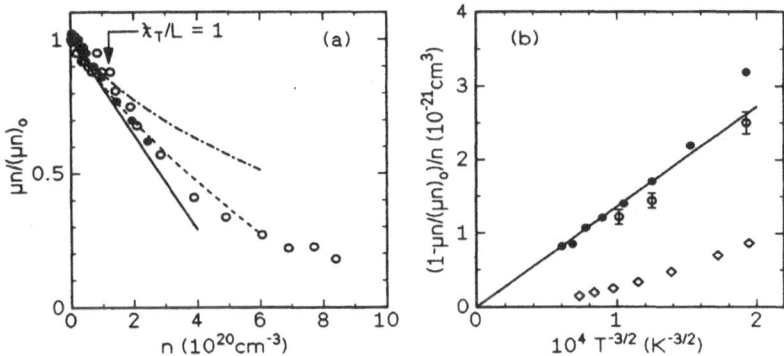

Figure 1. (a) $\mu n/(\mu n)_0$ of electrons in ammonia at T=400 K as a function of n: ●, Ref. 7; ○, Ref. 5,6. Solid line: Theoretical prediction given by Eq. (6) with $\langle\sigma_m^{av}(E)\rangle^{exp} = (7.85\pm0.65)\times10^{-14}\,cm^2$. Dash-pointed line: Prediction by Eq. (9), only valid for $\lambda_T/L \ll 1$. Dashed line: First order correction due to correlation in positions of the scatterers, Eq. (11). (b) $(1-\mu n/(\mu n)_0)/n$ as a function of $T^{-3/2}$ for electrons in NH₃ in the temperature range $300 \leq T \leq 650K$: ●, Ref. 7; ○, Ref. 5,6. Solid line: Eq. (6) with $\langle\sigma_m^{av}(E)\rangle^{exp} = (6.45\pm0.56)\times10^{-14}cm^2$ for T=460 K using a 1/T dependence of the mean scattering cross section. ◊ : Contribution -2B(T), see the text.

where $\varepsilon = E/k_B T$ and $\xi = (0.5\pi - 0.6)\hbar Am^2(2mk_BT)^{-3/2}$. The result of Eq. (9) is given in Figure 1a. Notice the small density range where Eq. (9) can be replaced by its approximation (6). Obviously, there is some other density correction to the mobility not involved in Eqs. (6) and (9). Such correction is due to correlation in position of the molecules as well as to correlation of orientation of the dipoles of the scatterers in a nonideal gas. Polischuks calculations just neglect these correlation effects. To first order, correlation in position give rise to a modified mobility for purely s wave electron-gas atom scattering:[11]

$$\mu^{imperf}/\mu_L = 1+2B(T)\,n \tag{10}$$

where B(T) is the second virial coefficient of the gas. Due to the lack of a suitable expression for correlated dipole scatterers we use Eq. (10) as a rough estimate for imperfect polar gases, too. One obtains immediately:

$$\mu n / (\mu n)_0 = 0.5 \int_{\epsilon_c}^{\infty} d\epsilon \exp(-\epsilon)\epsilon^2 \left[1 + \xi n/\epsilon^{3/2}\right]^{-1} + 2B(T)n \approx$$

$$\approx 1 - \frac{\sqrt{\pi}}{8}(0.5\pi - 0.6)\lambda_T \langle\sigma_m^{av}(E)\rangle n + 2B(T)n \qquad (11)$$

For electrons in ammonia (B(T=400K)= -120 cm^{3}mol^{-1})[12] the result of Eq. (11) is presented in Fig. 1a. It demonstrates in principle an excellent agreement between theory and experiment if one takes into account the error of $\langle\sigma_m^{av}(E)\rangle^{exp}$. This means that the properties of the imperfect gas have to be considered when studying multiple scattering processes in polar media. In Fig. 1b we have collected the results of the analysis of all experimental data in the temperature range $300 \leq T \leq 650K$ in context with Eq. (6). The correction -2B(T) presented separately in Fig. 1b demonstrates that at least at T=300 K the experimental slope $(1 - \mu n/(\mu n)_0)/n$ contains such a contribution.

ELECTRONS IN METHANOL

We have performed electron mobility measurements in sub- and supercritical CH$_3$OH (D=1.70 Debye) in the density range $1.7 \times 10^{19} \leq n \leq 4 \times 10^{21}$cm^{-3} and in the temperature range $393 \leq T \leq 573$ K.[13] The low density results are presented in Fig. 2a. The linear μn vs. n plots have been extrapolated to zero density in order to obtain $(\mu n)_0$ and finally $\langle\sigma_m^{av}(E)\rangle^{exp}$ as a function of T. The mean scattering cross sections can be described almost perfectly by Altshulers relationship (Eqs. (4) and (6)).[13] With these cross sections the first order multiple scattering correction given by Polischuk has been calculated and is compared in Fig. 2b with the corresponding experimental results. One should add a term -2B(T) to the normalized slope $(1 - \mu n/(\mu n)_0)/n$, a correction which is obviously of the right order to account for the observed deviation from a strict T$^{-3/2}$ dependence (Fig. 2b). It should be emphasized that at T=300 K the first order contribution due to this correlation in positions of the methanol molecules (6.55×10^{-21} cm^3) exceeds by far the multiple scattering contribution.

ELECTRONS IN ACETONITRILE

Due to the high dipole moment (D=3.96 Debye) the Born approximation should not be valid for electrons in CH$_3$CN. The experimental scattering cross section $\langle\sigma_m^{av}(E)\rangle^{exp} = 4.36 \times 10^{-13}$cm^2 determined by Christophorou et al.[1] at 298 K has, however, a smaller deviation from $\langle\sigma_m^{av}(E)\rangle^{theor} = 5.97 \times 10^{-13}$cm^2 calculated by Altshulers formula compared to the case of electrons in ammonia. Therefore, the following analysis of the mobilty data is justified. In Fig. 3 we have plotted a part of our experimental data in the density range for which $\lambda_T \leq L$. The different behaviour of electrons in CH$_3$CN is striking. First order multiple scattering and correlation effects underestimate by far the observed density dependence of the electron mobility in the *saturated* vapour at these low densities (being aware that B(T) does not describe well the imperfect gas near the coexistence curve).

As long as we have no multiple scattering theory including correlations in position and orientation of the scatterers the following interpretation of the strong density dependence of $\mu n / (\mu n)_0$ remains somewhat speculative. But we can show that with the assumption of an electron attachment/detachment equilibrium

$$e_f^- + CH_3CN \xleftrightarrow{K} (CH_3CN...e^-)_l \qquad \text{with } K = k_{fl}/k_{lf} \qquad (12)$$

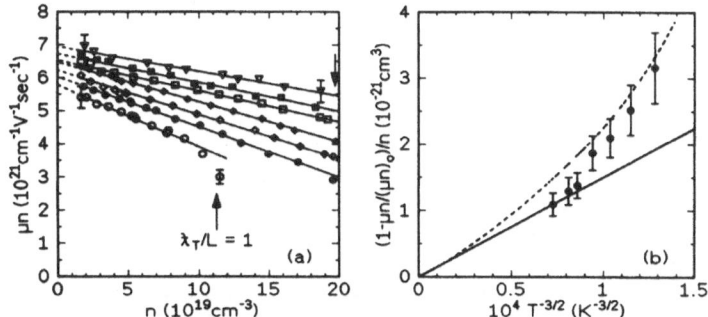

Figure 2. (a) Density normalized mobility μn of electrons in CH_3OH as a function of n at low gas densities: O , 393 K; ● , 423 K; ◊ , 453 K; ◆ , 483 K; □ ,513 K; ■ , 533 K; ∇ , 573 K. Solid lines: Linear least squares fit to the experimental data. Dashed lines: Extrapolation of these linear fits to zero density. **(b)** $(1-\mu n/(\mu n)_0)/n$ as a function of $T^{-3/2}$ for electrons in CH_3OH in the temperature range $393 \leq T \leq 573K$. Solid line: Theoretical prediction according to Eq. (6) with A = 5.02 cm^4sec^{-2}. Dashed line: Correction due to correlation in position (Eq. (11)).

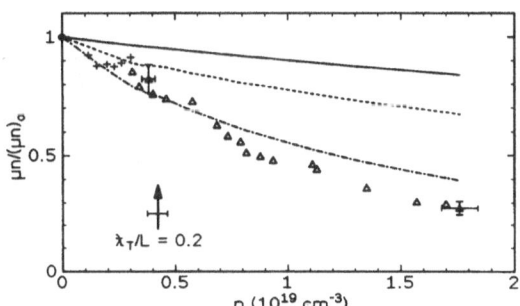

Figure. 3 $\mu n/(\mu n)_0$ of electrons in CH_3CN as a function of n ($n \leq 1/\lambda_T \langle \sigma_m^{av}(E) \rangle^{exp}$): ●, Ref. 1; +, T=298 K; Δ , $298 \leq T \leq 347K$ (saturated vapour). Solid line: Polischuks theoretical prediction, Eq. (6). Dashed line: Multiple scattering and first order correction due to correlation in position, Eq. (11). Dash-dotted line: Density normalised effective mobility $\mu_{eff}\, n/(\mu n)_0$ calculated by Eq. (13) with $\mu_f n$ from Eq. (11) and $K=4.0 \times 10^{-20} cm^3$ (assumed to be constant with respect to temperature).

we are able to describe the observed behaviour of the electrons. In this case the density normalized effective mobility is given by

$$\mu_{eff} n / \mu_f \, n = \left[1 + nK\right]^{-1} \tag{13}$$

μ_f is the mobility of multiple scattered electrons in the disordered medium of dipole scatterers correlated in position (Eq. (11)). $\tau_{fl} = 1/nk_{fl}$ is the mean life time of the electron in this state f. $\tau_{lf} = 1/k_{lf}$ is its mean life time in the temporarily occupied localized state l represented by $(CH_3CN...e^-)_l$. Eq. (13) is valid only if the mobility μ_l of the electron in the state $(CH_3CN...e^-)_l$ is much smaller than μ_f and if both, τ_{fl} and τ_{lf} are much smaller than the drift time (for details see Ref. 10). Then with a reasonable value for the equilibrium constant, i.e. $K = 4.0 \times 10^{-20} cm^3$, we can reproduce qualitatively the density dependence of the electron mobility, at least where $\lambda_T < L$ (see Fig. 3).

What is the nature of this electron state $(CH_3CN...e^-)_l$? More than twenty years ago a so-called dipole-bound electron state has been predicted:[14] Molecules with dipole moments greater than about 2 Debye should form an electronic state with a very diffuse and weakly bound electron which is sufficiently stable with respect to thermal energy collisional processes or stripping by electric fields. The dipole moment of CH_3CN is well beyond this critical value. However it was pointed out already by Garrett[15] that measurements on the very fragile dipole-bound system $(CH_3CN...e^-)$ (anion of a closed shell system) present difficult challenges to the experimentalist since an *ab initio* calculation of the electron affinity yields E.A.$= 1 \times 10^{-4} eV$. The experimental evidence for the existence of a dipole-bound state in CH_3CN is rather conflicting.[16] Recently, however, Hashemi and Illenberger[16] observed obviously dipole-bound $(CH_3CN...e^-)$ via resonant attachment of free electrons to CH_3CN clusters.

We suggest that the electron mobility in CH_3CN is controlled by such dipole-bound states which have, however, a finite lifetime. Using the attachment rate constant $k_{fl} = 7.2 \times 10^{-12} cm^3 sec^{-1}$ determined by Mothes et al.[17] for $n \leq 10^{12} cm^{-3}$ we obtain from K $\tau_{lf} = 5.6 \times 10^{-9} sec$ which is much smaller than the observed drift times of several hundred nanoseconds to a few microseconds in our present experimental set-up.

In order to learn more about such dipole-bound electron states, and also about the influence of the orientational correlation of the dipoles on multiple scattering processes of electrons the experimental work is being extended to HCN (D=2.99 Debye).

Acknowledgment

This work was supported by the Deutsche Forschungsgemeinschaft and by the Fonds der Chemischen Industrie.

REFERENCES

1. L.G. Christophorou and A.A. Christodoulides, Scattering of thermal electrons by polar molecules, *J. Phys. B: At. Mol. Phys.* 2:71(1969).
2. A.Ya. Polischuk, Theory of electron mobility in moderately dense polar gases, *J. Phys. B: At. Mol. Phys.* 18:829(1985), and references therein.
3. S. Altshuler, Theory of low-energy electron scattering by polar molecules, *Phys. Rev.* 107:114(1957).
4. P. Krebs and St. Vautrin, Mobility of excess electrons in acetonitrile vapour and multiple scattering processes in dense polar vapours, *J. de Physique IV, Colloque C5, suppl. au J. de Physique I*, Vol.1:115 (1991).
5. P. Krebs, V. Giraud, and M. Wantschik, Electron localization in dense ammonia vapor, *Phys. Rev. Lett.* 44:211(1980)

6. P. Krebs, Localization of excess electrons in dense polar vapors, *J. Phys. Chem.* 88:3702(1984), and references therein.

7. L.G. Christophorou, J.G. Carter, and D.V. Maxey, Electron motion in high-pressure polar gases: NH_3, *J. Chem. Phys.* 76:2653(1982).

8. V. Giraud and P. Krebs, The onset of electron localization in subcritical water vapor, *Chem. Phys. Lett.* 86:85(1982).

9. A.F. Ioffe and A.R. Regel, Non-crystalline, amorphous, and liquid electronic semiconductors, *Progr. Semiconductors* 4:237(1960).

10. P. Krebs, manuscript in preparation.

11. J.L. Levine and T.M. Sanders Jr., Mobility of electrons in low-temperature helium gas, *Phys. Rev.* 154:138(1967).

12. P. Davies, Ammonia *in* "Thermodynamic Functions of Gases", Vol. 1, F. Din, ed., Butterworths, London (1962).

13. P. Krebs and U. Lang, to be published.

14. J.E. Turner, Electron capture by rotational excitation of polar molecules, *Phys. Rev.* 141:21(1966).

15. W.R. Garrett, Critical binding of electron-dipole rotor systems; electronically excited states, *J. Chem. Phys.* 73:5721(1980) and references therein.

16. R. Hashemi and E. Illenberger, Observation of dipole-bound CH_3CN^- following electron attachment to CH_3CN clusters, *J. Phys. Chem.* 95:6402(1991).

17. K.G. Mothes, E. Schulten, and R.N. Schindler, Eine Untersuchung der Elektroneneinfangreaktion durch einfache Cyanide im thermischen Energiebereich, *Ber. Bunsenges. Phys. Chemie* 76:1258(1972).

THERMODYNAMICS OF ELECTRON INJECTION

Robert Schiller[*], R.A. Holroyd[+],
and Masaru Nishikawa[#]

[*]Central Research Institute for Physics
Atomic Energy Research Institute
P.O.B. 49, Budapest H-1525, Hungary
[+]Chemistry 555, Brookhaven National Laboratory
Upton, NY 11973, USA
[#]Department of Pure and Applied Sciences
University of Tokyo, 3-8-1 Komaba,
Meguro-ko, Tokyo 153, Japan

INTRODUCTION

The energy of the quasifree excess electron, V_0, in an insulating fluid is regarded as an equilibrium property. However, the usual experimental determination of V_0 involves rapid perturbations. Some recent results bring out this difference between theoretical concept and experimental observations markedly. The aim of the forthcoming thermodynamic considerations is to reconcile some of the contradictions which stem from the different footings of theory and experiment.

A liquid which comprises a quasifree electron is thought to have the macroscopic properties and microscopic structure of the liquid at equilibrium. The electron state is determined by the repulsive pseudo-potential of the closed shells of the liquid atoms or molecules, and by the attractive potential due to polarization. The space and energy distribution of the molecules are those which prevail in the neat liquid[1-3].

Thus theory defines the energy change of the system due to the inclusion of one quasifree electron, V_0^{th}, as the partial derivative of the system energy, U, with respect to the mole number of the electrons, N, at constant temperature, T, and volume, V, i.e.

$$\left(\frac{\partial U}{\partial N}\right)_{T,V} = V_0^{th} = V_0^{TV} \tag{1}$$

Linking the Gaseous and Condensed Phases of Matter
Edited by L.G. Christophorou et al, Plenum Press, New York, 1994

A microscopic theory of V_0^{th} is based on the Wigner-Seitz liquid model by which electron kinetic energy, K, and potential energy, U_p, are determined separately[4]. Here K is calculated as

$$K = \frac{\hbar^2 k_0^2}{2m} \quad ; \qquad \tan[k_0(r_s - a)] = k_0 r_s \quad , \tag{2}$$

where k_0 is the lowest wavenumber and m is the rest mass of the electron, a is the hard core radius and r_s the Wigner-Seitz radius of the liquid the latter being obtained from the number density, ρ, as $r_s = (3/4\pi\rho)^{1/3}$.

The expression for U_p is of the form

$$U_p = \frac{3}{2} \frac{\alpha e^2}{r_s^4} \left(\frac{8}{7} + \frac{1}{1 + \frac{8\pi\alpha}{3\varrho}} \right) \quad , \tag{3}$$

where α is the polarizability of the liquid molecule. Finally V_0^{th} is found as

$$V_0^{th} = V_0^{TV} = K + U_p \quad . \tag{4}$$

A recent theoretical work[5] lends strong support to this approximation.

The experimental determination of V_0 is based on photoelectron injection from a photocathode or photoionization of a solute[6]. In view of their rapidity these processes are regarded as adiabatic ones[1].

The obvious difference, however, between isothermal theory and adiabatic experiment has not been brought out until recently. This was so because experimental conditions, like temperature or pressure, were not varied drastically and it was easy to fit experimental data to theoretical predictions by a proper choice of the hard core radius, a, a parameter which is usually not too well known from independent sources.

Recently Holroyd et al.[7] determined V_0 for a number of hydrocarbon liquids under different pressures between 1 and 2500 bar and found the increase with increasing pressure to be markedly less than expected by eqs.(2-4). This observation is being tried to understand in terms of the difference between the thermodynamic conditions of theory and experiment.

THERMODYNAMIC CONSIDERATIONS AND RESULTS

Injection, as the majority of the kinetic processes, is expected to be polytropic, i.e. none of the thermodynamic variables are held constant throughout the process. Four limiting cases, with different sets of constraints, will be considered, asking whether the real conditions of photoelectron injection are nearer to the case of (a) constant entropy and volume, (b) constant entropy and pressure, (c) constant temperature and volume, or (d) constant temperature and pressure. One finds the expressions :

$$\left(\frac{\partial U}{\partial N}\right)_{S,V} = V_0^{SV} = \mu \quad , \tag{5}$$

$$\left(\frac{\partial U}{\partial N}\right)_{S,p} = V_0^{Sp} = \mu - pv_{Sp} = \mu - p\left(\frac{\partial \mu}{\partial p}\right)_{S,N} \quad , \tag{6}$$

$$\left(\frac{\partial U}{\partial N}\right)_{T,V} = V_0^{TV} = \mu + Ts_{TV} = \mu - T\left(\frac{\partial \mu}{\partial T}\right)_{V,N} \quad , \tag{7}$$

$$\left(\frac{\partial U}{\partial N}\right)_{T,p} = V_0^{Tp} = \mu - pv_{Tp} + Ts_{Tp} = \mu - p\left(\frac{\partial \mu}{\partial p}\right)_{T,N} - T\left(\frac{\partial \mu}{\partial T}\right)_{p,N} \quad . \tag{8}$$

Here s and v denote the partial molar entropy and volume, μ the chemical potential of the electron, respectively, with the variables in the subscript held constant.

As usual, a is determined so as to make V_0^{TV} equal to V_0(experimental). Making use of these a values for four hydrocarbons, the pressure (i.e. density) dependence of V_0^{TV} was calculated by eqs. (2-4). The results are given in Fig. 1. together with the experimental points. Theoretical values are seen to be always larger than experimental data, the difference increases with relative density.

Since experiments were said to be nearer to adiabatic than to isothermal conditions it looked reasonable to evaluate V_0^{Sp}. By combining eqs. (5) and (6) one finds

$$V_0^{Sp} = V_0^{TV}(p) - p\varrho \frac{dV_0^{TV}}{d\varrho} \kappa_s \quad , \tag{9}$$

where κ_s denotes adiabatic compressibility. V_0^{Sp} values by eq. (9) are also plotted in Fig. 1. Experimental data are seen to fall invariably between the $T,V=const.$ and the $S,p=const.$ curves.

An inspection and numerical estimates of eqs. (5-8) show the series of inequalities $\Delta V_0^{SV} > \Delta V_0^{TV} > \Delta V_0^{Sp} > \Delta V_0^{Tp}$ to prevail where ΔV_0 denotes $V_0(p) - V_0(1\ bar)$. Hence, in view of these inequalities and of Fig. 1. one can conclude that **the polytrop which describes electron injection lies between the isothermal-isochoric and isoentropic-isobaric curves.**

The data in Fig. 1. indicate that experimental and theoretical V_0 values coincide at low pressures, only because the hard-core radii, a, are properly fitted. Let now the low pressure limit of the systematic difference between theory and experiment be estimated. Approximating V_0(experimental) by $V_0^{Sp}(p=0) = V_0^0$ one finds the relationship

$$V_0^{th} = V_0^{TV} = V_0^0 - T\left[\frac{\alpha_p}{\kappa_T}\left(\frac{\partial V_0^0}{\partial p}\right)_{N,T} + \left(\frac{\partial V_0^0}{\partial T}\right)_{N,p}\right] \quad , \tag{10}$$

where α_p denotes isobaric thermal expansivity and κ_T isothermal compressibility. The derivative $(\partial V_0^0 / \partial T)_{N,p}$ was estimated[8] to be about -10^{-3} eV/K which, together with average liquid properties, results in some ± 0.03 eV difference between theory and experiment. (The magnitudes of the two terms in the RHS bracket of eq. (10) are commensurable but of

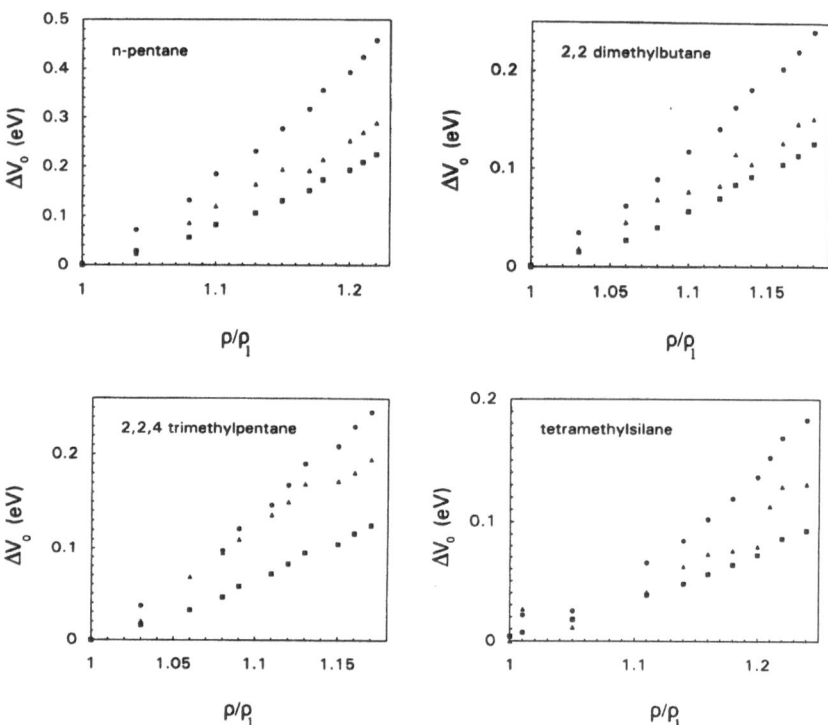

Figure 1. $\Delta V_o = V_o(p)\text{-}V_o(1\ bar)$ as a function of relative density $\rho(p)/\rho_1$ with ρ_1 denoting density at 1 bar. ● $T, V=const.$: ■ $S, p=const.$; ▲ experimental

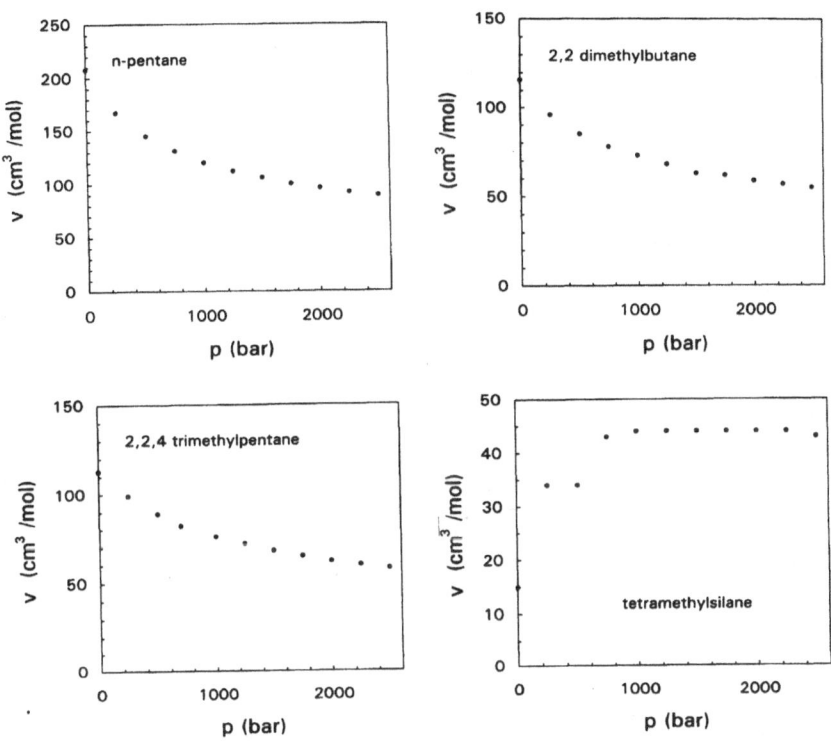

Figure 2. Isothermal-isobaric partial molar volume of quasifree electrons as a function of density, evaluated by eq.(11)

opposite sign, thus more specific data are needed for the predicition of even the sign of the small difference.) What is important from a practical point of view is that **low pressure injection measurements are reasonably well described by the generally accepted isothermal-isochoric model.**

The pressure derivative of V_0^{TV}, which can be evaluated by eqs. (2-4) and (9), renders the isothermal-isobaric partial molar volume of the quasifree electron as

$$(\frac{\partial V_0^{TV}}{\partial p})_{T,N} = v_{Tp} \quad . \tag{11}$$

The partial molar volumes in four neat hydrocarbons as a function of pressure are given in Fig. 2. The values are seen to vary in the range of 40 to 200 cm^3/mol. Molar volumes decrease with increasing pressure, with the exception of tetramethylsilane. No speculations are being made about these findings.

ACKNOWLEDGEMENT

Help and advice of Drs. L. Balázs, G. Jancsó, G. Nagy, L. Nyikos and T. Pajkossy is gratefully acknowledged. The work was supported by the National Research Fund of Hungary, under Contract No. OTKA 2981

REFERENCES

1. J.Jortner, Excess electron states in liquids, *in:* "Actions Chimiques et Biologiques des Radiations," M. Haissinsky, ed., Masson, Paris (1970)
2. H.T.Davis and R.G.Brown, Low-energy electrons in nonpolar fluids, *in:* "Advances in Chemical Physics Vol. XXXI," I.Prigogine and S.A.Rice, eds., Wiley, New York (1975)
3. W.F.Schmidt, Electronic energy levels in nonpolar dielectric liquids, *in:* "Excess Electrons in Dielectric Media", C.Ferradini and J.-P.Jay-Gerin, eds., CRC Press, Boca Raton (1991)
4. B.E.Springett, J.Jortner and M.H.Cohen, Stability criterion for the localization of an excess electron in a nonpolar fluid, *J. Chem. Phys.* 48:2720(1968)
5. J.-M.Lopez-Castillo, Y.Frongillo, B.Plenkiewicz and J.-P.Jay-Gerin, Path-integral molecular dynamics calculation of the conduction-band energy minimum V_0 of excess electrons in fluid argon, *J. Chem. Phys.* 96:9092 (1992)
6. R.A.Holroyd, The electron: its properties and reactions, *in:* "Radiation Chemistry, Principles and Applications," Farhataziz and M.A.J.Rodgers, eds., VCH Publ., Weinheim (1987)
7. R.Holroyd, M.Nishikawa, K.Nakagawa and N.Kato, Pressure dependence of the conduction-band energy of nonpolar liquids, *Phys. Rev. B* 45:3215(1992)
8. R.A.Holroyd, S.Tames and A.Kennedy, Effect of temperature on conduction band energies of electrons in nonpolar liquids, *J. Phys. Chem.* 79:2857(1975)

SECTION V. ELECTRON ATTACHMENT IN THE GASEOUS AND THE CONDENSED PHASES OF MATTER

ELECTRON ATTACHMENT TO MOLECULES

Eugen Illenberger

Institut für Physikalische und Theoretische Chemie
Freie Universität Berlin
Takustrasse 3
D-14195 Berlin, Germany

INTRODUCTION

STABLE ANIONS AND TRANSIENT NEGATIVE IONS

This contribution focuses on anion formation resulting from low energy (0-20 eV) electron impact to molecules. In the gas phase the interaction of free electrons with neutral molecules can directly generate a molecular anion via resonant electron capture [1-3]

$$e^- + M \rightarrow M^{(*)-} \tag{1}$$

An anion such formed contains more energy than the neutral and is thus unstable with respect to loss of the extra electron (autodetachment). During its lifetime, however, it may lose energy by different processes like emission of radiation or collisions with other molecules and thus relaxing into a stable configuration. The question is, whether such relaxation mechanisms can compete with autodetachment under the particular experimental conditions. Finally, the ion may also decompose into stable fragments (see below).

Due to its finite lifetime such a molecular ion is called a "temporary negative ion" (TNI). Synonymously the term "resonance" is frequently used since its formation is a resonant process. Before discussing the fate of such an unstable molecular ion, we shall recall some general facts on the stability of negative ions.

The possibility of an atom or a molecule to form a *thermodynamically* stable anion is expressed by the electron affinity. The *electron affinity* of a molecule is formally defined as the energy difference between the neutral (M) and the anion (M⁻) in their respective ground states. By convention, the electron affinity of M is considered *positive* if the ground state of M⁻ lies *below* that of M, and *negative* if the ground state energy of M⁻ is *higher* than M. A positive value for the electron affinity indicates the existence of a stable anion in which the extra electron exists in a bound state. The electron affinity of a neutral particle thus corresponds to the ionization energy (or detachment energy) of the anion. (Ionization is commonly defined as a transition associated with the removal of one or more electrons.)

In molecules which are characterized by considerable geometry change between the neutral and the anion, one has to distinguish between the (adiabatic) electron affinity (EA) of M and the vertical detachment energy (VDE) of M⁻ (Fig. 1). The latter is the number which is experimentally obtained in a Franck-Condon transition in photodetachment from molecular anions. The problem is analogue to adiabatic and vertical ionization energies in photoionization.

Thermodynamically stable molecular anions can thus not directly be generated by free electron attachment. They can e.g. be formed in charge transfer reactions from anions or neutrals

$$A + M \rightarrow A^+ + M^- \qquad (2a)$$
$$A^- + M \rightarrow A + M^- \qquad (2b)$$

Since the binding energy of the extra electron to M is usually less than the ionization energy of A, reaction (2a) is generally endothermic and can only occur if A and M contain sufficient (kinetic and/or internal) energy.

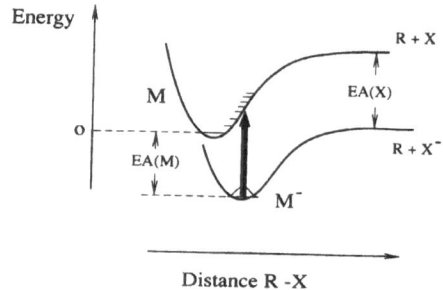

Distance R -X

Figure 1. Potential energy curves illustrating electron affinity (EA) and vertical detachment energy (VDE) in a molecule. From ref. 3.

A great many molecules, among them such ubiquitous compounds as N_2, CO_2, CH_4, ethylene, benzene, etc., are not able to form *thermodynamically stable anions*. By capture of free electrons, however, they can form temporary negative ions (TNIs).

As mentioned above free electron attachment is an electronic transition which, in any case, initially creates an unstable (or metastable) negative ion with respect to ejection of the extra electron. This is true, regardless of the sign of the electron affinity. If it is positive, electron attachment generates an (electronically) excited state of the anion. A subsequent transition to the thermodynamically stable ground state is possible, provided that stabilization mechanisms are operative within the lifetime of the TNI. This is, however, generally not the case under single collision conditions (see below).

While the ionization energy of most organic molecules lies around 10 eV, the electron affinity scale ranges to about 3.8 eV.[4] The comparatively low electron binding energy and a very restricted number of bound electronic states (if existent at all) are the salient features of negative ions as opposed to neutral molecules or positive ions. This can easily be rationalized by considering the interaction of an electron with the rest of the molecule: in a neutral molecule or in a positive ion an excited electron may occupy a valence type molecular orbital (MO) or a Rydberg type MO. In the latter case the excited electron and the positive core are subjected to the long range Coulomb interaction

$$V(r) = -ze^2/r \tag{3}$$

For the Rydberg series converging to the first ionization limit, this leads to an infinite number of bound, electronically excited states.

The situation changes completely on going to the anion. Here the extra electron interacts with a neutral molecule. At larger distances, this interaction can be approximated by the electron induced dipole potential

$$V(r) = -\alpha e^2/2r^4 \tag{4}$$

(α is the polarizability of the neutral molecule) which falls off much faster than the Coulomb potential leading to a comparatively low binding energy for the extra electron - if a bound state exists at all.

FREE ELECTRON ATTACHMENT TO MOLECULES ELECTRONIC CONFIGURATION OF RESONANCES

To discuss the mechanism responsible for the trapping of an extra electron we will briefly describe the interaction of electrons with molecules more generally. Consider the interaction of an electron with a molecule M in the gas phase under single collision conditions. This interaction may be divided into two classes, namely, *direct scattering* and *resonant scattering*. In *direct scattering* the incident electron collides with the target molecule and will eventually be deviated from its original direction (Fig. 2). If the energy of the electron is unaffected by the scattering process, we have *elastic scattering*. Strictly speaking, due to momentum and energy conservation, the electron always loses an amount of the order m/M of its initial energy. Since this fraction is generally less than 10^{-5}, it is negligible in most cases. If the electron loses some energy due to exciation of internal degrees of freedom in the target molecule, we speak of *inelastic scattering*. According to the argument above (m « M), excitation of rotational and vibrational energy is unlikely, so direct inelastic scattering predominantly causes electronic excitation of the target molecule. The electronic excitation, of course, may subsequently result in vibrational excitation of the ground electronic state through internal conversion, etc..

Resonant scattering is considered to occur if the incoming electron is trapped for a certain time (considerably longer than the direct transit time) in the vicinity of the molecule to form a temporary negative ion (TNI). Electron capture can only happen if the energy of the incident electron fits that of the TNI. The process thus represents an electronic transition from a continuum state (M+e⁻) to a discrete (quasi bound) state of the anion. The lifetime of a TNI varies on a large scale depending on the energy of the resonance and the size of the molecule. It ranges from less than a few vibrational periods (10^{-14} s), as in N_2^- [5] to the ms range for larger polyatomic molecules such as SF_6^- and perfluorinated) compounds .[6] For comparison, an electron of 1 eV takes 5×10^{-16} s to travel a distance of 3 Å, as is the case in direct scattering. According to Heisenberg's uncertainty principle, the lifetime of a TNI is connected with an energy width given by

$$\Gamma \approx \frac{\hbar}{\tau},$$

with $\hbar = h/2\pi = 6.6 \times 10^{-16}$ eV·s, the Planck constant. A resonance with a mean lifetime of $\tau = 10^{-14}$ s has thus a natural linewidth of $\Gamma \approx 66$ meV. We will see below that, in molecules, the width of a resonance is generally controlled by the Franck-Condon transition rather than by its lifetime.

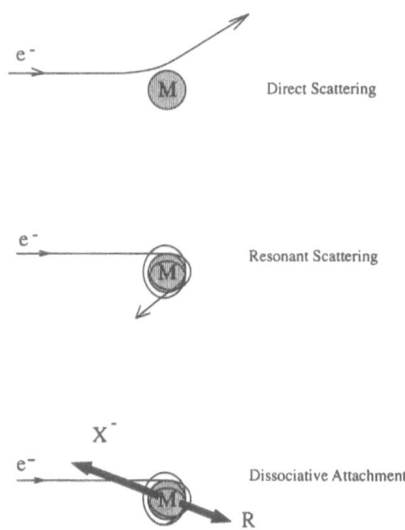

Figure 2. Schematic representation of direct and resonant electron scattering.

If the energy of the detached electron equals that of the incident electron, we speak of *resonant elastic scattering*; otherwise, we refer to *resonant inelastic scattering*. Due to the longer residence time of the extra electron, the neutral molecule may be left in a vibrationally excited state. This is the case when the TNI is characterized by a substantial geometry change, electron capture thereby providing a strong coupling between electronic and nuclear motion.

We will now briefly consider the *electronic configuration* of temporary negative ions. If the electron is attached without changing the electronic configuration of the target molecule, i.e., if the state of the TNI is formally obtained by adding the extra electron into one of the empty (virtual) MOs, we speak of a *single particle (1p) resonance*. On the other hand, if electron attachment is accompanied by an electronic excitation of the molecule, we have a *two particle - one hole (2p-1h) resonance*, with two electrons in normally unoccupied MOs. Such electronic states are also called *core excited resonances*. Note that formation of a 2p-1h resonance represents a two electron transition, similar to the 2h-1p electronically excited ("non Koopman's") states in positive ions.

If the energy of the core-excited resonance lies *above* that of the associated electronically excited neutral molecule (the parent), we speak of an *open channel core excited resonance*; if it lies below, we refer to a *closed channel* or *Feshbach resonance*. A Feshbach resonance cannot decay into the associated excited neutral molecule via a one electron process. Its decay into the ground state of the neutral requires a change of the electronic configuration which often results in a longer lifetime.

Finally, if the incoming electron induces a strong coupling between electronic and nuclear motion, the transfer of electronic to vibrational energy (EVT) can prevent the extra electron from autodetaching. Such states are called *nuclear-excited Feshbach resonances*.

What is the mechanism responsible for trapping an incoming electron for times considerably longer than the direct transit time through the molecule's dimension? In the case when the incoming electron induces an electronic excitation in the target molecule that results in a Feshbach resonance, the extra electron is then simply trapped in the field of the excited molecule; its emission via a one electron process is energetically not possible.

The other mechanism which applies for single particle and open-channel core-excited resonances describes the trapping by the *shape* of the effective interaction potential between the incoming electron and the molecule. Therefore, such resonances are called *shape resonances*.

The trapping mechanism can be pictured using the (spherically symmetric) polarization potential introduced above for the long-range interaction. The effective potential is then

$$V(r) = - \frac{\alpha e^2}{2r^4} + \frac{\hbar^2 l(l+1)}{2mr^2} ,$$

(5)

with l being the angular momentum quantum number of the incoming electron. Combination of the attractive polarization and the repulsive centrifugal potential results in a potential barrier. If the molecule possesses an energetically accessible unfilled MO, characterized through its symmetry by a particular value of l, then an incoming electron of that particular angular momentum quantum number can temporarily be captured within the barrier. The resulting TNI represents a discrete electronic state lying in the (ionization or detachment) continuum. Such states are called *quasi-bound states*; they only exist for $l \neq 0$

THE FATE OF A TEMPORARY NEGATIVE ION

We will now consider the evolution of a polyatomic temporary anion formed by resonant electron capture. The molecular anion can react via the following channels

$$M^{-(*)} \xrightarrow{\tau_a} M^{(*)} + e^-$$

(6)

$$M^{-(*)} \xrightarrow{\tau_r} M^- + h\omega$$

(7)

$$M^{-(*)} \xrightarrow{\tau_d} R + X^- \text{ or } R^- + X \text{ etc.}$$

(8)

Reaction (6) represents autodetachment, reaction (7) is radiative stabilization to the stable ground state which is only possible for compounds having a positive electron affinity, and (8) finally is the unimolecular decomposition into stable neutral and negatively charged fragments (dissociative electron attachment).

As mentioned above, the autodetachment lifetime (τ_a) may extend up to the ms scale for larger molecules. Such metastable anion states are typically generated within a narrow resonance near 0 eV where dissociation channels into stable fragments are not yet accessible. Owing to their long lifetime, they can easily be observed by mass spectrometric techniques.

Although radiative stabilization (Eq. 7) has been studied for a number of atoms[7], data for molecules are virtually non existent. Since radiative lifetimes are of the order of 10^{-9}-10^{-8} s, it is indeed likely that radiative stabilization is slow compared with the competing channels (6) and (8). In addition, metastable anions may undergo internal conversion rather than radiative stabilization.

Finally, dissociative attachment typically proceeds in the time domain between 10^{-14} and 10^{-12} s, depending on the mechanism of the reaction. The decomposition can occur if: (a) thermodynamically stable, negatively charged fragments exist for the respective compound, and (b) these channels are energetically available at the energy of the TNI. If the dissociative attachment channel is accessible, it will generally strongly compete with autodetachment.

Let us consider the simple reaction leading to two fragments

$$e^-(\varepsilon) + M \rightarrow M^-(\varepsilon) \rightarrow R + X^- \tag{9}$$

with ε is the electron energy. The process can be pictured in a Born-Oppenheimer potential energy diagram, as depicted in Fig.3. Such curves are rigorously applicable only to diatomic molecules. For polyatomics they represent two-dimensional cuts through

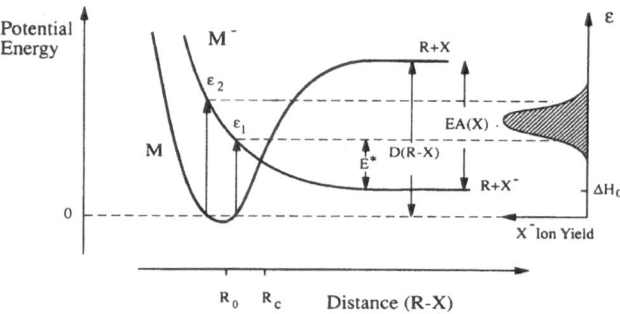

Figure 3. Born-Oppenheimer Potential energy curves illustrating electron attachment and direct electronic dissociation via a purely repulsive potential.

hyperdimensional surfaces along a reaction coordinate Q. However, they retain their simple meaning (even for polyatomic molecules) for a *direct electronic dissociation*. In this case, the extra electron has a strong and specific influence on the R-X bond, or, in terms of MOs, the electron is captured into an MO with a significant R-X antibonding character. In accordance with the Franck-Condon principle, transitions from the continuum state ($M + e^-$ ($r \rightarrow \infty$)) to M^- are only possible within the energy region between ε_1 and ε_2.

The molecular anion once formed may either dissociate into $R + X^-$ or lose the extra electron to regenerate the neutral molecule. Autodetachment can occur for $R \lesssim R_c$, where R_c is the crossing point of the two potential energy curves. For $R > R_c$ the additional electron is localized and bound to fragment X to an extent that autodetachment of M^- is no longer possible.

The line shape of the ion yield curve of fragment X^- is explained by the reflection principle.[8] Here, the continuum wave function is approximated by a delta function at the classical turning point, so that the vibrational wave function of the neutral ground state is *reflected* at the repulsive potential of the anionic state. The width of the DA resonance is then directly proportional to the slope of the anionic potential curve within the Franck-Condon region.

Of course, in a polyatomic molecule dissociative attachment may generally not proceed directly via a purely repulsive energy surface as in Fig. 3, but rather through more indirect processes such as electronic or vibrational predissociation or rearrangement in the precursor ion prior to dissociation.[3]

Figure 3 directly gives the energy balance for dissociative electron attachment

$$\varepsilon = D(R\text{-}X) - EA(X) + E^* \tag{10}$$

D is the bond dissociation energy in the neutral molecule, and EA is the electron affinity of X and $\varepsilon_1 \le \varepsilon \le \varepsilon_2$. The minimum heat of reaction is

$$\Delta H_O = D(R\text{-}X) - EA(X) . \tag{11}$$

Combining Eqs. (10) and (11) we can write for the total excess energy:

$$E^* = \varepsilon - \Delta H_O . \tag{12}$$

For reactions where ΔH_O is known, Eq. (12) gives the total excess energy for a given incident electron energy. In polyatomic molecules, E^* is generally shared among the different degrees of freedom (translational and internal energies of the fragments). If the kinetic energy of the ionic fragment is measured, the principle of conservation of linear momentum will give the total translational energy (E_T) imparted to both fragments. Furthermore, in the case where X^- is an atomic ion, energy amounting to ($E^*\text{-}E_T$) must be deposited as internal energy in the remaining neutral fragment. Measurement of the ion kinetic energy as a function of the incident electron energy yields detailed information on the distribution of excess energy in the unimolecular decomposition of the TNI.

Conversely, in cases where ΔH_O is not known, experimental determination of the appearance energy and the kinetic energy release of the ionic fragment will help to provide more detailed information on thermodynamic quantities such as bond dissociation energies and electron affinities. By that, the method represents a simple and powerful way of studying energy distributions in unimolecular decompositions. From photoionization e.g., analogous information is only accessible through coincidence techniques between photo-ions and photoelectrons.

EXPERIMENTAL CONSIDERATIONS

In electron attachment spectroscopy (EAS) a beam of electrons produced by an electron monochromator is crossed with the target gas beam and the anions are recorded with a mass spectrometer. Figure. 4 shows a configuration[3,19] which uses a trochoidal electron monochromator (TEM)[9] and a quadrupole mass filter. The monochromator is made of molybdenum in order to reduce surface problems. Non-magnetic stainless steel is used for the other components and for the entire ultrahigh vacuum system.

The operating principle of the TEM is based on the dispersion of the electrons from the heated filament according to their velocity.[9] This occurs in a region between two parallel plates (C_1,C_2) where they are subjected to a crossed electric and magnetic field. The energy-selected electrons are then accelerated or decelerated by the electrodes S_2 and S_3 and enter the reaction chamber which is kept at a fixed electrical potential.

Ions formed in the reaction volume (as defined by the crossing of the electron beam with the target gas beam effusing from a capillary) are extracted by a small electric field (less than 1 V cm^{-1}) and accelerated by a series of parallel electrodes onto the entrance hole of a commercial quadrupole mass filter.

After passing through the reaction chamber the electrons are accelerated towards the electron collector (A) which serves as a beam intensity monitor. The pressure in the reaction chamber is kept low (less than 10^{-4} mbar) in order to ensure single-collision conditions. This establishes that the electron intensity measured at the collector (A) is essentially equal to the primary intensity of the beam.

The electron energy distribution can be obtained either by applying a stopping voltage

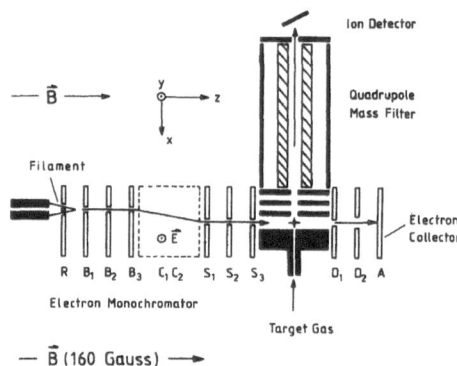

Figure 4. Schematic representation of the electron attachment spectrometer. From ref. 19.

to electrode S_3 and recording the electron retarding curve or by measuring the associative electron attachment to SF_6, which is known to be sharply peaked.[2] The SF_6 resonance is also used to calibrate the energy scale.

The ion yield curves presented here were measured with an energy resolution between 0.1 and 0.2 eV (and at times even worse!). Despite the limited energy resolution, the present configuration - the combination of a TEM with a quadrupole mass filter - is particularly suited to the study of electron attachment reactions in the very low energy region since it combines several important necessary properties: (1) the axial magnetic field prevents spreading of the electron beam at low energies so that reasonable intensities can be achieved even near 0 eV. Beam spreading is a major problem when electrostatic devices are used, (2) the TEM can be operated in a continuous mode. Because of the presence of the magnetic field, the energy of the electrons entering the reaction chamber is not influenced by the presence of the (continuous) ion draw-out field, and (3) the alignment of the electron beam by the magnetic field allows a separation of electrons and negative ions at the secondary electron mulitplier.

The translational energy analysis of the product ions uses a time of flight (TOF) technique in combination with a quadrupole mass filter as elsewhere described in detail.[10] In brief, the pulsed electron beam (pulse width < 1 μs) interacts with the molecular beam and the flight time of the ions (generated within the short time of the electron pulse) from the reaction volume through the quadrupole to the detector is measured. The TOF arrangement, in principle, corresponds to a two-field TOF spectrometer, consisting of a draw-out region, an acceleration region, and the quadrupole mass filter which acts as the drift tube. Since the quadrupole guarantees the mass separation, one can use a low draw-out field (4 Vcm^{-1}) in order to visualize excess translational energies on the time scale.

Ions formed with low kinetic energy (e.g. thermal ions) exhibit one peak in the TOF spectrum while ions generated with considerable kinetic energy produce two peaks due to "direct" and "turn around" ions. In the latter case discrimination against velocity components perpendicular to the flight tube axis results in a separation (in time) of ions ejected parallel (direct) or antiparallel (turn around) to the axis. The turn around ions are decelerated, reversed and then accelerated and reach the detector by some time delay (ΔT) with respect to the direct ions. From the experimentally determined flight time difference (ΔT), the initial kinetic energy release of the product ion can be calculated as

$$E_T^i = \frac{(\Delta T q \varepsilon_1)^2}{8 m_i} \tag{13}$$

where ε_1 is the draw-out field introduced above, q the elementary charge and m_i the mass of the ion. Conversely, for ions with low kinetic energy (< 0.1 eV for the present configuration) there is no discrimination against perpendicular velocity components and all ions are transmitted to the quadrupole. For a Maxwell-Boltzmann velocity distribution, the corresponding TOF peak has a Gaussian shape. Its width $\Delta T_{1/2}$ (fwhm) is related to the average kinetic energy by

$$\overline{E_T^i} = \frac{(\Delta T_{1/2} q \varepsilon_1)^2}{3.7 m_i} \quad . \tag{14}$$

Recording of the flight times is performed by a time-to-pulse height converter (TPHC) followed by A/D conversion.

RESULTS AND DISCUSSION

THE HALOGENATED METHANES CF₃I AND CF₄

Figure 5 shows the energy profiles of fragment ions observed from gas phase CF₃I under single collision conditions. We observe two prominent resonances near 0 eV and around 3.8 eV due to dissociative electron attachment. The continuous increase in the F⁻ and CF₃⁻ion signal above ≈ 12 eV is the result of the (non resonant) ion pair formation process

$$e^- + M \quad \rightarrow \quad M^{**} + e^-$$
$$\longmapsto \quad R^+ + X^- \tag{15}$$

The spectra in Fig. 5 were recorded at different target gas pressures for the different ions to ensure *single collision conditions*. The pressure indicated in the figure corresponds to a range where it was established that the signal of the respective ion varies linearly with the pressure. From Fig. 5 one obtains the pressure normalized relative intensities (peak values) of the different ions as I^-: F^-: FI^-: $CF_3^- \approx 10^5$: 30 : 10 : 0.5. We have thus a huge number of I^- ions formed near zero eV with respect to the other ions formed at higher

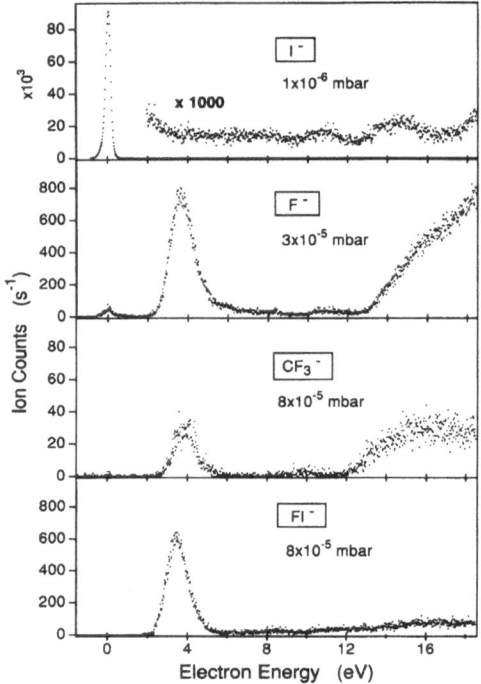

Figure 5. Negative ion formation from CF_3I under single collision conditions. The different products are recorded under different target gas pressure to ensure single collision conditions.

energy. With the exception of FI^- the unimolecular decompositions involve the cleavage of a single bond. The asymptotic limits (ΔH_0) for the corresponding reactions can be calculated as

$$
\begin{array}{llll}
e^- + CF_3I & \rightarrow & I^- + CF_3 \; ; & \Delta H_0 = -0.67 \pm 0.1 \text{ eV} & (16) \\
& \rightarrow & I + CF_3^- \; ; & \Delta H_0 = 0.7 \pm 0.3 \text{ eV} & (17) \\
& \rightarrow & F^- + CF_2I \; ; & \Delta H_0 \approx 1.5 \pm 0.3 \text{ eV} & (18)
\end{array}
$$

These numbers are derived from well established thermochemical data.[11]

The weak contribution of F^- near 0 eV can thus not arise from the direct reaction (18). Its intensity increases more than linearly with pressure but less than the square as expected for a pure secondary reaction. Due to the extreme overflow of I^- at zero eV we suggest that an ion-molecule reaction of the form

$$
I^- + CF_3I \quad \rightarrow \quad F^- + CF_2I_2
$$

may contribute to the formation of F⁻ near zero eV. This reaction is probably endothermic with some additional activation energy. On the other hand, I⁻ is formed with some excess energy (see below) which could help to make this reaction operative under the present experimental conditions.

Under single collision conditions, the parent molecular ion CF_3I^- is not formed. This ion was observed in charge transfer experiments with alkali atoms.[12-14] The electron affinity values of CF_3I derived from these experiments range from 1.4 to 2.2 eV.

The unimolecular reactions (16) and (17) exhibit a particular behavior in that they proceed via the release of remarkable amounts of excess energy as evident from the TOF measurements (Fig. 6).

Both I⁻ and CF_3^- show a separated TOF doublet due to direct and turn around ions indicating a fairly discrete initial energy of the ions. Note that the TOF spectrum of CF_3^- was recorded with a higher ion draw out field (4.0 Vcm⁻¹ instead of 1.6 Vcm⁻¹). This was necessary in order to prevent the energetic CF_3^- turn around ions from striking the back electrode of the reaction chamber. From the experimentally obtained time difference (ΔT) and Eq. (13) we obtain E_T^i (CF_3^-) = 1.7 ± 0.1 eV and E_T^i (I⁻) = 0.2 ± 0.02 eV.

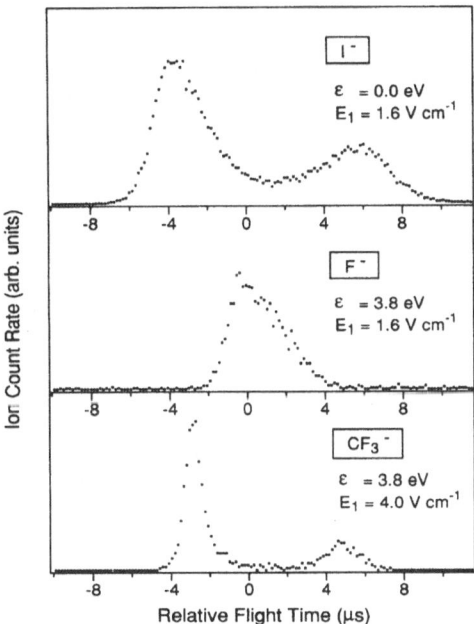

Figure 6. TOF spectra of I⁻, F⁻ and CF_3^-. The CF_3^- spectrum is recorded at a higher draw out field in order to prevent the turn around ions from striking the back electrode. From ref. 23.

In both cases energetics dictates that we have a decomposition into *two* fragments and the principle of linear momentum conservation determines the amount of translational energy imparted to the two fragments

$$E_T = E_T^i \cdot \frac{M}{m} \tag{19}$$

(M is the mass of the parent molecule and m that of the neutral fragment) which leads to $E_T = 0.57 \pm 0.06$ eV for reaction (16) and $E_T = 2.6 \pm 0.2$ eV for reaction (17). The analysis shows that in both cases most of the available excess energy appears as translational energy of the decomposition products.

In contrast to that, the TOF analysis of FI^- and F^- indicates low kinetic energy. FI^- exhibits one single and narrow Gaussian shaped peak (not shown here) and F^- shows one single and slightly asymmetric peak (Fig. 6). This shape is either caused by an unresolved doublet (incomplete discrimination of velocity components perpendicular to the flight tube axis due to low kinetic energy) or a metastable decay of the precursor ion in the ion draw out region. In any case, the reactions leading to $F^- + CF_2I$ and $FI^- + CF_2$ are associated with low translational energy ($E_T \lesssim 0.25$ eV).

The dissociation dynamics indicates that both precursor states (at 0.0 eV *and* at 3.8 eV) possess a significant F_3C-I repulsive nature. It is quite remarkable that the 3.8 eV resonance decomposes into $CF_3^- + I$, while fragment I^- shows no particular contribution at that energy (Fig. 5).

From photodissociation studies it is well known that excitation of CF_3I in the A band (350-200 nm) results in prompt dissociation along the C-I bond due to the repulsive nature of the excited state.[15] The A band of molecules of the type CX_3I arises from a $\sigma^* \leftarrow n$ transition localized in the C-I bond, and the dissociation dynamics is then discussed[16] in the simple model of "colinear pseudotriatomic dissociation": the molecule is considered as a triatomic A-B-C, with the A-B and B-C bonds being distinct normal coordinates. The electronic chromophore is localized at the A-B bond and the B-C bond is "spectator" during excitation. It eventually receives vibrational energy as a result of repulsion during dissociation. Due to the C_{3v} symmetry, vibrational excitation in CF_3 is confined to the v_2 umbrella vibration of CF_3.

The A band consists of three overlapping transitions from the ground N state to the $^3Q_1, ^3Q_0$, and $^1Q_1$, states with absorption maxima located around 4.1 eV, 4.7 eV and 5.2 eV, respectively.[17] Recent photodissociation studies in the 304 nm region have shown[18] the dominant formation of spin orbit excited $I^*(^2P_{1/2})$ due to the $^3Q_0 \leftarrow N$ transition which carries most of the A band oscillator strength and which directly correlates with spin orbit excited I^*. In this direct dissociation process it was found that only 5% of the available excess energy (0.91 eV) appears as vibrational energy of CF_3.

Returning to the present problem of electron attachment, we can describe the resonance at 0 eV as the ground state of CF_3I^- formed by the accommodation of the extra electron into the first normally unoccupied MO with a strongly localized $\sigma^*(C-I)$ character. The situation is thus similar to photoexcitation where a non-bonding electron is promoted to the $\sigma^*(C-I)$ MO.

The electronic state at 3.8 eV is most likely a *core excited* resonance: the incoming electron concomitantly excites one of the molecule's electrons and we have two electrons in normally unfilled MOs.

This core excited resonance shows quite unique features: As obvious from the $CF_3^- + I$ decomposition dynamics, it has a strong F_3C-I repulsive nature. From the TOF analysis of CF_3^- we can conclude that the neutral iodine atom must exclusively be formed in its ground state, $I(^2P_{3/2})$. The strong repulsive nature seen in the $CF_3^- + I$ channel suggests a two particle resonance associated with the first electronically excited state of the neutral molecule, i.e. associated with one of the components of the A band.

The observation that no particular I^- contribution appears near 3.8 eV can be rationalized by the fact that the resonance involves promotion of a 5p electron located principally on the iodine atom to the localized $\sigma^*(C-I)$ MO. This leaves a hole or a positive charge on I in the initially formed molecular anion thus making the decomposition into I^- unlikely.

The situation is summarized in the potential energy diagram (Fig. 7). The zero eV resonance is attributed to a one electron process with the extra electron captured into the lowest virtual MO with distinct σ^*(C-I) character. Although the associated potential energy surface has a minimum, CF_3I^- is not formed under single collision conditions.

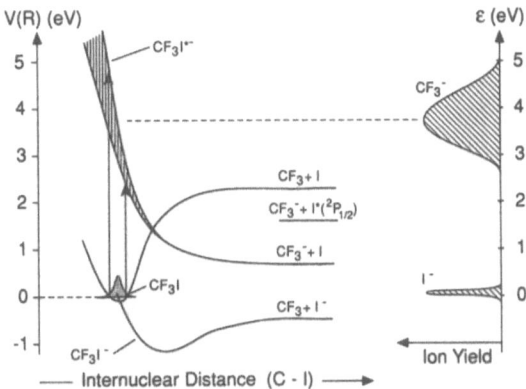

Figure 7. Schematic potential energy curves for the electronic states relevant in dissociative electron attachment to CF_3I. From ref. 23.

The electronic state at 3.8 eV is a core excited resonance where the incoming electron concomitantly induces a $\sigma^* \leftarrow n$ transition. This two particle resonance decomposes (among other channels) into $CF_3^- + I$ and can also decay by the emission of electrons (autodetachment) as indicated by the width of the potential energy curve in Fig. 7. The lowest excited state of the neutral is the $^3Q_1$ component of the first absorption band (A band) which has been resolved by magnetic circular dichroism (MCD)[17] and located at 4.1 eV. In the usual terminology of resonances the state at 3.8 eV would then be a Feshbach resonance which cannot autodetach into the associated $^3Q_1$ state.

In contrast to CF_3I, tetrafluoromethane captures electrons within a very broad resonance (Figs. 8 and 9), yielding the ions F^- and CF_3^-. From well established thermodynamic data the energetic threshold for reaction

$$e^- + CF_4 \rightarrow F^- + CF_3 \tag{20}$$

is calculated as $\Delta H_O = 2.26 \pm 0.2$ eV.[11] At the experimental threshold (≈ 4.5 eV) we thus have already more than 2 eV of excess energy.

The extraordinary large width of the observed resonance indicates strong repulsion of the anionic potential energy surface reached by the Franck-Condon transition. In the case of CF_3^-, we observe an incompletely separated TOF doublet at low energies which becomes more separated with increasing electron energy and, hence, total energy (Fig. 8). F^- formation, on the other hand, exhibits qualitatively different behavior: at low energies a doublet is visible signalizing F^- ions released with significant kinetic energy. Above 6 eV an additional peak at $T = 0$ appears which becomes dominant in the TOF spectrum on fur-

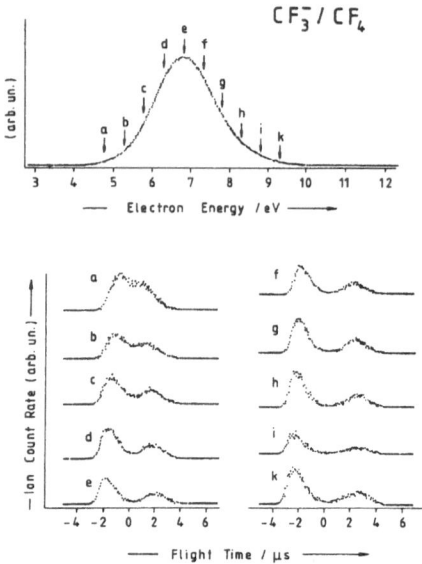

Figure 8. TOF spectra of CF3⁻ from CF4 recorded at different electron energies. From ref. 3.

ther increasing the electron energy (Fig. 9). This feature indicates the concomitant formation of thermal or near-thermal F⁻ ions. We suggest the following mechanism for this effect: Electron attachment to CF_4 occurs via two overlapping resonances associated with the parent anion in its electronic ground state and an electronically excited state.[19] The ground state ion decomposes into the complementary channels:

$$CF_4^- \rightarrow CF_3 + F^- \tag{20}$$
$$ \rightarrow CF_3^- + F \tag{21}$$

with high kinetic excess energy release in both cases. The CF_4^{*-} ion, in contrast, is only correlated with F⁻ ion formation. This reaction proceeds either according to

$$CF_4^{*-} \rightarrow F^- + CF_3^* \tag{22}$$
$$\phantom{CF_4^{*-} \rightarrow F^- + CF_3^*} \rightarrow CF_2 + F$$

or directly by

$$CF_4^{*-} \rightarrow F^- + CF_2 + F \tag{23}$$

The thermodynamic threshold for this multiple fragmentation process is 6.0 eV (as calculated using standard thermodynamic data) and hence above the energy of the first electronically excited state of CF_3 (6.41 eV[21]).

Returning to the channels (20) and (21) correlated with the ground state of the parent ion, we may now ask how much of the available excess energy E^* is released as kinetic energy of the two products. Since the precursor is a 5 atomic ion, F^- and CF_3^- formation is expected to be associated with a *distribution* of kinetic energy. The average energy of the ion is obtained by simply taking ΔT from the TOF doublet and applying eqn. (13) which yields the most probable kinetic energy.

If the final dissociation channel consists of two fragments, as in reaction (20) and (21), the principle of conservation of linear momentum (19) allows to calculate the translational energy imparted to both fragments.

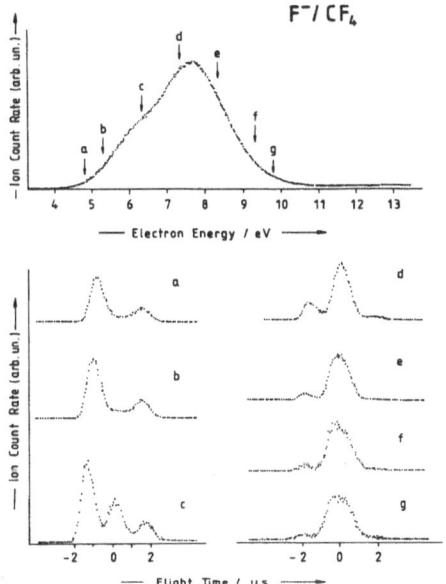

Figure 9. TOF spectra of F^- from CF_4 recorded at different electron energies. From ref. 3.

The evaluation reveals that the mean kinetic energy release to $CF_3^- + F$ and $CF_3 + F^-$, respectively, increases *linearly* with the electron energy and hence, with E^*. The ratio E_T/E^* is 0.33 for F^- ion formation and 0.62 for CF_3^- formation. In other words, in reaction (20), 33% of the available excess energy is released as kinetic energy while it is 62% for channel (21).

It is interesting to note that in electron stimulated desorption experiments via the 7 eV resonance from condensed CF_4 only F^- was observed.[22] Although in $CF_3^- + F$ formation nearly 2/3 of the total excess energy appears as translational energy, the unfavourable mass ratio between the products prevents CF_3^- from leaving the surface.

FLUOROETHYLENES

It is well known that low energy electron capture by ethylene and the fluoroethylenes occurs via accommodation of the extra electron into the b_{2g} (π^*) MO with the vertical attachment energy increasing from 1.8 eV in ethylene to 3.0 eV in perfluoroethylene.[24-26] Figures 10 and 11 illustrate the decomposition reactions of cis-1,2-difluoroethylene and 1,1-difluoroethylene. The hatched resonance is assigned to the $^2\Pi$ state of the anion.

The ion yield curves a show an additional, pronounced resonance in the region of 6-7 eV. A correlation of these energies with electronically excited states in the neutral fluoroethylenes makes it seem plausible that this is a core-excited resonance associated with the 3s Rydberg state converging to the first (π) ionization energy [26].

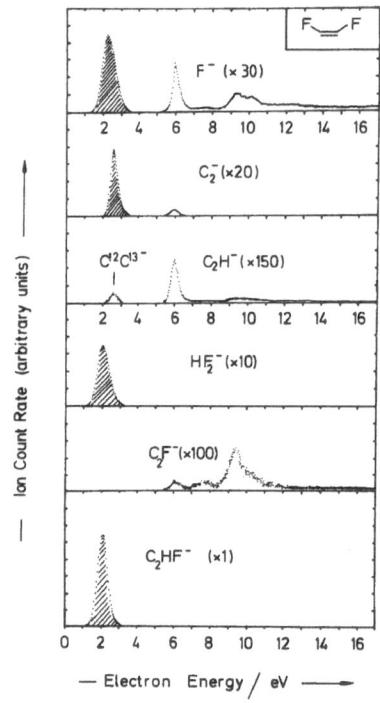

Figure 10. Dissociation poducts from cis-1,2-difluoroethylene. From ref. 26.

On inspecting the decomposition products and the intensity with which they are formed, some intriguing features of the unimolecular decomposition reactions become manifest: the C_2^- ion is formed from the $^2\Pi$ resonance in cis-1,2-difluoroethylene (Fig. 10) and trans-1,2-difluoroethylene (not shown here), but *not* in 1,1-difluoroethylene (Fig. 11). The energetically lowest lying dissociation channels for the formation of the C_2^- ion are

$$C_2H_2F_2^-(^2\Pi) \qquad \rightarrow \qquad C_2^- + 2HF \qquad\qquad (24)$$
$$\rightarrow \qquad C_2^- + H_2 + F_2 \qquad\qquad (25)$$

The energetic threshold for channel (24) is 2.5 eV in the case of cis- and trans-difluoroethylene, and 3.0 eV for 1,1-difluoroethylene (as calculated from established thermodynamic data). Channel (25) lies by 5.6 eV above channel (24), so the $^2\Pi$ resonance in the difluoroethylenes can decay into C_2^- ions only by formation of two HF molecules.

The capture of electrons into the π^* MO and the subsequent excitation of vibrational modes containing the C-C stretch and HCF scissoring component explains the observed preferential formation of HF when the hydrogen and fluorine atoms originate from the same carbon atom.

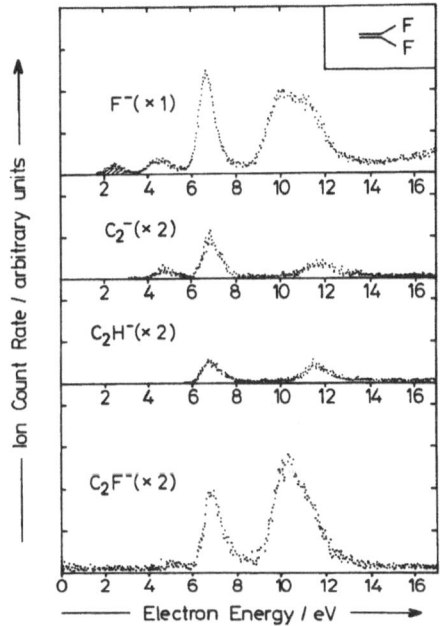

Figure 11. Dissociation products from 1,1-difluoroethylene. From ref. 26.

Similar arguments apply to the formation of the C_2HF^- ion which is, by far, the predominant fragment from cis-1,2-difluoroethylene (Fig.10) and also trans-1,2-difluoroethylene[26] (not shown here). This strongly suggests that C_2HF^- is a fluorovinylidene-like species. In fact, for the $C_2H_2^-$ ion it is well established experimentally[27] and theoretically[28] that the stable configuration is the vinylidene structure. Acetylene, on the other hand, possesses a strongly negative electron affinity, as shown by electron transmission spectroscopy.[29]

The bifluoride ion (FHF⁻) observed from 1,2-difluoroethylene (Fig. 10) is known to possess an unusually strong hydrogen bond, namely, D(FH-F) ≥ 1.5 eV.[27,28] Its observation from the $^2\Pi$ resonance in some fluoroethylenes suggests a remarkably high electron affinity for the neutral radical, EA(FHF) ≥ 4.8 eV.[31]

METHANOL AND DEUTERATED METHANOL

In contrast to the compounds discussed above, methanol shows unusual behavior in that negative ions only appear at energies far above their energetic thresholds with a rather complex feature of kinetic energy release. The ions observed are O⁻, OH⁻, and CH₃O⁻.[32] From well established thermodynamic data the minimum heat of reaction ΔH_0 for the lowest channels can be calculated as

$$e^- + CH_3OH \rightarrow OH^- + CH_3, \quad \Delta H_0 = 2.1\ eV \qquad (26)$$
$$\rightarrow O^- + CH_4, \quad \Delta H_0 = 2.4\ eV \qquad (27)$$
$$\rightarrow CH_3O^- + H, \quad \Delta H_0 = 2.9\ eV. \qquad (28)$$

We have studied[32] the deuterated species CH₃OD, CD₃OH, CD₃OD, and CH₂DOH to elucidate the decomposition mechanisms of the TNIs in some more detail.

As an example Fig. 12 shows the results for CH₃OD. While the CH₃O⁻ ion is formed in three resonances above 5.5 eV, OH⁻, OD⁻, and O⁻ ions only appear within a

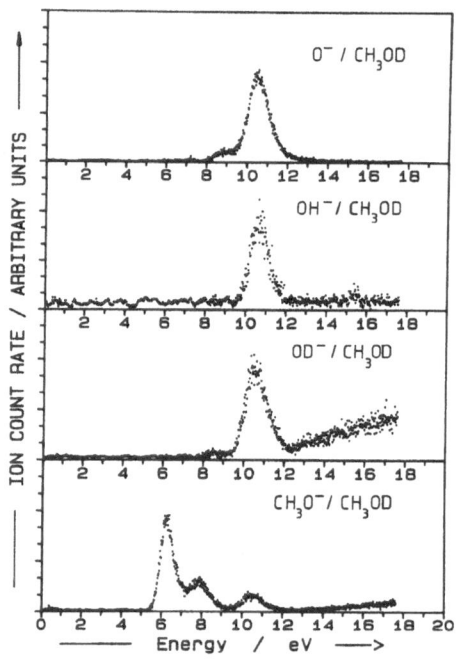

Figure 12. Negative Ion formation from CH₃OD. From ref. 32.

prominent resonance peaking at 10.5 eV. From Fig. 12 it is apparent that the high-energy resonance simultaneously yields OH$^-$ and OD$^-$ ions which points to "hydrogen scrambling" in the TNI prior to decomposition. In contrast, the observation that the CH$_3$O$^-$ ion, rather than the CH$_2$DO$^-$ ion is formed indicates a direct cleavage of the OD bond (without hydrogen scrambling). By measuring the other deuterated methanols (not shown here) it becomes apparent that OH$^-$ (or OD$^-$) ion formation is always accompanied by hydrogen scrambling, while formation of CH$_3$O$^-$ ions (and the deuterated analogues) proceeds directly (no hydrogen scrambling).

A TOF analysis of the different fragment ion shows that, in methanol and the deuterated methanols, only O$^-$ is formed with considerable translational energy. In the case of CH$_3$O$^-$ (and its deuterated analogues, channel (28)) the light hydrogen (or deuterium) atom carries away 97% (94%) of the total translational energy so that a TOF study of CH$_3$O$^-$ ions can not provide any further information about the energy distribution of this channel. The O$^-$ ion is formed at 9 eV with low kinetic energy and at 10.5 eV with a "quasi-discrete" translational energy (separated TOF doublet) corresponding to a mean energy of 1.0 ± 0.2 eV for the O$^-$ ion.

We thus have a rather complicated situation in methanol and the deuterated methanols: O$^-$ and OH$^-$ (OD$^-$) ions are formed from a resonance peaking at 10.5 eV which is roughly 8 eV above their energetic thresholds. It is very likely that the associated fragmentation channels consist of more than two fragments. In contrast, formation of CH$_3$O$^-$ ions (and the corresponding deuterated analogues) proceeds directly. The unfavorable mass ratio of hydrogen to CH$_3$O$^-$ ions, unfortunately precludes any detailed information on the excess energy distribution using TOF methods. Owing to the presence of the magnetic field in our experimental configuration, H$^-$ (and D$^-$) ions can not be detected, thus ruling out the study of the complementary reaction to (28).

HEXAFLUOROBENZENE

C$_6$F$_6$ belongs to the group of larger perfluoro compounds which are known to capture low energy electrons with large cross sections to form long lived molecular anions (see Introduction). In these molecular anions dissociation into R + F$^-$ is not possible since the (C-F) bond dissociation energy generally exceeds the electron affinity of fluorine by 1-2 eV.[11] Figure 13 shows that in C$_6$F$_6$ the parent molecular anion exclusively arises within a narrow resonance near 0 eV. At higher energies we additionally detect F$^-$ and C$_6$F$_5^-$ which arise from the complementary reactions

$$e^- + C_6F_6 \quad \rightarrow \quad F^- + C_6F_5^- \qquad (29)$$
$$\rightarrow \quad F^- + C_6F_5 \qquad (30)$$

with the thermodynamic limits of 1.54 eV and < 1.54 eV, respectively.[33] It is thus very likely that resonances above 8 eV which decompose into F$^-$ (see Fig. 13) yield more than one *neutral* fragment in the final channel. C$_6$F$_5^-$ formation, on the other hand, is uniquely associated with channel (30) consisting of only *two* fragments. At the maximum of the second resonance leading to C$_6$F$_5^-$ (ε = 8.5 eV) the large amount of excess energy (E* = (ε - ΔH$_0$) > 7 eV) must be shared between the two particles. Figure 14 indeed shows that C$_6$F$_5^-$ (inspite of its large mass) exhibits a doublet in the TOF spectrum at 8.5 eV indicating considerable translational energy. A detailed evaluation[33] leads to E$_T$ = 3.2 eV for channel (30). The rest of the total excess energy (>3.8 eV) must be internal energy of C$_6$F$_5^-$ since the lowest excited state of the F radical is at 12.7 eV.

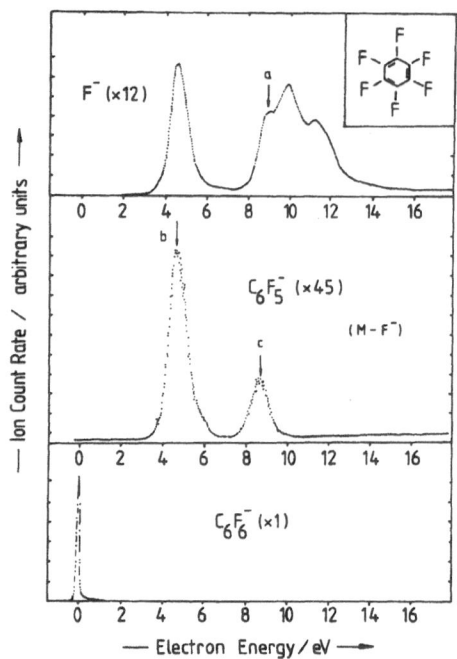

Figure 13. Negative ion formation in hexafluorobenzene. From ref. 33.

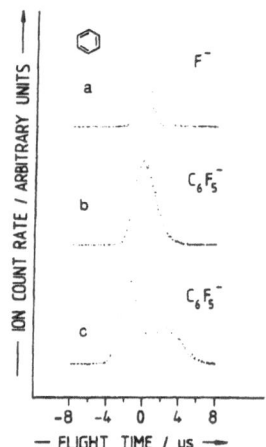

Figure 14. TOF spectra of F⁻ and C₆F₅⁻ at the energies indicated. From ref. 33.

Negative ion states in benzene and also hexafluorobenzene have extensively been studied by electron transmission techniques.[33,34] It it appears that the low energy resonance yielding the parent anion $C_6F_6^-$ is due to electron attachment into the lowest unoccupied MO, $e_{2u}(\pi^*)$ (doubly degenerate) while in the state at 4.5 eV the next virtual π^* MO (b_{2g}) is involved. The electronic states above 8 eV are most likely core excited resonances with two electrons in normally unfilled MOs. It should finally be noted that the $C_6F_6^-$ ground state anion is subject to a Jahn Teller distortion yielding an elongated and a compressed structure, both with D_{2h} symmetry.[33]

Acknowledgments

This work has been supported by the Deutsche Forschungsgemeinschaft, Stiftung Volkswagenwerk, Verband der Chemischen Industrie, and the NATO Scientific Affairs Division.

REFERENCES

1.   H.S.W. Massey, "Negative Ions", Cambridge University Press, Cambridge (1976).

2.   L.G. Christophorou, "Atomic and Molecular Radiation Physics", Wiley-Interscience, London (1971).

3.   E. Illenberger, J. Momigny, "Gaseous Molecular Ions. An Introduction to Elementary Processes Induced by Ionization", Steinkopff Verlag, Darmstadt - Springer -Verlag, New York (1992).

4.   K. Janousek and J.I. Brauman, Electron Affinities, in "Gas Phase Ion Chemistry" (M.T. Bowers Ed.), Academic Press, New York (1979).

5.   A. Herzenberg and D.T. Birtwistle, Vibrational Excitation of N_2 by Resonance Scattering of Electrons, *J. Phys. B: Atom. Molec. Phys. 4:53 (1971)*.

6.   L.G. Christophorou (Ed.), "Electron-Molecule Interactions and Their Applications", Vol. I, Academic Press, Orlando (1984).

7.   G. Mück and H.P. Popp, Quantitative Ausmessung des Chlor-Affinitätskontinuums, *Z. Naturforsch. A 23: 1213 (1968)*.

8.   H.S. Taylor, Models, Interpretations, and Calculations Concerning Resonant Electron Scattering Processes in Atoms and Molecules, in "Advances in Chemical Physics", Vol. XVIII. Interscience Publishers, New York (1970).

9.   A. Stamatovic and G.J. Schulz, Characteristics of the Trochoidal Electron Mono-chromator, *Rev. Sci. Instr. 41:423 (1970)*.

10.   E. Illenberger, A Method to Determine Excess Energies in Dissociative Electron Attachment, *Ber. Bunsenges. Phys. Chem. 86:247 (1982)*.

11. D.F. McMillen and D.M. Golden, Hydrocarbon Bond Dissociation Energies, *Annu. Rev. Phys. Chem. 33:493 (1982)*.

12. S.Y. Tang, B.P. Mathur, E.W. Rothe and G.P. Reck, Negative Ion Formation in Halocarbons by Charge Exchange with Cesium, *J. Chem. Phys. 64:1270 (1976)*.

13. R.N. Compton, D.W. Reinhard and C.D. Cooper, Collisional Ionization Between Alkali Atoms and some Methane Derivatives: Electron Affinities for CH_3NO_2, CF_3I and CF_3Br, *J. Chem. Phys. 68:43600 (1978)*.

14. P.E. McNamee, K. Lacmann and D.R. Herschbach, General Discussion, *Faraday Discuss. Chem. Soc. 55:318 (1973)*.

15. P. Felder, Photodissociation of CF_3I at 248 nm: Internal Energy Distribution of the CF_3 Fragments, *Chem. Phys. 143:141 (1990)*.

16. M. Shapiro and R. Bersohn, Vibrational Energy Distribution of the CH_3 Radical Photodissociated from CH_3I, *J. Chem. Phys. 73:3810 (1980)*.

17. A. Gedanken, The Circular Magnetic Dicroism of the *A* Band in CF_3I, C_2H_5I, and *t*-BuI, *Chem. Phys. Lett. 137:462 (1987)*.

18. H. Jin-Hwang and M.A. El-Sayed, Photodissociation of CF_3I at 304 nm: Effects of Photon Energy and Curve Crossing on the Internal Excitation of CF_3, *J. Phys. Chem. 96:8728 (1992)*.

19. T. Oster, A. Kühn and E. Illenberger, Gas Phase Negative Ion Chemistry, *Int. J. Mass Spectrom. Ion Proc. 89:1 (1989)* (Review Article).

20. M.B. Robin, "Higher Excited States of Polyatomic Molecules", Vol III, Academic Press, Orlando (1984).

21. M. Suto, N. Washida, H. Akimoto and M. Nakamura, Emission Spectra of CF_3 Radicals. II. Analysis of the UV Emission Spectrum of CF_3 Radicals, *J. Chem. Phys. 78:1019 (1983)*.

22. M. Meinke, L. Parenteau, P. Rowntree, L. Sanche and E. Illenberger, Low Energy Electron Stimulated Desorption of Anions from Condensed CF_4, *Chem. Phys. Lett. 205:213 (1993)*.

23. T. Oster, T. Jaffke, O. Ingolfsson, M. Meinke and E. Illenberger, Anion Formation from Gaseous and Condensed CF_3I on Low Energy Electron Impact, *J. Chem. Phys. 99: 5141 (1993)*.

24. N.S. Chiu, P.D. Burrow and K.D. Jordan, Temporary Anions of the Fluoro-ethylenes, *Chem. Phys. Lett. 68:121 (1979)*.

25. M. Heni, E. Illenberger, H. Baumgärtel and S. Süzer, The Dissociation of the $^2\Pi$ Fluoroethylene-Anions, *Chem. Phys. Lett. 87:244 (1982)*.

EFFECTS OF THE SOLID PHASE ON RESONANCE STABILIZATION, DISSOCIATIVE ATTACHMENT AND DIPOLAR DISSOCIATION

Léon Sanche

Groupe CRM en Sciences des radiations
Faculté de médecine, Université de Sherbrooke
Sherbrooke, QC, Canada J1H 5N4

INTRODUCTION

In a previous article of this book, it has been shown that the concept developed in the gas phase to explain the behavior of electrons scattered from isolated atoms and molecules could be utilized to explain several electron scattering phenomena occurring in the solid phase. The results presented clearly indicated that many of the electron resonances (i.e., transient anions) present in the gas-phase were also found in the various scattering cross sections of solids. Transient molecular anions are not only involved in electron scattering, in the sense that they do not necessarily reemit the captured electron They can also be stabilized into a permanent molecular anion or they can dissociate producing a stable anion fragment and a neutral atom, molecule or radical. In the present article, it is shown that these processes are present near the surface of solids where they can be observed by detecting stable anion emission in vacuum or charge accumulation on the surface. The mechanisms responsible for resonance stabilization and dissociative attachment (DA) in the condensed phase can be linked to those responsible for similar processes in the gas phase, by comparison between experimental results obtained in the two phases and, with those of the dipolar dissociation (DD) process which does not involve the initial formation of an anion resonance.

LINKING THE MECHANISMS OF ANION FORMATION IN THE GAS PHASE TO THOSE OCCURRING NEAR SURFACES

Anions on metallic, semi-conducting or dielectric surfaces may be formed as temporary or permanent charges. Their mechanisms of formation are in many respects similar to those found in the gas-phase. As shown in the previous article on electron scattering, a temporary anion may possess many decay channels depending on its energy.[1] The departing electron may leave the molecule in a rotationally, vibrationally or electronically excited state. If the resulting electronically excited neutral state is dissociative, ground state or excited fragments can be produced. If the lifetime of the resonance is, at least, of the order

Linking the Gaseous and Condensed Phases of Matter
Edited by L.G. Christophorou *et al*, Plenum Press, New York, 1994

of a vibrational period, if the intermediate state is dissociative in the Frank-Condon region, and if one of the possible fragments has a positive electron affinity, then the ion may dissociate into a stable anion and a neutral fragment in the ground or an excited state. This process is called dissociative attachment (DA). If during its lifetime, the transient anion transfers energy to another system (e.g., to another molecule by collisional interaction) it can become stable when the parent molecule has a positive electron affinity. We refer to this process as resonance stabilization. Finally, when the transient anion is formed at energies above the ionization potential, two-electron emission is possible. Thus, by DA or resonance stabilization a temporary anion may lead to the production of a stable anion. These processes require the addition of an electron and cannot therefore be induced with electromagnetic radiation, however dipolar dissociation (DD), which also produces stable anions from the dissociation of an excited electronic state of a molecule (e.g. for a molecule AB, $AB^* \to A^+ + B^-$), can be caused directly by either electron or photon impact since no extra charge is needed in this process.

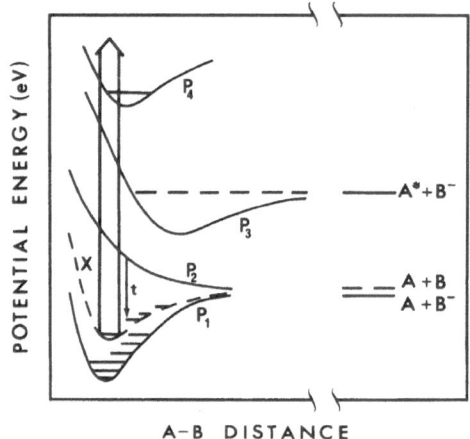

Figure 1. Hypothetical potential energy curves of a molecular anion AB^-. The dashed curve represents the gound state of the neutral molecule AB.

The decay of transient anions into the DA and stabilization channels may be better understood by considering the hypothetical potential energy diagram of a diatomic anion AB^- shown in Fig. 1. The dashed curve (X) represents the ground state of the molecule AB while the others are those of the anion system. Curve P_1 represents the ground state of AB^- which is stable for vibrational energy levels below the state X. Here, stabilization may be illustrated by noting that when an electron is captured in a vibrational level of AB^- in state P_1 lying above the ground state (X) of the molecule, vibrational energy transfer from the anion to another system can cause the electron to be captured in another vibrational state lying below the ground state X. The anion states P2 or P3 may dissociate into fragments $A + B^-$ or $A^* + B^-$, respectively, where A* represent an excited state, if the lifetime of the resonance is sufficiently long. If not, the electron will be reemitted (e.g., transition t in Fig. 1) leaving the target in vibrational and rotational excited states.

Although similar to the gas phase, the formation of a transient anion on a surface[2] or within a solid[3] is greatly perturbed from the isolated electron-molecule phenomenon. The main factors which modify isolated resonances include: the change in the symmetry of the scattering problem introduced by the presence of other nearby targets and the polarization of these targets by the charge of the electron and transient anion. The orientation of the molecule with respect to the solid surface or within the solid may cause serious perturbations when the target is a single crystal. The possible distortion and modification of the target molecule by the solid should only be considered as an important factor for chemisorbed molecules.

The change in symmetry in the condensed phase arises essentially from the fact that electrons interact with more than one scattering center. Near surfaces, it may be viewed to arise from multiple scattering of the electron waves between the molecule and the solid.[2] This modification may relax symmetry rules and change the partial wave content of the resonant electron. By acting on the isolated electron-molecule potential, the polarization force can modify the resonance lifetime, principally by promoting electron transfer to the substrate and/or the neighboring targets. In addition, the polarization force lowers the resonance energy by introducing an attractive potential. This perturbation may reduce the number of decay channels; thus, affecting in a drastic manner certain cross sections by increasing the survival probability or decreasing total decay width of the resonance. The dipolar process is also affected by condensation, but the initial state being neutral, the polarization field only affects the fragment ions with much less perturbation on the dissociating excited state. However, the dynamics of both the DA and DD mechanisms are perturbed by multiple electron scattering prior to formation of the transient state (i.e., AB^- or AB^*, respectively) and by reactive scattering of the stable ions produced with adjacent molecules. This latter perturbation includes the possible depletion of the stable anion species by associative detachment with nearby molecules (e.g., taking O^- formation from DA, an example of associative detachment is given by the reaction $O^- + CO \rightarrow (CO_2)^- \rightarrow e^- + CO_2$).[4] Other reactions which deplete anion currents from DD and DA include the reduction of the anion's kinetic energy in reactive exchange (e.g., $O^- + CO\ (v') \rightarrow O^- + CO\ (v > v')$,[5] or by secondary reaction processes (e.g., $O^- + C_nH_{2n} \rightarrow OH^- \rightarrow C_nH_{2n-1}$).[6] Such processes which occur after the initial anion has been produced are referred to as post-dissociation interactions (PDI).

EXPERIMENTS

For molecular adsorbates, the DA and DD process can be detected by measuring the electron stimulated desorption (ESD) of anions as a function of electron energy.[3,7] The presence of stable anions on surfaces produced by DA and resonance stabilization can be detected in surface charge measurements.[8] These experiments are briefly described in this section.

Target Preparation and Energy Calibration

Molecules can be condensed in different ways depending on the nature and degree of order of the solid surface, its coverage and temperature. At submonolayer coverages, molecules may be physisorbed or chemisorbed onto a substrate. Physisorbed molecules retain most of their molecular orbital characteristics, whereas in chemisorption exchange of electrons between the molecule and the substrate causes the formation of new bonds leading to modifications of the molecular identity. Thus, in order to link gas-phase behavior to that of the condensed phase, it is preferable to investigate physisorbed molecules. The results reported herein were recorded with molecules condensed as multilayers on a metal

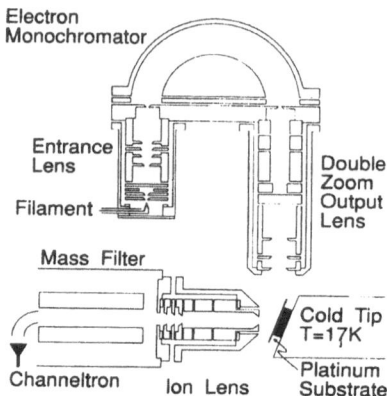

Electron
Monochromator

Entrance
Lens

Double
Zoom
Output
Lens

Filament

Mass Filter

Cold Tip
T=17K

Channeltron   Ion Lens   Platinum
Substrate

Figure 2. Schematic view of electron stimulated desorption (ESD) apparatus. The components shown are housed in a conventional UHV chamber within two concentric layers of magnetic shielding.

surface or as submonolayers on a dielectric surface; under these conditions they retain their molecular properties. In these experiments, a thin multilayer film is grown in an UHV system by the condensation of gases or organic vapors on a clean metal substrate held at cryogenic temperatures (e.g., target in Fig. 2). If it is desirable to exclude the effect of the substrate, the film thickness is made larger than the total electron mean free path.

The absolute electron energy scale can be calibrated with respect to the vacuum level by simply measuring the onset of the current arriving at the substrate.[8] The amount of particles condensed on the substrate can be estimated by expanding a given volume at a given pressure of a gas or a vapor in vacuum in front of the target using a gas-volume expansion dosing procedure.[9] Absolute calibration of the thickness is provided by monitoring the quantum interference oscillations present in the electron energy dependence of the target current of thin films.[10] With this calibration, the film thicknesses (10-500 Å) can usually be estimated with an accuracy equal or better than 30%. Prior to deposition, the metal substrate is usually cleaned by ion sputtering and resistive heating. Data acquisition and energy scale calibration are particularly difficult, when degradation and/or charging of condensed multilayer films occur. In this case, thermalized electron trapping within the film or near its surface can be observed by measuring the potential created by the trapped charges as explained in subsection C. *Unless otherwise stated the results presented in this chapter were obtained by condensing molecules onto a polycrystalline platinum substrate cooled to temperatures of about 20 K.* At these temperatures adsorbates are in the physisorbed state.

Electron Stimulated Desorption of Anions

A typical apparatus[11] used to measure ESD of anions is shown schematically in Fig. 2. It consists essentially of a monoenergetic electron source, a mass spectrometer, and a closed-cycle refrigerated cryostat. These components are housed in an ultra-high-vacuum (UHV) system reaching working pressures lying in the 10^{-11} Torr range. The electron source consists of a hemispherical electron monochromator equipped with double zoom lens, and two perpendicular pairs of parallel deflector plates. The electron beam is incident at 20° from the target surface. It has an energy resolution of about 80 meV full width at half-maximum (FWHM) and produces currents at the sample of the order of nanoamperes. The beam can be scanned on the target and aligned with respect to the axis of the mass spectrometer by applying proper voltages across the deflector plates. The condensed films are grown on a polycrystalline Pt surface fitted to the cold tip of the cryostat and isolated electrically from the Pt substrate by a ceramic sheet. A portion of the negative ions desorbed by electrons impinging on the film are focused, by the ion lens, at the entrance of the mass spectrometer whose axis is oriented at 70° with respect to the surface. The ions are mass selected by a quadrupole mass filter, then deflected at 90° from the primary axis by a pair of cylindrical plates and finally detected by a channeltron electron multiplier. The resulting pulses are stored in a computer as a function of electron energy.

Charge Trapping Measurements

The number of charges accumulated on a dielectric film resulting from bombardment

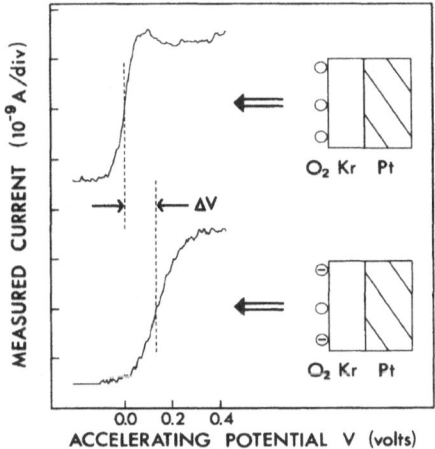

Figure 3. Current transmitted through an uncharged (top) and a charged (bottom) Kr film covered with 0.1 monolayer (ML) of O_2 as a function of the accelerating potential V of the incident electron beam.

with a low-energy electron beam of a specific energy can be measured with an apparatus of the type shown in Fig. 2 in which the electron beam is incident perpendicular to the surface.[8] When electrons from the monochromator have just enough energy to enter a multilayer film deposited on the substrate, a sharp rise, termed the "injection curve" (IC), is seen in the electron energy dependence of the target current spectrum. The IC for an uncharged film is represented by the upper curve of Fig. 3. When the same film is charged by the electron beam, the IC is shifted by ΔV to a higher accelerating potential (bottom curve, Fig. 3) since the incoming electrons must then possess additional kinetic energy to overcome the negative potential barrier. The IC is also broadened due to the effect of the spatial charges and current density distributions. If the film is purposely allowed to charge by a significant potential ΔV, this latter can be related to the trapping cross section by treating the dielectric film as a charged capacitor.[8] The potential barrier ΔV is related to the charge density $\sigma(t)$, which has accumulated after bombardment time t, by the relation $\Delta V(t) = \sigma(t)$ h/ε where $\sigma(t) = \sigma_0 (1 - e^{-\beta t})$ and $\beta = \mu J_0/e$; ε is the dielectric constant of the film, h its thickness, σ_0 the initial (t=0) trap density, μ the trapping cross section, J_0 the incident current density and e the unit charge. In the limit $t \to 0$ a charging coefficient $A_s = d\Delta V/dt$, directly proportional to the trapping cross section, can be expressed as $d\Delta/Vdt|_{t=0} = (h\sigma_0 J_0/\varepsilon\ e)\ \mu = A_s$.

The IC of a freshly deposited multilayer film is first recorded rapidly (i.e., 0.1 sec) to avoid any significant charging. The film is then bombarded at a given potential V for a much longer period (i.e., 25 sec) with the same incident current (i.e., $I_0 \cong 5 \times 10^{-9}$ A). Afterwards, the IC is again rapidly recorded and the shift ΔV determined by comparison with the initial IC. Such a cycle can be repeated many times on the same film with the same V to obtain the time dependence of the process. To measure the thickness or energy dependence, a new film has to be deposited for each data point.

ANION ELECTRON STIMULATED DESORPTION

In both the gas and solid phase, DA and DD processes can be identified by measuring the yield of anion produced as a function of the energy of an electron beam hitting the target. The DD process produces a smooth continuous signal which, beyond threshold, increases monotonically with electron energy. The threshold energy for DD within a solid or at its surface corresponds to the dissociation energy of the isolated positive and negative species screened by the polarization induced by the products on the adjacent molecules and the substrate. Below the energy threshold for DD, negative ions can be produced only via DA. This process constitutes a particular channel for the decay of molecular transient anions (Fig. 1) formed in the gas-phase or within molecules condensed on surfaces. Since a given molecular configuration of a transient anion appears at a well-defined   energy   each

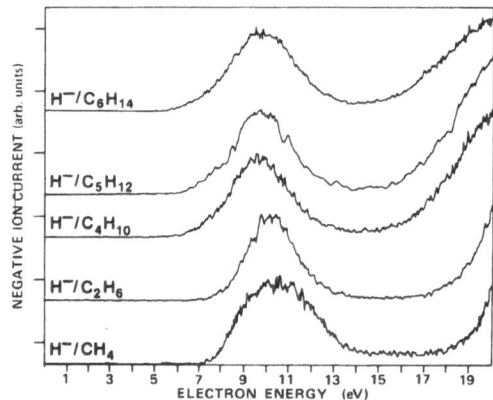

Figure 4. Energy dependence of H⁻ signal desorbed by electrons impinging on a 6-ML film of C_nH_{2n+2} molecules deposited on a Pt substrate.

peak in the electron-energy dependence of the anion yield identifies the energy of a particular resonant state.

Anion ESD by both DA and DD processes is exemplified in Fig. 4 by the energy dependence of H^- yields recorded with multilayer films of condensed C_nH_{2n+2} (n = 1,2,4-6) molecules.[12] All curves exhibit a single peak, whose maximum is located around 10 eV, and a rise which onsets at higher energy (16-18 eV); this latter is due to DD. The 10-eV peak is caused by the DA reaction

$$e^- + C_nH_{2n+2} \rightarrow (C_nH_{2n+2})^- \rightarrow H^- + C_nH_{2n+1}$$

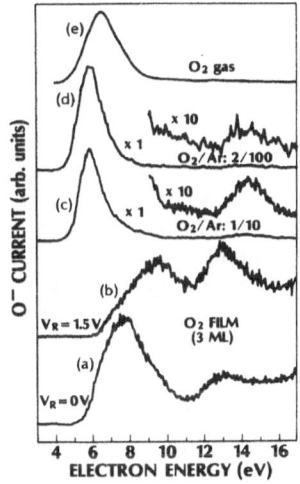

Figure 5. Incident electron energy dependence of O^- stimulated desorption yields from (a) and (b) 3-ML O_2, films; (c) and (d) 20-ML Ar films containing 10 and 2% volume O_2, respectively. Curve (e) represents the gas-phase yields (ref. 19). V_R is a potential retarding the desorbing anions.

DA features have also been observed in the O^-, C^-, Cl^- and H^- ESD yield functions measured from molecular solids formed by condensing O_2, NO, CO, Cl_2, N_2O, H_2O and hydrocarbon molecules on a metal substrate.[13] More recently, the anion-complexes $Ar.O^-$, $Ar.Cl^-$ were produced by ESD from Ar matrices containing N_2O and Cl_2, respectively.[13] O^-, $Ar.O^-$, $Kr.O^-$, O_3^- (or $O_2.O^-$) and O_2^- were also produced by ESD from solid rare-gas matrices containing O_2. Many of the DA reactions in the condensed state were found to be similar to the processes found in gas-phase[1] and cluster experiments.[14]

Analysis of the electron energy dependence of these anion yields[13] revealed that the gas-phase DA mechanism was perturbed in the condensed phase by: multiple electron scattering prior to electron attachment to a molecule;[15] post-dissociation interaction (PDI) of the resulting anion with the solid;[4,5] the induced polarization[7] and changes in the symmetry of the single-electron-molecule potential introduced by the presence of neighboring targets.[15] In addition, coherent scattering within the solid substrate was found to enhance ESD yields.[16] By analyzing the energy distributions of the desorbing anions as a function of incident electron energy,[13] it has been possible to distinguish between the contribution to the total yield arising directly from DA at a molecular site and the other two portions of the signal which involve multiple scattering and PDI. It was concluded that the lattice was not involved in the dissociation dynamics and that, therefore, the energy and momentum conservation laws of the free molecule could be applied to solid state DA reactions.[17]

Taking DA and DD in O_2 as an example, the effect of condensation on the symmetry of the scattering problem and the effects of the image charge of a metal surface and polarization of neighboring atoms and molecules is discussed in the following sections. It is also shown that surface DA can occur via simultaneous charge and energy exchange from other anion states formed within the substrate.

Symmetry of the Scattering Problem

The DA mechanism may differ in the condensed state from that of the isolated phenomenon, owing to the change in the symmetry of the problem involved in initial electron attachment. This is illustrated in Fig. 5, which displays the electron energy dependencies of the O^- ESD yield from a 3-ML O_2 film (a and b)[15] and a 20-monolayer (ML) Ar film containing 10% and 2% volume O_2 (c and d, respectively).[18] These results are compared with the O^- yields obtained by electron impact in gaseous O_2.[19] Curve (b) was recorded by retarding the ions with a potential $V_R = 1.5$ V at the entrance of the mass spectrometer. The results of Fig. 5 can be explained by considering the potential energy curves of the O_2^- gas-phase system lying below 15 eV. Although a multitude of molecular O_2^- states can be formed from the interaction of O and O^- fragments [e.g., 24 from the ground state $O(^3P)$ and $O^-(^2P)$],[20] only a few of these possess the proper characteristics to produce substantial decay into the DA channels.[21] Briefly, the state involved must have a multiplicity of 2 or 4, be repulsive in the Frank-Condon region, possess a lifetime of the order or longer than the O_2 vibrational period and be accessible in the energy range of interest. The multiplicity condition is dictated by spin conservation and the long lifetime is necessary to avoid electron detachment or autoionization before electron stabilization on one of the separating oxygen atoms.

Molecular orbital analysis[21] of the O_2^- states joining the two lowest separated-atom limits [i.e., $O(^3P) + O^-(^2P)$ and $O(^1D) + O^-(^2P)$] indicate that below 15 eV only the three states, shown in the potential energy diagram of Fig. 6, have the proper characteristics to produce O^- via DA with a considerable cross section. The dissociation limits of the molecular states of O_2^- are known from Wigner-Witmer rules.[22] According to these latter, the lowest $^2\Sigma_g^+$ state, shown in Fig. 6, correlates with the $O(^1D) + O^-(^2P)$ dissociation limit. Any correlation of that state with the $O(^3P) + O^-(^2P)$ limit requires a non-adiabatic curve crossing with another O_2^- state of the same symmetry. Experimental evidence from ESD measurements from pure O_2 multilayer films[15] and in CO and N_2 matrices containing O_2[5] corroborate the theoretical analysis and indicate the existence of three O_2^- states located around 6, 9 and 13 eV, which contribute to the O$^-$ signal in the 4-16 eV range. Analysis of ion energies at various incident electron energies within this range[15] indicates that the $^2\Sigma_g^+$ states dissociate predominantly into the lowest limit whereas the upper state dissociates predominantly into the second dissociation limit.

Out of the states shown in Fig. 6, the Σ^+ configurations cannot be formed by electron scattering from a single O$_2$ molecule in the $^3\Sigma_g^-$ ground state,[23] since in the single-target-electron frame of reference, the one-electron wavefunction must be σ^+. This selection rule excludes $\Sigma^- \leftrightarrow \Sigma^+$ transitions. Thus, it is found that in gas phase experiments only the $^2\Pi_u$

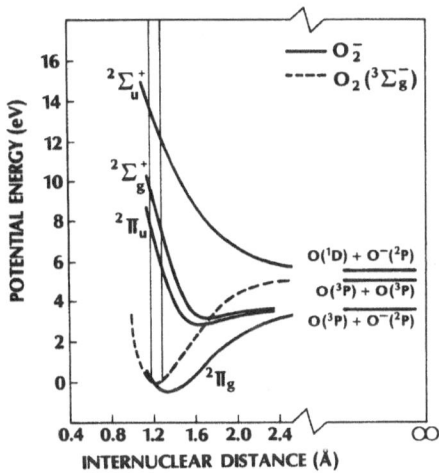

Figure 6. Potential energy diagram of the O$_2^-$ system (solid curves) and ground state O$_2$ (dashed curve). The states shown are only those which are expected to produce significant O$^-$ yields via dissociative attachment. The curves are correlated to the dissociation limit giving the most abundant signal.

state produces O$^-$ with considerable amplitude. This can be seen from curve (e) in Fig. 5 which exhibits only one broad peak with a maximum at 6.5 eV arising from the reaction

$$e + O_2(^3\Sigma_g^-) \rightarrow O_2^-(^2\Pi_u) \rightarrow O^-(^2P) + O(^3P)$$

The same reaction dominates ESD from O$_2$-doped rare gas matrices (e.g., curves c and d in Fig. 5) where, as in the gas phase, the yield function consists essentially of a single peak having a maximum at a slightly lower energy (i.e., 6.2 eV). This is expected since in matrices electrons attach principally to O$_2$ molecules which are isolated from each other by rare gas atoms as in the gas-phase. However, as the concentration of O$_2$ molecules increases from 2% (curve d) to 100% (curve a) another prominent structure appears at 13 eV in the yield function and the 6.2 eV peak distorts, broadens significantly and shifts to 7.6 eV. These modifications are amplified when only higher energy ions (i.e., >1.5 eV, curve b) are allowed to reach the mass spectrometer. In both curves a and b the appearance of the new features is attributable to breakdown of the $\Sigma^- \leftarrow|\rightarrow \Sigma^+$ selection rule, due to the change in electron-target symmetry in condensed O_2. The peaks at 9.5 and 13 eV were therefore ascribed, respectively, to the $^2\Sigma_g^+$ and $^2\Sigma_u^+$ states shown in Fig. 6.

Image Charge and Polarization Effects

The image charge induced in the metal substrate affects the DA and DD processes. This perturbation can be studied by inserting rare gas (RG) layers between the condensed molecules and metal surface. The effect on the O^- yield of isolating the O_2 molecule from

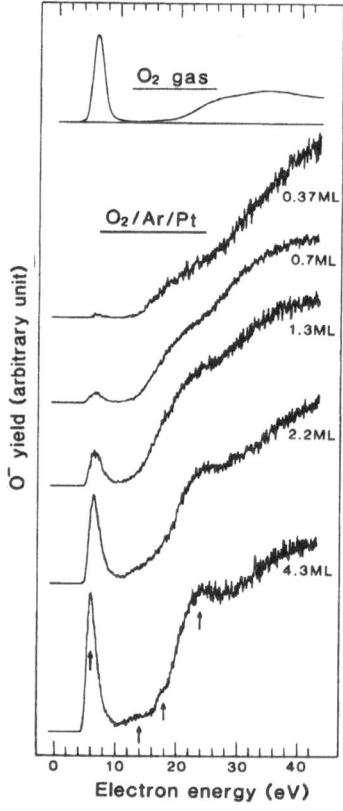

Figure 7. Comparison of O^- yields produced by ESD from O_2 gas (ref. 19) with those from $O_2/Ar/Pt$ samples with a constant (0.1 ML) O_2 coverage and variable (0.37–4.3 ML) Ar thicknesses. The $^2\Pi_u$ state of O_2 is indicated by the arrow at 6 eV; the others indicate features involving multiple electron scattering.

a polycrystalline Pt substrate by 1.3, 2.2 and 4.3 monolayers (ML) of Argon,[7] respectively, is shown in the lowest three curves of Fig. 7. For all curves but the upper one, which represents the gas-phase cross section for O^- production,[19] the O_2 coverage is 0.1 ML. The curves labeled 0.37 ML and 0.7 ML are the O^- ESD yields when these amounts of Ar are coadsorbed with 0.1 ML of O_2. The magnitude of the O^- DA peak around 8 eV increases by two orders of magnitude as the distance of the transient O_2^- anion from the metal is increased by spacer Ar layers up to 32 ML.[16] The DD contribution to the O^- yield is the smooth background signal above $\simeq$ 16 eV, over which multiple electron scattering features (indicated by arrows) are superimposed. These features arise from electrons having lost energy to electronic excitation of O_2 prior to formation of the O_2^- temporary states. Notice that the DD signal is much less affected by the proximity of the metal surface.

The effect of the image charge can be explained by considering the perturbation it induces on potential energy curves of the intermediate states O_2^- and O_2^* involved in the DA and DD processes, respectively. The solid curves in Fig. 8 correspond to the gas-phase

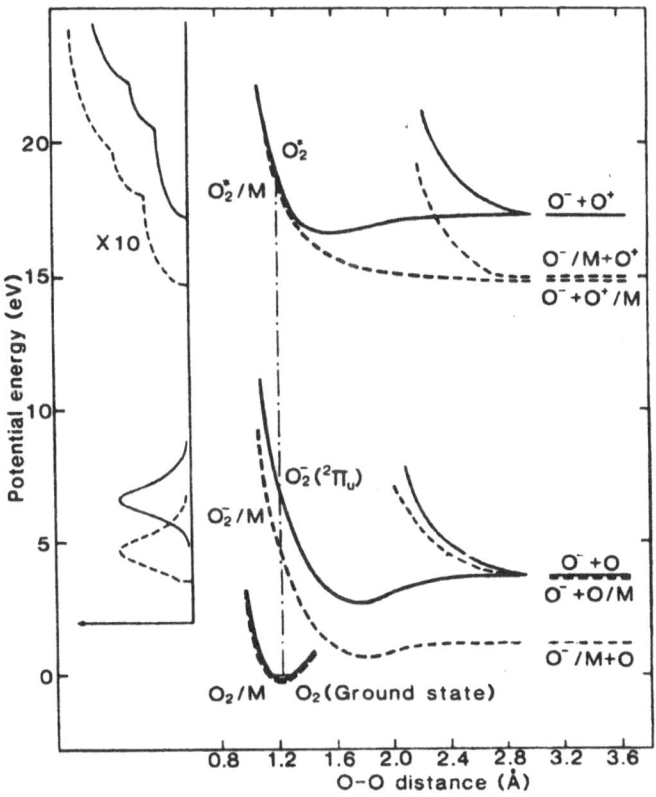

Figure 8. Potential-energy curves of O_2 gas (solid curves) and the corresponding potential-energy curves under the influence of an image charge (dashed curves). Spectra on the left side show schematically the relative number of vibrational and/or electronic states which have sufficient energy to yield O^- ions, for O_2 gas (solid lines) and for O_2/M (dashed lines).

O_2^- and O_2^* potential energy curves; the curves under the influence of an image charge are represented by broken lines. In these curves, species on the surface are identified by the symbol /M where M designate the metal surface (e.g., if O^- remains on the surface it is represented by O^-/M). The image charge due to the Pt lowers the energies of the ionic species O^+/M, O^-/M, and O_2^-/M, while the energies of the neutrals O/M, O_2^*/M, and O_2/M remain essentially the same. The magnitude of this lowering depends on the distance between the ion and the metal surface, and hence on the thickness of the rare-gas spacer layer. The image charge also splits the $O + O^-$ dissociation limit (i.e., O_2^-/M has essentially two dissociation limits $O^- + O$/M and O^-/M $+ O$ depending on whether or not the O^- ion escapes in vacuum). Thus, the branching ratio between the trapped O^-/M and desorbed (i.e., free O^-) states can vary considerably depending on the energy difference between the two dissociation limits. The free O^- yield seen in Fig. 7 is expected to increase as the energy of the O_2^- $^2\Pi_u$ state and the energy of the O^-/M $+ O$ limit rise with increasing Ar thickness. In contrast, the separation between the O^-/M $+ O^+$ and $O^- + O^+$/M DD dissociation limits is negligibly small since the polarization energies of O^+/M and O^-/M are similar. In this case, the competition is therefore expected to remain essentially the same with increasing Ar thickness. In conclusion, the DA contribution should increase rapidly and the DD contribution should not change appreciably as the thickness of the rare-gas layer increases, as far as branching ratios are concerned.

Another effect of the image charge involves the kinetic energy of the escaping O^- ions. In comparison with the corresponding O^- kinetic energy from O_2 gas, that of the O^- ions via O_2^-/M is decreased while that via O_2^*/M is increased. As can be seen in Fig. 8, the DD dissociation limit being lower for the O_2^* surface state than for the same state in vacuum, the kinetic energy of either O^+ or O^- escaping the surface is larger. However, the $O^- + O$/M limit is *essentially the same* for a surface or free anion dissociating via DA. Thus, the O_2^-/M state produces ESD of O^- ions having a greater chance to be recaptured by the metal; this decreases the DA O^- yield. The O_2^*/M produces faster O^- ions having a greater chance for escape, thus increasing the DD O^- yield.

The fourth effect of the image charge involves the quenching rate of the intermediate state. Near a metal surface, some of the O_2^- or O_2^* states may be neutralized or deexcited before dissociation. Since the O_2^- state can be destroyed by either charge or energy transfer to the metal, intermediate-state quenching has a stronger effect on the DA than on the DD process. Finally, the lowering of the intermediate anion energy by the polarization force induced in the metal, as well as in the rare gas layers, changes the probability for decay of the O_2^- state with respect to the isolated anion. For O_2 condensed on a rare gas solid, it has been shown that the lower energy of the $^2\Pi_u$ O_2^- state reduces its gas-phase autoionization probability due to a change in the crossing points between the potential energy curves of the anion and neutral states.[24] This modification increases considerably its survival probability of the $^2\Pi_u$ state which causes the DA cross section to increase by at least an order of magnitude in comparison with the gas-phase value.[24]

Environmental Effects on DA

In the previous subsection a it has been shown that the proximity of a metal surface may have a drastic effect on the O^- ESD yield arising from DA in condensed O_2, but me –

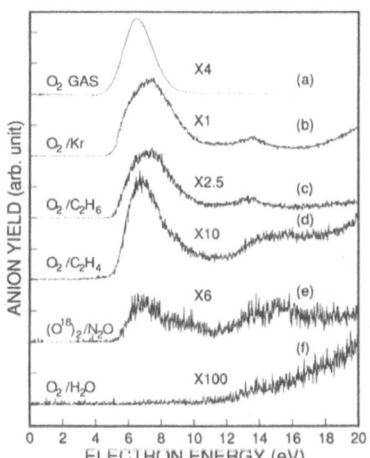

Figure 9. O⁻ yields produced by electron bombardment of (a) gaseous O_2 molecules, and (b)-(f) 0.15 ML of O_2 molecules condensed on 4 ML Kr, C_2H_6, C_2H_4, N_2O and H_2O substrates, respectively. The number over each curve indicates the gain or amplification factor relative to curve (b). The relative magnitude of the O⁻ intensity in curves (b) to (f) is highly sensitive to the nature of the substrate; it is reliable within 5%.

tals are not the only materials which can alter the DA process. Taking again as a reference DA in O_2, the O^- ESD yields from O_2 under different environmental conditions are shown in Fig. 9. The yield functions in this figure were produced by monoenergetic electrons impinging on 0.15 ML of O_2 condensed on different substrates. These latter were prepared by condensing 4 ML of (b) Kr, (c) C_2H_6, (d) C_2H_4, (e) N_2O and (f) H_2O on a clean Pt surface held at a temperature of 20 K in an ultra-high-vacuum system. Curves (b) to (f) were recorded by keeping all experimental parameters the same, so that the relative magnitude of each curve is reliable within a 5% error.

Comparison of the O^- yields in Fig. 9 clearly shows that dissociation of O_2 by DA, is highly dependent on the environment of the O_2 molecule. For example from curves (b) and (d), one can estimate that ESD via DA is about an order of magnitude stronger when O_2 is surrounded by Kr atoms rather than by C_2H_4 molecules. Even between hydrocarbon substrates composed of molecules of similar molecular weight and size (i.e., C_2H_6 and C_2H_4), an almost threefold difference is seen in O^- yields. Within the limits of detectability, *the DEA process is completely absent for O_2 condensed on amorphous ice* (curve f)! The very small signal monotonically increasing from about 12 eV in (f) is due to DD of O_2 (e + $O_2 \rightarrow O^+ + O^- + e$). From the noise level, the O^- signal from DEA is estimated to be *at least three orders of magnitude smaller* than that obtained with the Kr substrate. This translates into a cross section of the order of 10^{-20} cm^2 for O^- production below 12 eV and lower than 10^{-19} cm^2 below 20 eV. The combinations of mechanisms responsible for the changes in the magnitude of DA in O_2 are presently being investigated. In case of O_2 on the H_2O substrate, it has been suggested[25] that the major mechanism, responsible for the disappearance of the DA signal is a shortening of the anion lifetime due to its interaction with the H_2O substrate. It is possible that the additional electron of O_2^- is transfered to adjacent H_2O molecules before dissociation of the anion.

Transient Anions as Charge and Energy Transfer States

When a transient anion is formed in a low pressure gas it can only decay by DA or by reemitting the captured electron in vacuum; however, when formed at a surface or inside a solid composed of similar atoms or molecules, there exists a finite probability for the additional electron to hop from one site to another. If electron trapping occurs via a core-excited resonance, then simultaneous energy and charge transfer is expected (i.e., the additional electron can move with the electronic excitation energy from site to site). Such a simultaneous transfer of excitation energy and extra charge from one site to another is difficult to demonstrate for interactions between similar species, but recently it has been shown[11] that such an "electron-exciton complex" formed in a rare gas solid multilayer film can migrate to the film's surface ands under certain conditions, transfer its energy and charge to a molecular adsorbate. To illustrate this phenomenon, D^- ESD yield functions[11] are shown in Fig. 5 for (a) a pure C_2D_6 8-ML film, (b) 0.2 ML C_2D_6 deposited on a 80 ML Kr spacer film and, (c) 0.05 ML C_2D_6 deposited on a 50 ML Xe spacer film. The D^- yield distribution measured from pure multilayer C_2D_6 shown in (a) has a single broad peak at 9.9 eV which arises from the DA process e + $C_2D_6 \rightarrow C_2D_6^- \rightarrow C_2D_5 + D^-$. When deposited on Kr and Xe substrates, strong and narrow enhancements are observed at energies slightly below those needed to create an exciton in the rare gas solid (Fig. 10(b) and (c)). The direct ESD yield from the C_2D_6 is significantly reduced at incident electron energies well above and below that of the enhancement. For fixed quantities of C_2D_6, the absolute enhancement intensity increases as the rare-gas substrate is made thicker; the intensity of the sharp peak relative to the broad D^- distribution also increases as the C_2D_6 adsorbate coverage is decreased for a fixed substrate thickness. These trends suggest that the initial process occurs in the bulk rather than at the surface, and that it has significant mobility. Enhanced anion yields have also been obtained with an Ar substrate and with other molecular adsorbates.[11] In each of these cases, the enhanced ESD yields are related to electronic excitations of the rare-gas; however, the energies of the observed enhancements are *below* the energy of the lowest optically accessible exciton states of the respective rare gas solid. The sharp enhancements, whose energies were found to be independent of the incidence angle of the impinging electrons, arise from the formation of a core-excited resonance below the energy of the lowest exciton in the rare gas substrates (i.e., a transient anion formed by trapping an electron in the positive electron affinity well of the lowest excited state of the rare gas solid). Such states are described in a previous article in this volume by the author. They are well known for isolated rare-gas atoms (e.g., $Kr^-[4p^55s^2]\,^2P_{3/2}$ at 9.52 eV and $Xe^-[5p^56s^2]\,^2P_{3/2}$ at 7.90 eV), where they are observed ~0.5 eV below the lowest excited neutral state of the atoms.[1] Briefly stated, the core excited anion state or "electron-exciton complex" moves to the surface of the rare gas, where it exchanges its charge and energy by forming a core-excited anion of the molecular adsorbate (e.g., $C_2D_6^-$) which afterwards dissociates, thus producing the sharp enhancement in the ESD yield function (e.g. in the D^- signal of Fig. 10(b) and (c)) precisely at the energy of formation of core-excited anions in the rare gas solid.

SURFACE CHARGING

In surface experiments, when an anion is produced by DA or DD, it does not necessarily escape in vacuum. In fact, due to the polarization force and PDI most of the ions produced are expected to remain at the surface and/or in its vicinity. If the surface on which the molecules are absorbed is conducting, the ions may be neutralized, whereas on a dielectric insulating surface, the products of dissociation or electron capture can remain charged for extremely long periods of time. The production of charges on dielectric surfaces by low-energy electron impact has recently been measured with the technique described in section III.C. The results for O_2 molecules deposited at the surface of a   20-ML

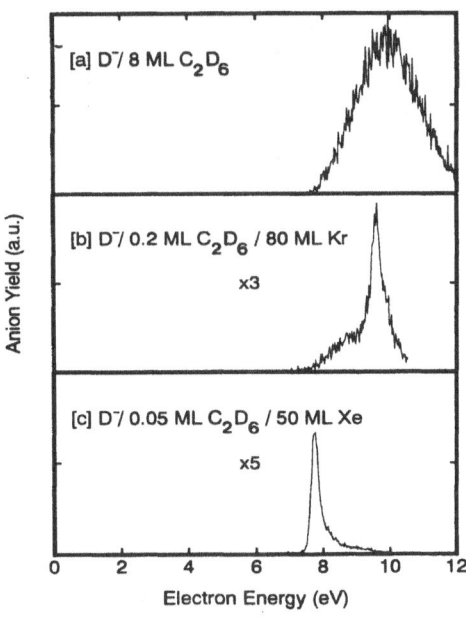

Figure 10. ESD yield obtained for (a) a 8 ML film of C_2D_6, (b) 0.2 ML C_2D_6 deposited on a 80 ML Kr spacer, and (c) 0.05 ML C_2D_6 deposited on 50 ML of Xe.

Kr film are shown in Fig. 11. The surface charging coefficient A_s, defined in section III.C for such a film, is represented by the middle curve in Fig. 11, as a function of the energy of the electron beam causing the charge.[26] This result [Fig. 11(c)] is comparable to the energy dependence of the anion yields derived from the gas-phase attachment rate coefficient for stable O_2^- production [Fig. 11(a)][27] and the gas-phase dissociative attachment (DA) cross section [Fig. 11(b)].[19] No signal has been reported between 1.2 and 4.5 eV in the gas-phase.[1] Curve (d) represents the ESD signal[16] from a Kr film covered with 0.1 ML of O_2. This coverage corresponds to the O_2 coverage of the charging experiment from which curve (c) was obtained.[26]

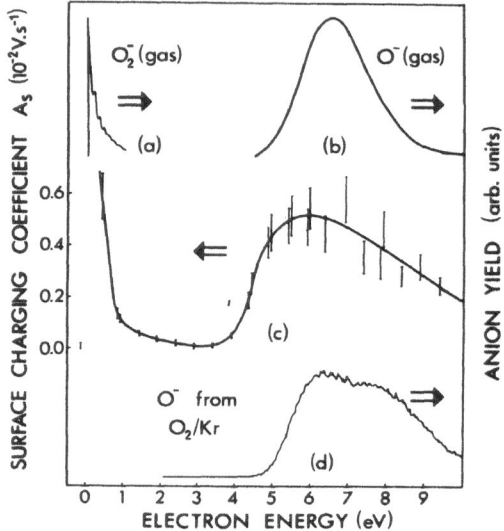

Figure 11. Anion yields produced by 0-10 eV electron impact on (a and b) gaseous O_2 and on (d) a 20-ML Kr film covered with 0.1 ML O_2. The electron energy dependence of the surface charging coefficient A_s for the same O_2/Kr film is shown in (c).

From the correspondence between curves (b), (c) and (d), charge trapping at the surface of the Kr film in the range 4-9 eV has been ascribed to dissociation of the $^2\Pi_u$ state of O_2^- into the limit $O(^3P) + O^-(^2P)$. According to the potential energy curves of the O_2^- system shown in Fig. 6, $O^-(^2P)$ ions forming the broad maximum, centered around 6 eV, could arise from both the $^2\Pi_u$ and the $^2\Sigma_g^+$ state of O_2^-. But, only a small fraction of the signal probably arises from the Σ state since this latter must be formed near another O_2

molecule in order to break the $\Sigma^- \leftrightarrow \Sigma^+$ selection rule. The maximum in the charging coefficient A_s is found at about 0.6 eV below the maximum in the gas-phase O^- yield due to lowering of the $^2\Pi_u$ state by the polarization potential of the Kr surface. For O^- remaining on the surface the dissociation limit corresponds to $O^-/M+O$ shown in Fig. 8. The DA

maximum in the ESD yield arises from the same O_2^- surface state as that in the charging signal. However, the DA maximum lies at higher energy, since in order to overcome the polarization force of the Kr film, (i.e., to reach the dissociation limit $O^- + O/M$ in Fig. 8), desorbing O^- ions must arise on the average, from a higher energy portion in the Frank-

Condon region of the O_2^- $^2\Pi_u$ state. According to the equation for A_s in section III.C, the maximum in Fig. 11(c) around 6 eV corresponds to a value of $(2.2 \pm 0.7) \times 10^{-17}$ cm^2 for the trapping cross section. This can be compared with the cross section in the gas phase[28] at 6.7 eV of $(1.3 \pm 0.2) \times 10^{-18}$. An enhancement of the DA cross section by at least a factor of 17 occurs on the surface. The origin of this enhancement has been explained[24] by

invoking a reduction in the autoionization of the $^2\Pi_u$ O_2^- state into the A $^3\Sigma_u^+$, C$^3\Delta_u$ and

C$^1\Sigma_u^-$ states of O_2 and into the ground state configurations due to the lower energy of the resonance at the Kr surface compared to the isolated anion. According to calculations[24] the greater survival probability of the $^2\Pi_u$ anion increases the dissociation probability into the limit $O(^3P) + O^-(^2P)$ by a factor of 21, thus accounting for the observed change.

In the 0-1.2 eV range, temporary electron attachment to gaseous O_2 leads to the

temporary formation of the $^2\Pi_g$ resonance state[1] of O_2^- whose potential energy curve is shown in Fig. 6. This occurs in vibrational levels v > 4, since the v < 4 levels lie below ground state O_2. However, when during the lifetime of the $^2\Pi_g$ anion vibrational energy is transferred to another molecule in a third body collision, the v < 4 levels can be reached and the electron becomes permanently attached to the molecule as explained by the stabilization process in Fig. 1. The comparison in Fig. 11 suggests that a similar stabilization process is effective at the Kr surface and responsible for surface charge accumulation in the range 0-2 eV. The electron can either lose energy to vibrationally excite O_2 via the $^2\Pi_g$

O_2^- state and afterwards stabilize at a Kr trapping site (including those containing O_2) or the $^2\Pi_g$ state may itself stabilize by energy transfer to phonons. The estimated lifetime of

the $^2\Pi_g$ O_2^- state within clusters is of the order of 10^{-12} s.[29] which is comparable to the phonon vibrational period of the Kr lattice (5.10^{-13} s.).[30] Hence, the additional electron resides a sufficiently long time at an O_2 site to polarize the phonon modes of the Kr lattice and thereby transfer energy to lattice vibrations.

ACKNOWLEDGMENTS

The author would like to thank Dr. Michael Huels for useful suggestions and corrections and Mrs Francine Lussier for the preparation of this manuscript. This work was supported by the Medical Research Council of Canada and the Canadian Center of Excellence in Molecular and Interfacial Dynamics.

REFERENCES

1. For a review of transient anion formation in gases see G.J. Schulz, *Rev. Mod. Phys.* 45:378 (1973); *ibid.*, 423 (1973); and Christophorou, L.G., "Electron-Molecule

Interactions and their Applications", vol. 1 and vol. 2, Academic Press, Orlando (1984).

2. For a review of anion formation with surface molecules see Sanche, L., Low-energy electron scattering from molecules on surfaces, *J. Phys. B*. 23:1597 (1990); and Palmer, R.E. and Rous, P.J., Resonance in electron-scattering by molecules on surfaces, *Rev. Mod. Phys*. 64:383 (1992).

3. For a review of electron interaction within dielectric solids see Sanche, L., Primary interactions of low-energy electrons in condensed matter, Chapt. 1 "Excess Electrons In Dielectric Media", Ferradini, C. and Jay-Gerin, J.-P. eds, CRC Press, Boca Raton (1991).

4. Azria, R., Parenteau, L. and Sanche, L., Post dissociation interaction in ESD: the $^{18}O^-$ - C ^{16}O surface reaction induced by 4-10 eV electrons, *Chem. Phys. Lett*. 171:229 (1990).

5. Azria, R., Parenteau, L., and Sanche, L., O^- electron stimulated desorption from O_2 in CO and N_2 matrices, *Chem. Phys. Lett*. 156: 606 (1989).

6. Sanche, L. and Parenteau, L., Ion-molecule surface reactions induced by slow (5-20 eV) electrons, *Phys. Rev. Lett*. 59:136 (1987); Sanche, L. and Parenteau, L., Surface reactions between O_2 and hydrocarbons induced by dissociative electron attachment, *J. Chem. Phys*. 93:7476 (1990).

7. Sambe, H., Ramaker, D.E., Parenteau, L., and Sanche, L., Image-charge effects in electron-stimulated desorption: O^- from O_2 condensed on Ar films grown on Pt, *Phys. Rev. Lett*. 59:236 (1987).

8. Marsolais, R.M., Deschênes M., and Sanche, L., Low energy electron transmission method for measuring charge trapping in dielectric films, *Rev. Sci. Instrum*. 60:2724 (1989).

9. L. Sanche, *J. Chem. Phys*. 71:4860 (1979); Sanche, L., Bader G., and Caron, L.G., Transmission of 0-15 eV monoenergetic electrons through aliphatic and alicyclic hydrocarbon films, *J. Chem. Phys*. 76:4016 (1982).

10. Perluzzo, G., Bader, G., Caron L.G., and Sanche, L., Direct determination of electron band energies by transmission interference in thin films, *Phys. Rev. Lett*. 55:545 (1985).

11. Rowntree, P., Sambe, H., Parenteau L., and Sanche, L., The formation of anionic excitatons in the rare gas solids, and their coupling to dissociative states of adsorbed molecules, *Phys. Rev. B* 47:4537 (1993).

12. Rowntree, P., Parenteau L., and Sanche, L., Anion yield produced by low-energy electron impact on condensed hydrocarbon films, *J. Phys. Chem*. 95:4902 (1991).

13. For a review of these results see Sanche, L., and Azria, R., Formation of negative ions at surfaces: Electron stimulated desorption, Chapt. 18 in "Negative ions", vol. 2, V. Esaulov, ed. in preparation.

14. For a review of anion formation by electron impact on cluster see Illenberger, E., Electron-attachment reactions in molecular clusters, *Chem. Rev*. 92:1589 (1992).

15. Azria, R., Parenteau L., and Sanche, L., Dissociative attachment from condensed O_2: Violation of the selection rule $\Sigma^- \leftrightarrow \Sigma^+$, *Phys. Rev. Lett*. 59:638 (1987).

16. Sambe, H., Ramaker, D.E., Parenteau L., and Sanche, L., Electron-stimulated desorption enhanced by coherent scattering, *Phys. Rev. Lett*. 59:505 (1987).

17. Azria, R., Parenteau L., and Sanche, L., Mechanisms for O^- electron stimulated desorption via dissociative attachment in condensed CO, *J. Chem. Phys*. 88:5166 (1988).

18. Sanche, L., Parenteau L., and Cloutier, P., Dissociative attachment reactions in electron stimulated desorption from condensed O_2 and O_2-doped rare-gas matrices, *J. Chem. Phys*. 91:2664 (1989).

19. Rapp D. and Briglia, D.D., Total cross sections for ionization and attachment in gases by electron impact. II. Negative-ion formation, *J. Chem. Phys.* 43:1480 (1965).

20. Krauss, M., Newmann, D., Wahl, C.A., Das G., and Zeinlse,W., Excited electronic states of O_2^-, *Phys. Rev. A* 7:69 (1973).

21. Sambe H. and Ramaker, D.E., Forbidden electron attachment in O_2, *Phys. Rev. A* 40:3651 (1989).

22. Krupenie, P.H., The band spectrum of molecular oxygen, *J. Phys. Chem. Ref. Data* 1:423 (1972).

23. Sambe H. and Ramaker, D.E., The σ^- selection rule in electron attachment and autoionization of diatomic molecules, *Chem. Phys. Lett.* 139:386 (1987).

24. Sambe, H., Ramaker, D.E., Deschênes, M., Bass, A.D., and Sanche, L., Enhancement of dissociative electron attachment cross section in O_2 condensed on a Kr film, *Phys. Rev. Lett.* 64:523 (1990).

25. Huels, M., Parenteau, L. and Sanche, L., Quenching of the dissociative electron attachment resonances of O_2 physisorbed on amorphous ice, *Chem. Phys. Lett.*, 210:340 (1993).

26. Sanche L. and Deschênes, M., Mechanisms of charge trapping at a dielectric surface: Resonance stabilization and dissociative attachment, *Phys. Rev. Lett.* 61:2096 (1988).

27. Spence D. and Schulz, G.J., Three-body attachment in O_2 using electron-beams, *Phys. Rev. A* 5:724 (1972).

28. Schulz, G.J., Cross sections and electron affinity for O^- ions from O_2, CO and CO_2 by electron impact, *Phys. Rev.* 128:178 (1962).

29. Hatano, Y., Electron attachment to van der Waals molecules, in "Electronic and Atomic Collisions", Lorentz D.C. et al., ed., Elsevier, New York.

30. Kern, K., Zeppenfield, P., David, R., and Comsa, G., Observation of adsorbate-substrate vibrational coupling in physisorbed Kr films on Pt(111), *Phys. Rev. B* 35:886 (1987).

PHOTOINDUCED DISSOCIATIVE ELECTRON CAPTURE PROCESSES IN BINARY ION-MOLECULE COMPLEXES

Donna M. Cyr and Mark A. Johnson

Department of Chemistry
Sterling Chemistry Laboratory
Yale University
New Haven, CT 06511

INTRODUCTION

Molecular negative ions are distinct from their neutral counter parts in that while their ground electronic states can be quite stable, their first excited electronic states often correspond to the electron detachment continuum, with the lowest valence excited states embedded in the continuum.[1,2] Thus, the absence of the long range coulomb attraction between the excess electron and the neutral core often completely eliminates the infinite number of bound Rydberg states present in all neutral molecules near their ionization potentials. There are, however, a few recently discovered cases in which the dipole moment of a neutral molecule is sufficiently large that the electrostatic (i.e. charge-dipole) interaction with the excess electron is sufficient to create an additional bound electronic state just below the continuum.[3] Water dimer anion, $(H_2O)_2^-$,[4] presents a spectacular case in which a dipole bound state actually forms the ground state since all the valence states are unstable with respect to autodetachment.

In this paper we are concerned with the electronically excited states which occur when a stable anion, A^-, possessing only one bound electronic state, is attached by electrostatic forces to a closed shell neutral molecule, M, to form the binary complex, $A^- \cdot M$. Let us begin by considering the degenerate case of an anion complexed to its corresponding neutral (i.e. $O_2^- \cdot O_2$ or $I^- \cdot I$). First we eliminate the cross-interactions between the molecular centers (i.e. electron transfer) to define a zero-order hamiltonian, H_o:

$$H = h_1 + h_2 + h_{12}(R) = H_o + h_{12}(R). \qquad [1]$$

where H_o is derived from the full electrostatic hamiltonian, H, in the limit of large separation between the molecular centers, R.[5] The zero order hamiltonian, $H_o = h_1 + h_2$, describes the separated neutral and anionic systems. The ground state of the pseudo two-particle system, under H_o, is described by the wavefunctions for the separated molecule and anion, $\phi(A_1^-)$ and $\phi(A_2)$ respectively:

$$\psi_{\pm}^o = 1/\sqrt{2}[\phi(A_1^-)\phi(A_2) \pm \phi(A_1)\phi(A_2^-)]. \qquad [2]$$

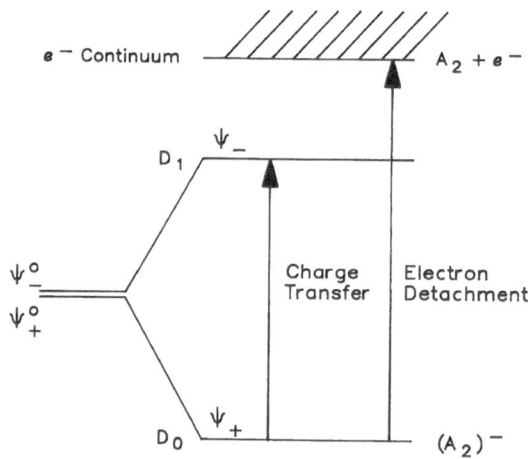

Figure 1. Schematic level structure of symmetrical binary complex $(A_2)^-$, where D_o and D_1 denote the first two doublet electronic states of the system.

ψ_+^o and ψ_-^o are degenerate according to whether the charge is on one molecule or the other. This degeneracy is broken when we re-introduce the cross terms in the hamiltonian and allow the electrons and nuclei on one molecular center to interact with the other. This effect amounts to allowing electron transfer between the charge-localized systems. After the interaction is turned on, ψ_- forms the first excited state, raised above the ground state by $E_{CT} = 2<\psi_+^o|h_{12}|\psi_-^o>^2$ (where E_{CT} is the charge transfer energy). This excited state can generally lie below the electron continuum even if the isolated anion does not have any bound excited states, i.e. $E_{CT} < EA(A \cdot A)$, where $EA(A \cdot A)$ denotes the electron affinity of the $A \cdot A$ complex. From the electronic structure point of view, therefore, we are discussing the situation shown in Figure 1:

For small systems, E_{CT} is on the order of 1 eV, and many charged van der Waals dimer systems (e.g. $(CO_2)_2^+$,[6] $(NO)_2^+$,[6] $(C_6H_6)_2^+$,[7] etc.) possess an intense absorption corresponding to a transition from ψ_+ to ψ_-. In the case of molecular dimers, note that the situation is complicated by the possibility of intra-molecular distortions upon oxidation and reduction,[8] allowing the ground state (ψ_+) to remain charge-localized owing to the different structures of A and A^- (e.g. linear CO_2 vs bent CO_2^-). In this case the degeneracy occurs only along a specific displacement of the internal coordinates corresponding to a simultaneous deformation of each component between the equilibrium geometries of the oxidized and reduced forms of A. These charge-transfer coupled dimers typically form polarons which trap the excess charge in condensed gases and form "core ions" in charged molecular clusters.[9]

These types of CT bands have now been found for several anionic systems [$(NO)_2^-$,[10] $(SO_2)_2^-$,[11] quinone$_2^-$[12]], but have been more difficult to identify in the heterocomplexes. If we now consider an excess charge on a binary complex composed of dissimilar species (i.e. $I^- \cdot O_2$), then the ground and excited states are principally charge-localized on the I^- and O_2^-, respectively, with the excited state correlating to the endoergic $O_2^- + I$ asymptote. If the electron affinity of the complexing neutral molecule (e.g. O_2) is large enough, the CT excited state will again lie below the electron affinity of the complex, forming one bound state below the detachment continuum. We emphasize that by "bound state" we mean that the electronic excited state is stable with respect to autodetachment; the CT excited states are typically unstable with respect to dissociation of the constituent molecules owing to the repulsion induced by the $\langle\psi_+^o|h_{12}|\psi_-^o\rangle$ interaction. As was the case in the degenerate complexes, there will again be a charge-transfer excitation in the asymmetric case resulting from pumping the electron off of the I^- anion and onto the O_2 molecule at an energy corresponding roughly to the difference between the electron affinity of the two species plus a remnant of the $\langle\psi_+^o|h_{12}|\psi_-^o\rangle$ interaction mixing these non-degenerate wavefunctions. In the case of cations, there is ample documentation of such transitions, for example, excitation of $Kr \cdot O_2^+$:

$$Kr \cdot O_2^+ + h\nu \rightarrow Kr^+ + O_2 \qquad \Delta E = 2.27 \text{ eV} \qquad [3]$$

yields the Kr^+ product ion[13] corresponding to endoergic charge transfer. The transition moment for excitation of the Charge Transfer (CT) state is dependent on the h_{12} matrix element such that the absorption coefficient to this state vanishes as $\langle\psi_+^o|h_{12}|\psi_-^o\rangle \rightarrow 0$. The purpose of this paper is to present the photophysics of one such class of asymmetric binary complexes in which an alkyl halide, RY, is complexed to an atomic halide ion, X^-. We shall see that the $X^- \cdot RY$ class of clusters indeed display rich photochemistry, and this behavior raises interesting issues concerning the photophysics at play when the CT band lies very close to the detachment continuum.

EXPERIMENTAL PROCEDURES

Binary complexes of the type $X^- \cdot RY$ are synthesized and cooled in an ionized, pulsed free jet and mass selected using time-of-flight.[9,14] The basic apparatus is illustrated in Figure 2. The complexes are formed by a two-step process in which the halide anion is first formed by dissociative electron attachment (DA):

$$e^- + CH_3I \rightarrow I^- + CH_3 \qquad\qquad [4]$$

followed by three-body association onto the alkyl halide:

$$I^- + RY + Ar \rightarrow I^- \cdot RY + Ar \qquad\qquad [5]$$

It is well known that the cross section for reaction [4] is strongly peaked near zero collision energy,[15] and that the electrons effecting reaction [4] are actually the secondary electrons generated upon high energy (1 keV) electron impact ionization at the throat of the expansion. In this high density region (about 10^{19} molecules/cm^3) near the nozzle, the secondary electrons are efficiently cooled and the jet evolves as an expanding neutral plasma. Reaction [5] also occurs in the relatively high density region of the expansion, allowing some cooling[16] of the ion-molecule complexes as they drift for 15 cm before being pulsed into the mass spectrometer.

Two experiments are carried out after mass selection. First, photoelectron spectra of the complexes are taken by time-of-flight energy analysis[17] to establish the structure of the complexes and their electron binding energies. In the second experiment, the ionic photofragments are monitored using a second, reflecting time-of-flight ("reflectron") mass spectrometer. Both experiments utilize frequency-doubled, pulsed Nd:YAG-pumped dye lasers to achieve high power in the UV, a capability required by the high electron affinities of the halide complexes.

PROPERTIES OF THE $X^- \cdot RY$ ION-MOLECULE COMPLEXES

While essentially all *ab initio* calculations[18-21] verify that the $X^- \cdot RY$ species are indeed charge-localized on the halide ion with an essentially unperturbed alkyl halide moiety attached this fact has not been verified experimentally. One of the simplest ways

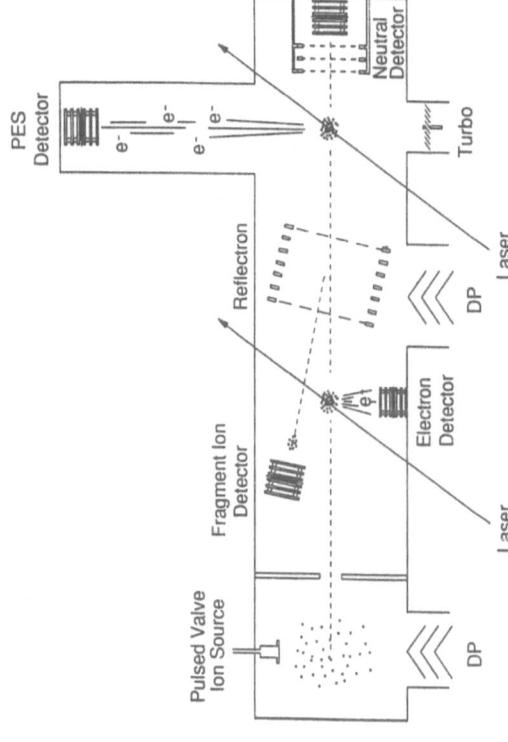

Figure 2. Schematic diagram of pulsed ion spectrometer at Yale.

PES = Photoelectron Spectrometer, DP = Diffusion Pump, and Turbo = Turbomolecular Pump

to established the chemical identity of such complexes is to break them apart in a high energy collision to reveal the ion core.[22] Such experiments on $X^- \cdot RY$ (X,Y = Br, I) are ambiguous since halide exchange ($X^- + RY \rightarrow Y^- + RX$) has been observed following collisional activation.[23] Photoelectron spectroscopy, on the other hand, should be a more powerful diagnostic of the complex structure since the spectroscopic fingerprint of the halide anion should persist in the complex if it is indeed an "ion-molecule" complex. We illustrate the situation in Figure 3, where the upper trace shows the two spin-orbit components of the ^2P ground state term of the iodine atom and the lower trace shows that

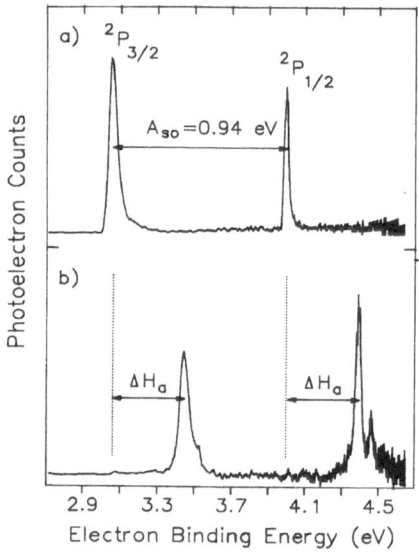

Figure 3. Photoelectron spectra (4.66 eV) of a) I^- and b) the $I^- \cdot CH_3I$ binary complex. ΔH_a and A_{so} indicate the gas phase association energy of the $I^- \cdot CH_3I$ complex and the spin orbit splitting of I, respectively.

this pattern is present in the species synthesized and mass-selected as $I^- \cdot CH_3I$.[24] The splitting of these features in the complex is identical, within experimental error, to the spin-orbit splitting in the I^- spectrum (0.94eV),[25] indicating that the neutral iodine atom formed upon photodetachment is distant from the closed-shell CH_3I molecule. The $I^- \cdot CH_3I$ spectrum is obviously shifted relative to I^-, however, which is easily understood when we consider the neutral and anionic potential curves describing the approach of I toward CH_3I,

shown schematically in Figure 4. The magnitude of the shift (0.37±0.01 eV) of the I⁻·
CH₃I bands relative to the bare I⁻ is almost identical to the heat of association of I⁻ onto
CH₃I measured by high pressure mass spectrometry ($\Delta H_a = 0.39 \pm 0.01$ eV),[26] indicating that
the neutral surface is near the asymptotic (i.e. flat) region over the geometry of the ionic
complex. The ground state in Figure 4 is drawn as a double minimum surface separated

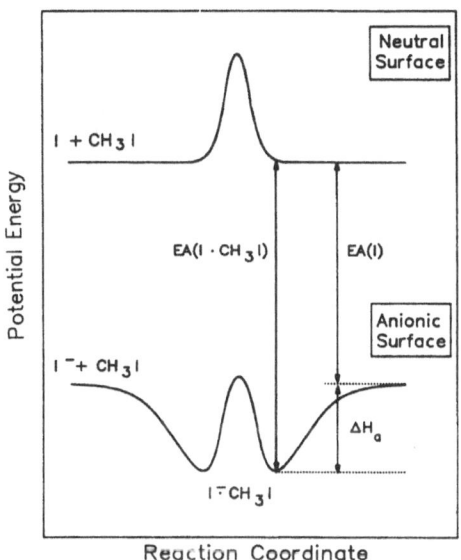

Figure 4. Schematic potential energy curves for I⁻ + CH₃I and I + CH₃I, where EA(I) and EA(I· CH₃· I)
are the electron affinities of atomic I and I· CH₃· I, respectively and ΔH_a is the gas phase association energy
of the I⁻· CH₃I complex.

by a barrier, since the methyl group can exchange iodine atoms with inversion of the
hydrogen atoms in the classic "S$_N$2" (nucleophilic bimolecular) reaction.[27] All of I⁻· RY
complexes we have investigated display sharp photoelectron spectra characteristic of an
unperturbed iodide ion shifted by ~0.4-0.5 eV toward higher electron binding energy in the
complex relative to bare I⁻. Figure 5 presents a survey of the spectra for the series I⁻,
I⁻· CH₃I, I⁻· CH₃Br, and I⁻· CH₂Br₂ to illustrate this point. Thus, the photoelectron
spectroscopy results reinforce the picture that the I⁻· RY complexes can indeed be regarded
as charge-localized species consisting of essentially unperturbed ions bound to neutral
molecules.

the sharp iodine line is broadened in the complexes, sometimes with the appearance of resolvable fine-structure as in the case of $I^- \cdot CH_3I$. One possible cause for such structure is the splitting of the I $^2P_{3/2}$ term into $\Omega = 1/2$ and $3/2$ angular momentum projection components along the ion-molecule axis; however, this fine structure persists in the band arising from the $^2P_{1/2}$ spin-orbit component (with only $\Omega = 1/2$), ruling out this hypothesis. Another possibility is that the fine structure corresponds to vibrational excitation of the

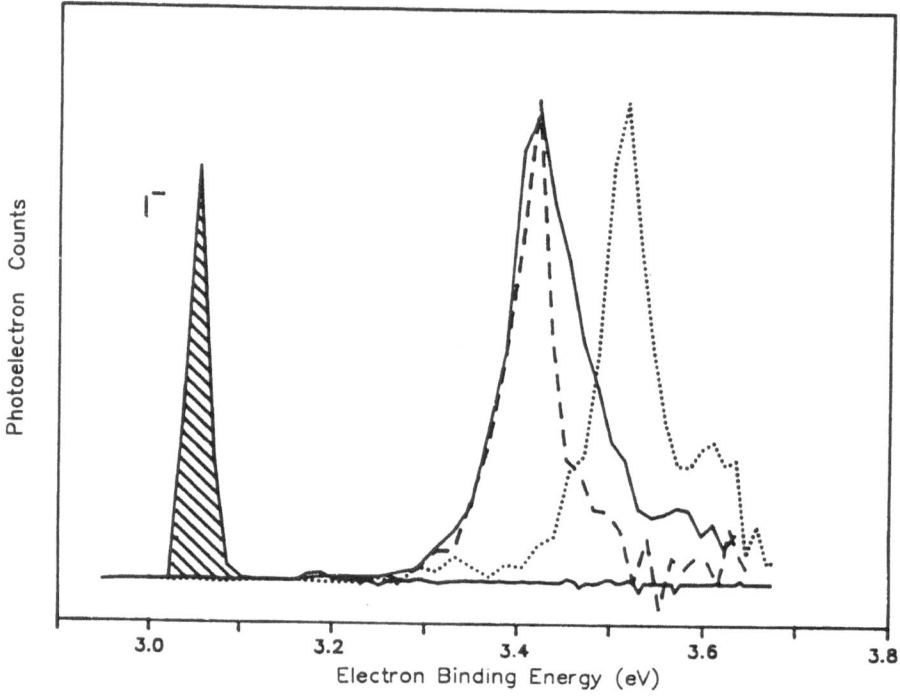

Figure 5. Photoelectron spectra (3.875 eV) of $I^- \cdot CH_3I$ ($-$), $I^- \cdot CH_3Br$ (---) and $I^- \cdot CH_2Br_2$ (···) in addition to the iodine $^2P_{3/2}$ line (shaded).

CH_3I neutral partner in the complex,[24,28] which is resolvable precisely because the neutral surface is so flat along the $I \cdot \cdot CH_3I$ bond that the CH_3I group can execute several vibrations before it retreats from the neutral iodine atom. This is indeed the case as verified by the high resolution photoelectron spectra[24] of the $I^- \cdot CH_3I$ complex shown in Figure 6. The dominant vibrational sequence in Figure 6 corresponds to a frequency of 550 ± 40 cm^{-1}, quite close to v_3 in neutral CH_3I (532 cm^{-1}) and possessing the correct isotope shift for this assignment (5% reduction in the frequency for CD_3I *vs* CH_3I). Since we have argued that the iodide ion is rather far removed from CH_3I in the complex, it is plausible that the envelope of the vibrational band can be modelled as due to a transition from a distorted CH_3I in the complex to a nearly undistorted CH_3I in the neutral. A Franck-Condon analysis[29] of the vibrational intensities indicates that CH_3I (nominally neutral)

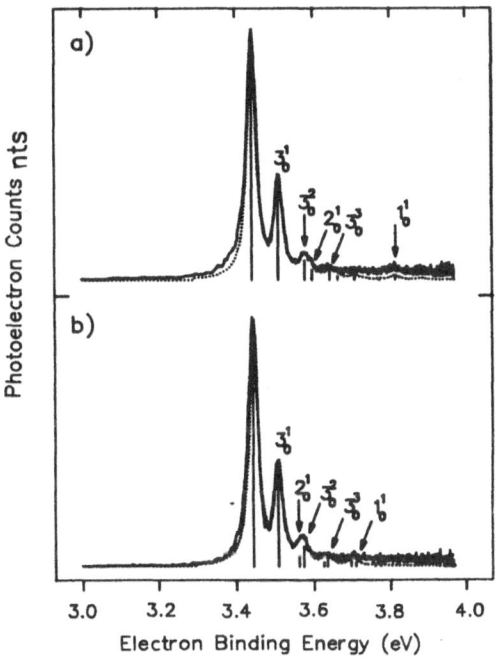

Figure 6. Isotope dependence of the vibrational fine structure in the high resolution photoelectron spectrum of the a) $I^- \cdot CH_3I$ and b) $I^- \cdot CD_3I$ bands arising from the $I\ ^2P_{3/2}$ state. Dotted lines through the spectra are the results of a Franck-Condon fit assuming that CH_3I is distorted upon complexation with I^-. The labels (e.g. 3_0^1) denote the vibronic assignments of the band.

moiety in the complex is distorted by a few percent along the C-I stretch, the CH_3 umbrella, and the C-H stretch totally symmetric coordinates.[30]

The magnitude of the CH_3I distortion is anticipated by CT interactions in the asymmetric complex. Recalling the remarks in the introduction, the ground state wavefunction, ψ, of the asymmetric complex can be approximated as an admixture of the charge-localized states:

$$\psi = c_1\phi(I^-)\phi(CH_3I) + c_2\phi(I)\phi(CH_3I^-) \qquad [6]$$

where the mixing matrix elements, c_i, are given by the usual 2X2 diagonalization of h_{12}. Shaik and co-workers[18] have considered the CT model for such complexes at length, and suggest that the intrinsic matrix element for charge transfer is on the order of 0.6 eV. This can be used along with the proximity of the zero order CT state (3.6 eV) to conclude that

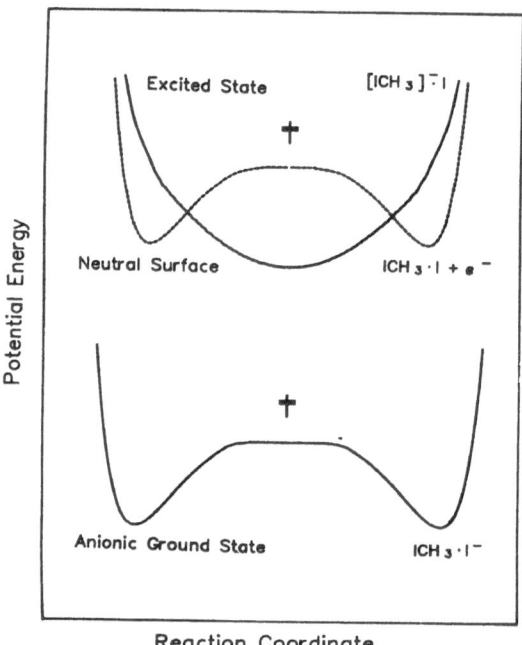

Figure 7. Schematic potential curves for the mixed CT states of the anion invoked to explain the behavior of the $I^- \cdot CH_3I$ complex along with the neutral surface (dotted line). The location of the neutral surface is unknown in the vicinity of the transition state ($\ddagger$).

about 3% of the $I \cdot [CH_3I]^-$ wavefunction is mixed with the ground state (e.g. $c_2^2 \cong 0.03$). This mixing nicely explains the distortions of the CH_3I in the complex, since the excess electron must reside in the C-I σ^* orbital, which is antibonding along the C-I bond and causes most of the distortion along is coordinate.[31] The underlying potential curves which hold the key to the $I^- \cdot CH_3I$ complex are schematically shown in Figure 7, displayed along the antisymmetric I-C-I stretch normal coordinate.

CHARGE-TRANSFER PHOTOCHEMISTRY OF THE COMPLEXES

Anticipating a discussion later in this paper, we also include the location of the neutral surface in Figure 7, indicating that the CT excited states of the $I^- \cdot RY$ complexes are generally quite close to, or are embedded in, the electron continuum. This arises because the relatively large distance between I^- and RY in the ground state results in an excited state which is approximately $[CH_3I]^-$ perturbed by the iodine atom. The valence ground state of $[CH_3I]^-$, however, is an electron-molecule scattering resonance, repulsive along the C-I$^-$ bond and becoming stable with respect to autodetachment as the C-I$^-$ bond ruptures.

The search for the CT excited states is therefore tantamount to detecting RY$^-$ since this transient negative ion is the signature of the excited state. The behavior of RY$^-$ is well documented in the literature of dissociative electron attachment:[32]

$$e^- + RY \rightarrow RY^- \qquad \text{[7a]}$$
$$\rightarrow Y^- + R \qquad \text{[7b]}$$
$$\rightarrow R^- + Y \qquad \text{[7c]}$$

where all channels are observed depending on the degree of halogenation of the alkyl halide. For example, halide-rich molecules such as CCl_4 exist as a stable negative ion and also display two classes of products in (electron transfer) reactive collisions which proceed via a transient CCl_4^-:

$$
\begin{array}{lll}
K + CCl_4 \rightarrow CCl_4^- + K^+ & 0.001 & \text{[8a]} \\
\rightarrow Cl^- + CCl_3 + K^+ & 0.906 & \text{[8b]} \\
\rightarrow CCl_3^- + Cl + K^+ & 0.074 & \text{[8c]} \\
\rightarrow Cl_2^- + CCl_2 + K^+ & 0.014 & \text{[8d]} \\
\rightarrow Cl_2 + CCl_2^- + K^+ & 0.004 & \text{[8e]}
\end{array}
$$

where the numbers in equation [8] indicate the branching fraction of each product channel.[33] The photofragments resulting from photoexcitation of the $Cl^- \cdot CCl_4$ complex are shown in Figure 8. Several of the products expected from DA are found in the photofragments, including the formation of CCl_4^-, with a much larger cross section that reaction 8a. The enhanced branching ratio may be due to stabilization of the ion by ejection of the Cl atom after photoexcitation. The mixed IBr$^-$ diatomic anion is also formed photofragmentation of $I^- \cdot CH_2Br_2$, presumably due to the recombination of a nascent Br$^-$ with the (now) neutral I atom. Interestingly, this recombination product becomes more important with increasing halogenation on the neutral partner, possibly indicating that it arises from the physical proximity of the two halogens in the ground state of the ionic complex. Also note that I$^-$ is not formed from $I^- \cdot CH_2Br_2$, indicating that the fragment ions arise from RY$^-$ and not from shake off of the neutral.

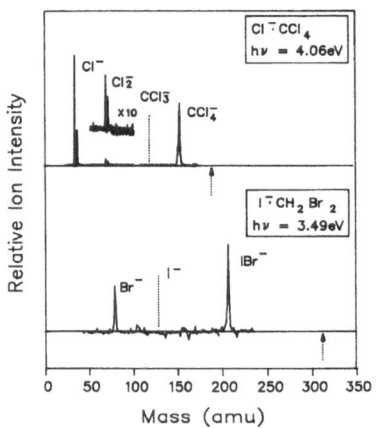

Figure 8. Ionic photoproduct distribution from excitation of $Cl^{-} \cdot CCl_4$ and $I^{-} \cdot CH_2Br_2$ at 4.06 and 3.49 eV respectively. The dotted lines indicate the absence of the anion photoproduct, and the dotted arrows indicate the position of the parent complexes relative to the photoproducts.

Table 1. Photofragmentation Branching Ratios for S_N2 Reaction Intermediates.

System	Excitation Wavelength	Observed Products	Average Product Ratios
$I^- \cdot CH_3Br$	355nm (3.55eV)	Br^-	
$I^- \cdot CH_2Br_2$	362-330nm (3.43-3.76eV)	Br^-, IBr^-	1:1
$Br^- \cdot CH_2Br_2$	315-290nm (3.94-4.28eV)	Br^-, Br_2^-	1.4:1
$Br^- \cdot CHBr_3$	320nm (3.87eV)	Br^-, Br_2^-, CBr_3^-	2:1:0.02
$Cl^- \cdot CH_3Cl$	305nm (4.07eV)	detachment only	-
$Cl^- \cdot CH_2Cl_2$	305nm (4.07eV) 279nm (4.44eV) 275nm (4.51eV)	no photofragmentation Cl^- detachment only	- - -
$Cl^- \cdot CHCl_3$	305nm (4.07eV) 279-266nm (4.44-4.66eV)	no photofragmentation $Cl^-:Cl_2^-:CCl_3^-$	- 1:0.2:0.4
$Cl^- \cdot CCl_4$	305nm	$Cl^-:Cl_2^-:CCl_4^-$	1:0.01:0.4
$Cl^- \cdot CHBr_3$	305nm	$Br^-:Cl^-:ClBr^-:Br_2^-$ CBr_3^- not observed	1:0.5:0.5:0.02
$I^- \cdot CH_3I$	355nm (3.55eV)	I^-: I_2^-	10:1
$I^- \cdot CH_3CH_2I$	355nm (3.55eV)	I^-, I_2^-	1:1
$I^- \cdot CH_3CH_2CH_2I$	355nm (3.55eV)	I^-, I_2^-	1:2
$I^- \cdot CH_3CHICH_3$	355nm (3.55eV)	I^-, I_2^-	1:6
$I^- \cdot CH_3CH_2CH_2CH_2I$	355nm (3.55eV)	I^-, I_2^-	1:3.5
$I^- \cdot (CH_3)_3CI$	355nm (3.55eV)	I^-, I_2^-	15:1

We have investigated the photoproducts from a number of the $X^- \cdot RY$ complexes, with the results summarized in Table I. All the observed products can be rationalized in terms of a transient $X \cdot (RY)^-$ species which decays by break-up of RY^- (as in DA) with the additional possibility of recombination of these products with X.

Assignment of the $X^- \cdot RY$ photochemistry as due to CT excitation must also be consistent with the known location of RY^- from DA studies. For many alkyl halides (especially for the heavy halides and polyhalogenated molecules), the RY^- state lies just above the RY neutral (plus free electron) curve, and the CT band location should be crudely (within the two state approximation) given by:

$$E_{CT} = EA(X) + VEA(RY) + \Delta E(complex) + \beta \qquad [9]$$

where VEA(RY) denotes the vertical electron attachment energy at the equilibrium geometry of RY, ΔE(complex) is binding energy of the ground state of the complex, and β includes correction factors arising from the CT matrix element repelling the zero-order states. For RY=CH$_3$I and CH$_2$Br$_2$, the VEA is very nearly zero, as evidenced by the propensity of these systems to capture very slow electrons with high cross section. Thus, for these systems, we expect the CT transition to occur at:

$$E_{CT} \approx EA(X) + \Delta E(\text{complex}) + \beta \qquad [10]$$

The right hand side of Eq. [10] is close to the electron binding energy of the complex I$^-\cdot$ CH$_3$I, since so little repulsive energy is created in the loosely bound neutral upon photodetachment (either in the I$\cdot\cdot$ CH$_3$I atom-molecule coordinate or in intramolecular modes of CH$_3$I, as we discussed in the context of Figure 6). Since the binding energy of the complex is easily measured using negative ion photoelectron spectroscopy, the validity of Eq. [10] can be directly tested by comparing the absorption (i.e.

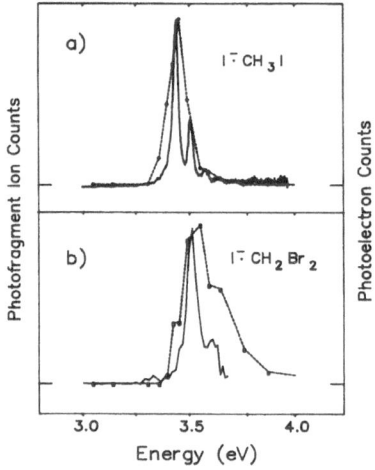

Figure 9. Relative photofragmentation cross sections (dotted lines) and photoelectron spectra (solid lines, displayed as binding energy distribution) for a) I$^-\cdot$ CH$_3$I and b) I$^-\cdot$ CH$_2$Br$_2$.

photofragment action) spectrum against the electron binding energy distribution extracted from the photoelectron spectra of the complexes.[34,35] Figure 9 presents the relative photofragment cross section *vs* photon energy in addition to the photoelectron binding energy distribution for the I$^-\cdot$ CH$_3$I and I$^-\cdot$ CH$_2$Br$_2$ complexes, respectively. The action spectra are remarkably close to the vertical detachment energies in each case, indicating that Eq. [10] correctly captures the essential physics of the process with β≈0. It is important to emphasize that Eq. [10] also anticipates the location of spectra in which RY$^-$ lies significantly above neutral (equilibrium) RY, such as in the case of CHCl$_3$. In that case, DA studies[36,37] indicate that RY$^-$ lies about 0.2eV above CHCl$_3$, and indeed no fragments are observed for CHCl$_3$ complexes at the detachment threshold, but are observed at 4.66 eV. Using VEA data from dissociative attachment data, Eq. [9] indicates that the CT band should occur at 4.42 eV, close to the observed value.

Another indication that the CT model is qualitatively correct can be inferred from the relative fragmentation cross sections depending on the nature of RY. In general, species with large DA cross sections give the largest photofragmentation cross sections and *vice versa*. I$^-$· CH$_3$Br, for example, has a fragmentation cross section about one order of magnitude lower than I$^-$· CH$_2$Br$_2$, which is similar to the relative DA cross sections for low energy electron attachment[36,37]. A similar trend is found for the X$^-$· CH$_{4-n}$Cl$_n$ series where no fragments are found for n=1, and the cross section increases to the largest value for n=4 (see Table 1). Thus, absorption in the I$^-$· RY complexes can be viewed as a kind of intra-cluster, photoexcited dissociative capture process in which the low energy electron is created by detachment from the halide ion.

AN INTRA-CLUSTER ELECTRON SCATTERING MODEL FOR THE PHOTOCHEMISTRY OF X$^-$· RY

The similarity of the photofragmentation spectrum to the electron binding energy spectrum (see Figure 9) indicates that in the iodide anion complexes, fragmentation occurs in the same vicinity as threshold electrons. This implies that we have the potential of exciting *both* the detachment continuum as well as the CT state, which is itself coupled to the continuum. The situation is quite reminiscent of the photoionization/electron capture experiments of Chutjian[15] in his very high resolution studies of DA in systems such as CH$_3$I and CH$_2$Br$_2$. In those experiments, an alkali atom is ionized with a tunable laser to produce a nearly monoenergetic electron which is then scattered off of molecules to monitor DA processes near threshold. In our case, we would have an exactly equivalent circumstance in the limit that I$^-$ is far removed from RY; photodetachment would then yield a monoenergetic electron which scatters off of RY. As the two components of the binary cluster are brought closer together, this scattering process should persist while a new, direct CT band begins to compete for oscillator strength. When the CT excited state lies below the electron continuum it is obvious that only CT absorption can occur to this state. When the CT state lies in the continuum, however, the continuum is mixed with the CT state, giving rise to the DA process in scattering, as well as autodetachment of the CT state once it is formed. The binary complexes give us a unique medium in which to explore direct optical excitation into these mixed states. The situation is analogous to preparing molecules in the act of dissociating by vertical excitation from a stable electronic state to a reactive excited state, as for example has been demonstrated in the case of CH$_3$I photodissociation.[38] The negative ion photochemistry represents a different case, however, in that the reaction which is initiated by photoexcitation is the dissociative attachment, electron-molecule scattering reaction.

A key question in the photoexcitation of a DA resonance is whether the mixed, electron-molecule scattering wavefunction is excited *via* the negative ion zero-order (i.e. CT) state or the continuum contribution. The two alternatives are diagrammed in Figure 10. Both cases should occur in nature depending on the relative transition moments for photodetachment and CT absorption. For example, as the I$^-$ and RY constituents of the complex are separated, the CT band must disappear as $\langle\psi_+^0|h_{12}|\psi_-^0\rangle\rightarrow 0$ (recall $h_{12}\rightarrow 0$ as R$\rightarrow\infty$), but the photodetachment from I$^-$ continues to occur out to infinite separation.

Coherent excitation of the continuum would then be manifested as "intra-cluster electron scattering," with a very fast but significant rise time for formation of RY⁻, while coherent excitation of the CT band would lead to prompt formation of RY⁻ with immediate decay. It would appear that one way to identify which process dominates the photodissociation of I⁻· RY would be to monitor the absolute absorption cross section of the CT band. If this absorption is much stronger than detachment (which continues above the Franck-Condon

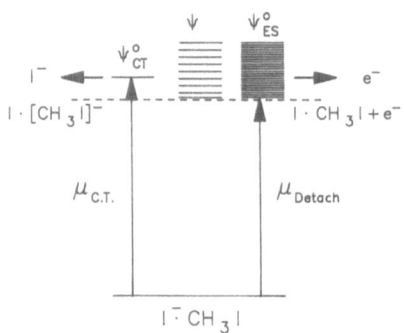

Figure 10. Illustration of two limiting cases when the transition moment for photoexcitation of a dissociative attachment resonance is dominated by a) CT absorption in the ion, and b) excitation of the continuum (e.g. photodetachment). ψ^o_{CT}, ψ^o_{ES} and ψ represent the CT excited state, electron scattering and mixed states respectively.

region of the CT band) at higher energy, then CT states transition moment, $\mu_{C.T.}$, must dominate. A cursory investigation of the absorption of I⁻· CH₃I above the fragmentation band indicates that absorption near the peak cross section for I⁻ production is not significantly stronger than that of bare at higher energy (e.g. 4.66 eV). This indicates that we might be in an intermediate regime or in a condition in which the continuum transition moment dominates the absorption.

CONCLUDING REMARKS

The binary ion-molecule complexes X⁻· CH₃Y are well described as charge-localized on the halide ion in their ground states. Photoexcitation of these complexes generally results in photofragmentation upon excitation near their electron affinities with an array of fragment ions consistent with a transition to a state with X· (RY)⁻ character, suggesting a Charge-Transfer (CT) transition. The shape and location of the bands can be reproduced with a simple model invoking a photoinitiated, intra-cluster dissociative electron attachment reaction. These experiments promise to provide access to the dissociative attachment reaction at the "half collision" between the electron and the RY molecule, with an iodine atom acting as a chaperon.

ACKNOWLEDGEMENTS

We thank the National Science Foundation for support of this work under CHE-9207894.

REFERENCES

1. H. Massey. "Negative Ions," Cambridge University Press, Cambridge (1976).
2. P.S. Drzaic, J. Marks, and J.I. Brauman, Electron Photodetachment from Gas Phase Molecular Anions in: "Gas Phase Ion Chemistry Vol. 3," M.T. Bowers, ed., Academic Press, New York (1984).
3. R.D. Mead, A.E. Stevens, and W.C. Lineberger, Photodetachment in Negative Ion Beams in: "Gas Phase Ion Chemistry Vol. 3," M.T. Bowers,ed., Academic Press, New York (1984).
4. H. Haberland, C. Ludenidgt, H.-G. Schindler, and D.R. Worsnop, Experimental Observation of Negatively Charged Water Dimer and Other Small $(H_2O)_n^-$ Clusters, J. Chem. Phys. 81:3742 (1984).
5. R.L. Fulton, and M. Gouterman, Vibronic Coupling. I. Mathematical Treatment of Two Electronic States, J. Chem. Phys. 35:1059 (1961).
6. G.P. Smith, P.C. Crosby, and J.T. Moseley, Photodissociation of Atmospheric Positive Ions I. 5300-6700 Å, J. Chem. Phys. 67:3818 (1977).
7. F. Misaizu, H. Shinohara, N. Nishi, T. Kondow, and M. Kinoshita, Two-Color 2+2 Photon Resonance Enhanced Ionization of Benzene-Carbon Tetrachloride Binary Clusters, Int. J. Mass Spec. Ion Proc. 102:99 (1990).
8. S.B. Peipho, E.R. Krausz, and P.N. Schatz, Vibronic Coupling Model for Calculation of Mixed Valence Absorption Profiles, J. Am. Chem. Soc. 100:2996 (1978).
9. M.A. Johnson, and W.C. Lineberger, Pulsed Methods for Cluster Ion Spectroscopy in: "Techniques for the Study of Ion Molecule Reactions," J.M. Farrar, and W. Saunders, Jr., eds., John Wiley & Sons, Inc. (1988).
10. L.A. Posey, and M.A. Johnson, Pulsed Photoelectron Spectroscopy of Negative Cluster Ions: Isolation of Three Distinguishable Forms of $N_2O_2^-$, J. Chem. Phys. 88:5383 (1988).
11. T. Dresch, H. Kramer, Y. Thurner, R. Weber, Photodissociation of Sulfur Dioxide Cluster Anions, Z. Phys. D.: At. Mol. Clust. 18:391 (1991).
12. P.B. Comita, and J.I. Brauman, Electronic Structure of Electron Transfer Intermediates. Photodissociation Spectroscopy of the Negative Ion Dimer of Toluquinone, J. Am. Chem. Soc. 109:7591 (1987).
13. M.F. Jarrold, L. Misev, and M.T. Bowers, Charge Transfer Half Collisions: Photodissociation of the Kr· O_2^+ Cluster Ion with Resolution of the O_2 Product Vibrational States, J. Chem. Phys. 81:4369 (1984).
14. M.J. DeLuca, Solvent Induced Effects in Small Homogeneous Cluster Ions Systems, Ph.D. Thesis, Yale University, 1990.
15. S.H. Alajajian, M.T. Bernius, and A. Chutjian, Electron Attachment Lineshapes, Cross Sections and Rate Constants at Ultra-Low Energies in Several Halomethyl and Haloethyl Molecules, J. Phys. B.: At. Mol. Opt. Phys 21:4021 (1988).
16. P.J. Campagnola, L.A. Posey, and M.A. Johnson, Controlling the Internal Energy Content of Size-Selected Cluster Ions: An Experimental Comparison of the Metastable Decay Rate and Photofragmentation methods of Quantifying the Internal Excitation of $(H_2O)_n^-$, J. Chem. Phys. 95:7998 (1991).
17. L.A. Posey, M.J. DeLuca, and M.A. Johnson, Demonstration of a Pulsed Photoelectron Spectrometer on Mass-Selected Negative Ions: O^-, O_2^-, and O_4^-, Chem Phys. Lett. 131:170 (1986).
18. S.S. Shaik, H.B. Schegel, and S. Wolfe. "Theoretical Aspects of Physical Organic Chemistry: the S_N2 Mechanism," Wiley-Interscience, New York (1992).
19. S.C. Tucker, and D.G. Truhlar, Ab Initio Calculation of the Transition State Geometry and Vibrational Frequencies of the S_N2 Reaction of Cl^- with CH_3Cl, J. Phys. Chem. 93:8138 (1989).
20. K. Hirao, and P. Kebarle, S_N2 Reactions in the Gas Phase. Transition States for the Reaction: Cl^- + RBr → ClR + Br^-, where R = CH_3, C_2H_5, and iso-C_3H_7 from Ab Initio Calculation and Comparison with Experiment. Solvent Effects, Can. J. Chem. 67:1261 (1989).
21. S.R. Vande Linde and W.L. Hase, Complete Multidimensional Analytic Potential Energy Surface for Cl^- + CH_3Cl S_N2 Nucleophilic Substitution, J. Phys. Chem. 94:2778 (1990).

22. R.G. Cooks. "Collision Spectroscopy," Plenum Press, New York (1978).

23. D.M. Cyr, L.A. Posey, G.A. Bishea, C.-C. Han, and M.A. Johnson, Collisional Activation of Captured Intermediates in the Gas-Phase S_N2 Reaction $Cl^- + CH_3Br \rightarrow Br^- + CH_3Cl$, *J. Am. Chem. Soc.* 113:9697 (1991). Collisional Activation of the entrance channel complexes can drive the S_N2 reaction.

24. D.M. Cyr, M.G. Scarton, and M.A. Johnson, Photoelectron Spectroscopy of the Gas-Phase S_N2 Reaction Intermediates $I^- \cdot CH_3I$ and $I^- \cdot CD_3I$: Distortion of the CH_3I at the 'Ion-Dipole' Complex, *J. Chem. Phys.* accepted July 1993.

25. C.E. Moore. "Atomic Energy Levels, Vol. III," National Bureau of Standards, Washington, D.C. (1958).

26. R.C. Dougherty, and J.D. Roberts, S_N2 Reaction in the Gas Phase: Nucleophilicity Effects, *Org. Mass Spec.* 8:81 (1974).

27. W.N. Olmstead, and J.I. Brauman, Gas-Phase Nucleophilic Displacement Reactions, *J. Am. Chem. Soc.* 99:4219 (1977).

28. D.W. Arnold, S.E. Bradforth, E.H. Kim, and D.M. Neumark, Anionic Photoelectron Spectroscopy of Iodine-Carbon Dioxide Clusters, *J. Chem. Phys.* 97:9268 (1992).

29. A Franck-Condon analysis of the photoelectron spectra was performed with a program (PESCAL) written by K.M. Ervin using the methods described in: K.M. Ervin, J. Ho, and W.C. Lineberger, Ultraviolet Photoelectron Spectrum of NO_2^-, *J. Phys. Chem.* 92:5405 (1988).

30. W.T. King, I.M. Mills, and B. Crawford, Normal Coordinates in Methyl Halides, *J. Chem. Phys.* 27:455 (1957).

31. W.A. Chupka, A.M. Woodward, and S.D. Colson, Electron Photodetachment from transient negative ions in the multiphoton ionization of CH_3I, *J. Chem. Phys.* 82:4880 (1985).

32. L.G. Christophorou. "Electron-Molecule Interactions and Their Applications," Academic Press, Orlando, Fla. (1984).

33. H. Dispert, and K. Lacmann, Negative Ion Formation in Collisions Between Potassium and Fluoro- and Chloromethanes: Electron Affinities and Bond Dissociation Energies, *Int. J. Mass Spec. and I. Phys.* 28:49 (1978)

34. D.M. Cyr, G.A. Bishea, M.G. Scarton, and M.A. Johnson, Observation of Charge-Transfer Excited States in the $I^- \cdot CH_3I$, $I^- \cdot CH_3Br$, and $I^- \cdot CH_2Br_2$ S_N2 Reaction Intermediates Using Photofragmentation and Photoelectron Spectroscopy, *J. Chem. Phys.* 97:5911 (1992).

35. D.M. Cyr, G.A. Bishea, C.-C. Han, L.A. Posey, M.A. Johnson, Photoinduced Intra-Cluster Electron Transfer Reaction of Captured Intermediates in Gas Phase S_N2 Reactions, *in:* "SPIE Proceedings-Optical Methods for Time- and State-Resolved Chemistry," Society of Photo-Optical Instrumentation Engineers, Bellingham, Washington (1992).

36. D. Spence, and G.J. Schulz, Temperature Dependence of Electron Attachment at Low Energies for Polyatomic Molecules, *J. Chem. Phys.* 58:1800 (1973).

37. R.P. Blaustein, and L.G. Christophorou, Electron Attachment to Halogenated Aliphatic Hydrocarbons, *J. Chem. Phys.* 49:1526 (1968).

38. D. Irme, L. Kinsey, A. Sinha, and J. Krenos, Chemical Dynamics Studied by Emission Spectroscopy of Dissociating Molecules, *J. Phys. Chem.* 88:3956 (1984).

ELECTRON ATTACHMENT TO EXCITED MOLECULES[1]

Loucas G. Christophorou, Lal A. Pinnaduwage, and Panos G. Datskos

Atomic, Molecular, and High Voltage Physics Group, Health and Safety Research Division, Oak Ridge National Laboratory, Post Office Box 2008, Oak Ridge, Tennessee 37831-6122, and Department of Physics, The University of Tennessee, Knoxville, Tennessee, 37996

ABSTRACT

Studies on electron attachment to molecules rotationally/vibrationally excited thermally or via infrared-laser excitation showed that the effect of internal energy of a molecule on its electron attachment properties depends on the mode--dissociative or nondissociative--of electron attachment. They quantified the effect of the internal energy of the molecule on the rate of destruction (by autodissociation or by autodetachment) of its parent transient anion. Generally, increases in ro-vibrational molecular energy increase the cross section for dissociative electron attachment and decrease the effective cross section for parent anion formation due mainly to increased autodetachment. These findings and their understanding are discussed. A discussion is given, also, of recent investigations of electron attachment to electronically excited molecules, especially photoenhanced dissociative electron attachment to long- and short-lived excited electronic states of molecules produced directly or indirectly by laser irradiation. These studies showed that the cross sections for dissociative electron attachment to electronically excited molecules usually are many orders of magnitude larger than those for the ground-state molecules. The new techniques that have been developed for such studies are briefly described also.

[1]Research sponsored by the Office of Health and Environmental Research, U.S. Department of Energy, under Contract DE-AC05-84OR21400 with Martin Marietta Energy Systems, Inc., and by the Wright Laboratory, U.S. Department of the Air Force under Contract AF33615-92-C-2221 with the University of Tennessee, Knoxville, Tennessee, 37996.

INTRODUCTION

Studies of electron attachment to ground-state molecules trace back many decades; they led to significant knowledge on and to an understanding of the various electron attachment processes and their dependence on molecular structure and the electron energy[1-6]. In contrast, studies of electron attachment to "hot" (ro-vibrationally excited) molecules (e.g., see Refs. 4, 7, and 8 and subsequent discussion in this chapter) and especially to electronically excited molecules (e.g., see Ref. 8, and later in this chapter) are more recent and more limited; they have shown that the cross sections for electron attachment to molecules depend rather strongly on the internal energy content of the molecules and in many instances are exceedingly large (almost macroscopic!).

Studies of electron attachment to excited molecules are of fundamental significance (e.g., theory of collision processes and the structure of matter; energy loss mechanisms and cross sections; plasma, electron, ion, and laser physics and chemistry; radiation and life sciences) and of applied/technological significance (e.g., plasmas, lasers, gas discharges, pulsed power switches, optogalvanic effects). Excited molecules are more reactive than are unexcited molecules, making their effect on the properties of many systems significant even when their number densities in such systems are small. The reason for the limited studies on electron attachment to (and, in general, electron interactions with[8]) excited molecules is largely due to the difficulty in producing sufficient numbers of excited molecules to study under controlled experimental conditions, and to the often rapid dissociation of the excited molecules.

ELECTRON ATTACHMENT TO VIBRATIONALLY/ROTATIONALLY EXCITED MOLECULES AND THERMALLY-ENHANCED DETACHMENT FROM POLYATOMIC NEGATIVE IONS

General Considerations

Electron attachment reactions depend strongly on both the kinetic energy, ε, of the attached electron and the structure and internal energy, $<\varepsilon>_{int}$ of the electron attaching molecule.[1-4] As $<\varepsilon>_{int}$ is increased, delicate and often profound changes occur in the electron attaching/detaching properties of the molecule which crucially depend on the molecule itself and the mode (dissociative or nondissociative) of electron attachment (see Refs. 4 and 8 and sources cited later in this chapter).

Within the resonance scattering theory of electron attachment, the electron e, of energy ε, is initially captured by the molecule AX--with a cross section $\sigma_0(\varepsilon)$--forming a transient negative ion AX^{-*} which, then, decays by attachment ($A + X^-$ or AX^-) or autodetachment ($AX^{(*)} + e$), viz.,

$$e(\varepsilon) + AX \xrightarrow{\sigma_0(\varepsilon)} AX^{-*} \begin{array}{c} \xrightarrow{p} A + X^- \quad (1a) \\ \\ \searrow AX^{(*)} + e(\varepsilon') \quad (1b) \end{array}$$

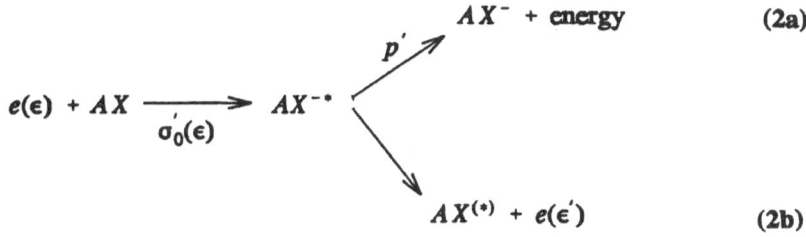

$$e(\epsilon) + AX \xrightarrow[\sigma_0'(\epsilon)]{} AX^{-*} \begin{cases} \xrightarrow{p'} & AX^- + \text{energy} \qquad (2a) \\ \\ \searrow & AX^{(*)} + e(\epsilon') \qquad (2b) \end{cases}$$

where $\epsilon'(\leq\epsilon)$ is the energy of the autodetached electron. The cross section for dissociative electron attachment, σ_{da}, and for nondissociative electron attachment, σ_{nda}, can be written, respectively, as $\sigma_o p$ and $\sigma_o' p'$, where p and p' are the probabilities, respectively, for AX^{-*} to decay producing $A + X^-$, or to become stabilized producing AX^-.

For ground state molecules, the cross sections σ_{da} and σ_{nda} vary by over ten orders of magnitude (from $\sim 10^{-24}$ to $> 10^{-14}$ cm^2) depending on the molecule and the energy position of the negative ion resonance[1,4]. Similarly, the autodetachment lifetimes of the isolated AX^{-*} ions vary by over thirteen orders of magnitude (from $\sim 10^{-15}$ to $> 10^{-2}$ s) (e.g., see Refs. 1, 4, 7, and 9). As the internal energy of the molecule is increased, changes may occur in σ_o and σ_o' due to variations in the Fanck-Condon factors and the fact that as $<\epsilon>_{int}$ is increased electrons with lower energy can reach the negative ion state. These changes, however, cannot explain the reported dependencies of the attachment cross section on T (e.g., see Refs. 4, 7, 8, 10-24).

For the case of dissociative electron attachment, the reported increases in σ_{da} with increasing T can be largely attributed to the increase of p (Eq. (1)) as T is increased;[4,7,8,10-18] at elevated temperatures electron capture occurs over larger internuclear distances and larger excited-vibrational-state amplitudes.

For the case of nondissociative electron attachment, the reported decreases in the cross section σ_{nda} (or the attachment rate constant, k_{nda}) have been attributed to a decrease in p' (i.e., increased autodetachment) as the anion's internal energy is increased[7,8,19-22]. This, however, requires further scrutiny (see later).

The autodetachment lifetime τ_a of an isolated AX^{-*} metastable negative ion depends on the negative ion state, the internal energy of AX^{-*}, the electron affinity, EA, of AX, and the electronic selection rules and Franck-Condon factors for the transition[7,9]

$$AX^{-*} \xrightarrow{\tau_a} AX^{(*)} + e \qquad (3)$$

Direct measurements of τ_a for reaction (3) as a function of the energy of the captured electron that led to the formation of AX^{-*} have shown that τ_a decreases with ϵ (e.g., see Ref. 9). In many high-pressure (swarm) experiments, however, AX^{-*} is effectively stabilized by collisions yielding AX^- (reaction (2a)) predominantly in its lowest state of excitation. In such cases the observed apparent decrease in σ_{nda} (or k_{nda}) with T is due to the enhanced probability of autodetachment via the reaction

$$AX^- + heat \longrightarrow AX + e \qquad (4)$$

The internal energy of polyatomic negative ions is appreciable even at temperatures within a few hundred degrees above ambient and reaction (4) becomes significant--especially when the EA is small (< 0.5 eV)--influencing the measured apparent electron attachment (see later and Refs. 19, 21, and 22).

The techniques that have been employed for such studies are conventional electron swarm methods with provisions for heating and temperature control and measurement[11,14,22], and electron impact mass spectrometers with similar provisions[10,12]; measurements dealing only with thermal electron attachment were also made using the flowing afterglow/Langmuir probe technique (e.g., see Refs. 20 and 23). Heating is, of course, the easiest way to increase the internal energy of AX (and AX⁻), but the resultant ro-vibrational excitation is nonselective. There have been only very few studies which employed selective vibrational excitation of molecules prior to electron attachment, using almost exclusively infrared CO_2 lasers[8,24]. A swarm technique which allows simultaneous determination of the effect of T on both the electron attachment and the electron detachment process has recently been developed and is discussed below.

Dissociative Electron Attachment to "Hot" Molecules

Techniques. As we mentioned above, conventional electron impact mass spectrometers and electron swarm techniques have been employed for the study of dissociative electron attachment to "hot" molecules. The principle of the former is illustrated in Fig. 1. In this particular arrangement, a trochoidal monochrometer is used to produce the monoenergetic electron beam which, subsequently, collides with the excited ("hot") molecules in the iridium chamber. The resulting negative ions were extracted at right angles into a quadrupole mass spectrometer where they are mass analyzed and detected. Temperatures of up to ~ 1200 K were reached in these studies.

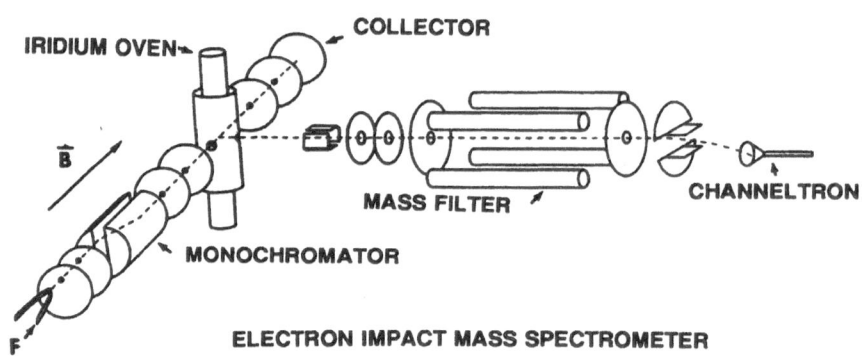

ELECTRON IMPACT MASS SPECTROMETER

Figure 1. Schematic of an electron impact mass spectrometer[10].

In Fig. 2 is depicted the principle of two electron swarm techniques which have been employed in studies of electron attachment to hot molecules. As is well known[1,4], in electron swarm experiments one deals with high pressures. Electrons are generated in various ways in a plane parallel to the two electrodes (see Fig. 2); they quickly (within a few ns, depending on the gas and its density) attain an equilibrium energy distribution $f(\varepsilon, E/N, T)$

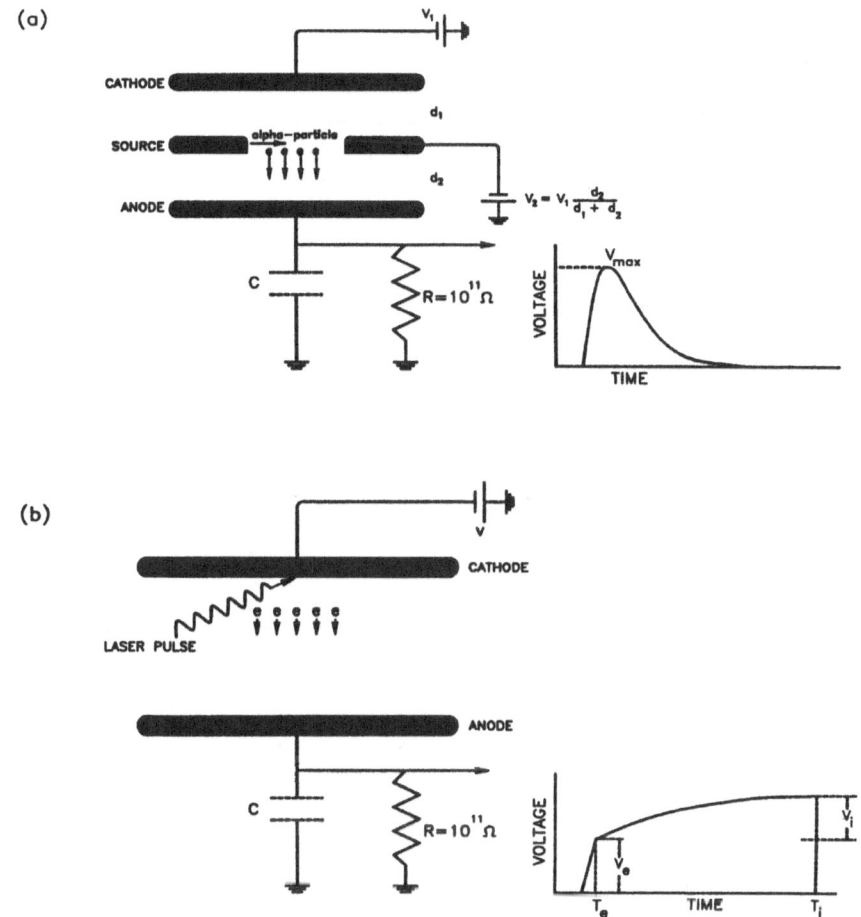

Figure 2. Schematic illustrating the principles of two electron swarm techniques (see the text).

(which depends on the gas, the applied density-reduced electric field E/N and the temperature) and drift, as a swarm, to the anode. As they drift, a fraction is removed from the electron swarm by attachment. The number of electrons removed is measured as a function of E/N (or the mean energy $<\varepsilon>$) and T, and is used to determine $k_a(<\varepsilon>,T)$ from which the attachment cross section σ_a (ε,T) is calculated[1,4]. The various swarm methods differ in the way the initial electrons are produced and the way in which their number is monitored as the swarm drifts to and is collected at the anode[1,4].

In Fig. 2a is shown the principle of the "pulse-shape swarm technique" used extensively by the Oak Ridge National Laboratory group. It is based on the fact that electron attachment modifies appreciably the shape of pulses obtained in ionization chambers. The electron swarm[1,25] (of about 5×10^5 electrons) is produced at a distance d_1 from the cathode every time an α-particle passes through the gas. The α-particle is emitted from a radioactive source (e.g., Cf^{232}) contained in the inner edge of a ring-shaped electrode situated at a distance d_1 from the cathode and at an equipotential. The electron swarm drifts a distance d_2 before reaching the anode. The peak in the <u>induced voltage</u> in the anode electrode is observed through a linear pulse amplifier (which has a known response to a step function) with a multichannel analyzer. At each value of E/N and T this is done for the pure buffer gas (normally Ar or N_2)[1,4] and then again for a mixture of a small quantity of the attaching gas and the buffer gas. From the ratio of the maxima of the pulses due to the induced voltages (see schematic in Fig. 2a) for the two cases, k_a (E/N,T) is determined and from an independent knowledge of $f(\varepsilon,E/N,T)$, $\sigma_a(\varepsilon,T)$ is calculated[4,11,14]. The measured rates and cross sections are total, since no mass analysis is possible in these experiments.

The schematic diagram in Fig. 2b shows the principle of the pulsed Townsend technique which has also been used by the ORNL group for electron-"hot" molecule studies[22,25]. Here a small number of electrons ($<10^6$ electrons/cm^3) is generated photoelectrically at the cathode with a short ($\leq$ 1 ns) UV laser pulse. The photoelectrons quickly reach a steady state and drift--as a swarm--to the anode under E/N. In the presence of electron attachment, the voltage induced in the anode circuit as a function of time (t = 0 $\equiv$ firing of the laser pulse) is as shown schematically in Fig. 2b, i.e., it consists of an initial fast rise due to the unattached electrons and a subsequent slow rise due to the slow-moving negative ions. If N_e and N_i are, respectively, the number of electrons and negative ions reaching the anode and V_e and V_i, are, respectively, the saturated values of the induced voltage due to the motion of electrons and negative ions, then

$$R_v(E/N,T) = \frac{N_e + N_i}{N_e} = \frac{V_e + V_i}{V_e} = \frac{\eta d}{1 - \exp(-\eta d)} \tag{5}$$

where d is the drift gap and η(E/N,T) is the unnormalized electron attachment coefficient. From the measured R_v(E/N,T), the η(E/N,T) is determined by an iterative procedure. The total electron attachment rate constant is, then, determined from

$$k_a (E/N,T) = \frac{\eta (E/N,T) w_e(E/N,T)}{N_a} \tag{6}$$

where w_e(E/N,T) is the electron drift velocity and N_a is the attaching gas number density.

Representative recent measurements. The rate constant/cross section for dissociative electron attachment generally increases with increasing $<\varepsilon>_{int}$ of the molecule, but the relative increase varies depending on the particular molecule and the position of the negative ion state above the v = 0 level of the neutral molecule. This has been shown for both diatomic (e.g., see Refs. 4, 10, 15-18) and polyatomic (e.g., see Refs. 4, 8, 11-14) molecules. As T increases the probability, p, for reaction (1b) increases profoundly; the transient ion AX^{-*} dissociates faster into $A+X^-$. In Figs. 3 and 4 are shown two recent examples for polyatomic molecules. The variation of k_{da} and σ_{da} with $<\varepsilon>_{int}$ for these two molecules and for the diatomic molecules HCl and DCl are shown in Fig. 5. The $<\varepsilon>_{int}$ for the polyatomic molecules was determined[13,14] by using the expression

$$<\varepsilon>_{int} \approx <\varepsilon>_{vib}(T) = \sum_{i=1}^{N} \frac{\hbar\omega_i}{e^{\hbar\omega_i /kT} - 1} \tag{7}$$

where $\hbar\omega_i$ are the vibrational frequencies of the neutral molecule. The use of Eq. (7) to determine $<\varepsilon>_{int}$ reglects the contribution from rotational excitation and takes the internal energy of the molecule to be mainly due to the vibrational excitation above the zero-point energy. The effect of rotational excitation can be seen from the data in Fig. 5c.

420

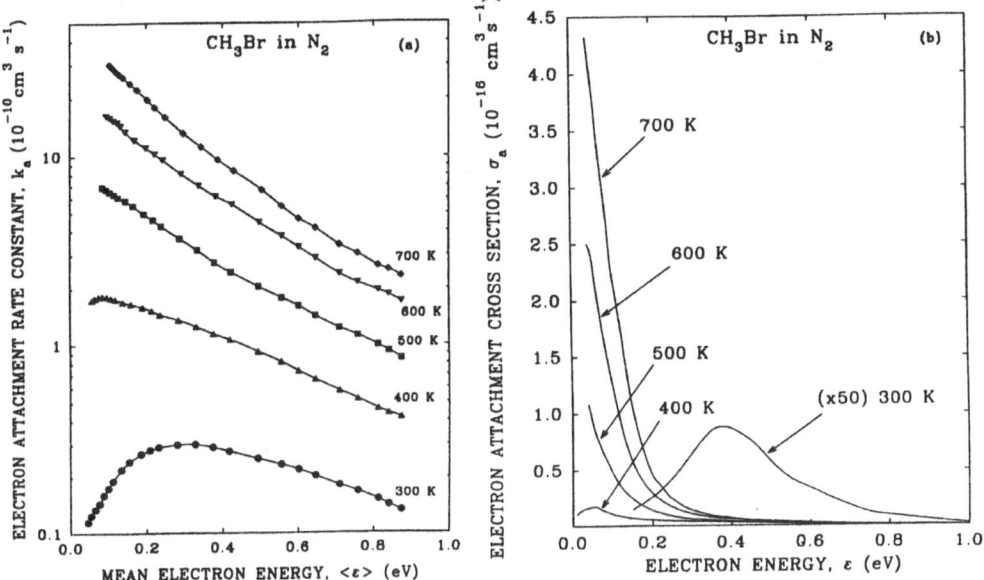

Figure 3. Dissociative electron attachment to CH_3Br at various temperatures. (a) total dissociative electron attachment rate constant k_a as a function of the mean electron energy $<\varepsilon>$. (b) Total dissociative electron attachment cross section σ_a as a function of the electron energy ε (from Ref. 14b).

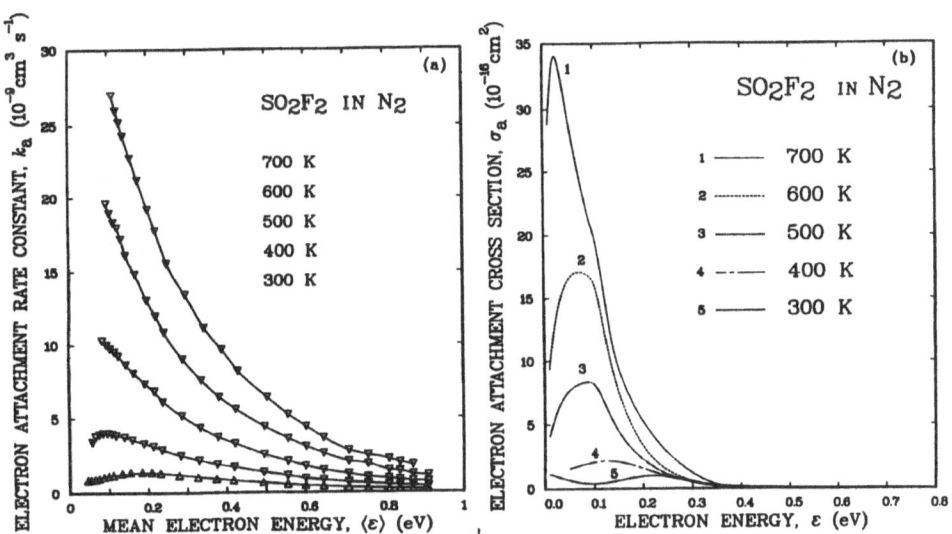

Figure 4. Dissociative electron attachment to SO_2F_2 at various temperatures. (a) k_a vs $<\varepsilon>$; (b) σ_a vs ε (from Ref. 13).

Nondissociative Electron Attachment to "Hot" Molecules and Thermally-Enhanced Detachment From Polyatomic Negative Ions

While the rate constants/cross sections for dissociative electron attachment to molecules, as a rule, increase with increasing T, just the opposite behavior has been reported for the nondissociative electron attachment reaction (2). The rate constant/cross section for nondissociative electron attachment has been found to decrease with increasing T for a number of molecules (e.g., see Refs. 4, 7, 8, 19-22a). An example of this type of behavior is shown in Fig. 6 for the C_6F_6 molecule which at low electron energies (< 1 eV) forms long-lived ($\tau_a > 10^{-5}$ s) parent negative ions (see Ref. 19a). The decrease in k_a (<ε>,T) with increasing T has been attributed to an increase in p' and/or a decrease in σ'_o as T increases. As will be discussed below, however, at the high pressures these swarm measurements were made, the long-lived transient anion AX^{-*} is stabilized by collision, and subsequent studies[22]

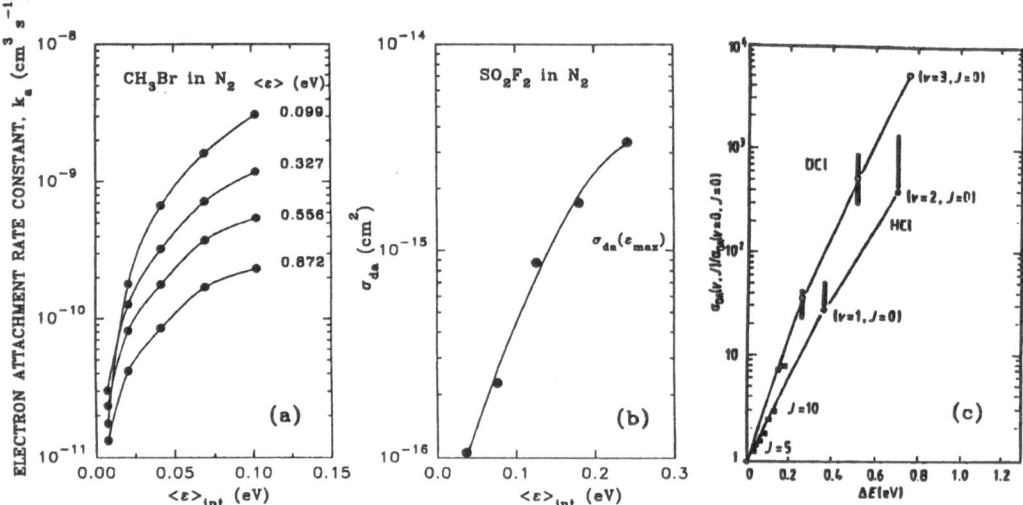

Figure 5. (a) k_a vs $<\varepsilon>_{int}$ for CH_3Br at a number of mean electron energies <ε> (from Ref. 14b); (b) Variation of the peak cross section σ_{da} (ε_{max}) for SO_2F_2 with $<\varepsilon>_{int}$ (from Ref. 13); (c) σ_{da} (v,j)/σ_{da}(v=0, j=0) versus internal energy for HCl/DCl: error bars (experiment)[10b]; • (HCl, v, j=0); o (DCl; v, j=0); ■ (v=0, j) (calculation for 25 meV above the respective thermodynamic threshold[17]).

showed that the detachment process affects the measurement of k_a as it is conducted in the traditional swarm experiments[19a]. For this reason the data in Fig. 6--and similar other work-- represent apparent values of k_a which are influenced by the measurement technique itself.

Recently we employed a new technique via which the effect of T on both the electron attachment and the electron detachment processes can be studied and quantified. This technique is decribed below.

The time-resolved electron swarm technique. This technique allows information on electron attachment and detachment processes to be obtained simultaneously from an analysis of transient electron waveforms. Its principle is shown in Fig. 7. The electron swarm is produced by a narrow (~ 6 x 10^{-10} s) N_2 laser pulse which strikes the cathode electrode through a hole in the anode electrode. The electrons drift to the anode under the influence of an applied electric field. As they drift a fraction is removed by attachment forming unstable negative ions which are quickly stabilized, by collision with the buffer gas, forming stable negative ions (M⁻ in Fig. 7). These stabilized anions are subsequently thermally autodetached giving rise to delayed electrons. The motion of those electrons which reach the anode without ever been attached ("prompt" electrons) and those electrons which have been captured and then thermally released ("delayed" electrons) induces a current in the anode circuit which is observed through a 50 Ω resistor to the ground. (For the μs time scales of interest here, the current is solely from the electrons; the contribution from the slowly moving negative ions is negligible and can be neglected). The electron current is given by[26]

$$i_e(t) = \frac{ew_e}{d} \int_{w_i t}^{\min(w_e t, d)} \rho_e(x,t)dx \tag{8}$$

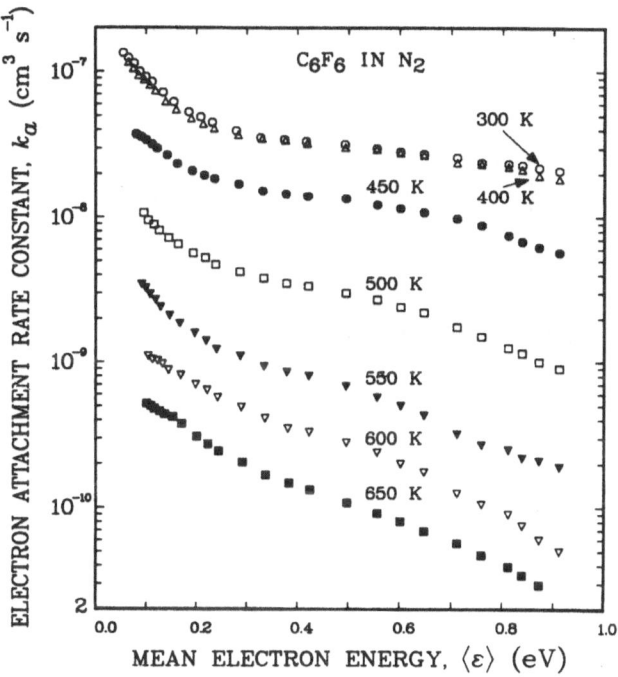

Figure 6. Electron attachment rate constant for C_6F_6 in N_2 buffer gas as a function of the mean electron energy at various temperatures (from Ref. 19a).

where w_e and w_i are the electron and ion drift velocities, d is the drift distance and ρ_e (x,t) is the electron number density given by[26]

$$\rho_e(x,t) = \frac{n_o}{w_e - w_i} \exp\left[-\frac{1}{t_a}\left(\frac{x - w_i t}{w_e - w_i}\right) + \frac{1}{t_d}\left(\frac{x - w_e t}{w_e - w_i}\right)\right]$$

$$x \left\{ \delta\left(\frac{w_e t - x}{w_e - w_i}\right) + \sqrt{\frac{1}{t_a t_d} \frac{x - w_i t}{w_e t - x}} \, I_1\left[\frac{2}{(w_e - w_i)}\sqrt{\frac{1}{t_a t_d}(w_e\, t - x)\,(x - w_i\, t)}\right]\right\} \tag{9}$$

423

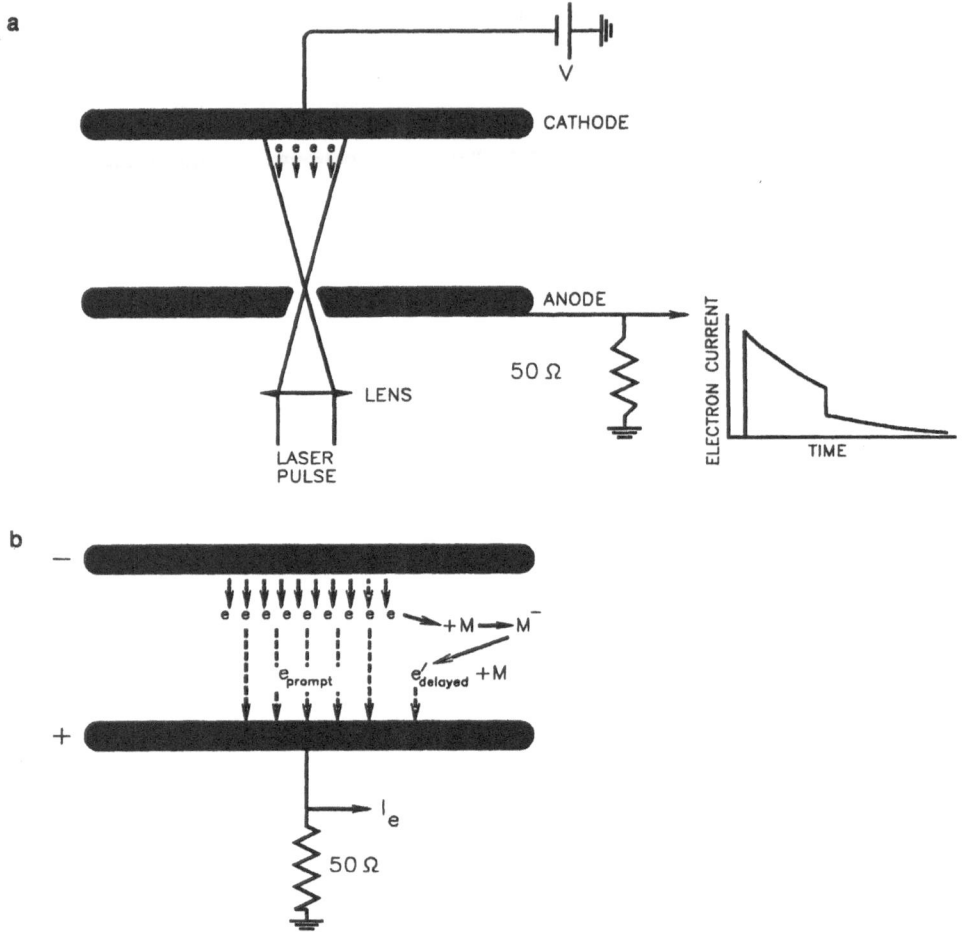

Figure 7. Schematic diagram of the experimental set-up of the time-resolved electron swarm technique (see the text and Ref. 22).

where n_o is the initial electron number density (at $x=0$), t_a^{-1} and t_d^{-1} are, respectively, the electron attachment and detachment frequencies, $\delta[(w_e t-x)/(w_e-w_l)]$ is a delta function and I_1 is the first order modified Bessel function. Figure 8 shows an example of the recorded waveforms as they were obtained for $C_6F_6^-$. The evolution of the autodetachment process as T is increased is clearly evident. The solid curves (curve 2 in Fig. 8f) are the measured waveforms. The dotted curves (curve 3 in Fig. 8f) represent the contribution to the total electron current when no autodetachment occurs, i,e., when all electrons reaching the anode are prompt. The difference between the total electron current (solid curves) and the dotted curves represents the contribution of the autodetached ("delayed") electrons; this is represented by the dashed curves (curve 4 in Fig. 8f). As T is increased this contribution becomes increasingly more significant; the parent anions autodetach faster.

From such recorded electron current waveforms the t_a^{-1} and t_d^{-1} are obtained--at each T-- using a nonlinear least squares fit. The k_{nda} is, then, obtained from

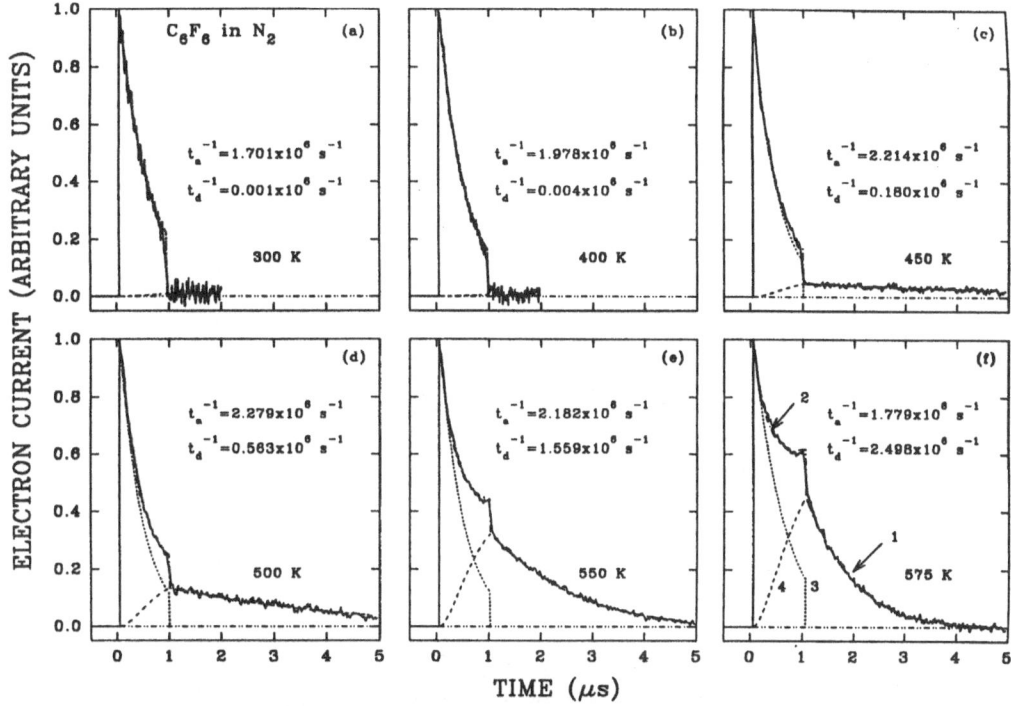

Figure 8. Electron current waveforms for C_6F_6 at T = 300, 400, 450, 500, 550, and 575 K. These waveforms correspond to an E/N = 1.09 x 10^{-17} V cm^2, a mean electron energy <ε> ~ 0.38 eV, a total gas number density N_T = 9.66 x 10^{19} molecules cm^{-3}, an attaching gas number density N_a = 5.64 x 10^{13} molecules cm^{-3} and an electrode gap d = 0.4183 cm. The solid curves (---) are the experimentally measured electron current waveforms, the dash-dot (-•-•) curves are the calculated electron current waveforms using t_a^{-1} and t_d^{-1} values obtained from the nonlinear least squares fit, the dotted (•••) curves represent the contribution to the total electron current of the initial ("prompt") electrons when only electron attachment is present, and the broken (---) curves are the contribution of the "delayed" (autodetached) electrons.

$$k_{nda} = \frac{1}{t_d N_a} \qquad (10)$$

Nondissociative electron attachment to "hot" molecules/thermally-induced detachment from polyatomic negative ions. The technique described in the preceding section has been applied to the study of C_6F_6 (Ref. 22a) and c-C_4F_6 (Ref. 22b). In Fig. 9 is given the measured k_a(<ε>) for C_6F_6 at various T. Indeed while increases in T affect the k_a(<ε>), the changes are not as profound as indicated by the earlier studies (Fig. 6). The k_a(<ε>) first increases slighly with increasing T and then decreases. This behavior is seen better in Fig. 10a where k_a is plotted as a function of <ε>$_{int}$ for fixed values of the mean electron energy. The <ε>$_{int}$ (T) was determined[22a] by assuming that the vibrational frequencies of the anion and the neutral molecule are the same and using published vibrational frequencies for the neutral molecule and Eq. (7). The rather weak dependence of k_a on T suggests that σ_o' (as σ_o) is not a strong function of T. In contrast to this small effect, temperature enhances dramatically the autodetachment of $C_6F_6^-$ (reaction (4)). This is clearly shown in Fig. 10b where the autodetachment frequency t_d^{-1} is seen to increase sharply with increasing <ε>$_{int}$ (the corresponding range of T is 450-575 K).

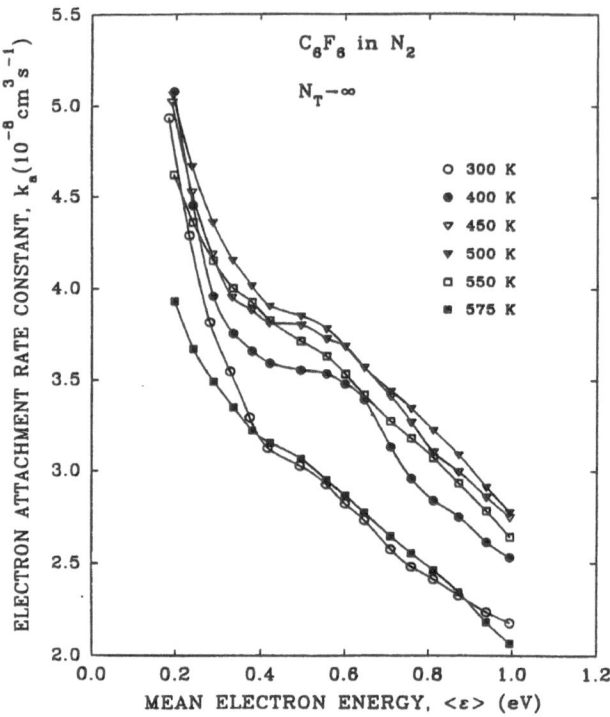

Figure 9. Electron attachment rate constant for the formation of $C_6F_6^-$ as a function of the mean electron energy at various temperatures (from Ref. 22a).

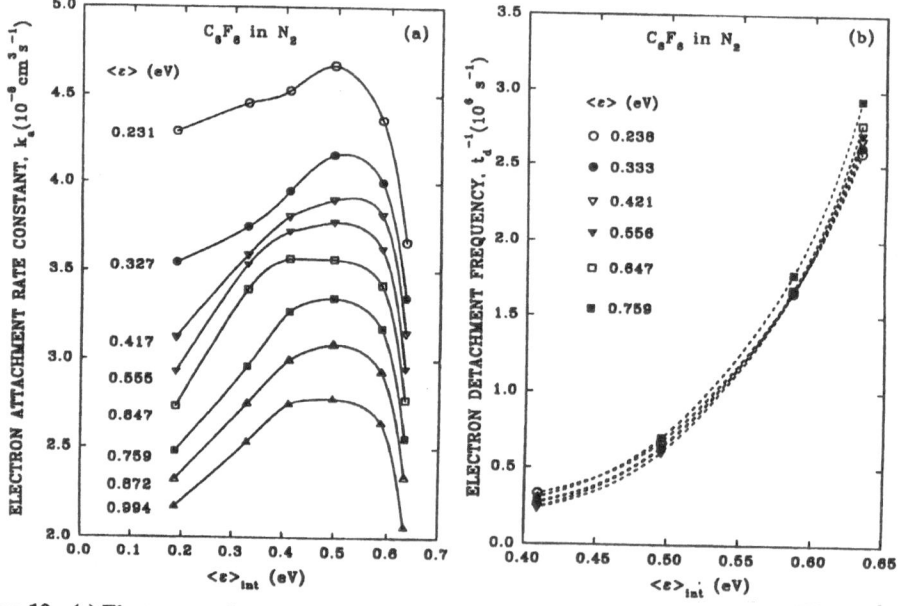

Figure 10. (a) Electron attachment rate constant, k_a (for $N_T \rightarrow \infty$) for the formation of $C_6F_6^-$ as a function of $\langle \varepsilon \rangle_{int}$ for various values of the mean electron energy. (b) Autodetachment frequency, t_d^{-1} (for $N_T \rightarrow \infty$) for the formation of $C_6F_6^-$ as a function of $\langle \varepsilon \rangle_{int}$ for various values of the mean electron energy.

Datskos et al.[22a] assumed that the autodetachment process has an activation energy E^* and that t_d^{-1} is related to T by

$$t_d^{-1} = A\, e^{-E^*/kT} \qquad (11)$$

where A is a constant. From a plot of $\ln(t_d^{-1})$ as a function of T^{-1} (Fig. 11) they obtained a value of E^* equal to 0.477 eV for thermal electrons ($<\varepsilon> \rightarrow 3/2\ kT$) and for completely

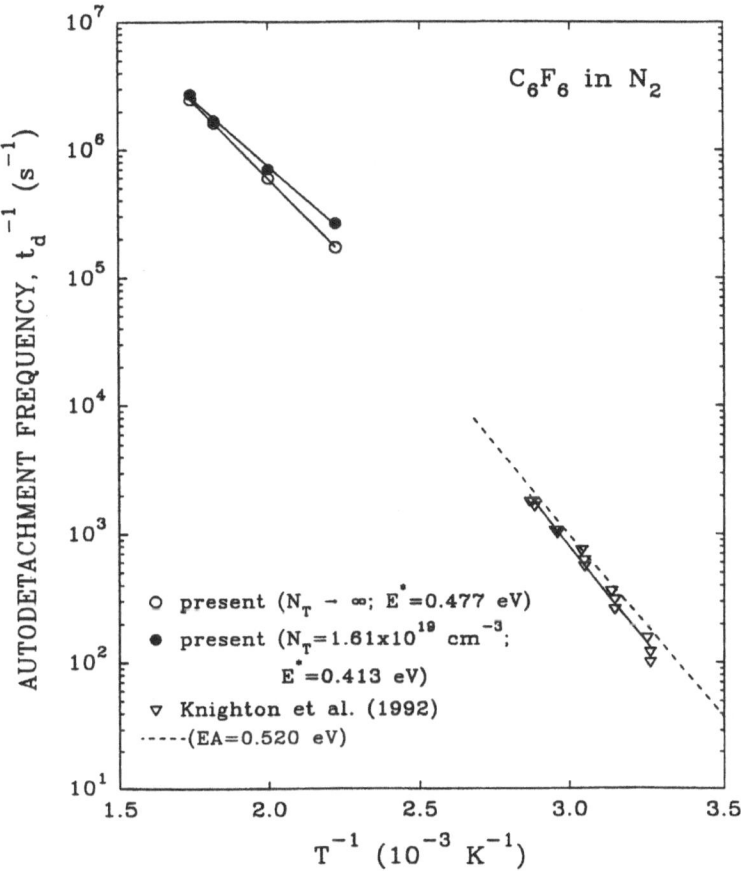

Figure 11. Autodetachment frequency, t_d^{-1} (for $<\varepsilon> \rightarrow 3/2\ kT$) for $C_6F_6^-$ as a function of 1/T. The open circles are for $N_T \rightarrow \infty$ and the closed circles for $N_T = 1.61 \times 10^{19}$ molecules cm^{-3}. Also plotted in the figure are the t_d^{-1} data of Ref. 28 extracted from Fig. 9 of that paper. The slope of the solid line through their data corresponds to an E^* of 0.603 eV. The broken line through their data corresponds to an E^* value of 0.52 eV which is the value reported by Ref. 27 for the EA of C_6F_6 [from Ref. 22a].

stabilized $C_6F_6^-$ (i.e., for total pressure $N_T \rightarrow \infty$). This value of E^* would correspond to the case where virtually all negative ions reach the ground electronic state of the neutral molecule adiabatically from the lowest internal energy state of $C_6F_6^-$. It should, thus, compare well--as indeed it does-- with the electron affinity (EA) value of 0.52 eV reported for C_6F_6 by Chowdhury et al.[27] Both of these values were basically determined by looking at the autodetachment reaction (4): that of Ref. 22a by monitoring the autodetached electrons at microsecond and sub-microsecond times and that of Ref. 27 by monitoring the decline of the negative ion number density as a function of time at long (ms) time scales.

ELECTRON ATTACHMENT TO ELECTRONICALLY EXCITED MOLECULES

In this section we shall describe briefly recent work on the <u>dissociative</u> attachment of slow electrons to electronically excited molecules prepared by laser light prior to or concomitantly with the generation of the attaching electrons. For these studies new techniques have been devised and are briefly discussed later in this section.

We shall separate these studies into two groups. The first group deals with metastable, long-lived (lifetimes $> 10^{-5}$ s) electronic states produced directly or indirectly (via internal conversion and intersystem crossing from higher-lying excited states initially reached by photoexcitation) by single or multiphoton light absorption, viz.

$$nh\nu\,(n{\geq}1) \,+\, AX \,\rightarrow\, AX^* \xrightarrow{+e} AX^{*-} \;\begin{array}{c} \nearrow\; AX^{(*)} + e \\[2ex] \searrow\; A + X^- \end{array} \qquad (12)$$

The second group involves short-lived ($< 10^{-6}$ s) or very short-lived (superexcited) electronic states

$$nh\nu\,(n{>}1) \,+\, AX \,\rightarrow\, AX^{**} \xrightarrow{+e} AX^{**-} \;\begin{array}{c} \nearrow\; AX^{(*)} + e \\[2ex] \searrow\; A + X^- \end{array} \qquad (13)$$

The excited AX^* and superexcited AX^{**} molecules may dissociate into vibrationally or electronically excited neutral fragments. Studies of electron attachment to vibrationally excited photofragments have been reported,[8,29] but, to our knowledge, electron attachment to electronically excited photofragments has not been observed to date.

Prior to discussing the results of these recent studies, it is important to note that electron capture accompanied by simultaneous electronic excitation of the molecule--thus leading to two electrons in normally unfilled molecular orbitals (MOs)--has long been known to manifest itself as resonances in dissociative electron attachment and in electron scattering[1-4,30]. They can be represented by the reaction

$$e_f \,+\, AX \,\longrightarrow\, [AX^* \text{----} e_s] \,\longrightarrow\, AX^{*-} \;\begin{array}{c} \nearrow\; AX^{(*)} + e \\[2ex] \searrow\; A + X^- \end{array} \qquad (14)$$

That is, their production involves concomitant electronic excitation of the molecule AX by the "fast" electron e_f which--having lost virtually all of its kinetic energy to exciting electronically the molecule--is "thermalized" in the very vicinity of the electronically excited molecule it produced and is quickly captured by it with a large cross section. The unstable anion AX^{*-} normally decays very quickly (within $< 10^{-12}$ s) by autodetachment and/or autodissociation. It has, however, been found that in certain cases [e.g., p-benzoquinone[31], aromatic hydrocarbon molecules with electron affinities > 0.5 eV (Ref. 32)] the intermediate AX^{*-} lives long enough ($> 10^{-5}$ s) to be detected directly as a parent anion in conventional mass spectrometers. This accounts[31,32] for the observation of negative ion resonances due to long-lived parent negative ions at electron energies well above thermal and with much larger cross sections than at thermal energies.

Experimental Techniques

The observation of electron attachment to electronically excited molecules requires production of an appreciable number of excited molecules and conditions such that electron attachment to the excited molecules occurs within their lifetimes. Since the electronically excited states of most molecules lie a few eV above the ground state, the excimer laser lines are ideal for these studies. With the availability of pulse energies of ~ 100 mJ in 10-20 ns pulses at the XeF (355 nm, 3.5 eV), XeCl (308 nm, 4.0 eV), KrF (248 nm, 5.0 eV), and ArF (193 nm, 6.4 eV), these lines can produce appreciable quantities of low-lying excited states with monophotonic excitation or high-lying excited states with 2- or 3-photon excitation. With the availability of a dye laser pumped by an excimer laser at the XeCl line, wavelength tunability can be obtained from 320 to ~ 1000 nm with pulse energies of ~ 10 mJ.

Depending on the lifetime of the excited state under study, a number of techniques have been developed[33,34,40,42] for electron attachment studies using pulsed lasers. Those techniques that have been developed by the authors are briefly described below.

<u>Measurements on long-lived ($\tau > 10^{-5}$ s) excited states using high-pressure swarm experiments with pulsed lasers.</u> If the lifetime of the excited state under relatively high ambient pressures (1 - 100 kPa) is longer than ~ 10^{-5} s, the experimental arrangement[33,34] shown in Fig. 12 can be employed. The technique is based on the pulsed Townsend method described earlier[22,25]. Depending on the absorption coefficient at the particular laser

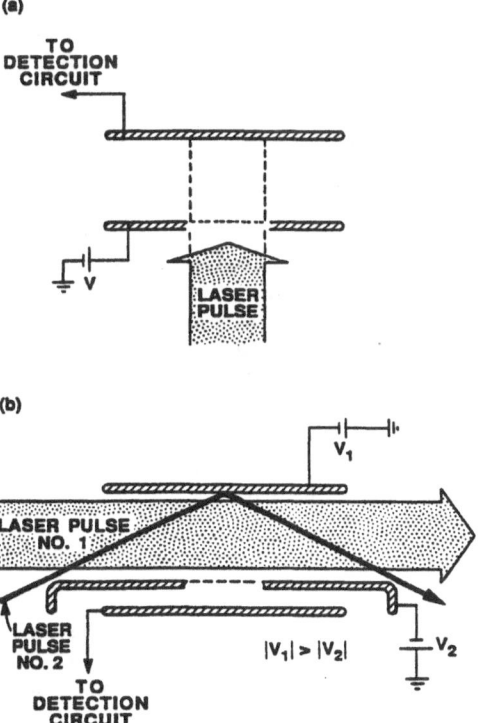

Figure 12. Schematic electrode-laser pulse(s) arrangements used in studies of photoenhanced electron attachment to long-lived (lifetime > 10^{-5} s) electronically excited molecules (see the text and Ref. 34).

wavelength involved, 0.1 to 100 Pa of the gas under study is mixed with a suitable buffer gas (usually N_2 or Ar) of high pressure (1-100 kPa). In Fig. 12a, a laser pulse enters the interaction region through the gridded bottom electrode, excites the molecules in the interation region, and produces a pulse (or swarm) of electrons at the top electrode. The electron swarm reaches a known steady-state energy distribution within $< 10^{-8}$ s, and drifts through the partially excited gas. The drift time taken by the electrons to reach the bottom electrode is $< 10^{-5}$ s, and thus electron attachment to the excited molecules can take place if the lifetime of the excited molecules is $> 10^{-5}$ s.

The arrangement of Fig. 12b is an improved version of that in Fig. 12a. The production of excited species is decoupled from that of the attaching electrons by using two lasers as shown; thus the time delay between the production of the excited species and the arrival of the attaching electrons in the interaction region can be varied. Furthermore, the use of three electrodes for separating the interaction and the detection regions allows the detection of negatively charged particles unambiguously; by applying suitably oriented electric fields, only the negative charges can be extracted to the bottom detection region through a grid in the middle electrode. This is crucial if positive ions are simultaneously produced (via laser photoionization in the interaction region). However, no direct identification of the negative ions can be made in these experiments although indirect anion identification is possible under certain conditions[35].

Measurements on long-lived ($\tau > 10^{-5}$ s) excited states using electron beam experiments with pulsed lasers. Most excited molecular electronic states undergo collisional destruction in a high pressure environment. If the lifetime of such excited states becomes $< 10^{-5}$ s, the experimental arrangement described in the previous section would not be suitable. Instead, an electron beam apparatus such as shown in Fig. 13 (Ref. 36) can be used to conduct measurements under isolated conditions (pressure $< 10^{-2}$ Pa) where the excited-state lifetime is essentially the radiative lifetime. However, the number of excited species in the collision region is determined by the diffusion of excited molecules out of the collision region (see below and Ref. 36). In the experimental arrangement in Fig. 13 a continuous effusive molecular beam is crossed at right angles with a continuous electron beam (energy resolution ~ 0.1 eV) and a pulsed laser beam, inside a vacuum chamber of base pressure ~ 10^{-6} Pa. The negative ions produced in the interaction region are drawn out by a weak electric field (~ 2 V cm^{-1}) into a quadrupole mass filter and the mass analysed ions are detected with a secondary electron multiplier.

Since the excimer lasers normally used for these experiments have pulse repetition rates of ~ 100 Hz (time delay between consecutive pulses ~ 10^4 μs), it is important to selectively detect negative ions arriving at the detector within a particular gate time after a preset delay from each laser pulse. The gate delay is associated with the time taken by the (laser-initiated) negative ions to arrive at the detector, and the gate time should be $\leq$ minimum $\{\tau, \tau_d\}$, where τ is the lifetime of the excited molecules and τ_d is the time taken by the excited molecules to diffuse out of the interaction region [36].

Measurements on short-lived ($\tau < 10^{-8}$ s) excited states. In order to measure electron attachment to short-lived excited species of lifetime τ produced via a laser pulse of duration τ_L, the electron attachment time, τ_c must be[33]

$$\tau_c < max\ \{\tau, \tau_L\} \tag{15}$$

i.e., electron attachment must occur before the decay of the excited species. For $\tau_L > \tau$, this upper limit is dictated by τ_L since excited species are continuously being produced within the duration of the laser pulse. Therefore, in order to observe electron attachment to short-lived species ($\tau < 10^{-8}$ s) produced by excimer lasers ($\tau_L \sim 10^{-8}$ s), the τ_c must be $< 10^{-8}$ s. Fortunately, the electron attachment cross sections are orders of magnitude larger for the excited states compared to the corresponding ground states making τ_c short.

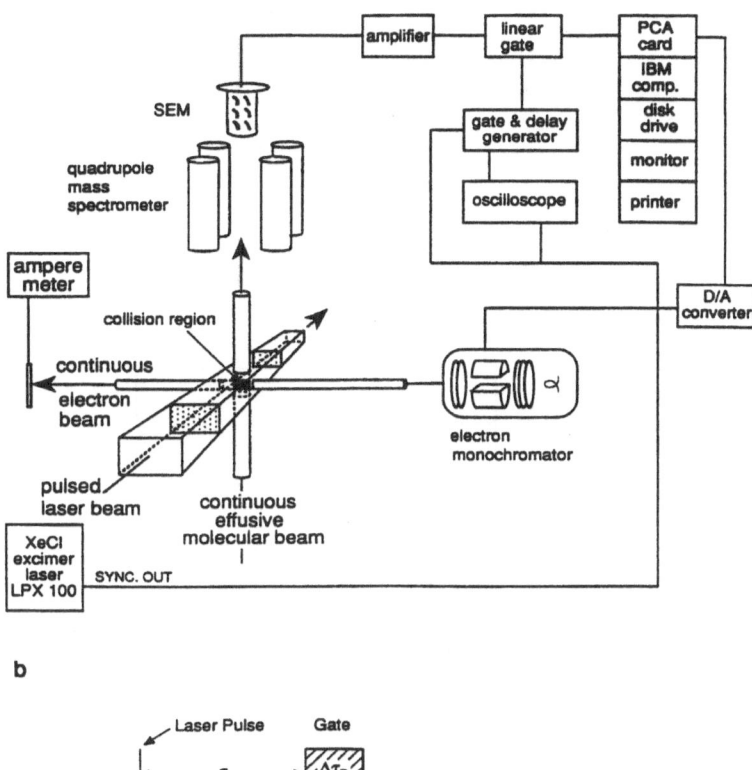

a

b

Figure 13. Schematic diagram of an electron spectrometer/mass spectrometer/laser experimental set up and the gate/delay arrangement employed to study electron attachment to electronically excited SO_2 molecules (see the text and Ref. 36).

A technique has been developed[33] to measure electron attachment to short-lived species in a high-pressure (1-50 kPa) environment. The basic idea behind this technique is to produce the excited species and the attaching electrons (via photoionization of the same gas under study or a suitable additive gas) by a single laser pulse (see Fig. 14). Since the excited species and the electrons are produced in close proximity,--depending on the number density and the electron attachment cross section of the excited species--electron attachment can occur in spite of the short τ.

Since in these arrangements positive ions are also produced by photoionization, an inherent requirement of the method is the ability to separate and to detect negative charges (electrons and negative ions) unambiguously. As shown in Fig. 14, this is accomplished by separating the detection region (located between electrodes 2 and 3) from the interaction region (located between electrodes 1 and 2) in a three-electrode arrangement. Charge transmission between the two regions is through a fine grid, which, also, electrically shields the two regions from each other. By applying suitable electric fields in the two regions either positive or negative charges can be extracted to the detection region. In the case of negative charges, the signal components due to electrons and negative ions are distinguished due to the different drift velocities; a "break" in the signal waveform could be easily seen[33,37] when the total pressure in the chamber exceeded ~ 0.1 kPa.

431

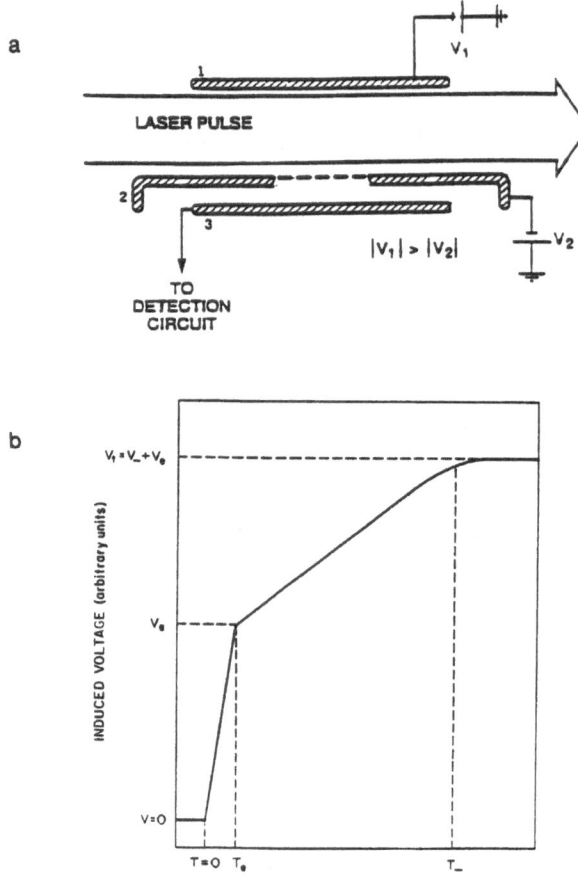

Figure 14. (a) Schematic diagram showing the principle of the technique used to measure electron attachment to short-lived (lifetimes < 10^{-8} s) excited states, and (b) schematic drawing of a typical signal waveform (see the text and Ref. 33).

Dissociative Electron Attachment to Metastable ($\tau \gtrsim 10^{-6}$ s) States

Singlet Oxygen.   Electron attachment measurements on pre-prepared electronically-excited molecules was first carried out[38] on the low-lying (excitation energy ~ 1 eV), extremely long-lived (lifetime ~ 2700 s) a $^1\Delta_g$ state of O_2. An electron beam apparatus was employed and the excited molecules produced in a microwave discharge were introduced to the interaction region via a Pyrex tubing; due to the long lifetime of the a $^1\Delta_g$ state, the other aspects of the experiment were the same as for ground state molecules. The cross section for O^- production via the O_2^{-*} ($^2\Pi_u$) state was shown to be ~ 3-4 times larger for the a $^1\Delta_g$ state compared to the $^3\Sigma_g^-$ ground state. Two additional O^- production channels [via the O_2^- $^2\Sigma_g^+$) state] involving the a $^1\Delta_g$ state were found and studied in two subsequent studies[39,40]. In Fig. 15 are shown the measured[40] cross sections for O^- from the ground state $^3\Sigma_g^-$ of O_2 via the negative ion state $^2\Pi_u$ and from the excited state a $^1\Delta_g$ of O_2 via the negative ion states $^2\Pi_u$ and $^2\Sigma_g^+$.

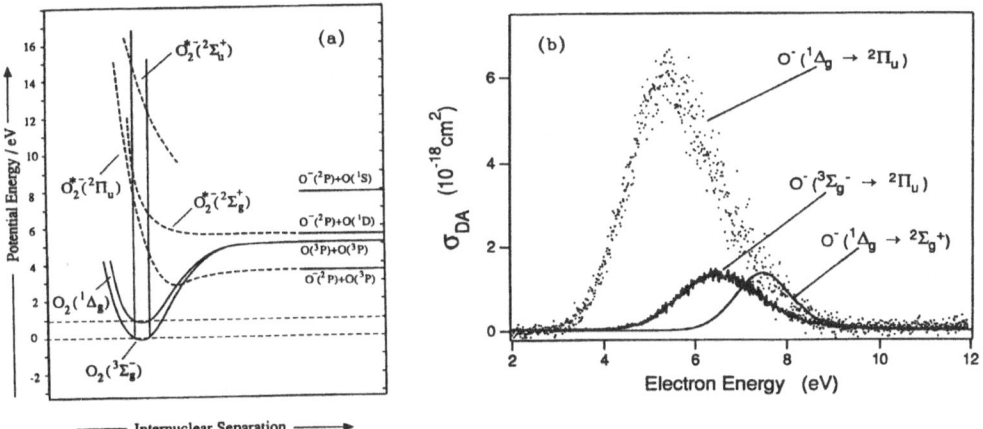

Figure 15. (a) Schematic potential energy diagrams for $O_2(^3\Sigma_g^-)$, $O_2(^1\Delta_g)$, and O_2^{-*} ($^2\Pi_u$, $^2\Sigma_u^+$). The broken curves refer to the negative ion states relevant to dissociative attachment. (b) Comparison of the cross section for the production of O^- from singlet O_2 via the $^2\Pi_u$ and $^2\Sigma_g^+$ negative ion states with that for O^- from the ground state O_2 via the $^2\Pi_u$ negative ion state (from Ref. 40).

First excited triplet state of thiophenol. The first observation of optically enhanced electron attachment to electronically excited states was reported[34,41] for the first excited triplet state, T_1, of thiophenol (PhSH) using the technique described earlier (Fig. 12a). This state is located ~ 4 eV above the ground state and has a lifetime of ~ 6 ms; it was excited indirectly (via internal conversion (IC) and intersystem crossing (ISC) from higher optically allowed states, S_n) using the 248 nm KrF excimer laser line:

$$PhSH + h\nu(KrF) \longrightarrow PhSH^*(S_n) \quad (optically-allowed)$$

$$PhSH^*(S_n) \xrightarrow{\quad IC \quad} PhSH^*(S_1) \xrightarrow{\quad ISC \quad} PhSH^*(T_1) \tag{16}$$

$$\sim10^{-13} \text{ s} \qquad\qquad \sim10^{-11}\text{-}10^{-8} \text{ s}$$

The measured electron attachment coefficient, η/N_a, for the ground and the excited state is shown in Fig. 16: data for curve 1 are for the ground state molecule and were obtained in a separate experiment without using lasers; curve 2 was obtained with 308 nm XeCl excimer laser line and yielded ground state attachment data (the photon energy is 4 eV at the XeCl line and lies below the first excited singlet state S_1 of PhSH); curve 3 was obtained using the KrF line and shows a large enhancement in electron attachment especially at low E/N values (low electron energy). Since only ~ 1 percent of the molecules are excited by a laser pulse, the actual enhancement in electron attachment at thermal energies was shown[34] to be ~ 10^5. This enhancement was attributed to the dissociative electron attachment to the $^-$PhSH(T_1) state produced indirectly via (16).

The curves 4-7 of Fig. 16 show the measured η/N_a for a double laser pulse experiment where the gas was irradiated with one pulse and electron attachment was measured 12.5 ms later when the attaching electrons were produced by a second similar laser pulse[41]. This

433

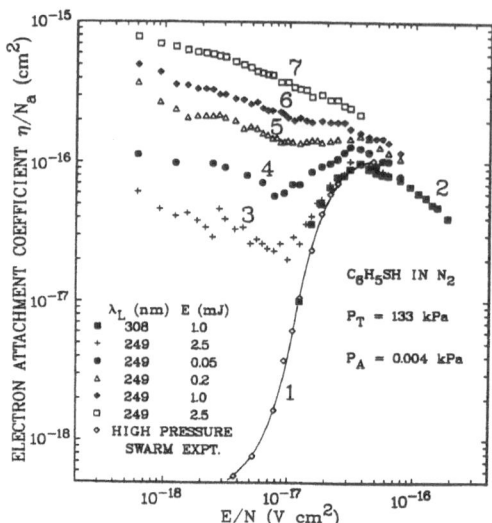

Figure 16. Electron attachment coefficient η/N_a versus E/N for C_6H_5SH in N_2 for the ground-state (curves 1 and 2) and the first-excited triplet state (curve 3) molecules. Curves 4-7 are photoenhanced attachment due to stable photoproducts formed in a double laser-pulse experiment [see the text and Refs. 34 and 41).

delayed photoenhanced attachment was studied in detail[34] and was shown to be due to electron attachment to diphenyl disulfide ($C_6H_5SSC_6H_5$) produced by the interaction of thiophenoxy radicals ($C_6H_5S^{\cdot}$) formed directly or indirectly via the first laser pulse.

The $A^2\Sigma^+$ state of NO. In this study a pulsed molecular beam was allowed to interact simultaneously with a beam of low energy electrons and a pulsed tunable laser beam[42]. The detector signal was measured with a gated integrator. As the laser wavelength was scanned several peaks appeared for the mass-resolved O^- signal in the wavelength range 226-227 nm. The wavelengths at which this enhancement occurred was shown to correspond to the (0,0) band of the $A^2\Sigma^+$ - $X^2\Pi$ band system of NO; thus, the photonenhanced signal was attributed to dissociative electron attachment to the $A^2\Sigma^+$ ($\upsilon = 0$) state of NO. The cross section for this process was estimated[42] to be at least a factor of ten greater than the maximum cross section for the ground state, which is 10^{-18} cm^2 at 8.1 eV.

First excited triplet states of p-benzoquinone and its methylated derivatives. This study was conducted[43] using an electron capture detector (ECD) with a xenon arc lamp as the excitation light source; a monochrometer of 20 nm bandwidth was used for wavelength selection. The ECD current was monitored during the gas chromatographic (GC) introduction of the compound of interest with and without light irradiation. The ECD response was simply the decrease in current due to electron attachment.

Similar to the study on PhSH(T$_1$) described earlier, the first excited triplet states of p-benzoquinone and its methylated derivatives were populated indirectly via higher-lying, optically-allowed singlet states. The lifetimes of the first excited triplet states of these molecules were ~ 30 μs and were independent of the pressure of the buffer gases argon or nitrogen employed. The estimated enhancement[43] in electron attachment was 10^5 to 10^7 which is quite similar to the enhancement estimated for PhSH(T$_1$) compared to the ground state.

Low-lying electronically-excited states of SO_2. Recently an electron attachment study[36] on XeCl-laser-irradiated SO_2 was conducted using the electron beam technique outlined earlier (Fig. 13). This investigation identified many of the experimental difficulties associated with electron beam studies of electron attachment to electronically-excited molecules, and illustrated how to overcome such difficulties and extract the relevant information on electron attachment to excited species[36].

Previous electron beam experiments[1,4,44] on the <u>ground</u> electronic state of SO_2 have established the following dissociative attachment channels:

ε_{max} (eV)

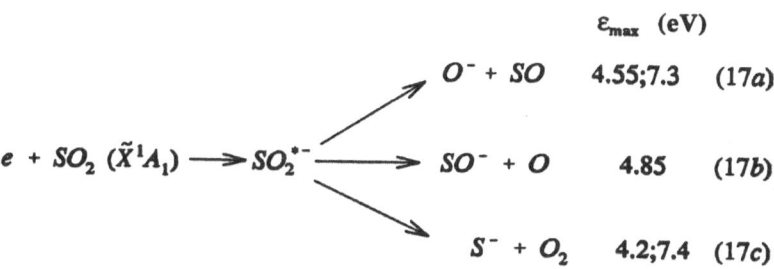

$$e + SO_2\ (\tilde{X}^1A_1) \longrightarrow SO_2^{\cdot-} \quad\nearrow\quad O^- + SO \qquad 4.55;7.3 \quad (17a)$$

$$\longrightarrow SO^- + O \qquad\quad 4.85 \qquad (17b)$$

$$\searrow\quad S^- + O_2 \qquad 4.2;7.4 \quad (17c)$$

where ε_{max} indicates the electron energies at the peak values of electron attachment cross sections. The maximum cross section for O^- formation (17a) was 2.46×10^{-18} cm^2 at 4.55 eV.

In the experiments of Ref. 36, enhanced O^- formation due to XeCl laser irradiation was studied in detail (enhanced SO^- and S^- formation was also observed). Figure 17a shows the O^- ion yield (monitored continuously) <u>in the absence of laser irradiation</u>, where the peaks due to the ground state process (17a above) can be clearly seen. Figure 17b shows the O^- signal with the laser on (laser repetition rate ~ 150 Hz) and with other conditions kept the same as for Fig. 17a; while the ground state process has not significantly changed, the laser-induced signal at electron energy of < 0.5 eV can hardly be seen without 100-fold amplification. This is due to the discrimination of the photoenhanced signal against the ground state signal by more than a factor of 1000: the ground state process generated O^- continuously while the photoenhanced process contributed only during a few μs per each laser pulse. This short time window available for electrons to interact with excited molecules is due to two factors, (i) decay of excited states, and (ii) the escape of excited states from the interaction region; the actual time available for collisions is the smaller of these two times. This was confirmed in an experiment[36] where the photoenhanced O^- signal was detected only during a 1 μs gate time following different time delays from the firing of the laser. Even though the lifetimes of the relevant excited states were > 50 μs (see below), the photoenhanced signal prevailed only for ~ 4.5 μs following the firing of the laser; this compared well with the estimated[36] "escape time" of ~ 4 μs.

When gated experiments were conducted where the O^- signal was detected only for a few μs following each laser pulse (with laser beam blocked and unblocked) the photoenhanced signal became more prominent, see the solid line of Fig. 17c; when this experimental curve was corrected for the variation of the electron beam current with electron energy, the dotted line of Fig. 17c was obtained. However, since only a fraction of SO_2 molecules are excited by each laser pulse, the photoenhanced signal shown by the dotted line is still not a good presentation of the actual enhancement. The peak cross section for the photoenhanced O^- formation was estimated[36] to be at least 2 orders of magnitude larger than the peak cross section (2.46×10^{-18} cm^2) for O^- formation from the ground state.

This observed photoenhanced O^- signal from XeCl-laser-irradiated SO_2 was attributed to the reaction:

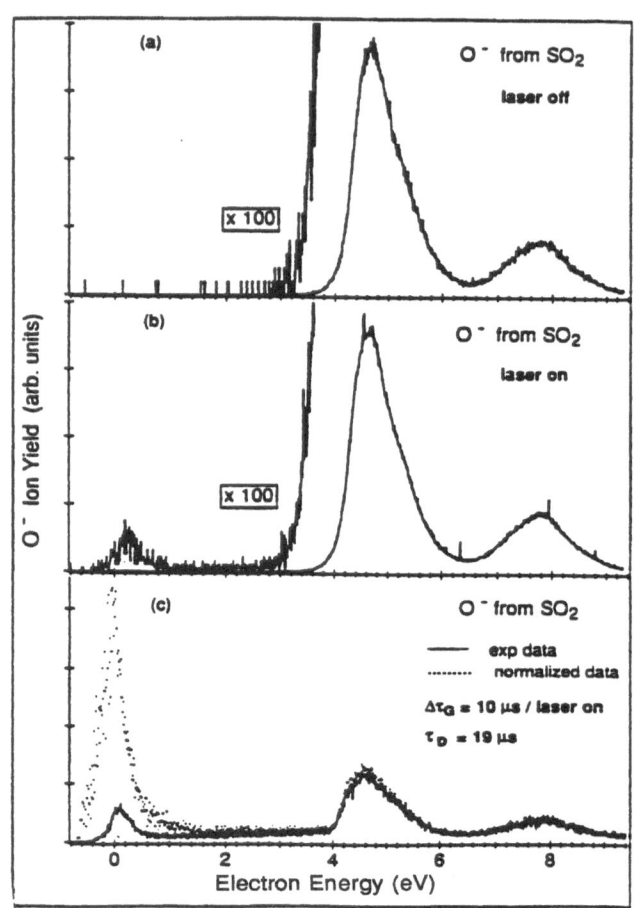

Figure 17. Relative cross section for the production of O⁻ from SO₂ as a function of the electron energy. (a) Ground state SO₂ (laser off). (b) Ground and excited SO₂; laser on (308 nm XeCl excimer laser) but no gating; (c) as in (b), but with a gate of 10 μs and a time delay (between the firing of the laser and the negative ion detection) of 3 μs: (---) experimental data; (--•-•) experimental data corrected for the variation of the electron current with energy and normalized at 8 eV (see the text and Ref. 36).

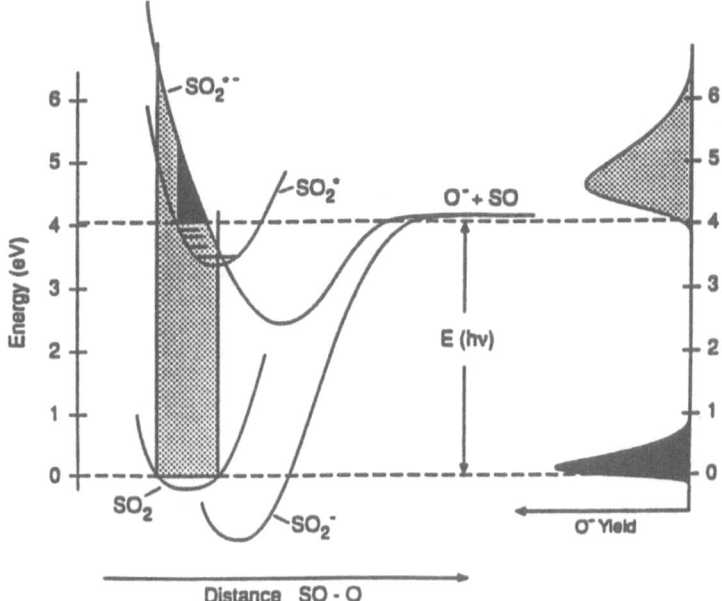

Figure 18. Schematic potential energy diagrams illustrating electron attachment to ground SO_2 and electronically excited SO_2^* (see the text and Ref. 36).

$$hv(308nm) + SO_2 \longrightarrow SO_2^*(^1B_1 \; or \, ^1A_2) \xrightarrow{+e(<0.5eV)} SO_2^{*-} \longrightarrow O^- + SO \quad (18)$$

i.e., dissociative electron attachment to the $^1B_1$ or $^1A_2$ states of SO_2 with lifetimes 50 and 80-530 μs respectively[36,45]. Figure 18 shows a schematic potential energy diagram illustrating electron attachment to SO_2 and electronically-excited SO_2^*. The electron energy required for the transition from SO_2^* to SO_2^{*-} is small, and thus the photoenhanced signal (shaded region of Fig. 18) appears at close to zero energy. The photoenhanced resonance is also narrower, due to the change in the equilibrium distance for SO_2^* compared to SO_2[36].

SO_2 is a "Douglas molecule,"[45] and along with other such molecules (e.g., CS_2, NO_2) are good candidates to study since they have long-lived electronically-excited states which do not readily dissociate when excited by easily available laser wavelengths.

Dissociative Electron Attachment to Short-Lived ($\tau < 10^{-8}$ s) States

Very efficient electron attachment to superexcited states, SESs, (electronically excited states lying above the first ionization threshold of molecules) was also reported[33]. Even though these highly energy rich states are short-lived, the presence of low-energy electrons in their close vicinity together with their implied extraordinarily large cross sections for electron attachment make it possible for electron attachment to occur within the short lifetimes of the SESs.

Triethylamine. This and other methylated amine compounds do not attach electrons significantly in their ground electronic states (electron attachment rate constants $< 10^{-11}$ cm^3 s^{-1} at electron energies < 1 eV). However, with excimer laser irradiation (at the KrF, KrCl, and ArF lines) efficient negative ion formation was observed in these compounds. Based on extensive experimental studies[33] on the dependencies of the total (electron plus negative ion) and negative ion signals on the laser fluence, amine partial pressure, buffer gas pressure and the applied electric field this laser enhanced electron attachment was attributed to the laser-excited superexcited states of these molecules.

Due to the space limitations we only wish to point out one unique aspect of these studies involving short-lived excited species. The dependence of the negative ion signal on the applied electric field at the KrF line is shown in Fig. 19 for triethylamine; these data were taken with two different buffer gases (Ar and N_2) and over a range of pressure (1.33 to 66.7 kPa). It is clear that the negative ion signal depends only on E, and does not depend significantly on either the identity of the buffer gas or the E/P (i.e., E/N) value. In a conventional electron swarm experiment, one would expect a different behavior, since an

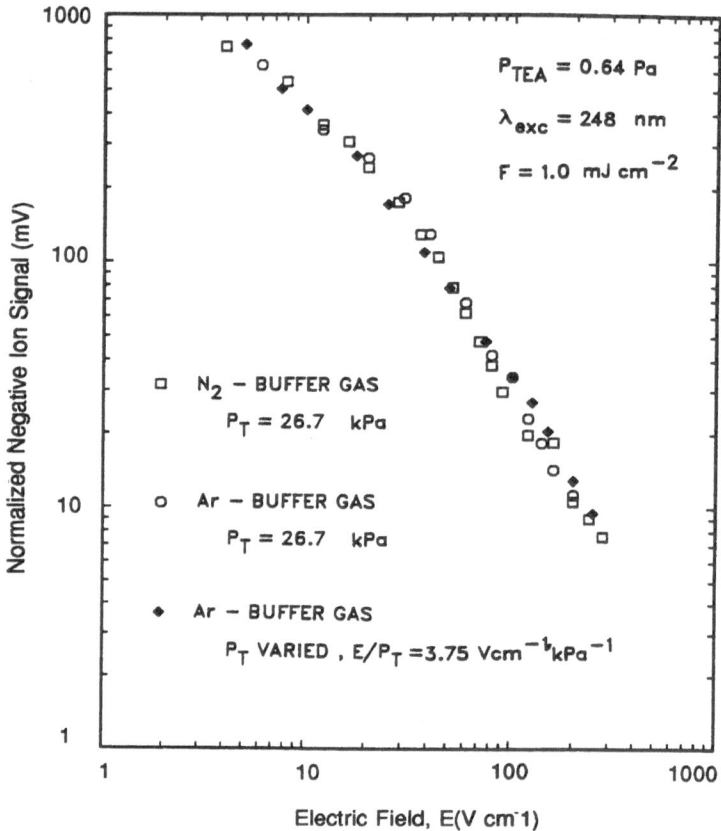

Figure 19. Electric field dependence of the normalized negative ion signal in laser irradiated triethylamine for the experimental parameters shown in the figure (see the text and Ref. 33).

electron allowed to reach steady-state conditions (within ~ 10^{-6} - 10^{-8} s) would have very different energies under these different conditions. In the present case, the electrons are attached just after they are produced (via photoionization) and well before any significant energy exchange occurs via collisions with the buffer gas. The electron attachment time must be less than the duration of the laser pulse which is ~ 10^{-8} s.

Nitric Oxide. In NO, electron attachment to the superexcited states was observed using the KrF excimer laser line and also at several wavelengths from a dye laser pumped by a

XeCl excimer laser[37]. In the case of dye-laser irradiation, resonance enhanced multiphoton ionization (REMPI) experiments were conducted first to determine the resonance wavelengths for optimum ionization signal. Since the attaching electrons in this technique are produced via photoionization, it is essential to have a large enough number of electrons (~ 10^8) per laser pulse[33] to quantify the measurements.

Similar to the amine compounds, no significant buffer gas pressure dependence was observed at 454.433 nm dye laser radiation, and the laser fluence dependence for negative ion formation was consistent with electron attachment to SESs. However, in the case of the KrF line the negative ion signal increased with the increase of buffer gas pressure, even though the total signal (and hence photoionization signal) did not depend on the buffer gas pressure. At the KrF line, the excess energy of the ionized electron is ~ 0.7 eV, and the collision time with N_2 at the pressures employed is < 10^{-10} s (Ref. 37). Thus, there was enough time for the electron to lose some energy via collisions and therefore to attach more efficiently. Previous photoelectron energy measurements had shown[46] that while near-zero-

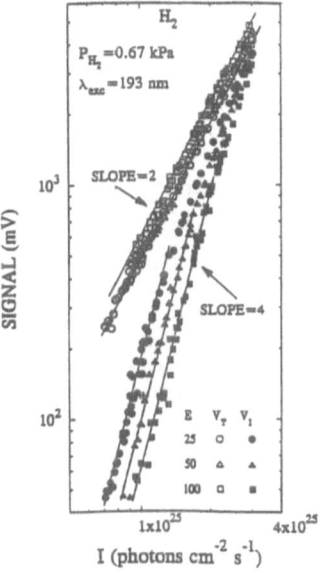

Figure 20. Laser intensity, I dependence of the measured total, V_T, and negative ion, I_I, signal for H_2 for the experimental parameters shown in the figure (see the text and Ref. 47).

energy electrons were produced at the dye-laser wavelength,--and hence energy loss was not needed--they were not produced at the KrF line.

Hydrogen and Deuterium. Due to their comparatively high ionization thresholds (~ 15.4 eV) it was not possible to obtain high ionization yield for H_2 and D_2 via frequency-doubled dye laser radiation. However, due to the serendipitous coincidence of the 2-photon energy of the ArF excimer laser line with the E,F $^1\Sigma_g^+$ (υ = 6) state, it was possible to carry out qualitative measurements at that laser line.

The laser intensity I dependence of the total signal and the negative ion signal are shown in Fig. 20 for H_2 irradiated by the ArF laser. The apparent I^2 dependence observed for the

3-photon photoionization process is due to the strong photon absorption from the E,F state[47]; since the excitation of a SES in turn should be an apparent 2-photon process, the observed I^4 dependence for negative ion formation is consistent with the attachment of an electron produced via photoionization to a molecule in SES, or an attaching species produced via a SES. Qualitatively similar results were obtained for D_2.

Nitrogen. No photoenhanced electron attachment was observed for N_2 (Ref. 47). This is to be expected since neither N_2 nor N can bind an extra electron.

CONCLUDING REMARKS

The internal energy of molecules plays a crucial role in determining their electron attachment properties. Dissociative electron attachment to "hot" (rotationally/vibrationally excited) molecules, and especially to electronically excited molecules, can be orders of magnitude larger than for ground state molecules. Nondissociative electron attachment to "hot" molecules decreases due to enhanced autodetachment. The experimental techniques that are now available for the systematic study of electron-energy rich molecule collisions will undoubtedly enrich our understanding[2] and increase the potential of this new knowledge for applications such as those involving optical switching, control or modulation of the impedance characteristics of gaseous matter.

REFERENCES

1. Christophorou, L. G., "Atomic and Molecular Radiation Physics," *Wiley-Interscience*, New York (1971).

2. Massey, H.S.W., "Negative Ions," *Cambridge University Press*, London (1976).

3. Smirnov, B. M., "Negative Ions," *McGraw-Hill*, New York (1982).

4. Christophorou, L. G., McCorkle, D. L., and Christodoulides, A. A., Electron Attachment Processes, in "Electron Molecule Interactions and Their Applications," L. G. Christophorou (Ed.), *Academic Press*, New York, 1984, Vol. 1, Chapt. 6.

5. Klar, D., Ruf, M.-W., and Hotop, H., Attachment of Electrons to Molecules at meV Resolution, *Austr. J. Phys.* 45:263 (1992).

6. Illenberger, E., Electron-Attachment Reactions in Molecular Clusters, *Chem. Rev.* 92:1589 (1992).

7. Christophorou, L.G., Electron Attachment and Detachment Processes in Electronegative Gases, *Cotrib. Plasma Phys.* 27:237 (1987).

8. Christophorou, L. G., Electron-Excited Molecule Interactions, in "Invited Papers, Proc. XXth Intern. Conf. Ionization Phenomena in Gases," V. Palleschi, D. P. Singh, and M. Vaselli (Eds.), Institute of Atomic and Molecular Physics, CNR, Pisa, Italy, July 8-12, 1991, pp. 3-13.

9. Christophorou, L. G., The Lifetimes of Metastable Negative Ions, *Adv. Electr. Electron Phys.*, 46:55 (1978).

10. Allan, M. and Wong, S. F., (a) Effect of Vibrational and Rotational Excitation on Dissociative Attachment to Hydrogen, *Phys. Rev. Lett.* 41:1791 (1978); (b) Dissociative Attachment from Vibrationally and Rotationally Excited HCl and HF, *J. Chem. Phys.* 74:1687 (1981).

11. Spyrou, S. M., and Christophorou, L. G., Effect of Temperature on the Dissociative Electron Attachment to $CClF_3$ and C_2F_6, *J. Chem. Phys.* 82:2620 (1985).

---

[2]The scattering of slow electrons from excited atoms is greatly enhanced compared to ground state atoms principally because of the larger polarizabilities of the excited states[48].

12. Chantry, P. J. and Chen, C. L., Ionization and Temperature Dependent Attachment Cross Section Measurements in C_3F_8 and C_2H_3Cl, *J. Chem. Phys.* 90:2585 (1989).

13. Datskos, P. G. and Christophorou, L. G., Variation with Temperature of the Electron Attachment to SO_2F_2, *J. Chem. Phys.* 90:2626 (1989).

14. Datskos, P. G., Christophorou, L. G., and Carter, J. G., (a) Temperature Enhanced Electron Attachment to CH_3Cl, *Chem. Phys. Lett.* 168:324 (1990); (b) Effect of Temperature on the Attachment of Slow (< 1 eV) Electrons to CH_3Br, *J. Chem. Phys.* 97:9031 (1992).

15. O'Malley, T. F., Calculation of Dissociative Attachment to Hot O_2, *Phys. Rev.* 155:59 (1967).

16. Bardsley, J. N. and Wadehra, J. M., (a) Dissociative Attachment and Vibrational Excitation in Low-Energy Collisions of Electrons with H_2 and D_2, *Phys. Rev. A* 20:1398 (1979); (b) Dissociative Attachment to HCl, DCl, and F_2, *J.Chem. Phys.* 78:7227 (1983).

17. Teillet-Billy, D. and Gauyacq, J. P., Dissociative Attachment in e-HCl, DCl Collisions, *J. Phys. B* 17:4041 (1984).

18. Hickman, A.P., Dissociative Attachment of Electrons to Vibrationally Excited H_2, *Phys. Rev. A* 43:3495 (1991).

19. Spyrou, S. M. and Christophorou, L. G., (a) Effect of Temperature on Nondissociative Electron Attachment to Perfluorobenzene, *J. Chem. Phys.* 82:1048 (1985); (b) Effect of Temperature on the Dissociative and Nondissociative Electron Attachment to C_3F_8, *J. Chem. Phys.* 83:2829 (1985).

20. Adams, N. G., Smith, D., Alge, E., and Burdon, J., Anomalous Temperature Dependence of the Coefficient of Electron Attachment to Hexafluorobenzene, *Chem. Phys. Lett.* 116:460 (1985).

21. Christodoulides, A.A., Christophorou, L. G., and McCorkle, D. L., Effect of Temperature on the Low-Energy (< 1 eV) Electron Attachment to Perfluorobutane (c-C_4F_8), *Chem. Phys. Lett.* 139:350 (1987).

22. Datskos, P. G., Christophorou, L. G., and Carter, J. G., (a) Temperature Enhanced Electron Detachment From $C_6F_6^-$ Negative Ions, *J. Chem. Phys.* 98:7875 (1993); (b) Temperature Dependence of Electron Attachment and Detachment in SF_6 and c-C_4F_6, *J. Chem. Phys.* (submitted).

23. (a) Alge, E., Adams, N. G., and Smith, D., Rate Coefficients for the Attachment Reactions of Electrons with c-C_7F_{14}, CH_3Br, CF_3Br, CH_2Br_2, and CH_3I Determined Between 200 and 600 K Using the FALP Technique, *J. Phys. B* 17:3827 (1984).

24. Chen, C. L. and Chantry, P. J., Photon-Enhanced Dissociative Electron Attachment in SF_6 and Its Isotopic Selectivity, *J. Chem. Phys.* 71:3897 (1979).

25. Hunter, S. R., Carter, J. G., and Christophorou, L. G., Electron Transport Measurements in Methane Using an Improved Townsend Technique, *J. Appl. Phys.* 60:24 (1986).

26. Wen, C. and Wetzer, J.M., (a) Electron Avalanches Influenced by Detachment and Conversion Processes, *IEEE Trans. Electr. Insul.* 23:999 (1988); (b) Time-Resolved Avalanches Current Waveforms in Octafluorocyclobutane, *IEEE Trans. Electr. Insul.* 24:143 (1989).

27. Chowdhury, S., Grimsrud, E. P., Heinis, T., and Kebarle, P., Electron Affinities of Perfluorobenzene and Perfluorophenyl Compounds, *J.Am. Chem. Soc.* 108: 3630 (1986).

28. Knighton, W. B., Bognar, J. A., and Grimsrud, E. P., Thermal Electron Detachment Rate Constants for the Molecular Anion of Perfluorobenzene, *Chem. Phys.Lett.* 192:522 (1992).

29. Rossi, M. J., Helm, H., and Lorents, D. C., Photoenhanced Electron Attachment to Vinylchloride and Trifluoroethylene at 193 nm, *Appl. Phys. Lett.* 47:576 (1985).

30. Schulz, G. J., Resonances in Electron Impact on Diatomic Molecules, *Rev. Mod. Phys.* 45:423 (1973).

441

31. (a) Christophorou, L. G., Carter, J. G., and Christodoulides, A. A., Long-Lived Parent Negative Ions in p-Benzoquinone Formed by Electron Capture in the Field of the Ground and Excited States, *Chem. Phys.Lett.* 3:237 (1969); (b) Collins, P. M., Christophorou, L.G., Chaney, E. L., and Carter, J. G., Energy Dependence of the Electron Attachment Cross Section and the Transient Negative Ion Lifetime for p-Benzoquinone and 1,4-Naphthoquinone, *Chem. Phys. Lett.* 4:646 (1970).

32. Tobita, S., Meinke, M., Illenberger, E., Christophorou, L. G., Baumgärtel, H., and Leach, S., Polycyclic Aromatic Hydrocarbons: Negative Ion Formation Following Low Energy (0-15 eV) Electron Impact, *Chem. Phys.* 161:501 (1992).

33. Pinnaduwage, L. A., Christophorou, L. G., and Bitouni, A. P., Enhanced Electron Attachment to Superexcited States of Saturated Tertiary Amines, *J. Chem. Phys.* 95:274 (1991).

34. Pinnaduwage, L. A., Christophorou, L. G., and Hunter, S. R., Optically Enhanced Electron Attachment to Thiophenol, *J. Chem. Phys.* 90:6275 (1989).

35. Pinnaduwage, L. A. and Christophorou, L. G., Verification of H⁻ Formation in UV Laser Irradiated Hydrogen; Implication for Negative Ion and Neutral Beam Technology, *J. Appl. Phys.* (submitted, 1993).

36. Jaffke, T., Hashemi, R., Christophorou, L. G., Illenberger, E., Baumgärtel, H., and Pinnaduwage, L. A., Photoenhanced Dissociative Electron Attachment to SO_2, *Chem. Phys. Lett.* 203:21 (1993).

37. Pinnaduwage, L. A. and Christophorou, L. G., Enhanced Electron Attachment to Superexcited States of Nitric Oxide, *Chem. Phys. Lett.* 186:4 (1991); erratum 189:486 (1992).

38. Burrow, P. D., Dissociative Attachment From the $O_2(a\ ^1\Delta_g)$ State, *J. Chem. Phys.* 59:4922 (1973).

39. Belić, D. S. and Hall, R.I., Dissociative Electron Attachment to Metastable Oxygen (a $^1\Delta_g$), *J. Phys. B* 14:365 (1981).

40. Jaffke, T., Meinke, M., Hashemi, R., Christophorou, L. G., and Illenberger, E., Dissociative Electron Attachment to Singlet Oxygen, *Chem. Phys. Lett.* 193:62 (1992).

41. Christophorou, L. G., Hunter, S. R., Pinnaduwage, L. A., Carter, J. G., Christodoulides, A. A., and Spyrou, S. M., Optically Enhanced Electron Attachment, *Phys. Rev. Lett.* 58:1316 (1987).

42. Kuo, C. T., Ono, Y., Hardwick, J. L., and Moseley, J. T., Dissociative Attachment of Electrons to the A $^2\Sigma^+$ State of Nitric Oxide, *J. Phys. Chem.* 92:5072 (1988).

43. Mock, R. S. and Grimsrud, E. P., Optically Enhanced Electron Capture by p-Benzoquinone and Its Methylated Derivatives, *J.Phys. Chem.* 94:3550 (1990).

44. Spyrou, S. M., Sauers, I., and Christophorou, L. G., Dissociative Electron Attachment to SO_2, *J. Chem. Phys.* 84:239 (1986).

45. Okabe, H., "Photochemistry of Small Molecules," *Wiley*, New York, (1978).

46. Miller, J. C. and Compton, R. N., Multiphoton Ionization Studies of Ultracold Nitric Oxide, *J.Chem. Phys.* 84:675 (1986).

47. Pinnaduwage, L. A. and Christophorou, L. G., H⁻ Formation in Laser-Excited Molecular Hydrogen, *Phys. Rev. Lett.* 70:754 (1993).

48. Christophorou, L. G. and Illenberger, E., Scattering of Slow Electrons From Excited Atoms: The Dominant Role of the Polarization Potential, *Phys. Lett. A.* 173:78 (1993).

ELECTRON REACTIONS IN NONPOLAR LIQUIDS - PRESSURE EFFECTS

Richard A. Holroyd

Chemistry Department
Brookhaven National Laboratory
Upton, NY 11973 USA

INTRODUCTION

Drift mobilities of excess electrons in nonpolar molecular hydrocarbons range from values as low as 0.016 cm^2/Vs for long chain alkanes[1] to 400 cm^2/Vs for methane[2] Similarly the rates of electron attachment reactions differ considerably from one nonpolar solvent to another.[3] To explain the first observation, the existence of two types of states in these liquids is assumed: a conducting state where the electron is quasi-free (qf) and mobile, and a trapped state (tr) where the electron is localized and relatively immobile. The range in mobilities is then attributed to differences in the probability of trapping. Trapping is more likely in n-alkanes and less likely in methane and neopentane, which are symmetrical molecules.

The explanation of the difference in electron reactivity is related to differences in the energy of these electron states in various liquids. The energy of the lowest conducting (qf) state in denoted V_O and ranges from -0.55 eV in tetramethylsilane to 0.18 eV in n-decane[3]. (V_O is defined relative to the energy of the electron in vacuum as zero). As the energy level of the electron changes, rates of electron reactions change. Sometimes maxima in rates are observed as a function of V_O as in electron attachment to fluorocarbons[3], N_2O and ethylbromide[4]. Apparent negative activation energies are even observed because of this dependence[3]. Energetics of reactions in solution differ from the gas phase. A major factor is that the product anion polarizes the solvent. Because of this, electron attachment reactions are energetically more favorable in liquids than in the gas phase.

Yet our understanding of electrons in these liquids is incomplete in regard to, for example, the quasi-free mobility as well as the nature of trapped electrons and ions. Some new insight is provided by pressure studies. For example, theory predicts [5,6] that the electron energy levels, V_O and E_{tr}, should increase in energy with density, or consequently with pressure increase. Experiment has verified the increase in V_O.[7] Therefore electron attachment rates in solution should change with pressure and conversely studies of the pressure dependence of electron reaction rates should elucidate their relation to the energy level of the electron in liquids. Further, since such studies reveal the magnitude of reaction volume changes, information is obtained about electrostriction around charged species in solution. Studies of the pressure effect on electron mobility show that some contraction occurs around trapped electrons (the observed volume change is about -20 cm^3/mol, but a much larger volume contraction occurs around ions in hydrocarbons.

Linking the Gaseous and Condensed Phases of Matter
Edited by L.G. Christophorou *et al*, Plenum Press, New York, 1994

Although the main focus here is on electron reactions, a brief discussion of pressure effects on electron energy levels and electron mobility is included in order to provide some relevant background information.

ELECTRON ENERGY LEVELS

The electron in the conduction band is said to be quasi-free, by which is meant it moves freely, as in the gas phase, but is restricted in its movement to the space between molecules. The density dependence of the conduction band energy (V_0) has been measured for several liquids.[1,8] Typically V_0 is close to zero at low densities and as the density increases it decreases, reaches a minimum value and increases up to the density corresponding to the normal liquid. Theory[5,9] predicts that $V_0 = T + U$ where U is the electron-molecule potential and T is $\hbar k^2/2m$, the kinetic energy term. The Springett, Jortner Cohen (SJC) theory[5] predicted the general density dependence that is observed. More recent efforts, using a more accurate molecular potential, reproduce the experimental V_0 data for methane quite well.[9]

Measurements of V_0 for liquids at high pressure provide an additional test of theory. The density increases with pressure and consequently for most liquids V_0 is expected to increase with pressure. The conduction band energy has been measured as a function of pressure for several liquids.[7] The results for n-pentane and 2,2-dimethylbutane are shown in figure 1. For both, V_0 increases with pressure; the increase is largest for n-pentane. However, the observed increase is about 50% less than predicted by the SJC model (ref 5). Another model by Schiller,[10] presented at this meeting, assumes the V_0 measurement is a constant entropy rather than constant volume process. This model underestimates the experimental points.

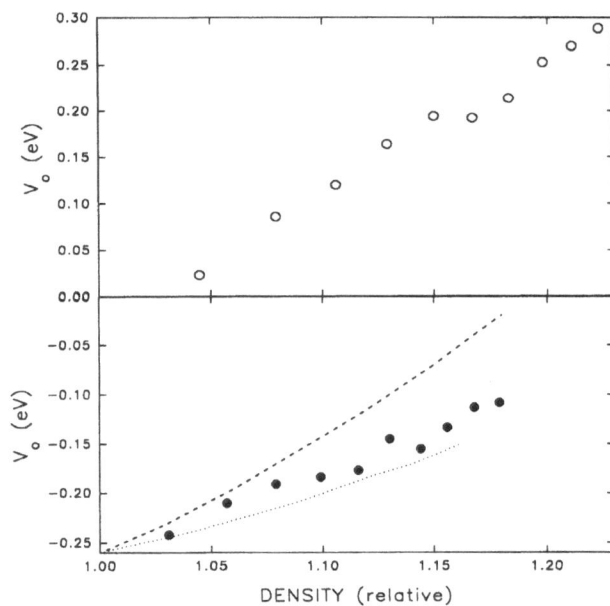

Fig. 1. Effect of pressure on the energy level of the quasi-free electron in a) n-pentane and b) 2,2-dimethylbutane. Theory:---Ref5,...Ref10

ELECTRON MOBILITY

The effect of pressure on the drift mobility of excess electrons has been studied for several hydrocarbon liquids. Some illustrative results are shown in figure 2 for 60°C. For branched alkanes exhibiting high mobility, like 2,2-dimethylbutane and 2,2,4-trimethylpentane, the mobility increases somewhat with pressure at 60°C and below and decreases with pressure at higher temperatures. These effects are believed to reflect the variation of the quasi-free mobility with pressure since trapping is less important in these liquids.

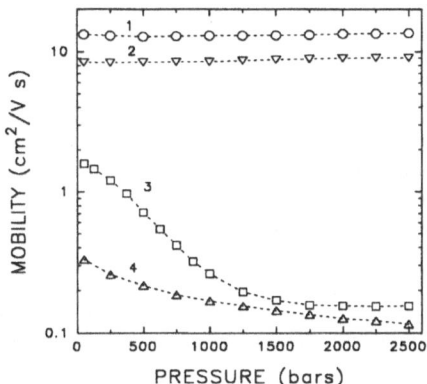

Fig. 2 Electron mobili. ersus pressure for: 1) 2,2-dimethylbutane at 60°C; 2) 2,2,4-trimethylpentane at 60°C; 3) benzene at 100°C; 4) n-pentane at 60°C.

Various theories of quasi-free mobility have been proposed. Lekner[11] proposed that μ_{qf} be given by:

$$\mu_{qf} = (2e/3) \, (2/\pi m \, k_BT)^{1/2} \, \Lambda \tag{1}$$

where

$$\Lambda = (\sigma N^2 \, k_BT \, \chi_T)^{-1}. \tag{2}$$

In equations (1) and (2), Λ is the mean free path, S(0) is the structure factor replaced by $Nk_BT \, \chi_T$, σ is a scattering cross section, N the number density and χ_T the isothermal compressibility. The compressibility of liquids decreases significantly over the pressure range investigated and consequently equation (1) predicts the quasi-free mobility will increase. For example, the calculated increase (over the pressure range in Fig 2) for 2,2-dimethylbutane at 60°C is a factor of five, much greater than the observed increase of a few percent.

In recent theories [12,13] scattering is assumed to arise from density fluctuations and the mean free path includes the density derivative of V_0.

$$\Lambda = c\hbar^4 \, (dV_0/dN \, N \, m)^{-2} \, (k_BT\chi_T)^{-1} \tag{3}$$

Since dV_o/dN increases with pressure (see figure 1 and ref 7), this compensates for the decrease in χ_T. The predicted increase is μ_{qf} for 2,2-dimethylbutane at 60° with Eq 3 is 75%, which is in better agreement with experiment.

For alkanes exhibiting low mobility, pressure increase to 2.5 kbar causes a reduction in the mobility of about 60% at 60°C; this is illustrated for n-pentane in figure 2. This decrease is attributed to a shift in the equilibrium:

$$e_{qf}^- \rightleftharpoons e_{tr}^- \qquad (4)$$

towards the right at high pressure. Pressure favors the trapped state because a volume change of -20 to -30 cm³/mol is associated with reaction (4). This volume change is due to electrostriction of the solvent around the electron which is localized in a cavity in the liquid. The classical theory of electrostriction,[14] equation 5, accounts reasonably well for this volume change if r, the radius of the cavity is about 0.4 nm and the fraction of the electron

$$V_{el} = \frac{-z^2 e^2}{2r} \frac{1}{\varepsilon^2} \frac{d\varepsilon}{dP} \qquad (5)$$

in the cavity is taken to be 0.7. Experimental volume changes, derived from the mobility, are linearly dependent on the parameter $(1/\varepsilon^2) d\varepsilon/dP$ for n-alkanes and 3-methylpentane.[15] The parameter $d\varepsilon/dP$ is proportional to the isothermal compressibility, which decreases with increasing pressure. Thus the largest volume changes are observed near 1 bar.

The effect of pressure on the drift mobility of electrons in aromatics[16] is illustrated in Fig 2 for benzene at 80°C. The mobility decrease with increasing pressure has a sigmoid shape, dropping off at an intermediate pressure and then is pressure independent at high pressure. Although benzene has an electron affinity of - 1.11 eV, in solution the polarization energy of the anion compensates and reaction 6 is expected to be favored especially

$$e^- + benzene \rightleftharpoons benzene^- \qquad (6)$$

at high pressure. Therefore transport involves both electron and benzene anions. The electron mobility at low pressure, which is 1.6 cm²/Vs, is due to the trapping equilibrium (4) because of which only a small fraction of the electrons are in the mobile quasi-free state. At pressures above 1. 5 kbar the mobility is a constant 0.15 cm²/Vs, which is the mobility of the benzene anion since the equilibrium (6) is shifted to the right. This mobility is, however, at least two orders of magnitude greater than that expected for ions moving diffusively at these pressures. The explanation is that electrons hop, or undergo self-exchange (reaction 7), and this accounts for the relatively high ion mobility. The pressure independence of the mobility is expected

$$benzene^- + benzene \rightleftharpoons benzene + benzene^- \qquad (7)$$

since the products and reactants of eq 7 are identical and thus there is no volume change.

The equilbrium constant for reaction 6 is 0.05 kg/mol at 250 bar in benzene at 80° but increases a factor of 50 by 1000 bar. The volume change in reaction 6 is evaluated from such data using the equation:

$$\Delta V_r = -RT \, d \ln K_6 / dP \qquad (8)$$

For electron attachment to benzene ΔV_r is -170 cm³/mol at 80°C and 375 bar. This result is typical of the large volume decreases observed in electron attachment reactions, more examples of which are considered below. Classical electrostriction theory (eq 5) accounts for about half of the observed volume change. The source of the additional volume change is also discussed below.

ELECTRON ATTACHMENT REACTIONS

Very significant effects of pressure are observed for electron attachment reactions in nonpolar solvents. Some data for attachment to N_2O, CO_2 and $n-C_5F_{12}$ are shown for tetramethylsilane as solvent in figure 3. In these cases the rates are accelerated about an order of magnitude by pressure increase to 2.5 kbar. This effect can be associated with a change in the energy level V_O. The energy of the V_O state in tetramethylsilane increases by 0.13 eV as the density increases 24 percent.[7] As was shown earlier for electron attachment to N_2O^4 and to some fluorocarbons[3] the rates increase with V_O and reach a maximum at a characteristic value of V_O. This behavior can be related to the known cross section data for such reactions in the gas phase. The increase in V_O in the liquid phase is equivalent to increasing the kinetic energy of the electron in the gas. The attachement rate to $n-C_5F_{12}$

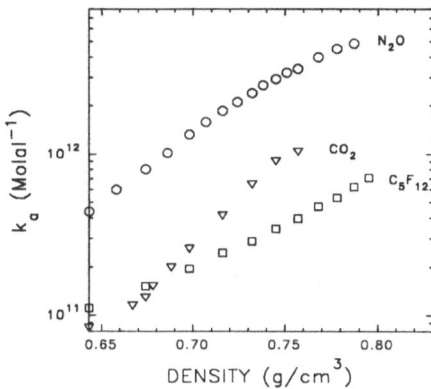

Fig. 3 Pressure dependence of the rate of electron attachment to: O, N_2O; ∇, CO_2; and $\square$ C_5F_{12} in tetramethylsilane.

reaches a maximum at 0.6 eV in the gas.[17] The energetics of attachment change in solution by the difference $V_O- P^-$, where P^- is the polarization energy of the $C_5F_{12}^-$ ion in solution. The difference V_O-P^- is roughly 0.5 eV at one bar in this case and increases with pressure.

Not all electron attachment reactions are accelerated by pressure. The rate constant for $e^-+1,3$-butadiene in n-hexane is about 5×10^{11} molal^{-1}s^{-1} independent of pressure from 0.5 to 2.5 kbar. The rate of attachment to CO_2 in 3-methylpentane decreases with pressure above 500 bar. The electron mobility in 3-methylpentane decreases similarly with pressure and the rate is diffusion controlled.

For reactions which increase in rate with pressure the activation volume, $\Delta V^{\ddagger}$, is negative and may be calculated from Eq 9.

$$\Delta V^{\ddagger} = -RT\frac{d \ln k_a}{dP} \qquad (9)$$

Typically plots of $\ln k_a$ versus pressure are curved, unlike the plot of $\ln k_a$ versus density (Fig 3) which is nearly linear. However, from the slopes of these plots in the 1 to 500 bar range the values of $\Delta V^{\ddagger}$ for attachment to N_2O, CO_2 and $n-C_5F_{12}$ can be determined and are

- 54, -68, and -28 cm^3/mole, respectively. The meaning of this decrease in volume at the transition state has been discussed.[18] The electrostriction around the trapped electron discussed earlier is not involved here as the electron is essentially quasi-free in tetramethylsilane. The overall reaction volumes for electron attachment are even greater (and negative) as discussed below and the contraction of the solvent occurring at the transition state is most likely part of the overall contraction which occurs.

ELECTRON EQUILIBRIA

Electron attachment to certain solutes is reversible in hydrocarbon solvents. This occurs for solutes with low or negative electron affinities. The use of pressure increases the range of solutes that can be studied. Those with quite negative electron affinities, which do not attach electrons at 1 bar in solution, do so at high pressure. This is illustrated by electron attachment to toluene (reaction 10)

$$e^- + toluene \rightleftharpoons toluene^- \tag{10}$$

which occurs at high pressure in n-pentane[19] and n-hexane.[20] The occurence of this equilibrium was detected by changes in the observed mobility. At low pressure, below 1 kbar, the electron mobility is unaffected by the presence of toluene at concentrations up to 1%, indicating reaction 10 is shifted to the left; that is, there is no reaction. At higher pressures the mobility decreases below that of the electron in the alkane and this effect increases with pressure. The decrease is due to temporary attachment of the electron to toluene. The contribution of the electron to the mobility depends on the fraction of time it is not on toluene, because of the low mobility of toluene anions in dilute solutions. Thus the mobility (μ_{obs}) is essentially:

$$\mu_{obs} = \mu_e/(1 + K_{10}[Tol]), \tag{11}$$

where μ_e is the mobility of the electron in the neat alkane and K_{10} the equilibrium constant of reaction 10. Values of K_{10} derivied from equation 11 and experimental mobility values increase a factor of 50 for a 1 kbar pressure increase. This effect is reflected in values of ΔG_{rx} for this reaction shown in Table 1. Even though the electron affinity of toluene is -1.1 eV, in solution ΔG_{rx} is negative at 2 kbar pressure and decreases further with increasing pressure.

The free energy change in solution, ΔG_{rx}, is given by the thermodynamic cycle:

$$\Delta G_{rx} = \Delta G_{rx}(gas) + \Delta G_{soln} (Tol^-) - \Delta G_{soln}(Tol) - \Delta G_s(e^-) \tag{12}$$

where ΔG_{rx}(gas) is 1.07 eV for toluene when the entropy of the electron in the gas phase is taken into account. The difference in energy of solution of the toluene anion and toluene neutral is given by the polarization energy of the negative ion, P^-, and thus eq 12 becomes:

$$\Delta G_{rx} = \Delta G_{rx}(gas) + P^- - \Delta G_s (e), \tag{13}$$

from which the energy of solution of the electron can be estimated. We approximate P^- from the Born equation as:

$$P^- = -\left(\frac{e^2}{2r}\right)\left(1 - 1/\varepsilon\right) \tag{14}$$

This procedure leads to the values of $\Delta G_s(e^-)$ given in the last column of Table 1. These values are close to 0.1 eV at high pressure and represent the energy of the ground state of the electron, which for these liquids is the trapped state. The value of the trapped state energy is higher than it is at 1 bar; thus like the energy V_0, the energy E_t also increases with pressure.

Benzene is similar to toluene and attaches electrons in n-pentane at pressures above 2 kbar where the electron energy is high (see Table 1). Benzene does not react with electrons in tetramethysilane, however, where the electron energy level is much lower (see Table 1). Although, as said earlier no reaction is observed at 1 bar, the results suggest that attachment to these aromatic hydrocarbons occurs but that detachment is rapid. That is, the equilibrium constants are too small and the mobility is unaffected (see equation 11).

Table 1. Thermodynamic data for electron equilibria $e^- + S \rightleftharpoons S^-$

Solute(S) EA in eV	Solvent	Pressure (bar)	$\Delta G_{rx}°$ (eV)	ΔH_{rx} (eV)	P^- (eV)	$\Delta G_S(e)$ (eV)
Toluene	n-hexane	2000	-0.107	-0.52	-1.06	0.12
(-1.11)	"	2750	-0.175	-0.59	–	–
"	n-pentane	2000	-0.077	-0.48	-1.06	0.10
"	"	2750	-0.141	-0.44	–	–
Benzene	n-pentane	2000	-0.093	-0.46	-1.12	0.09
(-1.15)						
CO$_2$	3-methylpentane	200	-0.608	-0.99	-1.50	-0.33
(-0.61)	2-methylbutane	100	-0.532	-1.01	-1.43	-0.34
"	2,2,4-trimethylpentane	150	-0.569	-0.98	-1.53	-0.40
"	2,2-dimethybutane	1	-0.469	-1.07	-1.46	-0.43
"	tetramethylisilane	1	-0.306	-0.92	-1.42	-0.55
	"	500	-0.402	-0.95	-1.51	-0.55
1,3-butadiene						
(-0.62)	n-hexane	700	-0.387	–	-1.09	-0.13

Another method of studying electron equilbria involves measuring both attachment (k_a) and detachment (k_d) reaction rates and calculating K_{eq} as k_a/k_d. The individual rates are obtained from analysis of the current trace following a short pulse. The existence of an equilbrium is indicated by an equilibrium current lasting long after the pulse. This method was used to study electron attachment to CO_2[18] and butadiene[21] in various solvents. These

$$e^- + CO_2 \rightleftharpoons CO_2^- \qquad (15)$$

reactions can be observed at very low ($\leq 1\mu M$) concentrations of solute because values of k_a are large (10^{11} to 10^{13} M^{-1}s^{-1}).

The equilbrium constants (K_{15}) for reaction of the electron with CO_2 in 2-methylbutane are shown in figure 4 as a function of density. Values of K_{15} typically increase a factor of 50 for an increase in pressure of 400 bar. Because the electron affinity of CO_2 is considerably higher than toluene, this reaction is observed even at 1 bar. Values of the free energy of reaction at 298 K, as shown in Table 1, change with conditions from -0.61 to -0.31 eV for the five solvents studied. This effect of solvent is attributed to the energy level of the electron in the liquid. The reaction is most exoergic for 3-methylpentane which has the highest $\Delta G_S(e)$ and least exoergic for tetramethylsilane where $\Delta G_S(e)$ is -.55 eV. The

449

observed increase in K_{15} by a factor of 50 corresponds to a decrease in ΔG_{rx} of 0.1eV. For the hydrocarbons this decrease is found to be due to a decrease of about 0.06 eV in P⁻ and an increase of 0.04 eV in the value of $\Delta G_s(e)$. For tetramethylsilane, there is a comparable decrease in ΔG_r for a 500 bar pressure change which, however, is attributed entirely to a decrease in P⁻; that is a stabilization of the CO_2^- ion.

Another reversible reaction is that of the electron with 1,3-butadiene which can be observed in n-hexane at pressures between 1bar and 1.5 kbar[21]. No reaction is observed at 1 bar at room temperature but the equilbrium can be observed at low pressure at -7°C. This reaction is less sensitive to pressure than the reaction with CO_2; K_{16} changes

$$e^- + 1,3\text{-butadiene} \rightleftharpoons 1,3\text{-butadiene}^- \tag{16}$$

a factor of 50 in 800 bar in n-hexane for temperatures from 23 to 60°C. Butadiene and CO_2

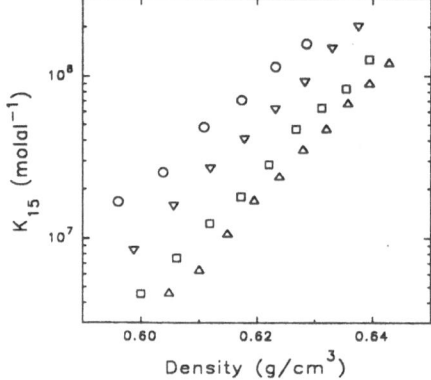

Fig. 4. Plot of equilibrium constant for e⁻ + CO_2 ⇌ CO_2^- versus density of the 2-methylbutane solvent at 0 55°C; ∇ 70°C, ☐ 85°C and △ 100°C.

have similar electron affinities but butadiene is less reactive because it is a larger molecule which means less polarization energy by the anion. The energy level of the electron is quite high in n-hexane at 700 bar (-.13 eV) which makes n-hexane a good solvent to observe this reaction. In 2,2-dimethylbutane, which has a lower electron level, no reaction with butadiene can be observed below 1.5 kbar. The shift of reaction 16 to the right with increasing pressure means the free energy of reaction decreases. As in the other reactions this is due to a decrease of the polarization energy of butadiene anion as well as an increase in the energy level of the electron as the pressure increases.

VOLUME CHANGES

Pressure causes equilbria to shift in the direction of a volume decrease. Thus since K_{eq}

increases with pressure for the attachment reactions we have studied, the reaction volume change is negative. The actual volume changes are calculated from eq 17, which is

$$\Delta V_{rx} = -RT \, d \ln K_{eq}/dP \tag{17}$$

analogous to equation 9. Some typical values of ΔV_{rx} for these reactions are given in Table 2. Since plots of $\ln K_{eq}$ versus pressure are slightly curved[18-20] (concave downward) the values given are from the slopes of such plots at the average pressure indicated. As the pressure increases the magnitude of ΔV_{rx} decreases (see e.g. the 1,3-butadiene data). The values range from -72 to -298 cm^3/mol. In the following we consider the meaning of these large volume changes.

Table 2. Volume changes for electron equilibria $e^- + S \rightleftharpoons S^-$

Solute S	Solvent	Temp °C	(Ave) Press bar	ΔV_{rx} (cm^3/mol)	ΔV_{el}[a] (cm^3/mol)
Toluene	n-hexane	23	2500	-85	-24
"	n-pentane	23	2500	-72	-27
Benzene	n-pentane	23	2000	-89	-33
1,3-Butadiene	n-hexane	23	525	-188	-63
"	"	"	1050	-101	-45
CO$_2$	3-methylpentane	80	125	-290	-237
"	2-methylbutane	47	125	-298	-207
"	2,2,4-trimethylpentane	51	150	-174	-143
"	2,2-dimethybutane	44	175	-254	-156
"	tetramethylisilane	23	350	-174	-137

a) From eq 5

In general the overall volume change in a reaction is the difference in partial molar volume of the products and reactants. The molar

$$\Delta V_{rx} = V_{el}(S^-) + V_i(S^-) - V(S) - V(e) \tag{18}$$

volume of the product anion (S$^-$) includes both an intrinsic volume of the ion V_i (S$^-$) and a solvent term V_{el} (S$^-$), which involves electrostriction. Because there are not very large structural changes when a molecule like benzene attaches an electron, the difference $V_i(S^-) - V(S)$ is small at least relative to the large volume changes observed. Thus ΔV_{rx} consists mainly of the terms $V_{el}(S^-)$ and $V(e)$, the partial molar volume of the electron.

Electrostriction of the solvent around the product anion is of major importance. The classical model of Drude and Nernst[14] (eq 5) considers electrostriction of the solvent around a sphere of radius R and charge ze immersed in a continuum dielectric. If the Clausius-Mosotti equation is assumed to correctly relate the dielectric constant (ε) to the density (ρ) at high pressure then eq 5 for a singly charged ion becomes:

$$V_{el} = -(e^2/2r) \, X_T \, (\varepsilon + 2)(\varepsilon - 1)/3\varepsilon^2 \tag{19}$$

451

The isothermal compressibility (χ_T) is obtained from density data. Experimental values of ΔV_{rx} for three reactions are shown in figure 5 plotted versus $(1/\varepsilon^2)$ dε/dP, which is evaluated using equation.19 Thus the slope of the plot depends inversely on the radius of S$^-$. The dashed lines in Figure 5 show how calculated values of V_{el} vary with pressure; the values of r used, indicated on the figure, are from molecular volumes of the solutes. These calculated values are significantly less than the observed volume changes (see also the last two columns of Table 2).

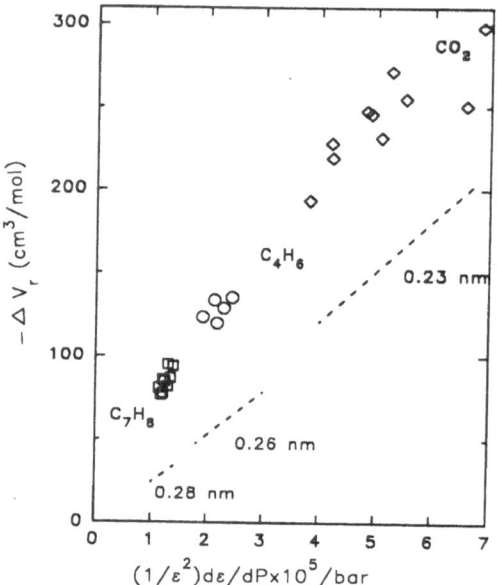

Fig. 5. Reaction volume for electron attachment to various solutes at various temperatures and pressures; □ toluene in n-hexane, ○ ,1,3-butadiene in n-hexane, ◊ CO_2 in several solvents.

In the first high pressure paper[18] of the series on electron attachment to CO_2 the discrepancy between calculated and experimental volumes was removed by making the radius r a parameter. Reasonably good agreement with experiment was obtained by assuming a radius for CO_2 of 0.16 nm. This radius is however considerably less than the hard core radius of CO_2.

In the paper on attachment to benzene and toluene in n-pentane[19] a similar fit of equation 19 to the reaction volume data led to radii of 0.11 nm for these solutes. This value is quite unrealistic. Alternatively when a radius dervived from the molecular volume is assumed, the theory accounts for only about 30% of the observed volumes as shown in Table 2. In a more recent paper dealing with attachment to toluene in n-hexane the classical theory of electrostriction was assumed to be valid and the rest of the volume change (about 70 cm^3/mole) attributed to a positive volume of the trapped electron in n-hexane. A positive volume of the electron is inconsistent with the earlier result that there is a negative volume change of -20 to -30 cm^3/mol associated with trapping of the electron in n-hexane. This enigma led to us to question the validity of the continuum model. Since it is clear there is constriction of the solvent around the ion, the density and dielectric constant are greater than the bulk values in the region surrounding the anion.

A recent paper by Schwarz[22] reexamines the situation. He points out that if the increase in density of the solvent near an ion is taken into account then an even larger discrepancy is found because the more dense solvent is less compressible. Instead he suggests that the increase in electrostatic energy causes the first solvent shell around an ion in a nonpolar liquid to freeze. This phase transition contributes an additional volume decrease. when this volume change is included, the calculation accounts very well for the experimental volume changes.

ACKNOWLEDGEMENT

The assistance of E. Stradowska in preparation of the figures is gratefully acknowledged. This research was carried out at Brookhaven National Laboratory under contract DE-AC02-76CH00016 with the U.S. Department of Energy and supported by its Division of Chemical Sciences, Office of Basic Energy Sciences.

REFERENCES

1. R. A. Holroyd and W. F. Schmidt, Transport of electrons in nonpolar fluids, Annual Rev. of Phys. Chem. **40**:439 (1989).

2. M. A. Floriano, G. R. Freeman, Electron transport in liquids: effect of unbalancing the sphere-like methane molecules by deuteration, and comparison with argon, krypton and xenon, J. Chem. Phys. **85**:1603 (1986).

3. R. Holroyd, Electron kinetics on nonpolar liquids energy and pressure effects in NATO-ASI series B Physics **193** "The Liquid State and its Electrical Properties" "E.E. Kunhardt, L.G. Christophorou and L.H. Luessen, ed. Plenum Press NY p221 (1987).

4. A.O. Allen, T.E. Gangwer and R. A. Holroyd, Chemical reaction rates of quasifree electrons in nonpolar liquids. II, J. Phys Chem. **79**:25 (1975).

5. B. E. Springett, J. Jortner, M.H. Cohen, Stability criterion for the localization of an excess electron in a nonpolar fluid, J. Chem. Phys. **48**:2720 (1968).

6. T. Kimura, K. Fueki, , Effect of density on excess electron localization in ethane and propane J. Chem. Phys. **66**:366 (1977).

7. R. Holroyd, M. Nishikawa, K. Nakagawa, N. Kato "Pressure dependence of the conduction band energy of nonpolar liquids, Phys. Rev B **45**:3215 (1992).

8. K. Nakagawa, K. Itoh, M. Nishikawa, The effect of molecular structure on the density dependence of electron mobility and conduction band energy in nonpolar fluids, IEEE Trans Electr. Insul **23**:509 (1988).

9. B. Plenkiewicz, P. Plenkiewicz, J. Gerin, Density dependence of the ground state energy of excess electrons in liquid methane, J. Chem. Phys. **90**:4907 (1989).

10. R. Schiller, Thermodynamics of electron injection, presented at this NATO-ASI.

11. J. Lekner, Motion of electrons in liquid argon, Phys. Rev. **158**:130 (1967).

12. Y.A. Berlin, L. Nyikos, R. Schiller Mobility of localized and quasi-free electrons in liquid hydrocarbons, J. Chem. Phys. **69**:2401 (1978).

13. S. Basak, M.H. Cohen, Deformation potential theory for the excess electrons in liquid argon, Phys. Rev. B **20**:3404 (1979).

14. P. Drude, W. Z. Nernst, Uber electrostricktion durch freie ionen, Z. Phys. Chem. **15**:79 (1894).

15. R. C. Munoz, R. A. Holroyd, K. Itoh, K. Nakagawa, M. Nishikawa K. Fueki, Excess electron mobility in hydrocarbon liquids at high pressure, J. Phys. Chem. **91**:4639 (1987).

16. K. Itoh, R. A. Holroyd, Effect of pressure on the electron mobility in liquid benzene and toluene, J. Phys. Chem. **94**:8850 (1990).

17. S.M. Spyrou, I. Sauers, L.G. Christophorou, Electron attachment to the perfluroalkanes n-C_NF_{2N+2} (N = 1 – 6) and i - C_4F_{10}, J. Chem. Phys. 78:7200 (1983).

18. M. Nishikawa, K. Itoh, R. Holroyd, Effect of pressure on the reaction of electrons with CO_2 in nonpolar solvents, J. Phys. Chem. 92:5262 (1988).

19. K. Itoh, R. Holroyd, Electron attachment to benzene and toluene in nonpolar solvents at high pressure, J. Phys. Chem. 94:8854 (1990).

20. K. Itoh, M. Nishikawa, R. Holroyd, Electron attachment to toluene in n-hexane and 2,2-dimethylbutane at high pressure, J. Phys. Chem. 97:503 (1993).

21. R.A Holroyd , E. Stradowska, S. Ninomiya, K. Itoh, M. Nishikawa, to be published.

22. H. A. Schwarz, Partial molar volumes of ions in non-polar solvents, J. Phys. Chem. (submitted).

THERMODYNAMIC PROPERTIES OF THE ELECTRON

Michael Henchman

Chemistry Department
Brandeis University
Waltham, MA 02254-9110

While the business of this institute is to study the dynamical properties of the electron, a fundamental constraint is imposed on these dynamical properties by the thermodynamic properties of the electron. It may be surprising to learn that in 1993 the thermodynamic properties of the electron are not well established and remain a matter of some controversy.

How can that be? One general answer is that the subject of thermodynamics is subtler than the quality of the minds addressing it. There are, however, more specific answers. Accurate measurement is not simple — even for physicists, even for the best of physicists, as Branscomb has noted.[1] When making measurements, we tend to report not our measurements but our perception of those measurements: even for the most able investigators, disturbingly subjective irregularities may interpose, for example from *intellectual phase locking*.[1] The historical development of gas-phase ion thermochemistry is peppered with instances of this sort; and the thermodynamic properties of the electron are far from being the most blatant example.

THE PROBLEM

The problem is readily apparent from consulting the standard thermodynamic tables for the thermodynamic properties of the electron. Some tables list one set of values:[2] others list a distinctly different set of values.[3] This difference derives from the use of two different sets of conventions. Some workers assert that the conventions are arbitrary and that any choice between them is also arbitrary. Pick either convention and stick with it and most thermodynamic quantities will be independent of the choice of convention. Other workers use theoretical arguments to favor one convention over the other. My own approach has been to

Linking the Gaseous and Condensed Phases of Matter
Edited by L.G. Christophorou *et al*, Plenum Press, New York, 1994

compare *experimentally measured quantities* with the values predicted according to the various conventions.[4] Through observing the natural world, we may evaluate our perceptions of it.

The Two Conventions

There are two principal conventions, the so-called "thermal electron" convention and the so-called "ion" convention.

The "thermal electron" convention is used by thermodynamicists, for example in the JANAF Tables,[2] and defines (i) the standard enthalpy of formation of the electron to be zero at all temperatures, and (ii) the heat capacity of the electron to equal that of all monatomic species, $C_p = 5/2\ R$.

The "ion" convention, used most extensively by gas-phase ion chemists, sets the standard enthalpy of formation of the electron and its heat capacity to be zero at all temperatures.[3]

It follows that the standard enthalpies of formation for the two conventions — $\Delta_f H_{298}(M^+)$ for the positive ions and $\Delta_f H_{298}(M^-)$ for the negative ions — are related as follows:

$$\Delta_f H_{298}(M^+)\ (\text{ion}) = \Delta_f H_{298}(M^+)\ (\text{thermal electron})\ - 6.197\ \text{kJ/mol} \tag{1}$$

$$\Delta_f H_{298}(M^-)\ (\text{ion}) = \Delta_f H_{298}(M^-)\ (\text{thermal electron})\ + 6.197\ \text{kJ/mol} \tag{2}$$

The "thermal electron" convention treats the electron as a boltzon, obeying Boltzmann statistics. There has been ongoing concern asking if the electron actually behaves as a fermion, obeying Fermi Dirac statistics.[3,5] This concern is simply answered.[6] In a metal, where the electron number density n_e is typically 10^{23} cm^{-3}, specific heat measurements show the electron to be behaving as a fermion. The situation is very different in ion/electron plasmas where typically $n_e < 10^{10}$ cm^{-3}. There the de Broglie wavelength of the electron at 300 K, ~4 x 10^{-7} cm, is very much less than the average electron/electron separation, > 5 x 10^{-4} cm, establishing the validity of Boltzmann statistics for such ion/electron plasmas.

The Conventions: Where Do They Differ?

The thermodynamic properties of the electron differ according to the different conventions. Therefore, by (3) and (4),

$$X = X^+ + e^- \tag{3}$$

$$Y + e^- = Y^- \tag{4}$$

absolute values for the standard thermodynamic properties of all positive and negative ions X^+ and Y^-, vary according to the convention selected.

The fact that the absolute values differ according to the convention chosen, presents fewer problems than might be supposed. Ion thermochemistry tables are most often used to compute thermodynamic properties of ionic *reactions*. Where these reactions involve ion-molecule reactions (5) or ion/ion recombination (6), the thermodynamic properties of the reactions are independent of the choice of the convention.

$$A^{+/-} + B = C^{+/-} + D \tag{5}$$
$$A^+ + B^- = C + D \tag{6}$$

In contrast, for all reaction types which feature free electrons, the thermodynamic properties of the reaction <u>do</u> depend on the choice of convention. Such reaction types comprise attachment (7), dissociative attachment (8) and ion/electron recombination (9).

$$e^- + AB = AB^- \tag{7}$$
$$e^- + AB = A + B^- \tag{8}$$
$$e^- + AB^+ = A + B \tag{9}$$

It follows that an experimental measurement of the thermodynamic properties for a reaction of the type (7), (8) or (9) should be able to determine which convention is correct and which incorrect.

AN EXPERIMENTAL ANSWER

It has been noted that a choice between conventions could be made if only data were available for ionization enthalpies (3) or electron-attachment enthalpies (4) as a function of temperature. These data are not available, because the reaction enthalpies are too large to permit measurement of the equilibrium composition at accessible temperatures. To follow this approach one must select a reaction which is closely thermoneutral and this limits the choice to type (8) dissociative electron attachment. A suitable choice is reaction (10)

$$e^- + HI \underset{k_r}{\overset{k_f}{\rightleftharpoons}} H + I^- \tag{10}$$

which has been carefully studied by Smith and Adams, at 300 K and 510 K.[7,8] Reaction (10) can be viewed as the sum of two parts

$$e^- + I = I^- \tag{11}$$
$$HI = H + I \tag{12}$$

the endothermic reaction (12) offsetting the exothermic reaction of interest (11) — it is an example of (4) — bringing thereby the overall reaction (10) into a measureable temperature range. These experimental measurements are very hard, the rate constant k_f for the dissociative electron attachment being measured in a FALP (flowing afterglow Langmuir probe) and k_r for the asssociative detachment involving atoms, being measured in a SIFT (selected ion flow tube).

Table 1: Rate constants and equilibrium constants for reaction (10)

Temperature (K)	k_f	k_r	$K = k_f/k_r$
300	3 ± 1 (-7)[a]	3 (-10)	1.0 ± 0.3 (3)
515	3 ± 1 (-7)	6 (-10)	0.5 ± 0.2 (3)

[a] (-7) implies x 10^{-7}. Rate constants k are in units of cm^3 /molec s.

These data are now analyzed using predictions for $\Delta H°_T$ for reaction (10) as a function of temperature, according to the two different conventions. At absolute zero, $\Delta H°_0$ is the same for both conventions.

$$\Delta H°_0 \ (10) = \Delta H°_0 \ (11) + \Delta H°_0 \ (12) = - EA(I) + D°_0(HI)$$
$$= - 3.0591 + 3.0541 = - 0.0050 \ eV = - 0.48 \ kJ/mol$$

The general expression for $\Delta H°_T$ at all temperatures is given by (13)

$$\Delta H°_T = \Delta_f H°_T(H) + \Delta_f H°_T(I^-) - \Delta_f H°_T(e^-) - \Delta_f H°_T(HI) \tag{13}$$

For the "thermal electron" convention, $\Delta_f H°_T(e^-) = 0$ for all T, reducing (13) to (14).

$$\Delta H°_T \ [\text{thermal electron}] = \Delta_f H°_T(H) + \Delta_f H°_T(I^-) - \Delta_f H°_T(HI) \tag{14}$$

Values for $\Delta_f H°_T(H)$, $\Delta_f H°_T(I^-)$ and $\Delta_f H°_T(HI)$ are taken from the JANAF Tables (which are based on the "thermal electron" convention)[2] and yield from (14) numerical predictions for $\Delta H°_T$ according to the "thermal electron" convention. Values for $\Delta H°_T$ are tabulated in Table 2 for the temperature range $0 < T \ (K) < 600$.

For the "ion" convention, corresponding values for $\Delta H°_T$ are evaluated using (15)

$$\Delta H°_T \ [\text{ion}] = \Delta H°_T \ [\text{thermal electron}] + C_p \ (e^-). \ T \tag{15}$$

which follows from applying (2) to (11). Corresponding values for $\Delta H°_T$ according to the "ion" convention, are also tabulated in Table 2 for the temperature range $0 < T \ (K) < 600$.

Table 2: $\Delta H°_T$ (kJ/mol) for (10) predicted for the "ion" and "thermal electron" conventions

Temperature (K)	0	300	350	400	450	500	600
$\Delta H°_T$ [thermal electron]	-0.48	-2.97	-3.39	-3.81	-4.24	-4.68	-5.61
$\Delta H°_T$ [ion]	-0.48	3.27	3.89	4.50	5.11	5.71	6.86

Table 2 shows two very different temperature dependences predicted for $\Delta H°_T$ according to the two conventions. With increasing T, (10) is either increasingly exothermic ("thermal electron" convention) or increasingly endothermic ("ion" convention).

So what do the data in Table 1 reveal about the validity of the two conventions? The data for (10) show an equilibrium constant decreasing with increasing temperature. That is compatible with (10) being an exothermic reaction ("thermal electron" convention) and incompatible with it being an endothermic reaction ("ion" convention). In an earlier treatment,[4] the data points at 300 K and 515 K were treated as a "two-point" van't Hoff plot to yield $\Delta H°_{300<T<515} = - 4 \pm 1$ kJ/mol. This was compared with the value of $\Delta H°_T$ predicted according to the "thermal electron" convention: $\Delta H°_{400} = - 3.81$ kJ/mol. The quantitative exactness of the comparison is however misleading because the van't Hoff analysis requires that changes in $\Delta S°_T$ for (10) $\Delta \Delta S°_T$ be considerably less than $\Delta H°_T/T$ throughout the temperature range $300 < T < 515$. This condition is not satisfied for (10) where $\Delta H°_T \sim 0$. More detailed analysis reveals that the data are still compatible with the predictions of the "thermal electron" convention within the experimental accuracy of the data.

WHERE WE STAND

It is possible to obtain experimental data for dissociative attachment reactions to choose between the alternative values for the thermodynamic properties of ions, according to the "thermal electron" asnd "ion" conventions. These data has been acquired using the SIFT and FALP techniques at variable temperatures. The limited data available for $e^- + HI = H + I^-$ suggest the "thermal electron" convention to be correct and the "ion" convention to be incorrect. More data and analysis are needed for this and for other systems.

There are significant consequences. (i) If tables based on the "ion" convention, such as *Gas-Phase and Neutral Thermochemistry*,[3] are used to compute thermodynamic properties for attachment, dissociative attachment, ion/electron recombination, detachment, associative detachment and ionization, the results will only be correct at 0 K. (ii) This emphasizes yet again that ionization energies and electron affinities are spectroscopic, state-to-state properties and they may only be equated with thermodynamic properties at 0 K. Thus the enthalpy of ionization of the H atom at 298 K is 6.197 kJ/mol larger than the ionization potential of the hydrogen atom. This has obvious implications for radiation chemistry.

ACKNOWLEDGMENTS

This study was made possible by the unique experimental data measured by David Smith,

Nigel Adams and Patrik Spanel. For these and for ongoing collaboration, I am most grateful. Partial funding was provided by the Air Force Geophysical Laboratory of the U. S. Air Force. I thank Professors Lindinger, Märk and Smith and the University of Innsbruck for a Visiting Professorship during the summer of 1993.

REFERENCES

1. L.M. Branscomb, Integrity in science, *Am. Sci.* 73:421 (1985).

2. M.W. Chase, Jr., C.A. Davies, J.R. Downey, Jr., D.J. Frurip, R.A. McDonald, and A.N. Syverud, "JANAF Thermochemical Tables", third edition, *J. Phys. Chem. Ref. Data* 14: Suppl. 1 (1985).

3. S.G. Lias, J.E. Bartmess, J.F. Liebman, J.L. Holmes, R.D. Levin, and W.G. Mallard, "Gas-Phase Ion and Neutral Thermochemistry", *J. Phys. Chem. Ref. Data* 17: Suppl. 1 (1988).

4. M. Henchman, Which is the proper thermodynamic convention for the enthalpy of ions? The "stationary electron" convention or the "thermal electron convention"? An experimental answer using $e^- + HI = H + I^-$, *in*: "Symposium on Atomic and Surface Physics '90", T.D. Märk and F. Howorka, eds., STUDIA Studentenförderungs Ges. m.b.H., Innsbruck (1990).

5. J.E. Bartmess,The thermodynamics of the electron and the proton, *J. Phys. Chem.* in press.

6. C. Kittel, "Elementary Statistical Physics," Wiley, New York (1958).

7. N.G. Adams, D. Smith, A.A. Viggiano, J.F. Paulson, and M.J. Henchman, Dissociative attachment reactions of electrons with strong acid molecules, *J. Chem. Phys.* 84:6728 (1986).

8. D. Smith and N.G. Adams, Studies of the reactions $HBr(HI) + e \rightleftharpoons Br^-(I^-) + H$ using the FALP and SIFT techniques, *J. Phys. B: At. Mol. Phys.* 20:4903 (1987).

THE THEORY OF ELECTRON ATTACHMENT TO MOLECULES

J. NORMAN BARDSLEY

LAWRENCE LIVERMORE NATIONAL LABORATORY,
LIVERMORE, CALIFORNIA 94550, USA

In analogy to the lecture on electron-ion recombination, we will focus on the processes of dissociative attachment (DA),

$$e + AB \; \text{->} \; AB^- \; \text{->} \; A + B^- \tag{1}$$

and three body attachment

$$e + AB + C \; \text{->} \; AB^- + C \tag{2}$$

where the third body C is usually another molecule.

The standard theory of dissociative attachment, first expounded by Bardsley et al (1966), Chen (1966), and O'Malley (1966) assumes the existence of a resonant state of the negative ion AB^- with a repulsive potential energy curve. Resonance formation increases the duration of the collision, leading to a larger probability of energy transfer from electronic to nuclear motion. The application of resonance theory provides a means to describe the process within the Born-Oppenheimer approximation. Recent studies of electron-molecule scattering at low energies have shown that whenever the matrix elements for electron scattering by molecules with fixed nuclei vary rapidly with internuclear separation, the transfer of energy to nuclear motion may be efficient and dissociation may result from non-adiabatic effects. Domcke (1991) has provided an excellent review of resonance and threshold effects in electron-molecule collisions, discussing both the available theoretical formalisms and the physics of the scattering processes. This lecture will concentrate on the latter.

Dissociative attachment of low energy electrons to H_2 has received most attention by theorists. The ground state $^2\Sigma_u$ of H_2^- is stable against electron emission only for values of the nuclear separation, R, greater than about 3 a_0. Since the equilibrium nuclear separation in H_2 is 1.4 a_0, the H_2 and H_2^- curves cross well outside the Franck-Condon region of the ground vibrational state. For R < 3 the H_2^- resonance has a large width, since the only restraint to electron emission is the weak p-wave centrifugal barrier. This means that the DA cross section is very small for the ground vibrational state, but rises rapidly with increasing vibrational quantum number, v. For low values of v, there is a strong isotope effect, since dissociation of D_2 takes more time and thus the competition from electron detachment is more severe. Ab initio calculations have been reported recently by Robicheaux (1991), using the frame-transformation method (Greene and Jungen, 1985), and by Hickman (1991) using resonant theory based upon parameters of Mundel et al (1985). Calculations on DA and the reverse process of associative detachment have been performed by Launay et al (1991) based upon semi-empirical resonance parameters.

Linking the Gaseous and Condensed Phases of Matter
Edited by L.G. Christophorou *et al*, Plenum Press, New York, 1994

Atems and Wadehra (1990) also used semi-empirical parameters in their study of non-local effects in resonant DA.

There has been little theoretical progress on attachment via higher resonances, or on attachment to electronically excited H_2 molecules, since the pioneering work of Bottcher and Buckley (1979). The observation by Pinnaduwage and Christophorou (1993) of efficient attachment to laser-excited H_2 is as yet unexplained. Their experiments on H_2, NO and other molecules will be discussed in the lecture.

Analysis of the high-resolution measurements on e-O_2 scattering below 1 eV by Ziesel et al (1993) lead to widths that are larger than those of previous calculations (Parlant and Fiquet-Fayard, 1976). The dependence of the widths upon v provides interesting information on the decay mechanism. The situation will be reviewed and implications for 3-body attachment will be discussed. Jaffke et al (1992) have deduced cross sections for DA to the $^1\Delta_g$ excited state of O_2 from measurements in a discharge. They suggest that the $^2\Pi_u$ resonance of O_2^- is coupled more strongly to the excited state than to the ground $^3\Sigma_g^-$ state of O_2. The implications of their result for electronic excitation should be examined, as was done for the $^2\Pi_g$ resonance by Teillet-Billy et al (1987).

Attachment of slow electrons to polar molecules, such as the hydrogen halides, has attracted much recent attention. The cross sections peak near threshold, and for HF and HCl are very sensitive to the vibrational quantum number. Ab initio R-matrix calculations for HCl were performed by Morgan et al (1990). The connection between the R-matrix and non-local resonance theories was discussed by Fabrikant (1990). Semi-classical methods to treat the nuclear motion have been developed by Kalin and Kazansky (1990) and Fabrikant (1991).

The presence of prominent zero energy peaks in attachment to molecules such as F_2 and SF_6 has shown that a permanent dipole moment is not necessary for the efficient capture of s-wave electrons. In F_2, attachment at around 1 eV can be explained in terms of a p-wave resonance associated with the lowest $^2\Sigma_u$ state of F_2^-, but at lower energies there is a strong s-wave component, behaving asymptotically as $E^{-0.5}$ (Chutjian and Alajajian, 1987). For SF_6, Klar et al (1992) have noted that the zero energy limit is close to that computed in a strong absorption model by Vogt and Wannier (1954), which is given solely by the polarizability of the target. The success of a polarizability model is surprising when the critical angular momentum for spiralling is less than h, but the same model also seems to work well for F_2. There does not seem to be a large zero-energy peak in the DA cross-section for CH_3Br at room temperature, although this feature does appear as the temperature is raised (Datskos et al, 1992). Speculations about the nature of the s-wave capture mechanism will be discussed in the lecture, and further examples will be cited (Shimamori et al, 1992).

ACKNOWLEDGEMENTS

This report was prepared under the auspices of the U. S. Department of Energy at the Lawrence Livermore National Laboratory under Contract No. W-7405-Eng-48.

REFERENCES

Atems, D. E., and Wadehra, J. M., 1990, Phys. Rev. 42, 5201.
Bardsley, J. N., Herzenberg, A., and Mandl F., 1966, Proc. Phys. Soc. 89, 305, 321.
Bottcher, C., and Buckley, 1979, J. Phys. B 12, L497.
Chen, J. C. Y., 1966, Phys. Rev. 148, 66.
Chutjian, A., and Alajajian, S. H., 1987, Phys. Rev. A 35, 4512.
Datskos, P. G., Christophorou, L. G., and Carter, J. G., 1992, J. Chem. Phys. 97, 9031.

Domcke, W., 1991, Phys. Rep. **208**, 97.

Fabrikant, I. I., 1990, Comments At. Mol. Phys. **24**, 37.

Fabrikant, I. I., 1991, Phys. Rev A **43**, 3478.

Greene, C. H., and Jungen, Ch., 1985, Adv. At. Mol. Phys. **21**, 51.

Hickman, A. P., 1991, Phys. Rev. A **43**, 3495.

Jaffke, T., Meinke, M., Hashemi, R., Christophorou, L., and Illenberger, E., 1992, Chem. Phys. Lett. **193**, 62.

Kalin, S. A., and Kalinsky A. K., 1990, J. Phys. B **23**, 3017, 4377.

Klar, D., Ruf, M.-W., and Hotop, H., 1992, Aust J. Phys. **45**, 263; Chem. Phys. Lett. **189**, 448.

Launay, J. M., Le Dourneuf M., and Zeippen, C. J., 1991, Astron. Astrophys. **252**, 842.

Mundel, C., Berman M., and Domcke, W., 1985, Phys. Rev. A **31**, 641; **32**,181.

Morgan, L. A., Burke, P. G., and Gillan, C. J., 1990, J. Phys. B **23**, 99.

O'Malley, 1966, Phys. Rev. **150**, 14.

Parlant, G., and Fiquet-Fayard, F., 1976, J. Phys. B **9**, 1617.

Pinnaduwage, L., and Christophorou, L. G., 1993, Phys. Rev. Lett. **70**, 754.

Robichaux, F. , 1991, Phys. Rev. A **43**, 5946.

Shimamori, H., Tatsumi, Y., Ogawa, Y., and Sunagawa, T.A., 1992, J. Chem. Phys. **97**, 6335.

Teillet-Billy, D., Malegat, L., and Gauyacq, J. P., 1987, J. Phys. B. **20**, 3201.

Vogt, E., and Wannier, G. H., Phys. Rev. **95**, 1190.

Ziesel, J. P., Randell, J., Field, D., Lunt, S. L., Mrotzek, G., and Martin, P., J. Phys. B **26**, 527.

SECTION VI. ELECTRON-ION RECOMBINATION IN GASES AND LIQUIDS

ELECTRON-ION RECOMBINATION
IN DENSE MOLECULAR MEDIA

Yoshihiko Hatano

Department of Chemistry
Tokyo Institute of Technology
Meguro-ku, Tokyo 152, Japan

INTRODUCTION

The investigation of the dynamic behavior of excess electrons such as electron transport and reactivities in dense molecular media is of essential importance in both fundamental and applied sciences.[1-12] Excess electrons here mean electrons produced by the interaction of ionizing radiation or laser and synchrotron radiation photons with molecules in the medium. These electrons initially have higher energies than the thermal energy of the medium and quickly lose their energies via cascading electron-molecule collisions down to a thermal equilibrium with the medium.

The values of an excess electron mobility μ_e have been measured extensively in a variety of dense molecular media and several theoretical models have been proposed for the transport mechanism. Excess electron reactivities with solute molecules or positive ions in these media have been also extensively investigated.[1-12] The former reaction is electron attachment and the latter electron-ion recombination. The investigation of these electron reactivities has provided a new insight into that of the behavior of slow electrons in matter, i.e., the entitled subject of the present NATO Advanced Study Institute. Since we already had opportunities of writing review articles on the attachment in dense media,[1,3] we will focus our present review in this chapter on the recombination in dense media. In the following, however, a brief survey is given of the recent investigation of these electron reactivities as divided into attachment and recombination to clarify our viewpoint of the investigation and to summarize distinctive features of these reactivities in dense molecular media as compared with those in low density gases.

Electron attachment is a process in which electrons are captured by atoms or molecules to form negative ions and classified into two types; dissociative and non-dissociative processes as shown in the following reaction scheme,[1,3,13]

$$e^- + AX \xrightarrow[1/\tau]{\sigma \text{ or } k} AX^{-*} \begin{cases} A + X^- & (1) \\ AX^- + \text{energy} & (2) \end{cases}$$

Interaction of low-energy electrons with molecules, AX, produces unstable negative ions, AX^{-*}, with a cross section σ or a rate constant k. The autodetachment of electrons from AX^{-*} with a lifetime τ may compete with the dissociation of AX^{-*} or with the formation of stable molecular negative ions, AX^-, which requires the release of the excess energy from AX^{-*}. The lifetime τ is related to the electron-energy width of the attachment resonance. The value of $1/\tau$ is a rate constant for the autodetachment process. In the presence of third-body molecules or in bulk media, AX^{-*} is collisionally stabilized to form stable AX^-. The branching ratios among the unimolecular processes of decaying AX^{-*} depend on the interrelationship of the potential energy curves between AX and AX^-, and also on electron energies. The relative importance of the collisional stabilization process in the overall decaying processes of electrons depends largely on these unimolecular processes particularly the lifetime τ and on the number density and character of third-body molecules, in which one may expect some environmental effects on the over-all scheme of electron attachment processes. In addition to the determination of cross sections or rate constants for electron attachment or negative-ion formation and their electron energy dependences, it has been of prime importance in electron attachment studies to clarify the attachment mechanism, not only the two-body mechanism but also the overall mechanism and how environmental conditions affect the mechanism.[1,3]

Electron attachment processes have been extensively studied theoretically and experimentally, and these are comprehensively summarized in recent review papers[1,3,13] and also in some lectures at the present NATO-ASI. Experimentally, the low-energy electron attachment to molecules has been studied using both beam and swarm methods, and each of them is consisted of several different techniques dependent on the method of detecting low energy electrons or formed negative ions.[14] In the investigation of low energy electron attachment in bulk gases particularly in dense gases, a microwave cavity technique combined with a pulse-radiolysis method has shown a distinct advantage over the other techniques because a sensitive detection of the time-resolved density of thermal and epithermal electrons is practicable with very fast response in a wide range of the density of an environmental gas which is chosen with virtually no limitation,[1,3] and therefore providing the absolute values of k and τ and also the quantitative information on the effect' of the density of environmental third-body molecules on these values, i.e., the absolute values of third-body attachment rate constants. Thus, the mechanism of low-energy electron attachment to molecules has been discussed primarily in terms of the interaction of electrons with molecules even in a multiple collision system. Recent studies of thermal electron attachment to O_2, N_2O and other molecules using this technique have revealed interesting features of the electron attachment to van der Waals (vdW) molecules,[1,3,15] i.e., the effect of the vdW potential on the electron attachment resonance. This experiment has motivated a recent development in beam experiments of electron collisions with vdW molecules.[16]

The rate constants for the electron attachment to vdW molecules (e.g., (O_2.M) and (N_2O.M) where M is O_2, N_2O or any other molecules) are much larger than those for the electron attachment to isolated O_2 or N_2O.[1,3,15] This result clearly shows that the electron attachment to an ordinary isolated single molecule is greatly enhanced by the presence of a surrounding guest molecule M. The modification of the interaction potential between an electron and an isolated single molecule by the formation of the vdW molecule reflects the departure from the gas-phase characteristics toward those of the condensed phase. The effect of the vdW potential or of the presence of a surrounding molecule on electron attachment has been summarized as follows:[3]

(1) The lowering of the attachment resonance energy due to a deeper ion-neutral potential in comparison with neutral-neutral potential of the vdW molecule,
(2) The additional vibrational structures of the vdW molecule,
(3) The symmetry breaking due to the vdW interaction which allows the molecule to attach the electron with additional partial waves,
(4) The deformation of the molecular structure or the change of the vibrational modes due to the surrounding molecules, and
(5) The effective vibrational relaxation of the formed negative ion with excess energies due to the presence of a built-in third body molecule in a vdW molecule.

The distinct features of the electron attachment to vdW molecules as summarized above may become a substantial clue to the understanding of the fundamental nature of electron attachment not only in dense gases but also in the condensed phase. It is also apparent that most of the electron attachment in bulk systems is no longer a simple process consisting of the interaction of an electron with an isolated molecule and is affected by the presence of a surrounding molecule.

It is of great interest to investigate also the electron-ion recombination in dense molecular media in the density range from dense gases to liquids and solids[11,12] because the Coulombic interaction distance between an electron and an ion as expressed by the Onsager length r_c is much bigger than the mean intermolecular distance of the nearest neighbor molecules in the medium. The Onsager length is given by,

$$ r_c = \frac{e^2}{\varepsilon k_B T} , \tag{3} $$

where e, k_B, ε, and T are the electron charge, the Boltzmann constant, the dielectric constant of the medium, and the absolute temperature of the medium, respectively.

Electron-ion recombination processes in isolated two-body collisions or in low pressure gases have been extensively studied mainly with a merged beam method,[17] a pulsed afterglow method,[18] and a flowing afterglow method.[19] A survey of recombination studies has been given very recently also from a theoretical viewpoint.[20] In the gas phase at one to several atmospheric pressures, these processes have been studied with a pulse-radiolysis method[21-23] as summarized in a recent review article,[24] and observed recombination rate constants k_r are expressed by

$$k_r = k_2 + k_3 n, \qquad (4)$$

where k_2 and k_3 are the two-body and the three-body coefficients, respectively, and n is the density of a third-body molecule. In dense media such as high pressure gases near the critical points, liquids, and solids, however, experimental and theoretical studies have been relatively very few and surveyed just recently.[11,12]

In the liquid phase, it has been shown[25] that observed electron-ion recombination rate constants k_r in a variety of nonpolar media are in good agreement with the calculated values k_D by the reduced Debye equation,

$$k_D = \frac{4\pi e}{\varepsilon} \mu_e, \qquad (5)$$

in the range of μ_e below about 150 cm²/Vs, and thus the reaction in these media was concluded to be diffusion controlled.

In liquid and high pressure gaseous methane in which most of μ_e values are larger than 150 cm²/Vs, however, it has been found recently[26] that the observed k_r values are much lower than k_D. This experiment has stimulated the several theoretical investigations[27-40] of the recombination process in dense media. Since the magnitudes of the electron mobilities in dense Ar, Kr, and Xe media are as high as in dense methane media, the values of k_r and μ_e in these media have been also measured[41-44] and compared further with theoretical results. The observed k_r values in both liquid and gas phases are again found to be much smaller than those calculated by the reduced Debye equation. In the solid phase, however, the observed k_r values are almost in agreement with those calculated by Eq.(5). The effect of an external dc electric field strength on both k_r and μ_e values has been also measured. A detailed comparison has been made further between the experimental results and the theoretical ones, from which it has been concluded that the recombination in liquid and dense gaseous Ar, Kr, and Xe as well as in methane is not a usual diffusion-controlled reaction. The experimental method of these investigations, the obtained results and discussion are described in the following sections.

EXPERIMENTAL

For the measurement of electron-ion recombination rate constants k_r in high pressure gases or in condensed nonpolar media, pulse radiolysis has been effectively adopted with a variety of detection techniques.[12,45]

(1) In a pulse-radiolysis dc-conductivity method,[46] k_r/μ_e is measured and then k_r is determined from the measured k_r/μ_e and a separately known μ_e.

(2) In a pulse-radiolysis charge-clearing method,[47] a residual amount of charge which is left unrecombined is collected and measured. Since the collected charge is expressed as a function of k_r, the value of

$k_r/(\beta\mu_e)$, where β is a correction factor due to a space charge is obtained. The k_r value is thus determined by using the value of μ_e known seperately and the value of β which is apporoximately assumed.

(3) In a pulse-radiolysis optical-absorption method,[46] the value of k_r/ε, where ε is a molar absorption coefficient, is measured by the time-resolved measurement of the optical absorption of solvated electrons and then the k_r value is determined by the observed value of k_r/ε and the value of ε known seperately.

(4) In a pulse-radiolysis dc-conductivity method with a long pulse-width[49] the resulting steady state current during the long pulse is considered by the assumption that the rate of excess electron generation becomes equal to the rate of electron loss because of the reaction with positive ions and impurities.

(5) In a pulse-radiolysis dc-conductivity method as called the decay-curve analysis method,[11,12,25,26,41-45,50-52] the k_r value is determined by analyzing transient current decay curves both at the high pulse dose of x-rays where the recombination is effective and at the low pulse dose where the recombination is negligible.

(6) In a pulse-radiolysis microwave-conductivity method,[22] the k_r value is determined by the time-resolved measurement of microwave conductivity.

(7) There is another way in which a radioisotope such as [207]Bi is used as an internal radiation-source for ionizing rare-gas liquids.[53] In this method the difference in a fluorescence intensity with and without a dc electric-field is assumed to be due to electron-ion recombination and the k_r value is determined by the time-resolved measurement of the fluorescence with the known number density of ionization which is determined separately.

Although the decay-curve analysis method is inadequate for the media in which μ_e values are less than about 5 cm²/Vs because of the too low signal, this method is useful because the k_r value can be determined without knowing the pulse dose of x-rays or the free-ion yield of the medium.[11,43,45] The following is a description of the decay-curve analysis method which has been adopted in this experiment.[11,12,25,26,41-44]

A research grade methane, argon, krypton, or xenon was purified in a grease-free vacuum-system which was pumped down to a pressure less than 10^{-6} Torr, passed through columns of "Gas Clean" which was an oxygen absorber, KOH pellets, and molecular seives 5A in this order, bubbled through Na-K liquid alloy, transferred to a conductivity cell as shown in Fig.1, and finally sealed in the cell.

The conductivity cell made of stainless steel can be operated up to the inner pressure of 130 atm. The cell has two parellel rectangular-electrodes with a rectangular guard-electrode.

The experimental setup for the measurement of a transient electron current is shown in Fig.2. An x-ray pulse of a few nanosecond was obtained by impinging an electron-beam pulse of 0.6 MeV from a Febetron 706 on a 0.1 mm thick tungsten-foil at the front of an electron-pulse emission window. The pulse dose was controlled by setting lead plates of various thickness behind the tungsten foil. The dose was calibrated by a thermoluminescent dosimeter.

The lower half of the conductivity cell containing the sample was set

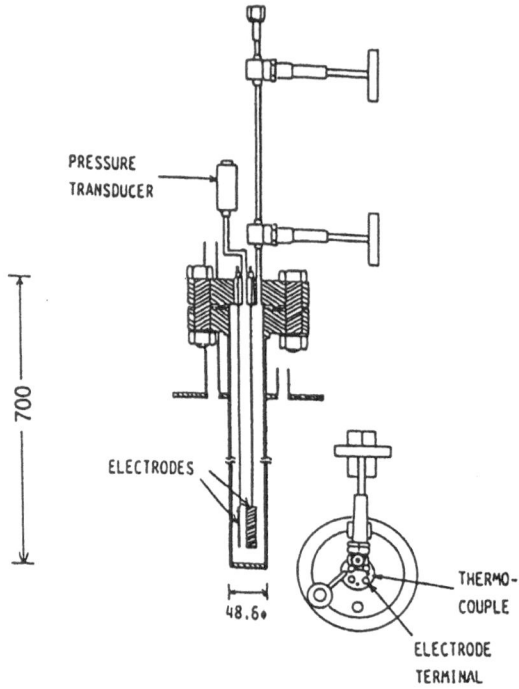

Figure 1. Conductivity cell.[41] (a) Cross sectional view, (b) top view.

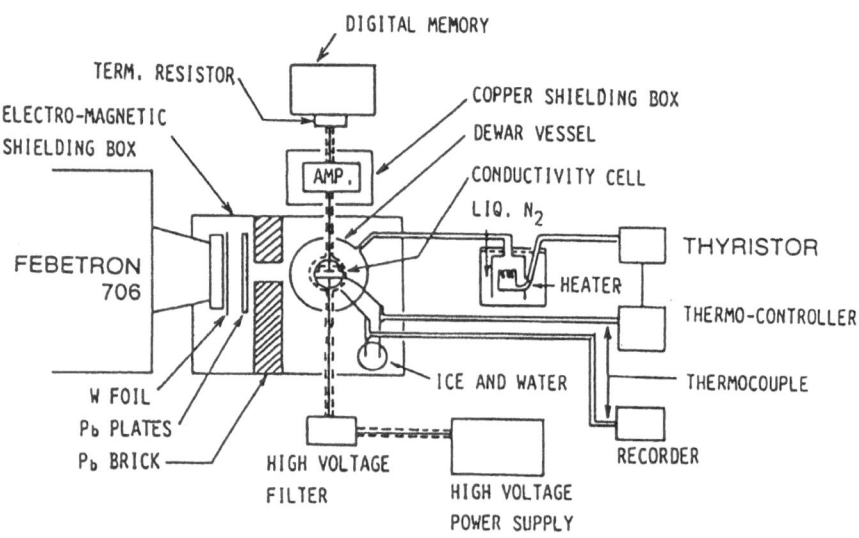

Figure 2. Setup for measurement of transient electron current.[41]

in a Dewar vessel in an electromagnetic-shielding box. Temperature was controlled by controlling the flow rate of a cold nitrogen-gas into the Dewar vessel using a temperature-control system. Temperature of the sample was monitored by the two copper-constantan thermocouples; one of them was set near the upper side of the guard electrode and the other was set near the lower side of the guard electrode. The difference in both temperature indications was less than one degree.

When a thermal equilibrium was established, a negative high voltage from a dc power-supply through a low-pass filter (8kHz) was applied to the high-voltage electrode of the cell, and then the sample was irradiated with an x-ray pulse. Under the influence of an electric field, a transient electron-current was induced in the external circuit of the cell.

The values of μ_e and k_r were determined by the analysis of observed decay-curves of transient electron-current as follows. The values of μ_e and k_{tr} were determined by fitting Eq.(6) to the observed decay-curve of an electron current under the condition of a low x-ray pulse-dose where the contribution of electron-ion recombination could be neglected,

$$I(t) = I_0(1-\mu_e Et/d)\exp(-k_{tr}t), \qquad (6)$$

where k_{tr} is a rate constant for electron trapping due to the medium itself or residual trace of electron attaching impurity even after the rigorous purification of a sample as described above. Using the obtained μ_e and k_{tr}, the value of k_r was determined by fitting Eq.(7) to the observed decay of electron current at a high x-ray pulse-dose where the contribution of electron-ion recombination was taken into account,

$$I(t) = \frac{I_0(1-\mu_e Et/d)}{1 + n_0 k_r t} \exp(-k_{tr}t). \qquad (7)$$

In Eqs.(6) and (7), I(t) is the electron current at the time t after the x-ray pulse, I_0 the peak current at the end of the pulse, E the electric field strength, d the distance between the electrodes, and n_0 the initial concentration of excess electrons. The error limits of the obtained values of μ_e and k_r were 10% and 15%, respectively.

The densities of a sample were evaluated by the measured pressure and temperature. The temperature dependence of the dielectric constant of a sample was calculated from the Clausius-Mossotti equation using the evaluated density.

RESULTS AND DISCUSSION

The values of k_r and μ_e which have been experimentally obtained in nonpolar liquids and solids are tabulated in Tables 1 and 2, respectively. These values are 2-7 orders of magnitude larger than the acid-base neutralization rate constant for the well-known recombination as a fast reaction of H_3O^+ with OH^- in aqueous solution (2.3×10^{-10} cm^3/s at 298K), reflecting much higher mobilities of electrons than those of solvated ions. When the values of μ_e are lower than ~ 150 cm^2/Vs, k_r values are well proportional to μ_e values as shown in Fig.3 and in good agreement with the values k_D calculated by Eq.(5). It has been concluded, therefore, that the electron-ion recombination in these media is diffusion-controlled.[25]

Table 1. Electron-ion recombination rate constants k_r and electron mobilities μ_e of nonpolar liquids.[12]

Liquid	$T[K]$	$k_r[cm^3/s]$	$\mu_e[cm^2/Vs]$	ref.
n-pentane	296	1.3×10^{-7}	0.14	47
n-hexane	293	1.2×10^{-7}		54
	293	1.0×10^{-7}	0.09	55
	296	7.8×10^{-8}	0.08	47
cyclohexane	293	3.2×10^{-7}	0.35	55
tetramethylsilane	296	8.0×10^{-5}	100	47
	298	8.5×10^{-5}	104	51
neopentane	283	4.7×10^{-5}	55	50
methane	120	2.7×10^{-4}	373	26
argon	87	7.0×10^{-5}	490	41
krypton	200	3.7×10^{-4}	357	43

Table 2. Electron-ion recombination rate constants k_r and electron mobilities μ_e of nonpolar solids.[12]

Solid	$T[K]$	$k_r[cm^3/s]$	$\mu_e[cm^2/Vs]$	ref.
neopentane	263	1.4×10^{-4}	152	50
tetramethylsilane (α-phase)	162	1.1×10^{-4}	161	51
methane	77	1.9×10^{-3}	1490	26
argon	80	1.5×10^{-3}	1690	41

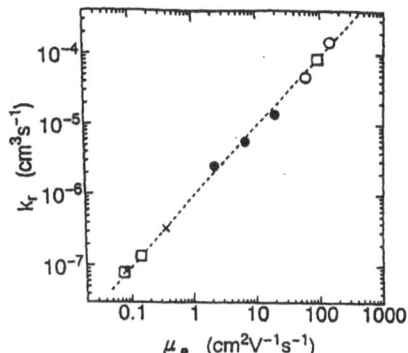

Figure 3. Variation of k_r with μ_e less than about 150 cm²/Vs in condensed nonpolar media.[25] n-Pentane, n-hexane, TMS (□) , ref.[47]; n-hexane, cyclohexane (×) , ref.[55]; neopentane-n-hexane mixtures (●) , ref.[45]; liquid and solid neopentane (○) , ref.[50].

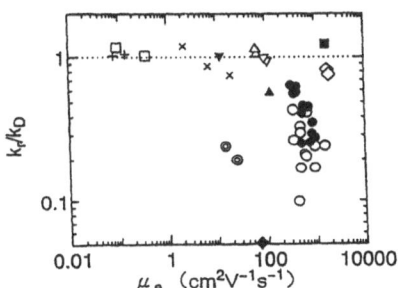

Figure 4. Variation of k_r/k_D with μ_e in condensed nonpolar media. Neopentane, liquid (△) and solid (▲) , ref.[50]; neohexane (▼) ; TMS (▽) , ref.[47]; n-pentane, n-hexane (+) , ref.[47]; n-hexane, cyclohexane (□) , ref.[55]; neopentane-n-hexane mixures (×) , ref.[45]; methane, liquid (●) and solid (■) , ref.[26]; argon, liquid (○) and solid (◇) , ref.[41]; krypton, gas (◆) ; and xenon, gas (◎) . For liquid krypton see Tables 3-5.

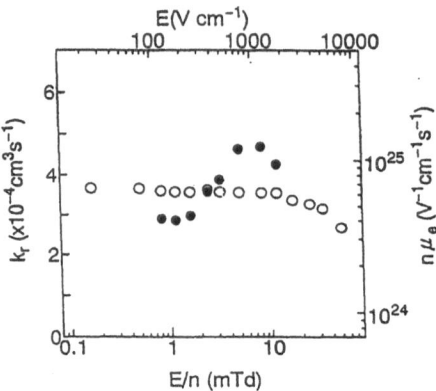

Figure 5. Electric field dependence of k_r (●) and $n\mu_e$ (○) in liquid methane (T = 123 K, n = 1.5× 10^{22}/cm^3).[26]

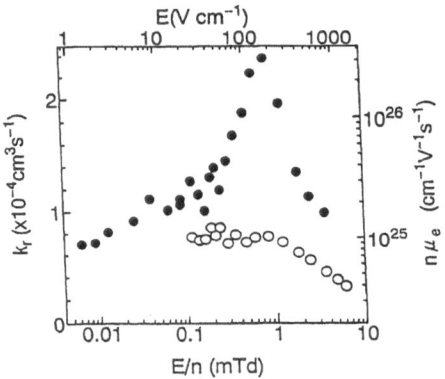

Figure 6. Electric field dependence of k_r (●) and $n\mu_e$ (○) in liquid argon (T = 87 K, n = 2.1× 10^{22}/cm^3).[41]

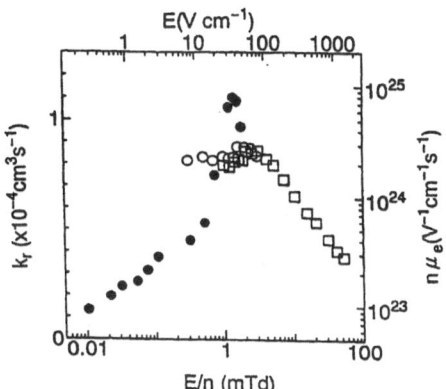

Figure 7. Electric field dependence of k_r ($\bullet$) and $n\mu_e$ ($\bigcirc$, $\square$) in high pressure argon gas ($\bullet$, $\bigcirc$; $T = 299$ K, $n = 2.6 \times 10^{21}/\text{cm}^3$; $\square$; $T = 296$ K, $n = 2.4 \times 10^{21}/\text{cm}^3$).[41]

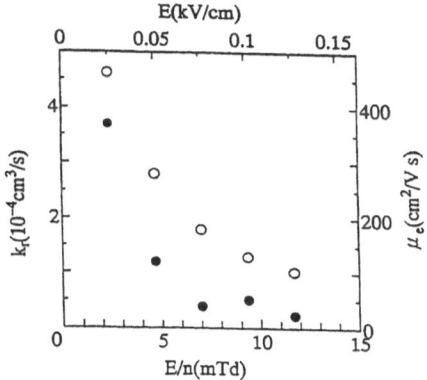

Figure 8. Electric field dependence of k_r ($\bullet$) and μ_e ($\bigcirc$) in liquid krypton ($T = 200$ K, $n = 1.08 \times 10^{22}/\text{cm}^3$).[43]

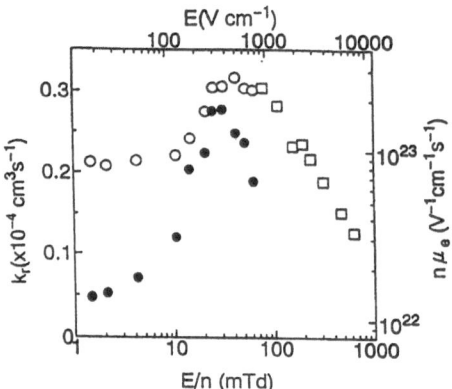

Figure 9. Electric field dependence of k_r (●) and $n\mu_e$ (○, □) in high pressure krypton gas (T = 291 K; ●, ○; n = 1.2× 10^{21}/cm³; □; n = 1.6× 10^{21}/cm³).[43]

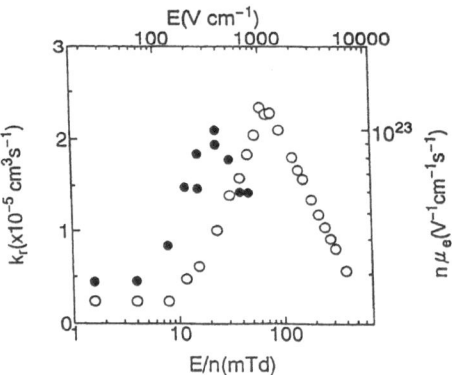

Figure 10. Electric field dependence of k_r (●) and $n\mu_e$ (○) in high pressure xenon gas (T = 293 K, n = 1.6× 10^{21}/cm³).

Table 3. Comparison of Warman's[27] theoretical ratio k_r/k_D with the experimental results in gaseous and liquid krypton,[43] argon,[41] and methane.[42]

	n $(10^{21}/cm^3)$	μ_{th} (cm^2/Vs)	T (K)	ε	E_{10} (V/cm)	k_r/k_D Expt.	Warman
Liquid							
CH_4	15	420	123	1.60	3750	0.6	8.1×10^{-1}
Kr	10.8	357	200	1.39	70	0.61	2.9×10^{-4}
Ar	21.1	490	87	1.49	250	0.11	8.2×10^{-2}
Gas							
CH_4	1.17	260	285	1.04	400	0.18	4.1×10^{-3}
Kr	1.2	70	291	1.04	120	0.051	3.4×10^{-4}
Ar	2.41	860	296	1.06	2.16	0.0088	5.7×10^{-5}

Table 4. Comparison of Tachiya's[31] and Sceat's[36] theoretical ratio k_r/k_D with the experimental results in gaseous and liquid krypton.[43]

	n $(10^{21}/cm^3)$	μ_e (cm^2/Vs)	T (K)	ε	k_r/k_D Expt.	Tachiya	Sceats	Λ (nm)/r_c(nm)
Liquid								
Kr	10.8	357	200	1.39	0.61	0.46	0.51	21.0/60.2
Gas								
Kr	1.2	70	291	1.04	0.051	0.87	0.94	4.96/55.3

Table 5. Comparison of Mozumder's[37] and Lopez-Quintela's[35] theoretical ratio k_r/k_D with the experimental results in gaseous and liquid krypton,[43] argon,[37] and methane.[42]

	T (K)	Λ (nm)	r_c (nm)	k_r/k_D Expt.	Mozumder	Lopez-Quintela
Gas						
Kr	291	4.96	55	0.051	0.75	0.75
Ar	296	61.6	53	0.010	0.19	0.19
CH_4	285	18.3	56	0.19	0.45	0.45
Liquid						
Ar	87	19.0	128	0.11	0.64	0.64
CH_4	123	19.4	85	0.6	0.54	0.54

However, in liquid and dense gaseous methane, argon and krypton, in which most of μ_e values are larger than 300 cm²/Vs, it has been found[26,41-43] that the observed k_r values are much lower than the k_D values except for solid methane and argon as shown in Fig.4. In dense xenon media the obtained results are qualitatively the same as those in dense methane, argon, and krypton media. Such deviation of k_r from k_D has stimulated several theoretical investigations of the recombination process in dense media.[27-40] Because the external electric field gives the change in the kinetic energy of excess electrons, recent experiments have been focussed on the effect of the external electric field strength on k_r values.[41-43] Experimental results of the effect of external electric field on k_r values are shown in Figs.5-10 for dense methane, argon, krypton, and xenon media, respectively, together with the variation of density normalized electron mobility $n\mu_e$.

Theoretical investigations trying to explain the k_r deviation from k_D are summarized as follows:

(1) A semi-empirical treatment taking both the diffusion controlled and electron-energy exchange controlled recombination processes into account,[27]
(2) A Monte Carlo simulation with the parameter Λ/a where Λ is the mean free path of electrons and a is the reaction radius,[29,31]
(3) A molecular dynamics simulation,[30]
(4) A fractal treatment,[35,37,39]
(5) A gas kinetic approach,[33,34] and
(6) An approach based on the Fokker-Planck equation.[38]

The followings are some comparisons between the experimental results and theoretical ones of k_r/k_D. Theoretical values of k_r/k_D shown in Tables 3-5 include not only the values reported by each theoretician but also those estimated by us using each theoretical method.

Warman proposed[27] a theoretical formula Eq.(8) for the recombination rate-constant taking the diffusion-controlled recombination rate-constant and the energy-exchange controlled recombination rate-constant into account:

$$k_r/k_D = [1+(55T^4\varepsilon^2/E^2{}_{10})]^{-1}. \tag{8}$$

Here, E_{10} is the critical field strength at which the drift velocity deviates 10% from $\mu_{th}E$, where μ_{th} is the thermal electron mobility and E is an external electric field strength. Table 3 shows k_r/k_D values calculated by Eq.(8) together with the experimental ones. In spite of many assumptions in the derivation of Eq.(8), the calculated values for liquid argon and methane are in rather good agreement with the experimental ones. One of the important assumptions seems to be the momentum and energy transfer cross sections for electrons in these media to be energy independent. In liquid krypton, the value calculated by Eq.(8) is much smaller than the experimental one, which may be partly attributed to the experimental k_r/k_D value not measured at a sufficiently low electric field. For dense gaseous methane, argon, and krypton, Warman's theoretical values are also far smaller than the experimental ones.

Taohiya pointed out[31] using a Monte Carlo simulation that if the magnitude of the ratio $\Lambda/\underline{a}$, where Λ is the mean free path of excess electrons and $\underline{a}$ the reaction radius at which the recombination occurs, is not negligible, then the rate constant of electron-ion recombination deviates from k_D. We do not know the values of reaction radius $\underline{a}$ and therefore take the Onsager length as a measure of $\underline{a}$. Our experimental values of k_r/k_D in liquid methane[26] vary on the curve obtained by the simulation,[31] which uses only $\Lambda/\underline{a}$ as a parameter (Fig.4 in ref.31).

Sceats[36] also developed the dependence of the steady-state electron-ion pair recombination rate-constant on $\Lambda/\underline{a}$. In the Sceats's theory, k_r/k_D is given by

$$k_r/k_D = 2s[(1+s^2)^{1/2}-s]/\{1+(\lambda/f)[(1+s^2)^{1/2}-s]\}, \qquad (9)$$

where $s=e^2/4\pi\lambda$ with the base of the natural logarithm e, λ is the ratio of the mean free path to the Onsager length taken as the reaction radius $\underline{a}$, and f is the function of the ratio of the Coulomb dissociation energy to k_BT. Table 4 shows the k_r/k_D values obtained by the Tachiya's Monte Carlo simulation[31] and by Eq.(9), and those experimentally obtained.[43] The k_r/k_D values for liquid krypton calculated with both Tachiya's and Sceats's theories are relatively in good agreement with the experimental value. The experimental value of k_r/k_D for gaseous krypton, however, is far lower than the theoretical values obtained by the Tachiya's and Sceats's theories, which use only Λ/r_c as a parameter. This discrepancy may be partly due to inadequate comparison not using a true reaction radius which has not been known yet. The reaction radius is probably affected by the density and other chemical and physical factors of a medium.

Lopez-Quintela et al.[35] analyzed the influence of the differential fractal dimension of reagent trajectories on the rate of diffusion-controlled bulk ion-recombination to explain our experimental results in liquid methane.[26] In their theory, k_r/k_D is given by

$$k_r/k_D = \{(1+d/r_c)-[1+d(R^{-1}+r_c^{-1})]\exp(-r_c/R)\}^{-1}, \qquad (10)$$

where d is a parameter related to Λ, R, and r_c. In their paper, d is treated as an optimizable parameter and is taken to be 3.7Λ by fitting Eq. (10) to experimental results.[26] Mozumder[37] has extended the theory of Lopez-Quintela et al. in order to treat the influence of a fractal dimension on bulk ion-recombination. In his theory,[37] k_r/k_D is given by

$$k_r/k_D = (1+d/r_c)^{-1}, \qquad (11)$$

where d is a parameter related to Λ. Mozumder also takes 3.7Λ for d following Lopez-Quintela et al. Table 5 shows the results obtained by Lopez-Quintela et al. and Mozumder together with the experimental values for krypton, methane, and argon. The results calculated by Eq.(10) agree with the results calculated by Eq.(11). The experimental results except for liquid methane, however, are not in good agreement with those calculated values.

The initial increase in k_r with increase in E/n as shown in Figs.5, 6, 7, 9, and 10 seems to be mainly due to the increase in the drift velocity of excess electrons because the increase in k_r is roughly proportional to the external electric field strength. According to this model, the increase in k_r with the external electric field strength is caused by the increase in the effective volume of the medium that an excess electron sweeps per unit time with a certain reaction radius. Tachiya has derived[32] theoretically that the rate constant of diffusion-controlled reactions k_d between a neutral particle and a charged particle increases with increasing the external electric field strength E as $k_d/4\pi D\underline{a} = 1+eE\underline{a}/2k_B T$ for low electric fields where $4\pi D\underline{a}$ is the rate constant in the absence of an electric field, D is the diffusion constant, and $\underline{a}$ is the reaction radius. Tachiya has extended the theory[32] to the case where Coulombic attraction acts between the reactants, $\Lambda/\underline{a}$ is negligible, and charge recombination is diffusion-controlled. According to his model, the increase in k_r with a small external electric field is given by

$$k_r = k_o[1+b(eEr_c/2k_B T)], \qquad (12)$$

where b is a function of both real reaction-radius and r_c. In the case of ion-ion recombination, b is negligible ($\ll 1$) so that the effect of the external electric field is very small. Consequently, as long as the recombination process obeys the usual diffusion-controlled kinetics, the effect of an external electric field is very small. In the present electron-ion recombination, however, the field dependence is so large as shown in Figs.5, 6, 7, 9, and 10 that the b value is estimated to be 160 if Eq.(12) is applicable to the result shown, e.g., in Fig.9. The difference in the b values for the above-mentioned two cases is at least partly caused by the assumption in the Tachiya's extended theory that $\Lambda/\underline{a}$ is negligible and the charge recombination is diffusion-controlled. In the present electron-ion recombination, however, $\Lambda/\underline{a}$ value is not negligible as shown by the Λ/r_c values in Table 4 and the electron-ion recombination is not a usual diffusion-controlled process. In the Tachiya's model[32] of the recombination between a neutral particle and a charged particle, where there is a much shorter-range force between them as compared with the force between charged particles, the electric field effect is suppressed even if $\Lambda/\underline{a}$ increases. Since, however, in the case where a long-range force acts between reactants, i.e., an electron and an ion, the situation does not seem so simple, then Tachiya and Isoda now try to extend further the Tachiya's extended theory to the case of the electric field effect on the electron-ion recombination rate constant.[56] Their idea that the external electric field makes the entangled trajectories of charged particles disentangle into the direction of the field and increases the effective volume which the charged particle sweeps per unit time may be partly applicable to our case.

The recombination rate constant decreases with increasing external electric field in the high electric-field region as shown in Figs.5-10. A tentative model, as described above, for the k_r value increasing with the external field in the low elctric-field region may be applicable also in the region of higher electric fields where the excess electrons are heated by the electric field. The heating of the excess electrons by the external electric field makes the electron-ion recombination more unfavorable. Of these opposite field effects on k_r the latter effect may become gradually predominant with increasing electric field, thus overcoming the increase in the value of k_r due to the increase in the drift velocity, and decreasing the net value of k_r in the higher electric field region.

The comparison, as described above, between theoretical values and experimental ones in different media and in different phases seems to show the requirement of further new theoretical treatments which take into account of other parameters indicating subtle differences in media such as a molecular structure, a molecular size, a density, and a reaction radius, in addition to an electron mobility, the mean free-path of excess electrons, etc. in the present theories.

A real reason why in solid argon and methane k_r/k_D ratio is nearly one as shown in Fig.4 in spite of much higher electron mobilities than in liquid has not been known yet. We tentatively consider that it could be due to a larger excess-energy loss of an electron per collision in the solid phase than that in the liquid phase as indicated by the shorter thermalization times in solid than in liquid.[27]

ACKNOWLEDGMENTS

The author thanks K.Shinsaka, T.Wada-Yamazaki, H.Namba, Y.Nakamura, T. Tezuka, M.Chiba, S.Yano, M.Yamamoto, M.Codama, T.Srithanratana, K.Serizawa, K.Endou, K.Honda, H.Yamada, K.Isoda, and T.Odaka for their excellent collaboration, and also thanks Drs. M.Tachiya, K.Kitahara, K.Kaneko, J.M. Warman, M.Hayashi, N.Kouchi, and M.Ukai for helpful discussions.

REFERENCES

1. Y.Hatano and H.Shimamori, Electron Attachment in Dense Gases, in: "Electron and Ion Swarms", ed., L.G.Christophorou, Pergamon Press, Oxford (1981), p.103.
2. J.M.Warman, The Dynamics of Electrons and Ions in Nonpolar Liquids, in: "The Study of Fast Processes and Transient Species by Electron Pulse Radiolysis", NATO Advanced Study Institute Series C86, eds., J.H. Baxendale and F.Busi, D.Reidel Publ. Co., Dordrecht (1982), p.433.
3. Y.Hatano, Electron Attachment to van der Waals Molecules, in: Electronic and Atomic Collisions", eds., D.C.Lorents, W.F.Meyerhof, and J.R. Peterson, Elsevier, New York (1986), p.153.
4. G.R.Freeman, Ionization and Charge Separation in Irradiated Materials, in: "Kinetics of Nonhomogeneous Processes", ed., G.R.Freeman, John Wiley and Sons Inc., New York (1987), p.19.
5. R.A.Holroyd, The Electron: Its Properties and Reactions, in: "Radiation Chemistry", eds., Farhataziz and M.A.J.Rodgers, VCH Publ. Inc., New York (1987), p.201.
6. R.A.Holroyd and W.F.Schmidt, Transport of Electrons in Nonpolar Fluids, Annu.Rev.Phys.Chem., 40, 439 (1989).
7. L.G.Christophorou and K.Siomos, Interphase Physics: Linking Knowledge on Electron-Molecule Interactions in Gases to Knowledge on Such Processes in Condensed Matter, in: "Electron-Molecule Interactions and Their Applications", Vol.2, ed., L.G.Christophorou, Academic Press, New York (1984), p.221.
8. R.Johnsen and H.S.Lee, Recent Advances in High-Pressure Swarms, in: "Swarm Studies and Inelastic Electron-Molecule Collisions", eds., L.C. Pitchford, B.V.Mckoy, A.Chutjian, and S.Trajmar, Springer-Verlag, New York (1985), p.23.

9. W.L.Morgan, Electron-Ion Recombination in High-Pressure Gases, ibid., p.43.

10. L.Sanche, Primary Interactions of Low Energy Electrons in Condensed Matter, in: "Excess Electrons in Dielectric Media", eds., C.Ferradini and J.-P. Jay-Gerin, CRC Press, Boca Raton (1991), p.1.

11. K.Shinsaka and Y.Hatano, Electron-Ion Recombination in Dense Molecular Media, in: "Non-Equilibrium Effects in Ion and Electron Transport", eds., J.W.Gallagher, D.F.Hudson, E.E.Kunhardt, and R.J.Van Brunt, Plenum Press, New York (1990), p.275.

12. K.Shinsaka and Y.Hatano, Electron-Ion Recombination in Condensed Non-polar Media, Nucl.Instr.Methods, $\underline{A327}$, 7 (1993).

13. L.G.Christophorou, D.L.MaCorkle, and A.A.Christodoulides, Electron Attachment Processes, in: "Electron-Molecule Interactions and Their Applications", Vol.1, ed., L.G.Christophorou, Academic Press, New York (1984), p.478.

14. Y.Hatano, Pulse Radiolysis in the Gas Phase, in: "Pulse Radiolysis", ed. Y.Tabata, CRC Press, Boca Raton (1991), p.199.

15. Y.Itikawa, A.Ichimura, K.Onda, K.Sakimoto, K.Takayanagi, Y.Hatano, M. Hayashi, H.Nishimura, and S.Tsurubuchi, Cross Sections for Collisions of Electrons and Photons with Oxygen Molecules, J.Phys.Chem.Ref.Data, $\underline{18}$, 23 (1989).

16. T.D.Mark, K.Leiter, W.Ritter, and A.Stamatovic, Low-Energy-Electron Attachment to Oxygen Clusters Produced by Nozzle Expansion, Phys.Rev. Lett., $\underline{55}$, 2559 (1985); A.Stamatovic, Electron Attachment to van der Waals Clusters: the Zero Energy Resonance, in: "Electronic and Atomic Collisions", eds., H.B.Gilbody, W.R.Newell, F.H.Read, and A.C.H.Smith, North Holland, Amsterdam (1988), p.729.

17. J.B.A.Mitchell and F.B.Yousif, Marged Beam Studies of Dissociative Recombination — Recent Results, in : "Dissociative Recombination: Theory, Experiment, and Applications", eds., J.B.A.Mitchell and S.L. Guberman, World Scientific, Singapore (1989), p.109.

18. Y.-J.Shiu, M.A.Biondi, and D.P.Sipler, Dissociative Recombination in Xenon: Variation of the Total Rate Coefficient and Excited-State Production with Electron Temperature, Phys.Rev., $\underline{A15}$, 494 (1977).

19. D.Smith and N.G.Adams, SIFT and FALP Determinations of Ionic Reactions Rate Coefficients, in: "The Physics of Electronic and Atomic Collisions XVI International Conference, New York, NY 1989", AIP Conference Proceedings 205, eds., A.Dalgarno, R.S.Freund, P.M.Koch, M.S.Lubell, and T.B.Lucatorto, AIP, New York (1990), p.325.

20. M.R.Flannery, Electron-Ion and Ion-Ion Recombination Processes, in: Adv.Atom.Molec.Opt.Phys., $\underline{32}$ (1993), in press.

21. E.S.Sennhauser, D.A.Armstrong, and J.M.Warman, The Temperature Dependence of Three-Body Electron Ion Recombination in Gaseous H_2O, NH_3 and CO_2, Radiat.Phys.Chem., $\underline{15}$, 479 (1980).

22. J.M.Warman, E.S.Sennhauser, and D.A.Armstrong, Three Body Electron-Ion Recombination in Molecular Gases, J.Chem.Phys., $\underline{70}$, 995 (1979).

23. R.J. Van Sonsbeek, R.Cooper, and R.N.Bhave, Pulse Radiolysis Studies of Ion-Electron Recombination in Helium. Pressure and Temperature Effects, J.Chem.Phys., $\underline{97}$, 1800 (1992).

24. R.Cooper, M.Burgers, R.Bhave, R.Van Sonsbeek, K.Caulfield, and J.Lowke, Mesurement of Three Body Ion-Electron Recombination Rate Constants in

Rare Gases at Moderate Pressures, in: "Gaseous Electronics and Its Applications", eds., R.W.Crompton, M.Hayashi, D.E.Boyd, and T.Makabe, KTK Sci.Publ., Tokyo (1991), p.73.

25. T.Tezuka, H.Namba, Y.Nakamura, M.Chiba, K.Shinsaka, and Y.Hatano, Free-Ion Yields, Electron Mobilities and Electron-Ion Recombination Rate Constants in Liquid and Solid Isooctane, Radiat.Phys.Chem., 21, 197 (1983).

26. Y.Nakamura, K.Shinsaka, and Y.Hatano, Electron Mobilities and Electron-Ion Recombination Rate constants in Solid, Liquid, and Gaseous Methane, J.Chem.Phys., 78, 5820 (1983).

27. J.M.Warman, Concerning Electron-Ion Recombination and Electron Thermalization in Liquid and Solid Methane, J.Phys.Chem., 87, 4353 (1983).

28. W.L.Morgan and J.N.Bardsley, Monte Carlo Simulation of Electron-Ion Recombination at High Pressure, Chem.Phys.Lett., 96, 93 (1983).

29. M.Tachiya, Influence of the Mean Free Path of Reactant Particles on the Kinetics of Diffusion-Controlled Reactions. II. Rate of Bulk Recombination, J.Chem.Phys., 84, 6178 (1986).

30. W.L.Morgan, Molecular Dynamics Simulation of Geminate Recombination by Electrons in liquid Methane, J.Chem.Phys., 84, 2298 (1986).

31. M.Tachiya, Breakdown of the Debye Theory of Bulk Ion Recombination, J. Chem.Phys., 87, 4108 (1987).

32. M.Tachiya, Effect of an External Electric Field on the Rate of Diffusion-Controlled Reactions, J.Chem.Phys., 87, 4622 (1987).

33. K.Kaneko, Y.Usami, and K.Kitahara, Gas Kinetic Approach for Electron Mobility in Dense Media, J.Chem.Phys., 89, 6420 (1988).

34. K.Kaneko, J.Takimoto, Y.Usami, and K.Kitahara, Gas Kinetic Approach for Electron-Ion Recombination Rate Constant in Dense Media, J.Phys.Soc. Jpn., 59, 56 (1990).

35. M.A.Lopez-Quintela, M.C.Bujan-Nunes, and J.C.Perez-Moure, Influence of the Fractal Geometry of Trajectories on the Rate of Diffusion-Controlled Bulk Ion Recombination, J.Chem.Phys., 88, 7478 (1988).

36. M.Sceats, Ion-Ion Recombination in Electronegative Gases, J.Chem.Phys., 90, 2666 (1989).

37. A.Mozumder, Influence of Fractal Geometry on Geminate Escape Probability, Mean Recombination Time, and Homogeneous Reaction Rate, J.Chem. Phys., 92, 1015 (1990).

38. M.Tachiya, Effect of an External Electric Field on Ion-Ion Recombination, Abstracts of Japanese Annual Symposium on Molecular Structure, 487 (1990).

39. M.A.Lopez-Quintela and M.C.Bujan-Nunes, Computer Simulation of Partially Diffusion-Controlled Reactions, Chem.Phys., 157, 307 (1991).

40. A.Mozumder, Comment on: Electron-Ion Recombination Rate Constants in Gaseous and Liquid Krypton, J.Chem.Phys., 98, 8347 (1993).

41. K.Shinsaka, M.Codama, T.Srithanratana, M.Yamamoto, and Y.Hatano, Electron-Ion Recombination Rate Constants in Gaseous, Liquid, and Solid Argon, J.Chem.Phys., 88, 7529 (1988).

42. K.Shinsaka, M.Codama, Y.Nakamura, K.Serizawa, and Y.Hatano, Electron-Ion Recombination Rate Constants in Liquid and Dense Gaseous Methane, Radiat.Phys.Chem., 34, 519 (1989).

43. K.Honda, K.Endou, H.Yamada, K.Shinsaka, M.Ukai, N.Kouchi, and Y.Hatano, Electron-Ion Recombination Rate Constants in Gaseous and Liquid Krypton, J.Chem.Phys., 97, 2386 (1992).

44. K.Honda, K.Endou, H.Yamada, K.Isoda, K.Shinsaka, M.Ukai, N.Kouchi, and Y.Hatano, Reply to Comment on: Electron-Ion Recombination Rate Constants in Gaseous and Liquid Krypton, J.Chem.Phys., $\underline{98}$, 8348 (1993).

45. T.Wada, K.Shinsaka, H.Namba, and Y.Hatano, Electron Reactivity in Liquid Hydrocarbon Mixtures, Can.J.Chem., $\underline{55}$, 2144 (1977).

46. P.G.Fuochi and G.R.Freeman, Molecular Structure Effects on the Free-Ion Yields and Reaction Kinetics in the Radiolysis of the Methyl-Substituted Propanes and Liquid Argon: Electron and Ion Mobilities, J.Chem.Phys., $\underline{56}$, 2333 (1972).

47. A.O.Allen and R.A.Holroyd, Chemical Reaction Rates of Quasi-free Electrons with Oxygen Molecules, J.Phys.Chem., $\underline{78}$, 796 (1974).

48. J.H.Baxendale, C.Bell, and P.Wardman, Observations on Solvated Electrons in Aliphatic Hydrocarbons at Room Temperature by Pulse Radiolysis, J.Chem.Soc. Faraday Trans. I, $\underline{69}$, 776 (1973).

49. B.S.Yakovlev, I.A.Boriev, L.I.Novikova, and E.I.Frankevich, Capture and Recombination Reactions of Electrons in Liquid Cyclohexane, Int.J. Radiat.Phys.Chem., $\underline{4}$, 395 (1972).

50. H.Namba, K.Shinsaka, and Y.Hatano, Effect of n-Butane Impurity on Electron Mobility and Electron-Ion Recombination Rate Constant in Solid Neopentane, J.Chem.Phys., $\underline{70}$, 5331 (1979).

51. Y.Nakamura, H.Namba, K.Shinsaka, and Y.Hatano, Electron Mobilities in Solid Tetrametylsilane: Effect of Liquid-Solid Phase Change, Chem.Phys. Lett., $\underline{76}$, 311 (1980).

52. K.Shinsaka, Y.Nakamura, K.Endou, K.Honda, H.Yamada, K.Isoda, M.Ukai, N. Kouchi, and Y.Hatano, Free-Ion Yields, Electron Mobilities and Electron-Ion Recombination Coefficients in Liquid Tetramethysilane, Nucl.Instr.Methods, $\underline{A327}$, 15 (1993).

53. S.Kubota, M.Hishida, M.Suzuki, and J.Ruan (Gen), Dynamical Behavior of Free Electrons in the Recombination Process in Liquid Argon, Krypton, and Xenon, Phys.Rev., $\underline{B20}$, 3486 (1979).

54. J.H.Baxendale and E.J.Rasburn, Pulse Radiolysis Study of Kinetics of Electron Reactions in Liquid n-Hexane at Room Temperature, J.Chem.Soc. Faraday Trans. I, $\underline{70}$, 705 (1974).

55. J.H.Baxendale, J.P.Keen, and E.J.Rasburn, Conductimetric Study of the Kinetics of Electrons in Pulse-Irradiated n-Hexane and Cyclohexane at Room Temperature, J.Chem.Soc. Faraday Trans. I, $\underline{70}$, 718 (1974).

56. K.Isoda, N.Kouchi, Y.Hatano, and M.Tachiya, Effect of an External Electric Field on Diffusion-Controlled Ion Recombination, Chem.Phys., to be published.

FALP STUDIES OF ELECTRON-ION RECOMBINATION

AND ELECTRON ATTACHMENT

David Smith and Patrik Spanel

Institut für Ionenphysik
Universität Innsbruck
Technikerstraße 25
A-6020 Innsbruck, Austria

THE FALP TECHNIQUE AND SOME PREVIOUS RESULTS

The flowing afterglow/Langmuir probe (FALP) technique has become a standard method for studying plasma reaction processes under thermal conditions, i.e. when the temperatures, T, of the composite electron, ion and neutral gases T_e, T_i and T_g, respectively are equal[1,2]. This is the situation that can readily be established in flowing afterglow plasmas in helium carrier gas because momentum transfer in electron-helium atom collisions is very efficient (this is not so for electron-argon collisions; see below). The concept of the method is simple. A microwave cavity discharge is created in the upstream region of the flowing helium (pressure ~1 Torr) and, due to the flow (velocity ~10^4 cms^{-1}), an afterglow plasma is established downstream, in which $T=T_g=T_e=T_i$ when the helium metastable atoms have been removed. (In practice, this is achieved by reacting the metastable atoms with argon which is introduced into the carrier gas upstream at low partial pressure (~1 mtorr)). T can be varied between 80 and 600 K by cooling and heating the whole flow tube (~1 metre long and ~8 cm diameter). A small cylindrical Langmuir probe can be positioned at any point along the axis of the flow tube to measure the electron and ion number densities, n_e and n_i, and T_e (and in a recent development, the electron energy distribution function, $f(E)$; see below). A small orifice in a nosecone positioned downstream samples the positive and negative ions from the plasma whence they pass into a mass spectrometer/detector system.

Specially-shaped gas entry ports are located at several positions along the flow tube to allow the introduction of measured amounts of reactant gases into the afterglow plasma. Hence the initial non-recombining atomic ions He$^+$ and Ar$^+$ (which are lost to the walls of the flow tube only by the relatively slow process of ambipolar diffusion) can be totally converted by ion-molecule reactions to molecular ions which usually recombine rapidly with electrons. So, for example, the addition of O_2 and NO to the afterglow results in the production of O_2^+ and NO$^+$ plasmas and then a measurement of the enhanced gradient of n_e along the flow tube (coordinate z), i.e. dn_e/dz, readily provides values for the dissociative recombination coefficients $\alpha_e(O_2^+)$ and $\alpha_e(NO^+)$. Thus these α_e and those for many other ions have been determined, including those for several interstellar ions[3,4] and ionospheric ions[5], some over the accessible T range from 80 to 600 K. Our study of $\alpha_e(H_3^+)$ has been particularly thorough (see below).

The addition of an electron attaching gas to the afterglow also results in an increase of dn_e/dz as electron attachment occurs and so attachment coefficients, β, can readily be determined. In this way, β have been determined over a range of T from 200 to 600 K for several species[6,7] most of which are dissociative attachment reactions (e.g. the Freons CCl_4, CCl_2F_2, CF_3Br etc.) and some are direct attachment reactions in which the parent negative ion is produced (e.g. SF_6^- and $C_7F_{14}^-$). The product ions are identified using the downstream mass spectrometer. Some of these reactions are slow at 300 K but become faster with increasing T. The measurements of β as a function of T then provides values for the "activation energies", E_a. An exceptional attachment reaction is that of C_6F_6[8], since β reduces rapidly with increasing T from a value at 300 K near to the upper limit value β_{max} (according to s-wave capture theory[9]). This is attributed[10] to an increase in the autodetachment rate with increasing T. Other notable FALP studies have included the first measurements of dissociative attachment to the molecular radicals CCl_3 and CCl_2Br[11], the observation of Br_2^- as a product of attachment to some dibromo organic molecules[12], measurements of the dissociative attachment coefficients for some acids and superacids[13] and dissociative attachment (and the reverse associative detachment) reactions of HBr and HI[14]. Our recognition that β for some reactions obtained using the thermal FALP method and two non-thermal methods were very different[15] was a stimulation for our new developments outlined in the next section. A review of these FALP studies of electron attachment is presented in our very recent paper[16].

If sufficient electron attaching gas is added to the afterglow then a positive ion/negative ion plasma can be created at high n_i and recombination (mutual neutralisation) of the positive and negative ions can be studied. In this situation, the positive and negative ion number density gradients are measured using the Langmuir probe and the ionic recombination coefficients, α_i, are obtained[17]. The determination of α_i for many reactions including those for 'simple ions' (e.g. $NO^+ + NO_2^-$) and 'cluster' ions (e.g. $H_3O^+ \cdot (H_2O)_3 + NO_3^- \cdot (HNO_3)$) were our first FALP studies of ionic recombination[18] yet they still remain the most thorough study of binary ionic recombination reactions. For one reaction ($NO^+ + NO_2^-$), the neutral products were identified by observing the radiation emmision in the reaction[19].

RECENT RESULTS AND NEW DEVELOPMENTS OF THE FALP TECHNIQUE

(a) Recombination of H_3^+ under Thermalised Conditions (Helium Carrier Gas)

Before outlining our new developments of the FALP technique and some interesting subsequential results, we describe our further studies of $\alpha_e(H_3^+)$ under thermalised conditions ($T_e = T_i$). This has become a controversial topic in recent years, but measurements of $\alpha_e(H_3^+)$ have been attempted for decades using various techniques. The first measurement[20] at 300 K indicated $\alpha_e(H_3^+)$ to be small ($\sim 3 \times 10^{-8} cm^3 s^{-1}$); the next[21] indicated a larger value ($\sim 3 \times 10^{-7} cm^3 s^{-1}$). The first FALP study[22] indicated a small value ($\leq 2 \times 10^{-8}$) $cm^3 s^{-1}$. Theoretical work has consistently predicted a small value for α_e for ground vibrational state ions, $\alpha_e(H_3^+(v=0))$. This apparent conflict is most unsatisfactory in view of the importance of H_3^+ in interstellar chemistry, and so we have been constrained to carry out a further thorough investigation of the H_3^+ recombination reaction. The detailed results and conclusions are reported in two papers[23,24] and so only the final results will be given here.

The essential requirements for the experiment are that the gases used in the experiment, especially the helium carrier gas, are ultra-pure because of the efficient proton transfer that occurs from H_3^+ to almost any gas. If this occurs it inevitably produces rapidly recombining ions (N_2H^+ and H_3O^+ are especially troublesome). Also observations of n_e must be made late into the afterglow when vibrationally-excited H_3^+ ions, i.e. $H_3^+(v)$, have been removed from the plasma either by rapid recombination or by vibrational relaxation in collisions with H_2 molecules (ultimately producing $H_3^+(v=0)$). Now when n_e is measured along z in the H_3^+/electron plasma generated from the reactions $Ar^+ + H_2 \rightarrow ArH^+ + H$, and $ArH^+ + H_2 \rightarrow H_3^+ + Ar$, an initial rapid decline of n_e is observed which we attribute to the fast recombination of $H_3^+(v)$ as predicted by theory[25], followed by an increasingly slower decline, that we attribute to the very slow recombination of the remaining $H_3^+(v=0)$ ions. This unusual form of the n_e versus z decay contrasts with that obtained in an O_2^+ plasma from which a z-independent value of $\alpha_e(O_2^+)$ is obtained. The α_e can be obtained from the n_e versus z data by analytically solving the equation describing dn_e/dz, i.e. $dn_i/dt = D_a \nabla^2 n_e - \alpha_e n_e^2$, for finite steps of z. Thus α_e is obtained as a function of z. In Fig. 1 are the data thus obtained in the O_2^+ and H_3^+ plasmas at T=300 K. Note the rapid rise (in the O_2 mixing zone) of $\alpha_e(O_2^+)$ to a constant value of $1.9 \times 10^{-7} cm^3 s^{-1}$, a value which is in excellent agreement with that obtained by other techniques[5]. However, $\alpha_e(H_3^+)$ first rises to a smaller value than $\alpha_e(O_2^+)$ but then falls to a much smaller value. Our final value of $(1-2) \times 10^{-8} cm^3 s^{-1}$ at 300 K which is consistent with our much earlier upper-limit value of $\leq 2 \times 10^{-8} cm^3 s^{-1}$ obtained in the FALP[22] has been obtained from many experiments. We have presented more data and argued forcibly in our recent papers[23,24] that this low value for $\alpha_e(H_3^+)$ is appropriate to ground state $H_3^+(v=0)$ ions and that the previously reported higher values are due variously to the presence of $H_3^+(v)$ ions and the failure to account for the rapid loss of $H_3^+(v=0)$ ions by high-order diffusion modes.

(b) Studies under Non-Thermal Conditions (Argon Carrier Gas)

All the work described above was carried out using helium carrier gas in which $T_e = T_i = T_g$, and the T_e were measured using the laborious and slow method of constructing semilogarithmic plots of the electron current to the probe as a function of the probe potential. Now we have developed a sophisticated technique by which the current to the probe is accumulated into a computer (via an A/D converter) and by the rapid procesing of the data (using a noise reduction technique to obtain the second derivatives) the electron energy distribution function, $f(E)$, in the afterglow plasmas can be obtained in a few seconds[26]. Hence in a Maxwellian plasma, T_e, can rapidly be obtained. Thus we have shown by numerous measurements that in the helium afterglow T_e is closely equal to T_g, but in the afterglow created in argon carrier gas T_e can greatly exceed T_g (and therefore T_i), this being due to the deep Ramsauer minimum in the cross section for electron-argon atom collisions. Further, we have shown that $f(E)$ is close to Maxwellian in both the helium and argon afterglows provided that n_e exceeds $10^7 cm^{-3}$ when electron-electron collisions efficiently randomise the velocity distribution[27]. So, using argon carrier gas, dissociative recombination and electron attachment can now be studied using the FALP apparatus over wide ranges of both T_g and T_e, the latter within the range from T_g to ~5000 K, and so the separate influences of T_g and T_e on the rates of these processes can be investigated.

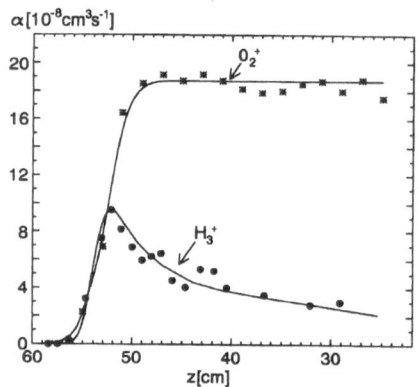

Figure 1. The recombination coefficients $\alpha(H_3^+)$ and $\alpha(O_2^+)$ determined as a function of axial position, z, along the flowing afterglow plasma (equivalent to the time variation). Note that $\alpha(O_2^+)$ is essentially independent of z (because only $O_2^+(v=0)$ ions are present), whereas $\alpha(H_3^+)$ varies with z (because a mixture of vibrationally excited H_3^+ and $H_3^+(v=0)$ ions is present). The data are taken from reference 24.

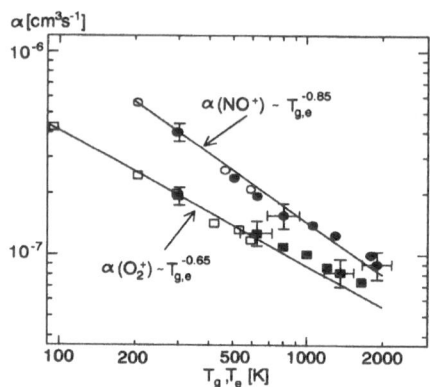

Figure 2. The recombination coefficients $\alpha(O_2^+)$ and $\alpha(NO^+)$ as a function of both T_g ($=T_e=T_i$) (open symbols) and T_e (for $T=T_i=300$ K) (filled symbols). Note that the power-law variations with T_g and T_e (indicated as $T_{g,e}$ for both $\alpha(O_2^+)$ and $\alpha(NO^+)$) are essentially the same. The data are taken from reference 28.

The following are summaries of our first experiments using these new features of the FALP apparatus.

(i) Dissociative recombination of O_2^+ and NO^+. $\alpha_e(O_2^+)$ has been measured using several techniques at 300 K and its value is well-established (see above); similarly, $\alpha_e(NO^+)$ is well established as 4×10^{-7}cm^{3}s^{-1} at 300 K. Both of these ions are major species in the terrestrial ionosphere and so it is important to know how $\alpha_e(O_2^+)$ and $\alpha_e(NO^+)$ vary with T_i and T_e. Previous FALP studies[5] over the T ($=T_e=T_i$) range of ~100 - 600 K established that both these coefficients varied in a power law fashion viz, $\alpha_e(O_2^+)\sim T^{-0.7}$ and $\alpha_e(NO^+)\sim T^{-0.9}$ in accordance with other measurements. Now we have determined $\alpha_e(O_2^+)$ and $\alpha_e(NO^+)$ over the T_e range from 300 to 2000 K at a T_g of 300 K. The maximum obtainable value of T_e (2000 K) is limited by the 'cooling' of the electron gas by the O_2 and NO which have to be introduced into the carrier gas to create the O_2^+ and NO^+ plasmas. $\alpha_e(O_2^+)$ and $\alpha_e(NO^+)$ are seen to vary with T_e in essentially the same power-law fashion as for T as can be seen in Fig. 2. The implication of these results is that T_e (not T_i) primarily controls $\alpha_e(O_2^+)$ and $\alpha_e(NO^+)$ in this T_i regimé in accordance with theoretical expectations. More experimental details and further discussion are given in our very recent paper[28].

(ii) Dissociative Electron Attachment to CF_3Br. As alluded to above, our FALP measurements of β for some halomethanes and haloethanes revealed astonishing differences between the variation of β with T and the β derived from the non-thermal swarm and Kr photoionisation methods[15]. We attributed these differences to the fact that the attaching molecule temperature was held constant and the electron energy was varied in the latter

490

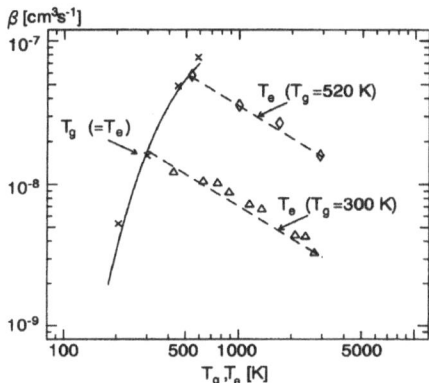

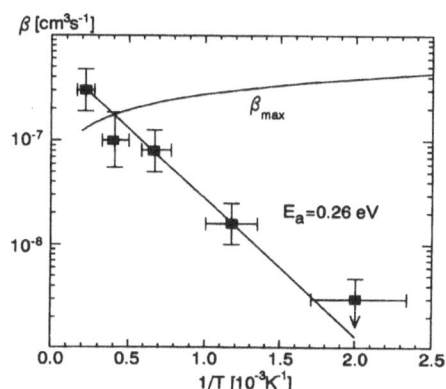

Figure 3. The dissociative electron attachment coefficient, β, for CF_3Br as a function of T_g ($=T_e$) obtained in helium carrier gas, and as a fuction of T_e obtained in argon carrier gas at fixed T_g of 300 K and 520 K. The data are taken from reference 30.

Figure 4. The direct attachment coefficient, β, for C_{60} as a function of T_e. The data are plotted in the form of an Arrhenius plot from the slope of which the 'activation energy', E_a, is obtained. Also shown is β_{max}, the 'upper limit' value of β as calculated using s-wave capture theory. The data are taken from reference 33.

experiments whereas in the FALP the T_g and T_e were varied simultaneously. So it is clear in the many attachment reactions which require "activation energy" to promote them (to increase β), that this E_a is more readily supplied by 'heating' the molecule, and that heating the electrons actually decreases β. Clearly, it is desirable to study the separate influence of T_g and T_e on β and this we can now do using the FALP apparatus. An excellent example of such a study is the dissociative electron attachment reaction of CF_3Br, the β for which increases rapidly with T as our previous FALP study had shown[7]. However, as the data in Fig. 3 show, the β actually decreases with increasing T_e at two T_g values of 300 K and 520 K, and overall β is described by the relation $\beta=8\times10^{-7}cm^3s^{-1}(T_e / 300 \text{ K})^{-0.75} \exp(-1150 \text{ K} / T_g)$. A similar behaviour must be expected for other attachment reactions as other recent experiments have shown[29]. Further details of these FALP experiments are given in a very recent paper[30].

(iii) Electron Attachment to C_{60}. The properties and reactivity of the fullerene molecule C_{60} have received an enormous amount of attention in the last few years[31]. In our Institute, electron attachment has been studied in a crossed electron-C_{60} beam apparatus and has been observed to be facile for electron energies form ~0.3 eV to ~15 eV [32]. Thus we chose to determine β for the C_{60} reaction in the FALP apparatus, firstly in the helium afterglow at 300 K. Surprisingly, we found that β was very small ($\leq 10^{-9}cm^3s^{-1}$) and so we then studied the reaction in the argon afterglow at elevated T_e. Now at $T_e=4500$ K, β was very large ($3\times10^{-7}cm^3s^{-1}$). Further studies at lower T_e produced the β values given in the Arrhenius-type plot shown in Fig. 4. These data indicate that there exists an "activation energy barrier" of 0.26 eV for electron attachment to C_{60}. At a T_e =4500 K, the mean thermal electron attachment cross section is ~80 $Å^2$ which exceeds the maximum value expected on the basis of s-wave capture theory (i.e. 30 $Å^2$ at this mean electron energy) by an amount (~50 Å^2) equal to the cross section of the C_{60} molecule! Thus the C_{60} molecule is an "electron sponge" for electrons of energy ≥ 0.5 eV and up to about 10 eV. An interesting experimental point concerns the determination of the number density of C_{60} molecules in the plasma (which is required for a determination of β). This was obtained by observing the departure from first order kinetics in the n_e loss rate at different initial n_e. This technique was conceived from our previous FALP studies of electron attachment to molecular radicals[11]. Further details of this C_{60} work are given in our recent paper[33].

REFERENCES

1. D. Smith and A.G. Dean. Electron temperature relaxation rates in flowing afterglow plasmas containing molecular nitrogen and oxygen. *J. Phys. B* 8, 997 (1975).

2. D. Smith, N.G. Adams, A.G. Dean and M.J. Church. The application of Langmuir probes to the study of flowing afterglow plasmas. *J. Phys. D* 8, 141 (1975).

3. N.G. Adams, D. Smith and E. Alge. Measurements of the dissociative recombination coefficients of H_3^+, HCO^+, N_2H^+ and CH_5^+ at 95 and 300K using the FALP apparatus. *J. Chem. Phys.* 81, 1778 (1984).

4. N.G. Adams and D. Smith. Measurements of the dissociative recombination coefficients for several polyatomic ion species at 300K. *Chem. Phys. Letters* 144, 11 (1988).

5. E. Alge, N.G. Adams and D. Smith. Measurements of the dissociative recombination coefficients of O_2^+, NO^+ and NH_4^+ in the temperature range 200-600K. *J. Phys. B* 16, 1433 (1983).

6. D. Smith, N.G. Adams and E. Alge. Atttachment coefficients for the reactions of electrons with CCl_4, CCl_2F_2, $CHCl_3$, Cl_2 and SF_6 determined between 200 and 600K using the FALP technique. *J. Phys. B* 17, 461 (1984).

7. E. Alge, N.G. Adams and D. Smith. Rate coefficients for the attachment reactions of electrons with c-C_7F_{14}, CH_3Br, CH_2Br_2, and CH_3I determined between 200 and 600K using the FALP technique. *J. Phys. B* 17, 3827 (1984).

8. N.G. Adams, D. Smith, E. Alge and J. Burdon. Anomolous temperature dependence of the coefficient of electron attachment to hexafluorobenzene. *Chem. Phys. Letters* 116, 460 (1985).

9. C. E. Klots. Rate constants for unimolecular decomposition at threshold. *Chem. Phys. Letters* 38, 62 (1976).

10. P.G. Datskos, L.G. Christophorou and J.G. Carter. Temperature-enhanced autodetachment from c-$C_4F_6^-$ *Chem. Phys. Letters* 195, 329 (1992).

11. N.G. Adams, D. Smith and C.R. Herd. Studies of dissociative electron attachment to the free radicals CCl_3 and CCl_2Br using the FALP apparatus. *Int. J. Mass Spectrom. Ion Processes* 84, 243 (1988).

12. D. Smith, C.R. Herd, N.G. Adams and J.F. Paulson. Formation of Br_2^- in the reactions of thermal electrons with some bromomethanes and bromoethanes. *Int. J. Mass Spectrom. Ion Processes* 96, 341 (1990).

13. N.G. Adams, D. Smith, A.A. Viggiano, J.F. Paulson and M.J. Henchman. Dissociative attachment reactions of electrons with strong acid molecules. *J. Chem. Phys.* 84, 6728 (1986).

14. D. Smith and N.G. Adams. Studies of the reactions HBr(HI) + e == Br$^-$(I$^-$) + H using the FALP and SIFT techniques. *J. Phys. B* 20, 4903 (1987).

15. D. Smith, C.R. Herd and N.G. Adams. Studies of dissociative electron attachment to some haloethanes using the FALP apparatus: comparisons with data obtained using non-thermal techniques. *Int. J. Mass Spectrom. Ion Processes* 93, 15 (1989).

16. D. Smith and P. Španel. Studies of electron attachment at thermal energies using the FALP technique. *Adv. Atom. Mol. Phys.* 32, 307 (1993).

17. D. Smith and M.J. Church. Binary ion - ion recombination coefficients determined in a flowing afterglow plasma. *Int. J. Mass Spectrom. Ion Physics* 19, 185 (1976).

18. D. Smith and N.G. Adams. Studies of ion - ion recombination using flowing afterglow plasmas. *in*: "Physics of Ion-Ion and Electron-Ion Collisions," F. Brouillard and J.W. McGowan, eds., Plenum Press, New York, pp. 501-531 (1983)

19. D. Smith, N.G. Adams and M.J. Church. On the emission of radiation from the neutralisation reaction $NO^+ + NO_2^-$. *J. Phys. B* 11, 4041 (1978).

20. K-B. Persson and S.C. Brown. Electron loss processes in the hydrogen afterglow. *Phys. Rev.* 100, 729 (1955).

21. M. T. Leu, M.A. Biondi and R. Johnsen. Measurements of recombination of electrons with H_3^+ and H_5^+ ions. *Phys. Rev. A* 8, 413 (1973).

22. N.G. Adams, D. Smith and E. Alge. Measurements of the dissociative recombination coefficients of H_3^+, HCO^+, N_2H^+, and CH_5^+ at 95 and 300K using the FALP apparatus. *J. Chem. Phys.* 81, 1778 (1984).

23. D. Smith and P. Španel. Dissociative recombination of H_3^+. Experiment and theory reconciled. *Chem. Phys. Letters 211, 454* (1993).

24. D. Smith and P. Španel. Dissociative recombination of H_3^+ and some other interstellar ions: a controversy resolved. *Int. J. Mass Spectrom. Ion Processes* (1993). In Press.

25. H.H. Michels and R.H. Hobbs. Low temperature dissociative recombination of $e + H_3^+$. *Astrophys. J. Lett.* 286, L27 (1984).

26. P. Španel and D. Smith. A technique for the rapid determination of electron energy distribution functions in afterglow plasmas. Paper in preparation.

27. D. Trunec, P. Španel and D. Smith. Electron temperature relaxation in afterglow plasmas: diffusion cooling. *Contributions to Plasma Physics* (1993). In Press.

28. P. Španel, L. Dittrichová and D. Smith. FALP studies of the dissociative recombination coefficients for O_2^+ and NO^+ within the electron temperature range 300 to 2000K. *Int. J. Mass Spectrom. Ion Processes* (1993). In Press.

29. H. Shimamori, Y. Tatsumi, Y. Ogawa and T. Sunagawa. Low energy electron attachment to molecules studied by pulse-radiolysis microwave-cavity technique combined with microwave heating. *J. Chem. Phys.* 97, 6335 (1992).

30. P. Španel and D. Smith. FALP studies of electron attachment at elevated electron temperatures: the influence of attachment on electron energy distributions. *Int. J. Mass Spectrom. Ion Processes* (1993). In Press.

31. See the papers in the Special Issue of *Phil. Trans. Roy. Soc. London. A* 343 (1993).

32. M. Lezius, P. Scheier and T.D. Märk. Free electron attachment to C_{60} and C_{70}. *Chem. Phys. Letters* 203, 232 (1993).

33. D. Smith, P. Španel and T.D. Märk. Electron attachment to C_{60} at low energies. *Chem. Phys. Letters* 213, 202 (1993).

THE THEORY OF ELECTRON-ION RECOMBINATION

J. NORMAN BARDSLEY

LAWRENCE LIVERMORE NATIONAL LABORATORY
LIVERMORE, CALIFORNIA 94550, USA

This lecture will focus upon the recombination of electrons with molecular ions, since the interaction between electronic and nuclear motion is an important feature of molecular processes and condensed matter physics, but is not present in the recombination of electrons with atomic ions. Two processes will be considered, dissociative recombination,

$$e + AB^+ \; \rightarrow \; AB^{**} \; \rightarrow \; A + B \tag{1}$$

which has recently been reviewed by Mitchell (1990), and three body-recombination

$$e + AB^+ + C \; \rightarrow \; AB + C \tag{2}$$

where the third body C could be an electron or another molecule.

The theory of dissociative recombination was first presented by Bates (1950). In the simplest or "direct" process, the electron loses energy through electronic excitation of the target ion and falls into a bound orbital. The potential energy curve of the resulting molecular state AB^{**} is usually repulsive, the nuclei are forced apart and two neutral fragments are produced. Recombination is most rapid when there are intermediate states AB^{**} with potential curves that intersect that of the molecular ion near the equilibrium nuclear separation. This direct mechanism was described in terms of resonance scattering theory by Bardsley (1968), who also pointed out the possibility of an indirect mechanism involving Rydberg molecular states. At low energies, the incident electron can also lose energy through rotational or vibrational excitation of the ion. Because the rotational and vibrational energy spacings are small, the electron will fall into a Rydberg orbital. The intermediate state is then a rotationally or vibrationally excited Rydberg state. If this state decays by predissociation, the recombination is completed. However, the intermediate state can also decay by autoionization, giving back an electron-ion pair. O'Malley (1981) realized that, because of the competition between these two decay mechanisms, the presence of the Rydberg states usually reduces the recombination cross section, producing narrow window resonances. This prediction has been confirmed by high-resolution experiments and calculations using multi-channel quantum defect theory (Giusti,1980; Greene and Jungen 1985).

Three features will be stressed as we survey recent studies of dissociative recombination. The first is the magnitude of the cross section. At low energies, the cross section σ for the direct process varies as $1/E$, following the Wigner threshold law. For many diatomic molecules, such as the atmospheric or rare gases other than He, σE is of order 10^{-15} cm^2 eV. For these molecules, there appear to be several intermediate states that contribute to the process. The second feature is the sensitivity of the cross section to the vibrational state of the initial ion. This is not strong when there are several intermediate states with favorable potential curves.

Finally, since recombination often leads to the formation of excited states, determination of the dissociation products is of great interest.

For the simplest ion, H_2^+, the dominant intermediate state is the $(1\sigma_u)^2 \, ^1\Sigma_g$ state, with a potential curve which crosses the ionic curve near the extremity of the Franck-Condon region of the first excited vibrational level. The magnitude of the cross section for ground state ions is relatively small, with σE below 10^{-16} cm^2 eV (Nakashima et al. 1987, Van der Donk et al. 1991, Schneider et al. 1991). The experiments show three window resonances below 0.1 eV, but there is disagreement over the identification of the responsible Rydberg levels. The cross section for ions with $v = 1$ are higher, due to the larger Franck-Condon factors.

Other molecules with low recombination rates for ground state ions are HeH and H_3. For these molecules there appear to be no intermediate states suitable for the direct recombination process, and an alternative mechanism may be required to explain the observed rates (Mitchell,1990). Bates (1992) has argued that the key to the recombination is the exchange of energy between electronic and nuclear motion, and that several intermediate states may be involved. In these cases also, the recombination rate should be considerably higher for excited vibrational states of the ions.

Knowledge of the products of dissociative recombination is important for the analysis of rare gas and excimer lasers, and for atmospheric and astrophysical applications. For example, dissociative recombination provides an important source of $O(^1S)$ atoms in the ionosphere. However, the branching ratio for $O(^1S)$ production is calculated to be very small for ground state ions (Guberman and Giusti-Suzor, 1991). In contrast, the production of $O(^1D)$ is efficient for all ionic vibrational states (Guberman,1988). For N_2^+, Guberman (1991) argues that the dominant dissociation channel leads to $N(^4S) + N(^2D)$. Each atom emerges with kinetic energy of over 1.72 eV, which is very close to the 1.74 eV necessary to escape from the gravitational force in the Martian atmosphere.

Studies of the branching ratios of recombination products is particularly interesting for polyatomic molecules. Although few complete calculations have been performed, Bates (1991) and Galloway and Herbst (1991) have suggested guiding principles, based upon valence structure and phase space considerations, respectively.

Measurements of molecular densities in interstellar clouds have shown that the HCO^+/CO density ratio is two orders of magnitude smaller than the corresponding ratio for HCS^+/CS. It has been suggested that this may be partially due to a very small recombination rate for HCS^+. The data on ground state ions is conflicting, with a relatively small rate of 5×10^{-8} cm^3 s^{-1} reported by Millar et al (1985) and a larger value of 2×10^{-7} cm^3 s^{-1} obtained by Abouelaziz et al (1992). A theoretical comparison of recombination in HCO^+ and HCS^+ has been made by Talbi et al (1989), but the results are not conclusive. Abouelaziz et al (1992) also suggest that the branching ratios for H atom production may be different in the two cases.

For ions to be neutralized by electron collisions without fragmentation, it is necessary that the excess energy be dissipated through photon emission, which is very unlikely, or in a collision with a third body (Bates and Khare,1965; Bates et al 1971). In 1982, Bates pointed out that dissociative recombination can also be enhanced through third-body collisions. Experiments and computer simulations (Morgan and Bardsley, 1983; Morgan, 1984) have confirmed earlier experiments showing that the recombination rate can reach peak values between 10^{-5} and 10^{-4} cm^3 s^{-1} at pressures of around 10 atmospheres. As the pressure is increased further, the electron mobility is decreased and the recombination rate falls. A summary of the theoretical and experiment results was provided by Morgan (1987).

ACKNOWLEDGEMENTS

This report was prepared under the auspices of the U. S. Department of Energy at the Lawrence Livermore National Laboratory under Contract No. W-7405-Eng-48.

REFERENCES

Abouelaziz, H., Queffelec, J.L., Rebrion, C., Rowe, B. R., Gomet, J. C., and Canosa, A., 1992, Chem. Phys. Lett. 194, 263.

Bardsley, J. N., 1968, J. Phys B 1, 349, 365.

Bates, D. R., 1950, Phys Rev. 78, 492.

Bates, D. R., 1982, Chem. Phys. Lett. 89, 294.

Bates, D. R., 1991, J. Phys. B 24, 3267

Bates, D. R., 1992, J. Phys. B 25, 5479.

Bates, D. R., and Khare, S. P., 1965, Proc. Phys. Soc. 85, 231.

Bates, D. R., Malaviya, V., and Young, N. A., 1971, Proc. Roy. Soc. Lond. A 359, 437.

Galloway, E. T., and Herbst, E., 1991, Ap. J. 376, 531.

Giusti, A., 1980, J. Phys. B 13, 3867.

Greene, C. H., and Jungen, Ch., 1985, Adv. At. Mol. Phys. 21, 51.

Guberman, S. L., 1988, Planet. Spa. Sci. 36, 47.

Guberman, S. L., 1991, Geophys. Res. Lett. 18, 1051.

Guberman, S. L. and Giusti-Suzor, A., 1991, J. Chem Phys. 2602.

Millar, T. J., Adams, N. G. Smith D., and Clary D. C., 1985, M. N. R. A. S. 216, 1025.

Mitchell, J. B. A., 1990, Phys. Rep. 186, 215.

Morgan, W. L., 1984, J. Chem. Phys. 80, 4564.

Morgan, W. L., 1987, in Recent Studies in Atomic and Molecular Processes, ed. A. E. Kingston (Plenum).

Morgan, W. L., and Bardsley, J. N., 1983, Chem. Phys. Lett. 96, 93.

Nakashima, K., Takagi, H. and Nakamura, H., 1987, J. Chem. Phys. 86, 726.

O' Malley, T. F., 1981, J. Phys. B 14, 1229.

Schneider, I. F., Dulieu, O., and Giusti-Suzor, A.,1991, J. Phys B 24, L289.

Talbi, D., Hickman, A.P., Pauzat, F., Ellinger, Y., and Berthier, G., 1989, Ap. J. 339, 231.

Van der Donk, P., Yousif F. B., Mitchell, J. B. A., and Hickman, A. P., 1991, Phys. Rev. Lett. 67, 42.

SECTION VII. ELECTRON TRANSFER AT INTERFACES

LOW ENERGY ELECTRONS FOR THE INVESTIGATION OF LIQUID SURFACES

Harald Morgner
Institut für Experimentalphysik
Universität Witten/Herdecke
Stockumer Str.10, D-58448 Witten

INTRODUCTION

Studies of liquid surfaces with electron spectroscopy are comparetively scarce even though the microscopic understanding of these systems could benefit to the same extent as is the case for surfaces of solids. The short mean free path λ_e of slow electrons in condensed matter ensures that electrons leaving the surface must originate from a site which is separated from the very surface by at most a few times λ_e. This holds for XPS(=x-ray photoelectron spectroscopy), UPS(=ultraviolet photoelectron spectroscopy), AES(=Auger electron spectroscopy) and EELS(=electron energy loss spectroscopy). Indeed, these techniques have been used by a small number of groups for the investigation of liquid surfaces. The first such study was published by H. and K.Siegbahn on formamide using XPS [1], followed by a large number of works by H.Siegbahn and coworkers on several liquid surfaces, e.g. solutions of surface active salts [2,3,4]. Two groups have made use of the HeI resonance line to perform UPS on liquid surfaces [5,6]. Ballard showed that AES and EELS (with energy losses in the eV range) could reproducibly be carried out on these systems [7] even though there remains the need to understand the precise contribution of these methods to the diagnosis of liquid surfaces. The above mentioned spectroscopies have in common that their surface sensitivity is governed by the value of the mean free path λ_e of the emitted electrons. There is, however, one electron spectroscopy that is distinguished in that its observation depth is not related to a mean free electron path. It is MIES (=Metastable Induced Electron Spectroscopy), a technique which monitors exclusively the topmost layer of a surface [8]. The energy needed for electron emission is supplied by the excitation energy (19.819 eV) of metastable helium atoms $He^{*}(1s2s, {}^{3}S)$. The surface sensitivity of MIES goes so far that even the orientation

Linking the Gaseous and Condensed Phases of Matter
Edited by L.G. Christophorou *et al*, Plenum Press, New York, 1994

of the molecules in the topmost layer can be determined[8]. We have been able
demonstrate this for the pure liquids formamide (FA)[8] and benzylalcohol
(BA) [9,10]. Other liquids like hydroxipropionitrile (HPN) do not show a preferred
orientation at the surface [11].

In the present paper we concentrate on liquids which are composed of two
components, binary mixtures and salt solutions. In both cases it is of interest to
characterize the surface composition which usually deviates from the bulk value
due to segregation. The component which shows an enhanced concentration is
called the surface active species. Its excess concentration must necessarily fall off
to zero when followed into the bulk. The profile of this excess concentration along
the surface normal can be assessed by comparing several experimental methods
with different observation depths. This will be demonstrated for the case of the
0.5m solution of tetrabutylammoniumiodide (TBAI) in formamide by evaluating
data from MIES, UPS and XPS. Further, we address to the question to which
extent the spectra contain information on the lateral structure of the surface. Computer simulations by the Monte Carlo method turn out to be of great heuristic
value in this respect. In the last two sections of the paper we present data from
Auger electron spectroscopy (AES) and electron energy loss spectroscopy (EELS)
taken at liquid surfaces and discuss their possible merits as diagnostic tools.

THEORY

We consider a liquid consisting of two components A and B. The concentration of A as function of distance z from the surface is written as

$$n_A(z) = n_A^{bulk} + n_A^{exc}(z) \tag{1}$$

with n_A^{bulk} being the known number density of A in the bulk of the mixture and
$n_A^{exc}(z)$ describing the deviation of the actual density from the bulk value. Its
value is positive for surface active species, but negative for the other component.
Electron spectroscopy with an effective observation depth λ' of the liquid leads to
a contribution from component A which is equivalent to a signal from N_A molecules of species A if no weakening of the signal via a limited free path would
occur

$$N_A = \int dz \; n_A(z) \; e^{-z/\lambda'} = \lambda' \, n_A^{bulk} + \int dz \; n_A^{exc}(z) \; e^{-z/\lambda'} \tag{2}$$

The analoguous definition shall hold for component B.

Usually, the excess concentration $n_A^{exc}(z)$ is described by an analytical ansatz.
Commonly used shapes are exponential or step functions [2,3]. As discussed by Siegbahn[3], the determination of $n_A^{exc}(z)$ without a mathematical ansatz requires experimental data of extreme precision. Our own experience supports this statement.
In order to evaluate presently available data and still not to influence the out-

come of a profile by a rigid choice of a trial function we use the ansatz

$$n_A(z) = n_A^{bulk} + \Delta z \, n_A^{exc}(k) \qquad for \qquad (k-1)\Delta z < z < k\Delta z, \; k=1,..,10 \qquad (3)$$

This histogram ansatz is rather flexible, but requires large sets of data in order to determine the increased number of parameters.

The observed ratio of the contributions of both molecules to the electron energy spectrum is described in terms of eq.(2) as

$$R = c \, N_A / N_B \qquad (4)$$

Here c accounts for different ionization cross sections of A and B. If λ' is varied via observation angle α then c remains constant as long as effects of angular distributions can be excluded. This is the case if the bands considered for A and B in the electron spectra have the same angular distribution. In case of the valence electron spectra evaluated below we found c to depend on the photon energy but not on the emission angle. The C(1s) core electron spectra we evaluate assuming c=1.

In the above formulation of the problem, the parameters $n_A^{exc}(k)$, $n_B^{exc}(k)$ and - for any energy of emitted electrons - the mean free path λ_e and eventually c have to be fitted to experimental data. These data are given as ratios according to eq.(4). A reduction of parameters can achieved by assuming the liquid to be compact, i.e.

$$n_B^{exc}(z) = - n_A^{exc}(z) \; v_A / v_B \qquad (5)$$

with v_A, v_B being the volumes of the molecules. This is an approximation that appears necessary at the moment, but is not to severe and may be overcome in the future. In consequence of eq.(5) we have

$$n_B^{exc}(k) = - n_A^{exc}(k) \; v_A / v_B \qquad (6)$$

The width Δz of the 10 layers introduced in eq.(3) has been varied in the initial stages of the work but then set to the fixed value

$$\Delta z = 0.05 * \lambda_e(1000)$$

where $\lambda_e(1000)$ denotes the mean free path of electrons emitted with 1000eV kinetic energy. This refers to the situation of Siegbahn's XPS data on C(1s) [3]. $\lambda_e(1000)$ serves as scaling length in our treatment. Thus, not only Δz but as well the electron path λ_e for the smaller kinetic energies in our UPS data are measured in units of this quantity. $\lambda_e(1000)$ itself will be evaluated at the end by comparing with surface tension measurements.

We follow a simple strategy to determine the still large number of parameters: 10 values for the excess concentration $n_A^{exc}(k)$, $k = 1,..,10$ plus the energy depending parameters c and λ_e. All necessary parameters are selected via a random number generator. Then the quantities R from eq.(4) can be determined and be compared to their experimental counterparts. Combinations of parameters that lead to an averaged relative deviation

$$F = \left[\frac{1}{J} \sum_{j=1}^{J} \left(\frac{R_{exp}(j) - R_{calc}(j)}{R_{exp}(j)} \right)^2 \right]^{1/2} \qquad \text{(j numbers the experimental data)} \qquad (7)$$

which exceeds a preset limit, are rejected. Among the accepted sets of parameters we compute the mean values and the standard deviations. This method is simple and has the advantage that it is unlikely to end up in a local minimum of F in the multidimensional (i.e. 15-16 dimensional) parameter space. The acceptance rate depends on the bounds between which the random number generator selects trial parameter values. Its typical value has been adjusted to approximately 0.1%.

The observation depth λ' is varied in two ways. First, by tuning the photon energy and , second, by varying the detection angle of the electrons. If one accepts the simple model of a straight finite mean free path of the electrons in the bulk and conservation of the flight direction when passing through the surface the observation depth and the angle of detection α against the surface normal are related by

$$\lambda' = \lambda_e * \cos \alpha \qquad (8)$$

In a recent paper the method of reconstructing the concentration depth profile from electron spectrometric data has been considerably refined[12]. The description is similar to ours in that the range of depths studied is conceived as a set of layers with constant composition. However, by taking into account elastic scattering of the electrons on their way to the surface and by assuming a mean free path which in turn depends on the composition of the individual layer it introduces a much higher degree of sophistication into the problem. It is obvious that in the future this approach or another more advanced treatment will become standard - in particular, if data based on electrons with low kinetic energies of only a few eV like our UPS data rather than XPS results as in ref.12 are to be evaluated. In that case one may even prefer to propagate the slow electrons in the bulk quantum mechanically[13]. One must realise, of course, that these improved methods require input on electron scattering cross sections[12] or complex scattering phase shifts[13] that are not readily available for slow electrons moving in bulk material. Thus, we content ourselves for the moment to evaluate our data within the simple framework outlined above. In part, the shortcomings of the model are compensated by fitting the mean free path separately for every electron energy.

We address now to the absolute value of our scaling parameter $\lambda_e(1000)$. It can be determined in the following way. The Gibbs equation relates the surface excess Γ to the thermodynamic quantities surface tension γ and activity $a = f X$ with X being the bulk molar fraction and f the activity coefficient of component A

$$\Gamma = - \frac{v_B / v_A}{v_B / v_A + X(1-X)} * \frac{1}{RT} \frac{d\gamma}{d\ln a} \qquad (9)$$

If we assume that the bulk molar fraction of component A is sufficiently low that Henry's law holds[14] , i.e. that f can be treated as a constant, then the Gibbs equation simplifies to

$$\Gamma = - \frac{v_B / v_A}{v_B / v_A + X(1-X)} * \frac{X}{RT} \frac{d\gamma}{dX} \tag{10}$$

For the 0.5m solution of TBAI in formamide considered below we have $X_{TBAI}=0.022$. This low value makes eq.(10) a good approximation. Accordingly, surface tension measurements as carried out in our lab[10] are sufficient to determine the surface excess Γ. The integrated excess concentration, i.e. the sum of the $n_A^{exc}(k)$, is related to the surface excess via

$$\Gamma = \left\{ \sum_{k=1}^{10} n_A^{exc}(k) \frac{\Delta z}{\lambda_e(1000)} \right\} \lambda_e(1000) \tag{11}$$

since in the fitting procedure all lengths are consistently measured in units of $\lambda_e(1000)$. The above expression allows to determine

$$\lambda_e(1000) = 45 \text{ Å } (\pm 15\%) \tag{12}$$

THE SURFACE OF A SALT SOLUTION AS EVALUATED FROM MIES-, UPS- AND XPS- DATA

In this section we try to access surface properties of a solution of 0.5m TBAI in formamide. This system has been investigated before with MIES[15] and XPS[3]. A large set of UPS-data has been taken recently[16]. Using light from the Berlin Synchrotron Radiation Facility (BESSY) valence and core electron spectra have been measured for electron detection angles α against the surface normal between $0°$ and $75°$. For the valence electron spectra photon energies between 25 eV and 60 eV were employed, the C(1s)-core electron spectra were taken with $h\nu = 310 ... 390$ eV. As an example, valence spectra are shown in fig. 1 for several photon energies, the electrons measured in a direction normal to the surface. The peaks characteristic for the solvent FA and for the positive and negative salt ions TBA$^+$ and I$^-$ are indicated. The qualitative effect of decreasing the photon energy, and in consequence the kinetic energy of the emitted electrons, is clearly seen: the relative weight of the observation depth is restricted more and more to the surface where the salt segregates. A similar result is obtained if the photon energy is kept constant but the direction of detection of the electrons with respect to the surface normal is increased, cf. fig 2. The fact that the relative signal due to I$^-$ varies much less with angle α than that of the positive ion TBA$^+$ is discussed later. At the moment we make use of the TBA$^+$ signal normalized to the FA-peak as function of photon energy and electron detection angle. This quantity is to be identified with R as defined in eq.(4). We combine these data with the MIES result for the same concentration[15] and XPS-data[3]. The fitting procedure outlined in the previous section leads to the determination of the parameters c and λ_e for every electron energy and of the excess concentration $n^{exc}(z)$ of the salt TBAI. From existing results[3,15] we do not have any reason to assume a differential segregation between positive

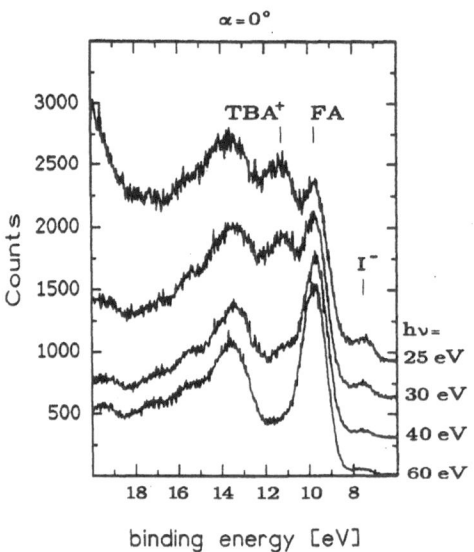

Fig.1 UPS valence electron spectra from the surface of a 0.5m solution of TBAI in formamide. The electrons are detected in a direction normal to the surface. The photon energies are hν = 25, 30, 40, 60 eV from top to bottom spectrum.

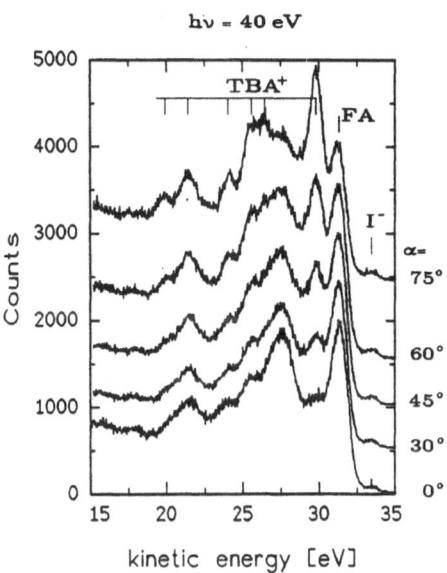

Fig. 2 ARUPS valence electron spectra from the surface of a 0.5 solution of TBAI in formamide. The photon energy is hν = 40 eV. The detection angle of the electrons takes on the values α = 75°, 60°, 45°, 30°, 0° from top to bottom spectrum.

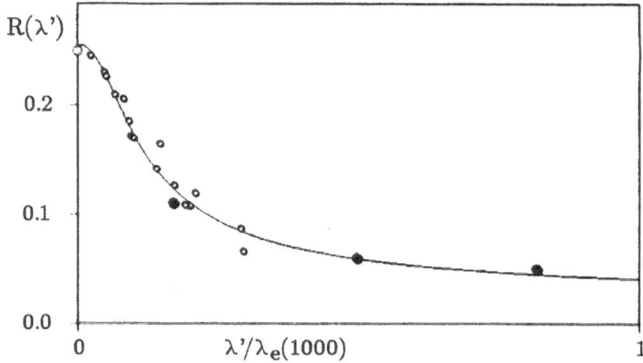

Fig. 3 Ratio $R(\lambda') = N_{TBAI}/N_{FA}$ as function of the observation depth λ' in units of $\lambda_e(1000)$. Small circles: ARUPS valence electron spectra; big circles: MIES; dots: XPS

and negative ions. Our data from fig. 1 support this: from $h\nu = 60$ eV to $h\nu = 25$ eV the relative increase of the I^- signal is the same as that of TBA^+. We will discuss below that the seeming discrepancy with the I^- signal in fig. 2 is not a counter argument. In fig. 3 the experimental values (fitted c and λ_e values taken into account) are plotted together with the quantity R as back calculated from the fitted

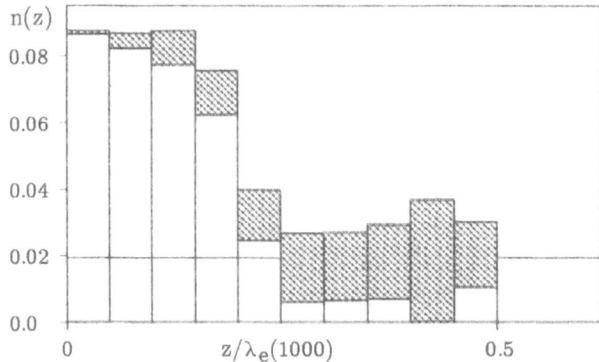

Fig. 4 Density profile of TBAI $n_{TBAI}(z)$ at the surface of a 0.5m solution in formamide. n is given in units of the bulk number density of pure formamide. Other explanations in the text.

concentration density profile. The latter is displayed in fig. 4. The abzissa is the distance from the surface, in units of $\lambda_e(1000)$. The shaded bars visualize the result of the fit. Their lower and upper values represent the mean minus and plus the standard deviation:

$$\langle n_{TBAI}^{exc}(k) \rangle - \sigma[n_{TBAI}^{exc}(k)] \quad \text{and} \quad \langle n_{TBAI}^{exc}(k) \rangle + \sigma[n_{TBAI}^{exc}(k)]$$

The standard deviation does not show the correlation between adjacent values of the histogram. One value being larger than the mean leads to a smaller values with its neighbors. This can be read off the fact that the integrated excess concentration

$$\sum_{k=1}^{10} n_{TBAI}^{exc}(k)$$

has only a small standard deviation of a few percent. The analogue to fig.3, but based on C(1s) core electron spectra, is shown in fig.5. Inspecting the data carefully one observes that the scatter of the data around the fitted curve R is a systematic one within every set of data. Whether this points out the limitations of our model presented in the previous section cannot be decided at present. It is interesting to note in this context, that the density profile fitted from the data in fig. 5 agrees with the profile in fig. 4 within the determined error bars.

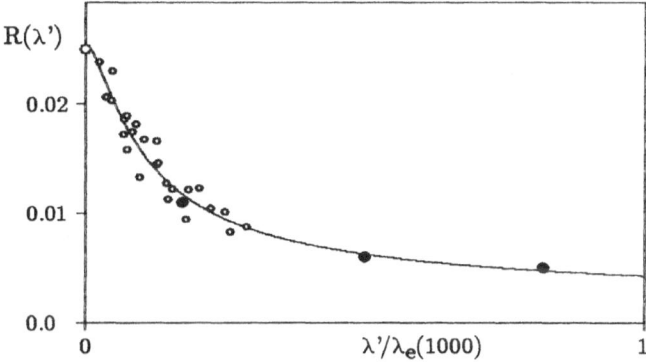

Fig. 5 Ratio $R(\lambda') = N_{TBAI}/N_{FA}$ as function of the observation depth λ' in units of $\lambda_e(1000)$. Small circles: ARUPS core electron spectra; big circles: MIES; dots: XPS

We will consider now the above mentioned observation that the relative intensity of the I^-- peak does not grow with increasing electron detection angle as does the TBA$^+$- signal, cf. fig. 2. Effects of an intrinsic angular distribution can be excluded since $I^-(^1S)$ is spherically symmetric and since the measurements are taken with a fixed angle between observed electron emission direction and the polarization vector of the light. A convincing explanation can be found in the outcome of computer simulations [10]. The underlying model is the following: only a two dimensional array of particles is used to simulate the surface of the salt solution. The different species are modelled by hard discs the radii of which are chosen to properly represent the size of the respective particle[15]. In case of the ions point charges are placed in the center of the disc whereas the disc representing the solvent molecule carries a point dipole. The repulsive hard core potentials and the electrostatic interaction within the layer are considered to dominate the surface structure rather than the interaction with deeper layers. This concept gets some support from the density profile shown in fig. 4 which suggests a sur-

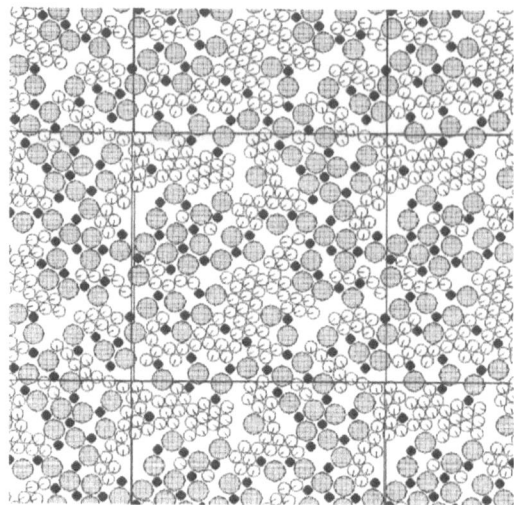

Fig. 6 Monte Carlo Simulation of a two dimensional layer of TBAI/FA. Big discs represent positive TBA$^+$ ions, small discs the negative I$^-$ ions. The solvent molecules are represented by the medium size discs the mark of which indicates the direction of the dipole moment. The ratio of salt to solvent molecules correspond to the surface composition of a 0.9m TBAI/FA solution. From ref. 10.

face layer of enhanced TBAI - concentration which is about 10Å thick. Since this value corresponds closely to the diameter of the TBA$^+$ - ion one feels inclined to conceive the surface as a salt enriched 'monolayer' on top of a homogeneous bulk solution. In fig. 6 a typical structure as obtained after equilibration of the Monte Carlo calculation is shown. The surface composition is chosen so as to correspond to the MIES result for a 0.9m TBAI/FA solution. Even though this bulk concen-

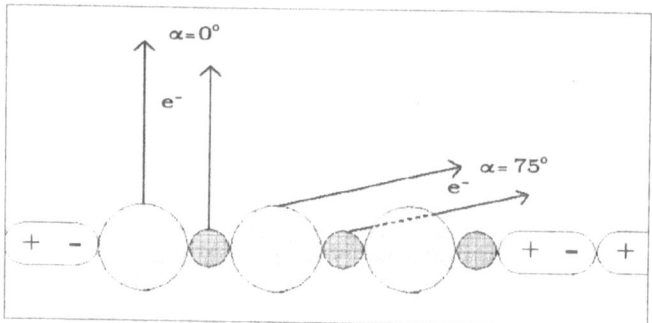

Fig.7 Schematic cross section of a TBAI/FA surface layer. The tetrahedral TBA$^+$ ions are drawn as large and the I$^-$ ions as small spheres. Solvent molecules are denoted by two signs indicating the orientation of their dipole moment. The straight lines are meant to illustrate the relative weakening of the I$^-$ - signal for large angle emission against the surface normal.

tration is larger than 0.5m the result has bearing to our case. This can be read off another calculation which simulates a 0.2m TBAI/FA solution which displays the same characteristic features. Fig. 6 shows that we do not only have a segregation of the salt to the surface, but that - within the model calculation - we find a segregation between salt and solvent within the surface layer. The impact of this finding is the following: the small iodide ions are usually in the close neighborhood of the large TBA$^+$- ions. If viewed under a large angle α their signal is shadowed by their big neighbors, cf. fig. 7. Even though the solvent molecules are small entities as well the above effect does not apply to them since they find themselves only rarely near a TBA$^+$ ion. It is obvious, that at normal observation such a shadowing effect would not occur, in agreement with the spectra in fig. 1. Thus, we may conclude that the electron energy spectra contain information not only on the composition of the surface but as well on the lateral surface structure.

AUGER ELECTRON SPECTROSCOPY

Auger electron spectroscopy is specific for the elemental composition of a probe. How this can be used to assess the chemical composition will be discussed again for the solution of TBAI/FA. Formamide contains carbon, nitrogen and oxygen in equal amounts:

$$\overset{H}{\underset{H}{\oplus}} > N = C \overset{\nearrow O^{\ominus}}{\underset{H}{}}$$

The positive salt ion TBA$^+$ has 16 carbon atoms per one nitrogen and does not contain oxygen at all:

$$
\begin{array}{c}
\text{CH}_3 \qquad\qquad\qquad \text{H}_3\text{C} \\
\text{H}_2\text{C}\text{--}\text{CH}_2 \quad \text{H}_2\text{C}\text{--}\text{CH}_2 \\
\text{H}_2\text{C}\underset{\text{N}^{\oplus}}{\diagdown}\text{--}\dot{\text{C}}\text{H}_2 \\
\text{H}_2\text{C}\diagup\diagdown\text{CH}_2 \\
\text{H}_2\text{C}\text{--}\dot{\text{C}}\text{H}_2 \quad \text{H}_2\text{C}\text{--}\text{CH}_2 \\
\dot{\text{C}}\text{H}_3 \qquad\qquad\qquad \text{H}_3\dot{\text{C}}
\end{array}
$$

Thus, for a given probe the total ratio of carbon to oxygen atoms would allow to calculate the relative amount of salt and solvent. Qualitatively, the effect is demonstrated by the differentiated Auger spectra from 0.4m TBAI/FA, shown in fig. 8. The primary e$^-$- beam is aimed under an angle of -22.5° against the surface normal of the probe. The detection angle is varied from $\alpha = 22.5°$ to 75°. The contributions from C,N,O are indicated. The spectra are normalized to the C KVV Auger structure. The decrease of the O-intensity with increasing angle α signals the depletion of the solvent near the surface, known from the preceding section.

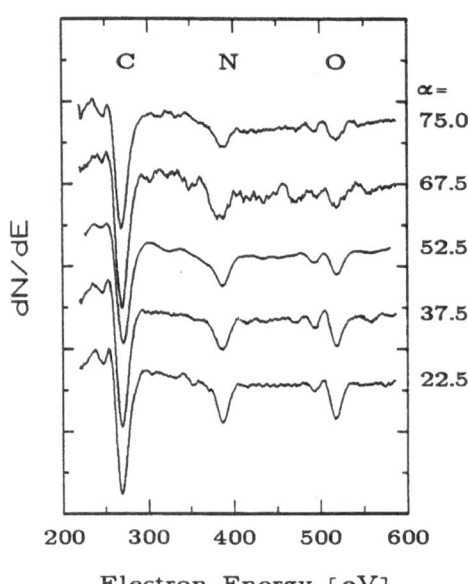

Fig. 8 Differentiated Auger electron spectra from the surface of a 0.4m TBAI solution in formamide. The detection angles α = 22.5°, 37.5°, 52.5°, 67.5°, 75° with respect to the surface normal. The exciting electron beam has an angle of incidence of -22.5°, the kinetic energy being 2 keV.

The fact that the N-signal is similarly reduced may indicate that the nitrogen atom of the TBA$^+$- ion does not contribute much to the signal. So far, a quantitative evaluation of the data has not been done. The full merit of the method is still to be explored. In principle, one has to vary: the kinetic energy and the angle of incidence of the primary electron beam and the detection angle of the Auger electrons. Which combinations of these parameters will prove useful to investigate composition and maybe structure of the surface will be tested in the future.

ELECTRON ENERGY LOSS SPECTROSCOPY

Electron energy loss spectroscopy can be used for two purposes. First, the investigation of electronic excitations which leads to energy losses of several eV. An energy resolution of $\Delta E = 1\,eV$ is often sufficient for this task. Second, it is used to perform a spectroscopy of vibrational frequencies, the required energy resolution being $\Delta E = 10$ meV or better. This latter application could be very fruitful for the characterization of molecules at a liquid surface as it is the case for molecules adsorbed on solid substrates.

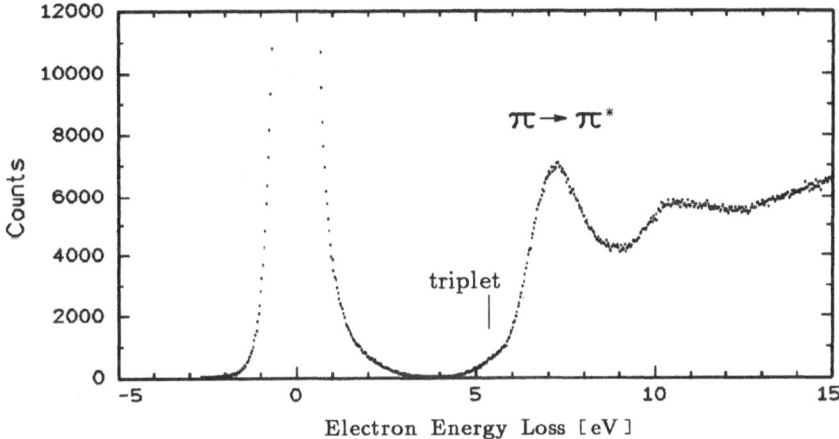

Fig. 9 Electron energy loss spectrum of pure formamide. The primary kinetic electron energy is 70 eV.

Measurements reported in the literature[7] were performed with a resolution of about 1eV. The equipment used for our own experiments[16] neither did allow a better resolution. Thus, the object of our investigation could only be the electronic excitation of the molecules in the liquid. In fig. 9 we show a loss spectrum obtained from the surface of pure formamide with electrons of a primary energy of 70 eV. The peak at ~7.5 eV corresponds to the excitation of the highest occupied level of FA to the lowest unoccupied level, a so called $\pi \rightarrow \pi^*$ transition. A weak, but interesting structure is found at ~5.5 eV energy loss. The underlying physical process is the excitation of a triplet state via electron exchange[17]. As expected,

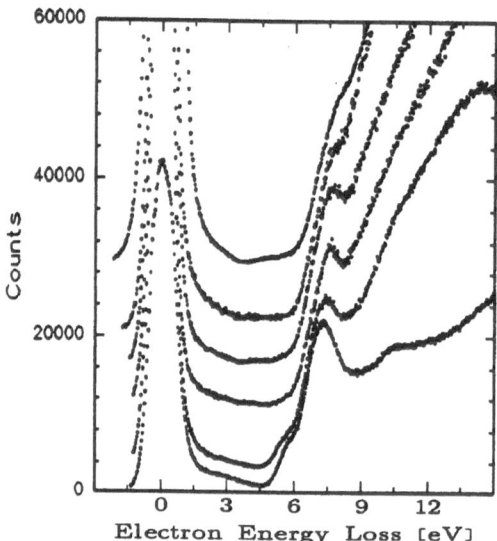

Fig.10 Electron energy loss spectra of a liquid surface with a primary electron energy of 20 eV. Spectrum at the bottom: pure formamide. The upper spectra are taken with increasing amounts of dissolved sodium oleate: 0.55, 1.1, 1.6, 4.1, 11 in units of 10^{-3} mol/l.

this latter process is more probable at lower excitation energies, cf. fig. 10. The surface sensitivity of this techniques shows if surface active species are dissolved in formamide and change the character of the surface. As seen in fig.10, the features characteristic for FA begin to vanish if small amounts of sodium oleate are present. Sodium oleate (NaOl) is a typical tenside

$$\text{Na}^+ \text{O}^- \diagdown \underset{\text{O}}{\overset{}{\diagup}} C - (CH_2)_7 - CH = CH - (CH_2)_7 - CH_3$$

the behaviour of which at the surface of formamide has been investigated before by MIES and UPS[18]. At the concentrations used in the experiments reported in fig. 10, NaOl tends to cover the surface fully within milliseconds. The EEL-spectrum of NaOl is yet to be established in order to derive quantitative information from EELS.

ACKNOWLEDGMENTS

The author is grateful to M.Wulf for preparing the major part of the figures. The experimental and theoretical work has been financially supported by Deutsche Forschungsgemeinschaft (DFG), Bundesminister für Forschung und Technologie (BMFT) and Land Nordrhein-Westfalen.

REFERENCES

1. H.Siegbahn and K.Siegbahn, ESCA applied to liquids, *J.Electr. Spectr. Rel. Phen.* 2:319–25 (1973)
2. S.Holmberg, R.Moberg, C.Y.Zhong and H.Siegbahn, Angle resolved electron spectroscopy for measurement of surface segregation phenomena in liquids and solutions, *J.Electr. Spectr. Rel. Phen.* 41:337–42 (1986)
3. S.Holmberg, C.Y.Zhong, R.Moberg and H.Siegbahn, Surface segregation of Tetra-N-Alkylamonium halides in formamide investigated by angle resolved electron spectroscopy, *J.Electr. Spectr. Rel. Phen.* 47:27–38 (1988)
4. R.Moberg,F.Bökman,O.Bohman and H.Siegbahn, ESCA studies of phase transfer catalysts in solution; ion pairing and surface activity, *J. Am. Chem. Soc.* 113:3663 (1991)
5. L.Nemec, H.J.Gaehrs, L.Chia and P.Delahay, *J.Chem. Phys.* 66:4450 (1977)
6. R.E.Ballard, J.Jones and E.Sutherland, He(I) photoelectron spectrum of Tetra-N-Butylammonium iodide and bromide in solution. The effects of surface activity,*Chem. Phys. Lett.* 112:310–2 (1984)
7. R.E.Ballard, J.Jones, D.Read, A. Inchley and M.Cranmer, Auger and electron energy loss studies on liquid surfaces, *Chem. Phys. Lett.* 147:629–31 (1988)
8. W.Keller, H.Morgner and W.A.Müller, Probing the outermost layer of a free liquid surface. Electron spectroscopy of formamide under He(2^3S) impact, *Mol. Phys.* 57:623–36 (1986)
9. K. Roth, PhD Thesis 1990, University Witten/Herdecke

10. J.Oberbrodhage, PhD Thesis 1992, University Witten/Herdecke

11. K.Richter, PhD Thesis 1992, University Witten/Herdecke

12. O.A.Baschenko, F.Bökman, O.Bohman and H.O.G. Siegbahn, Distribution of ions in subsurface layers of liquid solutions studied by ARXPS, *J.Electr.Spectr. Rel. Phen.* 62:317 (1993)

13. J.V.Peetz and W.Schattke, 5th *Int. Conf. Electr. Spectr.*, Kiev (1993) and *J.Electr.Spectr. Rel. Phen.* (1993) submitted

14. R.A.Alberty. "Physical Chemistry", John Wiley&Sons, New York (1987)

15. H.Morgner, J.Oberbrodhage, K.Richter and K.Roth, The solution of tetrabutyl-ammoniumiodide in formamide: investigation of the surface by MIES, *J.Phys.:Condens. Matter* 3:5639-55 (1991)

16. F.Eschen, M.Heyerhoff, H.Morgner and M.Wulf (1993) unpublished results

17. A.Kuppermann, W.M.Flicker and O.A.Mosher, Electronic spectroscopy of polyatomic molecules by low-energy, variable-angle electron impact, *Chem.Reviews* 79:77-90 (1979)

18. H.Morgner, K.Richter and M.Wulf, The first stages of tenside layer formation. A metastable impact electron spectroscopy study of sodium oleate dissolved in formamide, *Mol. Phys.* 79:169-77 (1993)

PHOTOELECTRON SPECTROSCOPY AT LIQUID WATER SURFACES

M.Faubel and B.Steiner

Max-Planck-Institut für Strömungsforschung
Bunsenstraße 10, D 37073 Göttingen

ABSTRACT

A new method for surface studies of volatile liquids is employed for first measurements of photoelectron spectra of liquid water using $h\nu$ = 21.21eV HeI radiation. Also studied are surface segregation phenomena in formamide/benzyl alcohol mixtures which occur on a ten microsecond time scale. A further example are observations of surface adsorbed salt ions in aqueous solutions. These show the presence of surface active PF_6^-ions at the water surface while common ions such as Cl^- appear to be screened by a solvation shell.

INTRODUCTION

Electron spectroscopy and analogous molecular beam type experiments are particularly difficult at the surfaces of common laboratory liquid solvents for three major reasons: (1)standard solvents, such as water, are volatile substances which are not readily compatible with the high vacuum conditions required for an unperturbed sampling and analysis of the electrons, or ions and molecules, respectively. (2) It is problematic to prepare a clean surface of a liquid which can not be heated to high temperatures. And (3) the liquids are frequently poor conductors, or isolators, and charge neutralization is here of crucial importance for any electron energy analysis experiment.

These notorious liquid target problems we solve simultaneously by employing in the present experiment a very thin and fast flowing liquid jet as the probe surface. Using this technique for liquid water we have demonstrated, by velocity distribution measurements of the vapor molecules, that a free vacuum surface is obtained for 6.5µm diameter liquid jets [1]. In the employed high purity water electrokinetic charging phenomena produce flow dependent additional surface potentials in the order of up to several eV which can be controlled and corrected for [2]. This charging has a beneficial side effect because it generates an electric potential gradient which is removing spurious vapor phase contributions to the photoelectron spectra.

This liquid target is used in a slightly modified standard photoelectron spectrometer arrangement for obtaining photoelectron spectra from the surface of pure liquid water. The employed HeI lamp illuminates an area of 10^{-2} mm^2 with approximately 10^9 photons per second and produces less than a total of 10^5 photoelectrons per second at the liquid surface. Only a small fraction is entering the 100 mm radius hemispherical electron energy analyser through a 30μm wide protection slit. With a single electron multiplier detector the data recording time for one photoelectron spectrum for a 6.5μm jet is at present in the order of 24 hours.

THE ELECTRON SPECTRUM OF PURE LIQUID WATER

In Fig.1a and 1b we show for comparison spectra measured with this apparatus in the gas phase and at the liquid surface of pure water. The gas phase spectrum is in quantitative agreement with the well known literature results [3]. And, for a subsequent discussion we identify in Fig.1a the three principal electronic peak structures with a sketch of the dominant molecular orbitals of the H_2O ground state which, respectively, have been depleted by the photoionization process. The sharp peak structure at 8.6 eV kinetic electron energy originates from the oxygen "lone pair orbital" with a molecular binding energy of ~12.6 eV [3]. The other two peaks are broadened by molecular vibration structure, and, their formal electronic assignment is indicated in the first column of table 1.

Table 1. The observed liquid water ionization potentials

formal MO assignment	I_{vapor}	I_{liquid}	ΔI
$1b_1$	12.6eV (26%)*	10.9eV (16%)*	1.7eV
$3a_1$	14.7eV (49%)	13.5eV (54%)	1.2eV
$1b_2$	18.5eV (25%)	17.0eV (30%)	1.5eV

* (exp. peak area ratio)

The here for the first time observable HeI photoelecton spectrum of the liquid water surface, in Fig.1b, shows three major features when it is compared versus the gas phase results: (1) the gas phase peak structures are considerably broadened, (2) as usual for condensed phases a large background electron peak appears at small kinetic energies. And, (3) the principal three major peak structures of the gas phase spectroscopy appear here alltogether shifted by approximately 1.5 eV toward lower vertical electron work functions in the liquid state.

The broadening (1) may in part be attributed to inelastic electron scattering. To a not yet quantifiable extent it, also, will result from a random environment of the neutral dielectric molecules which had not yet time to relax into an ordered hydration shell around the newly formed ion. The

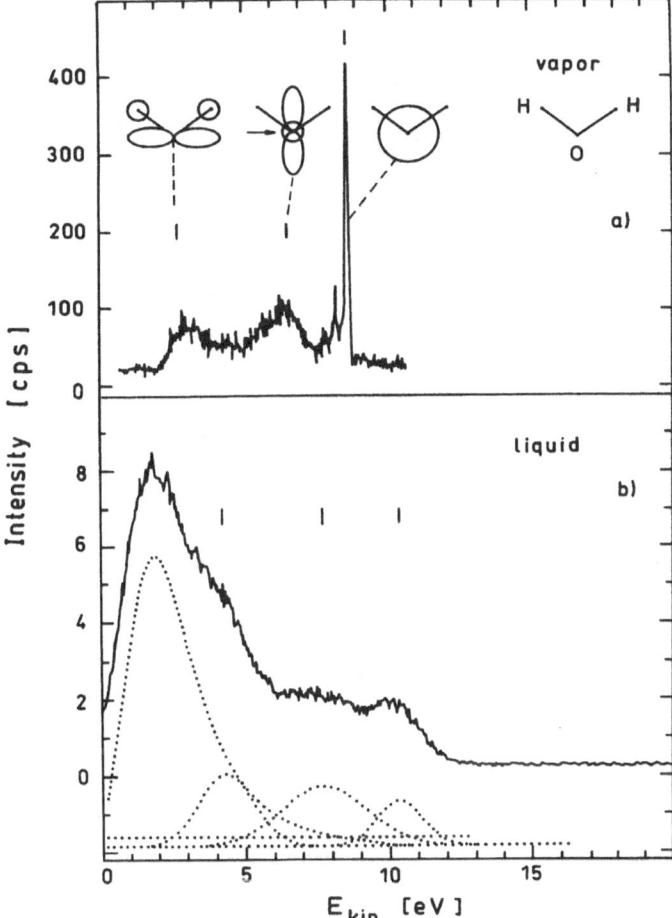

Figure 1. HeI (h =21.8eV) photoelectron spectra of water vapor (a) and of a liquid water surface (b). In (a) the three principal peak structures are labeled by a sketch of the prevailing MO orbitals in the neutral H_2O which, respectively, are removed by the photoionization process. (b) At the liquid surface these peaks are considerably broadened and shifted. For the deconvolution (- - - -) see text.

average energy shift (3) of the gas phase peak structures re-flects the Born solvation process of ions in a dielectric. For a model sphere with "ion radius" r_0 and charge Z the electric field energy changes by [4]:

$$\Delta G_{Born} = -\frac{Z^2}{8\pi\varepsilon_0 r_0} (1/1 - 1/\varepsilon)$$

when it is transferred from vacuum into a (continuous) di-electric medium with dielectric constant ε. For water, assum-ing a locally relaxed $\varepsilon \sim 80$, this gives an "ion radius" esti-mate of $r_0=0.47$nm for the observed energy shift in the order of 1.5 eV. Alternatively, a realistic value $r_0=0.29$nm [4] for the maximum of the liquid water radial distribution function provides a value in the order of $\varepsilon_{eff} \sim 2.5$ for the instantaneous polarization of the environment of a newly formed H_2O^+-ion.

For further analysis the experimental photoelectron spectrum of liquid water was evaluated with an automatic fitting routine. Allowing for one exponentially modified Gaussian function for the asymmetric low energy electron tail and for three Gaussian functions we obtain the deconvolution shown by the intermittent smooth curves in Fig.1. The numerical values for the centerline ionization potentials are listed in table 1 for a comparison versus the respective gas phase energy levels. Estimates of the relative peak intensities are given in brackets. From gas to liquid they change at maximum by less than 40% of their relative amplitudes. The lowest ionization energy of $I_{av}=10.9eV$ is in perfect agreement with Delahay's earlier determination of a threshhold energy of 10.06eV for the appearance of photoelectrons at the liquid surface of water [5]. The second observed level at $I_{av}=13.5eV$ seems to correspond to the region where for liquid water maximum UV absorption at 13.7eV has been reported in optical absorption experiments [6]. In principle this agrees with our finding of an approximately 54% contribution of this second transition peak to our HeI photoelectron spectrum.

The relative peak shifts between gas phase and liquid differ markedly. As given in the last column of the table they vary from $\Delta I=1.7eV$ to $\Delta I=1.2eV$ for the three different electronic states of the H_2O molecule. For a first order qualitative explanation of this observation it seems we can employ a simple generalization of the Born solvation energy model: the insert contour illustrations in Fig.1a show that the spatial distributions of the respectively ionized neutral H_2O molecular orbitals differ considerably for the three ionic states; for the oxygen lone pair orbital, which is responsible for the peak at the right hand side of the photoelectron spectra in Fig.I, the electron charge distribution is spherical. Thus, it extends into a larger environmental regime of the H_2O molecule and can interact with a comparably large volume of polarized solvent molecules. For the two other ionic states larger parts of the relevant charge clouds extend into the interior of the H_2O molecule and are to a lesser extent available for the build up of a Born solvation energy change. Correspondingly, one can expect a largest peak shift for the right hand peak in Fig.1, as is actually observed. In contrast the orbital which is responsible for the middle peak appears to be the comparably best shielded of the sketched electronic states and shows actually the smallest experimental shift in the liquid state of 1.2 eV, only.

LIQUID MIXTURES AND SALT SOLUTIONS

For mixtures of liquids and for solutions the high flow velocity of the microjet offers also a unique opportunity for studies of the equilibrium formation of the free jet surface on a microsecond time scale. A first experimental example are here measurements of photoelectron spectra at several different downstream positions at a $16\mu m$ thick jet of a mixture of formamide and 25% volume of benzyl alcohol. At the present

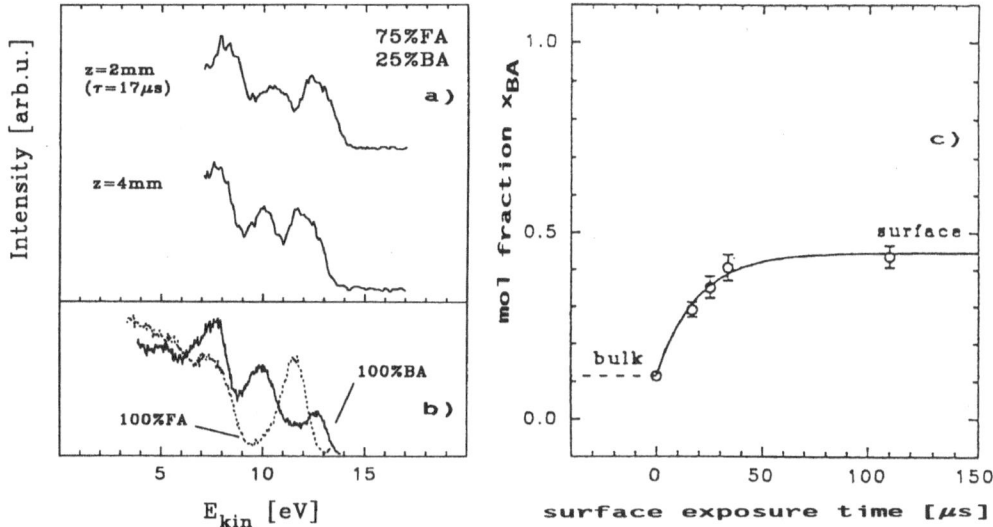

Figure 2. Molecular segregation of benzylalcohol (BA) and formamide (FA) at the surface of a 16μm diameter liquid jet of 25% BA in FA.

experimental nozzle temperature of 8°C the vapor pressure for formamide is near 10^{-2} mbar and for benzyl alcohol it is in the order of 0.1 mbar. This low vapor pressure system has recently been studied by Morgner's group [7], and we have used it for a cross-check of the UPS performance of our quite different new apparatus. The two spectra of this mixture, which are shown in Fig.2a, are measured at a distance of 2mm and at 4mm downstream of the nozzle exit. This corresponds to flow times of 17μsec and of 34μsec from the nozzle exit. For their interpretation we use the HeI photoelectron spectra of the pure benzyl alcohol (BA) and for pure formamide (FA) which are shown for reference in Fig.2b. The spectra at the mixture surface show, thus, from z=2mm to z=4mm a visible increase of the benzyl alcohol component as is exemplified by the relative increase of the middle peak at z=4mm distance.

From several measurements of the benzyl alcohol surface mol fraction versus the elapsed segregation time at the free jet surface we obtain, in Fig.2c, a best fit exponential time constant of τ =19.2μsec for the asymptotic formation of the equilibrium at the FA-BA free vacuum surface. With a typical liquid diffusion constant in the order of D ~ 10^{-5} cm^{2}sec^{-1} the diffusion relation x ~ $(D \cdot \tau)^{-1/2}$ ~ 0.1μm provides an estimate for the surface skin depth. For a future, accurate evaluation we note that the benzyl alcohol has a non negligible vapor pressure of ~0.1mbar and does actually evaporate with a rate of one monolayer in the 10μsec to 20μsec observation time period.

For a second example we show, in Fig.3, results of a surface adsorption study of salt ions in aqueous electrolyte solutions. The simple Born electrostatic solvation theory suggests that ions are removed from the free liquid surface by one or two ion hydration shell layers. In agreement with this expectation we observe no change in the shape of the photoelectron energy spectrum for aqueous KCl solutions with-

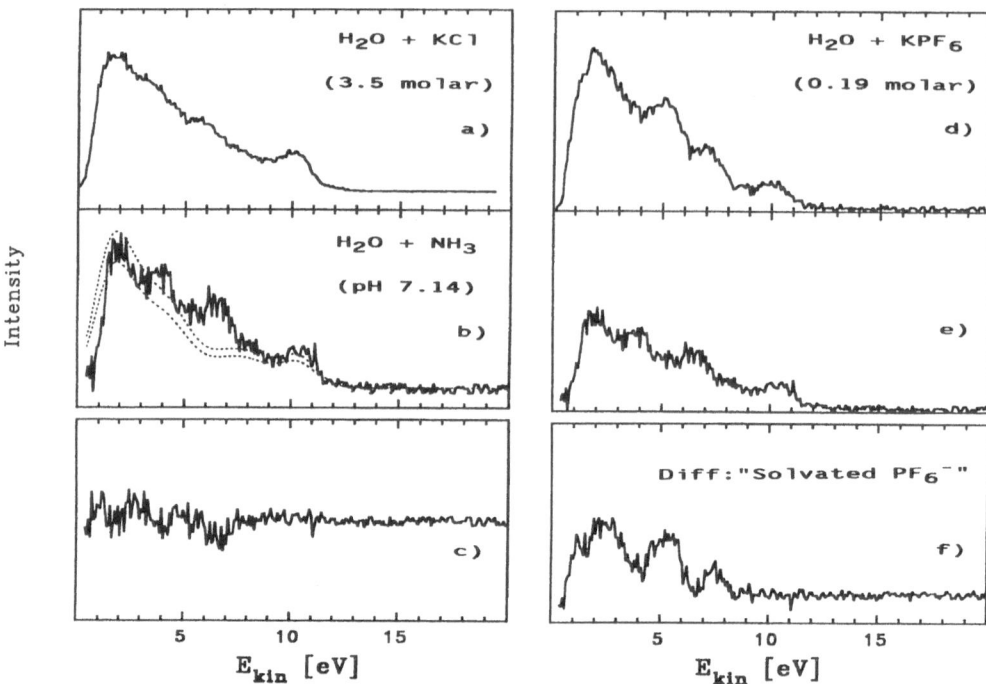

Figure 3. Absence of ionic surface adsorption in water-KCl solutions (a) versus PF$_6^-$ surface adsorption in water-KPF$_6$ electrolytes (d). (b) shows the measured gas phase contribution in photoelectron spectra of a pH 7.14 with NH$_3$ neutralized, liquid water jet. (f) gives the isolated spectrum of PF$_6^-$ at the liquid water surface.

in a wide range of KCl concentrations. The actual experimental appearance of the salt-in-water spectrum, shown for a 3.5 molar KCl concentration in Fig.3a, however, differs significantly from the allready discussed spectrum for the pure liquid surface of water because the electrically conductive jet can not generate automatically the electrokinetic potential which had removed before the gas phase vapor spectrum. The gas phase contribution could here only be partially suppressed by application of an inhomogeneous electric potential of 18 V between nozzle and spectrometer entrance. We can experimentally simulate the non-charging jet sampling condition for (almost) pure water by adding a minimum quantity of NH$_3$ to the pure water. As shown previously [2], we obtain zero electrokinetic charging at a pH-value of 7.14, i.e. at a 10^{-10} impurity level of NH$_3$-derived ions. The measured photoelectron spectrum for this pH 7.14-water is shown in fig.3b. The previously (in Fig.1) obtained pure liquid water photoelectron spectrum is shown here as the smooth intermittent lines and reveals clearly the H$_2$O-gas phase origin of the new structures in the non charging, pH 7.14, H$_2$O-NH$_3$ solution. In the difference spectrum, Fig.3c, of the "noncharging" pH 7.14 liquid water (of Fig.3b) and the 3.5 molar H$_2$O-KCl solution (of Fig.3a) remains no significant structure above the noise level. This provides a quantitative confirmation of the absence of any ions or neutral salt components, or of any significant H_2O restructuring at the top-most one or two monolayers of the aqueous KCl solution surface.

This microscopic result for KCl-H_2O solutions had long been anticipated from macroscopic measurements of the surface tension and of changes in the electric surface potential. Similar macroscopic measurements have also shown the existence of some ions where chemical forces are stronger than the electrostatic image forces and drive ions to the free liquid surface. A well known example for this phenomenon are aqueous solutions of KPF_6 where the PF_6^- ion is considered to be adsorbed at the free electrolyte surface. In agreement with this macroscopic expectation we find significant changes in the shape of the photoelectron spectrum when KPF_6 is dissolved in water, as is shown by Fig.3d for a 0.19 molar solution. This spectrum is decomposed by substraction of the in Fig.3e shown scaled partial spectrum of a neutral water surface (i.e. one which is proportional to the just discussed H_2O+NH_3 spectrum at pH 7.14) which has been fitted in amplitude to the water surface peak at 10 eV kinetic energy. The resulting difference spectrum, in Fig.3f, shows then the salt induced surface change of liquid water. It reflects very likely the photoelectron spectrum of the solvated PF_6^-ion at the liquid water surface.

CONCLUSIONS

This first study of photoelectron spectra of volatile liquid solvents can only demonstrate the versatility and the broad application range of this newly developed technique. Most questions of a detailed analysis of these new, and pre-liminary, experimental data remain naturally left to future work.

We gratefully acknowledge Professor J.P.Toennies for continuous interest, encouragement and support for this work.

REFERENCES

1. M.Faubel, S.Schlemmer and J.P.Toennies, A molecular beam study of the evaporation of water from a liquid jet, Z. Phys. D, 10:269 (1988)

2. M.Faubel and B.Steiner, Strong bipolar electrokinetic charging of thin liquid jets emerging from 10μm Pt Ir nozzles, Ber. Bunsenges. Phys. Chem. 96:1167 (1992)

3. K.Kimura, S.Katsumata, Y.Achiba, T.Yamazaki, S.Iwata, "Handbook of HeI Photoelectron Spectra", Halsted Press, N.Y. (1981)

4. W.J.Moore, "Physical Chemistry", 5th ed., Longman, London (1972)

5. P.Delahay and K.von Burg, Photoemission spectroscopy of liquid water, Chem. Phys. Lett. 83:250 (1981)

6. J.M.Heller, R.N.Hamm, R.D.Birkhoff and L.R.Painter, Collective oscillation in liquid water, J.Chem. Phys. 60:3483 (1974)

7. H.Morgner, J.Oberbrodhage, K.Richter and K.Roth, Surface segregation of a binary liquid mixture as studied by metastable impact electron spectroscopy, Mol. Phys. 73:1295 (1991)

LIGHT-INDUCED ELECTRON EMISSION
FROM SURFACES OF ORGANIC LIQUIDS

Klaus Lacmann, Hitoshi Koizumi and Werner F. Schmidt

Hahn-Meitner-Institut
Abteilung Strahlenchemie
14109 Berlin, Germany

ABSTRACT

The method of Ultraviolet Photoelectron Emission (UPE) is described and has been applied to determine the total quantum yield of photoelectrons by excitation from the energy threshold up to several electron volts above it in the energy range between 6 and 11 eV. Several high-molecular organic liquids were studied, which have a vapor pressure smaller than 10^{-4} Pa at room temperature . The results of several phthalates and long chain hydrocarbons such as squalane and squalene are discussed. σ- and π-bonds show different threshold energies. The total quantum yields approach a saturation value of 10^{-2} to 10^{-1} at three electron volts above the threshold.

The electron energy levels and the photo emission process are discussed by taking into account the electronic properties of the isolated molecules, the influence of charge polarization, and the transport of hot electrons through the liquid/vapor interface.

INTRODUCTION

The studies of electron emission from liquid surfaces utilize in most cases two different methods. Either the wavelength of the exciting light is fixed and the energy spectrum of the emitted electrons is determined or the spectrum of the exciting light is varied and the total yield of photoelectrons is determined without further energy analysis.

The first method is described in detail in this volume in the contributions by Morgner and Faubel (see index of this volume). This method is called **Ultraviolet** excited

Linking the Gaseous and Condensed Phases of Matter
Edited by L.G. Christophorou *et al*, Plenum Press, New York, 1994

Photoelectron Spectroscopy to liquids, abbreviated by **UPS**. In contrast, we vary the energy of the exciting photons and measure the total yield of electrons without analysis of the energy distribution and call it accordingly **UPE, Ultraviolet Photoelectron Emission.**

UPE has been applied in the past by e.g. Birkhoff et al. They determined absolute quantum yields for formamide and hexamethylphosphoric triamide[1,2] in the energy range of 16 to 25 eV, which is well above the emission threshold. Delahay, Watanabe et al.[3,4] on the other hand determined the threshold energies for several solutes in water and other polar solvents, but could not get reliable values of the absolute quantum yields.

From the experimental point of view, measurements on insulating or weakly conducting liquids, as compared to solid insulators, have the advantage that they are much less perturbed by the accumulation of positive charges, since the positive ions will eventually migrate to and discharge at the cathode. A disadvantage of most liquids is their higher vapor pressure as compared to solids. This makes it difficult to carry out experiments in the vacuum ultraviolet region (VUV) due to the optical absorption of the molecules in the vapor phase. Using very low vapor pressure liquids, however, we could investigate the photo emission from surfaces of liquids.

In this work we will compare the photoelectron yield and the threshold energies for squalane and squalene and for three different phthalates.

From a fundamental point of view, knowledge of the energetics of the emission process is required in order to develop a better understanding of the change of energy levels upon condensation. Measurement of the absolute quantum yield gives insight into the transport of hot electrons through the liquid and across the liquid/vapor interface.

EXPERIMENTAL

In the literature one finds different experimental techniques applied to the study of liquids.

The main experimental differences between the study of a static surface, as we do, and the liquid beam techniques are: The use of a static surface has the advantage that one needs less pumping speed and the temperature of the liquid is given by the containment / apparatus. There is no temperature change by adiabatic expansion as in a supersonically expanding beam. The disadvantage is a possible contamination of the liquid surface with time by other gases adsorbed on the surface while in a beam the surface is permanently renewed. Nevertheless we observed no change in the photoelectron emission even when a measurement ran over an hour.

The advantage of a deuterium discharge lamp as light source over synchrotron radiation is its easy setup in any laboratory; its disadvantage is its limited energy range from 6 to 11 eV. The relative photon intensity distribution of the monochromatized light was measured by guiding the light onto a sodium salicylate coated window and detecting its fluorescence by a photo multiplier. The quantum yield of the fluorescence is independent of the exciting wavelength in our experimental region[5]. The result is shown in Fig. 1.

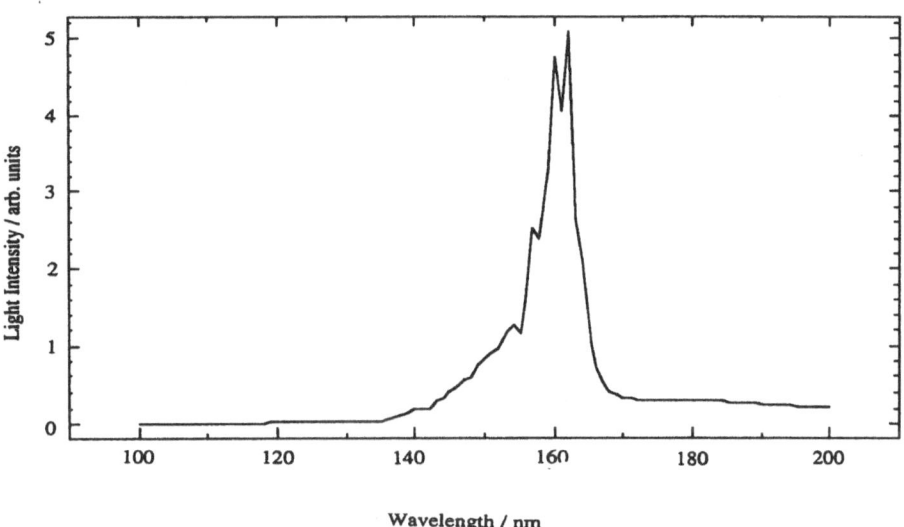

Figure 1.   Spectral distribution of the relative intensity of the light of a deuterium lamp

The deuterium lamp could either be used for preliminary measurements, which could later be extended over a wider energy range if useful with synchrotron light. Since the thresholds of most of these molecules with rather big molecular weights lie mostly in the emission range of the D_2-lamp, we used no synchrotron light.

The disadvantage of a helium lamp is that the experiments are limited to only two different wavelengths of excitation.

The applied experimental setup is shown schematically in Fig. 2.

A deuterium discharge lamp with a MgF_2 window   followed by a vacuum ultraviolet monochromator was used as light source. The bandwidth of the monochromatized light was 1nm. In the photo emission experiment, the light fell on the

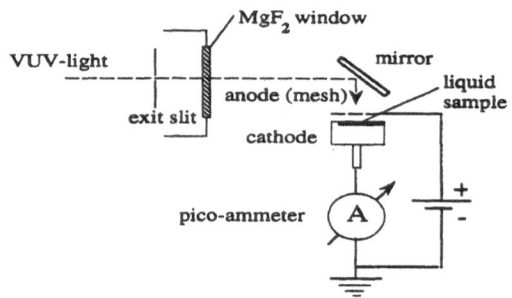

Figure 2.   Experimental setup to measure the photoelectron emission from a liquid

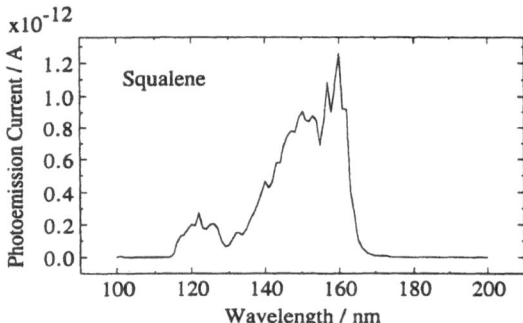

Figure 3.   Photoelectron current as function of the wavelength of the irradiated light for squalene

cathode after passing through a MgF$_2$ window, a MgF$_2$ coated aluminum mirror, and a gold coated mesh. A thin liquid film of about 150-200 μm thickness was put on the cathode in air, and then it was evacuated to 10^{-4} Pa. The emitted electron current was measured between the cathode and anode mesh as a function of wavelength. The separation of cathode and anode was about 2 mm. At an applied voltage of 10 V practically all electrons were detected. The actual electron emission current as a function of the wavelength of photons obtained with squalene is shown in Fig. 3.

Division of the emission current by the relative intensity (Fig. 2) gives the relative quantum yield as a function of photon energy. To obtain the absolute quantum yield, Y , the absolute photon-intensity must be known. It was determined by measurement of the photoelectron emission from a thin gold film and by the measurement of the photo ionization current of gaseous nitric oxide, NO, in the ionization chamber. For both systems, absolute quantum yields have been reported in the literature[6-8]. The photoelectron yield from gold measured by Cairns and Samson[6] differs considerably from the value reported by Krolikowski and Spicer[7]. This is probably due to the different vacuum conditions employed. We adopted the former value since the absolute intensity obtained with the ionization current of NO corrected for the transmittance of the MgF$_2$ window, agreed within 25 % with that from the photo emission current of Au using the value of Cairns and Samson.

In the following, we will show the results of our studies for two types of molecules, one group with squalane and squalene and the group consisting of three different phthalates.

All chemicals were used as supplied by the companies (Aldrich and Merck, respectively) with a purity of >98%.

The structural formula of squalane and squalene is:

(a)   Squalane:

(b)   Squalene:

The other examples are three different phthalates. Their structural formula is:

$$CH_2CH_3$$

Bis(2-ethylhexyl)phthalate: $R = R' = CH_2CHCH_2CH_2CH_2CH_3$

Dibutylphthalate: $R = R' = CH_2CH_2CH_2CH_3$

Benzylbutylphthalate: $R = CH_2-$⬡

$R = CH_2CH_2CH_2CH_3$

RESULTS and DISCUSSION

Photoelectron emission yields

For many organic crystals, a power function for the dependence of Y on the energy difference (hv-E_{th}) was observed[9-11]. We applied this power function to obtain the threshold energy for electron emission, E_{th}, of the liquids studied by us:

$$Y = A \cdot (hv - E_{th})^n \qquad (1)$$

A is a constant, and the exponent was taken as n=3[10,11]. Sometimes, other power functions are proposed in the literature[12,13]. The threshold values obtained with n = 2.5, n = 3.5, or n = 4, agree with those of n = 3 within ±0.1 eV. We estimate the maximum error of our experimentally determined threshold energy to be ±0.2 eV.

In Fig. 5 is shown the quantum yield Y of squalane and squalene as a function of the photon energy. The cube roots of these values are plotted in Fig. 4. The threshold energies obtained are 8.4 eV for squalane and 6.9 eV for squalene, respectively. The threshold for squalane is comparable to that of solid polyethylene (8.5 eV[14] or 8.8 eV[15]) or polypropylene (8.5 eV[16]).For the isolated molecule, the lowest binding energy of the electron is designated as IP. For alkanes, IP is determined by the σ-electrons of the single bonds, while for alkenes the first IP is determined by the π-electrons of double bonds. The second ionization energy IP* is determined by σ-electrons. The difference of IP for alkanes and alkenes is about 1.0 eV to 2.5 eV[17,18]. The corresponding difference in the emission threshold energy, ΔE_{th} =1.5 eV, is ascribed to the different binding energies of σ- and π-electrons. The absolute quantum yield shows the influence of the π-electrons, too.

As indicated in Fig. 4, the quantum yield of squalene near the threshold energy is due to π-electrons, only. At (E_{th}*) = 8.9 eV the onset of a second component is observed . This is interpreted as being due to the additional ionization of σ-electrons. In squalane, only σ-electrons can be emitted. As the photon energy is increased, the contribution

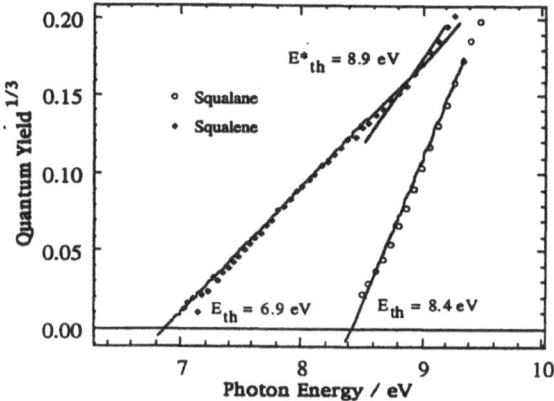

Figure 4. Photon energy dependence of squalane and squalene on quantum yield near the threshold energy

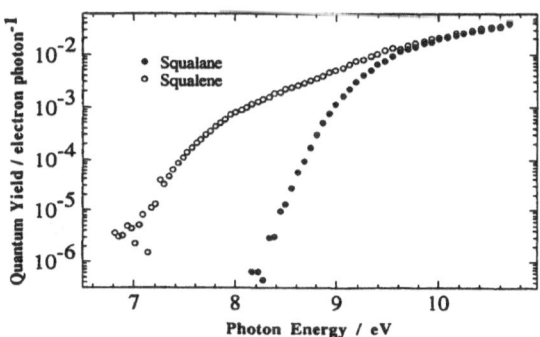

Figure 5. Quantum Yield versus exciting photon energy

of the π-electrons to the emission current of squalene decreases, and the quantum yields for squalane and squalene approach each other (Fig. 5).

Fig. 7 shows the quantum yields of phthalate esters, while Fig. 6 shows the extrapolation of the data according to Eq. 1. The threshold energies are indicated in Fig. 6.

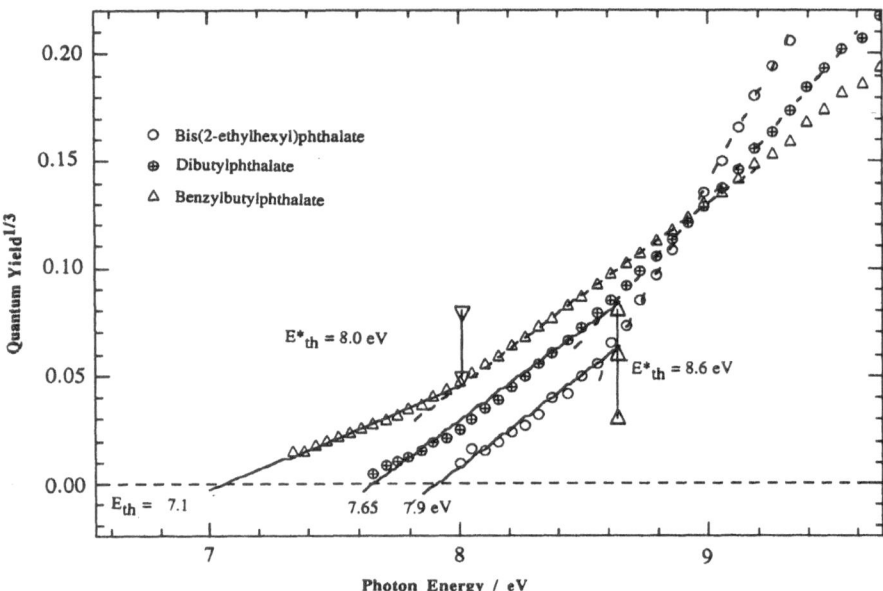

Figure 6.   Power law behavior of different phthalates near the threshold energy

Photo emission in the region lower than about 8.5 eV is due to ionization from π-orbitals of the benzene rings. The lowest value of E_{th} was observed for benzylbutylphthalate. From gas phase electron spectra of substituted benzenes[17,19] it may be inferred that the π-electrons of the benzyl-group have the lowest binding energy. Electrons from π-orbitals on the benzene ring bound to the C=O groups have a higher binding energy[20-22]. These conditions are reflected in the values of E_{th} and $E_{th}*$ of the benzylbutyl phthalate (see Fig. 7). In the other two phthalates, E_{th} is due to electrons from the benzene ring bound to the C=O groups while $E_{th}*$ is either due to electrons from non-bonding orbitals of oxygen atoms or from σ-bonds. Above 8.5 eV, in all three phthalates, the electron emission from σ-bonds sets in as in the case of the structurally similar dicarboxylic esters. A description of the results of a group dibutylsebacate and bis(2-ethylhexyl)adipate will be published together with some amines and phosphates[23].

Fig. 7 reflects again, as already shown for squalane and squalene (Fig. 5) the bigger quantum yield out of σ-bonds than from electrons out of π-orbitals. The larger number of σ-bonds in bis(2-ethylhexyl)phthalate than in dibutyl- and benzylbutylphthalate leads to a cross-over of the quantum yield with increasing excitation energies.

Electronic energy levels in the liquids

The emission threshold energies of the different organic compounds are tabulated in table 1. The following equation relates the photoelectron emission threshold in the liquid (E_{th}) to the ionization potential in the gas phase (IP):

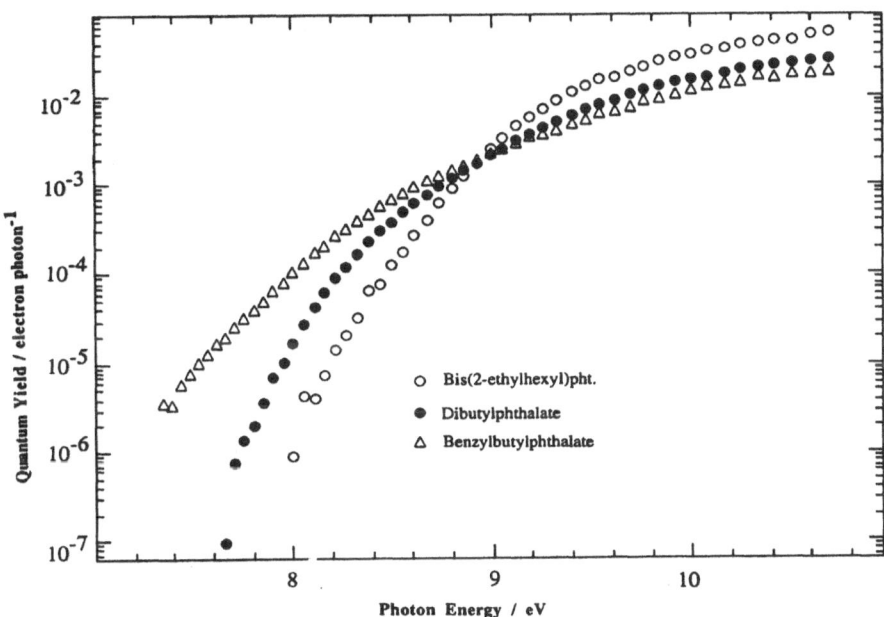

Figure 7. Quantum yield of photoelectron current versus exciting photon energy

$$E_{th} = IP + P_+ + E_{val}$$

where P_+ denotes the polarization energy of the positive ion, and E_{val} is the broadening of the valence levels of the isolated molecule upon condensation. E_{val} is between -0.1 and -0.2 eV for solid hydrocarbons[24]. A similar value can be expected for squalene and squalane and also the phthalates. Neglecting E_{val}, we can estimate P_+ if the values of (IP) are available. As far as we know, no (IP) values in the gas phase are available for the compounds investigated here. With a slightly changed cell we could heat up a drop of dimethylphthalate to 120°C. With a wire as cathode we could collect all electrons at the walls, connected as anode. Determinations of (IP) for other compounds could not be performed due to the higher temperatures required for a suitable vapor pressure. Though dimethylphthalate is different from the phthalates we measured, its (IP) might be a good approximation for the phthalates investigated here. The (IP) is due to the photoelectron ejection from the π-orbital of the phenyl group. The alkyl group has little effect on this orbital, since the orbital is separated from the alkyl group by a carboxyl group (-COO-).

The polarization energy, P_+, can be interpreted by Born's equation[25],

$$ P_+ = \frac{-e^2}{8\pi\varepsilon_0 R}\left[1 - \frac{1}{\varepsilon_r}\right] \tag{3} $$

where R is the radius of the positive ion, e is the electronic charge, ε_0 is the permittivity of the vacuum, and ε_r denotes the relative dielectric constant of the liquid. The polarization

Table 1.   1st. photo emission threshold liquid to vacuum, $E_{th,}$; 2nd. threshold, $E_{th}*$

Compounds	E_{th} / eV	E_{th}^* / eV
Bis(2-ethylhexyl)phthalate	7.9	8.6
Dibutyl phthalate	7.65	8.6
Benzylbutyl phthalate	7.1	8.0
Squalane	8.4	
Squalene	6.9	8.9

energy is measured from the vacuum level, E_{vac}, which is set to zero. Since photo ionization is a much faster process than orientation of molecules in a liquid, only the electronic polarization contributes to P_+. The dielectric constant is then given by,

$$ \varepsilon_r = n^2 \tag{4} $$

where n denotes the well known refractive index. The ε_r values calculated from n vary between 2.05 and 2.3. Using Eqs. 3 and 4, we calculated R. Since the values of n do not differ much from one liquid to another, the differences of P_+ reflect those of R. It might be interesting to compare the ionic volume with that of the neutral molecule, as shown in Table 2.

Table 2.   Photo emission thresholds 1st. photo emission threshold liquid to vacuum, E_{th}, 2nd. threshold, $E_{th}*$; ionization potentials in the gas phase, IP; polarization energies of ion, P_+; ionic radius, R; ion volume, V_i; average radius of molecule, r; average volume of molecule, V.

Compounds	E_{th}	IP	P_+	R	r	V_i	V	V_i/V
	eV	eV	eV	nm	nm	nm^3	nm^3	
Bis(2-ethylhexyl)- phthalate	7.9	9.1[a]	-1.2	0.33	0.54	0.15	0.66	0.22
Dibutyl phthalate	7.65	9.1[a]	-1.5	0.27	0.47	0.08	0.44	0.18

a) from value of dimethylphthalate (this work).

The molecules of the liquids investigated here interact by weak van der Waals forces. The factors governing the photo ionization process of a molecule in the liquid are almost the same as for the isolated molecule. The energy levels are shifted by the contribution of the polarization energy. From some calculations performed by Morgner et al.[26], it is apparent that over 90 percent of the polarization energy stems from two to three layers of neutral molecules around the ion. The apparent emission yield is influenced by the transport of the hot electrons through the liquid and across the liquid/vapor interface.

The absolute quantum yield

A three step model is commonly used for the explanation of the photo emission process[1,7,9]. The three steps are schematically depicted in Fig. 8. The first step consists of the optical excitation of an electron; the second step describes the electron tansport through the sample towards the surface; and the third step represents the escape through the surface into the vacuum. The optical excitation of the electron is described by an initial quantum yield η, in analogy to liquid phase photo ionization[27]. The hot electron undergoes a random walk losing its excess energy by elastic or inelastic collisions till it becomes thermalized; at a distance r in Fig. 8. If the energy of the bottom of the electronic conduction band V_0 is below the vacuum level, then only electrons with a kinetic energy > $|V_0|$ can leave the liquid (process A). Thermalized electrons will arrive at the surface and encounter a barrier given by V_0 (process B). Some delayed emission may occur if rearrangement of the surface due to thermal motion takes place. If $V_0 > 0$, emission into vacuum is not hampered by the liquid/vacuum interface. A factor of minor importance is the existence of an image barrier due to the change in dielectric constant from liquid to vacuum. Most organic molecules have a V_0 value between 0 and -1 eV.

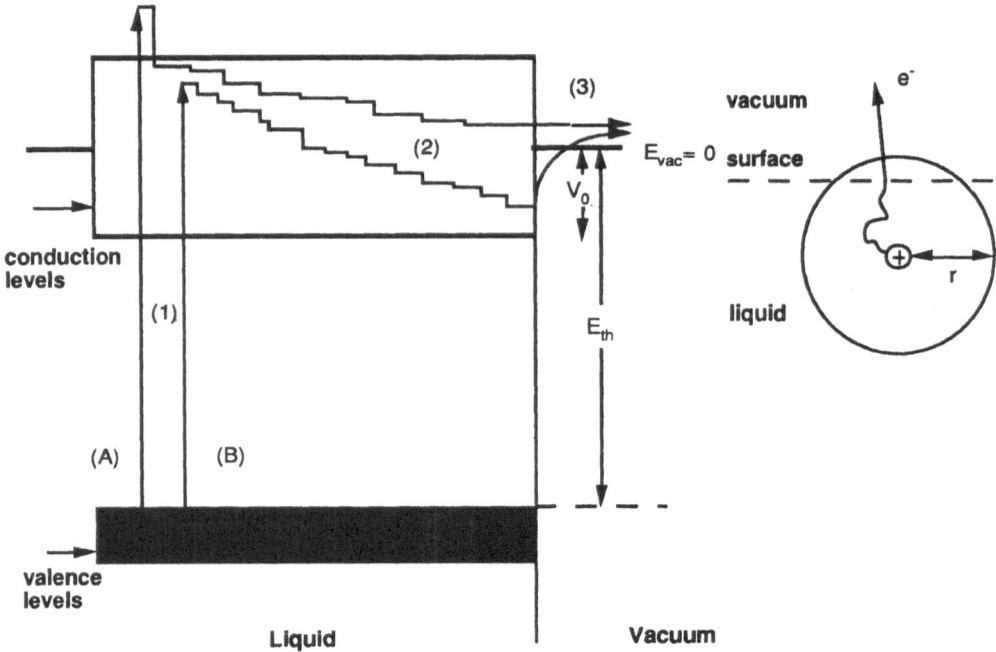

Figure 8. Three step model of the electron emission process

Birkhoff et al.[1,2] derived an equation relating the photo emission yield Y to the photo absorption coefficient μ and to the photoelectron escape depth λ_e. They employed the three step model of electron emission which was developed by Spicer et al. for metals[28].

The photon intensity decreases exponentially with distance x from the surface. The number of ionization events between x - x+dx is then given by,

$$\eta \times \mu \cdot \exp(-\mu \cdot x) \cdot dx \qquad (5)$$

where η and μ are initial photo ionization quantum yield and the photo absorption coefficient, respectively. The electron escape probability P depends exponentially on the escape length λ_e according

$$P \propto \exp\left(-\frac{x}{\lambda_e}\right) \qquad (6)$$

Birkhoff et al. neglected non-ionizing processes and assumed that η is unity[1,2]. An ejected photoelectron undergoes a random walk, losing its excess kinetic energy by elastic or inelastic collisions until it become thermalized. If the electron reaches the liquid/vapor interface while it still has a kinetic energy

$$\langle \varepsilon \rangle > |V_0| \qquad\qquad \text{for } V_0 < 0 \qquad\qquad (7)$$

it will escape into the vapor space. If the electron is thermalized inside the liquid, it will encounter the potential barrier V_0 at the interface. If $V_0 < 0$ emission into the vapor space is hampered. We estimate the emission yield by assuming that it is mainly due to hot electrons traversing the liquid/vapor interface. The quantum yield Y is then obtained by integration of the product of Eqs. 5 and 6 over all separation distances between electrons and parent cations in the liquid.

Y increases with increasing η, μ, and λ_e. Although few data on the absorption coefficient in the VUV exist, an estimate of Y is possible to give an estimation. The absorption coefficient of poly-dimethylsiloxane was measured by Sowers et al.[29]. At a photon energy of 10 eV it is $7 \cdot 10^5$ cm^{-1}. At the same time, values of the thermalization length r are available from measurements of the γ-radiation induced conductivity by Schmidt and Holroyd[30] and from the VUV photo conductivity by Baron, Casanovas et al.[31]. A value of 7 nm seems to be a good estimate for the average thermalization distance. λ_e might be somewhat smaller, depending on the V_0 value in Eq. 7.

Inserting the numbers, we estimate $Y \approx 0.05$ at 10 eV which is close to most measured values in Figs. 5 and 7.

A short description of the calculation of the photoelectron emission yield is given by Koizumi et al.[33]. A more detailed calculation will follow by Koizumi[32].

CONCLUDING REMARKS

The investigation of the VUV-light-induced electron emission from low vapor pressure non-polar liquids yields insight into the energetics of the ionization process in the condensed liquid phase. At the same time, absolute quantum yield measurements provide additional information on the transport of hot electrons in these liquids and through the liquid/vapor interface.

REFERENCES

1. Birkhoff, R. D., J. M. Heller, J., Painter, L. R., Ashley, J. C., and H. H. Hubbell, J., Photo emission and electron mean free paths in liquid formamide in the vacuum UV, J. Chem. Phys. 76:5208 (1982).

2. Birkhoff, R. D., H. H. Hubbell, J., Ashley, J. C., and Painter, L. R., Yields and mean free paths of photoelectrons from liquid hexamethyl phosphoric triamide, J. Chem. Phys. 77:4350 (1982).

3. Delahay, P., Photoelectron Emission Spectroscopy of Liquids and Solutions, in "Electron Spectroscopy: Theory, Tech. Appl.", ed. by C. R. Brundle and A. D. Baker, Academic, London, 1984, Vol. 5, pp.123.

4. Watanabe, I., Ono, K., and Ikeda, S., Photoelectron emission study of iron(II) and cobalt(II) complexes in aqueous solution. Reorganization energies., Bull. Chem. Soc. Jpn. 64:352 (1991).

5. Samson, J. A. R., "Techniques of Vacuum Ultraviolet Spectroscopy", Wiley, New York, (1967).

6. Cairns, R. B., and Samson, J. A. R., Metal Photo cathodes as Secondary Standards for Absolute Intensity measurements in the Vacuum Ultraviolet, J. Opt. Soc. Am. 56:1568 (1966).

7. Krolikowski, W. F., and Spicer, W. E., Photo emission Studies of the Noble Metals. II. Gold, Phys. Rev. B 1:478 (1970).

8. Watanabe, K., Matsunaga, F. M., and Sakai, H., Absorption coefficient and photo ionization yield of NO in the region 580-1350 Å, Appl. Opt. 6:391 (1967)

9. Pope, M., and Swenberg, C. E., "Electronic Processes in Organic Crystals", Clarendon, Oxford, (1982).

10. Kochi, M., Harada, Y., Hirooka, T., and Inokuchi, H., Photo emission from organic crystal in vacuum ultraviolet region. IV, Bull. Chem. Soc. Jpn. 43:2690 (1970).

11. Belkind, A. I., Brodskii, A. M., and Grechov, V. V., Theory of Photo emission from Molecular Crystals, Phys. Stat. Sol. (b) 85:465 (1978).

12. Watanabe, I., Flanagan, J. B., and Delahay, P., Vacuum ultraviolet photoelectron emission spectroscopy of water and aqueous solutions, J. Chem. Phys. 73:2057 (1980).

13. Brodsky, A. M., An investigation of the photoelectron emission from solutions containing solvated electrons, and the physical nature of the solvated electron, J. Phys. Chem. 84:1856 (1980).

14. Fujihira, M., and Inokuchi, H., Photo emission from Polyethylene, Chem. Phys. Lett. 17:554 (1972).

15. Less, K. J., and Wilson, E. G., Intrinsic photo conduction and photo emission in polyethylene, J. Phys. C 6:3110 (1973).

16. Chen, S. X., Seki, K., Inokuchi, H., Hashimoto, S., Ueno, N., and Sugita, K., Ultraviolet photoelectron spectroscopy of some fundamental vinyl polymers and the evolution of their electronic structures, Bull. Chem. Soc. Jpn. 58:890 (1985).

17. Kimura, K., Katsumata, S., Achiba, Y., Yamazaki, T., and Iwata, S., "Handbook of HeI Photoelectron Spectra of Fundamental Organic Molecules", Japan Scientific Societies Press, Tokyo, (1981).

18. Ashmore, F. S., and Burgess, A. R., Photoelectron Spectra of the Unbranched C5-C7 Alkenes, Aldehydes, and Ketones, J. Chem. Soc. Faraday Trans. II 74:485 (1978).

19. Selim, E. T. H., and Helal, A. I., The study of C1-C3 monosubstituted alkyl benzenes by the inverse convolution of first differential ionization efficiency curves, Org. Mass Spectrom. 17:539 (1982).

20. McLoughlin, R. G., and Traeger, J. C., A photoionization study of some benzoyl compounds-Thermochemistry of [C7H5O]+ Formation, Org. Mass Spectrom. 14:434 (1979).

21. Gal, J.-F., Geribaldi, S., Pfister-Guillouzo, G., and Morris, D. G., Basicity of the carbonyl group. Part 12. Correlation between Ionization Potentials and Lewis Basicities in Aromatic Carbonyl Compounds, J. Chem. Soc. Perkin Trans. II :103 (1985).

22. McLoughlin, R. G., Morrison, J. D., and Traeger, J. C., Org. Mass Spectrom. 14:104 (1979).

23. Koizumi, H., Lacmann, K., and Schmidt, W. F., VUV Light-Induced Electron Emission from Organic Liquids, J. Electron Spectrosc. Relat. Phenom. (accepted for publication).

24. Pireaux, J. J., and Caudano, R., X-ray photoemission study of core-electron relaxation energies and valence-band formation of the linear alkanes. II. Solid-phase measurements, Phys. Rev. B 15:2242 (1977).

25. Born, M., Z. Physik 1:45 (1920).

26. Morgner, H., Oberbrodhage, J., Richter, K., and Roth, K., The gas-liquid phase transition shift at surfaces: experimental method and interpretation, J. Electron Spectrosc. Rel. Phenom. 57:61 (1991).

27. Böttcher, E. H., and Schmidt, W. F., Photoconductivity of nonpolar liquids induced by vacuum-ultraviolet light, J. Chem. Phys. 80:1353 (1984).

28. Spicer, W. E., Optical Density of States Ultraviolet Photoelectric Spectroscopy, J. Res. Nat. Bur. Stand. 74A:397 (1970).

29. Sowers, B. L., Williams, M. W., Hamm, R. N., and Arakawa, E. T., Optical properties of some silicone diffusion-pump oils in the vacuum ultraviolet using a closed-cell technique, J. Appl. Phys. 42:4252 (1971).

30. Schmidt, W. F., and Holroyd, R. A., Ion Mobilities and Yields in X-irradiated Polymethylsiloxane Oils, Radiat. Phys. Chem. 39:349 (1992).

31. Baron, P. L., Casanovas, J., Guelfucci, J. P., and Hoi, R. L. S., Photoconductivity Induced By VUV Photons In Polydimethylsiloxane and Polymethylphenylsiloxane Oils, IEEE Trans. Electr. Insul. 23:563 (1988).

32. Koizumi, H., (to be published).

33. Koizumi, H., Lacmann, K., and Schmidt, W. F., Light-induced emission from low vapor pressure organic liquids, Nucl. Instr. and Meth. A327:75 (1993).

SECTION VIII. APPLICATIONS

PHYSICS OF NOBLE GAS X-RAY DETECTORS:
A MONTE CARLO SIMULATION STUDY

Teresa H. V. T. Dias

Departamento de Física
Universidade de Coimbra
3000 Coimbra
Portugal

INTRODUCTION

As it is well known, x-rays interact easily with matter, through multiple atomic processes involving the production of electrons. Most x-ray detection devices are based on this effect, and the absorption of x-ray photons with energy E_{xr} translates into a number N of primary electrons (i.e. subionization electrons). The filling of these detectors is often a gas, chosen to have a high cross section for photoelectric effect. The measurement of the energy E_{xr} is basically obtained in terms of the number of primary electrons produced: the position of a peak corresponding to monoenergetic x-rays is related to the average number of electrons, $\overline{N}$, while its width is related to the fluctuations in N. In principle E_{xr} is proportional to $\overline{N}$; however, energy linearity does not always hold, as it will be shown later.

Basically, there are two types of gaseous detectors for x-rays (Knoll 1989): the gas proportional *scintillation* counter (GPSC) and the gas proportional *ionization* counter (GPIC), see Fig.1. In a GPSC, the primary electrons are first allowed to drift along a region of low electric field, and are then accelerated in a region of higher field where they gain enough energy to excite (but not ionize) the atoms in the gas, the deexcitation leading to the production of scintillation. In this way, each primary electron can produce a large number of photons in this amplification stage. Xenon is often the filling for GPSC's as it has a high efficiency for photoelectric effect and for scintillation. In a GPIC, although the field is increasing as electrons approach the anode, the x-ray absorption and electron drift stages are basically the same as in a GPSC, but each primary electron does now produce an electron avalanche near the anode: a large number of secondary electrons is produced per each primary electron in this amplification stage. In a GPSC the scintillation photons are collected by a PMT, in a GPIC the secondary electrons are collected instead. GPSC's have a very good energy resolution - ~2 times better than GPIC's - actually approaching the intrinsic energy resolution (which is defined in terms of the fluctuations of N alone), and have been used in many applications, namely x-ray astronomy, medical imaging, x-ray fluorescence analysis (Butler and Scarsi 1990, Perez-Mendez 1987, dos Santos et al. 1993b).

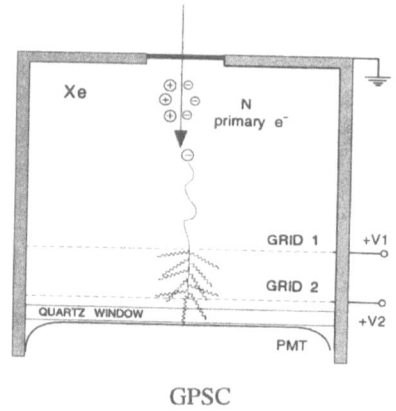

GPSC

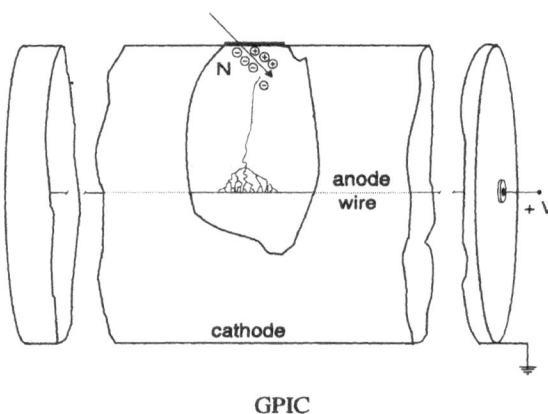

GPIC

Figure 1. Schematic representation of a gas proportional scintillation counter (GPSC) and a gas proportional ionization counter (GPIC).

GPSC's have been developed for a long time in Coimbra (Conde et al. 1977, Policarpo 1981, dos Santos 1993b). Recently, a thorough investigation of the physics involved in gaseous xenon filled detectors has been developed using a detailed Monte Carlo simulation study at P=760Torr, T=293K, as will be described in the next sections.

MONTE CARLO SIMULATION

Details of the Monte Carlo method developed can be found in Dias et al. 1993, but a brief outline of the relevant physical processes considered in the simulation will be made here.

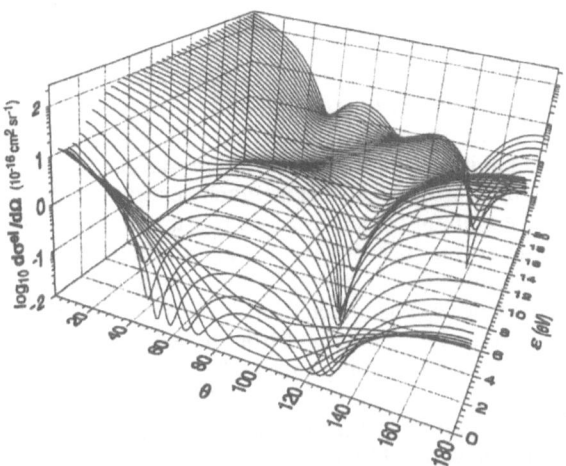

Figure 2. Electron impact angular differential elastic cross section $\sigma^{el}(\varepsilon,\theta) = d\sigma^{el}(\varepsilon)/d\Omega$ for electrons in Xe (McEachran and Stauffer 1986, 1987, Stauffer et al. 1986).

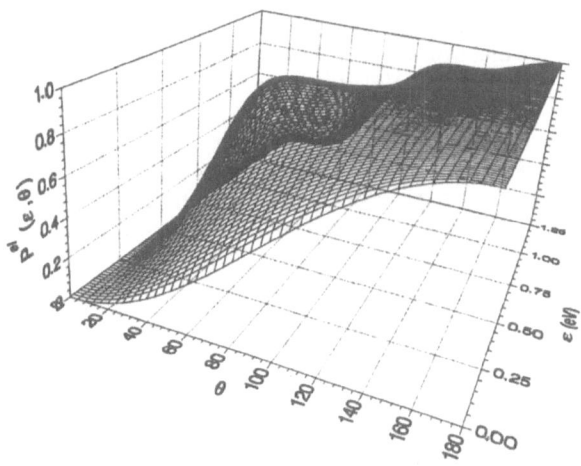

Figure 3. Cumulative probability $P^{el}(\varepsilon,\theta)$ for angular elastic scattering of electrons

in xenon ($\varepsilon \leq 1.25$ eV): $\quad P^{el}(\varepsilon,\theta) = \int_0^\theta \sigma^{el}(\varepsilon,\theta)\sin\theta\,d\theta \Big/ \int_0^\pi \sigma^{el}(\varepsilon,\theta)\sin\theta\,d\theta.$

When one x-ray photon interacts with a Xe atom in the gas, the atom is photoionized and the resulting xenon excited ion is left with a vacancy in the photoionized shell. This vacancy can decay through Auger/Coster-Kronig electron emission or x-ray fluorescence emission. Together with the photoelectron or Auger emission, a second electron can be ejected through shakeoff effect. In each case a new vacancy is created, which may again decay by the above processes. The Xe excited ion thus decays through a vacancy cascade process which results in the emission of a number n of cascade electrons and a multiple charged Xe^{n+} ion. The number of cascade electrons and their energy distribution are related with which subshell is photoionized and which transitions are involved in the subsequent vacancy decay.

The n cascade electrons are then followed along their free parabolic paths and elastic, excitation or ionization collisions with xenon atoms, until all electrons are slowed down to an energy below the Xe ionization threshold. In the end, for each x-ray photon absorbed in the gas, N primary (i.e. subionization) electrons are produced.

To reproduce all these processes with as much accuracy as possible along the Monte Carlo simulation, a large amount of data was put together, as fully described in Dias et al. 1993, in order to enable correct decisions to be taken at any stage. This data includes Xe photoionization cross sections (total, partial, angular differential), transition rates (for all subshells) for the vacancy decay phenomena involved in the Xe ion vacancy cascade decay, and also elastic, excitation and ionization electron impact scattering cross sections in xenon (again integral and angular differential, as well as energy partition for the case of ionization). As an example, Fig.2 and Fig.3 illustrate the electron impact angular differential elastic cross section and the corresponding cumulative probability surface for angular elastic scattering in Xe used in the Monte Carlo simulation.

RESULTS

The Monte Carlo simulation developed allowed the investigation of all three stages involved in a GPSC, namely the absorption of x-rays and the formation of the primary electron cloud, the drift of the primary electrons, and the electroluminescence stage. Results have been obtained for gaseous Xe on ion charge distributions (Santos et al. 1992), the size of the primary electron cloud (Dias et al. 1991a), the intrinsic energy resolution and the Fano factor (Dias et al. 1991b), energy linearity and the w-value i.e. the mean energy necessary to produce a primary electron (Santos et al. 1991), distortion effects of soft x-ray energy spectra (Dias et al. 1992), electron drift parameters -mean energy, drift velocities, diffusion coefficients, mobilities- (Dias et al. 1993, Santos et al. 1993) and electroluminescence (Santos et al. 1993). Although more extensive results are presented and discussed in all the above mentioned papers, representative results will be shown and discussed here.

Xenon Ion Charge Distributions

From the interaction of each x-ray photon with a xenon atom, n cascade electrons and a multiple charged ion Xe^{n+} are formed. It is important to get the right number and energy distribution of these cascade electrons, as they are the source for the N primary electrons. As the xenon ion charge distribution is sensitive to the way the photoionization and the vacancy cascade decay are modelled in the simulation, the Xe^{n+} charge distributions were calculated and compared with all available results from the literature, in order to assess the accuracy of the simulation. Two illustrative examples are shown in Fig.4 and Fig.5, and agreement was in general found to be good.

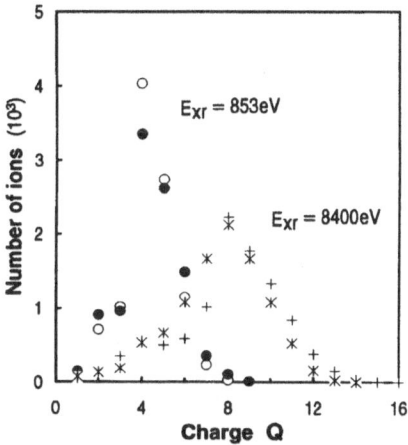

Figure 4. Xenon ion charge distribution produced by the absorption of 10^4 x-ray photons with E_{xr}= 853 and 8400eV: •, * Carlson et al. 1966; ○, + this work.

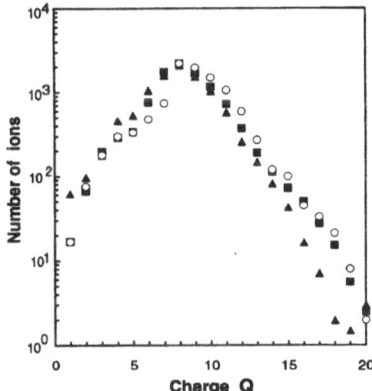

Figure 5. Xenon ion charge distribution produced by internal conversion following the decay of $^{131}Xe^m$ (10^4 ions): ▲ Pleasonton and Snell 1957, experimental; ■ Mukoyama 1986, Monte Carlo; ○ this work.

Radial Distribution of the Primary Electron Cloud

Xenon gaseous detectors are often used in position sensitive devices (Culhane 1991), so it is important to know the size of the primary electron cloud, as this is an intrinsic limit to the spatial resolution that can be achieved. Fig.6 shows examples of calculated primary electron densities as a function of radial distance to the x-ray incidence direction, for various x-ray energies.

We observe that there is an increasing spread in the radial distribution with increasing E_{xr}: a broad maximum gradually appears and shifts to higher distances with x-ray energy. It was found that this maximum corresponds to the contribution to the primary electrons arising from photoelectrons which, as far as the same subshells are photoionized, increase its energy for increasing E_{xr}. In the same circumstances, the contribution from Auger and shakeoff emission remains approximately unchanged, originating mainly the central part of the electron cloud.

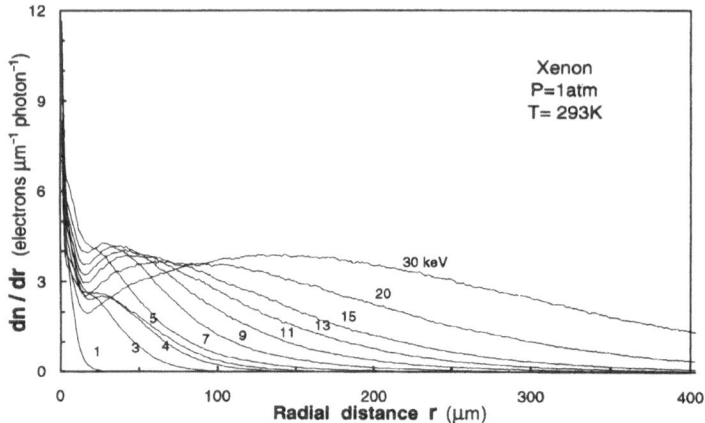

Figure 6. Calculated radial distribution of primary electrons, for the range $E_{xr} = 1$ to 30 keV.

Energy Resolution, Fano Factor and W-value

A very important characteristic of any energy detector is the ultimate energy resolution that can be achieved.

Fig.7 represents an x-ray energy peak, calculated as a frequency plot of the number N of primary electrons produced per absorbed x-ray photon. These peaks are obtained in the Monte Carlo simulation and allow, for each x-ray energy, a direct calculation of i) the relative variance of N, the Fano factor $F = \sigma^2/\overline{N}$, ii) the mean energy necessary to produce one primary electron, $w = E_{xr}/\overline{N}$, and iii) the intrinsic energy resolution $R_{int} = 2.35\sigma/\overline{N} = 2.35(Fw/E_{xr})^{1/2}$ (full width half maximum).

The Fano factor - a measure of the fluctuations in N - and the w-value depend in general on the absorbing medium and on the ionizing radiation or particle (Doke et al. 1992, Inokuti et al. 1992). In principle, if F and w are constant, the energy resolution is expected to vary continuously with E_{xr}, but this may not always be true for soft x-rays, as it is illustrated in Fig.8 where the energy resolution is represented as a function of E_{xr} in xenon. The Monte Carlo results exhibit clear discontinuities which reproduce well the unexpected experimental behaviour observed with xenon GPSC's by Kowalski et al. 1989, which was unexplained at the time.

However, the Monte Carlo investigation found a clear dependence of the Fano factor and w-value on x-ray energy which is consistent with this discontinuous behaviour of the energy resolution. Both the calculated Fano factor, represented in Fig.9, and the w-value, represented in Fig.10, exhibit in fact important discontinuities, which are located near the xenon absorption edges. The experimentally derived Fano factor values from Kowalski et al. 1989 which are included in Fig.9 were obtained by making $F \propto R^2 E_{xr}/w$ (R as in Fig.8) and adopting a constant value for w (21,9eV).

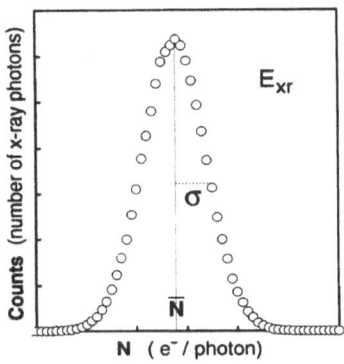

Figure 7. Typical peak for a given x-ray energy E_{xr}.

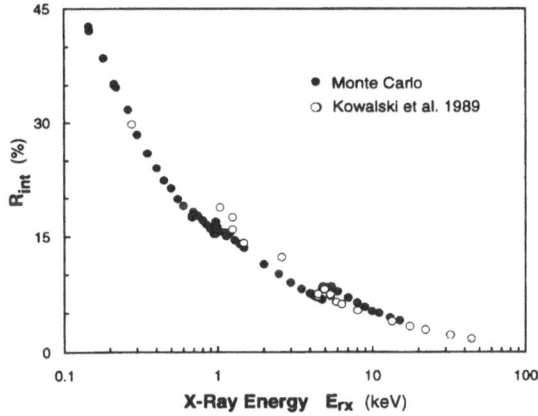

Figure 8. Energy resolution as a function of absorbed x-ray energy in xenon.

The discontinuous behaviour of the Monte Carlo calculated Fano factor shown in Fig.9 was found to be caused basically by the fact that, whenever a new (inner) atomic subshell can be photoionized, the energy distribution of the cascade electrons is different (photoelectrons with near zero energy, new Auger transitions) and this translates into larger fluctuations in N. On the other hand, the discontinuous behaviour of the calculated w-value observed in Fig.10 is related mainly with the extra energy retained in the larger number of vacancies with which xenon atoms are left whenever a new (inner) atomic subshell can be photoionized, and partly related with the different history the corresponding cascade electrons will follow while slowing down to subionization energies.

The x-ray energy dependence and discontinuous behaviour of the w-value has implications on energy linearity: strictly speaking, linearity will only hold if w is constant. In Fig.11 the calculated mean number $\overline{N}$ of primary electrons (i.e., the position of the x-ray energy peaks, see Fig.7) is represented as a function of x-ray energy, in the region of the

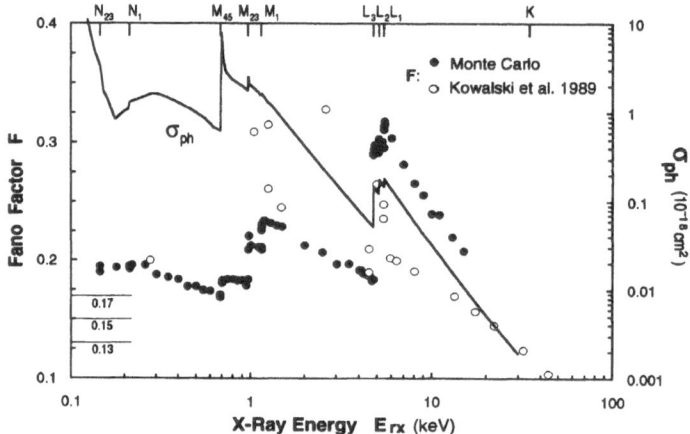

Figure 9. Fano factor F and photoionization cross section σ_{ph} as a function of x-ray energy in xenon.

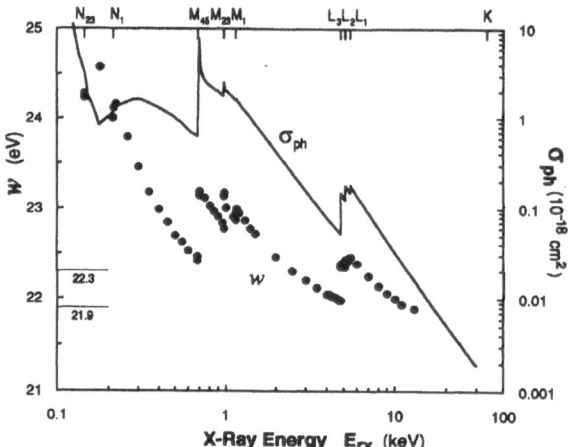

Figure 10. Calculated w-value in xenon as a function of absorbed x-ray energy.

xenon L absorption edges. The behaviour is approximately linear, but visible discontinuities arise near the absorption edges, and so energy linearity does not hold. These Monte Carlo calculated results were able to reproduce very well similar effects observed experimentally (Lamb et al. 1987, dos Santos et al. 1993a), and only the simulation was capable of giving an explanation for this phenomenon. Furthermore, we can see on Fig.11 that distinct x-ray energies may produce the same number of electrons, which means that in a real detector the absorption of these x-ray energies will generate pulses with equal

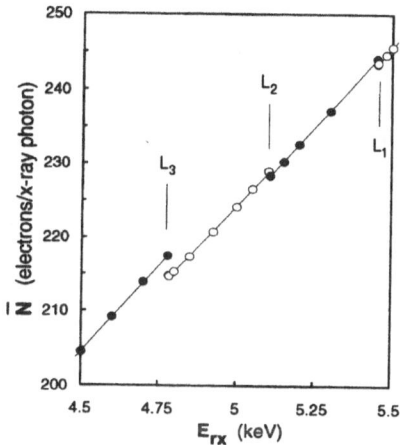

Figure 11. Calculated average number of primary electrons produced per absorbed x-ray photon in Xe as a function of x-ray energy, in the range 4500 to 5500 eV.

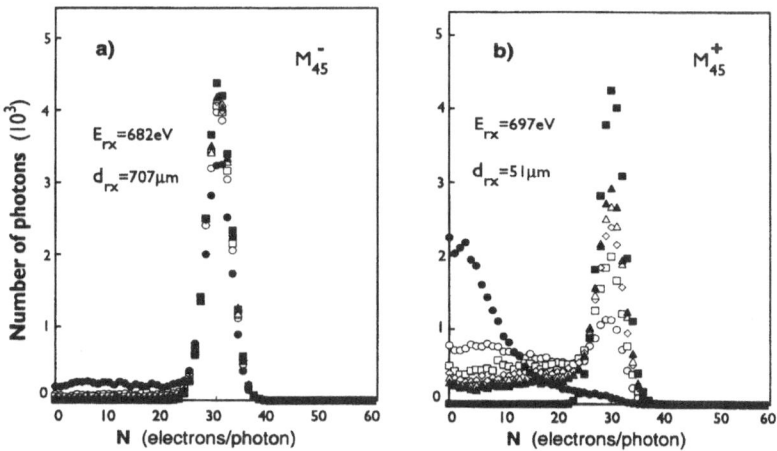

Figure 12. Calculated x-ray energy peaks for 25000 photons absorbed in xenon, corresponding to E_{xr}=682eV(M_{45}^-) and 697eV(M_{45}^+), for six different applied electric fields: ● 100V/cm, ○ 500V/cm, □ 1000V/cm, ◊ 1500V/cm, △ 2000V/cm, ▲ 2500V/cm. Reference peak (electron losses neglected): ■.

amplitude, thus falling in the same energy channel. This effect has originated unexpected features in experimental x-ray spectra (Peacock et al. 1985, Lamb et al. 1987), only now fully explained.

Distortion Effects in Soft X-Ray Energy Spectra

Another important experimental problem that was investigated was the cause of the distortion observed in soft x-ray energy spectra (Inoue et al. 1978, Hamilton et al. 1980, Simons and de Korte 1989), an effect that the Monte Carlo simulation was able to relate with the loss of electrons to a detector's entrance window. For 697eV x-rays - an energy just above the M_{45} absorption edge -, Fig.12b shows a series of six energy peaks calculated taking into account electron losses and at increasing applied electric fields in the absorption region, together with the "pure" peak obtained when this effect was neglected. Comparison with this reference peak clearly shows important tail enhancement and deterioration of resolution on all "real" spectra.

On the other hand, in Fig.12a we can observe that in a similar series of peaks for 682eV x-rays - an energy just below the M_{45} absorption edge -, the distortion effects are negligible. This is explained by the fact that the absorption length d_{rx} for $682eV(M_{45}^-)$ is much larger than for $697eV(M_{45}^+)$, as the photoionization cross section exhibits a large increase at this absorption edge.

An exhaustive study of these effects was made for soft x-rays up to 10keV, which

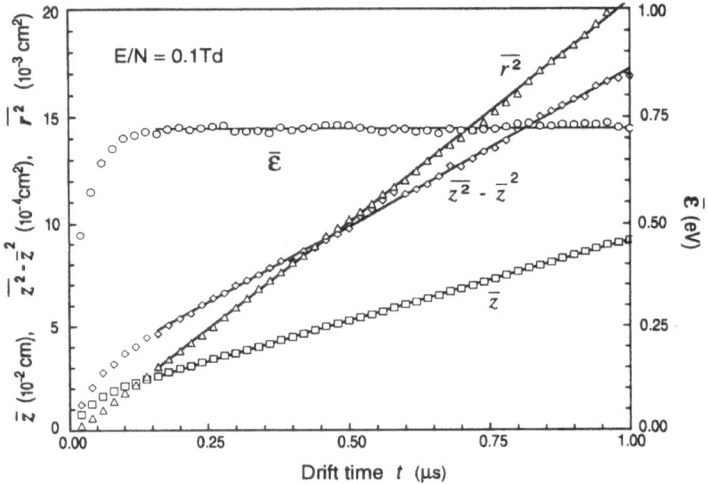

Figure 13. Calculated mean electron energy $\bar{\varepsilon}$ and mean displacements $\bar{z}$, $\overline{z^2}$ and $\overline{r^2} = \overline{x^2 + y^2}$ in Xe (760Torr, 293K) as a function of drift time t, for E/N=0.1Td.

fully assessed the role of the electric field in minimizing distortion by improving the collection of electrons near the entrance window of a detector.

Electron Drift Parameters

All results described until now are essentially related with the stage of x-ray absorption and formation of the primary electron cloud. However, the Monte Carlo simulation method developed for this purpose could also be used to calculate electron drift parameters in xenon: a sample of thermal electrons is in this case "injected" directly into the gas, and allowed to drift under various E/N values, where E is the electric field intensity and N is now the gas number density. The E/N range investigated was the range of interest for GPSC work, which is E/N<18Td, as ionization threshold for gaseous xenon is near this value, and electron multiplication is avoided in GPSC's.

Fig.13 illustrates how drift velocities and diffusion coefficients are calculated in the Monte Carlo simulation, taking as an example the case of E/N=0.1Td. The sample of electrons is allowed to drift for a time long enough to guarantee that equilibrium with the field is reached, i.e., that average electron energy $\bar{\varepsilon}$ stays approximately constant. Electron drift velocity, and longitudinal (//$\vec{E}$) and transversal ($\perp\vec{E}$) diffusion coefficients $D_l=d\overline{z^2}/dt$ and $D_t=d\overline{r^2}/dt$ are then calculated from the slopes of the linear increase with drift time t of the mean displacements $\bar{z}$, $\overline{z^2}$ and $\overline{r^2}=\overline{x^2+y^2}$ respectively. The initial non-equilibrium transient region is always neglected (~0.15µs in the example shown in Fig.13).

Fig.14 represents the calculated mean electron energy $\bar{\varepsilon}$ vs. E/N, together with the calculated mean collision frequency υ (the inverse mean time between collisions), which displays the minimum caused by the Ramsauer effect in the scattering cross section. The behaviour of $\bar{\varepsilon}$ reflects the way the collision frequency varies with E/N: a sharp increase when υ is decreasing, a slower increase when υ is increasing after the minimum. The flattening of $\bar{\varepsilon}$ observed from 3 to 16 Td can be attributed to the contribution of the excitation collisions which are occuring in this range.

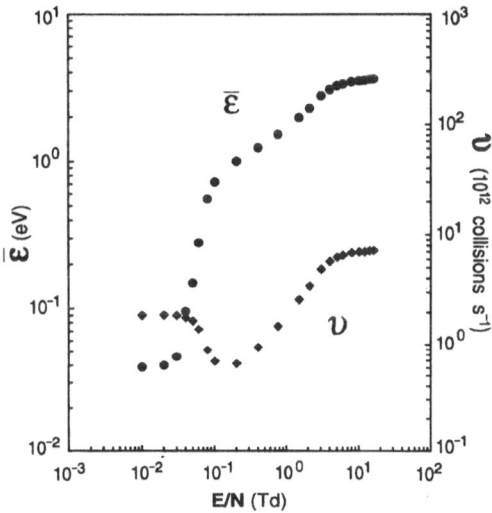

Figure 14. Calculated mean electron energy $\bar{\varepsilon}$ (●) and collision frequency υ (◆) as a function of E/N in gaseous xenon.

Calculated electron drift velocities, v_d, and reduced mobilities, $\mu N=v_d/(E/N)$, are shown in Fig.15. The agreement with experimental results is found to be good. As it is known, the behaviour of v_d and μ reflect the behaviour of the collision frequency υ: drift velocity displays a shoulder, and mobility displays a maximum, which are consistent with the existence of the minimum in υ.

Figs. 16 and 17 represent the computed characteristic energies $\varepsilon_{kl}=eD_l/\mu$ and $\varepsilon_{kt}=eD_t/\mu$ vs. E/N, shown together once more with the collision frequency υ. The process

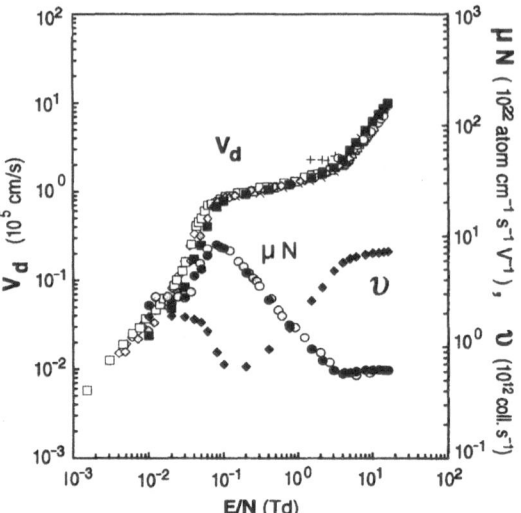

Figure 15. Electron drift velocity v_d, mobility μ and collision frequency υ as a function of E/N in gaseous xenon: △ Bowe 1960; □ Pack et al. 1962, Dutton 1975; ◈,◉ Huang and Freeman 1978; ○ Cumpstey et al. 1980; + Brooks et al. 1982; ◊ Hunter et al. 1988; × Hashimoto and Nakamura 1990; ■, ●, ◆ this work.

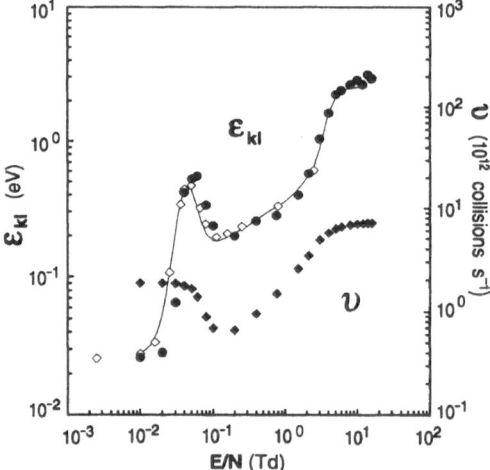

Figure 16. Characteristic energy $ε_{kl}$ and collision frequency υ as a function of E/N in gaseous xenon: ◊ Lowke and Parker 1969, Dutton 1975; — Suzuki et al. 1992; ●, ◆ this work.

of diffusion is known to be anisotropic in the presence of an applied electric field, and D_l tends to be lower than D_t when $\partial \upsilon / \partial \varepsilon > 0$, so that the behaviour of D_l, D_t, and ε_{kl}, ε_{kt}, is again related with the presence of a minimum in υ. Agreement of the Monte Carlo calculated ε_{kl} and ε_{kt} with experimental or Boltzmann analysis results found in the literature is good. The data from Lowke and Parker 1964, and from Suzuki et al. 1992 were calculated through a Boltzmann analysis using an electron scattering cross section derived in both cases from experimental drift velocities: this can explain the discrepancy found with the Monte Carlo ε_{kt} results, as it is known that drift velocities are much less sensitive to the scattering cross section than characteristic energies are. In fact, Koizumi et al. 1986 have derived from their experimental results a scattering cross section that is actually similar over the region of the Ramsauer minimum to the one we adopt in the Monte Carlo simulation, and so our characteristic results show a better agreement with his ε_{kt} values.

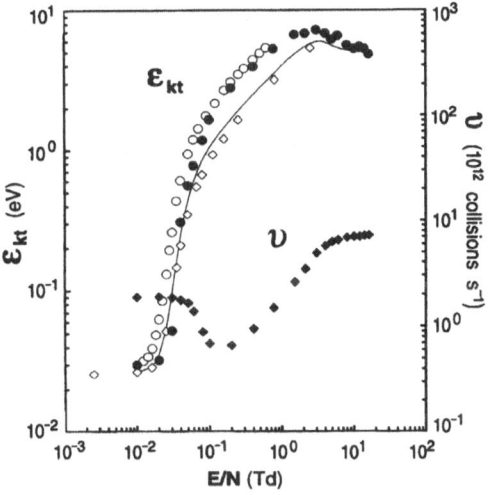

Figure 17. Characteristic energy ε_{kt} and collision frequency υ as a function of E/N in gaseous xenon: ◊ Lowke and Parker 1969, Dutton 1975; ○ Koizumi et al. 1986; — Suzuki et al. 1992; ●, ◆ this work.

The minimum in our cross section and in the one belonging to Koizumi's experimental results, is actually sharper than in the cross sections used in the case of Lowke and Parker 1964 (from Frost and Phelps 1964) and Suzuki et al. 1992.

Electroluminescence

The number of primary electrons produced per absorbed x-ray photon is usually small, and so, as we have seen, gas proportional scintillation counters have an amplification stage where the primary electrons produce electroluminescence: in the scintillation region, electrons are accelerated by an electric field so that they acquire enough energy to excite, but not ionize, the xenon atoms. At the pressure generally used in GPSC's (~1atm) the formation of excited dimers in three body collisions is favoured, which deexcite to a repulsive ground state with the emission of a VUV photon from a continuum: this way each primary electron can produce in the scintillation region a large number of these VUV photons. In the Monte Carlo simulation it is assumed that for every excitation collision, one scintillation VUV photon is emitted.

The calculated reduced number of photons produced per electron per unit length, i.e., the reduced electroluminescence yield, is represented in Fig.18 as a function of reduced electric field. We must bring to the reader´s attention that the electric field in the scintillation region of a GPSC is usually chosen to be close to (but below) the ionization threshold in order to optimize the working conditions of the detector in terms of energy resolution: a maximum number of scintillation photons with the least fluctuations can be obtained, achieving in these conditions an energy resolution which approaches the intrinsic value.

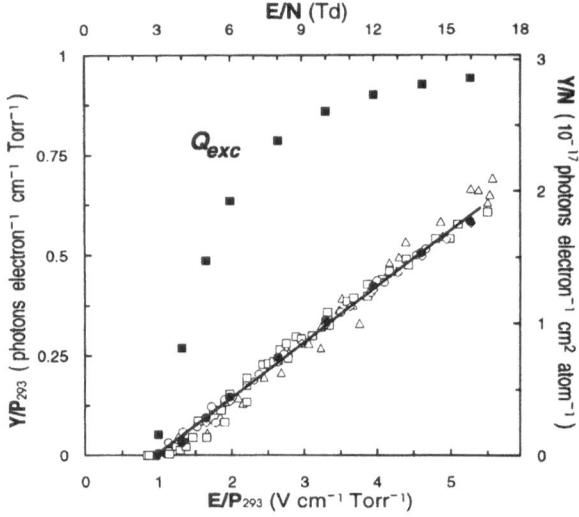

Figure 18. Reduced electroluminescence yield in Xe as a function of E/N: □ Conde et al. 1977, △ Andresen et al. 1977, ○ Favata et al. 1990, ◊ Fraga et al. 1990, ● this work, — linear fitting to ●. Excitation efficiency Q_{exc}: ■ this work.

In Fig.18, the Monte Carlo values are compared with three groups of experimental results and with one group of Boltzmann analysis results. The experimental values, always relative, have been scaled to the Monte Carlo results in the region of higher excitation efficiency. The excitation efficiency, also represented in Fig.18, was calculated as the ratio of the energy which electrons gain from the field during drift to the energy which is used up in excitation collisions with xenon atoms. We see that the electroluminescence yield shows an approximately linear behaviour with E/N, with a threshold at ~3Td, a behaviour that is in very good agreement with the experimental results. Below this scintillation threshold, no VUV photons are ever produced as electrons never acquire enough energy to produce excitation (onset at 8.3eV), loosing a significant amount of energy in elastic collisions through recoil to the xenon atoms, a loss which is never neglected in this Monte Carlo simulation. For $E/P_{293}=0.5$ Vcm⁻¹Torr⁻¹ (E/N~1.5Td), a field which does not yet allow for excitation to occur, this situation is clearly illustrated in Fig.19a, where the position of an electron (recorded every 200 collisions) is represented as a function of the

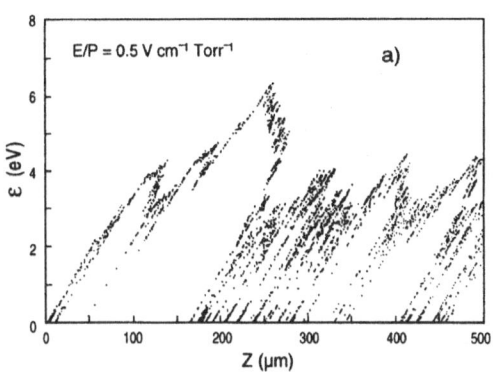

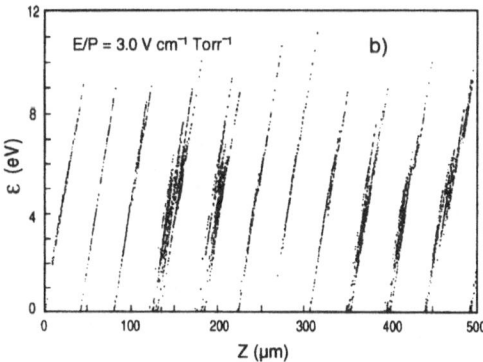

Figure 19. Electron energy ε in gaseous xenon (760Torr, 293K) as a function of position z along the electric field, for E/P = 0.5 and 3.0 V cm^{-1} Torr^{-1}.

distance travelled along the electric field (earlier results, from an unidimensional model in Dias et al. 1986). This can be compared with Fig.19b (every 100 collisions) for the case of $E/P_{293}=3.0$ Vcm^{-1}Torr^{-1} (E/N~9Td) where, along the same distance, several scintillation VUV photons are now produced: the large periodic discontinuities in ε for this higher field correspond to the losses on excitation collisions with xenon atoms.

CONCLUSION

A detailed study by a Monte Carlo simulation of the absorption of x-rays and electron drift in gaseous xenon was described, which yielded results in very good agreement with experimental work and achieved an important insight in the physics involved in xenon gas filled x-ray detectors.

Acknowledgements

Support is acknowledged from JNICT (Junta Nacional de Investigação Científica e Tecnológica, Portugal) and LNETI (Laboratório Nacional de Engenharia e Tecnologia Industrial, Portugal) through Eureka Project EU-39.

REFERENCES

Andresen, R.D., Leimann, E.A. and Peacock, A., 1977, The nature of the light produced inside a gas scintillation proportional counter, *Nucl. Instrum. Methods* 140:371.

Bowe, J.C., 1960, Drift velocity of electrons in nitrogen, helium, neon, argon, krypton and xenon, *Phys. Rev.* 117:1411.

Brooks, H.L., Cornell, M.C., Fletcher, J., Littlewood, I.M. and Nygaard, K.J., 1982, Electron drift velocities in xenon, *J. Phys. D: Appl. Phys.* 15:L51.

Butler, R.C. and Scarsi, L., 1990, The SAX mission, *SPIE* 1344:465.

Carlson, T.A., Hunt, W.E. and Krause, M.O., 1966, Relative abundances of ions formed as the result of inner shell vacancies in atoms, *Phys. Rev.* 151:41.

Conde, C.A.N., Requicha-Ferreira, L. and Ferreira, M.F., 1977, The secondary scintillation output of xenon in a uniform field gas proportional scintillation counter, *IEEE Trans. Nucl. Sci.* NS24:221.

Culhane, J.L., 1991, Position sensitive detectors in X-ray astronomy, *Nucl. Instrum. Methods A* 310:1.

Cumpstey, D.E. and Vass, D.G., 1980, The scintillation process in gas proportional scintillation detectors with uniform electric fields, *Nucl. Instrum. Methods* 171:473.

Dias, T.H.V.T., Santos, F.P., and Conde, C.A.N., 1991a, The primary electron cloud in xenon for x-rays in the 0.1 to 10 keV range, *Nucl. Instrum. Methods A* 310:137.

Dias, T.H.V.T., Santos, F.P., Stauffer, A.D., and Conde, C.A.N., 1991b, The Fano factor in gaseous xenon: a Monte Carlo calculation for x-rays in the 0.1 to 25 keV energy range, *Nucl. Instrum. Methods A* 307:341.

Dias, T.H.V.T., Santos, F.P., Stauffer, A.D., and Conde, C.A.N., 1992, Distortion effects in soft x-ray energy spectra for xenon gaseous detectors: a Monte Carlo simulation study, *Phys. Rev. A* 46:237.

Dias, T.H.V.T., Santos, F.P., Stauffer, A.D., and Conde, C.A.N., 1993, A Monte Carlo simulation of x-ray absorption and electron drift in gaseous xenon, *Phys. Rev. A* 48:2887.

Dias, T.H.V.T., Stauffer, A.D. and Conde, C.A.N., 1986, A unidimensional Monte Carlo simulation of electron drift velocities and electroluminescence in argon, krypton and xenon, *J. Phys. D: Appl. Phys.* 19:527.

Doke, T., Ishida, N. and Kase, M., 1992, Fano factors in rare gases and their mixtures, *Nucl. Instrum. Methods B* 63:373.

Dutton, J., 1975, A survey of electron swarm data, *J. Phys. Chem. Ref. Data* 4:577.

dos Santos, J.M.F, Conde, C.A.N. and Bento, A.C., 1993a, The energy linearity of gaseous xenon radiation detectors for x-rays with energies between 2 and 60 keV: experimental results, *Nucl. Instrum. Methods A* 324:611.

dos Santos, J.M.F, Conde, C.A.N. and Bento, A.C., 1993b, A simple, inexpensive gas proportional scintillation counter for x-ray fluorescence analysis, *X-Ray Spectrom.* 22:328.

Favata, F., Smith, A., Bavdaz, M. and Kowalski, T.Z., 1990, Light yield as a function of gas pressure and electric field in gas scintillation proportional counters, *Nucl. Instrum. Methods A* 294:595.

Fraga, M.M.F.R., Ferreira, C.M., Loureiro, J. and Leite, M.S.P., Excitation cross sections in Kr and Xe, *ESCAMPIG 1990 (Orleans, France, 28-31 Aug)*.

Frost, L.S. and Phelps, A.V., 1964, Momentum transfer cross sections for slow electrons in He, Ar, Kr and Xe from transport coefficients, *Phys. Rev. A* 136:1538.

Hamilton, T.T., Hailey, C.J., Ku, W.H.-M. and Novick, R., 1980, A high resolution gas scintillation proportional counter for studying low energy cosmic x-ray sources, *IEEE Trans. Nucl. Sci.* NS27:190.

Hashimoto, T. and Nakamura, Y., 1990, Electron swarm parameters in xenon and its momentum transfer cross section, *Papers of Gas Discharge Technical Committee* ED-90-61 (Japan:IEE).

Huang, S.S.S. and Freeman, G.R., 1978, Electron mobilities in gaseous, critical and liquid xenon: density, electric field and temperature effects: quasilocalization, *J. Chem. Phys.* 68:1355.

Hunter, S.R., Carter, J.G. and Christophorou, L.G., 1988, Low-energy electron drift and scattering in krypton and xenon, *Phys. Rev. A* 38:5539.

Inokuti, M., Kowari, K. and Kimura M., 1992, Statistical fluctuations in the yield of ionization due to protons or α particles, *Phys. Rev. A* 45:4499.

Inoue, H., Koyama, K., Matsuoka, M., Ohashi, T., Tanaka, Y. and Tsunemi, H., 1978, Properties of gas scintillation proportional counters, *Nucl. Instrum. Methods* 157:295.

Knoll, G.F., 1989, "Radiation Detection and Measurement", Wiley & Sons, New York.

Koizumi, T., Shirakawa, E. and Ogawa, I., 1986, Momentum transfer cross sections for low energy electrons in krypton and xenon from characteristic energies, *J. Phys. B: At. Mol. Phys.* 19:2331.

Kowalski, T.Z., Smith, A. and Peacock, A., 1989, Fano factor implications from gas scintillation proportional counter measurements, *Nucl. Instrum. Methods A* 279:567.

Lamb, P., Manzo, G., Re, S., Boella, G., Villa, G., Andresen, R., Sims, M.R. and Clark, G.F., 1987, The gas scintillation proportional counter in the spacelab environment: in-flight performance and post-flight calibration, *Astrophys. & Space Sci.* 136:369.

Lowke, J.J. and Parker Jr., J.H., 1969, Theory of electron diffusion parallel to electric fields. II. Application to real gases, *Phys. Rev.* 181:302.

McEachran, R.P. and Stauffer, A.D., 1986, Spin polarisation of electrons elastically scattered from xenon, *J. Phys. B: At. Mol. Phys.* 19:3523.

McEachran, R.P. and Stauffer, A.D., 1987, Relativistic low energy elastic and momentum transfer cross sections for electron scattering from xenon, *J. Phys. B: At. Mol. Phys.* 20:3483.

Mukoyama, T., 1986, Monte Carlo simulation of vacancy cascade in xenon, *J. Phys. Soc. Japan* 55:3054.

Pack, J.L., Voshall, R.E. and Phelps, A.V., 1962, Drift velocities of slow electrons in krypton, xenon, deuterium, carbon monoxide, carbon dioxide, water vapour, nitrous oxide and ammonia, *Phys. Rev.* 127:2084.

Peacock, A., Taylor, B.G., White, N., Courvoisier, T. and Manzo, G., 1985, The in-orbit performance of the Exosat gas scintillation proportional counter, *IEEE Trans. Nucl. Sci.* NS32:108.

Perez-Mendez, L.B.L.V., 1987, Radiation detectors in medicine, *in* "Instrumentation in Elementary Particle Physics", C.W. Fabjan and J.E. Pilcher eds., World Scientific, New Jersey, p.419.

Pleasonton, F. and Snell, A.H., 1957, Ionization following internal conversion in xenon, *Proc.Royal Soc. A* 241:141.

Policarpo, A.J.P.L., 1981, Light production and gaseous detectors, *Physica Scripta* 23:539.

Santos, F.P., Dias, T.H.V.T., Stauffer, A.D. and Conde, C.A.N., 1991, Variation of energy linearity and w value in gaseous xenon radiation detectors for x-rays in the 0.1 to 25 keV energy range: a Monte Carlo simulation study, *Nucl. Instrum. Methods A* 307:347.

Santos, F.P., Dias, T.H.V.T., Stauffer, A.D. and Conde, C.A.N., 1992, Xenon charge distribution resulting from x-ray absorption in xenon: a Monte Carlo simulation, *Proc. of SASP 92*, D. Bassi, M. Scotoni, and P. Tosi, ed., Universita di Trento, p. 2-97.

Santos, F.P., Dias, T.H.V.T., Stauffer, A.D. and Conde, C.A.N., 1993, Tridimensional Monte Carlo calculation of the VUV electroluminescence and other electron transport parameters in xenon, to be published in *J. Phys. D: Appl. Phys.*

Simons, D.G. and de Korte, P.A.J., 1989, Soft x-ray energy resolution and background rejection in a driftless gas scintillation proportional counter, *Nucl. Instrum. Methods A* 277:642.

Stauffer, A.D., Dias, T.H.V.T. and Conde, C.A.N, 1986, Analytical expressions for phase shifts and cross sections for low energy electron-atom scattering in noble gases, *Nucl. Instrum. Methods* 242:327, errata *Nucl. Instrum. Methods A* 256:406, 261:610.

Suzuki, M., Taniguchi, T., Yoshimura, N. and Tagashira, H., Momentum transfer cross section of xenon deduced from electron drift velocity data, 1992, *J. Phys. D: Appl. Phys.* 25:50.

A CARCINOGEN-SCREENING TEST BASED ON ELECTRONS

George Bakale

Case Western Reserve University
Department of Radiology
2054 Abington Road
Cleveland, Ohio 44106

INTRODUCTION

The 60,000 chemicals that are commonly used commercially[1] and the continual generation of new chemicals at the approximate rate of three/hour[2] underscore the need to develop efficacious means of screening chemicals for potentially biohazardous effects. In
addition to this exogenous component of potential human exposure to primarily synthetic
chemicals, exposures to "natural carcinogens"[3] and a lifestyle component[4] also contribute
to our burden of environmental exposures to carcinogens. These multiple routes of exposure to a myriad of chemicals underscore the need to identify chemical carcinogens.

The classical means of identifying chemical carcinogens have been well documented[5] and
were recently summarized by the author.[6] Among these methods, epidemiology is the only
means to identify <u>human</u> carcinogens; however, the ~20-year latency period of many
carcinogens diminishes the value of this method which is also insensitive if multiple
exposures to the host occurs. Another means to identify chemical carcinogens is the long-
term rodent (generally mice and/or rats) bioassay; however, as shown in Table 1, the cost
and time required to conduct these bioassays severely limit the number of chemicals that can
be screened. Also, the efforts of animal-rights activists during the past decade have placed
added pressure to seek alternatives to animal bioassays to identify chemical carcinogens.[7]
Another means of identifying potential carcinogens is the short-term test (STT) of which
more than 200 different types have been developed.[8] The most dominant of these STTs is
the Ames *Salmonella*/microsome test which has been used worldwide in more than 3000
laboratories to screen 20,000 chemicals.[8] Initial validation studies of the Ames test in the
mid-'70s indicated that this test based on bacterial mutagenicity identifies chemical
carcinogens with an accuracy of ~90%.[9] This accuracy declined to <60%, however, as

Linking the Gaseous and Condensed Phases of Matter
Edited by L.G. Christophorou *et al*, Plenum Press, New York, 1994

Table 1. Compasion of key characteristics of rodent bioasssays and the Ames *Salmonella* and k_e short-term tests.

Characteristic	Rodents	Ames *Salmonella*	k_e test
Cost (US $)	$> 10^6$	~ 1,000	~ 400
Time	> 2 years	> 2 days	~ 1 hour
Basis	rodent carcinogens are human carcinogens	bacterial mutagens are carcinogens	electrophiles are carcinogens

more chemicals that were less potent carcinogens were tested in the '80s by the National Toxicology Program (NTP).[10] The problem of low accuracy of the Ames test was exacerbated by the finding that other *in vitro* STTs yield the same incorrect responses as the Ames test which means that these tests are not complementary,[11] i.e. predictivity is not increased by using two or more *in vitro* STTs in a screening battery. Thus, the need to identify chemical carcinogens which the Ames test appeared to fulfill in the mid-'70s[9] was again an unrealized goal a decade later.[5,10]

During this decade of the rise and fall of the Ames and other *in vitro* tests, non-biological means of identifying chemical carcinogens were also developed. One of these methods is the third test listed in Table 1, which is the k_e test that is based on k_e, the rate constant of attachment of electrons to the test chemical in a nonpolar solvent. This STT can be viewed as having a similar basis as the Ames test in that both tests probe the electrophilicity or "electron-loving" potential of the test chemical; however, contrasts between these tests are also evident. For example, in the Ames test the DNA of *Salmonella* serve as the "reagent" that provides electrons with which the electrophilicity of chemicals is sampled, whereas freely drifting excess electrons in cyclohexane measure this property in the k_e test. In view of differences in the <u>nucleophilicity</u> of electrons delocalized in the aromatic bases of DNA relative to excess electrons in cyclohexane, it is not surprising that the Ames test and k_e test provide very different measures of the electrophilicity of a test chemical. This difference contributes to two unique advantages of the k_e test: the k_e test complements the Ames test and it yields positive responses to many unactivated procarcinogens. Both of these topics are discussed in the concluding section.

EXPERIMENTAL

A full description of the pulse-conductivity technique used to measure k_e has been reported[12] and is now briefly reiterated. A Van de Graaff generator that produces a 15 ns pulse of 1 MeV electrons is used to produce excess electrons in purified cyclohexane that is contained in a parallel-plate ion chamber. A storage oscilloscope (100 MHz) is used to monitor the decay of electrons that drift in an applied electric field of 30 kV/cm over a 2-μs time range. A sufficient concentration [S] of the test chemical is added to the solvent to reduce the electron half life $t_{1/2}$ to <200 ns. From the pseudo-first-order kinetics that apply,

k_e is calculated from $k_e = \ln 2/\{[S]\ t_{1/2}\}$. If the k_e of the test chemical that is measured is equal to or greater than 3.0×10^{12} M^{-1}s^{-1}, which is the k_e of the nonpolar carcinogen carbon tetrachloride,[12] the test chemical is clearly electrophilic, the k_e test response is regarded as positive, and the test chemical is predicted to be a carcinogen.

RESULTS

The results to be considered are those reported in four major validation studies of the k_e test which are: (i) a study[12] of 34 carcinogens and 51 noncarcinogens for which the rodent carcinogenicity and Ames *Salmonella* mutagenicity were reported by Kawachi et al.,[13] (ii) a study[14] of 46 nonmutagenic carcinogens and 20 mutagenic noncarcinogens tested by the NTP and reported by Zeiger,[15] (iii) a study[16] of 61 mutagenic carcinogens and 44 nonmutagenic noncarcinogens also tested by the NTP and reported by Zeiger,[15] and (iv) a study[17] of 26 chemicals for which k_es were measured while the NTP rodent bioassays were in progress.[18] The responses of the k_e and Ames tests for these 282 chemicals are compared in Table 2 for three measures of predictivity of STTs.

Table 2. Comparison of three measures of predictivity for the Ames and k_e tests.

Predictivity Criterion	Ames *Salmonella*	k_e test
Sensitivity[1]	84/156; 54%	100/156; 64%
Specificity[2]	94/126; 75%	89/126; 71%
Accuracy[3]	178/282; 63%	189/282; 67%

[1]Number of positive responses obtained/number of carcinogens tested.
[2]Number of negative responses obtained/number of noncarcinogens tested.
[3]Number of correct responses obtained/number of chemicals tested.

The three predictivity criteria listed in Table 2 include sensitivity and specificity, which are the percentages of carcinogens and noncarcinogens, respectively, that a test correctly identifies, and accuracy, which is a combination of the other two criteria. The values of the sensitivities listed in Table 2 can be compared with an estimate of the sensitivity of the mouse bioassay which is obtained from the percentage of rat carcinogens for which a mouse bioassay yields a positive response. For the NTP-screened rat carcinogens considered in this work, that percentage is 69% which can be viewed as the upper limit of sensitivity that could be attained with a STT having a surrogate target for mouse tissue such as *Salmonella* DNA in the Ames test or freely drifting electrons in the k_e test. The latter target yields a k_e test sensitivity of 59% for the same NTP-screened chemicals which although satisfactory is of less significance than the result pertaining to the complementary nature of the k_e and Ames tests. The sensitivity of the Ames/k_e test battery is 84% for the 156 carcinogens considered in this study which is similar to the value of 82% reported earlier for the same battery for 107 NTP-screened carcinogens.[19] This comparison of the sensitivities of the Ames/k_e battery and the mouse bioassay illustrates that the Ames/k_e battery identified more rat carcinogens correctly than did the NTP mouse bioassays that cost >$130 million. This comparison vividly illustrates the cost effectiveness of the Ames/k_e test battery which has been discussed elsewhere.[6,19]

DISCUSSION

One goal of this NATO Advanced Study Institute is to elucidate the unifying role that electrons play in linking gases and liquids by serving as sensitive probes of atomic and molecular interactions in both phases. The k_e test illustrates that this linking role of electrons can also be extended to biological systems where electrons participate in a myriad of life-sustaining and potentially life-threatening processes. In a recent review of structure-activity relationships, Hansch focused attention on "chemical⇔life interaction," an unnamed science on which he estimated 40 billion dollars is spent annually.[20] The k_e test described herein may be viewed as one tool with which "chemical⇔life interaction" can be probed by chemical physicists in unraveling the role that electrons play in carcinogenesis. The k_e test has already been used to probe chemical⇔life interaction by mathematicians who developed techniques to calculate k_e for a variety of chemicals.[21-22]

The basis of the k_e test has common roots with that of the Ames test in the somatic mutation theory of carcinogenesis which states that the one physico-chemical property shared by most carcinogens is electrophilicity and that electrophiles react with electron-rich biotargets such as DNA.[6,9] The k_e test markedly deviates from the Ames test and the somatic mutation theory, however, in yielding positive responses to many procarcinogens that have not been metabolically activated to the ultimate form that induces mutagenesis in bacteria and carcinogenesis in rodents.[23] The rationale offered to reconcile this apparent paradox, which is also consistent with the k_e test complementing the Ames test, is simply that freely drifting excess electrons in cyclohexane are more <u>nucleophilic</u> than are bound electrons in the target DNA of *Salmonella*, rodents and humans.[24] Consequently, electrons in cyclohexane attach to a procarcinogen at every encounter whereas the same procarcinogen is unreactive with a biotarget unless the procarcinogen is converted to a stronger electrophile. An implication of this rationale is that a "pre-chemical" electron transfer occurs which initiates carcinogen-biotarget interaction that culminates in adduct formation, intercalation, alkylation, etc., each of which disrupts subsequent transcription of the target if not excised by the host's repair facilities. The proposed pre-chemical electron transfer step may in some cases be "single-electron transfer" or SET reactions which have been found in the last two decades to be more ubiquitous than had been previously assumed.[25]

A final note on the k_e test pertains to its implications to radiation-induced biological damage. Ionizing radiation is considered to be the most thoroughly characterized human carcinogen,[26] and carcinogens are known to be electrophiles, therefore, some component of ionizing radiation that induces biological damage should have electrophilic properties.

Although such a radiolytic electrophile has been proposed as a precursor of biological damage,[27] no concerted effort to identify this species has been made. In view of the k_e test results presented in this work, a more rigorous search for an electrophilic precursor of radiation-induced biological damage appears warranted.

ACKNOWLEDGMENTS

The author thanks the organizers of this Advanced Study Institute and NATO for providing an opportunity to present this work which was supported by the US. Department of Energy; Grant DE-FG02-88ER60617.

REFERENCES

1. L.B. Lave and F.K. Ennever, Toxic substances control in the 1990s: Are we poisoning ourselves with low-level exposures? *Annu. Rev. Public Health* 11:69 (1990).
2. W. Hammer. "Occupational Safety Management and Engineering," 3rd ed. p.392, Prentice Hall, Englewood Cliffs, NJ (1985).
3. B.N. Ames, "Mutagenesis and carcinogenesis: Endogenous and exogenous factors", *Environ. Mol. Mutagen.* 14, Suppl. 16:66 (1989).
4. R. Doll and R. Peto, "The causes of cancer: Quantitative estimates of avoidable risk of cancer in the United States today", *J. Natl. Cancer Inst.*66:1191 (1981).
5. R.W. Hart and 19 co-authors, "Chemical carcinogens: A review of the science and its associated principles," *Environ. Health Persp.*67:201 (1986).
6. G. Bakale, Detection of mutagens and carcinogens by physiochemical techniques. *in*: "Hazard Assessment of Chemicals -Current Developments," vol. 6, J. Saxena, ed., Hemisphere, Washington , DC. (1989).
7. J. Kaplan, The use of animals in research, *Science* 242:839 (1988).
8. J. Ashby, Origins of common uncertainties in carcinogen/mutagen screening, *Environ. Mol. Mutagen.* 14, Suppl. 16:51 (1989).
9. B.A. Bridges, Short-term screening tests for carcinogens, *Nature* 261:195 (1976).
10. J. Ashby and I.F.H. Purchase, Reflections on the declining ability of the *Salmonella* assay to detect rodent carcinogens as positive, *Mutat. Res.* 205:51 (1988).
11. R.W. Tennant, B.H. Margolin, M.D. Shelby, E. Zeiger , J.K. Haseman, J. Spalding, W. Caspary, M. Resnick, S. Stasiewicz, B. Anderson, and R. Minor, Prediction of chemical carcinogenicity in rodents from *in vitro* genetic toxicity assays. *Science*, 236:933 (1987).
12. G. Bakale and R.D. McCreary, A physico-chemical screening test for chemical carcinogens: the k_e test, *Carcinogenesis* 8:253 (1987).
13. T. Kawachi, T. Yahagi, T. Kada, Y Tazima, M. Ishidate, M. Sasaki, and T. Sugiyama, Cooperative program on short-term assays for carcinogenicity in Japan, *in*:

"Molecular and Cellular Aspects of Carcinogen Screening Tests," R. Montesano, H. Bartsch, L. Tomatis, and W. Davis, eds. IARC, Lyon (1980).

14. G. Bakale and R.D. McCreary, Response of the k_e test to NCI/NTP-screened chemicals. I. Nongenotoxic carcinogens and genotoxic noncarcinogens, *Carcinogenesis* 11:1811 (1990).

15. E. Zeiger, Carcinogenicity of mutagens: Predictive capability of the *Salmonella* mutagenesis assay for rodent carcinogenicity, *Cancer Res.* 47:1287 (1987).

16. G. Bakale and R.D. McCreary, Response of the k_e test to NCI/NTP-screened chemicals. II. Genotoxic carcinogens and nongenotoxic noncarcinogens, *Carcinogenesis* 13:1437 (1992).

17. G. Bakale and R.D. McCreary, Prospective k_e screening of potential carcinogens being tested in rodent bioassays by the NTP, *Mutagenesis*, 7:91 (1992).

18 R.W. Tennant, J. Spalding, S. Stasiewicz, and J. Ashby, Prediction of the outcome of rodent carcinogenicity bioassays currently being conducted on 44 chemicals by the National Toxicology Program. *Mutagenesis*, 5:3 (1990).

19. F.K. Ennever and G. Bakale, Response of the k_e test to NCI/NTP-screened chemicals. III. Complementary value of k_e in screening for carcinogens, *Carcinogenesis*, 13:2059 (1992)

20. C. Hansch, Quantitative structure-activity relationships and the unnamed science, *Acc. Chem. Res.* 26:147 (1993).

21. R. Benigni, C. Andreoli, and A. Giuliani, Structure-activity studies of chemical carcinogens: use of an electrophilic reactivity parameter in a new QSAR model, *Carcinogenesis* 10:55 (1989).

22. R.Benigni, M. Cotta-Ramusino, C. Andreoli, and A. Giuliani, Electrophilicity as measured by k_e: molecular determinants, relationship with other physical chemical and quantum mechanical parameters, and ability to predict rodent carcinogenicity. *Carcinogenesis*,13:547 (1992).

23. E.C. Miller and J.A. Miller, The metabolism of chemical carcinogens to reactive electrophiles and their possible mechanisms of action in carcinogenesis, *in*: "Chemical Carcinogens," C.E. Searle, ed., American Chemical Society, Washington, D.C. (1976).

24. G. Bakale, Theoretical implications of the k_e carcinogen-screening test, *in*: "Chemical Carcinogens--Activation Mechanisms, Structural and Electronic Factors, and Reactivity," P. Politzer and F.J. Martin, eds., Elsevier, Amsterdam (1988).

25. E.C. Ashby, Single-electron transfer, a major reaction pathway in organic chemistry. An answer to recent criticisms, *Acc. Chem. Res.* 21:414 (1988).

26. J.D. Boice, Cancer following medical irradiation, *Cancer* 47:1081 (1981).

27. G. Bakale and E.C. Gregg, Conjecture on the role of dry charges in radiosensitization, *Br. J. Cancer*, 37, Suppl. III:24 (1978).

SECTION IX. SUMMARY OF DISCUSSION PANEL

Summary of the Discussion Panel on Experimental Techniques*

The slow electron as a probe to study the properties of matter in the gaseous and condensed phases was discussed during the Institute in the case of a variety of different experimental methods and techniques. The gas density and stage of aggregation ranged from isolated molecules under single collision conditions ($p < 10^{-5}$ mbar), beyond single collision conditions ($>10^{-3}$ mbar), gases at high pressure (>10 bar), stationary liquids, liquid beams, molecules adsorbed and condensed on solid surfaces, and finally, free clusters which are considered to represent a link between the gaseous and the condensed phase.

We can categorize (though somewhat arbitrarily) the experimental methods according to the processes which were studied:

(a)   the *removal* of electrons *from* matter (i.e. ionization),
(b)   the *transport* of electrons *through* matter, and
(c)   the *attachment* of electrons *to* matter.

Ionization experiments were presented using *photons* (21.21 eV from the He discharge lamp, synchrotron radiation and multiphoton ionization with tunable lasers), *electrons* or *He** *metastable* atoms. The latter probe is particularly surface sensitive: the energy is slowly carried as excitation energy by the atom and ionization occurs immediately upon the first contact of the excited atom with the surface of a liquid or solid. Ionization of gas-phase particles can either be observed via the ejected electrons or the generated cations. While the photoelectron spectrum is an image of the electronic structure of the neutral particle from which it is ejected, the observation of ions (parents and eventually fragment ions) contains information on the fate of the target ion created in the primary ionization event. Combination of both, i. e. recording photoelectrons and cations *in coincidence* yields detailed information on the decomposition of the precursor ion in dependence of its internal energy. In contrast to the gas phase (molecules, clusters) ionization of condensed molecules can directly be followed only by recording the emitted electrons while information on the evolution of the cations is much more difficult.

Transport of electrons *through* matter can be studied by observing the transmitted current of free electrons as a function of energy through a sample of gas or condensed material (electron transmission spectroscopy). Sharp structures in the gas phase transmission spectrum indicate the formation of transient negative ions, while a transmission spectrum from condensed molecules is much more complicated as it represents a multi scattering problem involving scattering at the interface vacuum/film in the film and at the interface film/substrate. Another more traditional method creates the excess electrons directly in the gas, liquid or solid by ionizing radiation. The transport of electrons through the material of interest can then be recorded as a function of different parameters like the external electric field, pressure, temperature, etc..

---

*prepared by Eugen Illenberger

Electron attachment to gas phase-particles is usually performed by swarm techniques and beam techniques. In the first case the attachment coefficient is derived from the attenuation of the intensity of the electrons drifting through the attaching gas. The average energy of the drifting electrons is determined by the electric field and the gas pressure. In beam experiments the resulting ions (parent and fragment anions) are usually recorded as a function of the primary electron energy. Under *single collision conditions* electron capture to isolated molecules usually yields fragment ions since stabilization mechanisms producing the stable parent ion are not operative under such conditions. The situation changes completely in clusters where intermolecular energy transfer can easily generate stable molecular anions thereby evaporating the target cluster (evaporative attachment). Such van der Waals clusters are commonly produced in supersonic beams under fairly controlled conditions. In the condensed phase, transitory negative ions can also be formed by resonant electron capture, and the decomposition of a negative ion resonance at the surface can also lead to desorption of fragment ions. However, energetic and structural constraints at the surface may in many cases prevent the fragment anion from desorption. The alternative reaction path of the transient molecular anion is its immediate stabilization (facilitated by the environment) or decomposition at the surface and stabilization of the fragment ion at or near the surface. The quantity of charging can then be measured by the shift of the injection curve in a transmission experiment .

The Panel Discussion was accompanied by short contributions from Y. Hatano, E. Illenberger, H. Morgner and L. Sanche.

H. Hatano illustrated the considerable improvement in measuring attachment rates by the introduction of the microwave conductivity method combined with pulse radiolysis. This technique allows the measurement of absolute attachment rates for two body and three body processes with electrons almost completely thermalized having a Maxwell-Boltzmann energy distribution, it also eliminates completely the effect of ambipolar diffusion. In addition, the density of a third body molecule can be chosen with virtually no limitation which makes this technique most interesting for environmental application.

Some aspects concerning experiments with clusters were discussed by E. Illenberger. In most of the contributions presented during the Institute, the clusters were produced by adiabatic expansion of a gas (or gas mixture) using the seeded beam technique. Such a beam always contains clusters of different size. Although the average size of the distribution can be varied in a broad range by the experimental conditions (stagnation pressure, nozzle diameter, temperature, etc.) any process in the target cluster (ionization, electron attachment and subsequent dissociation reactions) are most likely accompanied by cleavage of the much weaker intermolecular van der Waals bonds (evaporation). Thus, just by measuring a product arising from a reaction in a van der Waals cluster it is not possible to *directly* conclude on the exact size of the initial target cluster. This is the cruicial point in most experiments dealing with neutral custers. The only method so far to perform experiments on size sected custers was introduced by the Göttingen group (U. Buck et al.[1]). In this technique the neutral cluster beam is crossed with a neutral rare gas beam. Due to momentum conservation, small clusters scatter into large angles and *vice versa*. Although some energy is also transferred to the cluster by the scattering process, this is in most cases below the intermolecular dissociation energy and the method has sucessfully been applied to studies on small size selected clusters of organic molecules.

The situation changes completely in the case of ionic clusters, like in photo-detachment from cluster anions as discussed during the Institute by M.A Johnson. Cluster ions (which can be created by some technique like gas discharge, electron impact, etc.) can easily be separated according to e/m in space or time by applying appropriate electric and/or magnetic fields, i.e. by conventional mass spectrometric techniques.

A further point concerned electron attachment to electronically excited molecules. So far the material is virtually restricted to results from swarm experiments as was pointed out by L.G. Christophorou during the Institute. The extension to corresponding experiments in molecular beams using pulsed laser systems suffer from the simple shortcoming of an inherently low "duty cycle": in the most favourable case (e.g. formation of a long lived electronic states by photoexcitation and subsequent singlet-triplet conversion) the presence of the excited species in the interaction volume is limited by their thermal velocity. This makes the duty cycle generally less than 10^{-4}, i. e. the average ratio between excited and ground state targets present in the interaction region is in any case below 10^{-4}.

H. Morgner considered some aspects related to electron spectroscopy of liquid beams. Electron spectroscopy studies on liquids are surprisingly scarce although the microscopic understanding of liquid surfaces could benefit to the same extend as in the case of solid surfaces. One problem here is the vapour pressure of the liquid which restricts the method to systems with low vapour pressure. The beam is maintained in a closed cycle system driven by a pump and exposed to "vacuum" and the ionizing probe (photons or He* metastables) within a distance of a few cm. The exposure of the beam is short enough to avoid contamination of the liquid surface from the background gas and charging of the beam due to ionization. The surface is probed by MIES (Metastable Induced Electron Spectroscopy) and conventional PES (using 21.22 eV photons from the He discharge lamp and photons of variable energy from synchrotron radiation). As mentioned above, the MIES technique is extremely surface sensitive while photons have a much larger free path. In PES the observation depth is thus virtually controlled by the escape depth of the ionized electrons. From a clever combination of different techniques using synchrotron radiation depth profiles of the liquid beam composition can be derived. This has been applied to binary mixtures where the surface composition often deviates considerably from that in the bulk.

Finally, electron scattering (L. Sanche) and electron spectroscopy from condensed and adsorbed molecules definitely requires UHV conditions ($p < 10^{-9}$ mb) in order to avoid surface contamination. In contrast to a liquid beam the stationary molecular film is usually charged by trapped positive or negative charges from ionization or electron attachment, respectively. This complicates any electron spectroscopy, but opens the possibility to study microscopic aspects of surface charging.

In the general discussion it became evident that the variety of subjects and problems within this rather diversified field range from very fundamental questions of basic research to application like in environmental problems, surface preparation and modification, etc.. The Institute thus established a profound background for an increasing activity of interdisciplinary research in that field.

[1] U. Buck and H. Meyer, Scattering Analysis of Cluster Beams: Formation and Fragmentation of Small Ar_n Clusters, *Phys. Rev. Letters* 52: 109 (1984).

THEORY: INTERACTIONS OF ELECTRONS WITH DENSE MEDIA[1]

The gargantuan feast of ideas and information, offered by the lecturers and discussants during this strenuous fortnight of intellectual exercise, both requires and deserves an overview from certain theoretical aspects. It is, however, sadly true that it is practically impossible to bring such a summary to any level of perfection, particularly not in the immediate wake of the meeting. Hence I feel compelled to waive completeness either in width or depth and, instead of giving a balanced review, I can only summarize my personal views and understanding. Nevertheless, I hope to be able to articulate some of the thoughts and opinions which appeared to me to be generally accepted by the majority of the participants. No credit will be given to any individual author or discussant their personal results or ideas being mainly printed in the previous pages.

The *wrong* approach to the problem of electrons in dense media would be: solve the Schrödinger equation, or perhaps the von Neumann equation, for all the electrons and nuclei of the medium plus the excess electron. A calculation of this type, *pace* for supercomputers, cannot be carried through and fortunately so, because if once obtained, one could not handle the data if only due to their sheer amount. Thus mathematical and/or physical approximations are to be employed and talking about theories of excess electrons in dense media necessarily means that one refers to the concept or application of one approximation or another.

Each approximation rests on the knowledge of the interaction potential between the excess electron and the atoms or molecules of the medium. This is quite similar to the outset concerning the theories of liquids: also there one must start out with the knowledge of the intermolecular potentials. Electron-in-liquid theories, in their most complete versions, are solution theories, where both the state of the electron and the structure of the liquid are subjected to calculation. The obvious difficulty lies with the much differing masses of molecule and electron. Whereas molecules can be regarded as classical particles, the electron must be treated in terms of quantum mechanics.

The electron-to-liquid potentials are given as a sum of a repulsive and an attractive term. Repulsion, characterized by a pseudopotential, accounts for the effect of closed atomic shells in view of the exclusion principle. Attraction is due to polarization interactions, hence its determination amounts to some theory of the dielectric properties

---

[1]Prepared by Robert Schiller

Linking the Gaseous and Condensed Phases of Matter
Edited by L.G. Christophorou *et al*, Plenum Press, New York, 1994

of the medium. The difficulty in all these calculations lies with the electron states which, on the one hand, depend on the delicate balance of repulsion and attraction and, on the other hand, also influence this balance greatly.

Usually it seems reasonable to start the calculations with a system which consists of the medium, whose structure is that of the neat liquid, and of the electron, completely delocalized. This electron is called quasifree. The electron energy in such a system, usually denoted by V_0 can be measured by standard techniques and, at least for simple liquids like argon or methane, some contemporary theoretical data are also at hand. Assuming a band structure of electron energies to prevail in liquids V_0 corresponds to the bottom of the conduction band.

Theories of V_0 were only referred to throughout the meeting whereas experimental data and their thermodynamic meaning were discussed in some detail. Critical thinking was given to the validity of any energy band model for systems without long range order. The analogy between liquids and amorphous solids was taken as a qualitative argument in favour of the idea.

By knowing the potentials the structure and energetics of the coupled classical-cum-quantum mechanical system must be evaluated, a task nowadays mostly done by Feynman's path integral method. Here the electron is represented by a large ring of classical particles. The method, also applied to the analogous cases of positron and positronium in liquids, yields encouraging results. It is clear, however, that much work is still to be done, e.g. on the optical absorption spectra.

An earlier approximation was based on the Wigner-Seitz model which attributes average long range order to the liquid phase. The approximation treated kinetic and potential energies of the electron separately, a simplification which made analytical solutions feasible. Although the mathematical simplicity and easy viusalization have made this treatment appealing its validity has been questioned. Even so, recent results based on a more rigorous approach have shown the old approximation to be reasonably sound.

Theories show that in liquids where V_0 is negative the electron is delocalized and the liquid structure in the vicinity of the electron is very similar to that of the neat medium, whereas if V_0 is positive the electron gets localized in a much rearranged, low density portion of the liquid (in a "bubble"). These simple statements hold only if the absolute value of V_0 is much larger than kT. Whereas the above predictions on complete localization or delocalization are well supported by electron mobility and positronium lifetime measurements, new evidence has emerged in both theory and experiment on the prevalence of intermediary states in liquids of lower absolute value V_0 This is an old idea, much cherished by this reviewer. Now the question arose whether the "partially localized states" correspond to a two-state model, i.e. to a mixture of completely localized and delocalized electrons, in other words to localization and delocalization as often recurring, random events throughout the life of an electron. Or, otherwise, the electron wave function is only partially localized around the bubble and this single state does not vary in time. Dynamic calculations and statistical mechanical arguments, based on the ergodic nature of the system, seem to support strongly the two-state model.

Feynman's method is a mathematical approximation, and certainly not the only one even if it seems to be one of the most powerful treatments. The problem, however, can be approached from the physical side by investigating liquid-like systems which are simpler than real liquids. Generally speaking one can decrease either the dimensionality or the size of the systems.

Amorphous thin layers or liquid surfaces can be regarded as laminar difformed systems, i.e. liquids of essentially two dimensions. The experimental information gained on such systems is usually very enlightening. There are, however, at least two limitations which impede direct translation of the results to three dimensional systems. Thin layers exhibit quantum size effects owing to electron wave interferences; and relaxation phenomena at or near to the surface might be much different from those in the bulk.

The experimental realization of small systems is nowadays best done by producing atomic or molecular clusters. Reports on the production and nature of metallic clusters, such as Cu_n, covalent clusters, the archetype of which is C_{60}, and van der Waals clusters, the simplest of them being Ar_n, are given in the previous articles. It has been made clear that van der Waals clusters may be regarded the best as model liquids.

Cluster theories, by virtue of a conceptual order, rest on the enumeration of configurations this being followed by the construction of potential surfaces for each configuration. The result, even if n is small, is a large number of configurations and complicated potential maps with numerous metastable states. Only after these conditions are elucidated can one determine the configurations and corresponding energetics of an ion- or electron-containing cluster. Properties, like optical spectra or chemical reaction rates depend largely on the size and structure of the cluster. From the point of view of the electron-in-liquid problem the most important question is how one can conclude from cluster data to electron properties in a macroscopic liquid.

It has been observed that certain electron properties, taken as a function of $n^{-1/3}$, i.e. of the reciprocal of the cluster radius, can be justifiably extrapolated to infinitely large radii by taking $n^{-1/3} \to 0$, thus obtaining the electron properties in the bulk liquid. However, whether this simple contrivance can find general application is still a matter of future research.

The effect of cluster size cannot be ascribed to individual atom-electron interactions. As has been expressed it is the phonon bath of the cluster bonds that affects the electron states. Interatomic or intermolecular vibrations play the most important role.

Going over from clusters to bulk macroscopic liquids the need to account for the effect of interatomic motions on electron states remains with us. The enumeration of configurations and vibration modes is in this case, however, impossible. The most appropriate way to meet this need, in my opinion, is to consider density and energy fluctuations in the liquid as an average measure of intermolecular displacements around the equilibrium states. Local and momentary deviations of density and energy must be evaluated on the length- and time-scales of the pair correlation and temporal auto-correlation functions, respectively. If a deviation occurs in the vicinity of an electron, this will influence its state, hence its spectral and transport properties.

Thermodynamic fluctuations are ubiquitous but only rarely perceptible in macroscopic systems. A solution of conventional ions, e.g. NaCl in water, reflects but poorly if at all the local deviations from equilibrium. The excess electron in a dense medium differs, however, from any heavy ion in the strong non-linearity of its properties as a function of external conditions. As said previously, repulsive and attractive forces maintain a very delicate balance in most liquids, hence a very small change in local energy or density might change the net attraction to repulsion or *vice versa*. This, in turn, makes the quasifree electron localized or sets the localized electron free, a variation which is accompanied by huge changes in electron mobility, reactivity, spectra, etc. These strong non-linearities amplify the effects of fluctuations so as to make the small changes perceptible.

Before finishing this short review it might be apposite to mention two areas of research which I regard as relevant to the main subjects of the Institute; these, however, were not tackled at the meeting.

One of them is non-linearity; one aspect I mentioned above and I feel sure that there are several other facets of that group of phenomena which would deserve interest from the students of our field.

The other item is periodic or chaotic behaviour. The necessary conditions for a system to behave in such a way are (a) a number of metastable states, (b) a far-from-equilibrium steady condition, and (c) non-linear feedback. Given the large number of metastable states e.g. of clusters, the strong non-linearities of electron properties, and the distance from thermodynamic equilibrium if mobile and reactive electrons are in dense media, one might reasonably expect some success in the search for periodic or chaotic processes.

The two-week meeting has left me with the impression that recent experimental results render a good number of problems for theoretical work. In order to get further impetus theoreticians asked the experimentalists to introduce new methods of perturbing the media. Experimentalists, however, would expect more than sheer explanations. They would be most pleased to have predictions from theories, statements which would govern their future experimental work.

The support of NATO ASI, the National Research Fund (Hungary) under Contract No. OTKA 2981 and the National Committee for Technical Development (Hungary) is gratefully acknowledged.

THE BEHAVIOR OF SLOW ELECTRONS IN MOLECULAR SUBSTANCES AND ITS SIGNIFICANCE IN RADIATION AND LIFE SCIENCES

Mitio Inokuti

Argonne National Laboratory
Argonne, Illinois 60439 U.S.A.

ELEMENTS OF RADIATION PHYSICS

Any ionizing radiation passing through matter excites or liberates many electrons, which are the lightest constituent of any substance. Energetic electrons thus generated, as well as excited or ionized molecular species, are the major precursors of the physical and chemical and, hence, the biological effects of radiation actions. (As a qualification, radiation effects on crystalline solids, most notably metals, arise primarily from atomic displacements that are due to the transfer of radiation energy into the kinetic energies of atoms.) Consequently, electrons play a central role in radiation physics.

The electrons generated in matter have widely distributed energies, slow down through collisions with molecules, and also produce more electrons. This process, called electron degradation, is the subject of extensive studies.[1] Roughly speaking, electrons of every decade of energies are equally important in the production of ions and most of excited species; in other words, electrons of energies between 100 eV and 1 keV produce about the same number of ions as electrons of energies between 1 keV and 10 keV, provided that the first-generation electrons, either incident or produced by high-energy photons or other radiations, have energies higher than 10 keV. Exceptions to this rule of thumb arise in the production of triplet states, negative ions, or other species that result almost exclusively from electrons with energies lower than 100 eV. Thus, much of the discussion in the present NATO Institute is highly pertinent to fundamentals of radiation chemistry and biology, as discussed by Schmidt.[2]

SOME CURRENT ISSUES

A recent comprehensive survey of knowledge about physical and chemical mechanisms in radiation biology appeared in the proceedings of the Woods Hole Conference.[3] In what follows, I summarize some current issues in three topical areas.

First, it is of fundamental importance to characterize excited and ionized molecular species resulting from collisions of electrons with water (the dominant constituent of the biological cell), as well as DNA, proteins, and many other

Linking the Gaseous and Condensed Phases of Matter
Edited by L.G. Christophorou *et al*, Plenum Press, New York, 1994

biomolecules. What needs to be characterized includes the energy levels of these molecules, the probabilities of their excitation and ionization, and the decay processes of excited and ionized molecular species.

Second, the biological cell has highly complex and inhomogeneous structures, and the spatial distribution of excited and ionized molecular species is inhomogeneous on roughly the same scale as the structures. This spatial distribution, commonly referred to as track structure, is the subject of extensive studies.[4-6] Indeed, the track structure is usually considered a reason for the generally nonlinear dependence of the biological effect on the dose (i.e., the mean energy absorbed per unit mass). [In contrast, many of the physical and even chemical effects in homogeneous substances occur roughly proportionally to the dose (i.e., the total energy absorbed).]

Third, most radiation biologists hold the opinion that DNA is the main target of radiation actions. Ionizing radiations and the resulting electrons certainly cause excitation and ionization of DNA molecules; this is commonly called the direct effect. Ionizing radiations also produce hydroxyl radicals, hydrated electrons, hydronium ions, and other molecular species. Among them, hydroxyl radicals react with the sugar residues of DNA to cause damage; this is called the indirect effect. According to the majority opinion, the direct and indirect effects are both appreciable; in other words, neither of them is negligible except in special circumstances.

As a qualification, other ideas need to be cited. For instance, hydrated electrons, which are abundantly produced, react primarily with the base residues of the DNA; however, the biological consequences of chemical changes in the bases remain to be clarified. An important role of slow electrons is strongly suggested by the remarkable correlation between the electron-attachment rates of many molecules and their carcinogenicity, discussed by Bakale[7] at the present NATO Institute, although precise mechanisms leading to the correlation have not been delineated. Moreover, some radiation biologists emphasize the role of cell membranes as an important target; this idea is certainly consistent with physics in the sense that the cell membranes are compact and therefore can absorb more energy per volume than other parts of the cell. In addition, ionization of the DNA, produced either directly or indirectly should have serious consequences for the double-helix conformation of the DNA, which is ordinarily maintained by hydrogen bonds, van der Waals forces, and other dielectric properties. The influence of ionization on protein structure was first discussed by Platzman and Franck.[8]

ROLES OF PHYSICS AND CHEMISTRY

In what way do physics and chemistry contribute to the understanding of radiation effects on life? My brief answer to this question is as follows.

First, physics gives <u>instrumentation</u> necessary for research in life and radiation sciences. Instrumentation here means various devices related to radiation. These devices may be classified into three classes: (1) radiation sources, which produce radiations in a controllable way; (2) radiation dosimeters, which enable one to measure radiation fields; and (3) detectors and probes, which permit one to see material structures affected by radiation. In the development of instrumentation, especially of dosimeters, detectors, and probes, chemistry often plays an important role.

Second, both physics and chemistry are crucial in the clarification of mechanisms of processes leading to biological effects, such as cell death, mutation, and carciogenesis. Studies of the mechanisms of any macroscopic phenomenon are never straightforward but require many intellectual endeavors. To cite a simple example, to calculate the number of ions produced in a gas used in a radiation detector, one must fully understand all the major processes of the interactions of electrons and other particles with molecules in the gas, as seen in the lecture by Dias[9] in the present NATO Institute.

Materials of interest in radiation and life sciences include biomolecules such as DNA, proteins, and related molecules, as well as those used in detectors and dosimeters.

For these materials, broad spectroscopic work is required to characterize excited and ionized states and their behavior, such as the decay modes (e.g., molecular fragmentation, energy transfer, luminescence, and internal conversion). Despite the obvious need for and also the presence of various techniques, as we have seen in the present NATO Institute, the current knowledge of the spectroscopy of biomolecules is less than satisfactory.

After the spectroscopy, one must characterize the probabilities of producing excitation and ionization (i.e., cross sections for collisions of electrons and other particles). Current knowledge of the cross sections for biomolecules is sparce and fragmentary.

Only after a considerable amount of reliable data on cross sections is available will a truly trustworthy model of radiation track structures be feasible. Work to date[4,5] on the subject has taken the major component (water) explicitly into account but has omitted biomolecules because of the absence of usable cross section data.

ACKNOWLEDGMENT

The present work was supported in part by the U. S. Department of Energy, Office of Energy Research, Office of Health and Environmental Research, under Contract W-31-109-Eng-38.

REFERENCES

1. M. Kimura, M. Inokuti, and M.A. Dillon, Electron degradation in molecular substances, in: "Advances in Chemical Physics," Vol. 84, I. Prigogine and S.A. Rice, eds., John Wiley & Sons, Inc. (1993) p. 193.
2. W.F. Schmidt, Electron transport in gases and liquids, in the present volume.
3. W.A. Glass and M.N. Varma, eds., "Physical and Chemical Mechanisms in Molecular Radiation Biology," Plenum Press, New York (1991).
4. H.G. Paretzke, in: "Kinetics of Nonhomogeneous Processes," G.R. Freeman, ed., John Wiley & Sons, Inc. (1987) p. 89.
5. R.H. Ritchie, R.N. Hamm, J.E. Turner, H.A. Wright, and W.E. Bloch, Radiation interactions and energy transport in the condensed phase, p. 99 of Ref. 3.
6. M. Zaider, Charged particle transport in the condensed phase, p. 137 of Ref. 3.
7. G. Bakale, A carcinogen-screening test based on electrons, in the present volume.
8. R.L. Platzman and J. Franck, A physical mechanism for the inactivation of proteins by ionizing radiation, in: "Symposium on Information Theory in Biology," H.P. Yockey, R.L. Platzman, and H. Quastler, eds., Pergamon Press, Oxford (1958) p. 262.
9. Teresa H.V.T. Dias, Physics of noble gas X-ray detectors, in the present volume.

PARTICIPANTS

C. S. Askun
Middle East Technical University
Department of Chemistry
Ankara 06531
TURKEY

G. Bakale
Case Western Reserve University
Department of Radiology
Cleveland, OH 44106

N. J. Bardsley
Lawrence Livermore National Laboratory
P. O. Box 808, MS-L-296
Livermore, CA 94550

H. Baumgärtel
Freie Universität Berlin
Institut für Physikalische und
Theoretische Chemie
Takustr. 3
14195 Berlin
GERMANY

R. S. Berry
The University of Chicago
Department of Chemistry
5735 S. Ellis Avenue
Chicago, IL 60367

A. F. Borghesani
University of Padua
Low Temperature Laboratory
Via F. Marzolo 8
Padova 35131
ITALY

V. Byakov
Institute of Theoretical and
Experimental Physics (ITEP)
B. Cheremushkinskaya 25
Moskow 117259
RUSSIA

L. Carlomusto
Università degli Studi di Cassino
Dipartimento di Ingegneria Industriale
Via Zamosch, 43
Cassino (Fr) 03043
ITALY

R. Carlomusto
Università degli Studi di Cassino
Dipartimento di Ingegneria Industriale
Via Zamosch, 43
Cassino (Fr) 03043
ITALY

V. Chepel
Universidade de Coimbra
Dept. de Fisica LIP - Coimbra
Coimbra P-3000
PORTUGAL

A. A. Christodoulides
Department of Physics
University of Ioannina
Ioannina
GREECE

L. G. Christophorou
Oak Ridge National Laboratory
P. O. Box 2008
Oak Ridge, TN 37831-6122

C. A. N. Conde
University of Coimbra
Department of Physics
Rua Larga
Coimbra P-3000
PORTUGAL

P. G. Datskos
University of Tennessee
Department of Physics
401 Nielsen Physics Building
Knoxville, TN 37996-1200

C. Dedonder-Lardeux
Laboratoire Photophysique Moleculaire
CNRS
Orsay
FRANCE

T. H. V. T. Dias
University of Coimbra
Department of Physics
Coimbra P-3000
PORTUGAL

N. Elmaci
Middle East Technical University
Department of Chemistry
Ankara 06531
TURKEY

I. I. Fabrikant
University of Nebraska
Department of Physics & Astronomy
Lincoln, NE 68588

M. Faubel
Max-Planck-Institut für
  Strömungsforschung
Bunsenstr. 10
3400 Göttingen
GERMANY

J. Ferch
Universität Bielefeld
Fakultät für Physik
Postfach 100131
4800 Bielefeld 1
GERMANY

M. Hahn
Universität Hamburg
Institut für Physikalische Chemie
Bundesstraße 45
2000 Hamburg 13
GERMANY

R. N. Hamm
Oak Ridge National Laboratory
P. O. Box 2008
Oak Ridge, TN 37831-6123

R. Hashemi
Freie Universität Berlin
Institut für Physikalische und
  Theoretische Chemie
Takustraße 3
14195 Berlin
GERMANY

Y. Hatano
Tokyo Institute of Technology
Chemistry Department
12-1 Ohkayama, 2-Chome
Meguro-ku, Tokyo 152
JAPAN

M. Hayashi
503 Sakae High Home
15-14 Sakae 4-Chome
Nakaku, Nagoya 460
JAPAN

M. J. Henchman
Brandeis University
Department of Chemistry
Waltham, MA 02254-9110

O. Hilt
Hahn-Meitner-Institut
Abteilung Strahlenchemie
Glienicker Straße 100
14109 Berlin
GERMANY

R. A. Holroyd
Brookhaven National Laboratory
Department of Chemistry
Bldg. #555
Upton, Long Island, NY 11973

I. Iakubov
Institute for High Temperatures
Academy of Sciences
Korovinskoye Road
Moscow 127 412
RUSSIA

E. Illenberger
Freie Universität Berlin
Physikalische Chemie
Takustraße 3
14195 Berlin
GERMANY

O. Ingolfsson
Freie Universität Berlin
Institut für Physikalische und
Theoretische Chemie
Takustraße 3
14195 Berlin
GERMANY

M. Inokuti
Argonne National Laboratory
9700 S. Cass Avenue, Building 203
Argonne, IL 60439

K. Itoh
University of Tokyo
Department of Pure & Applied Sciences
3-8-1 Komaba, Meguro-ku
Tokyo 153
JAPAN

T. Jaffke
Freie Universität Berlin
Institut für Physikalische und
  Theoretische Chemie
Takustraße 3
14195 Berlin
GERMANY

M. A. Johnson
Department of Chemistry
Yale University
225 Prospect Street
New Haven, CT 06511

C. Jouvet
Université de Paris-Sud
Laboratoire de Photophysique
Moleculaire du CNRS, Bât 213
F-91405 Orsay Cedex
FRANCE

R. Katoh
Hahn-Meitner-Institut
Abteilung Strahlenchemie
Glienicker Straße 100
14109 Berlin
GERMANY

A. G. Khrapak
Hokkaido University
Department of Electrical Engineering
Sapporo 060
JAPAN

W. Klein
Hahn-Meitner-Institut
Abteilung Strahlenchemie
Glienicker Straße 100
14109 Berlin
GERMANY

A. Krasinsky
The Hebrew University
Racah Institute of Physics
Jerusalem 91904
ISRAEL

P. Krebs
Universität Karlsruhe
Institut für Physikalische Chemie und
  Elektrochemie
Kaiserstraße 12
7500 Karlsruhe
GERMANY

J. Kristiak
Slovak Academy of Sciences
Institute of Physics
Dúbravská cesta 9
Bratislava 842 28
SLOVAKIA

C. Krontiras
University of Patras
Department of Physics
Patras 25110
GREECE

K. Lacmann
Hahn-Meitner-Institut
Abteilung Strahlenchemie
Glienicker Straße 100
14109 Berlin
GERMANY

P. Lamp
Max-Planck-Institut für Physik
Werner Heisenberg Institut
Föhringer Ring 6
8000 München 40
GERMANY

F. Mafuné
The University of Tokyo
Department of Chemistry Faculty
  of Science
Hongo Bunkyo-Ku
Tokyo 113
JAPAN

T. D. Märk
Universität Innsbruck
Institut für Ionenphysik
Techniker Straße 25
A-6020 Innsbruck
AUSTRIA

N. J. Mason
University College London
Department of Physics and Astronomy
Gower Street
London WC1E 6BT
ENGLAND

M. Meinke
Freie Universität Berlin
Institut für Physikalische Chemie
Takustraße 3
14195 Berlin
GERMANY

B. N. Miller
Texas Christian University
Department of Physics
P. O. Box 32915
Fort Worth, TX 76129

J. C. Miller
Oak Ridge National Laboratory
Chemical Physics
P. O. Box 2008
Oak Ridge, TN 37831-6125

M. Momirlan
Institute of Physical Chemistry
Romanian Academy
Spl. Independentei 202
Sector 6
Bucharest 77208
ROMANIA

H. Morgner
Institut für Experimental Physik
Naturwissenschaftliche Facultät
Universität Witten/Herdecke
Stockumer str. 10
58448 Witten
GERMANY

D. Neri
University of Padua
Departimento di Fisica
Via F. Marzolo 8
Padova 35131
ITALY

584

I. M. Obodovski
Moscow Engineer Physical Institute
Kashirskoe sh., 31
Moscow 115409
RUSSIA

I. Oleinik
Volgograd State University
Volgograd
RUSSIA

T. M. Orlando
Molecular Science Research Center
Batelle
Pacific Northwest Laboratory
P. O. Box 999, Mailstop K2-14
902 Batelle Boulevard
Richland, WA 99352

V. D. Peskov
Fermi National Accelerator Laboratory
MS 220
P. O. Box 500
Batavia, IL 60510

M. N. Pisanias
University of Patras
Department of Physics
Solid State Physics Laboratory
Patras 26110
GREECE

P. J. B.M. H.V.T. Rachinhas
University of Coimbra
Department of Physics
Coimbra P-3000
PORTUGAL

E. Rühl
Freie Universität Berlin
Physikalische und Theoretische Chemie
Takustr. 3
14195 Berlin
GERMANY

Y. Sakai
Hokkaido University
Department of Electrical Engineering
Sapporo 060
JAPAN

L. Sanche
Université de Sherbrooke
Faculté des Médecine
Centre Hospitalier
Sherbrooke, Québec J1H 5N4
CANADA

M. F. P. Santos
University of Coimbra
Department of Physics
Rua Larga
Coimbra P-3000
PORTUGAL

R. Schiller
Hungarian Academy of Science
Central Research Institute for Physics
Atomic Energy Research Institute
P. O. Box 49
Budapest H-1525
HUNGARY

W. F. Schmidt
Hahn-Meitner-Institut
Abteilung Strahlenchemie
Glienicker Straße 100
14109 Berlin
GERMANY

D. Smith
Universität Innsbruck
Institut für Ionenphysik
Technikerstraße 25
A-6020 Innsbruck
AUSTRIA

D. Solgadi
Laboratoire Photophysique Moléculaire
CNRS
Bat. 213
Université Paris XI
91405 Orsay
FRANCE

P. Spanel
Innsbruck University
Innsbruck
AUSTRIA

S. M. Spyrou
The National Hellenic Research
Foundation, TPCI
Theor. Phys. Chemistry
48, Vassileos Constantinou Avenue
Athens 11635
GREECE

P. Stampfli
Freie Universität Berlin
Fachbereich Theoretische Physik
Arnimallee 14
14195 Berlin
GERMANY

S. V. Stepanov
Institute of Theoretical & Experimental
Physics, ITEP
Moscow 117259
RUSSIA

M. N. Varma
U.S. Department of Energy
Physical & Technological Research Division,
Office of Health and Environmental
Research
ER-74
Washington, D. C. 20545

T. Demetro
University of Patras
Department of Physics
Patras 26110
GREECE

F. Weik
Freie Universität Berlin
14195 Berlin
GERMANY

N. Xanthopoulos
University of Patras
Department of Physics
Patras 26110
GREECE

T. Yalcin
Middle East Technical University
Department of Chemistry
c-blok 207
Ankara 06531
TURKEY

P. Yianoulis
University of Patras
Department of Physics
Patras 26110
GREECE

V. Zengin
Middle East Technical University
Department of Chemistry
Anakara 06531
TURKEY

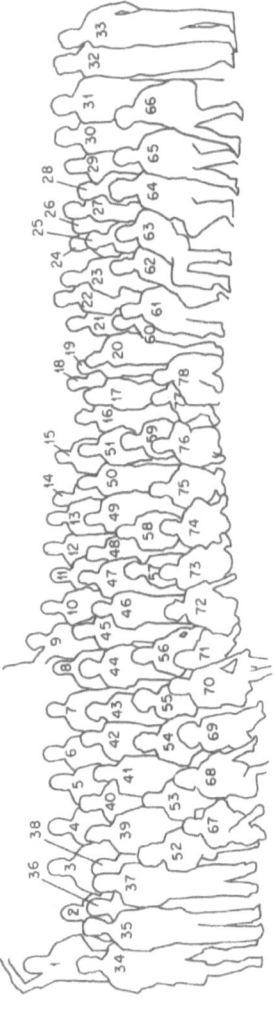

CONFERENCE PHOTO

1. Oliver Hilt
2. Vsevolod Byakov
3. Yoshihiko Hatano
4. Carlos Conde
5. Luigi Carlomusto
6. Bruce Miller
7. John C. Miller
8. Alexei G. Khrapak
9. Fritz Weik
10. Tilmann D. Märk
11. Talat Yalcin
12. Daniel Solgadi
13. Christophe Jouvet
14. Patrick Spanel
15. David A. Smith
16. Thomas Jaffke
17. Flavia Borghesani
18. A.Francesco Borghesani
19. Michael N. Pisanias
20. Robert Schiller
21. Paulo Rachinhas
22. Cengiz Savas Askun
23. Nigel J. Mason
24. Michael Henchman
25. Maria Iakubov
26. Ilya Obodowski
27. I. Iakubov

28. Oxana Obodowski
29. Norman J. Bardsley
30. Panayiotis Yiannoulis
31. Manfred Faubel
32. Peter Krebs
33. Wolfram Klein
34. Matthias Hahn
35. Reza Hashemi
36. Robert Neill Hamm
37. Jozef Kristiak
38. Yosuke Sakai
39. Eckart Rühl
40. Oddur Ingolfsson
41. Panos G. Datskos
42. Harald Morgner
43. Marc A. Johnson
44. Thomas Orlando
45. George Bakale
46. Matesh Varma
47. Vitaly Chepel
48. Joachim Ferch
49. Brigitte Lacmann
50. Klaus Lacmann
51. Peter Lamp
52. Léon Sanche
53. Eugen Illenberger
54. R. Stephen Berry

55. Lois Hamm
56. Toula Christophorou
57. Dwana Holroyd
58. Richard Holroyd
59. Vera Schiller
60. Kate Henchman
61. Mitio Inokuti
62. Makoto Hayaschi
63. Werner F. Schmidt
64. Loucas G. Christophorou
65. Traute Krebs
66. Ursula Schmidt-Heilbronn
67. Spyros Spyrou
68. Martina Meinke
69. Nuram Elmaci
70. Ivan Oleinik
71. Filomena P. dos Santos
72. Teresa H.V.T. Dias
73. Kengoh Itoh
74. Dino Neri
75. Veysel Zengin
76. Furmitaka Mafuné
77. Ryuzi Kato
78. Sergey Stepanov

INDEX

Actvation energy, 427
  volume, 447
  barrier, 491
Adsorbed molecules, 31, 569
Amorphous,
  ice, 8, 43–45
  solids, 46
Annihilation,
  positron, 121, 144
  positronium, 121, 131, 144
Aqueous electrolyte, 521
Ar,
  collision cross sections,
    gas, 264, 308
    liquid, 308
  electron drift velocity,
    gas, 312
    liquid, 312
    liquid mixtures, 315
  electron energy distribution,
    liquid, 314
  electron impact ionization of,
    Ar, 157, 170, 174
    Ar_2, 174
    Ar clusters, 160, 161
  electron-ion recombination,
    dense gas, 477, 479
    liquid, 475, 476, 479
    solid, 474, 475
  electron mobility,
    dense gas, 261, 264, 272
    273, 276
  energy loss spectrum,
    film, 40, 41
  ionization threshold of,
    clusters, 19
  mean electron energy,
    liquid, 312
  negative ion states,
    solid, 40, 43
  structure factor, 305–311
Auger,
  spectra, 512
  spectroscopy, 103, 501, 511–513

Auto-
  detachment, (see, also
      negative ions, negative ion
      states and electron attachment).
      33, 50, 355, 359, 416, 417, 425–427
  scavenging, 167
Azulene,
  ionization threshold in
  nonpolar liquids, 15, 16

Benzene,
  anions, 446
  clusters, 218–220
  mobility of electrons in, 445
Binary ion-molecule complexes, 397–413
  charge transfer photochemistry of
    407–411
  intra-cluster electron scattering in,
    411, 412
  of the form $X^-.RY$, 400–402
  photoelectron spectra of, 402–406
  photofragmentation of, 408–411
Boltzmann transport equation, 260
  for liquids, 303, 309–311
Born equation, 15, 448, 519, 534
Born-Oppenheimer approximation, 232, 259,
    360, 461
Br_2,
  metastable impact electron spectrum, 108
  ultraviolet photoelectron spectrum, 108
Bubble models, 286–290
C_{60},
  electron attachment, 491, 492
  partial ionization cross sections, 159, 160
Cage effects, 51
Carcinogen screening, 561
CF_4,
  electron attachment, 63, 64, 363, 367–369
    in condensed phase, 63
  electron impact ionization, 172
C_6F_6,
  electron attachment, 373–375, 423–427, 488
  electron detachment, 422–427
  photodetachment of $C_6F_6^-$ in liquids, 23–26

CF$_3$Br,
  electron attachment, 490-491
CF$_3$Cl,
  anions from clusters, 61-63
CF$_3$I,
  electron attachment,
    clusters, 56, 57, 61
    films, 59, 60
    single molecules, 56-58, 363-367
CH$_4$,
  electron-ion recombination,
    gas, 479
    liquid, 474-476, 479
    solid, 474, 479
CH$_3$Br,
  electron attachment, 421, 462
CH$_3$CN,
  electron drift in, 339, 342-344
CH$_3$OH,
  clusters of, 219, 221
  electron attachment, 372, 373
  electron drift in, 342
Clusters(s)
  anions (see also, individual species),
    49,50
  core level excitation, 217
  electron affinity, 27
  electron attachment (see, also, electron
    attachment), 21, 22, 167, 196-200
    electron impact ionization, 13, 14,
      155, 160-162, 174-176

  EXAFS in molecular, 217, 219
  excimer-induced fragmentation of cluster
    ions, 164, 167
  magic, numbers, 163, 204, 210
  multiply charged, 161, 217
  multiphoton ionization (see, also,
    ionization), 203
  NEXAFS in molecular, 217
  phase diagrams, 239
  phases of, 231
  photodetachment, 570
  photofragmentation, 183-201
  photoionization, 203-214
  photoionization energetics, 14-19
  size distribution, 52, 197, 570
  size-selected anions, 152, 183, 570
  surface melting, 243
  van der Waals, 49, 151, 155, 160,
    174, 203, 575
Compressibility, 445
Computer simulations, 114, 117, 497
CO$_2^-$.(H$_2$O)$_n$, 196, 197
Coulomb explosion, 217, 220
Cu$_n$, 27

Delta rays, 99
Density fluctuations, 275, 319, 445
Desorption, 51, 55, 59

Dipolar dissociation, 377
Dipole-bound states, 184, 339
Dissociative attachment (see, also,
  negative ions), 7, 32, 50, 377-379
  415-422, 428-440, 461
  environmental effects on, 33, 388-390
  in
    clusters, 20, 21, 49, 56-58, 61-63,
      65-70
    gases, 7, 20, 363-374, 415
    liquids, 20
    solids, 59, 60, 63, 64, 377-393
  photoinduced, 397-413
  time-of-flight spectra, 363-369
  to excited molecules (see excited
    molecules)
Dissociative recombination, of O$_2^+$,
  NO$^+$, 487, 488
DNA, 578

Effective
  charge, 123, 129
  mass, 333
Electroluminescence, 555, 556
Electron(s)
  collisions, 5
  conduction band energy
    minimum, V$_o$, 5, 15 16, 84, 127,
      128, 330-333, 347, 348, 443, 444,
      574
  emission from liquid surfaces, 525-537
  energy levels, 84, 444
  energy loss
    spectrometer, 34, 35
    spectroscopy, 103, 501, 513-515
    spectrum, 36, 40, 103
  equilibria, 448-453
  escape, 92, 194, 195
  excess, 259
  heat capacity, 456
  impact ionization, 12, 14, 155
  impact mass spectrometer, 418
  in liquids, 573
  injection thermodynamics, 347
  ion convention, 456, 458, 459
  localization, 141, 142, 183, 281
  quasi-free, 259, 443, 444
  self-trapping, 125
  slow, 3, 4, 569
  spectroscopy, 103
    of liquid beams, 571
  stimulated desorption (see, also,
    desorption), 380, 382-384
  swarm techniques, 419, 420, 422-425
  thermalization, 183
    distance, 91
  thermodynamic properties, 455-459
  transport (see electron drift
    velocity and electron mobility)
Electron affinity, 23-25, 50, 355

Electron attachment, 51, 355, 415
468, 487-492, 569
dissociative, 7, 19
evaporative, 51, 70
spectrometer, 362
temperature dependence of, 416-427
theory of, 461, 462
to
clusters, 21, 22, 53, 57, 58,
61-63, 65-70, 167
condensed molecules, 49
excited molecules, 415-440
van der Waals molecules, 469
to molecules in,
liquids, 19, 20, 447-452
solids, 377-393
Electron detachment, 416-418, 422-427
Electron drift velocity, 10, 11, 263, 311, 312
Electron impact ionization,
mechanisms of, 155-170
of atoms, 155-157
clusters, 160-162, 174, 175
molecules, 170-174
Electron-ion recombination, 467-483
coefficients, 490
diffusion controlled rates, 470,
480-483
dissociative, 487, 489, 490, 495
in condensed nonpolar media, 474-476
dense gases, 477-483
nonpolar liquids, 474-483
nonpolar solids, 474, 475
rate constants, 470, 474-483
theory of, 495, 496
Electron mobility,
bubble models for, 286-290
in dense rare gases, 259-261, 263
264, 268, 270, 272, 276, 281-286
nonpolar liquids, 445, 446, 474
nonpolar solids, 474
polar media, 339-344
maximum in dense argon, 273
mesoscopic model for, 296-298
perculation model for, 290-296
theories of density dependence, 265-
277, 285-298, 321-337
Electron scattering, 7, 31, 319
direct, 5, 6, 31, 355
elastic, 45
in clusters, 69
gases, 3, 31
liquids, 307, 326
solids, 31, 35-46
indirect, 5, 6, 31, 355
multiple, 39, 43-45, 259, 339
EXAFS, 217
Excited molecules (electronically), 35
dissociative attachment to, 428-440, 570
H$_2$, 439, 440
NO, 434, 438, 439

Excited molecules (electronically) (cont'd)
SO$_2$, 435-437
p-benzoquinone, 434
singlet oxygen, 432, 433
thiophenol, 433, 434
triethylamine, 437, 438
techniques for, 429-432
Excited molecules (vibrationally/rotationally),
36
dissociative attachment to, 416-427, 488, 490,
491
CF$_3$Br, 490, 491
CH$_3$Br, 421, 422
DCl, 422
HCl, 422
SO$_2$F$_2$, 421, 422
nondissociative attachment to, 417-427
C$_6$F$_6$, 422-427
techniques for, 418-420

FALP technique, 487
Fano factor, 548, 550
Field detachment, 192
Fluoroethylenes,
electron attachment to, 370, 371
Formaldehide,
Auger electron spectra, 512
electron energy loss spectrum, 513-515
metastable impact on gas, liquid, 110-112
phase transition shift, 116
photoelectron spectrum of gas, liquid,
115
valence electron spectrum, 506-508
Free ion yield, 14, 85, 91
density effects on, 95, 96, 99, 100
in liquids, 14, 85, 93
molecular structure effects on, 93
temperature effects on, 94
track effects on, 96-99
Freons (see, also, individual species)
electron attachment to, 7, 56-64, 363-372,
421, 422, 462, 488, 490, 491
Fullerene, 155

Gas proportional
ionization counter, 543, 544
scintillation counter, 543, 544
Geminate recombination, 80
Onsager theory of, 80-84
escape probability, 82-84, 92, 94
length, 469
Glasses, 254
G-value(s), 77, 78
for nonpolar liquids, 85, 86, 91-94

H$_2$,
dissociative attachment to, 439, 461, 462
electron impact ionization of clusters, 14,
174

H_3^+
  recombination of, 488, 489
H_2O,
  amorphous films, 40, 43, 44
    electron energy loss spectrum of, 43, 44
  differential oscillator strength, 8
  liquid
    differential oscillator strength, 8
    electron spectrum, 518
    ionization potentials, 518
    surface photoelectron studies, 517-523
  vapor
    photoelectron spectra, 519
He,
  momentum transfer cross section, 264
$(H_2O)_n^-$, 184-201

Image charge, 386, 387
Interphase physics, 3
Ionization (see, also, electron impact ionization, photoionization and multiphoton
  ionization)
  as a function of
    density, 15-17, 91
    phase, 75
  by electrons,
    of atoms, 155, 157, 170
    clusters, 155, 160, 174-176
    molecules, 155, 158, 159, 170
  by high-energy particles, 75, 77, 85-88
    in liquids, 85, 86, 91-100
    polymers, 87, 88
    solids, 86
    track effects, 96
  by photons (see photoionization), 75
  current, 79
  energetics, 84, 85
  mechanisms, 155, 156
  potential, 75
  spatial distribution, 78
  threshold energy, 85

Isomerization, 167
Isothermal compressibility, 445

Kinetic energy shift, 188, 271, 273
Kr,
  electron
    collision cross section, liquid, 308
    drift velocity, gas, 312
    -ion recombination,
      dense gas, 475, 478, 479
      liquid, 474, 477, 479
    mean electron energy, liquid, 312
  ionization potential
    of clusters, 19
    liquid, 85

Kr (cont'd)
  solid films of, 43, 389-393
  structure factor, 306

Langmuir probe, 487
LET, 77, 78, 86
Life sciences, 577
Linking gaseous and
  liquid phase, 3, 75, 103, 316
  solid phase, 45, 75, 377
Liquid
  jets, 517
  surfaces, 501

Magic numbers, 163, 204, 210
Mesoscopic model, 296-298
Metastable impact electron,
  spectra, 106
  spectroscopy, 105, 501
Molecular dynamics simulations, 113, 164, 24
Monte Carlo calculations, 142, 510, 543
Multilayer films, 34-46, 59
Multiphoton ionization, 203
  in dense gases, 14-17
    liquids, 14-16
  of clusters, 203-214
  nonresonant, 210-214
  resonant, 207-209

N_2,
  condensed on substrates, 37
  electron scattering, 6
  energy loss spectrum, solid film, 36
  excitation functions from condensed, 38, 39
  ionization, 106
  metastable impact electron spectrum, 106
  negative ion states, 6, 12
  positron annihilation in, 135
  solid films, 12, 13, 35, 36
  vibrational excitation, gas/solid, 36-39
  ultraviolet photoelectron spectrum, 106
Ne,
  electron localization in, dense gas, 281
  electron mobility in, dense gas, 263, 268, 270
    276, 281, 284-298
  momentum transfer cross section, 264
Negative ions (see, also, electron
  attachment and clusters), 355, 377
  autodetachment of, 416, 417
  desorbed, 51
  electron stimulated desorption of,
    380, 382-384
  fates of temporary, 359
Negative ion states
  decay channels, 12, 32, 359
  environmental effects on, 33
NEXAFS, clusters, 217

NH₃,
  electrons in, 340-342
N₂O,
  electron attachment, 19, 20, 447
(NO)ₙ,
  multiphoton ionization, 207-209, 212
NO-CH₄ clusters,
  multiphoton ionization, 213
NO₂-C₂H₄ complex,
  photoinduced reactions, 223
NO-rare gas molecules,
  multiphoton ionization, 205-211
Nucleophilicity, 562

O₂,
  dissociative attachment to,
    clusters, 21, 22, 65-70
    isolated O₂, 21, 389
    singlet O₂, 432, 433, 462
    solid films of, 21, 383-386, 388-393
  potential energy curves,
    for O₂ and O₂⁻, 21, 67, 385
Onsager
  escape probability, 80-84, 91, 92, 94
  length, 469
  theory of geminate recombination, 80
Optical potential approximation, 127
Ore-gap, 124
Orthopositronium, 131, 141
Oscillator strength, 7, 8, 191, 192
  condense phase effects on, 8
  for H₂O, 8

Path integral method, 141, 143
Penning ionization, 103
Phase,
  diagram, 239
  equilibrium, 234
  transition, 231
  transition shift, 116, 117
Photoconductivity,
  of nonpolar liquids, 77, 84
  method, 76
Photodetachment,
  in clusters, 22, 25-27, 200
    gases, 22, 23
    liquids, 22-26
  technique, 24-26
Photoelectron
  spectrometer, 401
  spectroscopy, 517
    of benzyl alcohol, 521
    liquid mixtures, 520-523
    salt solutions, 520-523
    solvated PF₆⁻, 523
Photoemission,
  absolute quantum yield, 535
  threshold, liquid to vacuum, 534, 535
Photofragmentation, 183, 189

Photoionization,
  as a function of density, 16, 17
    phase, 75-88
  energetics, 15-18
  of clusters (see, also, clusters), 16-19, 217
    molecules in liquids, 14
Physisorbed molecules, 379
Polar gases,
  attachment to, 462
  electrons in, 339-344
Polarization
  effects, 32, 386
  energy, 15, 42, 448, 534
Polaron theory, 145
Polymers, 87, 88
Positron,
  annihilation, 121
    clusters, 135-137
    cross sections, 122
    in liquids, 134
    rates, 122
    time spectrum, 122
  scattering cross sections, 124-127
Positronium, 121
  bubbles, 125, 130-135
  elastic scattering, 127
  pick-off rate, 130
Potential surface landscapes, 251
  topography of, 253
Protein folding, 254
Pulse radiolysis, 470-473
Pulsed-Townsend photoinjection technique, 261, 262

Quasi-equilibrium theory, 158

Radiation physics, 577
Ramsauer-Townsend minimum, 9, 124, 125, 260, 275, 489
Rare gas(es),
  clusters (see clusters and individual species)
  electron transport in liquid, 303, 311-316
  positron scattering from, 126
Reaction volume, 443, 451-453
Recombination, 77
  geminate (see geminate), 80
  volume, 92
Rydberg transitions, 35

Scattering length, 125
  potential, 304
Segregation time, 521
Self-trapping
  of electrons, 137, 141, 161
    positrons, 137, 141
SF₆,
  electron attachment, 167, 362, 462
Smoluchowski equation, 80
SO₂F₂,
  dissociative attachment, 421

Spectral moments, 185
Spur,
    distribution, 78
    model, 124
Structure factor, 9, 46, 305-307
    for liquid rare gases, 306
Supersonic molecular beams, 54
Surface,
    charging, 391
    melting, 243
    multidimensional potential, 251
    properties of solutions 505-511
    studies of volatile liquids, 517
    tension, 132

Tetramethylsilane
    electron drift in, 11
    electron ion-recombination, 474
    free ion yield, 93
    ionization threshold, 85
Thermalization distance, 91, 92
Thermionic emission,
    from $(H_2O)_n^-$, 193
Thermodynamic fluctuations, 575
Time spectrum, 122
Time-of-flight technique, 363
    spectra, 365-369
Time-resolved electron swarm technique,
    422-425
TMPD,
    photoionization, 16, 17

Ultraviolet photoelectron,
    emission, 525, 526
        yield, 530-533
    spectroscopy, 104
    valence electron spectrum, 506
Unimolecular reactions, cluster ions, 162

Van der Waals,
    clusters (see, also, clusters), 49, 160-162,
    203
    molecules, 203, 469
    solid, 43
Vertical detachment energy, 23, 24, 184, 194,
    356
V_o (see electron conduction band energy
    miminum)

Wigner-Seitz,
    energy, 291
    model, 128, 129, 269
W-values, 77, 78, 85

Xe,
    amorphous Xe solid, 46
    annihilation in Xe of
        orthopositronium, 131
        positron and positronium, 144
    electroluminescence yield in, 556
    electron,
        characteristic energy,
            gas, 554, 555
            liquid, 312
        collision frequency, gas, 553, 554
        drift velocity,
            gas, 10, 312, 554
            liquid, 10
        ionization coefficient, liquid, 313
        - ion recombination, dense gas, 475, 478
        mean energy,
            gas, 552, 553
            liquid, 312
        mobility, gas, 554
        scattering cross section,
            gas, 10, 308
            liquid, 10, 308
            solid, 11
    Fano factor of, 548, 550
    ion charge distribution in gas, 546, 547
    ionization threshold
        clusters, 19
        gas, 14
        liquid, 14, 85
    multiphoton ionization, clusters, 210
    photoionization cross section, gas, 550
    structure factor, 306
    W-value,
        gas, 14, 548, 550
        liquid, 14
X-ray,
    detectors, 543
    distortion in energy spectra, 551
    photoelectron spectroscopy, 104, 501